Teacher, Student One-Stop Internet Resources

Log on to blue.msscience.com

ONLINE STUDY TOOLS
- Section Self-Check Quizzes
- Interactive Tutor
- Chapter Review Tests
- Standardized Test Practice
- Vocabulary PuzzleMaker

ONLINE RESEARCH
- WebQuest Projects
- Prescreened Web Links
- Career Links
- Internet Labs

INTERACTIVE ONLINE STUDENT EDITION
- Complete Interactive Student Edition available at mhln.com

FOR TEACHERS
- Teacher Bulletin Board
- Teaching Today—Professional Development

SAFETY SYMBOLS

SAFETY SYMBOLS	HAZARD	EXAMPLES	PRECAUTION	REMEDY
DISPOSAL	Special disposal procedures need to be followed.	certain chemicals, living organisms	Do not dispose of these materials in the sink or trash can.	Dispose of wastes as directed by your teacher.
BIOLOGICAL	Organisms or other biological materials that might be harmful to humans	bacteria, fungi, blood, unpreserved tissues, plant materials	Avoid skin contact with these materials. Wear mask or gloves.	Notify your teacher if you suspect contact with material. Wash hands thoroughly.
EXTREME TEMPERATURE	Objects that can burn skin by being too cold or too hot	boiling liquids, hot plates, dry ice, liquid nitrogen	Use proper protection when handling.	Go to your teacher for first aid.
SHARP OBJECT	Use of tools or glassware that can easily puncture or slice skin	razor blades, pins, scalpels, pointed tools, dissecting probes, broken glass	Practice commonsense behavior and follow guidelines for use of the tool.	Go to your teacher for first aid.
FUME	Possible danger to respiratory tract from fumes	ammonia, acetone, nail polish remover, heated sulfur, moth balls	Make sure there is good ventilation. Never smell fumes directly. Wear a mask.	Leave foul area and notify your teacher immediately.
ELECTRICAL	Possible danger from electrical shock or burn	improper grounding, liquid spills, short circuits, exposed wires	Double-check setup with teacher. Check condition of wires and apparatus.	Do not attempt to fix electrical problems. Notify your teacher immediately.
IRRITANT	Substances that can irritate the skin or mucous membranes of the respiratory tract	pollen, moth balls, steel wool, fiberglass, potassium permanganate	Wear dust mask and gloves. Practice extra care when handling these materials.	Go to your teacher for first aid.
CHEMICAL	Chemicals can react with and destroy tissue and other materials	bleaches such as hydrogen peroxide; acids such as sulfuric acid, hydrochloric acid; bases such as ammonia, sodium hydroxide	Wear goggles, gloves, and an apron.	Immediately flush the affected area with water and notify your teacher.
TOXIC	Substance may be poisonous if touched, inhaled, or swallowed.	mercury, many metal compounds, iodine, poinsettia plant parts	Follow your teacher's instructions.	Always wash hands thoroughly after use. Go to your teacher for first aid.
FLAMMABLE	Flammable chemicals may be ignited by open flame, spark, or exposed heat.	alcohol, kerosene, potassium permanganate	Avoid open flames and heat when using flammable chemicals.	Notify your teacher immediately. Use fire safety equipment if applicable.
OPEN FLAME	Open flame in use, may cause fire.	hair, clothing, paper, synthetic materials	Tie back hair and loose clothing. Follow teacher's instruction on lighting and extinguishing flames.	Notify your teacher immediately. Use fire safety equipment if applicable.

 Eye Safety Proper eye protection should be worn at all times by anyone performing or observing science activities.

 Clothing Protection This symbol appears when substances could stain or burn clothing.

 Animal Safety This symbol appears when safety of animals and students must be ensured.

 Handwashing After the lab, wash hands with soap and water before removing goggles.

 Glencoe

New York, New York · Columbus, Ohio · Chicago, Illinois · Woodland Hills, California

Glencoe Science

Level Blue

This furnace is used in the production of steel. Steel is formed by heating molten iron and scrap steel at 1550°C. The Grand Canyon was formed by over one billion years of erosion. These tropical fish are found in waters that maintain a constant temperature between 70°C and 78°C. Color may give these fish a way to communicate with or attract mates.

The McGraw·Hill Companies

Copyright © 2008 by The McGraw-Hill Companies, Inc. All rights reserved. Except as permitted under the United States Copyright Act, no part of this publication may be reproduced or distributed in any form or by any means, or stored in a database or retrieval system, without prior written permission of the publisher.

The National Geographic features were designed and developed by the National Geographic Society's Education Division. Copyright © National Geographic Society. The name "National Geographic Society" and the Yellow Border Rectangle are trademarks of the Society, and their use, without prior written permission, is strictly prohibited.

The "Science and Society" and the "Science and History" features that appear in this book were designed and developed by TIME School Publishing, a division of TIME Magazine. TIME and the red border are trademarks of Time Inc. All rights reserved.

Send all inquiries to:
Glencoe/McGraw-Hill
8787 Orion Place
Columbus, OH 43240-4027

ISBN: 978-0-07-877810-0
MHID: 0-07-877810-7

Printed in the United States of America.

4 5 6 7 8 9 10 079/043 12 11 10 09

Contents In Brief

Unit 1 — Humans and Heredity 2
- **Chapter 1** The Nature of Science 4
- **Chapter 2** Traits and How They Change 36
- **Chapter 3** Interactions of Human Systems 62

Unit 2 — Ecology 90
- **Chapter 4** Interactions of Life 92
- **Chapter 5** The Nonliving Environment 120
- **Chapter 6** Ecosystems 148

Unit 3 — Earth's Changes over Time 178
- **Chapter 7** Plate Tectonics 180
- **Chapter 8** Earthquakes and Volcanoes 208
- **Chapter 9** Clues to Earth's Past 240
- **Chapter 10** Geologic Time 270

Unit 4 — Earth's Place in the Universe 302
- **Chapter 11** The Sun-Earth-Moon System 304
- **Chapter 12** The Solar System 334
- **Chapter 13** Stars and Galaxies 368

Unit 5 — Chemistry of Matter 400
- **Chapter 14** Inside the Atom 402
- **Chapter 15** The Periodic Table 432
- **Chapter 16** Atomic Structure and Chemical Bonds 462
- **Chapter 17** Chemical Reactions 490

Unit 6 — Motion, Forces, and Energy 518
- **Chapter 18** Motion and Momentum 520
- **Chapter 19** Force and Newton's Laws 548
- **Chapter 20** Work and Simple Machines 578
- **Chapter 21** Thermal Energy 606

Unit 7 — Physical Interactions 632
- **Chapter 22** Electricity 634
- **Chapter 23** Magnetism 664
- **Chapter 24** Waves, Sound, and Light 692

Authors

Education Division
Washington, D.C.

Alton Biggs
Retired Biology Teacher
Allen High School
Allen, TX

Ralph M. Feather Jr., PhD
Assistant Professor
Department of Educational Studies
and Secondary Education
Bloomsburg University
Bloomsburg, PA

Peter Rillero, PhD
Associate Professor of Science Education
Arizona State University West
Phoenix, AZ

Dinah Zike
Educational Consultant
Dinah-Might Activities, Inc.
San Antonio, TX

Series Consultants

CONTENT

Alton J. Banks, PhD
Director of the Faculty Center
for Teaching and Learning
North Carolina State University
Raleigh, NC

Jack Cooper
Ennis High School
Ennis, TX

Sandra K. Enger, PhD
Associate Director,
Associate Professor
UAH Institute for Science Education
Huntsville, AL

David G. Haase, PhD
North Carolina State University
Raleigh, NC

Michael A. Hoggarth, PhD
Department of Life and
Earth Sciences
Otterbein College
Westerville, OH

Jerome A. Jackson, PhD
Whitaker Eminent Scholar in Science
Program Director
Center for Science, Mathematics,
and Technology Education
Florida Gulf Coast University
Fort Meyers, FL

William C. Keel, PhD
Department of Physics
and Astronomy
University of Alabama
Tuscaloosa, AL

Linda McGaw
Science Program Coordinator
Advanced Placement Strategies, Inc.
Dallas, TX

Madelaine Meek
Physics Consultant Editor
Lebanon, OH

Robert Nierste
Science Department Head
Hendrick Middle School, Plano ISD
Plano, TX

Connie Rizzo, MD, PhD
Depatment of Science/Math
Marymount Manhattan College
New York, NY

Dominic Salinas, PhD
Middle School Science Supervisor
Caddo Parish Schools
Shreveport, LA

Cheryl Wistrom
St. Joseph's College
Rensselaer, IN

Carl Zorn, PhD
Staff Scientist
Jefferson Laboratory
Newport News, VA

MATH

Michael Hopper, DEng
Manager of Aircraft Certification
L-3 Communications
Greenville, TX

Teri Willard, EdD
Mathematics Curriculum Writer
Belgrade, MT

READING

Elizabeth Babich
Special Education Teacher
Mashpee Public Schools
Mashpee, MA

Barry Barto
Special Education Teacher
John F. Kennedy Elementary
Manistee, MI

Carol A. Senf, PhD
School of Literature,
Communication, and Culture
Georgia Institute of Technology
Atlanta, GA

Rachel Swaters-Kissinger
Science Teacher
John Boise Middle School
Warsaw, MO

SAFETY

Aileen Duc, PhD
Science 8 Teacher
Hendrick Middle School, Plano ISD
Plano, TX

Sandra West, PhD
Department of Biology
Texas State University-San Marcos
San Marcos, TX

ACTIVITY TESTERS

Nerma Coats Henderson
Pickerington Lakeview Jr. High School
Pickerington, OH

Mary Helen Mariscal-Cholka
William D. Slider Middle School
El Paso, TX

Science Kit and Boreal Laboratories
Tonawanda, NY

Reviewers

Deidre Adams
West Vigo Middle School
West Terre Haute, IN

Sharla Adams
IPC Teacher
Allen High School
Allen, TX

Maureen Barrett
Thomas E. Harrington Middle School
Mt. Laurel, NJ

John Barry
Seeger Jr.-Sr. High School
West Lebanon, IN

Desiree Bishop
Environmental Studies Center
Mobile County Public Schools
Mobile, AL

William Blair
Retired Teacher
J. Marshall Middle School
Billerica, MA

Tom Bright
Concord High School
Charlotte, NC

Lois Burdette
Green Bank Elementary-Middle School
Green Bank, WV

Marcia Chackan
Pine Crest School
Boca Raton, FL

Obioma Chukwu
J.H. Rose High School
Greenville, NC

Nerma Coats Henderson
Pickerington Lakeview Jr. High School
Pickerington, OH

Karen Curry
East Wake Middle School
Raleigh, NC

Joanne Davis
Murphy High School
Murphy, NC

Anthony J. DiSipio, Jr.
8th Grade Science
Octorana Middle School
Atglen, PA

Annette D'Urso Garcia
Kearney Middle School
Commerce City, CO

Sandra Everhart
Dauphin/Enterprise Jr. High Schools
Enterprise, AL

Mary Ferneau
Westview Middle School
Goose Creek, SC

Cory Fish
Burkholder Middle School
Henderson, NV

Linda V. Forsyth
Retired Teacher
Merrill Middle School
Denver, CO

George Gabb
Great Bridge Middle School
Chesapeake Public Schools
Chesapeake, VA

Lynne Huskey
Chase Middle School
Forest City, NC

Maria E. Kelly
Principal
Nativity School
Catholic Diocese of Arlington
Burke, VA

Michael Mansour
Board Member
National Middle Level Science
Teacher's Association
John Page Middle School
Madison Heights, MI

Mary Helen Mariscal-Cholka
William D. Slider Middle School
El Paso, TX

Michelle Mazeika-Simmons
Whiting Middle School
Whiting, IN

Joe McConnell
Speedway Jr. High School
Indianapolis, IN

Sharon Mitchell
William D. Slider Middle School
El Paso, TX

Amy Morgan
Berry Middle School
Hoover, AL

Norma Neely, EdD
Associate Director for Regional Projects
Texas Rural Systemic Initiative
Austin, TX

Annette Parrott
Lakeside High School
Atlanta, GA

Nora M. Prestinari Burchett
Saint Luke School
McLean, VA

Mark Sailer
Pioneer Jr.-Sr. High School
Royal Center, IN

Joanne Stickney
Monticello Middle School
Monticello, NY

Dee Stout
Penn State University
University Park, PA

Darcy Vetro-Ravndal
Hillsborough High School
Tampa, FL

Karen Watkins
Perry Meridian Middle School
Indianapolis, IN

Clabe Webb
Permian High School
Ector County ISD
Odessa, TX

Alison Welch
William D. Slider Middle School
El Paso, TX

Kim Wimpey
North Gwinnett High School
Suwanee, GA

Kate Ziegler
Durant Road Middle School
Raleigh, NC

Teacher Advisor Board

The Teacher Advisory Board gave the authors, editorial staff, and design team feedback on the content and design of the Student Edition. They provided valuable input in the development of the 2008 edition of *Glencoe Science Level Blue.*

John Gonzales
Challenger Middle School
Tucson, AZ

Rachel Shively
Aptakisic Jr. High School
Buffalo Grove, IL

Roger Pratt
Manistique High School
Manistique, MI

Kirtina Hile
Northmor Jr. High/High School
Galion, OH

Marie Renner
Diley Middle School
Pickerington, OH

Nelson Farrier
Hamlin Middle School
Springfield, OR

Jeff Remington
Palmyra Middle School
Palmyra, PA

Erin Peters
Williamsburg Middle School
Arlington, VA

Rubidel Peoples
Meacham Middle School
Fort Worth, TX

Kristi Ramsey
Navasota Jr. High School
Navasota, TX

Student Advisory Board

The Student Advisory Board gave the authors, editorial staff, and design team feedback on the design of the Student Edition. We thank these students for their hard work and creative suggestions in making the 2008 edition of *Glencoe Science Level Blue* student friendly.

Jack Andrews
Reynoldsburg Jr. High School
Reynoldsburg, OH

Peter Arnold
Hastings Middle School
Upper Arlington, OH

Emily Barbe
Perry Middle School
Worthington, OH

Kirsty Bateman
Hilliard Heritage Middle School
Hilliard, OH

Andre Brown
Spanish Emersion Academy
Columbus, OH

Chris Dundon
Heritage Middle School
Westerville, OH

Ryan Manafee
Monroe Middle School
Columbus, OH

Addison Owen
Davis Middle School
Dublin, OH

Teriana Patrick
Eastmoor Middle School
Columbus, OH

Ashley Ruz
Karrar Middle School
Dublin, OH

The Glencoe middle school science Student Advisory Board taking a timeout at COSI, a science museum in Columbus, Ohio.

HOW TO...
Use Your Science Book

Why do I need my science book?

Have you ever been in class and not understood all of what was presented? Or, you understood everything in class, but at home, got stuck on how to answer a question? Maybe you just wondered when you were ever going to use this stuff?

These next few pages are designed to help you understand everything your science book can be used for . . . besides a paperweight!

Before You Read

- **Chapter Opener** Science is occurring all around you, and the opening photo of each chapter will preview the science you will be learning about. The **Chapter Preview** will give you an idea of what you will be learning about, and you can try the **Launch Lab** to help get your brain headed in the right direction. The **Foldables** exercise is a fun way to keep you organized.

- **Section Opener** Chapters are divided into two to four sections. The **As You Read** in the margin of the first page of each section will let you know what is most important in the section. It is divided into four parts. **What You'll Learn** will tell you the major topics you will be covering. **Why It's Important** will remind you why you are studying this in the first place! The **Review Vocabulary** word is a word you already know, either from your science studies or your prior knowledge. The **New Vocabulary** words are words that you need to learn to understand this section. These words will be in **boldfaced** print and highlighted in the section. Make a note to yourself to recognize these words as you are reading the section.

As You Read

- **Headings** Each section has a title in large red letters, and is further divided into blue titles and small red titles at the beginnings of some paragraphs. To help you study, make an outline of the headings and subheadings.

- **Margins** In the margins of your text, you will find many helpful resources. The **Science Online** exercises and **Integrate** activities help you explore the topics you are studying. **MiniLabs** reinforce the science concepts you have learned.

- **Building Skills** You also will find an **Applying Math** or **Applying Science** activity in each chapter. This gives you extra practice using your new knowledge, and helps prepare you for standardized tests.

- **Student Resources** At the end of the book you will find **Student Resources** to help you throughout your studies. These include **Science, Technology,** and **Math Skill Handbooks,** an **English/Spanish Glossary,** and an **Index.** Also, use your **Foldables** as a resource. It will help you organize information, and review before a test.

- **In Class** Remember, you can always ask your teacher to explain anything you don't understand.

FOLDABLES Study Organizer

Science Vocabulary Make the following Foldable to help you understand the vocabulary terms in this chapter.

STEP 1 Fold a vertical sheet of notebook paper from side to side.

STEP 2 Cut along every third line of only the top layer to form tabs.

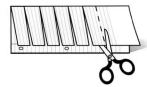

STEP 3 Label each tab with a vocabulary word from the chapter.

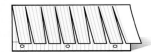

Build Vocabulary As you read the chapter, list the vocabulary words on the tabs. As you learn the definitions, write them under the tab for each vocabulary word.

Look For... **FOLDABLES** At the beginning of every section.

In Lab

Working in the laboratory is one of the best ways to understand the concepts you are studying. Your book will be your guide through your laboratory experiences, and help you begin to think like a scientist. In it, you not only will find the steps necessary to follow the investigations, but you also will find helpful tips to make the most of your time.

- Each lab provides you with a **Real-World Question** to remind you that science is something you use every day, not just in class. This may lead to many more questions about how things happen in your world.

- Remember, experiments do not always produce the result you expect. Scientists have made many discoveries based on investigations with unexpected results. You can try the experiment again to make sure your results were accurate, or perhaps form a new hypothesis to test.

- Keeping a **Science Journal** is how scientists keep accurate records of observations and data. In your journal, you also can write any questions that may arise during your investigation. This is a great method of reminding yourself to find the answers later.

Look For...
- **Launch Labs** start every chapter.
- **MiniLabs** in the margin of each chapter.
- **Two Full-Period Labs** in every chapter.
- **EXTRA Try at Home Labs** at the end of your book.
- the **Web site** with **laboratory demonstrations**.

Before a Test

Admit it! You don't like to take tests! However, there *are* ways to review that make them less painful. Your book will help you be more successful taking tests if you use the resources provided to you.

- Review all of the **New Vocabulary** words and be sure you understand their definitions.
- Review the notes you've taken on your **Foldables,** in class, and in lab. Write down any question that you still need answered.
- Review the **Summaries** and **Self Check questions** at the end of each section.
- Study the concepts presented in the chapter by reading the **Study Guide** and answering the questions in the **Chapter Review.**

Look For...

- **Reading Checks** and **caption questions** throughout the text.
- the **Summaries** and **Self Check questions** at the end of each section.
- the **Study Guide** and **Review** at the end of each chapter.
- the **Standardized Test Practice** after each chapter.

Let's Get Started

To help you find the information you need quickly, use the Scavenger Hunt below to learn where things are located in Chapter 1.

1. What is the title of this chapter?
2. What will you learn in Section 1?
3. Sometimes you may ask, "Why am I learning this?" State a reason why the concepts from Section 2 are important.
4. What is the main topic presented in Section 2?
5. How many reading checks are in Section 1?
6. What is the Web address where you can find extra information?
7. What is the main heading above the sixth paragraph in Section 2?
8. There is an integration with another subject mentioned in one of the margins of the chapter. What subject is it?
9. List the new vocabulary words presented in Section 2.
10. List the safety symbols presented in the first Lab.
11. Where would you find a Self Check to be sure you understand the section?
12. Suppose you're doing the Self Check and you have a question about concept mapping. Where could you find help?
13. On what pages are the Chapter Study Guide and Chapter Review?
14. Look in the Table of Contents to find out on which page Section 2 of the chapter begins.
15. You complete the Chapter Review to study for your chapter test. Where could you find another quiz for more practice?

Contents

unit 1
Humans and Heredity—2

chapter 1
The Nature of Science—4

Section 1	What is science?	6
	Lab Battle of the Beverage Mixes	12
Section 2	Doing Science	13
Section 3	Science and Technology	24
	Lab When is the Internet the busiest?	28

chapter 2
Traits and How They Change—36

Section 1	Traits and the Environment	38
	Lab Jelly Bean Hunt	43
Section 2	Genetics	44
Section 3	Environmental Impact over Time	49
	Lab Toothpick Fish	54

In each chapter, look for these opportunities for review and assessment:
- Reading Checks
- Caption Questions
- Section Review
- Chapter Study Guide
- Chapter Review
- Standardized Test Practice
- Online practice at blue.msscience.com

Get Ready to Read Strategies
- Preview 6A
- Identify the Main Idea 38A
- New Vocabulary 64A

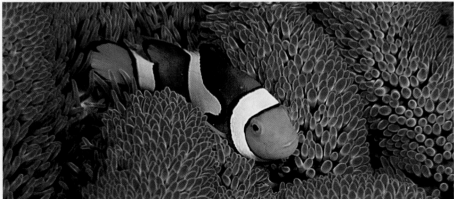

chapter 3
Interactions of Human Systems—62

Section 1	The Human Organism	64
	Lab Observing Cells	72
Section 2	How Your Body Works	73
	Lab Does exercise affect respiration?	82

xiii

Contents

unit 2 Ecology—90

Chapter 4 — Interactions of Life—92
Section 1	Living Earth	94
Section 2	Populations	98
Section 3	Interactions Within Communities	106
	Lab Feeding Habits of Planaria	111
	Lab Population Growth in Fruit Flies	112

Chapter 5 — The Nonliving Environment—120
Section 1	Abiotic Factors	122
	Lab Humus Farm	129
Section 2	Cycles in Nature	130
Section 3	Energy Flow	136
	Lab Where does the mass of a plant come from?	140

Chapter 6 — Ecosystems—148
Section 1	How Ecosystems Change	150
Section 2	Biomes	154
	Lab Studying a Land Ecosystem	162
Section 3	Aquatic Ecosystems	163
	Lab Exploring Wetlands	170

In each chapter, look for these opportunities for review and assessment:
- Reading Checks
- Caption Questions
- Section Review
- Chapter Study Guide
- Chapter Review
- Standardized Test Practice
- Online practice at blue.msscience.com

Contents

Unit 3 — Earth's Changes over Time—178

Chapter 7 — Plate Tectonics—180
- Section 1 Continental Drift 182
- Section 2 Seafloor Spreading 186
  - Lab Seafloor Spreading Rates 189
- Section 3 Theory of Plate Tectonics 190
  - Lab Predicting Tectonic Activity 200

Chapter 8 — Earthquakes and Volcanoes—208
- Section 1 Earthquakes 210
- Section 2 Volcanoes 219
  - Lab Disruptive Eruptions 225
- Section 3 Earthquakes, Volcanoes, and Plate Tectonics ...226
  - Lab Seismic Waves 232

Chapter 9 — Clues to Earth's Past—240
- Section 1 Fossils 242
- Section 2 Relative Ages of Rocks 250
  - Lab Relative Ages 256
- Section 3 Absolute Ages of Rocks 257
  - Lab Trace Fossils 262

Chapter 10 — Geologic Time—270
- Section 1 Life and Geologic Time 272
- Section 2 Early Earth History 280
  - Lab Changing Species 287
- Section 3 Middle and Recent Earth History 288
  - Lab Discovering the Past 294

Get Ready to Read Strategies
- Comparing and Contrasting 94A
- Make Inferences 122A
- Take Notes 150A
- Summarize 182A
- Monitor 210A
- Take Notes 242A
- Questions and Answers 272A

Contents

unit 4 Earth's Place in the Universe—302

chapter 11 The Sun-Earth-Moon System—304

Section 1	Earth	306
Section 2	The Moon—Earth's Satellite	312
	Lab Moon Phases and Eclipses	321
Section 3	Exploring Earth's Moon	322
	Lab Tilt and Temperature	326

chapter 12 The Solar System—334

Section 1	The Solar System	336
	Lab Planetary Orbits	341
Section 2	The Inner Planets	342
Section 3	The Outer Planets	348
Section 4	Other Objects in the Solar System	356
	Lab Solar System Distance Model	360

Contents

Stars and Galaxies—368

Section 1	Stars	370
Section 2	The Sun	375
	Lab Sunspots	379
Section 3	Evolution of Stars	380
Section 4	Galaxies and the Universe	386
	Lab Measuring Parallax	392

In each chapter, look for these opportunities for review and assessment:
- Reading Checks
- Caption Questions
- Section Review
- Chapter Study Guide
- Chapter Review
- Standardized Test Practice
- Online practice at blue.msscience.com

Get Ready to Read Strategies
- Summarize 306A
- Comparing and Contrasting 336A
- Make Inferences 370A
- Visualize 404A

unit 5 Chemistry of Matter—400

Inside the Atom—402

Section 1	Models of the Atom	404
	Lab Making a Model of the Invisible	414
Section 2	The Nucleus	415
	Lab Half-Life	424

Contents

chapter 15 — The Periodic Table—432

Section 1	Introduction to the Periodic Table	434
Section 2	Representative Elements	441
Section 3	Transition Elements	448
Lab	Metals and Nonmetals	453
Lab	Health Risks from Heavy Metals	454

chapter 16 — Atomic Structure and Chemical Bonds—462

Section 1	Why do atoms combine?	464
Section 2	How Elements Bond	472
Lab	Ionic Compounds	481
Lab	Atomic Structure	482

chapter 17 — Chemical Reactions—490

Section 1	Chemical Formulas and Equations	492
Section 2	Rates of Chemical Reactions	502
Lab	Physical or Chemical Change?	509
Lab	Exothermic or Endothermic?	510

unit 6 — Motion, Forces, and Energy—518

chapter 18 — Motion and Momentum—520

Section 1	What is motion?	522
Section 2	Acceleration	528
Section 3	Momentum	533
Lab	Collisions	539
Lab	Car Safety Testing	540

In each chapter, look for these opportunities for review and assessment:
- Reading Checks
- Caption Questions
- Section Review
- Chapter Study Guide
- Chapter Review
- Standardized Test Practice
- Online practice at blue.msscience.com

Contents

chapter 19
Force and Newton's Laws—548
Section 1	**Newton's First Law**	550
Section 2	**Newton's Second Law**	556
Section 3	**Newton's Third Law**	563
	Lab Balloon Races	569
	Lab Modeling Motion in Two Directions	570

chapter 20
Work and Simple Machines—578
Section 1	**Work and Power**	580
	Lab Building the Pyramids	585
Section 2	**Using Machines**	586
Section 3	**Simple Machines**	591
	Lab Pulley Power	598

chapter 21
Thermal Energy—606
Section 1	**Temperature and Thermal Energy**	608
Section 2	**Heat**	612
	Lab Heating Up and Cooling Down	618
Section 3	**Engines and Refrigerators**	619
	Lab Comparing Thermal Insulators	624

Get Ready to Read Strategies
- Make Connections 434A
- Questioning 464A
- Make Predictions 492A
- Summarize 522A
- Comparing and Contrasting 550A
- Questions and Answers 580A
- Identify the Main Idea 608A

xix

Contents

unit 7
Physical Interactions—632

chapter 22
Electricity—634

Section 1	Electric Charge	636
Section 2	Electric Current	643
Section 3	Electric Circuits	648
	Lab Current in a Parallel Circuit	655
	Lab A Model for Voltage and Current	656

chapter 23
Magnetism—664

Section 1	What is magnetism?	666
	Lab Make a Compass	672
Section 2	Electricity and Magnetism	673
	Lab How does an electric motor work?	684

chapter 24
Waves, Sound, and Light—692

Section 1	Waves	694
Section 2	Sound Waves	701
	Lab Sound Waves in Matter	706
Section 3	Light	707
	Lab Bending Light	714

In each chapter, look for these opportunities for review and assessment:
- Reading Checks
- Caption Questions
- Section Review
- Chapter Study Guide
- Chapter Review
- Standardized Test Practice
- Online practice at blue.msscience.com

Get Ready to Read Strategies
- Make Predictions 636A
- Identify Cause and Effect 666A
- Make Connections 694A

Contents

Student Resources—722

Science Skill Handbook—724
Scientific Methods 724
Safety Symbols 733
Safety in the Science Laboratory 734

Extra Try at Home Labs—736

Technology Skill Handbook—748
Computer Skills 748
Presentation Skills 751

Math Skill Handbook—752
Math Review 752
Science Applications 762

Reference Handbooks—767
Topographic Map Symbols 767
Physical Science Reference Tables 768
Periodic Table of the Elements 770

English/Spanish Glossary—772

Index—793

Credits—812

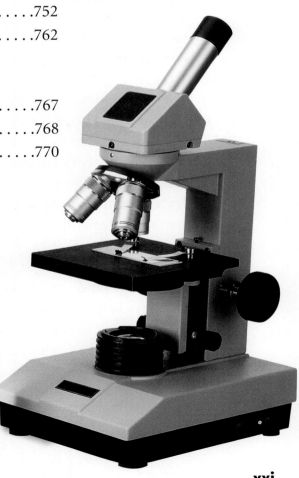

xxi

Cross-Curricular Readings

NATIONAL GEOGRAPHIC Unit Openers

- **Unit 1** How are Electricity and DNA Connected? 2
- **Unit 2** How are Beverages and Wildlife Connected? 90
- **Unit 3** How are Volcanoes and Fish Connected? 178
- **Unit 4** How are Thunderstorms and Neutron Stars Connected? 302
- **Unit 5** How are Charcoal and Celebrations Connected? 400
- **Unit 6** How are City Streets and Zebra Mussels Connected? 518
- **Unit 7** How are Radar and Popcorn Connected? 632

NATIONAL GEOGRAPHIC VISUALIZING

1. Descriptive and Experimental Research 20
2. Natural Selection 51
3. Human Cells 69
4. Population Growth 104
5. The Carbon Cycle 134
6. Secondary Succession 152
7. Plate Boundaries 193
8. Tsunamis 216
9. Unconformities 253
10. Unusual Life Forms 283
11. The Moon's Surface 318
12. The Solar System's Formation 339
13. The Big Bang Theory 390
14. Tracer Elements 422
15. Synthetic Elements 451
16. Crystal Structure 478
17. Chemical Reactions 493
18. The Conservation of Momentum 537
19. Newton's Laws in Sports 565
20. Levers 595
21. The Four-Stroke Cycle 621
22. Nerve Impulses 638
23. Voltmeters and Ammeters 675
24. Common Vision Problems 712

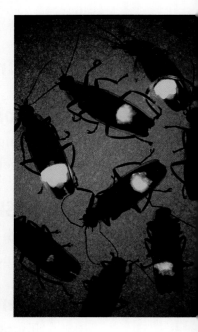

Cross-Curricular Readings

TIME Science and Society

- 2 How did life begin? ... 56
- 6 Creating Wetlands to Purify Wastewater ... 172
- 19 Air Bag Safety ... 572
- 20 Bionic People ... 600
- 21 The Heat Is On ... 626
- 22 Fire in the Forest ... 658

TIME Science and History

- 4 The Census Measures a Human Population ... 114
- 8 Quake ... 234
- 11 The Mayan Calendar ... 328
- 14 Pioneers in Radioactivity ... 426
- 17 Synthetic Diamonds ... 512

Oops! Accidents in Science

- 9 The World's Oldest Fish Story ... 264
- 12 It Came From Outer Space ... 362
- 18 What Goes Around Comes Around ... 542
- 24 Jansky's Merry-Go-Round ... 716

Science and Language Arts

- 1 The Everglades: River of Grass ... 30
- 7 Listening In ... 202
- 15 "Anansi Tries to Steal All the Wisdom in the World" ... 456
- 16 "Baring the Atom's Mother Heart" ... 484
- 23 "Aagjuuk and Sivulliit" ... 686

Science Stats

- 3 Astonishing Human Systems ... 84
- 5 Extreme Climates ... 142
- 10 Extinct! ... 296
- 13 Stars and Galaxies ... 394

DVD available as a video lab on DVD

DVD	1	Measure Using Tools	5
	2	How are people different?	37
	3	Model Blood Flow in Arteries and Veins	63
	4	How do lawn organisms survive?	93
	5	Earth Has Many Ecosystems	121
	6	What environment do houseplants need?	149
DVD	7	Reassemble an Image	181
DVD	8	Construct with Strength	209
	9	Clues to Life's Past	241
	10	Survival Through Time	271
	11	Model Rotation and Revolution	305
DVD	12	Model Crater Formation	335
	13	Why do clusters of galaxies move apart?	369
	14	Model the Unseen	403
DVD	15	Make a Model of a Periodic Pattern	433
DVD	16	Model the Energy of Electrons	463
	17	Identify a Chemical Reaction	491
	18	Motion After a Collision	521
	19	Forces and Motion	549
	20	Compare Forces	579
	21	Measuring Temperature	607
	22	Observing Electric Forces	635
	23	Magnetic Forces	665
	24	Wave Properties	693

Mini LAB

	1	Inferring from Pictures	9
	2	Observing Fruit Fly Phenotypes	46
	3	Observing a Chemical Reaction	76
DVD	4	Comparing Biotic Potential	103
	5	Comparing Fertilizers	133
	6	Modeling Freshwater Environments	164
	7	Modeling Convection Currents	195
	8	Observing Deformation	211
	9	Modeling Carbon-14 Dating	258
	10	Dating Rock Layers with Fossils	282
	11	Making Your Own Compass	308

12	Inferring Effects of Gravity	345
13	Measuring Distance in Space	388
14	Graphing Half-Life	418
15	Designing a Periodic Table	435
16	Drawing Electron Dot Diagrams	470
17	Observing the Law of Conservation of Mass	496
18	Modeling Acceleration	531
19	Measuring Force Pairs	567
20	Observing Pulleys	596
21	Observing Convection	615
22	Identifying Simple Circuits	650
23	Observing Magnetic Fields	670
24	Separating Wavelengths	710

Mini LAB — Try at Home

1	Comparing Paper Towels	18
2	Observing Gravity and Stem Growth	40
3	Observing the Gases That You Exhale	74
4	Observing Seedling Competition	99
5	Determining Soil Makeup	124
6	Modeling Rain Forest Leaves	158
7	Interpreting Fossil Data	184
8	Modeling an Eruption	220
9	Predicting Fossil Preservation	242
10	Calculating the Age of the Atlantic Ocean	292
11	Comparing the Sun and the Moon	313
12	Modeling Planets	350
13	Observing Star Patterns	371
14	Modeling the Nuclear Atom	411
16	Constructing a Model of Methane	475
17	Identifying Inhibitors	506
18	Measuring Average Speed	525
19	Observing Friction	554
20	Work and Power	583
21	Comparing Rates of Melting	614
22	Investigating the Electric Force	644
23	Assembling an Electromagnet	674
24	Refraction of Light	699

 available as a video lab on DVD

One-Page Labs

1. Battle of the Beverage Mixes 12
2. Jelly Bean Hunt 43
3. Observing Cells 72
4. Feeding Habits of Planaria 111
5. Humus Farm 129
6. Studying a Land Ecosystem 162
7. Seafloor Spreading Rates 189
8. Disruptive Eruptions 225
9. Relative Ages 256
10. Changing Species 287
11. Moon Phases and Eclipses 321
12. Planetary Orbits 341
13. Sunspots 379
14. Making a Model of the Invisible 414
15. Metals and Nonmetals 453
16. Ionic Compounds 481
17. Physical or Chemical Change? 509
18. Collisions 539
19. Balloon Races 569
20. Building the Pyramids 585
21. Heating Up and Cooling Down 618
22. Current in a Parallel Circuit 655
23. Make a Compass 672
24. Sound Waves in Matter 706

Two-Page Labs

2. Toothpick Fish 54–55
5. Where does the mass of a plant come from? 140–141
8. Seismic Waves 232–233
11. Tilt and Temperature 326–327
22. A Model for Voltage and Current 656–657
23. How does an electric motor work? 684–685
24. Bending Light 714–715

Design Your Own Labs

- **3** Does exercise affect respiration? 82–83
- **4** Population Growth in Fruit Flies 112–113
- **13** Measuring Parallax 392–393
- **14** Half-Life 424–425
- **17** Exothermic or Endothermic? 510–511
- **18** Car Safety Testing 540–541
- **19** Modeling Motion in Two Directions 570–571
- **20** Pulley Power 598–599
- **21** Comparing Thermal Insulators 624–625

Model and Invent Labs

- **9** Trace Fossils 262–263
- **12** Solar System Distance Model 360–361
- **16** Atomic Structure 482–483

Use the Internet Labs

- **1** When is the Internet the busiest? 28–29
- **6** Exploring Wetlands 170–171
- **7** Predicting Tectonic Activity 200–201
- **10** Discovering the Past 294–295
- **15** Health Risks from Heavy Metals 454–455

Activities

Applying Math

2	Percent of Offspring with Certain Traits	47
3	Lung Volume	77
5	Temperature Changes	126
6	Temperature	166
8	P-wave Travel Time	230
10	Calculating Extinction by Using Percentages	291
12	Diameter of Mars	346
14	Find Half-Lives	419
17	Conserving Mass	498
18	Speed of a Swimmer	524
18	Acceleration of a Bus	530
18	Momentum of a Bicycle	534
19	Acceleration of a Car	559
20	Calculating Work	582
20	Calculating Power	583
20	Calculating Mechanical Advantage	587
20	Calculating Efficiency	589
21	Converting to Celcius	610
22	Voltage from a Wall Outlet	649
22	Electric Power Used by a Lightbulb	652
24	Speed of Sound	698

Applying Science

1	Problem-Solving Skills	14
4	Do you have too many crickets?	101
7	How well do the continents fit together?	192
9	When did the Iceman die?	260
11	What will you use to survive on the Moon?	319
13	Are distance and brightness related?	372
15	What does *periodic* mean in the periodic table?	439
16	How does the periodic table help you identify properties of elements?	469
23	Finding the Magnetic Declination	669

Activities

INTEGRATE

Career: 50, 127, 165, 197, 317, 444, 467, 622
Chemistry: 39, 107, 187, 221, 228, 281, 383, 456, 645
Earth Science: 67, 128, 137, 160, 202, 420, 696
Environment: 15, 49, 168, 420
Health: 21, 452, 505, 682, 703
History: 75, 109, 384, 412, 507, 557, 589, 646, 681
Language Arts: 229, 352
Life Science: 30, 248, 274, 307, 423, 445, 473, 495, 551, 564, 589, 593, 616, 617
Physics: 187, 198, 257, 308, 317, 338, 340, 382, 450, 484
Social Studies: 7, 245, 534

Science Online

8, 25, 42, 52, 66, 80, 96, 102, 127, 135, 151, 167, 183, 192, 216, 221, 251, 254, 284, 289, 309, 311, 315, 323, 337, 346, 377, 382, 416, 421, 438, 452, 466, 477, 497, 503, 526, 536, 553, 564, 584, 587, 620, 640, 653, 671, 679, 708

Standardized Test Practice

34–35, 60–61, 88–89, 118–119, 146–147, 176–177, 206–207, 238–239, 268–269, 300–301, 332–333, 366–367, 398–399, 430–431, 460–461, 488–489, 516–517, 546–547, 576–577, 604–605, 630–631, 662–663, 690–691, 720–721

unit 1
Humans and Heredity

How Are Electricity & DNA Connected?

In 1831, a scientist produced electricity by passing a magnet through a coil of wire. This confirmed what hundreds of previous scientific investigations had attempted. That simple experiment began a new era—the electrical age. By 1892, it was realized that electricity could be used for heating. Two years later, electricity was first used for cooking. In 1918, electric washing machines became available, and the next year, the refrigerator appeared on the electrical scene. By this time, electricity had been accepted as the energy of the future. Gas lighting and steam engines were losing ground to the new electric lights and motors. With electricity came radios, televisions, power tools, and automated appliances. In the 1970s, a new use was discovered for electricity. Scientists discovered that an electrical current passed through a gel containing fragments of DNA could separate larger ones from smaller ones. This process, called electrophoresis (ih lek truh fuh REE sus), is now used in DNA fingerprinting.

unit projects

Visit blue.msscience.com/unit_project to find project ideas and resources. Projects include:

- **History** Become a genealogist looking for patterns of hereditary traits in your family tree.
- **Career** Research the process of DNA fingerprinting and how it has helped police find and prosecute criminals in a court of law.
- **Model** Using DNA codes, investigate a mock crime scene, and then present mock evidence to a judge and jury of your peers.

WebQuest Human Clone: Ethical Consideration provides an opportunity to explore cloning, to discover the ethical debate, and what laws govern cloning.

chapter 1

The BIG Idea
Science and technology can make lives healthier, more convenient, and safer.

SECTION 1
What is Science?
Main Idea Science is an organized way of studying things and finding answers to questions.

SECTION 2
Doing Science
Main Idea Scientists use different types of research to discover new information.

SECTION 3
Science and Technology
Main Idea The scientific discoveries often lead to new technologies and vice versa.

The Nature of Science

Science at Work

Science is going on all the time. You probably use science skills to investigate the world around you. In labs, such as the one shown, scientists use skills and tools to answer questions and solve problems.

Science Journal Describe the most interesting science activity you've done. Identify as many parts of the scientific process used in the activity as you can.

Start-Up Activities

Measure Using Tools

Ouch! That soup is hot. Your senses tell you a great deal of information about the world around you, but they can't answer every question. Scientists use tools, such as thermometers, to measure accurately. Learn more about the importance of using tools in the following lab.

1. Use three bowls. Fill one with cold water, one with lukewarm water, and the third with hot water. **WARNING:** *Make sure the hot water will not burn you.*
2. Use a thermometer to measure the temperature of the lukewarm water. Record the temperature.
3. Submerse one hand in the cold water and the other in the hot water for 2 min.
4. Put both hands into the bowl of lukewarm water. What do you sense with each hand? Record your response in your Science Journal.
5. **Think Critically** In your Science Journal, write a paragraph that explains why it is important to use tools to measure information.

Make the following foldable to help you stay focused and better understand scientists when you are reading the chapter.

STEP 1 **Draw** a mark at the midpoint of a sheet of paper along the side edge. Then **fold** the top and bottom edges in to touch the midpoint.

STEP 2 **Fold** in half from side to side.

STEP 3 **Turn** the paper vertically. **Open and cut** along the inside fold lines to form four tabs.

STEP 4 **Label** each tab.

Classify As you read the chapter, list the characteristics of the four major divisions of scientists under each tab.

 Preview this chapter's content and activities at blue.msscience.com

Get Ready to Read

Preview

① Learn It! If you know what to expect before reading, it will be easier to understand ideas and relationships presented in the text. Follow these steps to preview your reading assignments.

1. Look at the title and any illustrations that are included.
2. Read the headings, subheadings, and anything in bold letters.
3. Skim over the passage to see how it is organized. Is it divided into many parts?
4. Look at the graphics—pictures, maps, or diagrams. Read their titles, labels, and captions.
5. Set a purpose for your reading. Are you reading to learn something new? Are you reading to find specific information?

② Practice It! Take some time to preview this chapter. Skim all the main headings and subheadings. With a partner, discuss your answers to these questions.
- Which part of this chapter looks most interesting to you?
- Are there any words in the headings that are unfamiliar to you?
- Choose one of the lesson review questions to discuss with a partner.

③ Apply It! Now that you have skimmed the chapter, write a short paragraph describing one thing you want to learn from this chapter.

Target Your Reading

Reading Tip
Forming your own mental images will help you remember what you read.

Use this to focus on the main ideas as you read the chapter.

① **Before you read** the chapter, respond to the statements below on your worksheet or on a numbered sheet of paper.
- Write an **A** if you **agree** with the statement.
- Write a **D** if you **disagree** with the statement.

② **After you read** the chapter, look back to this page to see if you've changed your mind about any of the statements.
- If any of your answers changed, explain why.
- Change any false statements into true statements.
- Use your revised statements as a study guide.

Before You Read A or D		Statement	After You Read A or D
	1	Scientists often use their prior knowledge to predict an experiment's outcome.	
	2	Most scientists prefer that their discoveries remain secret.	
	3	There is just one way to approach a scientific problem.	
	4	Observation is one way to make scientific discoveries.	
	5	In a well-planned experiment, only one part is changed at a time.	
	6	Repeating an experiment is a waste of time for a scientist.	
	7	Only university graduates can be scientists.	
	8	SI units enable accurate communication among scientists.	
	9	A scientist learns nothing if an experiment does not support the hypothesis.	

Science Online
Print out a worksheet of this page at blue.msscience.com

section 1

What is science?

as you read

What You'll Learn
- **Identify** how science is a part of your everyday life.
- **Describe** what skills and tools are used in science.

Why It's Important
What and how you learn in science class can be applied to other areas of your life.

Review Vocabulary
observation: gathering information through the use of one or more senses

New Vocabulary
- science
- technology

Science in Society

When you hear the word *science*, do you think only of your science class, your teacher, and certain terms and facts? Is there any connection between what happens in science class and the rest of your life? You might have problems to solve or questions that need answers, as illustrated in **Figure 1. Science** is a way or a process used to investigate what is happening around you. It can provide possible answers to your questions.

Science Is Not New Throughout history, people have tried to find answers to questions about what was happening around them. Early scientists tried to explain things based on their observations. They used their senses of sight, touch, smell, taste, and hearing to make these observations. From the Launch Lab, you know that using only your senses can be misleading. What is cold or hot? How heavy is heavy? How much is a little? How close is nearby? Numbers can be used to describe observations. Tools, such as thermometers and metersticks, are used to give numbers to descriptions. Scientists observe, investigate, and experiment to find answers, and so can you.

Figure 1 You use scientific thinking every day to make decisions.

6 CHAPTER 1

Science as a Tool

As Luis and Midori walked into science class, they still were talking about their new history assignment. Mr. Johnson overheard them and asked what they were excited about.

"We have a special assignment—celebrating the founding of our town 200 years ago," answered Luis. "We need to do a project that demonstrates the similarities of and differences between a past event and something that is happening in our community now."

Mr. Johnson responded. "That sounds like a big undertaking. Have you chosen the two events yet?"

"We read some old newspaper articles and found several stories about a cholera epidemic here that killed ten people and made more than 50 others ill. It happened in 1871—soon after the Civil War. Midori and I think that it's like the E. coli outbreak going on now in our town," replied Luis.

"What do you know about an outbreak of cholera and problems caused by E. coli, Luis?"

"Well, Mr. Johnson, cholera is a disease caused by a bacterium that is found in contaminated water," Luis replied. "People who eat food from this water or drink this water have bad cases of diarrhea and can become dehydrated quickly. They might even die. E. coli is another type of bacterium. Some types of E. coli are harmless, but others cause intestinal problems when contaminated food and water are consumed."

"In fact," added Midori, "one of the workers at my dad's store is just getting over being sick from E. coli. Anyway, Mr. Johnson, we want to know if you can help us with the project. We want to compare how people tracked down the source of the cholera in 1871 with how they are tracking down the source of the E. coli now."

Using Science Every Day

"I'll be glad to help. This sounds like a great way to show how science is a part of everyone's life. In fact, you are acting like scientists right now," Mr. Johnson said proudly.

Luis had a puzzled look on his face, then he asked, "What do you mean? How can we be doing science? This is supposed to be a history project."

Figure 2 Newspapers, magazines, books, and the Internet are all good sources of information.

Science in Advertising
You can't prevent all illnesses. You can, however, take steps to reduce your chances of coming in contact with disease-causing organisms. Antibacterial soaps and cleansers claim to kill such organisms, but how do you know if they work? Read ads for or labels on such products. Do they include data to support their claims? Communicate what you learn to your class.

SECTION 1 What is science? **7**

Topic: Disease Control
Visit blue.msscience.com for Web links to information about disease control and the Centers for Disease Control and Prevention (CDC).

Activity Research two different diseases that the CDC have tracked down and identified in the past five years. Prepare a poster that includes the following information: symptoms, cause, cures or treatments, and locations.

Figure 3 When solving a problem, it is important to discover all background information. Different sources can provide such information.
Explain *how you would find information on a specific topic. What sources of information would you use?*

Scientists Use Clues "Well, you're acting like a detective right now. You have a problem to solve. You and Midori are looking for clues that show how the two events are similar and different. As you complete the project, you will use several skills and tools to find the clues." Mr. Johnson continued, "In many ways, scientists do the same thing. People in 1871 followed clues to track the source of the cholera epidemic and solve their problem. Today, scientists are doing the same thing by finding and following clues to track the source of the *E. coli*."

Using Prior Knowledge

Mr. Johnson asked, "Luis, how do you know what is needed to complete your project?"

Luis thought, then responded, "Our history teacher, Ms. Hernandez, said the report must be at least three pages long and have maps, pictures, or charts and graphs. We have to use information from different sources such as written articles, letters, videotapes, or the Internet. I also know that it must be handed in on time and that correct spelling and grammar count."

"Did Ms. Hernandez actually talk about correct spelling and grammar?" asked Mr. Johnson.

Midori quickly responded, "No, she didn't have to. Everyone knows that Ms. Hernandez takes points away for incorrect spelling or grammar. I forgot to check my spelling in my last report and she took off two points."

"Ah-ha! That's where your project is like science," exclaimed Mr. Johnson. "You know from experience what will happen. When you don't follow her rule, you lose points. You can predict, or make an educated guess, that Ms. Hernandez will react the same way with this report as she has with others."

Mr. Johnson continued, "Scientists also use prior experience to predict what will occur in investigations. Scientists form theories when their predictions have been well tested. A theory is an explanation that is supported by facts. Scientists also form laws, which are rules that describe a pattern in nature, like gravity."

Using Science and Technology

"Midori, you said that you want to compare how the two diseases were tracked. Like scientists, you will use skills and tools to find the similarities and differences." Mr. Johnson then pointed to Luis. "You need a variety of resource materials to find information. How will you know which materials will be useful?"

"We can use a computer to find books, magazines, newspapers, videos, and web pages that have information we need," said Luis.

"Exactly," said Mr. Johnson. "That's another way that you are thinking like scientists. The computer is one tool that modern scientists use to find and analyze data. The computer is an example of technology. **Technology** is the application of science to make products or tools that people can use. One of the big differences you will find between the way diseases were tracked in 1871 and how they are tracked now is the result of new technology."

Science Skills "Perhaps some of the skills used to track the two diseases will be one of the similarities between the two time periods," continued Mr. Johnson. "Today's doctors and scientists, like those in the late 1800s, use skills such as observing, classifying, and interpreting data. In fact, you might want to review the science skills we've talked about in class. That way, you'll be able to identify how they were used during the cholera outbreak and how they still are used today."

Luis and Midori began reviewing the science skills that Mr. Johnson had mentioned. Some of these skills used by scientists are described in the Science Skill Handbook at the back of this book. The more you practice these skills, the better you will become at using them.

Figure 4 Computers are one example of technology. Schools and libraries often provide computers for students to do research and word processing.

Inferring from Pictures

Procedure
1. Study the two pictures to the left. Write your observations in your **Science Journal**.
2. Make and record inferences based on your observations.
3. Share your inferences with others in your class.

Analysis
1. Analyze your inferences. Are there other explanations for what you observed?
2. Why must you be careful when making inferences?

Observation and Measurement Think about the Launch Lab at the beginning of this chapter. Observing, measuring, and comparing and contrasting are three skills you used to complete the activity. Scientists probably use these skills more than other people do. You will learn that sometimes observation alone does not provide a complete picture of what is happening. To ensure that your data are useful, accurate measurements must be taken, in addition to making careful observations.

Reading Check *What are three skills commonly used in science?*

Luis and Midori want to find the similarities and differences between the disease-tracking techniques used in the late 1800s and today. They will use the comparing and contrasting skill. When they look for similarities among available techniques, they compare them. Contrasting the available techniques is looking for differences.

Communication in Science

What do scientists do with their findings? The results of their observations, experiments, and investigations will not be of use to the rest of the world unless they are shared. Scientists use several methods to communicate their observations.

Results and conclusions of experiments often are reported in one of the thousands of scientific journals or magazines that are published each year. Some of these publications are shown in **Figure 5.** Scientists spend a large part of their time reading journal articles. Sometimes, scientists discover information in articles that might lead to new experiments.

Figure 5 Scientific publications enable scientists around the world to learn about the latest research. Papers are submitted to journals. Other scientists review them before they are published.
Explain *why other scientists review papers before they are published.*

Science Journal Another method to communicate scientific data and results is to keep a Science Journal. Observations and plans for investigations can be recorded, along with the step-by-step procedures that were followed. Listings of materials and drawings of how equipment was set up should be in a journal, along with the specific results of an investigation. You should record mathematical measurements or formulas that were used to analyze the data. Problems that occurred and questions that came up during the investigation should be noted, as well as any possible solutions. Your data might be summarized in the form of tables, charts, or graphs, or they might be recorded in a paragraph. Remember that it's always important to use correct spelling and grammar in your Science Journal.

Figure 6 Your Science Journal is used to record and communicate your findings. It might include graphs, tables, and illustrations.

Reading Check *What are some ways to summarize data from an investigation?*

You will be able to use your Science Journal, as illustrated in **Figure 6,** to communicate your observations, questions, thoughts, and ideas as you work in science class. You will practice many of the science skills and become better at identifying problems. You will learn to plan investigations and experiments that might solve these problems.

section 1 review

Summary

Science in Society
- People use their senses to observe their surroundings.
- Scientific process is used to solve problems and answer questions.

Using Prior Knowledge
- Scientists use prior knowledge to predict the outcome of investigations.
- After hypotheses have been tested many times, theories are formed.

Using Science and Technology
- Journals, newspapers, books, and the Internet can be useful sources of information.
- Observation, classification, and interpretation are important scientific skills.

Communication in Science
- Scientists communicate their observations, experiments, and results with others.

Self Check

1. **Infer** why scientists use tools, such as thermometers and metersticks, when they make observations.
2. **Determine** what some skills used in science are. Name one science skill that you have used today.
3. **Evaluate** one example of technology. How is technology different from science?
4. **Think Critically** Why is a Science Journal used to record data? What are three different ways you could record or summarize data in your Science Journal?

Applying Skills

5. **Compare and Contrast** Sometimes you use your senses to make observations to find the answer to a question. Other times you use tools and measurements to provide answers. Compare and contrast these two methods of answering scientific questions.
6. **Communicate** In your Science Journal, record five things you observe in or about your classroom.

Science Online blue.msscience.com/self_check_quiz

Battle of the Beverage Mixes

You can use science skills to answer everyday questions or to solve problems. For example, you might know that the cheapest brand of a product is not always the best value. In this lab, you will test one aspect, or quality, of a product.

◉ Real-World Question
Which brand of powdered beverage mix dissolves best?

Goals
- **Determine** which brand of powdered beverage mix dissolves best using science skills.

Materials
weighing paper
50-mL graduated cylinder
powdered beverage mixes (3 or 4)
triple-beam balance
*electronic balance
250-mL beaker
water
spoon

*Alternate materials

Safety Precautions

WARNING: *Never eat or drink anything during science experiments.*

◉ Procedure
1. Copy the following data table in your Science Journal.

Beverage-Mix Data	
Beverage Mix	Mass of Dissolved Powder (g)
Do not write in this book.	

2. Using the graduated cylinder, measure 50 mL of water and pour the water into the beaker.
3. **Measure** 20 g of powder from one of the brands of beverage powder.
4. Gradually add the powder to the water. Stir the mixture after each time that you add more powder. Stop adding powder when undissolved powder begins to accumulate at the bottom of the beaker.
5. **Measure** the mass of the remaining powder. Subtract this number from 20 g to find the amount of powder that was dissolved. Record your answer in your data table.
6. Empty the beverage mix into the sink, rinse out your beaker, and repeat steps 2 through 5 for the other brands of beverage mix.

◉ Conclude and Apply
1. **Identify** the beverage-mix powder that dissolved best in the water.
2. **Infer** which beverage-mix brand would taste the best, based on the data you collected.
3. **List** the science skills you used during this experiment to help you determine the best beverage mix. Which beverage-mix brand would you buy?
4. **Review** promotional pamphlets. Make a list of inferences about the claims presented.

𝒞ommunicating Your Data
Write and perfom a 15-second ad about why people should buy the best-dissolving beverage-mix brand. **For more help, refer to the** Science Skill Handbook.

section 2

Doing Science

Solving Problems

When Luis and Midori did their project, they were answering a question. However, there is more than one way to answer a question or solve a scientific problem. Every day, scientists work to solve scientific problems. Although the investigation of each problem is different, scientists use some steps in all investigations.

Identify the Problem Scientists first make sure that everyone working to solve the problem has a clear understanding of the problem. Sometimes, scientists find that the problem is easy to identify or that several problems need to be solved. For example, before a scientist can find the source of a disease, the disease must be identified correctly.

How can the problem be solved? Scientists know that scientific problems can be solved in different ways. Two of the methods used to answer questions are descriptive research and experimental research design. **Descriptive research** answers scientific questions through observation. When Luis and Midori gathered information to learn about cholera and *E. coli*, they performed descriptive research. **Experimental research design** is used to answer scientific questions by testing a hypothesis through the use of a series of carefully controlled steps. **Scientific methods,** like the one shown in **Figure 7,** are ways, or steps to follow, to try to solve problems. Different problems will require different scientific methods to solve them.

as you read

What **You'll Learn**
- **Examine** the steps used to solve a problem in a scientific way.
- **Explain** how a well-designed investigation is developed.

Why **It's Important**
Using scientific methods and carefully thought-out experiments can help you solve problems.

🔄 Review Vocabulary
experiment: a set of controlled steps carried out to discover, test, or demonstrate something

New Vocabulary
- descriptive research
- experimental research design
- scientific methods
- model
- hypothesis
- independent variable
- dependent variable
- constant
- control

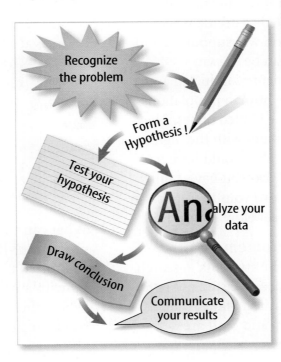

Figure 7 This poster shows one way to solve problems using scientific methods.

SECTION 2 Doing Science **13**

Descriptive Research

Some scientific problems can be solved, or questions answered, by using descriptive research. Descriptive research is based mostly on observations. What observations can you make about the objects in **Figure 8?** Descriptive research can be used in investigations when experiments would be impossible to perform. For example, a London doctor, Dr. John Snow, tracked the source of a cholera epidemic in the 1800s by using descriptive research. Descriptive research usually involves the following steps.

Figure 8 Items can be described by using words and numbers. **Describe** *these objects using both words and numbers.*

State the Research Objective This is the first step in solving a problem using descriptive research. A research objective is what you want to find out, or what question you would like to answer. Luis and Midori might have said that their research objective was "to find out how the sources of the cholera epidemic and *E. coli* epidemic were tracked." Dr. John Snow might have stated his research objective as "finding the source of the cholera epidemic in London."

Applying Science

Problem-Solving Skills

Drawing Conclusions from a Data Table

During an investigation, data tables often are used to record information. The data can be evaluated to decide whether or not the prediction was supported and then conclusions can be drawn.

A group of students conducted an investigation of the human populations of some states in the United States. They predicted that the states with the highest human population also would have the largest area of land. Do you have a different prediction? Record your prediction in your Science Journal before continuing.

Identifying the Problem

The results of the students' research are shown in this chart. Listed are several states in the United States, their human population, and land area.

State Population and Size

State	Human Population	Area (km^2)
New York	18,976,457	122,284
New Jersey	8,414,350	19,210
Massachusetts	6,349,097	20,306
Maine	1,274,923	79,932
Montana	902,195	376,978
North Dakota	642,200	178,647
Alaska	626,902	1,481,350

Source: United States Census Bureau, United States Census 2000

1. What can you conclude about your prediction? If your prediction is not supported by the data, can you come up with a new prediction? Explain.
2. What other research could be conducted to support your prediction?

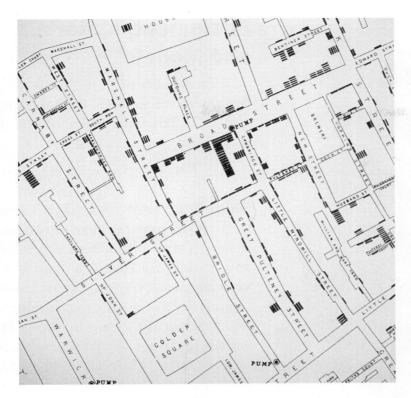

Describe the Research Design How will you carry out your investigation? What steps will you use? How will the data be recorded and analyzed? How will your research design answer your question? These are a few of the things scientists think about when they design an investigation using descriptive research. An important part of any research design is safety. Check with your teacher several times before beginning any investigation.

Reading Check *What are some questions to think about when planning an investigation?*

Dr. John Snow's research design included the map shown above. The map shows where people with cholera had lived, and where they obtained their water. He used these data to predict that the water from the Broad Street pump, shown in **Figure 9,** was the source of the contamination.

Eliminate Bias It's a Saturday afternoon. You want to see a certain movie, but your friends do not. To persuade them, you tell them about a part of the show that they will find interesting. You give only partial information so they will make the choice you want. Similarly, scientists might expect certain results. This is known as bias. Good investigations avoid bias. One way to avoid bias is to use careful numerical measurements for all data. Another type of bias can occur in surveys or groups that are chosen for investigations. To get an accurate result, you need to use a random sample.

Figure 9 Each mark on Dr. Snow's map shows where a cholera victim lived. Dr. Snow had the water-pump handle removed, and the cholera epidemic ended.

The Clean Water Act The U.S. Congress has passed several laws to reduce water pollution. The 1986 Safe Drinking Water Act is a law to ensure that drinking water in the United States is safe. The 1987 Clean Water Act gives money to the states for building sewage- and wastewater-treatment facilities. Find information about a state or local water quality law and share your findings with the class.

SECTION 2 Doing Science **15**

Equipment, Materials, and Models

When a scientific problem is solved by descriptive research, the equipment and materials used to carry out the investigation and analyze the data are important.

Selecting Your Materials Scientists try to use the most up-to-date materials available to them. If possible, you should use scientific equipment such as balances, spring scales, microscopes, and metric measurements when performing investigations and gathering data. Calculators and computers can be helpful in evaluating or displaying data. However, you don't have to have the latest or most expensive materials and tools to conduct good scientific investigations. Your investigations can be completed successfully and the data displayed with materials found in your home or classroom, like paper, colored pencils, or markers. An organized presentation of data, like the one shown in **Figure 10,** is as effective as a computer graphic or an extravagant display.

Figure 10 This presentation neatly and clearly shows experimental design and data.
List *the aspects of this display that make it easy to follow.*

Using Models One part of carrying out the investigative plan might include making or using scientific models. In science, a **model** represents things that happen too slowly, too quickly, or are too big or too small to observe directly. Models also are useful in situations in which direct observation would be too dangerous or expensive.

Dr. John Snow's map of the cholera epidemic was a model that allowed him to predict possible sources of the epidemic. Today, people in many professions use models. Many kinds of models are made on computers. Graphs, tables, and spreadsheets are models that display information. Computers can produce three-dimensional models of a microscopic bacterium, a huge asteroid, or an erupting volcano. They are used to design safer airplanes and office buildings. Models save time and money by testing ideas that otherwise are too small, too large, or take too long to build.

Table 1 Common SI Measurements

Measurement	Unit	Symbol	Equal to
Length	1 millimeter	mm	0.001 (1/1,000) m
	1 centimeter	cm	0.01 (1/100) m
	1 meter	m	100 cm
	1 kilometer	km	1,000 m
Liquid volume	1 milliliter	mL	0.001 L
	1 liter	L	1,000 mL
Mass	1 milligram	mg	0.001 g
	1 gram	g	1,000 mg
	1 kilogram	kg	1,000 g
	1 tonne	t	1,000 kg = 1 metric ton

Scientific Measurement Scientists around the world use a system of measurements called the International System of Units, or SI, to make observations. This allows them to understand each other's research and compare results. Most of the units you will use in science are shown in **Table 1.** Because SI uses certain metric units that are based on units of ten, multiplication and division are easy to do. Prefixes are used with units to change their names to larger or smaller units. See the Reference Handbook to help you convert English units to SI. **Figure 11** shows equipment you can use to measure in SI.

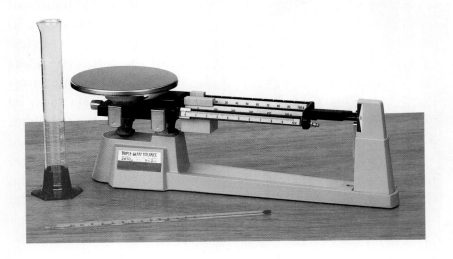

Figure 11 Some of the equipment used by scientists is shown here. A graduated cylinder is used to measure liquid volume. Mass is measured with a balance. A scientist would use a thermometer with the Celsius scale to measure temperature.

Mini LAB

Comparing Paper Towels

Procedure
1. Make a data table similar to the one in **Figure 12**.
2. Cut a 5-cm by 5-cm square from each of **three brands of paper towel.** Lay each piece on a level, smooth, waterproof surface.
3. Add one drop of water to each square.
4. Continue to add drops until the piece of paper towel no longer can absorb the water.
5. Tally your observations in your data table and graph your results.
6. Repeat steps 2 through 5 three more times.

Analysis
1. Did all the squares of paper towels absorb equal amounts of water?
2. If one brand of paper towel absorbs more water than the others, can you conclude that it is the towel you should buy? Explain.
3. Which scientific methods did you use to compare paper towel absorbency?

Try at Home

Figure 12 Data tables help you organize your observations and results.

Paper Towel Absorbency (Drops of Water Per Sheet)			
Trial	Brand A	Brand B	Brand C
1			
2		Do not write in this book.	
3			
4			

Data

In every type of scientific research, data must be collected and organized carefully. When data are well organized, they are easier to interpret and analyze.

Designing Your Data Tables A well-planned investigation includes ways to record results and observations accurately. Data tables, like the one shown in **Figure 12,** are one way to do this. Most tables have a title that tells you at a glance what the table is about. The table is divided into columns and rows. These are usually trials or characteristics to be compared. The first row contains the titles of the columns. The first column identifies what each row represents.

As you complete a data table, you will know that you have the information you need to analyze the results of the investigation accurately. It is wise to make all of your data tables before beginning the experiment. That way, you will have a place for all of your data as soon as they are available.

Analyze Your Data
Your investigation is over. You breathe a sigh of relief. Now you have to figure out what your results mean. To do this, you must review all of the recorded observations and measurements. Your data must be organized to analyze them. Charts and graphs are excellent ways to organize data. You can draw the charts and graphs, like the ones in **Figure 13,** or use a computer to make them.

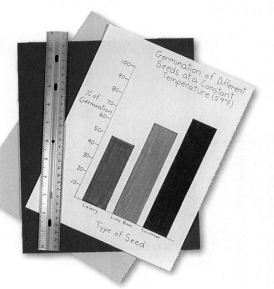

Figure 13 Charts and graphs can help you organize and analyze your data.

18 CHAPTER 1 The Nature of Science

Draw Conclusions

After you have organized your data, you are ready to draw a conclusion. Do the data answer your question? Was your prediction supported? You might be concerned if your data are not what you expected, but remember, scientists understand that it is important to know when something doesn't work. When looking for an antibiotic to kill a specific bacteria, scientists spend years finding out which antibiotics will work and which won't. Each time scientists find that a particular antibiotic doesn't work, they learn some new information. They use this information to help make other antibiotics that have a better chance of working. A successful investigation is not always the one that comes out the way you originally predicted.

Communicating Your Results Every investigation begins because a problem needs to be solved. Analyzing data and drawing conclusions are the end of the investigation. However, they are not the end of the work a scientist does. Usually, scientists communicate their results to other scientists, government agencies, private industries, or the public. They write reports and presentations that provide details on how experiments were carried out, summaries of the data, and final conclusions. They can include recommendations for further research. Scientists usually publish their most important findings.

Figure 14 Communicating experimental results is an important part of the laboratory experience.

Reading Check *Why is it important for scientists to communicate their data?*

Just as scientists communicate their findings, you will have the chance to communicate your data and conclusions to other members of your science class, as shown in **Figure 14.** You can give an oral presentation, create a poster, display your results on a bulletin board, prepare computer graphics, give a multimedia presentation, or talk with other students or your teacher. You will share with other groups the charts, tables, and graphs that show your data. Your teacher, or other students, might have questions about your investigation or your conclusions. Organized data and careful analysis will enable you to answer most questions and to discuss your work confidently. Analyzing and sharing data are important parts of descriptive and experimental research, as shown in **Figure 15.**

SECTION 2 Doing Science **19**

NATIONAL GEOGRAPHIC VISUALIZING DESCRIPTIVE AND EXPERIMENTAL RESEARCH

Figure 15

Scientists use a series of steps to solve scientific problems. Depending on the type of problem, they may use descriptive research or experimental research with controlled conditions. Several of the research steps involved in determining water quality at a wastewater treatment plant are shown here.

A Gathering background information is an important first step in descriptive and experimental research.

B Some questions can be answered by descriptive research. Here, the scientists make and record observations about the appearance of a water sample.

C Some questions can be answered by experimentation. These scientists collect a wastewater sample for testing under controlled conditions in the laboratory.

D Careful analysis of data is essential after completing experiments and observations. The technician at right uses computers and other instruments to analyze data.

20 CHAPTER 1 The Nature of Science

Experimental Research Design

Another way to solve scientific problems is through experimentation. Experimental research design answers scientific questions by observation of a controlled situation. Experimental research design includes several steps.

Form a Hypothesis A **hypothesis** (hi PAH thuh sus) is a prediction, or statement, that can be tested. You use your prior knowledge, new information, and any previous observations to form a hypothesis.

Variables In well-planned experiments, one factor, or variable, is changed at a time. This means that the variable is controlled. The variable that is changed is called the **independent variable.** In the experiment shown below, the independent variable is the amount or type of antibiotic applied to the bacteria. A **dependent variable** is the factor being measured. The dependent variable in this experiment is the growth of the bacteria, as shown in **Figure 16.**

To test which of two antibiotics will kill a type of bacterium, you must make sure that every variable remains the same but the type of antibiotic. The variables that stay the same are called **constants.** For example, you cannot run the experiments at two different room temperatures, for different lengths of time, or with different amounts of antibiotics.

Figure 16 In this experiment, the effect of two different antibiotics on bacterial growth was tested. The type of antibiotic is the independent variable.

At the beginning of the experiment, dishes A and B of bacteria were treated with different antibiotics. The control dish did not receive any antibiotic.

The results of the experiment are shown. All factors were constant except the type of antibiotic applied.
Draw a conclusion *about the effects of these antibiotics on bacteria based on these photographs.*

SECTION 2 Doing Science

Identify Controls Your experiment will not be valid unless a control is used. A **control** is a sample that is treated like the other experimental groups except that the independent variable is not applied to it. In the experiment with antibiotics, your control is a sample of bacteria that is not treated with either antibiotic. The control shows how the bacteria grow when left untreated by either antibiotic.

Figure 17 Check with your teacher several times as you plan your experiment. **Determine** *why you should check with your teacher several times.*

Reading Check *What is an experimental control?*

You have formed your hypothesis and planned your experiment. Before you begin, you must give a copy of it to your teacher, who must approve your materials and plans before you begin, as shown in **Figure 17.** This is also a good way to find out whether any problems exist in how you proposed to set up the experiment. Potential problems might include health and safety issues, length of time required to complete the experiment, and the cost and availability of materials.

Once you begin the experiment, make sure to carry it out as planned. Don't skip or change steps in the middle of the process. If you do, you will have to begin the experiment again. Also, you should record your observations and complete your data tables in a timely manner. Incomplete observations and reports result in data that are difficult to analyze and threaten the accuracy of your conclusions.

Number of Trials Experiments done the same way do not always have the same results. To make sure that your results are valid, you need to conduct several trials of your experiment. Multiple trials mean that an unusual outcome of the experiment won't be considered the true result. For example, if another substance is spilled accidentally on one of the containers with an antibiotic, that substance might kill the bacteria. Without results from other trials to use as comparisons, you might think that the antibiotic killed the bacteria. The more trials you do using the same methods, the more likely it is that your results will be reliable and repeatable. The number of trials you choose to do will be based on how much time, space, and material you have to complete the experiment.

Analyze Your Results After completing your experiment and obtaining all of your data, it is time to analyze your results. Now you can see if your data support your hypothesis. If the data do not support your original hypothesis, you can still learn from the experiment. Experiments that don't work out as you had planned can still provide valuable information. Perhaps your original hypothesis needs to be revised, or your experiment needs to be carried out in a different way. Maybe more background information is available that would help. In any case, remember that professional scientists, like those shown in **Figure 18,** rarely have results that support their hypothesis without completing numerous trials first.

After your results are analyzed, you can communicate them to your teacher and your class. Sharing the results of experiments allows you to hear new ideas from other students that might improve your research. Your results might contain information that will be helpful to other students.

In this section you learned the importance of scientific methods—steps used to solve a problem. Remember that some problems are solved using descriptive research, and others are solved through experimental research.

Figure 18 These scientists might work for months or years to find the best experimental design to test a hypothesis.

section 2 review

Summary

Solving Problems
- Scientific methods are the steps followed to solve a problem.
- Descriptive research is used when experiments are impossible to use.

Equipment, Materials, and Models
- Models are important tools in science.
- The International System of Units (SI) is used to take measurements.
- Data is collected, recorded, and organized.

Draw Conclusions
- Scientists look for trends in their data, then communicate their findings.

Experimental Research Design
- Experiments start with a hypothesis.
- Variables are factors that are changed. Controls are samples that are not changed.
- Conclusions are drawn. Research is communicated to other scientists.

Self Check

1. **Explain** why scientists use models. Give three examples of models.
2. **Define** the term *hypothesis.*
3. **List** the three steps scientists might use when designing an investigation to solve a problem.
4. **Determine** why it is important to identify carefully the problem to be solved.
5. **Measure** Use a meterstick to measure the length of your desktop in meters, centimeters, and millimeters.
6. **Think Critically** The data that you gathered and recorded during an experiment do not support your original hypothesis. Explain why your experiment is not a failure.

Applying Math

7. **Use Percentages** A town of 1,000 people is divided into five areas, each with the same number of people. Use the data below to make a bar graph showing the number of people ill with cholera in each area. *Area: A–50%; B–5%; C–10%; D–16%; E–35%.*

section 3
Science and Technology

as you read

What You'll Learn
- **Determine** how science and technology influence your life.
- **Analyze** how modern technology allows scientific discoveries to be communicated worldwide.

Why It's Important
Modern communication systems enable scientific discoveries and information to be shared with people all over the world.

Review Vocabulary
computer: an electrical device that can be programmed to store, retrieve, and process data

New Vocabulary
- information technology

Science in Your Daily Life

You have learned how science is useful in your daily life. Doing science means more than just completing a science activity, reading a science chapter, memorizing vocabulary words, or following a scientific method to find answers.

Scientific Discoveries

Science is meaningful in other ways in your everyday life. New discoveries constantly lead to new products that influence your lifestyle or standard of living, such as those shown in **Figure 19**. For example, in the last 100 years, technological advances have enabled entertainment to move from live stage shows to large movie screens. Now DVDs enable users to choose a variety of options while viewing a movie. Do you want to hear English dialogue with French subtitles or Spanish dialogue with English subtitles? Do you want to change the ending? You can do it all from your chair by using the remote control.

Figure 19 New technology has changed the way people work and relax.
Identify which of the technologies in the photo you have used.

24 CHAPTER 1 The Nature of Science

Technological Advances Technology also makes your life more convenient. Hand-held computers can be carried in a pocket. Foods can be prepared quickly in microwave ovens, and hydraulic tools make construction work easier and faster. A satellite tracking system in your car can give you verbal and visual directions to a destination in an unfamiliar city.

New discoveries influence other areas of your life as well, including your health. Technological advances, like the ones shown in **Figure 20,** help many people lead healthier lives. A disease might be controlled by a skin patch that releases a constant dose of medicine into your body. Miniature instruments enable doctors to operate on unborn children and save their lives. Bacteria also have been engineered to make important drugs such as insulin for people with diabetes.

Figure 20 Modern medical technology helps people have better health. The physician is studying a series of X rays. New, more complete ways of seeing internal problems helps to solve them.

Reading Check *What new scientific discoveries have you used?*

Science—The Product of Many

New scientific knowledge can mean that old ways of thinking or doing things are challenged. Aristotle, an ancient Greek philosopher, classified living organisms into plants and animals. This system worked until new tools, such as the microscope, enabled scientists to study organisms in greater detail. The new information changed how scientists viewed the living world. The current classification system will be used only as long as it continues to answer questions scientists have or until a new discovery enables them to look at information in a different way.

Topic: Student Scientists
Visit blue.msscience.com for Web links to information about students who have made scientific discoveries or invented new technologies.

Activity Select one of the student scientists you read about. Work with a partner and prepare an interview where one of you is the interviewer and the other is the student scientist.

SECTION 3 Science and Technology

Figure 21 Science and technology are the results of many people's efforts.

Who practices science? Scientific discoveries have never been limited to people of one race, sex, culture, or time period, or to professional scientists, as shown in **Figure 21**. In fact, students your age have made some important discoveries.

Sarita M. James was a teenager when she developed a system that enables computers to recognize human speech easily.

Stephen Hawking, a physicist, studies the universe and black holes.

Grace Murray Hopper, a mathematician and software developer, helped pioneer the computer field.

Fred Begay is a physicist who studies ways to produce heat energy without harming the environment.

Ellen Ochoa is an inventor and an astronaut in NASA's space shuttle program.

Daniel Hale Williams performed the first open-heart surgery and founded a hospital.

Use of Scientific Information Science provides new information every day that people use to make decisions. A new drug can be found or a new way to produce electricity can be developed. However, science cannot decide whether the new information is good or bad, moral or immoral. People decide whether the new information is used to help or harm the world and its inhabitants. The Internet quickly spreads word of new discoveries. New knowledge and technology brought about by these discoveries are shared by people in all countries. Any information gathered from the Internet must be checked carefully for accuracy.

Looking to the Future

Midori and Luis discovered that technology has changed how modern scientists track the source of a disease. New information about bacteria and modern tools, such as those shown in **Figure 22,** help identify specific types of these organisms. Computers are used to model how the bacteria kill healthy cells or which part of a population the bacteria will infect. Today's scientists use cellular phones and computers to communicate with each other. This **information technology** has led to the globalization, or worldwide distribution, of information.

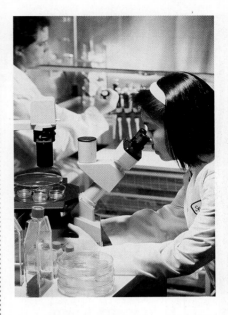

Figure 22 Modern laboratories enable scientists to track the source of a disease or solve many other scientific problems.

section 3 review

Summary

Science in Your Daily Life
- New discoveries lead to new technologies that make your life more convenient.
- Technological advances help many people lead healthier lives.

Science—The Product of Many
- New information and discoveries change how scientists view the world.
- Discoveries aren't limited to one gender, race, culture, or time period.
- The Internet enables information to spread quickly but it must be checked carefully for accuracy.
- Computers are used to model different situations in science.
- Information technology has led to the worldwide distribution of information.

Self Check

1. **Identify** one way that science or technology has improved your health.
2. **Infer** what might cause scientists to change a 100-year-old theory.
3. **List** five ways that scientists are able to communicate their discoveries.
4. **Describe** an advance in technology that makes your life more enjoyable. What discoveries contributed to this technology?
5. **Think Critically** Explain why modern communications systems are important to scientists worldwide.

Applying Skills

6. **Use a Word Processor** Research the life of a famous scientist. Find at least two sources for your information. Take notes on ten facts about the scientist and use a word processing program to write a short biography.

LAB Use the Internet

When is the Internet the busiest?

Goals
- **Observe** when you, your friends, or your family use the Internet.
- **Research** how to measure the speed of the Internet.
- **Identify** the times of day when the Internet is the busiest in different areas of the country.
- **Graph** your findings and communicate them to other students.

Data Source

Science Online

Visit **blue.msscience.com/ internet_lab** for more information on how to measure the speed of the Internet, when the Internet is busiest, and data from other students.

Real-World Question

Using the Internet, you can get information any time from practically anywhere in the world. It has been called the "information superhighway." But does the Internet ever get traffic jams like real highways? Is the Internet busier at certain times? How long does it take data to travel across the Internet at different times of the day?

Make a Plan

1. **Observe** when you, your family, and your friends use the Internet. Do you think that everyone in the world uses the Internet during the same times?
2. How are you going to measure the speed of the Internet? Research different factors that might affect the speed of the Internet. What are your variables?
3. How many times are you going to measure the speed of the Internet? What times of day are you going to gather your data?

28   **CHAPTER 1**   The Nature of Science

Using Scientific Methods

▶ Follow Your Plan

1. Make sure your teacher approves your plan before you start.
2. Visit the link shown below. Click on the Web Links button to view links that will help you do this activity.
3. Complete your investigation as planned.
4. **Record** all of your data in your Science Journal.
5. **Share** your data by posting it at the link shown below.

▶ Analyze Your Data

1. **Record** in your Science Journal what time of day you found it took the most time to send data over the Internet.
2. **Compare** your results with those of other students around the country. In which areas did data travel the most quickly?

▶ Conclude and Apply

1. **Compare** your findings to those of your classmates and other data that were posted at the link shown below. When is the Internet the busiest in your area? How does that compare to different areas of the country?
2. **Infer** what factors could cause different results in your class.
3. **Predict** how you think your data would be affected if you had performed this experiment during a different time of the year, like the winter holidays.

Communicating Your Data

Find this lab using the link below. **Post** your data in the table provided. Combine your data with those of other students and plot the combined data on a map to recognize patterns in internet traffic.

blue.msscience.com/internet_lab

Science and Language Arts

The Everglades: River of Grass
by Marjory Stoneman Douglas

In this passage, Douglas writes about Lake Okeechobee, the large freshwater lake that lies in the southern part of Florida, north of the Everglades. A dike is an earthen wall usually built to protect against floods.

Something had to be done about the control of Okeechobee waters in storms. . . . A vast dike was constructed from east to south to west of the lake, within its average rim.[1] Canal gates were opened in it. It rises now between the lake itself and all those busy towns. . . .

To see the vast pale water you climb the levee[2] and look out upon its emptiness, hear the limpkins[3] crying among the islands of reeds in the foreground, and watch the wheeling creaking sea gulls flying about a man cutting bait in a boat. . . .

From the lake the control project extended west, cutting a long ugly canal straight through the green curving jungle and the grove-covered banks of Caloosahatchee [River].

1 "Average rim" refers to the average location of the southern bank of the lake. Before the dike was built, heavy rains routinely caused Lake Okeechobee to overflow, emptying water over its southern banks into the Everglades. The overflowing water would carry silt and soil toward the southern banks of the lake, causing the southern banks to vary in size and location.

2 dike

3 waterbirds

Understanding Literature

Nonfiction Nonfiction stories are about real people, places, and events. Nonfiction includes autobiographies, biographies, and essays, as well as encyclopedias, history and science books, and newspaper and magazine articles. How can you judge the accuracy of the information?

Respond to the Reading

1. How would you verify facts contained in this passage such as the construction of the dike and its location?
2. What hints does the author give you about her opinion of the dike-building project?
3. **Linking Science and Writing** Write a one-page nonfiction account of your favorite outdoor place.

Because nonfiction is based upon real life, nonfiction writers must research their subjects thoroughly. Author Marjory Stoneman Douglas relied upon her own observations as a long-time resident of Florida. She also conducted scientific investigation when she thoroughly researched the history of the Florida Everglades. *The Everglades: River of Grass* brought the world's attention to the need to preserve the Everglades because of its unique ecosystems.

chapter 1 Study Guide

Reviewing Main Ideas

Section 1 What is science?

1. Science is a process that can be used to solve problems or answer questions. Communication is an important part of all aspects of science.
2. Scientists use tools to measure.
3. Technology is the application of science to make tools and products you use each day. Computers are a valuable technological tool.

Section 2 Doing Science

1. No one scientific method is used to solve all problems. Organization and careful planning are important when trying to solve any problem.
2. Scientific questions can be answered by descriptive research or experimental research.
3. Models save time and money by testing ideas that are too difficult to build or carry out. Models cannot completely replace experimentation.

4. A hypothesis is an idea that can be tested. Sometimes experiments don't support the original hypothesis, and a new hypothesis must be formed.
5. In a well-planned experiment, there is a control and only one variable is changed at a time. All other factors are kept constant.

Section 3 Science and Technology

1. Science is part of everyone's life. New discoveries lead to new technology and products.
2. Science continues to challenge old knowledge and ways of doing things. Old ideas are kept until new discoveries prove them wrong.
3. People of all races, ages, sexes, cultures, and professions practice science.
4. Modern communication assures that scientific information is spread around the world.

Visualizing Main Ideas

Copy and complete the following concept map with steps to solving a problem.

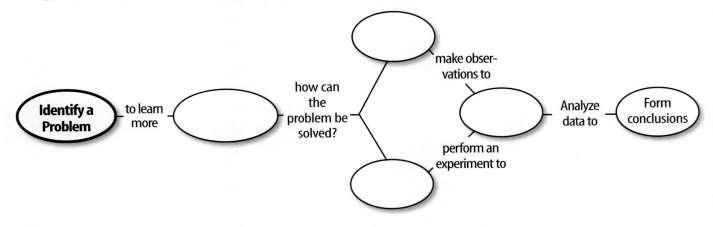

chapter 1 Review

Using Vocabulary

constant p. 21
control p. 22
dependent variable p. 21
descriptive research p. 13
experimental research design p. 13
hypothesis p. 21
independent variable p. 21
Information technology p. 27
model p. 16
science p. 6
scientific methods p. 13
technology p. 9

Match each phrase with the correct vocabulary word from the list.

1. the factor being measured in an experiment
2. a statement that can be tested
3. use of knowledge to make products
4. sample treated like other experimental groups except variable is not applied
5. steps to follow to solve a problem
6. a variable that stays the same during every trial of an experiment
7. the variable that is changed in an experiment

Checking Concepts

Choose the word or phrase that best answers the question.

8. To make sure experimental results are valid, which of these procedures must be followed?
   A) conduct multiple trials
   B) pick two hypotheses
   C) add bias
   D) communicate uncertain results

9. Predictions about what will happen can be based on which of the following?
   A) controls
   B) technology
   C) prior knowledge
   D) number of trials

10. Which of the following is the greatest concern for scientists using the Internet?
    A) speed
    B) availability
    C) language
    D) accuracy

11. In an experiment on bacteria, using different amounts of antibiotics is an example of which of the following?
    A) control
    B) hypothesis
    C) bias
    D) variable

12. Computers are used in science to do which of the following processes?
    A) analyze data
    B) make models
    C) communicate with other scientists
    D) all of the above

13. If you use a computer to make a three-dimensional picture of a building, it is an example of which of the following?
    A) model
    B) hypothesis
    C) control
    D) variable

14. When scientists make a prediction that can be tested, what skill is being used?
    A) hypothesizing
    B) inferring
    C) taking measurements
    D) making models

15. Which of the following is the first step toward finding a solution?
    A) analyze data
    B) draw a conclusion
    C) identify the problem
    D) test the hypothesis

16. Which of the following terms describes a variable that does not change in an experiment?
    A) hypothesis
    B) dependent
    C) constant
    D) independent

17. Carmen did an experiment to learn whether fish grew larger in cooler water. Once a week she weighed the fish and recorded the data. What could have improved her experiment?
    A) setting up a control tank
    B) weighing fish daily
    C) using a larger tank
    D) measuring the water temperature

chapter 1 Review

Thinking Critically

18. **Infer** why it is important to record data as they are collected.

19. **Compare and contrast** analyzing data and drawing conclusions.

20. **Explain** the advantage of eliminating bias in experiments.

21. **Determine** why scientists collect information about what is already known when trying to solve a problem.

22. **Recognize Cause and Effect** If three variables were changed at one time, what would happen to the accuracy of the conclusions made for an experiment?

Use the photo below to answer question 23.

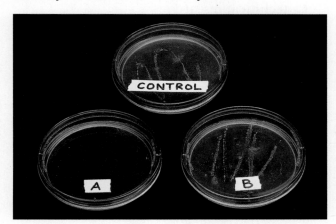

23. **Interpret** You applied two different antibiotics to two bacteria samples. The control bacteria sample did not receive any antibiotics. Two of the bacteria samples grew at the same rate. How could you interpret your results?

Performance Activities

24. **Poster** Create a poster showing steps in a scientific method. Use creative images to show the steps to solving a scientific problem.

Applying Math

Use the table below to answer question 25.

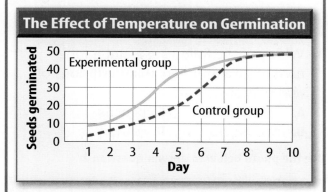

The Effect of Temperature on Germination

25. **Seed Germination** A team of student scientists measured the number of radish seeds that germinated over a 10-day period. The control group germinated at 20°C and the experimental group germinated at 25°C. According to the graph below, how many more experimental seeds than control seeds had germinated by day 5?

26. **SI Measurements** You have collected a sample of pond water to study in the lab. Your 1-L container is about half full. About how many milliliters of water have you collected? Refer to **Table 1** in this chapter for help.

Use the table below to answer question 27.

Disease Victims	
Age Group (years)	Number of People
0–5	37
6–10	20
11–15	2
16–20	1
over 20	0

27. **Disease Data** Prepare a bar graph of the data in this table. Which age group seems most likely to get the disease? Which age group seems unaffected by the disease?

Chapter 1 Standardized Test Practice

Part 1 Multiple Choice

Record your answers on the answer sheet provided by your teacher or on a sheet of paper.

1. Which of the following is not a tool used to give numbers to descriptions of observations?
   A. thermometers   C. pencils
   B. metersticks   D. scales

Use the photo below to answer question 2.

2. These students are doing an important step before beginning an investigation. They are
   A. drawing conclusions.
   B. analyzing data.
   C. controlling variables.
   D. collecting information.

3. What would be a good source of information when doing a science report on a bacterial epidemic that happened over a hundred years ago locally?
   A. pictures   C. television
   B. Internet   D. newspapers

Test-Taking Tip

Sleep Instead of Cramming Get plenty of sleep—at least eight hours every night—during test week and the week before the test.

4. Giving partial information to get the results you want is known as
   A. using a random sampling.
   B. having a bias.
   C. stating a theory.
   D. conducting a survey.

5. The application of science to make products or tools that people use is
   A. engineering.   C. industry.
   B. technology.   D. mechanics.

6. What name is given to scientific research that answers questions by observation?
   A. descriptive research
   B. experimental research
   C. technical research
   D. analytical research

Use the photo below to answer question 7.

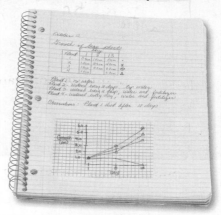

7. Which of the following would not go in this notebook?
   A. listings of materials
   B. drawings of equipment setups
   C. specific results of an investigation
   D. English assignments

8. What type of research answers scientific questions by testing a hypothesis?
   A. experimental   C. technical
   B. descriptive   D. analytical

34 STANDARDIZED TEST PRACTICE

Standardized Test Practice

Part 2 | Short Response/Grid In

Record your answers on the answer sheet provided by your teacher or on a sheet of paper.

9. Explain the basic steps to follow when solving a scientific problem.

10. Why do scientists around the world use the International System of Units (SI)?

Use the illustration below to answer question 11.

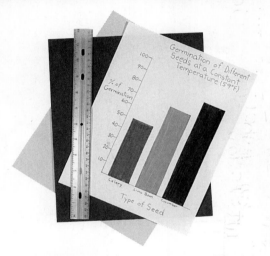

11. Why might you use this type of graphic in your scientific notebook?

12. Why is it important that you do multiple trials in an experiment?

13. Why is it important for scientists to search for new discoveries?

14. What is an experimental control?

15. Experiments are often conducted in secret. Why would you want to share your experiments and results with others?

16. Why are computers important to science? Describe three ways a scientist might use a computer.

17. Early scientists just used observation to explain things. What are some possible problems with just using observation in science?

Part 3 | Open Ended

Record your answers on a sheet of paper.

18. Some people, such as farmers, produce food, while others consume what is produced. What would be your hypothesis as to what would happen if all farmers suddenly decided not to produce vegetables anymore? Is there a way to check your hypothesis?

Use the figure below to answer questions 19 and 20.

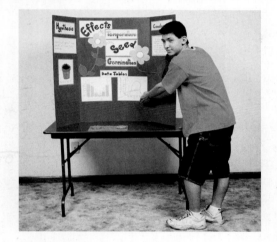

19. Describe how you would present the data in question 11 using the picture above.

20. The picture shows headlines labeled Data Tables, Effects, Conclusions, etc. What headlines would you put on your presentation board and what would go under each headline?

21. The black plague killed thousands of people in the Middle Ages. Explain how you would go about finding information about this disease. How was is spread? Is the disease still around today? If so, how is it treated?

22. How would you go about telling the world about observations you made in countries with drought and famine?

chapter 2

Traits and How They Change

The BIG Idea
The environment and changes over time can affect genetic traits.

SECTION 1
Traits and the Environment
Main Idea The interaction of the environment with alleles and genes affects phenotypes.

SECTION 2
Genetics
Main Idea Genetics is the science that studies how genetic traits pass from parents to offspring.

SECTION 3
Environmental Impact over Time
Main Idea If an environment changes over time, species adapt to the changes, move to a new environment, or become extinct.

Are these dogs related?
It hardly seems possible that these dogs are members of the same species. In this chapter, you will learn how the differences among members of a species develop over time, about the effects of the environment on traits, and how to predict traits of offspring.

Science Journal List two traits that a dog inherits, and two that are determined by the environment.

Start-Up Activities

How are people different?

Some of your unique qualities were present before you were born. You can observe one unique quality by studying your own fingerprints. Do the lab below to compare the fingerprint patterns labeled in the photograph with your fingerprints and those of your classmates.

Loop Arch Whorl

1. Press the thumb and fingertips of one hand on a washable ink pad.
2. Gently roll each fingertip on a sheet of blank paper to produce a set of fingerprints.
3. Use a magnifyng lens to observe your fingerprints and those of others in your class to find whorl, arch, and loop patterns.
4. **Think Critically** Write a paragraph in your Science Journal suggesting why no two people in your class have exactly the same fingerprints.

Traits Make the following Foldable to help you classify your traits into groups.

STEP 1 Fold a vertical sheet of paper from top to bottom.

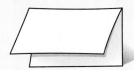

STEP 2 Fold it in half from side to side with the fold at the top.

STEP 3 Draw a picture of yourself and write your name and a short biography on the front of your Foldable.

Classify Traits Before you read the chapter, make a list on the inside of your Foldable of the traits that you think are inherited from your parents. Decide if each trait you listed is a phenotype or genotype and write your decision next to each.

Preview this chapter's content and activities at blue.msscience.com

Get Ready to Read

Identify the Main Idea

1 Learn It! Main ideas are the most important ideas in a paragraph, section, or chapter. Supporting details are facts or examples that explain the main idea. Understanding the main idea allows you to grasp the whole picture.

2 Picture It! Read the following paragraph. Draw a graphic organizer like the one below to show the main idea and supporting details.

> Many environmental factors other than seasonal temperatures and rainfall influence the survival of a species. Some environmental factors, like pollution, limit whether a species can survive in a habitat. Other factors can influence a species so that it changes in appearance. Fire, height of mountains, volcanic eruptions, and periodic flooding of rivers can have influences on the animals and plants in an area significantly.
>
> —*from page 49*

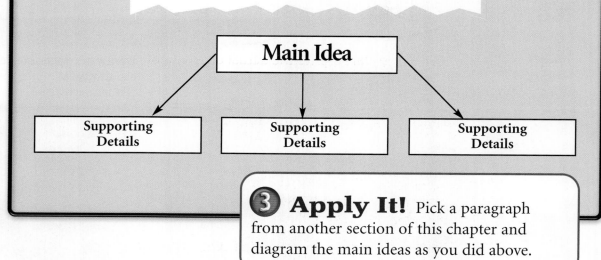

3 Apply It! Pick a paragraph from another section of this chapter and diagram the main ideas as you did above.

38 A   **CHAPTER 2**   Traits and How They Change

Target Your Reading

Reading Tip
The main idea is often the first sentence in a paragraph but not always.

Use this to focus on the main ideas as you read the chapter.

1 Before you read the chapter, respond to the statements below on your worksheet or on a numbered sheet of paper.
- Write an **A** if you **agree** with the statement.
- Write a **D** if you **disagree** with the statement.

2 After you read the chapter, look back to this page to see if you've changed your mind about any of the statements.
- If any of your answers changed, explain why.
- Change any false statements into true statements.
- Use your revised statements as a study guide.

Science Online
Print out a worksheet of this page at blue.msscience.com

Before You Read A or D		Statement	After You Read A or D
	1	Your genetic traits are the same as one of your parents.	
	2	An organism's internal and external environments can influence its genetic traits.	
	3	The two alleles for a gene can be the same or different.	
	4	A Punnett square shows the actual offspring of two parents.	
	5	The color markings of an adult Siamese cat result from environmental effects.	
	6	An organism's genotype depends on its phenotype.	
	7	The environment and other species affect the number of species present in an environment.	
	8	Each present-day species evolved from a different ancient species.	
	9	Adapted species are selected naturally and increase in number more than other species.	

38 B

section 1
Traits and the Environment

as you read

What You'll Learn
- **Compare and contrast** phenotype and genotype.
- **Describe** some effects the environment has on traits.
- **Explain** how traits are formed.

Why It's Important
The features of many organisms, including humans, are partly determined by the environment.

Review Vocabulary
variations: inherited trait that makes an individual different from other members of the same species and results from a mutation in the organism's genes

New Vocabulary
- trait
- gene
- genotype
- phenotype

What are traits?

Every living thing has many inherited features—characteristics that came from its parents. For example, if you have a cat, its coloration, length of hair, and many other features came from its parents. All of the features that an organism inherits are its **traits.** The color of your eyes and the shape of your ears are two of your traits.

Observing Traits People observed the inheritance of traits long before scientists understood how the inheritance occurred. Many breeds of domestic animals and crops were developed based on these observations. For example, over thousands of years, Native Americans developed maize (MAYZ) from a wild grass called teosinte (tay oh SIHN tee), shown in **Figure 1.** By carefully selecting and breeding individual plants with desired traits, modern corn was developed.

When people wanted to improve an existing plant or animal, they based breeding on observable traits. The Native Americans may have based their breeding of maize on the number and size of kernels each plant produced. But, what would have happened if environmental factors, like the amount of rain and temperature, determined the number and size of kernels instead of the traits from the parent plants? The breeding of maize would not have been successful. Sometimes it is obvious how the environment affects traits, but other times it is not.

Figure 1 Teosinte, below, an ancestor of modern corn, had few kernels on each plant. This modern corn, *Zea mays,* to the right, has many kernels on each ear.

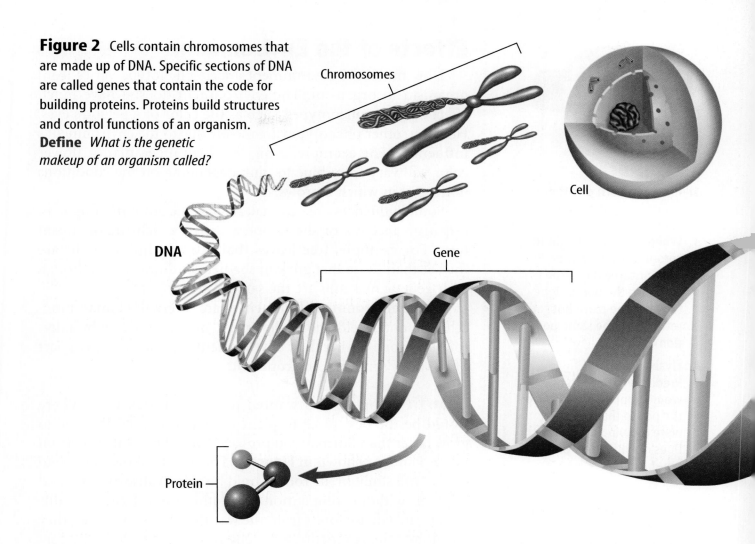

Figure 2 Cells contain chromosomes that are made up of DNA. Specific sections of DNA are called genes that contain the code for building proteins. Proteins build structures and control functions of an organism.
Define *What is the genetic makeup of an organism called?*

Phenotypes and Genotypes A cat, a plant, or even you can be thought of as a collection of thousands of traits. Each trait results from the coded information in the hereditary material called DNA, which is found in every cell. DNA is a complex molecule, shaped like a twisted ladder. It contains all of the information required to produce a living organism.

In cells that have a nucleus, DNA is found in chromosomes (KROH muh sohmz). A **gene** is a part of the DNA code on a chromosome. Humans have tens of thousands of genes on their chromosomes. The genes that an organism has—its genetic makeup—are called its **genotype** (JEE nuh tipe). Within each cell, the DNA code directs the production of specific proteins, as shown in **Figure 2**.

 *How are genes, DNA, and chromosomes related?*

When you look at an organism, you see the organism's phenotype (FEE nuh tipe). A **phenotype,** like hair color in humans, is the combination of genetic makeup and the environment's effect on that makeup.

INTEGRATE Chemistry

DNA Structure DNA is a unique molecule that has a twisted-ladder shape. The uprights of the ladder are made of repeating molecules of a phosphate and deoxyribose, a sugar. The rungs of the ladder are made of four bases that contain nitrogen. Use reference materials to find out what the four bases of DNA are called. List the four bases in your Science Journal.

SECTION 1 Traits and the Environment **39**

Mini LAB

Observing Gravity and Stem Growth

Procedure
1. Plant **popcorn or other seeds** in a **pot of soil**.
2. **Water** thoroughly.
3. When the plants begin to sprout, turn the pot on its side for three days. You may turn the pot upright to water the plant but return it to the same position each time.

Analysis
1. Predict what eventually would happen if you rolled the pot so that the stems were growing downward.
2. Does gravity affect phenotype? Explain.

Try at Home

Effects of the Environment

How much the environment affects phenotype varies from organism to organism. The environment doesn't have much effect on some phenotypes, such as the color of a person's eyes. However, other phenotypes are mostly due to the environment's influences. For example, a big-leaf hydrangea plant's flower color will vary from blue to pink depending on the conditions of the soil in which it grows.

Some influences are external, such as the amount of light an organism receives or the temperature in which the organism lives. For example, tree leaves that grow in full sunlight are thicker than those that grow in shadier conditions, even though their genetic makeups are the same.

Other environmental influences are internal. Human brain cells will not develop normally unless they are acted on by a thyroid hormone during their development. The hormone is a part of the body's internal environment.

Growth Suppose you wanted to plant an oak tree. Where would be the best place to plant it? If you wanted the tree to grow faster than normal, you probably would plant it away from other plants, as shown in **Figure 3**. This would allow the tree to receive full sunlight. Its roots would be able to absorb water and minerals without competition from other trees. Trees grow differently in a dense forest from the way that they grow when they are alone. The competition for environmental factors in a forest—light, water, soil minerals, and many others—have significant effects on the populations of trees in it.

Figure 3 Many plants, such as the oak tree shown here, grow faster when they are planted away from other plants.
Infer *In a forest, what environmental factors slow tree growth?*

Figure 4 Water buttercup leaves are finely divided below the water level. Leaves that grow above the water level are not as finely divided.
Explain what causes these differences.

Appearance The water buttercup shown in **Figure 4** has leaves that are shaped differently depending on where the leaves develop. Although the cells of the plant have the same genes, leaves that grow submerged in water are threadlike and those that grow above the water are broad. What environmental factor do you think determines the difference? The presence of water makes the difference because a leaf that grows halfway in the water is half threadlike and half broad.

Reading Check What causes water buttercup cells with the same genotype to have different phenotypes?

The color markings on a Siamese cat are another phenotype affected by the environment. Siamese kittens, like the one shown in **Figure 5,** are pure white at birth. Because the gene for colored fur is less active in heat, colored markings, as shown in **Figure 5,** develop more quickly on cooler parts of the cat's body, such as the ears. In warmer climates the fur color might not develop fully until the cat is more than a year old.

The arctic fox's fur color is a phenotype that changes with the seasons. During the winter months the arctic fox does not produce pigment that colors fur, so the fox's fur is white. As a result, the fox blends with the snowy ground helping it avoid predators. In warmer summer months, the arctic fox produces fur pigment. Then, the arctic fox's fur is brown, which is perfect for blending with the tundra.

Figure 5 This Siamese kitten has not developed markings. An adult Siamese cat has darker markings on the cooler parts of its body.

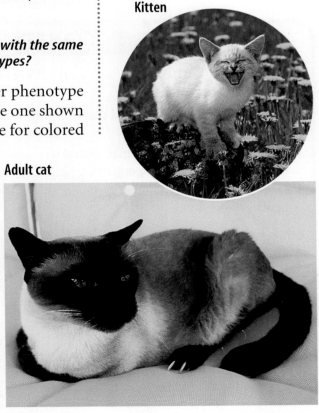

Kitten

Adult cat

SECTION 1 Traits and the Environment

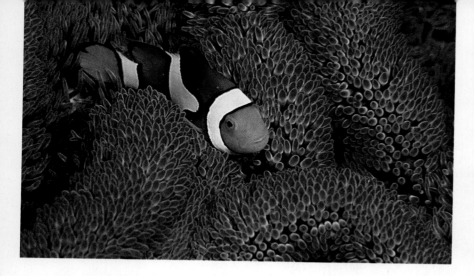

Figure 6 The large clownfish is female. This fish was a top-ranking male earlier in its life. It changed into a female when the female who occupied this territory died.

Gender Most living things are born male or female and remain that way for life. However, some species of fish, including many clownfish, parrot fish, wrasses, and sea bass, are born with the ability to change sex. This allows these species to maintain a desired male-to-female ratio in a group under different conditions. For example, as shown in **Figure 6,** one large female clownfish lives near a sea anemone. Several males, including one top-ranking male, might share the territory with the female. Only the top ranking male mates with the female to produce offspring. When the female dies, the top-ranking male changes phenotype to become a female. All of the other males then move up in the social order. In some of these species, sex changes are not reversible. A few fish have the ability to switch back and forth between sexes, depending on the number of males and females in the population.

Science Online

Topic: Changing Gender
Visit blue.msscience.com for Web links to recent news or magazine articles about animals that can change their gender.

Activity List two animals whose gender can change and describe the conditions under which it changes.

---

section 1 review

Summary

What are traits?
- All living things inherit features that come from its parents.
- DNA contains all of the information that is needed to produce a living organism.
- There are tens of thousands of genes on human chromosomes.

Effects of the Environment
- Some environmental influences are external and some are internal.
- The competition for environmental factors can affect growth.
- The appearance of an organism can be different depending on its environment.
- The gender of some living things can change under different conditions.

Self Check

1. **Describe** two factors that determine the phenotype of a trait.
2. **Identify** a phenotype that changes as seasons change.
3. **Explain** the difference between an organism's genotype and its phenotype.
4. **Think Critically** The environment can affect phenotypes in desirable and undesirable ways. Describe an example of each.

Applying Skills

5. **Record Observations** Observe a family pet or other organism. List five traits that could be influenced by the environment and explain why you think so for each one.
6. **Communicate** Sickle-cell disease is inherited. Research then explain in your Science Journal how the environment can affect this human disorder.

Science Online blue.msscience.com/self_check_quiz

Jelly Bean Hunt

The environment plays an important role in the development of some phenotypes. In this lab, you will observe how camouflaged animals are less likely to be captured by predators.

Jelly Bean Data

	Shade #1	Shade #2	Shade #3	Shade #4	Shade #5
Hunt #1					
Hunt #2		Do not write in this book.			
Hunt #3					
Hunt #4					
Hunt #5					

Real-World Question

How do differences in animal coloration camouflage some but expose others to predation?

Goals
- **Model** camouflage and predation.
- **Infer** how the effect of the environment on phenotype helps some animals survive.

Materials
five shades of green jelly beans (10 each)
*five shades of another color (10 each)
green poster board
*poster board that matches chosen color
*Alternative materials

Safety Precautions

WARNING: *Never eat or drink anything in the lab.*

Procedure

1. Copy the data table on this page in your Science Journal. Determine which shade is which number.
2. Put the poster board on the desk. Have your partner turn his or her back to the poster.
3. Arrange the 50 jelly beans on the poster. Mix up the different shades of jelly beans.
4. Have your partner, who is the hunter, turn and pick up one at a time, as many jelly beans as possible in 3 s.
5. **Count** the number of each shade of jelly bean the hunter caught. Record these numbers in the *Hunt #1* row.
6. Mix up the jelly beans and have the hunter make four more hunts.

Conclude and Apply

1. **Observe** Which shade of jelly bean did the hunter select most often? Least often?
2. **Explain** why the hunter caught more of certain shades of green jelly beans than others.
3. **Predict** your results with a different shade of poster board.
4. **Infer** how your experiment could explain the specific green color of tropical lizards.

Communicating Your Data

Describe how the environment's effect on some phenotypes can help animals survive. **For more help, refer to the** Science Skill Handbook.

section 2
Genetics

as you read

What You'll Learn
- **Differentiate** between genetics and heredity.
- **Explain** the results of Mendel's pea plant experiments.
- **Identify** the results shown by a Punnett square.

Why It's Important
The transfer of genetic information from one generation to the next allows traits to change over time.

Review Vocabulary
cloning: making copies of organisms, each of which is a clone that receives DNA from only one parent cell

New Vocabulary
- genetics
- allele
- dominant
- recessive
- Punnett square

Science of Genetics

Long before scientists understood genes, chromosomes, or how sex cells are produced, they tried to figure out heredity. Some early scientists proposed that the male parent contributed all of the traits and that the female parent was only a supplier of food for the new organism. While observing sperm through a microscope, they even imagined that they could see a tiny human curled up in the head of sperm. Other early scientists hypothesized that the traits of the parents blended to form those of the offspring. Only within the past 200 years have scientists begun to understand the true nature of how organisms inherit traits.

What is genetics? Heredity is the passing of traits from parents to offspring. Eventually, the study of heredity developed into a science called **genetics.** Researchers in the field of genetics, as shown in **Figure 7,** are rapidly providing more information about the genetics of humans and other organisms. Studies in genetics, combined with an understanding of chemical interactions and other cell processes, provide an explanation of how species can change through the generations.

Reading Check *Why is genetics important?*

Figure 7 The Human Genome Project is an international effort to provide information about the genetics of humans. In the photo to the right, a technician prepares DNA samples for sequencing trials. **Infer** *why medical doctors might be interested in results from the Human Genome Project.*

Beginning with Mendel

Gregor Mendel was the first researcher to use numbers to describe the results of genetics experiments, as shown in **Figure 8**. Mendel's work was presented in the 1860s, but its importance was not recognized for many years. Although Mendel did not know about chromosomes or genes, he was able to develop principles of genetics by experimenting with thousands of pea plants.

Dominant and Recessive Traits One of Mendel's conclusions from his experiments was that traits are determined by different factors. Mendel explained that each trait of an individual is determined by at least two of these factors, as shown in **Figure 9**. Today, Mendel's factors are called genes. The different forms of a gene are each called an **allele** (uh LEEL).

Mendel's principle of dominance explains why only one form of a trait is expressed even when both alleles are present. **Dominant** (DAH muh nunt) alleles will show their effect on the phenotype whenever they are present in the genotype. These traits often are seen in each generation. **Recessive** (rih SE sihv) alleles will show their effect on the phenotype only when two of them for a trait are present in the genotype.

Understanding how dominant and recessive alleles show their effects has helped scientists figure out how some genetic diseases are passed down through families.

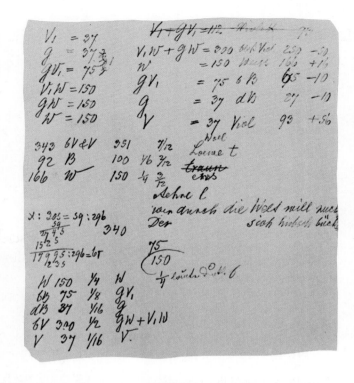

Figure 8 This page from Gregor Mendel's notebook shows some of his experimental data from more than 28,000 pea plants.

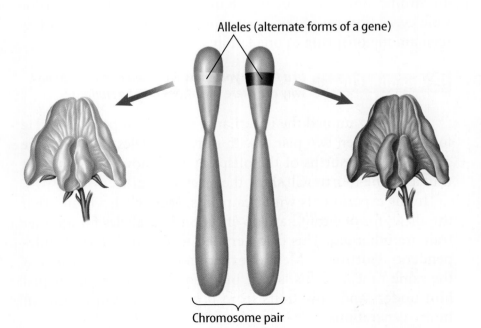

Figure 9 Mendel studied the factor (gene) that determines flower color in pea plants. One allele codes for purple flowers and another allele codes for white flowers. The purple allele is the dominant allele.
Determine *how many white alleles have to be present to produce a white flower.*

Figure 10 A parent with three traits on three different chromosomes could produce eight genetically different sex cells. **Explain** *how this creates variation among offspring.*

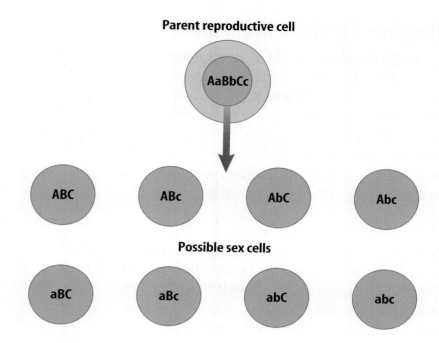

From Parents to Offspring Mendel also concluded that each parent passes only one of the alleles for a trait to its offspring. This is known as the principle of segregation. This explains why variation exists among the offspring of parents. Suppose that a parent has three pairs of chromosomes with a different trait on each pair. The traits can be called A, B, and C. Each trait has two different alleles—A and a, B and b, and C and c. When sex cells—eggs or sperm—form in the parent, each sex cell will have three chromosomes, each with one form of the A, B, or C trait. The three chromosomes and their alleles can combine in eight possible ways, as shown in **Figure 10**. Humans have 46 chromosomes, so more than 8 million combinations are possible every time an egg or sperm forms. That's why variation is seen among offspring in one family.

 *In a family you know with two or more siblings, why are there variations among siblings?*

Mendel examined the inheritance of traits in pea plants. He found that when two plants with different alleles for a trait are crossed, three fourths of the offspring will show the dominant trait and one fourth will show the recessive trait.

In his experiments with pea plants, Mendel also found that the alleles for one trait have no effect on how alleles for another trait are inherited. This discovery led to Mendel's law of independent assortment. Mendel experimented with two traits at the same time, and his law of independent assortment helped him understand how traits from both parents can appear in future generations.

Mini LAB

Observing Fruit Fly Phenotypes

Procedure
1. Obtain a container of **fruit flies** from your teacher.
2. Use a **magnifying lens** to observe the eyes and wings of the flies.

Analysis
1. What variations did you find in the eyes and wings of the flies?
2. Hypothesize what caused the variations.

Predicting Genetic Outcomes

Mendel's greatest achievement might have been his ability to apply mathematics to scientific problem solving. He stated predictions about his experiments in terms of probability. Almost 50 years after Mendel's work was published, Reginald C. Punnett developed a chart called a Punnett square. A **Punnett square** can help you understand and make genetic predictions.

Applying Math — Find a Percentage

PERCENT OF OFFSPRING WITH CERTAIN TRAITS In mice, black fur color is dominant and white fur color is recessive. If both parent mice have different alleles for the trait, what percent of their offspring would have white fur?

Solution

1. *This is what you know:*
   - Each parent has different alleles for fur color, Bb.
   - Offspring with BB or Bb will have black fur; offspring with bb will have white fur.

2. *This is what you need to find:* percent of white fur offspring

3. *This is the procedure you need to use:*
   - Complete the Punnett square by combining the letter in each column with the letter in each row.

	B	b
B	BB	Bb
b	Bb	bb

   - $\dfrac{\text{number of offspring with white fur}}{\text{total number of possible offspring}} = P/100$
   - Substitute the known values: $1/4 = P/100$.
   - Find the cross products: $1 \times 100 = P \times 4$.
   - Divide each side by 4: $25 = P$.

4. *Check your answer:* Substitute your answer into the proportion. You should get 1/4.

Practice Problems

1. Suppose a male mouse is Bb and a female is bb. What percent of the mice offspring from these two mice would you expect to have black fur?

2. One fruit fly is heterozygous for long wings, and another fruit fly is homozygous for short wings. Long wings are dominant to short wings. Using a Punnett square, find out what percent of the offspring are expected to have short wings.

For more practice, visit blue.msscience.com/math_practice

	Female (XX)	
	X	X
Male (XY) X	XX	XX
Y	XY	XY

Figure 11 Females have the genotype XX, and males have the genotype XY.

Punnett Square The hereditary traits in organisms can be predicted using a Punnett square. It is a model that is used to predict the possible offspring of crosses between different organisms of known genotypes. When an organism has two different alleles for a trait, the organism is called a hybrid. For example, a pea plant with an allele for purple flowers and an allele for white flowers is a hybrid. A monohybrid cross is one that includes one trait, such as flower color. The ability to predict the possible offspring becomes more complex as more traits are involved.

Understanding Results When you use a Punnett square, like the one shown in **Figure 11,** to predict the sex of one offspring, the results are one-half males and one-half females. Suppose a mother has given birth already to three boys and is expecting a fourth child. What are the chances that it will be a girl? You might expect that the chances would be increased that the next baby would be a girl, but the chances are still only one in two. Each result is independent of the others that came before or come after it.

When Mendel was studying heredity in garden peas, his results were close to his predicted outcomes. This was because he studied large numbers of pea plants in each experiment. When large numbers are studied, the probability increases that the predicted result will occur.

section 2 review

Summary

Science of Genetics
- Only within the last 200 years have scientists begun to understand the inheritance of traits.
- Genetic researchers are rapidly providing more information about the genetics of humans.

Beginning with Mendel
- Mendel developed genetic principles by experimenting with pea plants.
- Two recessive alleles are needed in the genotype to show an effect in the phenotype.
- Each parent passes only one allele for a trait to its offspring.

Predicting Genetic Outcomes
- When an organism has two different alleles for a trait, it is called a hybrid.
- Predicting the traits of an offspring becomes more complex as more traits are involved.

Self Check

1. **State** some of the early beliefs about human heredity.
2. **Contrast** heredity and genetics.
3. **Explain** why some people call Mendel the "Father of Genetics."
4. **Describe** the primary purpose of using a Punnett square.
5. **Think Critically** Which of Mendel's principles would apply to mating two organisms that have two different alleles for three different traits?

Applying Math

6. **Calculate a Ratio** Long wings (L) are dominant to short wings (l) in fruit flies. Make a Punnett square to show the predicted outcome of a cross between two parents that have both alleles for wing length. What is the ratio of flies produced?

section 3
Environmental Impact over Time

Survival and the Environment

Would you be more likely to see a blue jay or a cactus wren where you live? Depending on where you live, winters might be long and cold or they might be short and mild. More than 125 cm or less than 25 cm of rain might fall in a year. Over a long period of time, the environment influences which organisms can live in an area. You wouldn't expect to find many cacti where it rains a lot, and you would never see giant evergreen trees growing in a desert.

Nonliving Influences Many environmental factors other than temperature and rainfall influence the survival of a species. Some environmental factors, like pollution, limit whether a species can survive in a habitat. Other factors can influence a species so that it changes in appearance. Fire, height of mountains, volcanic eruptions, and periodic flooding of rivers can influence the animals and plants in an area significantly.

The chaparral shrub land of California, as shown in **Figure 12,** and forests of Yellowstone National Park require periodic fires to survive. Some of the plant species in these areas have seeds that can germinate only after fire. Some trees, such as aspen trees, can sprout from underground roots when fire has burned away competing plants.

If you travel up a mountain's slopes, you will notice that the environment gradually changes. Temperature decreases and wind usually increases. At high elevations, the trees can be short and stubby, and above certain elevations, trees don't grow at all. The types of animals found also vary at different elevations on a mountain.

as you read

What You'll Learn
- **Explain** how living and nonliving environmental factors impact evolution.
- **Describe** how natural selection occurs in a species.
- **Compare and contrast** selective breeding and natural selection.

Why It's Important
Species can evolve over time due to the impact of the environment.

Review Vocabulary
environment: the objects and conditions that surround an organism

New Vocabulary
- evolution
- natural selection
- mutation
- adaptive radiation
- extinction

Figure 12 In California, plants and animals are adapted for periodic, destructive fires.

49

Interactions with Other Organisms

Living factors in the environment also affect the species that are present. Predators, availability of food, and how many of the same species that live in an area have an effect. Predators, as shown in **Figure 13,** often limit the number of individuals. Over generations, groups can adapt to the presence of predators. They might evolve ways to escape detection, defenses against predators, or ways to increase their number.

Reading Check *What are some effects of predation?*

Figure 13 Raccoons prey on insects, fish, crayfish, and other small organisms.
Describe *what might happen to the population of crayfish as the population of raccoons increases.*

Species and the Environment

About the same time that Gregor Mendel was discovering the rules of genetics, Charles Darwin and Alfred Russell Wallace, two British biologists, were separately hypothesizing about how so many living things came to exist on Earth. Darwin studied the diversity of living things while sailing aboard the HMS *Beagle*. After visiting the Galápagos Islands off the coast of South America, Darwin began to hypothesize about reasons for the diversity he observed and recorded. Wallace came to the same conclusions as Darwin while studying in the East Indies. Darwin and Wallace concluded that different, long-term, environmental influences on populations produced the variety of species they observed.

Natural Selection According to Darwin and Wallace, changes happen from generation to generation that result in adaptations to the environment. This process is called evolution. **Evolution** is the change in the genetics of a species over time. Darwin's theory, the theory of evolution by natural selection, is an explanation of how, over time, several factors can act together and result in a new species. This theory provides an explanation of how organisms could have changed to produce the several million species that are alive today.

The big question for Darwin and Wallace was how evolution happens. They proposed that organisms that are better adapted to an environment survive and reproduce at a greater rate than organisms that are not. They called this **natural selection,** as shown in **Figure 14,** because the adapted organisms are selected naturally to survive and increase in number. Natural selection can produce new organisms or new species.

Volcanologist Volcanic activity changes the landscape. Scientists who study volcanoes are called volcanologists. To become a volcanologist it is necessary to study many areas of science. Research to find out what classes are required to become a volcanologist. In what parts of the United States would it be easiest to study volcanoes?

NATIONAL GEOGRAPHIC VISUALIZING NATURAL SELECTION

Figure 14

British naturalist Charles Darwin hypothesized that the 14 species of finches he found on the Galápagos Islands developed from a common ancestor through a process of natural selection. "Darwin's finches," as they became known, probably developed their different beak structures and feeding habits over time, as a result of the specific environment on each of the islands.

◀ **LARGE GROUND FINCH** Ground finches have short, stout "crushing" beaks, useful for breaking seeds. They spend much of their time foraging on the ground.

▼ **SMALL TREE FINCH** The beak of this tree-dwelling finch is sharper than that of the ground finch—and better suited to the tree finch's plant and insect diet.

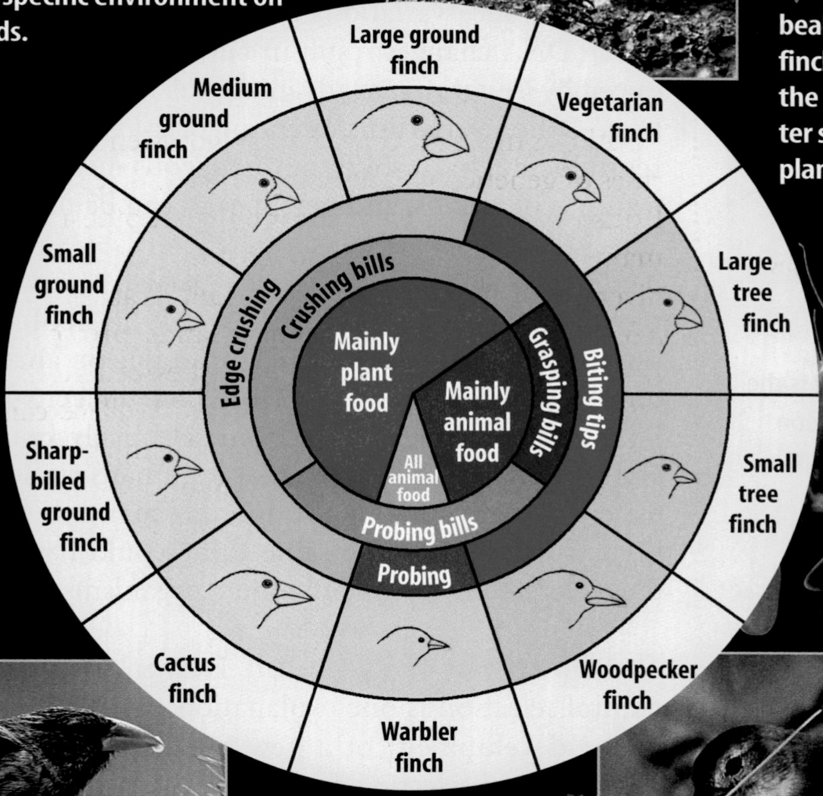

▼ **CACTUS FINCH** The long beak of the cactus finch allows it to eat the fruit of the prickly pear cactus.

▼ **WARBLER FINCH** The smallest of Darwin's finches, the warbler finch, has a long, narrow beak for insect eating.

▲ **WOODPECKER FINCH** This finch uses twigs or cactus spines to pry insects or their larvae out of small holes in cacti or from beneath bark.

SECTION 3 Environmental Impact over Time

Fantail pigeon

Frillback pigeon

Homing pigeon

Rock dove

Figure 15 The rock dove is the common ancestor of all pigeon breeds, such as the three shown here.

Mutation You read in the last section that different forms of alleles produce variations in traits. **Mutation** is the process in which DNA changes result in new alleles. Some variations produced by mutation are advantageous for survival and reproduction. Other variations keep an organism from surviving or reproducing. In this way, advantageous mutations are passed to future generations, and new species can be produced.

Selective Breeding Charles Darwin was well aware of the methods of selective breeding. One of his hobbies was breeding pigeons. From the rock dove, a wild pigeon ancestor that looks much like some pigeons you would see in a city park, many different pigeon breeds have been selectively bred, as shown in **Figure 15.** Darwin inferred that if humans could select so many different variations to produce so many different breeds of organisms, perhaps the same thing could happen naturally in the different environments where organisms live.

The Direction of Evolution Darwin's theory of evolution by natural selection is one explanation of how variations can lead to the development of a new species. New species can form when natural selection favors members of a population with a variation in a trait. In another way, more than one variation of a trait is favored. This can lead to two or more new species from one ancestral species.

The production of several species from one ancestral species is called **adaptive radiation.** Darwin observed many species of finches and tortoises when he visited the Galápagos Islands. He concluded that one ancestral population of finches and tortoises had reached the islands. Because they were isolated geographically from the same species on the mainland, they adapted to the various conditions of the islands. Eventually, each ancestral species produced several different species; each adapted to the different environments on each island.

Topic: Different Species
Visit blue.msscience.com for Web links to information about other theories of how different species came about.

Activity List and describe two theories on the formation of different species.

Extinction of Species

All indviduals experience a life cycle that includes birth and death. **Extinction** occurs when the last individual of a species dies. During Earth's history, millions of species have become extinct. Fossils are evidence of these species. The rate of extinction today of known species is as great or greater than at any time in the recent past, as shown in **Figure 16.** Extinction can occur for many reasons, including the destruction of habitat and the introduction of new species.

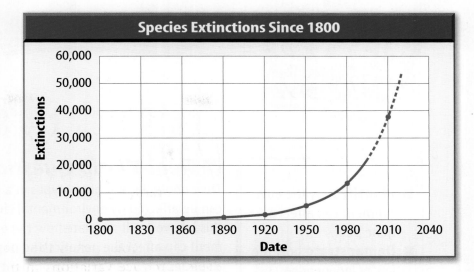

Figure 16 The rate at which species have become extinct has risen rapidly.

Humans impact environments when they construct buildings, recreational areas, or roads, and when they farm or mine land. Some species increase in number because of changes, but others cannot cope and either leave the area or die.

Sometimes newly introduced species prey on organisms that do not have defenses against them. The introduced species also might produce many offspring that crowd out other species. In either case, some species might become extinct. Zebra mussels were accidentally introduced into the Great Lakes. They have affected food webs and some species are disappearing.

section 3 review

Summary

Survival and the Environment

- Nonliving influences can affect species survival in a habitat.
- Predators, availability of food, and how many of the same species that live in an area can affect the species that are present.

Species and the Environment

- The process in which changes result in adaptations to the environment is called evolution.
- Organisms that are better adapted to the environment survive and reproduce at a greater rate than organisms that are not.
- Variations can result in new species.
- Millions of species have become extinct during the history of Earth.

Self Check

1. **List** some nonliving factors in the environment that can cause change in species over several generations.
2. **Differentiate** between selective breeding and natural selection.
3. **Describe** how evolution and extinction are related.
4. **Think Critically** What are some changes made by humans in the area where you live that might affect other species there?

Applying Skills

5. **Relate Cause and Effect** Members of a butterfly species are blown onto an island. Over many generations, they evolve into a new species. What environmental factors might explain the evolution?

Toothpick Fish

Goals
- **Identify** how the environment can affect a gene pool.
- **Demonstrate** how the law of independent assortment is random.
- **Distinguish** between the number of phenotypes and genotypes.

Materials
petri dish
toothpicks (24)
   (8 green, 8 red, 8 yellow)

Safety Precautions

Real-World Question
The genotypes and phenotypes in a population can be affected by environmental changes. In this lab, you will simulate how the environment can affect the genetics and population of a species. How can the environment affect a species' gene pool?

Procedure

1. The petri dish represents a fish gene pool and the colored toothpicks represent the alleles that control fish skin color. The green allele is dominant. The red and yellow alleles are recessive to green, but fish with a red allele and a yellow allele have orange skin. List all of the genotypes for the four fish skin colors in your Science Journal.

2. Copy Table A and Table B in your Science Journal. Select an allele pair (2 toothpicks) without looking and record the results in Table A. Continue selecting and recording pairs until the gene pool is empty. Do not mix up the pairs.

3. Count and record in Table B the numbers of each color of fish offspring in the first generation.

4. Predators easily spot yellow fish in the green seaweed. Remove the yellow fish and put the remaining alleles back in the gene pool. Select a second generation of fish without looking. Record your results in Table A. Repeat step 3.

Table A Allele Pairs and Fish Offspring Skin Colors										
Generations										
First		Second		Third		Fourth		Fifth	
Allele pair	Skin color	Allele pair	Skin color	Allele pair	Skin color	Allele pair	Skin color	Allele pair	Skin color
				Do not write in this book.					

54 CHAPTER 2 Traits and How They Change

Using Scientific Methods

Table B Number of Fish Offspring

Environment	Generation	Green	Red	Orange	Yellow
Green seaweed grows everywhere.	First				
	Second		Do not write in this book.		
	Third				
	Fourth				
The seaweed dies.	Fifth	0			

5. Remove yellow fish again and return the surviving fish alleles to the petri dish. Repeat step 4 two more times to model the third generation and fourth generation.

6. Draw a fifth generation from the gene pool. Record the data in Table A.

7. Factory wastes are dumped into the stream and kill the seaweed. The green fish now are easily seen by predators. Remove the green fish and record the number of surviving offspring in the last row of Table B.

Analyze Your Data

1. **Compare** the population in the fourth generation to the first, second, and third generation. Explain any differences.
2. **Determine** if any alleles have disappeared. Describe why it did or did not occur.

Conclude and Apply

1. **Explain** how the environment affected the fish population.
2. **Infer** how environmental changes could lead to the extinction of a species.

Combine the data in **Table B** from all students in your class. Calculate the average number of each fish color for each generation. How do your data compare to the class averages?

LAB 55

TIME SCIENCE AND Society

SCIENCE ISSUES THAT AFFECT YOU!

Genetic Engineering

Genetically modified "super" corn can resist heat, cold, drought, and insects.

What would happen if you crossed a cactus with a rose? Well, you'd either get an extra spiky flower, or a bush that didn't need to be watered very often. Until recently, this sort of mix was the stuff science fiction was made of. But now, with the help of genetic engineering, it may be possible.

Genetic engineering is a way of taking genes—sections of DNA that produce certain traits, like the color of a flower or the shape of a nose—from one species and giving them to another.

In 1983, the first plant was genetically modified, or changed. Since then, many crops in the U.S. have been modified in this way, including soybeans, potatoes, tomatoes, and corn.

One purpose of genetic engineering is to transfer an organism's traits. For example, scientists have changed lawn grass by adding to it the gene from another grass species. This gene makes lawn grass grow slowly, so it doesn't have to be mowed very often. Genetic engineering can also make plants that grow bigger and faster, repel insects, or resist herbicides. These changes could allow farmers to produce more crops with fewer chemicals. Scientists predict that genetic engineering soon will produce crops that are more nutritious and that can resist cold, heat, or even drought.

Genetic engineering is a relatively new process, and some people are worried about the long-term risks. One concern is that people might be allergic to modified foods and not realize it until it's too late. Other people say that genetic engineering is unnatural. Also, farmers must purchase the patented genetically modified seeds each growing season from the companies that make them, rather than saving and replanting the seeds from their current crops.

People in favor of genetic engineering reply that there are always risks with new technology, but proper precautions are being taken. Each new plant is tested and then approved by U.S. governmental agencies. And they say that most "natural" crops aren't really natural. They are really hybrid plants bred by agriculturists, and they couldn't survive on their own.

As genetic engineering continues, so does the debate.

Debate Research the pros and cons of genetic engineering at the link shown to the right. Decide whether you are for or against genetic engineering. Debate your decision with a classmate.

Science Online
For more information, visit
blue.msscience.com

Chapter 2 Study Guide

Reviewing Main Ideas

Section 1 Traits and the Environment

1. Traits are the features that are coded for in DNA and can be inherited.
2. Genes are segments of DNA on chromosomes.
3. Phenotypes are observed traits. Genotypes are the genes an individual has.
4. The environment can affect phenotypes.

Section 2 Genetics

1. Genetics is the study of heredity.
2. Mendel discovered and stated these important conclusions about genetics:
   - The principle of dominance explains how one allele can be responsible for producing a trait.
   - The law of segregation states that a parent can pass on only one allele for each trait to its offspring.
   - The law of independent assortment explains how traits from both parents can appear in their offspring.
3. A Punnett square is a model for predicting the offspring of known genotypes.

Section 3 Environmental Impact over Time

1. Environmental factors might cause species to change gradually over time and determine which species survive in an area.
2. Natural selection can lead to evolution. Mutations can result in new variations.
3. The death of an entire species is called extinction. Extinctions cause a loss of biological diversity.

Visualizing Main Ideas

Copy and complete the following concept map about traits.

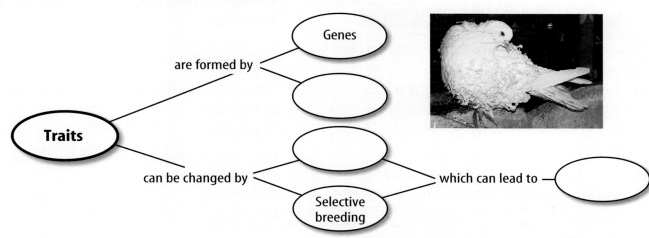

Chapter 2 Review

Using Vocabulary

adaptive radiation p. 52
allele p. 45
dominant p. 45
evolution p. 50
extinction p. 53
gene p. 39
genetics p. 44
genotype p. 39
mutation p. 52
natural selection p. 50
phenotype p. 39
Punnett square p. 47
recessive p. 45
trait p. 38

Fill in the blank with the correct vocabulary word or words.

1. Any feature of an organism that is inherited is a(n) _____.
2. The visible product of a gene and the environment's influences is a(n) _____.
3. An alternate form of a gene is a(n) _____.
4. _____ is the result of the death of the last member of a species.
5. _____ is the study of heredity.
6. The mechanism of evolution is _____.
7. A model used to predict the offspring of known genotypes is a(n) _____.
8. _____ is the change in traits over several generations.

Checking Concepts

Choose the word or phrase that best answers the question.

9. What is the term used for the different forms of a gene?
   A) phenotypes   C) genetics
   B) genotypes   D) alleles

10. What is a trait called that appears only when both alleles are present?
    A) dominant   C) genetic
    B) recessive   D) environmental

11. Which is a sequence of DNA that directs a cell to make a protein?
    A) chromosome   C) nucleus
    B) gene   D) recessive

12. What is the study of heredity called?
    A) genetics
    B) environmental
    C) evolution
    D) natural selection

13. How was the species of corn shown in the photo developed?
    A) by natural selection
    B) by selective breeding
    C) by evolution
    D) by extinction

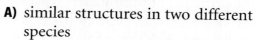

14. Which would NOT be evidence of evolution?
    A) similar structures in two different species
    B) variations in a species in two environments
    C) a bird that is the same in two different areas of its range
    D) a species that can avoid a predator

15. What is the production of several species from one ancestral species called?
    A) adaptive radiation
    B) heredity
    C) extinction
    D) phenotypes

16. What results in the formation of new alleles?
    A) mutation   C) dominant
    B) genetics   D) recessive

17. Which is NOT a way that humans influence the rate of species extinction?
    A) mining   C) volcanic eruptions
    B) farming   D) construction

58 CHAPTER REVIEW

chapter 2 Review

Thinking Critically

18. **Apply** Suppose you raise tropical fish that are black. How could you explain a white fish among the newly hatched young?

19. **Infer** why a mutation that results in larger-than-normal corn kernels might be helpful.

20. **Determine** which of Mendel's principles is most important in explaining the production of males and females.

21. **Communicate** Use an example to explain natural selection.

22. **Draw Conclusions** Huntington's disease is a human disorder caused by a dominant allele, *H*. If a person does not have Huntington's disease, what is his or her genotype for this trait?

23. **Predict** Red flowers are dominant to white flowers in a particular species of plants. Predict the possible genotypes of the parent plants that were crossed to produce a white-flowered plant.

Use the graph below to answer question 24.

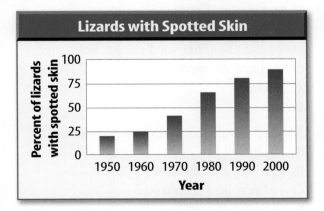

24. **Interpret Data** A population of lizards has been studied for the past 50 years. Study the graph above. Explain what might be happening and why.

Performance Activities

25. **Letter** Imagine that you were living in England and read a newspaper article that announced Darwin's return to England from his voyage on the HMS *Beagle*. Write a letter to a friend telling about Darwin's experiences.

Applying Math

Use the diagram below to answer question 26.

	R	r
r	Do not write	in this book.
r		

26. **Wrinkled Seeds** In pea plants, the trait for round seeds (R) is dominant to wrinkled seeds (r). Using the Punnett square above, predict the percent of offspring that will have wrinkled seeds when a parent with both alleles is crossed with a parent that has only the wrinkled allele.

Use the diagram below to answer question 27.

	T	t
T	Do not write	in this book.
T		

27. **Different Alleles** The trait for a tall plant (T) is dominant to short (t). What percent of offspring will have two different alleles when a parent with both alleles for plant height is crossed with a parent that has two tall alleles?

Chapter 2 Standardized Test Practice

Part 1 Multiple Choice

Record your answers on the answer sheet provided by your teacher or on a sheet of paper.

Use the photo below to answer question 1.

1. The fur color changes seen in the cat above are determined by body temperature. What is this an example of?
   A. environmental influences on gene expression
   B. expression of a dominant allel
   C. expression of a recessive allele
   D. climate control

2. If a parent has the genotype CcBB, how many different sex cells will be produced?
   A. 2
   B. 4
   C. 1
   D. 7

3. For which did Darwin propose natural selection as an explanation?
   A. selective breeding
   B. evolution
   C. predation
   D. mutation

4. What explains all of the different breeds of dogs and cats?
   A. phenotype
   B. genes and the environment
   C. dominant alleles
   D. selective breeding

5. What provides evidence of an extinct species?
   A. fossils
   B. genetics
   C. natural selection
   D. evolution

Use the table below to answer questions 6 and 7.

Table 1 Some of the Traits Compared by Mendel			
Traits	Shape of Seeds	Color of Seeds	Plant Height
Dominant Trait	Round	Yellow	Tall
Recessive Trait	Wrinkled	Green	Short

6. If you cross a tall plant with a dwarf plant and all of the offspring are tall, what was the genotype of the tall parent?
   A. T
   B. Tt
   C. tt
   D. TT

7. If one parent has two dominant alleles and the other parent has one dominant and one recessive allele for seed shape, what percentage of offspring will be wrinkled?
   A. 100%
   B. 0%
   C. 50%
   D. 25%

8. Mendel observed what types of alleles?
   A. high and low
   B. genotype and phenotype
   C. natural and unnatural
   D. dominant and recessive

9. What scientific field specializes in the study of heredity?
   A. botany
   B. cell biology
   C. genetics
   D. astronomy

60 STANDARDIZED TEST PRACTICE

Standardized Test Practice

Part 2 | Short Response/Grid In

Record your answers on the answer sheet provided by your teacher or on a sheet of paper.

Use the illustration below to answer questions 10 and 11.

10. Describe how mutation might have influenced the development of different species of finches that Darwin observed on the Galápagos Islands.

11. If fire destroyed the leafy vegetation on the islands, which of the finches would be most affected? Explain.

12. Genes that are inherited together or linked violate which of Mendel's laws?

13. How does predation limit the number of a species in a given area?

14. How can the introduction of a new species into a habitat help to cause the extinction of a species?

15. If a parent had the genotype XxYYZz, list the genotypes found in its sex cells.

16. What environmental factors might cause a species of insects to have mostly individuals that are light colored or dark colored, but few intermediate-colored individuals?

Part 3 | Open Ended

Record your answers on a sheet of paper.

17. How are genes, chromosomes, DNA, and proteins related to the genotype and phenotype of a trait?

18. List environmental factors that Mendel had to keep the same while growing his pea plants. How would these factors have effected growth or appearance of the plants?

19. In mice, brown fur is dominant over white fur. How can you determine the unknown genotype of a brown mouse by crossing it with a white mouse?

20. Explain why a variation that caused a fur color change in the arctic fox was favored over remaining the same color throughout the year.

Use the photo below to answer question 21.

21. Explain why there is so much variation among the kittens in this litter.

Test-Taking Tip

Practice Remember that test-taking skills can improve with practice. If possible, take at least one practice test and familiarize yourself with the test format and instuctions.

chapter 3

Interactions of Human Systems

The BIG Idea
Your health and survival depend on the interactions of your body systems.

SECTION 1
The Human Organism
Main Idea All organisms require certain elements that combine and form countless substances needed for life.

SECTION 2
How Your Body Works
Main Idea A cell can survive only if substances can move within the cell and pass through its cell membrane.

Will they win?
When a rowing team competes, individual team members work together to achieve a common goal. In a similar way, your body is made up of individual parts that interact and work together to help you perform the activities necessary for life.

Science Journal Write a paragraph describing why it is important for the members of a sports team, orchestra, or another group to work together.

Start-Up Activities

Model Blood Flow in Arteries and Veins

Your body's circulatory system is made up of a pumping heart and three types of blood vessels—arteries, veins, and capillaries. The blood flowing in your blood vessels is under constant pressure. However, the blood in your arteries has the highest pressure.

1. Fill a plastic sports bottle with water, leave the cap off, and squeeze it over a sink. Record the results in your Science Journal.
2. Refill the bottle, screw on the cap, and squeeze it over a sink. Record the results in your Science Journal.
3. Compare the results of step 1 and step 2.
4. **Think Critically** In your Science Journal, indicate which step modeled your veins and which step modeled your arteries. Explain.

Preview this chapter's content and activities at blue.msscience.com

Human Systems Make the following Foldable to ensure you have understood the content by defining the vocabulary terms on the interactions of human systems.

STEP 1 Fold a sheet of notebook paper vertically from left to right.

STEP 2 Cut along every third line of only the top layer to form tabs.

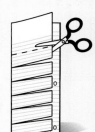

STEP 3 Label each tab.

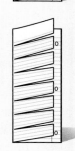

Read and Write Before you read the chapter, write the vocabulary words listed on the front page of each section on the front of the tabs. On the back of each tab, write what you think the word means. As you read the chapter, check your definitions and add to or change them as needed.

Get Ready to Read

New Vocabulary

① Learn It! What should you do if you find a word you don't know or understand? Here are some suggested strategies:

1. Use context clues (from the sentence or the paragraph) to help you define it.
2. Look for prefixes, suffixes, or root words that you already know.
3. Write it down and ask for help with the meaning.
4. Guess at its meaning.
5. Look it up in the glossary or a dictionary.

② Practice It! Look at the term *organic compounds* in the following passage. See how context clues can help you understand its meaning.

Context Clue
...compounds that contain carbon, with few exceptions.

Context Clue
...not considered organic are carbon dioxide and carbon monoxide.

Context Clue
The four groups of organic chemicals that make up all living things are carbohydrates, lipids, nucleic acids, and proteins.

> Living things are made of **organic compounds,** which are compounds that contain carbon, with few exceptions. Two of the carbon-containing compounds that are not considered organic are carbon dioxide and carbon monoxide. The four groups of organic chemicals that make up all living things are carbohydrates, lipids, nucleic acids, and proteins.
>
> —*from page 67*

③ Apply It! Make a vocabulary bookmark with a strip of paper. As you read, keep track of words you do not know or want to learn more about.

Target Your Reading

Use this to focus on the main ideas as you read the chapter.

1 Before you read the chapter, respond to the statements below on your worksheet or on a numbered sheet of paper.
- Write an **A** if you **agree** with the statement.
- Write a **D** if you **disagree** with the statement.

2 After you read the chapter, look back to this page to see if you've changed your mind about any of the statements.
- If any of your answers changed, explain why.
- Change any false statements into true statements.
- Use your revised statements as a study guide.

Reading Tip

Read a paragraph containing a vocabulary word from beginning to end. Then, go back to determine the meaning of the word.

Science Online
Print out a worksheet of this page at blue.msscience.com

Before You Read A or D		Statement	After You Read A or D
	1	The human body is organized into organ systems.	
	2	Proteins are inorganic substances needed by the body to build and repair tissues.	
	3	Nucleic acids are organic compounds that store genetic information.	
	4	The human body contains cells that function independently.	
	5	The nervous system does not function with digestive system.	
	6	In your body, cellular respiration can occur without help from your circulatory system.	
	7	You skin helps to eliminate wastes and excess substances from your body.	
	8	Your body responds to internal and external stimuli continually.	
	9	Negative feedback systems continue processes until homeostasis is restored.	

64 B

section 1
The Human Organism

as you read

What You'll Learn
- **Describe** the basic structure and function of a typical human cell.
- **Identify and describe** the five levels of organization in the body.

Why It's Important
Your body is a complex structure made up of many parts. Good health depends upon all of these parts functioning properly.

Review Vocabulary
matter: any substance that has mass and occupies space

New Vocabulary
- mineral
- organic compound
- cell
- tissue
- organ
- organ system

Organization in the Human Body

The next time you pass a building under construction, like the one in **Figure 1,** stop for a moment and think about how something so complex can be built. In the early stages of construction, the building lacks organization and looks different from the way it will look when it is finished. Some of the raw materials from which it will be built began as sand grains, clay particles, trees, and iron ore. Eventually, these materials are made into more useful structures, like concrete, plaster, wood beams, and steel supports. The separate pieces then are organized to form the basic framework of the building. When other finishing materials are added, the building is complete.

Although your body is not made of sand, clay, wood, and steel, it is composed of a series of building blocks that are different in size and complexity. As in a building, all of the things that your body is made up of are vital to your health.

Chemical Basis of Life

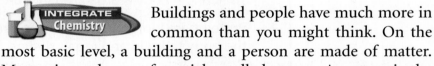 Buildings and people have much more in common than you might think. On the most basic level, a building and a person are made of matter. Matter is made up of particles called atoms. An atom is the smallest part of matter.

Elements and Compounds Matter that has the same composition and properties throughout is a substance. Substances are either elements or compounds. An element is made up of only one kind of atom. Scientists have discovered 90 natural elements on Earth.

Substances that are made of more than one element are called compounds. A compound is two or more elements that are chemically combined. What elements make up the compounds carbon dioxide (CO_2) and ammonia (NH_3)?

Figure 1 This building is in various stages of construction.

Inorganic Substances Chemicals in living things are classified as either inorganic or organic. Inorganic substances typically come from nonliving sources such as air, soil, or water.

Minerals are inorganic substances that are involved in many of the body's chemical reactions, as listed in **Table 1.** They are required in small amounts, yet your body must have them to maintain good health and fight disease.

Table 1 Important Minerals in Your Diet		
Mineral	**Use**	**Dietary Source**
Calcium	Formation of bones and teeth, blood clotting, muscle and nerve functions	Dairy products, leafy green vegetables, nuts, whole grains
Chlorine	Fluid balance, pH balance	Table salt
Cobalt	Formation of red blood cells	Meat, dairy products
Copper	Development of red blood cells and respiratory enzymes	Kidney, liver, beans, whole-meal flour, lentils, raisins
Fluorine	Formation of bones and teeth	Fluoridated water
Iodine	Part of thyroid hormone	Seafood, iodized table salt
Iron	Part of hemoglobin and some enzymes	Liver, egg yolk, peas, nuts, whole grains, red meat, raisins, leafy green vegetables
Magnesium	Muscle and nerve functions, bone formation, breakdown of proteins and carbohydrates, enzyme function	Potatoes, fruits, whole-grain cereal, vegetables, dairy products
Manganese	Growth of cartilage and bone; breakdown of carbohydrates, proteins, and fats	Wheat germ, nuts, bran, leafy green vegetables
Phosphorus	Bone formation, nerve function, regulation of blood pH, muscle contraction	Milk, whole-grain cereal, meats, vegetables, nuts
Potassium	Muscle and nerve function	Grains, fruits, vegetables, ketchup
Sodium	Muscle and nerve function, water balance, regulation of body fluid pH	Table salt, bacon, butter, vegetables
Sulphur	Builds hair, nails and skin; part of insulin	Nuts, dried fruits, oatmeal, eggs, beans
Zinc	Digestion, healing, taste, smell	Liver, seafood

Inorganic Chemicals in Your Body Inorganic chemicals play many important roles in your body. Salt, an important chemical in your blood, is an example of an inorganic substance in the body. Chemically speaking, blood is mostly water (H_2O), but it also contains sodium ions (Na^+) and chlorine ions (Cl^-) in the form of dissolved sodium chloride (NaCl), also known as table salt. Water, for instance, makes up more than 70 percent of the body's tissues. It also plays a role in nearly every bodily function, such as digesting foods, muscle function, delivering oxygen to cells, and removing wastes from the body. Over long periods of time, a lack of water can lead to problems in digestion, circulation, and kidney function. On a daily basis, not getting enough water can cause dry skin, headaches, and fatigue.

Reading Check *What are the roles of water in your body?*

Organic Substances You probably have heard the expression "You are what you eat." For most people, it means that a person's health is a reflection of the kinds of foods he or she eats. In another way, it means that your body is made up of the same kinds of chemicals that are found in the things that you eat every day.

Living things are made of **organic compounds,** which are compounds containing carbon, with a few exceptions. Two of the carbon-containing compounds that are not considered organic are carbon dioxide and carbon monoxide. The four groups of organic compounds are carbohydrates, lipids, nucleic acids, and proteins. Foods that contain these compounds are shown in **Figure 2.**

Topic: Water
Visit blue.msscience.com for Web links to information about the importance of water to living things.

Activity Design a label to place on the back of a bottle of water detailing the benefits of water to the human body.

Figure 2 Organic compounds needed by your body—carbohydrates, lipids, nucleic acids, and proteins—can be found in the foods you eat.
Identify *four carbohydrates in this picture.*

66 CHAPTER 3 Interactions of Human Systems

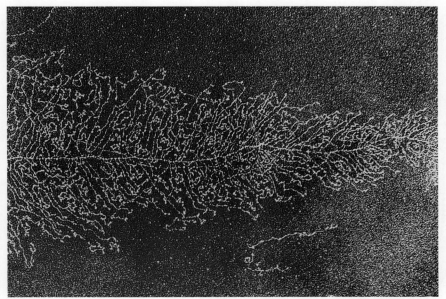

Figure 3 This scanning electron micrograph shows a nucleic acid. Nucleic acids are made of carbon, hydrogen, oxygen, nitrogen, and phosphorus. DNA and RNA are examples of nucleic acids found in your body.

SEM Magnification: 1500×

Life's Organic Compounds The next time you come home from school feeling tired and listless, do what many professional athletes do when they need an energy boost—eat some pasta. Maybe you would prefer a baked potato or a bowl of rice. These foods contain carbohydrates, which are the main source of energy for living things. Carbohydrates are made up of carbon (C), hydrogen (H), and oxygen (O).

Most of your energy comes from carbohydrates because they make up the largest part of your diet. But lipids contain more energy per molecule than carbohydrates do. Lipids, commonly called fats and oils, are stored in your body as energy reserves. When your supply of carbohydrates is low, your body turns to its fat reserves for energy.

Reading Check *Which organic compound contains more energy than carbohydrates?*

Nucleic acids, as shown in **Figure 3,** are large, complex organic compounds that store information in the form of a code. One type of nucleic acid called DNA often is called the blueprint for a living thing. DNA carries information that directs cell activities and instructions for making all proteins. Another nucleic acid called RNA is responsible for making proteins from amino acids.

Proteins serve many important functions in living things. Many of the structural parts of your body—hair, nails, skin, muscles, and blood vessels—are made, in part, of protein. Other proteins, such as enzymes, help your body carry out important processes, such as growth, repair, digestion, cellular respiration, and the transmission of nerve impulses.

Body Elements Three of the chemical elements that make up the human body also are some that make up Earth's crust. The common elements are oxygen, calcium, and potassium. Earth's crust is 46.4 percent oxygen, 3.6 percent calcium and 2.6 percent potassium. What percentages of these elements are present in a typical human body?

Figure 4 Cells are complex structures that perform the activities necessary for life. Cells are made up of various parts. Each part has a special job to do in the cell.
Name four cell parts that are inside of the cell membrane.

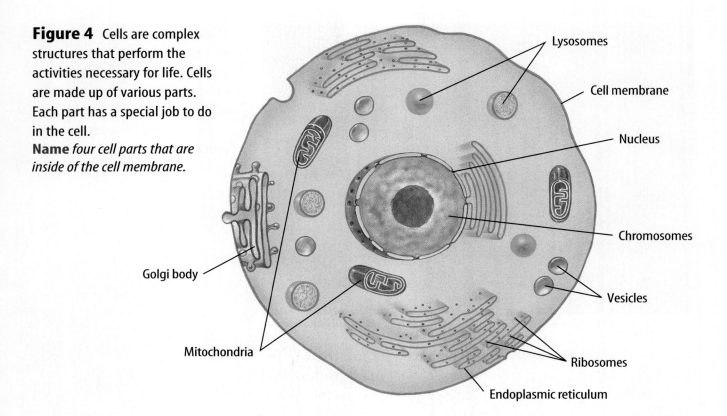

Cells—Living Factories

Organic and inorganic chemicals combine in an organized way to form cells. A **cell** is the smallest functional unit in an organism. Every living thing is made up of one or more cells. **Figure 4** illustrates the basic structure of a human cell.

Most cells can't be seen with the unaided eye. You need a microscope to see human cells. Despite its size, the cell is the reason that living things can carry on all of the important activities of life, such as digestion, growth, and reproduction. Cells contain different parts that perform these activities necessary for life. Human cells, like the one shown in **Figure 5,** have a variety of shapes and sizes, but more similarities than differences can be identified.

In some ways, a living cell can be compared to a busy factory. Many factories have separate locations inside the building to complete the tasks of the production process. Raw materials are brought in, assembled into products, packaged, and delivered to customers. In a similar way, living cells take in raw materials from their surroundings, such as food, oxygen, water, and minerals. Cells then use these materials in chemical reactions to make proteins and other products necessary for life. During some of these chemical reactions, energy is released.

Reading Check *How is a cell like a factory?*

NATIONAL GEOGRAPHIC VISUALIZING HUMAN CELLS

Figure 5

Throughout your body, many different types of cells carry out the complex processes that keep you alive and functioning at your best. A cell's size, shape, and structure are closely related to the tasks it performs, as shown here.

▼ **EPITHELIAL (ep uh THEE lee ul) CELLS** Humans have three types of epithelial cells, which protect or lubricate various parts of the body. Squamous (SKWAY mus) epithelial cells cover your body and line your mouth and esophagus. Cuboidal (kyew BOY dul) epithelial cells line your body cavities and glands. Columnar (kuh LUM nur) epithelial cells line your stomach, intestines, and respiratory passages.

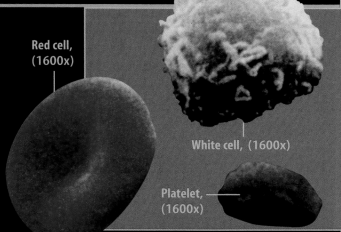

▲ **BLOOD CELLS** Red blood cells carry oxygen to your tissues and remove carbon dioxide. White blood cells protect your body from harmful, foreign substances or organisms. Platelets help form blood clots.

Red cell, (1600x)
White cell, (1600x)
Platelet, (1600x)

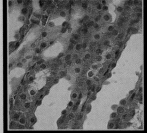
Squamous epithelium, (125x)

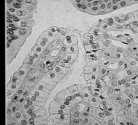
Cuboidal epithelium, (80x)

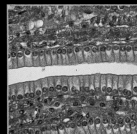
Columnar epithelium, (100x)

▶ **CONNECTIVE TISSUE CELLS** Like the cells in blood—a connective tissue—some connective tissue cells wander. Wandering connective tissue cells can protect, repair, and transport. Other connective tissue cells are fixed, like those in the photo to the right. These cells can insulate, maintain, store, support, or produce substances.

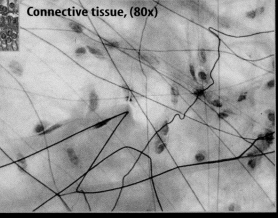
Connective tissue, (80x)

▶ **MUSCLE CELLS** Skeletal muscle cells move body parts such as arms and legs. Smooth muscle cells move substances within or through internal organs and vessels. Cardiac muscle cells are found only in the heart.

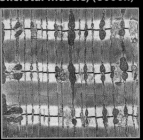
Skeletal muscle, (6000x)

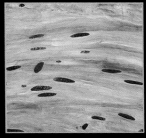
Smooth muscle, (100x)

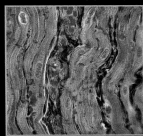
Cardiac muscle, (225x)

SECTION 1 The Human Organism

From Cell to Organism Your body is constructed from a variety of different types of cells. Are you no more than a collection of cells with each one doing its own thing?

Many-celled organisms are highly organized. Think about what it is like to be in a band. Each member plays the music written for his or her instrument. However, for the band to be successful, the members must work together. In a similar way, the cells in an organism need to perform their individual functions yet work together so the whole organism can function.

Levels of Organization The different levels of organization in the human body are shown in **Figure 6.** As stated earlier, cells represent the first level of organization in the body. Cells are organized into tissues. **Tissues** are groups of similar cells that do the same sort of work. In the human body, nerve tissue carries nerve impulses around the body. It helps different parts of the body communicate. Blood tissue transports oxygen and nutrients to and from cells.

Tissues are organized into larger structures called organs. An **organ** is a structure made up of different types of tissues that work together. For example, your heart is an organ made up of cardiac muscle tissue, nerve tissue, and blood tissue.

Figure 6 The human organism is made of small parts that work together and form larger, more complex parts.

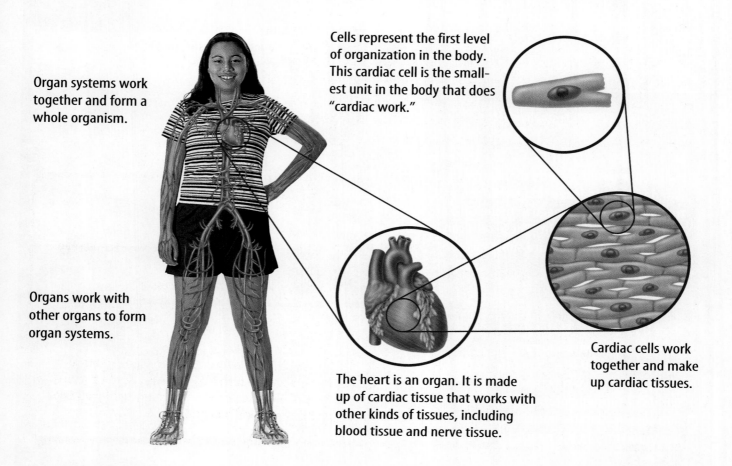

Organ systems work together and form a whole organism.

Organs work with other organs to form organ systems.

Cells represent the first level of organization in the body. This cardiac cell is the smallest unit in the body that does "cardiac work."

The heart is an organ. It is made up of cardiac tissue that works with other kinds of tissues, including blood tissue and nerve tissue.

Cardiac cells work together and make up cardiac tissues.

Table 2 The Systems of the Human Body

System	Main Organs	Function
Integumentary	Skin	Protects the body and prevents water loss
Muscular	Muscles	Movement of the body, attached to bones
Skeletal	Bones	Support and protection of soft body parts
Nervous	Brain, spinal cord, nerves	Controls mental and bodily functions
Endocrine	Pancreas, pituitary gland	Controls homeostasis by releasing hormones
Circulatory	Heart, blood vessels	Transport of materials to and from body cells
Lymphatic	Spleen, thymus, tonsils	Remove dead cells and foreign bodies from body fluids
Respiratory	Lungs, trachea	Exchange of gases between blood and the environment
Digestive	Stomach, small intestine	Break down food for absorption into the blood
Urinary	Kidneys, bladder	Control of water balance and chemical makeup of blood
Reproductive	Testes, ovaries	Production of sex cells

Systems Organs in your body involved in the circulation of blood include your heart, arteries, veins, and capillaries. A group of organs that work together to do a certain job is called an **organ system**. Human body systems, as listed in **Table 2**, are interdependent and work together to perform life functions.

section 1 review

Summary

Organization in the Human Body
- The human body is complex, but organized.

Chemical Basis of Life
- Inorganic substances in the human body include minerals, salt, and water.
- Organic substances in the human body are carbohydrates, lipids, nucleic acids, and proteins.

Cells—Living Factories
- Cells perform the activities necessary for life.
- The human body has five levels of organization.

Self Check

1. **List**, in order, the levels of organization in the human organism.
2. **Explain** why your body needs minerals.
3. **Think Critically** Explain why cells are vital for life.

Applying Math

4. **Solve One-Step Equations** Every milliliter of blood has five million red blood cells. On average, a person has 3.3 L of blood. Calculate how many red blood cells are found in a typical human body.

Observing Cells

Cells come in a variety of shapes and sizes. They might look different and perform different functions, but they have some common structures.

▶ Real-World Question

How do prepared human cheek cells and living onion cells compare?

Goals
- **Observe** the structure of human cheek cells.
- **Observe** the structure of onion cells.

Materials
compound light microscope
microscope slide
coverslip
prepared human cheek cell slide
small piece of onion
forceps
iodine solution
water
medicine dropper

Safety Precautions

WARNING: *These solutions can cause stains. Do not allow them to contact your skin or clothing.*

▶ Procedure

1. Obtain a prepared slide of human cheek cells from your teacher.
2. **Examine** the cheek cells under low power of the microscope and then under high power. In your Science Journal, draw several cheek cells as they appear under high power.
3. Using forceps, remove a small section of paper-thin tissue from the onion. Prepare a wet-mount slide of the onion tissue. Before putting on the coverslip, place a drop of iodine on the onion tissue.
4. **Examine** the onion-tissue slide under low power, then under high power. In your Science Journal, draw several onion cells as they appear under high power.

▶ Conclude and Apply

1. **Describe** the shape of human cheek cells.
2. **Describe** the shape of the onion cells.
3. **Name** the structure you saw in the onion cell but not in the human cheek cell.

Communicating Your Data

Make a colorful, poster-sized illustration of one of these cells. Label all visible parts of the cell. **For more help, refer to the** Science Skill Handbook.

Section 2: How Your Body Works

Body System Connections

The heart is the main organ of the circulatory system. Lungs carry out the most important role in the body's respiratory system. The heart and lungs are important organs working together to put oxygen into blood and take carbon dioxide out of it. This partnership of the heart and lungs, as shown in **Figure 7,** is one of the most important interactions in the body.

The heart also keeps blood moving so all the cells in the body can stay alive. Your heart and lungs work more quickly when you are active. When you rest or sleep, your heart and lungs slow down but never stop. In this section, you will read how other systems in your body interact and work together to carry out important life processes.

Feeding Cells

The foods you eat are filled with the chemicals that your body needs to grow strong and healthy. How are those chemicals distributed throughout your body for energy, growth, repair, and other important body functions?

as you read

What You'll Learn
- **Discuss** how body systems work together to carry out important life functions.
- **Explain** how negative feedback mechanisms in the body help maintain homeostasis.
- **Compare** negative feedback mechanisms and positive feedback mechanisms in the body.

Why It's Important
Systems in the body are interdependent—they depend on each other and work together.

Review Vocabulary
energy: the ability to cause change

New Vocabulary
- digestion
- enzyme
- villi
- absorption
- cellular respiration
- alveoli
- excretion
- homeostasis
- negative feedback

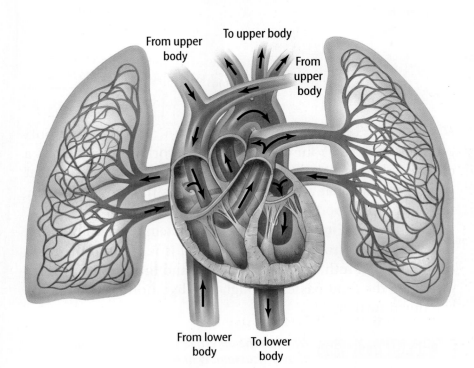

Figure 7 The arrows show the direction of blood flow through the heart, lungs, and body. **Explain** what occurs in the lungs.

SECTION 2 How Your Body Works **73**

Figure 8 During digestion, glucose, salt, vitamins, and water can be absorbed immediately into the bloodstream. All other foods must be broken down before they can pass through the walls of the small intestine and be absorbed by the blood.

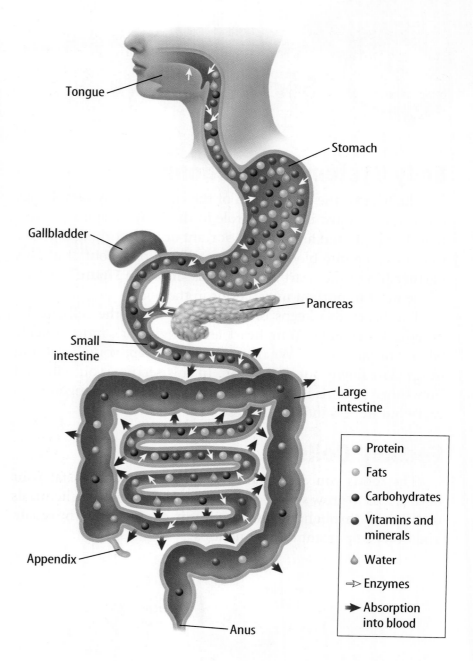

Observing the Gases That You Exhale

Procedure
1. Dry your hands with a **towel**.
2. Hold the palm of your hand up to your mouth, then exhale into it.
3. Feel your palm with your other hand.
4. Hold a **mirror** up to your mouth and exhale again. Observe what happens to the mirror.

Analysis
1. How did your hand feel after you breathed on it?
2. What happened to the mirror when you breathed on it?
3. Besides carbon dioxide, hypothesize what other things are exhaled.

Digestive System In the factory, an assembly line puts products together. Your digestive system works in reverse. It is a disassembly line. **Digestion** is the breakdown of the foods that you eat into smaller and simpler molecules that can be used by the cells in your body.

As shown in **Figure 8,** the digestive system is basically a long tube that runs through your body. Food is taken into the body through the mouth, where it is chewed and broken into smaller pieces. As the food passes through the digestion tube, it is broken down further.

Reading Check Why can the digestive system be thought of as a disassembly line?

74 CHAPTER 3 Interactions of Human Systems

Helping Digestion Other body systems help the digestive system with this work. For instance, your brain signals some cells in your mouth to produce saliva when you see, smell, taste, or sometimes just think of food. Saliva softens and lubricates food and also contains an enzyme that helps the digestive process. An **enzyme** (EN zime) is a protein that helps the body carry out chemical reactions. When food reaches other parts of the digestive tract, such as the stomach and small intestine, the brain again directs the body to produce other enzymes that help break down the chemicals in food even further.

Circulation and Digestion The body's circulatory system transports blood that provides nutrients for the cells of the digestive system, just as it does for all the cells in the body. However, the most important connection between the circulatory system and the digestive system occurs in the small intestine, as shown in **Figure 9.**

Your small intestine is 2.5 cm to 5 cm in diameter but is 4 m to 7 m in length. Imagine a 7-m garden hose coiled up inside your abdomen. Unlike a garden hose, the small intestine is not smooth. It is lined with tiny fingerlike projections called **villi.** The many blood vessels in the villi are a clue to their function. By the time food reaches the small intestine, it has been broken down into molecules that can pass through the walls of the villi, then into the bloodstream. This process is known as **absorption.** After they are absorbed into the bloodstream, nutrients are carried to cells throughout the body. Materials that are not absorbed pass out of the body as wastes.

Antibiotics In 1928, Alexander Fleming discovered the first antibiotic—penicillin. This led to the discovery of other antibiotics and changed the way infections are treated. However, antibiotics that are used to kill disease-causing bacteria also can kill some helpful ones, like the bacteria that live in your digestive tract and help with the process of digestion. Killing these helpful bacteria upsets the digestive process. What is a common side effect of taking antibiotics?

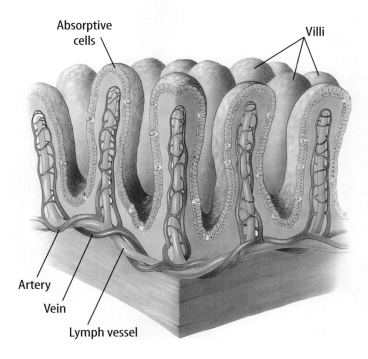

Figure 9 Villi make absorption more efficient because they increase the surface area of the small intestine by ten times.
Infer *what might happen to a person's weight if the number of villi were drastically increased. Why?*

Observing a Chemical Reaction

Procedure
1. Pour a small amount of **3% hydrogen peroxide solution** into a **petri dish.**
2. Use a **thermometer probe** to take the temperature of the peroxide solution.
3. Place a small chunk of fresh **liver** in the petri dish. Observe what happens.
4. Wait 1 min, then take the temperature of the solution a second time.

Analysis
1. Was energy released during this chemical reaction? How do you know?
2. How is this chemical reaction similar to the process of respiration in cells?

Energy for the Body

The body's cells release energy from food in the process of carrying out different activities. This important process occurs in the mitochondria of every cell. **Cellular respiration** is a series of chemical processes in which oxygen combines with food molecules and energy is released. A cell uses this released energy to perform all of its tasks. Carbon dioxide (CO_2) and water are given off as wastes.

For cellular respiration to occur, a constant supply of nutrients and oxygen must be available. When not enough oxygen is available, the cycle cannot be completed. In cells, hydrogen ions typically combine with oxygen molecules to produce water as a waste product. When the oxygen level is low, hydrogen combines with a compound that is produced when glucose is broken down to form lactic acid as a waste product. As lactic acid increases, muscle fatigue occurs, cellular respiration is stimulated, and rapid breathing results.

Respiratory System Several body systems work together to remove wastes. Other body systems help the raw materials of cellular respiration—food and oxygen—get to your cells. You've read how the digestive and circulatory systems interact to break down food and transport it to cells. As shown in **Figure 10,** oxygen enters your body and carbon dioxide is released through your respiratory system.

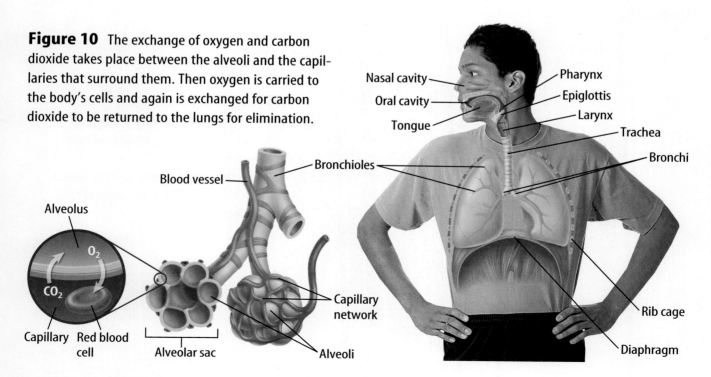

Figure 10 The exchange of oxygen and carbon dioxide takes place between the alveoli and the capillaries that surround them. Then oxygen is carried to the body's cells and again is exchanged for carbon dioxide to be returned to the lungs for elimination.

Oxygen Absorption Recall how food molecules move into the bloodstream through the villi of the small intestine. A similar process of oxygen absorption occurs in the lungs, where your respiratory and circulatory systems interact. At the lower end of the trachea are two short branches, called bronchi, which carry air into the lungs. Within the lungs, the bronchi branch into smaller and smaller passageways. At the ends of the narrowest tubes are clusters of tiny, thin-walled sacs called **alveoli** (al VEE uh li). Lungs are made up of millions of alveoli. A network of tiny blood vessels called capillaries surrounds each cluster of alveoli.

Reading Check *Compare food absorption and oxygen absorption in the body.*

Applying Math — Calculate Volume

LUNG VOLUME A person's lung capacity depends on their gender and overall physical condition. Using the table to the right, calculate the lung volume for a 5-ft-tall, nonathletic female.

Body-Size Factor	
Body	(cm^2)
Female	20
Female athlete	22
Male	25
Male athlete	29

Solution

1 *This is what you know:*
- height in inches = 12 × 5 = 60 inches
- height in centimeters = 60 × 2.54 cm = 152.4 cm
- body-size factor (from chart) = 20 cm^2

2 *This is what you need to find out:* What is the lung capacity of a 5-ft-tall, nonathletic female?

3 *This is the procedure you need to use:*
- Use this equation:
  height in centimeters × body-size factor = volume of lungs
- Substitute in the known values and solve:
  152.4 cm × 20 cm^2 = 3,048 cm^3 is the lung volume of a 5-ft-tall, nonathletic female

4 *Check your answer:* Divide 3,048 cm^3 by 20 cm^2 and you should get 152.4 cm.

Practice Problems

1. Calculate your lung volume.
2. Calculate the difference in the lung volumes of a male athlete who is 62 inches tall and a female athlete who is 56 inches tall.

For more practice, visit blue.msscience.com/math_practice

Getting Rid of Wastes Carbon dioxide gas is released from your body through the combined efforts of the circulatory and respiratory systems. But CO_2 isn't the only waste product of the body. Earlier you read that undigested material is eliminated by your digestive system. Some salts also are given off when you sweat. All of these systems function together to make up your excretory system. The removal of waste products, or **excretion**, is an important life process in all organisms. If wastes aren't removed, they can build up to toxic levels and damage cells and eventually the whole organism.

Another body system that removes wastes is the urinary system, as shown in **Figure 11**. The main organs of the urinary system are two bean-shaped kidneys. All of your blood passes through the kidneys many times each day. The kidneys remove cell wastes and help control the amount of water in the blood.

Figure 11 Waste products are removed from blood in the kidneys by nephrons. Three processes occur in nephrons—filtration, secretion, and reabsorption.

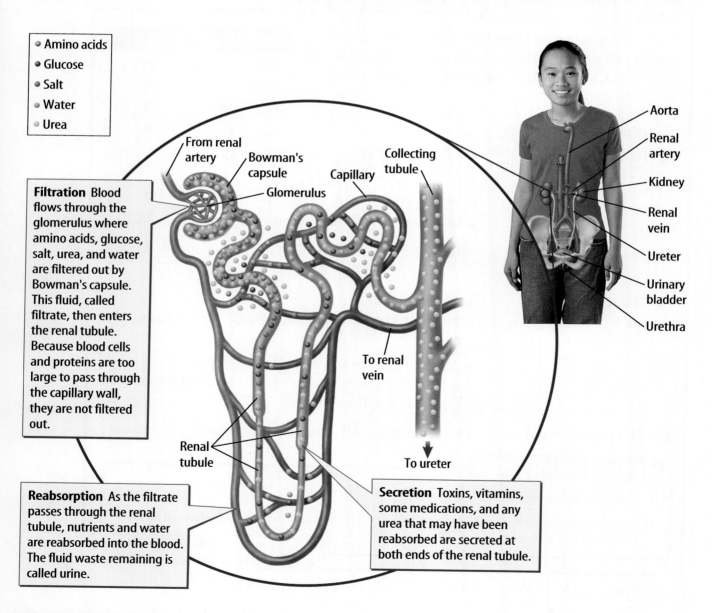

- Amino acids
- Glucose
- Salt
- Water
- Urea

Filtration Blood flows through the glomerulus where amino acids, glucose, salt, urea, and water are filtered out by Bowman's capsule. This fluid, called filtrate, then enters the renal tubule. Because blood cells and proteins are too large to pass through the capillary wall, they are not filtered out.

Reabsorption As the filtrate passes through the renal tubule, nutrients and water are reabsorbed into the blood. The fluid waste remaining is called urine.

Secretion Toxins, vitamins, some medications, and any urea that may have been reabsorbed are secreted at both ends of the renal tubule.

78 CHAPTER 3 Interactions of Human Systems

Nephrons The kidney is made up of millions of tiny units, called nephrons, where the blood is filtered. Each nephron has a cuplike structure and a snakelike tubule. Water, salt, sugar, and wastes from your blood first pass into the cuplike structure and then into the tubule. Capillaries surrounding the tubule reabsorb most of the water, sugar, and salt and return them to the blood. Anything left behind is waste. The waste liquid, urine, contains excess water, salts, and other wastes, and eventually is eliminated from the body.

Interdependence of Body Systems

The human body has the remarkable ability to sense changes in its environment and to respond by making changes in body functions. As a result of the changes in functions, your body's internal environment is kept stable. The process by which the body maintains a stable internal environment is called **homeostasis**.

Reading Check *What is homeostasis?*

You probably know how your heart rate and breathing increase as you exercise. These bodily changes are direct responses to the increased level of activity. During exercise, your leg muscle cells use more and more oxygen and produce lots of carbon dioxide waste. You are not even aware that your brain responds to these internal changes and directs your heart and lungs to work harder. This delivers oxygen to your muscle cells and eliminates carbon dioxide at a faster rate.

When the body becomes overheated, as shown in **Figure 12,** it responds in a coordinated way to maintain homeostasis. The brain senses the increase in internal temperature and directs the body to make changes that will keep it from becoming too hot. Sweating, for example, is one response to an increase in temperature. As sweat evaporates, it cools the skin. Another response is the widening or dilation of blood vessels in the skin. Dilation helps release the body's thermal energy.

Rapid Breathing Breathing gets faster to help move more heat out of the body through the lungs.

Flushing Blood vessels in your skin dilate. This action helps the body release internal body heat.

Sweating When sweat evaporates it helps cool the skin surface.

Figure 12 The regulation of body temperature is a familiar example of homeostasis.
Describe *what happens when the body becomes too hot.*

Topic: Negative Feedback
Visit blue.msscience.com for Web links to information about negative feedback.

Activity Make a model of a simple negative feedback system.

Negative Feedback Most body systems maintain homeostasis by the action of negative feedback mechanisms. In a **negative feedback** mechanism, the body changes an internal condition back to its normal state. The responses that occur when you exercise or your body becomes overheated are examples of negative feedback.

Reading Check *What type of mechanism is used by most body systems to maintain homeostasis?*

Blood pressure also is controlled by a negative feedback system. Specialized cells within the walls of major arteries are sensitive to changes in blood pressure. If blood pressure rises, a message is relayed to the brain signaling this internal change. The brain responds by sending a message to the heart to slow down—an action that decreases blood pressure.

Maintaining Chemical Balance Negative feedback is important for maintaining a proper chemical balance in the body. **Figure 13** shows what occurs when you eat and the glucose concentration in your blood rises above normal. Glucose is a type of sugar. When glucose levels are too high, a hormone called insulin is secreted by the pancreas. Insulin stimulates the absorption of glucose by cells and the liver's conversion of glucose into glycogen. Glycogen is a sugar that can be stored in the liver and muscle cells. As glucose levels decrease, less insulin is produced. A hormone called glucagon (GLEW kuh gahn) is produced when glucose levels are too low. Glucagon stimulates the liver's conversion of stored glycogen into glucose. Glucose is released into the blood. As glucose levels increase, less glucagon is produced.

Figure 13 A negative-feedback response within the pancreas regulates blood sugar levels.
Describe *what would happen to blood sugar levels if the pancreas did not secrete enough insulin.*

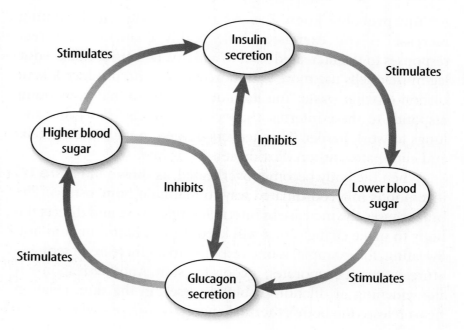

Positive Feedback In a healthy body, positive feedback mechanisms are rare. The term *positive* means that when a change from the normal state occurs, the positive feedback mechanism does not restore the body to a normal state. It causes an even greater change. One example of positive feedback is the contraction of the uterus during childbirth, as shown in **Figure 14.** Contractions of the uterus push the baby against the opening of the birth canal. When the brain senses this change, it signals the uterus to increase contractions. The positive feedback mechanism continues until the baby passes through the birth canal.

Blood Clotting Another example of a positive feedback response is blood clotting. When a blood vessel is cut or torn, the vessel constricts and chemicals are released. The walls of the vessel also become sticky and adhere to each other. Platelets attach to the vessel walls. This causes more of the chemicals to be released. Therefore, more platelets become sticky and adhere to one another. When the hole is plugged, the process stops. The clot becomes hard, white blood cells destroy any bacteria, and skin cells begin to repair the hole.

In this chapter, you've read that the human body is a complex, yet highly organized structure. It is a natural system made up of individual parts that carry out specific jobs. As in all large systems, proper functioning at all levels is necessary for the health of the entire system.

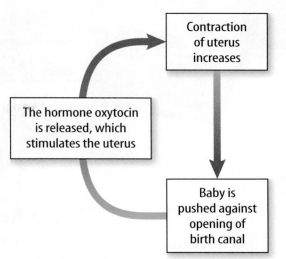

Figure 14 A positive feedback response within the uterus stimulates contractions during childbirth.

section 2 review

Summary

Feeding Cells
- Your digestive and circulatory systems work together to feed cells.

Energy for the Body
- The process of cellular respiration combines food and oxygen to release energy for use by the cell.
- The circulatory, respiratory, and excretory systems remove wastes from the body.

Interdependence of Body Systems
- Your body systems work together to maintain homeostasis.
- Negative feedback systems help the body maintain homeostasis.

Self Check

1. **Discuss** how the digestive and circulatory systems work together to get food to cells.
2. **Compare and contrast** negative and positive feedback mechanisms.
3. **Infer** why a circulatory system disorder might be harmful to other body systems.
4. **Think Critically** Discuss the similarities between the villi of the small intestine and alveoli of the lungs.

Applying Skills

5. **Communicate** In your Science Journal, write a short paragraph that compares the structure and organization of the human body to a sports team.

Design Your Own

Does exercise affect cellular respiration?

Goals
- **Observe** the effect of carbon dioxide on the bromthymol blue indicator solution.
- **Predict** how exercise will affect the amount of carbon dioxide that is exhaled by the body.

Possible Materials
graduated cylinder
large beakers (2)
straws (2)
balloons (2)
bromthymol blue indicator solution (200 mL)
stopwatch
*clock with a second hand
glass-marking pencil
*Alternate materials

Safety Precautions

● Real-World Question

When you exercise, muscle cells in your body use up a lot of energy. Carbon dioxide is a waste product of cellular respiration, the energy-releasing process in your cells. Does exercise affect the amount of carbon dioxide exhaled by the lungs?

● Form a Hypothesis

Make a hypothesis about how exercise affects cellular respiration. Will exercise increase or decrease the amount of carbon dioxide you exhale?

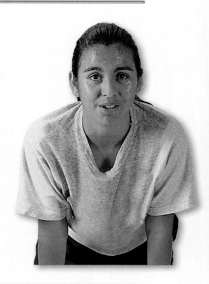

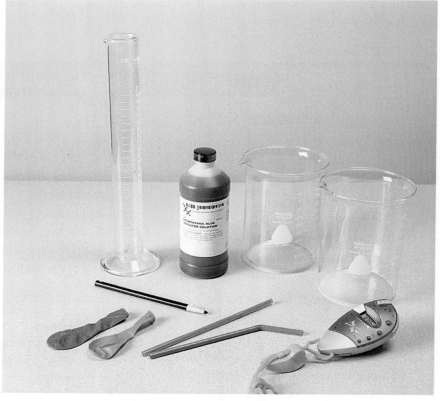

82 CHAPTER 3 Interactions of Human Systems

Using Scientific Methods

▶ Test Your Hypothesis

Make a Plan

1. As a group, predict how exercise will affect the amount of carbon dioxide exhaled by the lungs. Identify a way that you can test your hypothesis.
2. **List** the steps you will follow to test your hypothesis. Be sure to describe exactly what you will do in each step.
3. Make a data table in your Science Journal to record your observations.
4. Read over your entire experimental procedure. Do the steps make sense? Are they arranged in the correct order?

Follow Your Plan

1. Make sure your teacher approves your plan and your data table before you start.
2. Carry out the experiment according to the approved plan.
3. **Record** all of your observations in your data table.

▶ Analyze Your Data

1. What caused the indicator solution to change color? Describe the color change.
2. **Compare** the time it took the bromthymol blue indicator solution to change color before exercising and after exercising.
3. **Compare** your results with the results of other groups in your class.

▶ Conclude and Apply

1. **Explain** whether the results supported your hypothesis.
2. **Describe** how exercise affects the amount of carbon dioxide you exhale.

Communicating Your Data

Prepare a poster project showing how increasing levels of activity affect the production of carbon dioxide. Include photos, a data table, and a step-by-step procedure.

LAB 83

SCIENCE Stats

Astonishing Human Systems

Did you know...

...If all the capillaries in an adult's body were flattened, they would cover an area of about 4,815 m², or about 90 percent of the area of a football field. In comparison, the total surface area of the average adult's skin is only 2 m².

Football field: 5,390 square meters
Adult's capillaries: 4,815 square meters
Adult's skin: 2 square meters

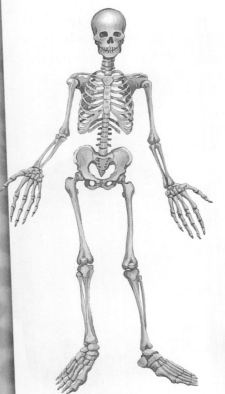

Average Brain Mass of Different Species

- Pig
- Cow
- Giraffe
- Human (adult)
- Elephant (African)

Mass in grams: 1,000 2,000 3,000 4,000 5,000

...More than half of the bones in your body are in your hands and feet. Your skeleton has 206 bones. There are 27 bones in each hand and 26 bones in each foot.

Applying Math Use a circle graph to show the percentages of bones in your hands, feet, and the rest of your body.

Find Out About It

Draw an events-chain concept map showing either the steps in the digestion process, excretory process, or process of respiration. For help, visit **blue.msscience.com/science_stats**.

chapter 3 Study Guide

Reviewing Main Ideas

Section 1 The Human Organism

1. Your body is a complex, organized structure.
2. Minerals are inorganic substances that your body needs for good health and to fight disease.
3. Water plays a role in nearly every body function.
4. All living things are made up of four groups of organic chemicals—carbohydrates, lipids, nucleic acids, and proteins.
5. Cells are composed of organic and inorganic chemicals.
6. Complex organisms are made of organ systems, organ systems are made of organs, organs are made of tissues, tissues are made of cells, and cells are made of chemicals.

Section 2 How Your Body Works

1. Organ systems in your body do not work alone. They are interdependent and work together to perform all of the body's functions.
2. Energy is transported to your cells by the combined efforts of the digestive and circulatory systems.
3. Waste products are removed from your body by the combined efforts of the circulatory, respiratory, and excretory systems.
4. Body systems work together to maintain homeostasis.
5. Homeostasis is maintained through negative feedback mechanisms, which help return the body to a normal state.

Visualizing Main Ideas

Copy and complete the following concept map.

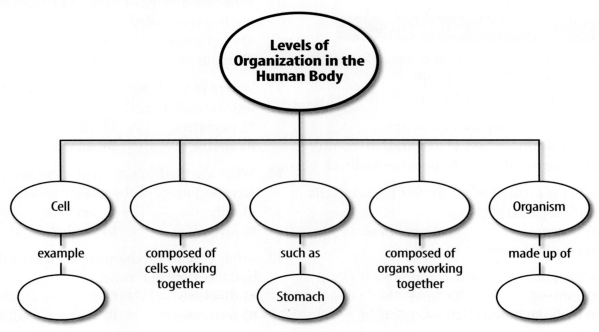

chapter 3 Review

Using Vocabulary

absorption p. 75	mineral p. 65
alveoli p. 77	negative feedback p. 80
cell p. 68	organ p. 70
cellular respiration p. 76	organic compound p. 66
digestion p. 74	organ system p. 71
enzyme p. 75	tissue p. 70
excretion p. 78	villi p. 75
homeostasis p. 79	

Use what you know about the vocabulary words to explain the differences in the following sets of words. Then explain how the words are related.

1. homeostasis—negative feedback
2. cell—tissue
3. absorption—excretion
4. alveoli—villi
5. digestion—cellular respiration
6. minerals—organic compounds
7. organ system—organ
8. organic compounds—enzymes
9. digestion—absorption

Checking Concepts

10. Which chemicals help you digest foods?
    A) enzymes   C) villi
    B) nephrons D) cells

11. What is the purpose of sweating?
    A) digestion   C) respiration
    B) homeostasis D) positive feedback

12. What is formed when similar types of cells work together?
    A) organs   C) organelles
    B) organ systems D) tissues

13. What kind of organic compound is DNA?
    A) mineral   C) lipid
    B) carbohydrate D) nucleic acid

Use the illustration below to answer questions 14 and 15.

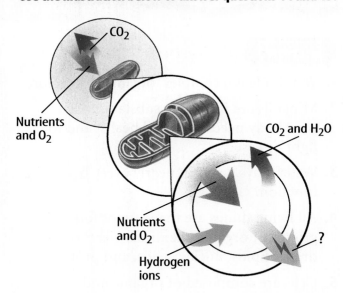

14. What cell process is illustrated above?
    A) excretion   C) cellular respiration
    B) absorption D) photosynthesis

15. What is released at the end of the chemical process shown in the illustration above?
    A) oxygen C) carbon dioxide
    B) energy D) water

16. What are groups of different tissues working together called?
    A) organelles   C) organs
    B) organ system D) organisms

17. Where in the lungs are oxygen and carbon dioxide exchanged?
    A) nephrons C) villi
    B) alveoli D) cells

18. What are the filtering, reabsorption, and secretion units of the kidney called?
    A) villi   C) neurons
    B) alveoli D) nephrons

19. What type of mechanism helps return the body to a normal state?
    A) digestive   C) negative feedback
    B) respiratory D) positive feedback

chapter 3 Review

Thinking Critically

20. **Explain** why a kidney disorder could disrupt the chemical balance of the body.

21. **Explain** why food needs to be digested before cells can use it for energy.

22. **Infer** why muscle cells have more mitochondria than skin cells.

23. **Describe** what might happen if your body had no way of detecting internal temperature changes.

24. **Discuss** an example, not given in this chapter, of how two of your body systems work together.

25. **Classify** You're stranded on an island and equipped with only a microscope. Describe what you could do to determine whether an unknown object is from a living or a nonliving thing.

26. **Compare and contrast** the levels of organization in the human body to the organization of a book.

27. **Concept Map** Study the cycle concept map in **Figure 13**. Copy and complete this cycle concept map, which shows how salt levels are maintained in the body.

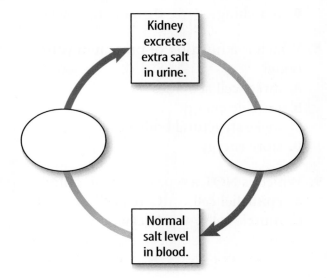

Performance Activities

28. **Experiment** Design an experiment to show that the process of respiration occurs in living yeast cells.

29. **Models** Using assorted household materials such as cardboard, foam, pipe cleaners, yarn, buttons, dry macaroni, or other small objects, construct a three-dimensional model of a human cell. On a separate sheet of paper, make a key to indicate which materials represent the different cell parts.

Applying Math

30. **New Skin** If the human body replaces the top layer of skin every 15 to 30 days, how many times per year does the body have a new top layer of skin?

Use the graph below to answer question 31.

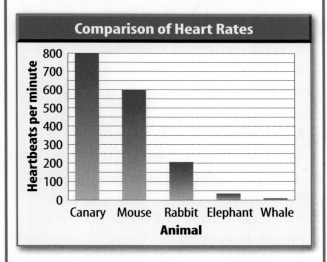

31. **Heart Rate** Smaller animals have faster heart rates than larger animals. Using the graph above, estimate how many times faster a mouse's heart beats when compared to an elephant's heart. How many times faster does a canary's heart beat when compared to a rabbit's heart?

chapter 3 Standardized Test Practice

Part 1 Multiple Choice

Record your answers on the answer sheet provided by your teacher or on a sheet of paper.

1. Which is a chemical compound?
   A. H_2
   B. CO_2
   C. O_2
   D. Cl

Use the illustration below to answer questions 2–4.

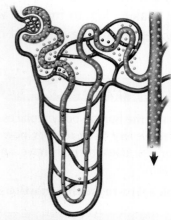

2. What structure is shown in the illustration?
   A. capillary
   B. alveolus
   C. nephron
   D. ureter

3. What is the function of the structure shown in the illustration?
   A. filtration
   B. digestion
   C. ingestion
   D. inspiration

4. If there are approximately one million of these structures in one kidney, approximately how many of these structures are present in your body?
   A. 1 million
   B. 0.5 million
   C. 2 million
   D. 10 million

> **Test-Taking Tip**
> **Study Aid** Do not "cram" the night before the test. It can hamper your memory and make you tired.

5. Which is the main energy source for your body?
   A. protein
   B. lipids
   C. carbohydrates
   D. minerals

Use the table below to answer questions 6 and 7.

Results from Ashley's Activities			
Activity	Pulse Rate (beats/min)	Body Temperature	Degree of Sweating
1	80	98.6°F	None
2	90	98.8°F	Minimal
3	100	98.9°F	Little
4	120	99.1°F	Moderate
5	150	99.5°F	Considerable

6. According to the information in this table, which activity caused Ashley's pulse to be less than 100 beats per minute?
   A. Activity 2
   B. Activity 3
   C. Activity 4
   D. Activity 5

7. Which is a reasonable hypothesis based on these data about what Ashley was doing during Activity 2?
   A. sprinting
   B. marching
   C. sitting down
   D. walking slowly

8. Which is a function of protein in your body?
   A. direct cell activities
   B. provide energy
   C. make structural body parts
   D. store energy

9. Which is NOT a type of human cell?
   A. epithelial cell
   B. muscle cell
   C. red cell
   D. cavity cell

Standardized Test Practice

Part 2 — Short Response/Grid In

Record your answers on the answer sheet provided by your teacher or on a sheet of paper.

10. If red blood cells are made at the rate of two million per second in the center of long bones, how many red blood cells are made in one hour?

11. If a cubic milliliter of blood has 10,000 white blood cells and 400,000 platelets, how many times more platelets than white blood cells are present in a cubic milliliter of blood?

Use the photo below to answer questions 12 and 13.

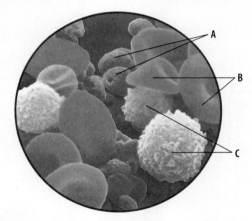

12. Identify the blood components labeled *A*. How do they help your body maintain homeostasis?

13. Thrombocytopenia is a condition in which the number of platelets in the blood is decreased. How would this affect the body?

14. At birth, your skeleton has approximately 300 bones. Some bones fused together as you developed. Now you have 206 bones. How many fewer bones do you have now?

15. If a cardiac muscle contracts about 70 times per minute, how many times does it contract in a day?

Part 3 — Open Ended

Record your answers on a sheet of paper.

16. Compare and contrast organic and inorganic substances. Give an example of each that is important for the human body.

17. Smooth muscle, found in the walls of blood vessels, controls the narrowing and widening of these vessels. What might happen to your body temperature if blood vessels in the skin did not contain smooth muscle?

18. Which body systems remove wastes? Discuss how these systems remove the wastes.

19. Urine can be tested for any signs of a urinary tract disease. Mrs. Chavez had a urine test that showed protein in the urine. Is protein normally present in the urine? Explain. What might the results of this urine test mean?

Use the illustration below to answer questions 20 and 21.

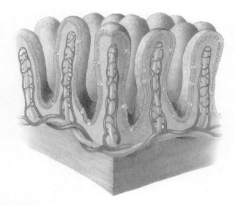

20. In which body system would you find these structures? Explain the function of these structures.

21. If a person had fewer of these structures than normally present, how would the person be affected? Explain.

unit 2 Ecology

How Are Beverages & Wildlife Connected?

In ancient times, people transported beverages in clay jars and animal skins. Around 100 B.C., hand-blown glass bottles began to be used to hold liquids. In 1903, the invention of the automatic glass-bottle-blowing machine made it possible to mass-produce bottles. They were used for everything from milk to soda. Consumers returned the empty bottles to be refilled. In 1929, companies began experimenting with cans for beverages. Cans were stackable, non-breakable, and fast cooling—and consumers didn't have to return them. The plastic six-pack yoke came along with the popular use of cans for beverages. This device bound cans together for easy carrying. Unfortunately, the yokes bound more than cans. Millions of yokes found their way into the environment where they entangled thousands of birds, fish, and marine animals. Today, animals are still being harmed—in some cases they are killed—by plastic six-pack yokes.

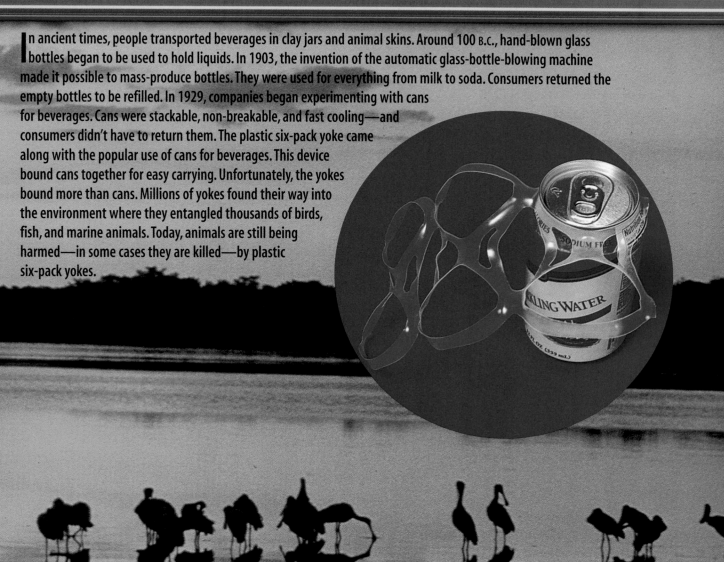

unit projects

Visit blue.msscience.com/unit_project to find project ideas and resources. Projects include:

- **Career** Design a concept map all about trash. Create new ways to reuse trash and limit waste disposal in your life.
- **Technology** Research microbial hydrocarbon bioremediation, and write a newspaper article describing these helpful organisms.
- **Model** Take another look at packaging, and design a "blueprint" for a more eco-friendly product and container.

WebQuest *Hybrid Vehicles* promotes understanding of the new vehicles being produced by car manufacturers. Analyze the advantages and disadvantages of hybrid electric vehicles.

chapter 4

Interactions of Life

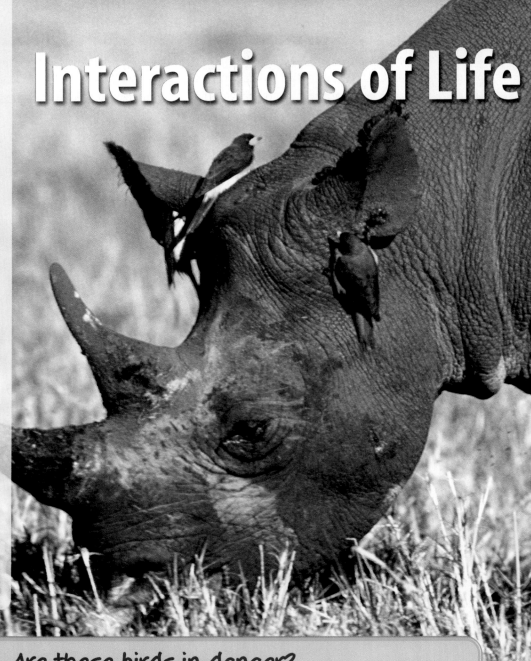

The BIG Idea
Living organisms interact with their environment and with one another in many ways.

SECTION 1
Living Earth
Main Idea All living and nonliving things on Earth are organized into levels, such as communities and ecosystems.

SECTION 2
Populations
Main Idea A population's size is affected by many things, including competition.

SECTION 3
Interactions Within Communities
Main Idea Every organism has a role in its environment.

Are these birds in danger?
The birds are a help to the rhinoceros. They feed on ticks and other parasites plucked from the rhino's hide. When the birds sense danger, they fly off, giving the rhino an early warning. Earth's living organisms supply one another with food, shelter, and other requirements for life.

Science Journal Describe how a familiar bird, insect, or other animal depends on other organisms.

Start-Up Activities

How do lawn organisms survive?

You probably have taken thousands of footsteps on grassy lawns or playing fields. If you look closely at the grass, you'll see that each blade is attached to roots in the soil. How do grass plants obtain everything they need to live and grow? What other kinds of organisms live in the grass? The following lab will give you a chance to take a closer look at the life in a lawn.

1. Examine a section of sod from a lawn.
2. How do the roots of the grass plants hold the soil?
3. Do you see signs of other living things besides grass?
4. **Think Critically** In your Science Journal, answer the above questions and describe any organisms that are present in your section of sod. Explain how these organisms might affect the growth of grass plants. Draw a picture of your section of sod.

Ecology Make the following Foldable to help organize information about one of your favorite wild animals and its role in an ecosystem.

STEP 1 Fold a vertical sheet of paper from side to side. Make the front edge 1.25 cm shorter than the back edge.

STEP 2 Turn lengthwise and fold into thirds.

STEP 3 Unfold and cut only the top layer along both folds to make three tabs. Label each tab.

Identify Questions Before you read the chapter, write what you already know about your favorite animal under the left tab of your Foldable. As you read the chapter, write how the animal is part of a population and a community under the appropriate tabs.

Preview this chapter's content and activities at blue.msscience.com

Get Ready to Read

Compare and Contrast

① Learn It! Good readers compare and contrast information as they read. This means they look for similarities and differences to help them to remember important ideas. Look for signal words in the text to let you know when the author is comparing or contrasting.

Compare and Contrast Signal Words	
Compare	**Contrast**
as	but
like	or
likewise	unlike
similarly	however
at the same time	although
in a similar way	on the other hand

② Practice It! Read the excerpt below and notice how the author uses contrast signal words to describe the differences between the biotic potentials of species.

> The highest rate of reproduction under ideal conditions is a population's biotic potential. The **larger** the number of offspring that are produced by parent organisms, the **higher** the biotic potential of the species will be. Compare an avocado tree to a tangerine tree.
>
> —from page 102

③ Apply It! Compare and contrast the different types of symbiotic relationships on page 108.

Target Your Reading

Use this to focus on the main ideas as you read the chapter.

1 Before you read the chapter, respond to the statements below on your worksheet or on a numbered sheet of paper.
- Write an **A** if you **agree** with the statement.
- Write a **D** if you **disagree** with the statement.

2 After you read the chapter, look back to this page to see if you've changed your mind about any of the statements.
- If any of your answers changed, explain why.
- Change any false statements into true statements.
- Use your revised statements as a study guide.

Reading Tip

As you read, use other skills, such as summarizing and connecting, to help you understand comparisons and contrasts.

Science Online
Print out a worksheet of this page at blue.msscience.com

Before You Read A or D		Statement	After You Read A or D
	1	A community is all the populations of species that live in an ecosystem.	
	2	All deserts are hot and dry environments.	
	3	An ecosystem is made up of only the living things in an area.	
	4	Organisms living in the wild always have enough food and living space.	
	5	The greatest competition in nature is among organisms of the same species.	
	6	Both nonliving and living parts of an ecosystem can limit the number of individuals in a population.	
	7	Living organisms do not need a constant supply of energy.	
	8	All consumers are predators.	
	9	Relationships between organisms of different species cannot benefit both organisms.	

section 1
Living Earth

as you read

What You'll Learn
- **Identify** places where life is found on Earth.
- **Define** ecology.
- **Observe** how the environment influences life.

Why It's Important
All living things on Earth depend on each other for survival.

Review Vocabulary
adaptation: any variation that makes an organism better suited to its environment

New Vocabulary
- biosphere
- ecosystem
- ecology
- population
- community
- habitat

The Biosphere

What makes Earth different from other planets in the solar system? One difference is Earth's abundance of living organisms. The part of Earth that supports life is the **biosphere** (BI uh sfihr). The biosphere includes the top portion of Earth's crust, all the waters that cover Earth's surface, and the atmosphere that surrounds Earth.

Reading Check *What three things make up the biosphere?*

As **Figure 1** shows, the biosphere is made up of different environments that are home to different kinds of organisms. For example, desert environments receive little rain. Cactus plants, coyotes, and lizards are included in the life of the desert. Tropical rain forest environments receive plenty of rain and warm weather. Parrots, monkeys, and tens of thousands of other organisms live in the rain forest. Coral reefs form in warm, shallow ocean waters. Arctic regions near the north pole are covered with ice and snow. Polar bears, seals, and walruses live in the arctic.

Figure 1 Earth's biosphere consists of many environments, including ocean waters, polar regions, and deserts.

Desert   Arctic   Coral reef

94 CHAPTER 4 Interactions of Life

INTEGRATE Astronomy

Life on Earth In our solar system, Earth is the third planet from the Sun. The amount of energy that reaches Earth from the Sun helps make the temperature just right for life. Mercury, the planet closest to the Sun, is too hot during the day and too cold at night to make life possible there. Venus, the second planet from the Sun, has a thick, carbon dioxide atmosphere and high temperatures. It is unlikely that life could survive there. Mars, the fourth planet, is much colder than Earth because it is farther from the Sun and has a thinner atmosphere. It might support microscopic life, but none has been found. The planets beyond Mars probably do not receive enough heat and light from the Sun to have the right conditions for life.

Ecosystems

On a visit to Yellowstone National Park in Wyoming, you might see a prairie scene like the one shown in **Figure 2.** Bison graze on prairie grass. Cowbirds follow the bison, catching grasshoppers that jump away from the bisons' hooves. This scene is part of an ecosystem. An **ecosystem** consists of all the organisms living in an area, as well as the nonliving parts of that environment. Bison, grass, birds, and insects are living organisms of this prairie ecosystem. Water, temperature, sunlight, soil, and air are nonliving features of this prairie ecosystem. **Ecology** is the study of interactions that occur among organisms and their environments. Ecologists are scientists who study these interactions.

Reading Check *What is an ecosystem?*

Figure 2 Ecosystems are made up of living organisms and the nonliving factors of their environment. In this prairie ecosystem, cowbirds eat insects and bison graze on grass.
List *other kinds of organisms that might live in this ecosystem.*

SECTION 1 Living Earth

Science Online

Topic: Human Population Data

Visit blue.msscience.com for Web links to information about the estimated human population size for the world today.

Activity Create a graph that shows how the human population has changed between the year 2000 and this year.

Populations

Suppose you meet an ecologist who studies how a herd of bison moves from place to place and how the female bison in the herd care for their young. This ecologist is studying the members of a population. A **population** is made up of all organisms of the same species that live in an area at the same time. For example, all the bison in a prairie ecosystem are one population. All the cowbirds in this ecosystem make up a different population. The grasshoppers make up yet another population.

Ecologists often study how populations interact. For example, an ecologist might try to answer questions about several prairie species. How does grazing by bison affect the growth of prairie grass? How does grazing influence the insects that live in the grass and the birds that eat those insects? This ecologist is studying a community. A **community** is all the populations of all species living in an ecosystem. The prairie community is made of populations of bison, grasshoppers, cowbirds, and all other species in the prairie ecosystem. An arctic community might include populations of fish, seals that eat fish, and polar bears that hunt and eat seals. **Figure 3** shows how organisms, populations, communities, and ecosystems are related.

Figure 3 The living world is arranged in several levels of organization.

Figure 4 The trees of the forest provide a habitat for woodpeckers and other birds. This salamander's habitat is the moist forest floor.

Habitats

Each organism in an ecosystem needs a place to live. The place in which an organism lives is called its **habitat.** The animals shown in **Figure 4** live in a forest ecosystem. Trees are the woodpecker's habitat. These birds use their strong beaks to pry insects from tree bark or break open acorns and nuts. Woodpeckers usually nest in holes in dead trees. The salamander's habitat is the forest floor, beneath fallen leaves and twigs. Salamanders avoid sunlight and seek damp, dark places. This animal eats small worms, insects, and slugs. An organism's habitat provides the kinds of food and shelter, the temperature, and the amount of moisture the organism needs to survive.

section 1 review

Summary

The Biosphere
- The biosphere is the portion of Earth that supports life.

Ecosystems
- An ecosystem is made up of the living organisms and nonliving parts of an area.

Populations
- A population is made up of all members of a species that live in the same ecosystem.
- A community consists of all the populations in an ecosystem.

Habitats
- A habitat is where an organism lives.

Self Check

1. **List** three parts of the Earth included in the biosphere.
2. **Define** the term *ecology*.
3. **Compare and contrast** the terms *habitat* and *biosphere*.
4. **Identify** the major difference between a community and a population, and give one example of each.
5. **Think Critically** Does the amount of rain that falls in an area determine which kinds of organisms can live there? Why or why not?

Applying Skills

6. **Form a hypothesis** about how a population of dandelion plants might be affected by a population of rabbits.

section 2
Populations

as you read

What You'll Learn
- **Identify** methods for estimating population sizes.
- **Explain** how competition limits population growth.
- **List** factors that influence changes in population size.

Why It's Important
Competition caused by population growth reduces the amount of food, living space, and other resources available to organisms, including humans.

Review Vocabulary
natural selection: hypothesis that states organisms with traits best suited to their environment are more likely to survive and reproduce

New Vocabulary
- limiting factor
- carrying capacity

Competition

Wild crickets feed on plant material at night. They hide under leaves or in dark damp places during the day. In some pet shops, crickets are raised in cages and fed to pet reptiles. Crickets require plenty of food, water, and hiding places. As a population of caged crickets grows, extra food and more hiding places are needed. To avoid crowding, some crickets might have to be moved to other cages.

Food and Space Organisms living in the wild do not always have enough food or living space. The Gila woodpecker, shown in **Figure 5,** lives in the Sonoran Desert of Arizona and Mexico. This woodpecker makes its nest by drilling a hole in a saguaro (suh GWAR oh) cactus. Woodpeckers must compete with each other for nesting spots. Competition occurs when two or more organisms seek the same resource at the same time.

Growth Limits Competition limits population size. If available nesting spaces are limited, some woodpeckers will not be able to raise young. Gila woodpeckers eat cactus fruit, berries, and insects. If food becomes scarce, some woodpeckers might not survive to reproduce. Competition for food, living space, or other resources can limit population growth.

In nature, the most intense competition is usually among individuals of the same species, because they need the same kinds of food and shelter. Competition also takes place among different species. For example, after a Gila woodpecker has abandoned its nest, owls, flycatchers, snakes, and lizards might compete for the shelter of the empty hole.

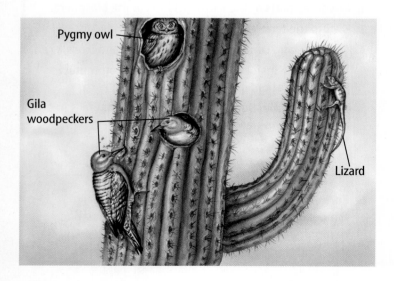

Figure 5 Gila woodpeckers make nesting holes in the saguaro cactus. Many animals compete for the shelter these holes provide.

Population Size

Ecologists often need to measure the size of a population. This information can indicate whether or not a population is healthy and growing. Population counts can help identify populations that could be in danger of disappearing.

Some populations are easy to measure. If you were raising crickets, you could measure the size of your cricket population simply by counting all the crickets in the container. What if you wanted to compare the cricket populations in two different containers? You would calculate the number of crickets per square meter (m^2) of your container. The number of individuals of one species per a specific area is called population density. **Figure 6** shows Earth's human population density.

Reading Check *What is population density?*

Measuring Populations Counting crickets can be tricky. They look alike, move a lot, and hide. The same cricket could be counted more than once, and others could be completely missed. Ecologists have similar problems when measuring wildlife populations. One of the methods they use is called trap-mark-release. Suppose you want to count wild rabbits. Rabbits live underground and come out at dawn and dusk to eat. Ecologists set traps that capture rabbits without injuring them. Each captured rabbit is marked and released. Later, another sample of rabbits is captured. Some of these rabbits will have marks, but many will not. By comparing the number of marked and unmarked rabbits in the second sample, ecologists can estimate the population size.

Mini LAB

Observing Seedling Competition

Procedure
1. Fill **two plant pots** with **moist potting soil**.
2. Plant **radish seeds** in one pot, following the spacing instructions on the seed packet. Label this pot *Recommended Spacing*.
3. Plant **radish seeds** in the second pot, spaced half the recommended distance apart. Label this pot *Densely Populated*. Wash your hands.
4. Keep the soil moist. When the seeds sprout, move them to a well-lit area.
5. Measure and record in your **Science Journal** the height of the seedlings every two days for two weeks.

Analysis
1. Which plants grew faster?
2. Which plants looked healthiest after two weeks?
3. How did competition influence the plants?

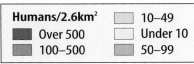

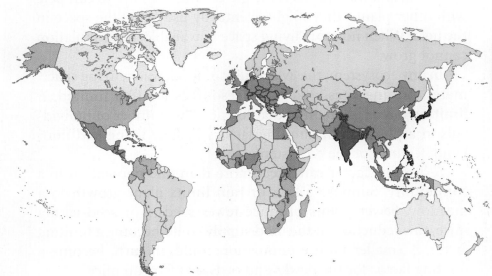

Figure 6 This map shows human population density. **Interpret Illustrations** *Which countries have the highest population density?*

SECTION 2 Populations **99**

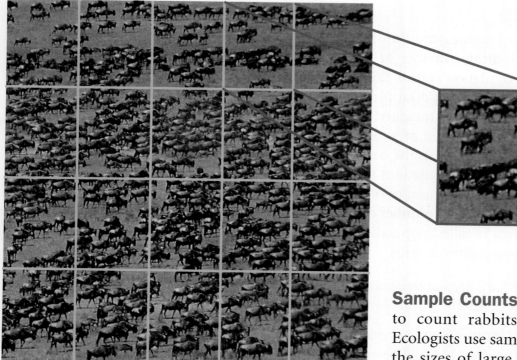

Figure 7 Ecologists can estimate population size by making a sample count. Wildebeests graze on the grassy plains of Africa.
Draw Conclusions How could you use the enlarged square to estimate the number of wildebeests in the entire photograph?

Sample Counts What if you wanted to count rabbits over a large area? Ecologists use sample counts to estimate the sizes of large populations. To estimate the number of rabbits in an area of 100 acres, for example, you could count the rabbits in one acre and multiply by 100 to estimate the population size. **Figure 7** shows another approach to sample counting.

Limiting Factors One grass plant can produce hundreds of seeds. Imagine those seeds drifting onto a vacant field. Many of the seeds sprout and grow into grass plants that produce hundreds more seeds. Soon the field is covered with grass. Can this grass population keep growing forever? Suppose the seeds of wildflowers or trees drift onto the field. If those seeds sprout, trees and flowers would compete with grasses for sunlight, soil, and water. Even if the grasses did not have to compete with other plants, they might eventually use up all the space in the field. When no more living space is available, the population cannot grow.

In any ecosystem, the availability of food, water, living space, mates, nesting sites, and other resources is often limited. A **limiting factor** is anything that restricts the number of individuals in a population. Limiting factors include living and nonliving features of the ecosystem.

A limiting factor can affect more than one population in a community. Suppose a lack of rain limits plant growth in a meadow. Fewer plants produce fewer seeds. For seed-eating mice, this reduction in the food supply could become a limiting factor. A smaller mouse population could, in turn, become a limiting factor for the hawks and owls that feed on mice.

Carrying Capacity A population of robins lives in a grove of trees in a park. Over several years, the number of robins increases and nesting space becomes scarce. Nesting space is a limiting factor that prevents the robin population from getting any larger. This ecosystem has reached its carrying capacity for robins. **Carrying capacity** is the largest number of individuals of one species that an ecosystem can support over time. If a population begins to exceed the environment's carrying capacity, some individuals will not have enough resources. They could die or be forced to move elsewhere, like the deer shown in **Figure 8**.

Figure 8 These deer might have moved into a residential area because a nearby forest's carrying capacity for deer has been reached.

 *How are limiting factors related to carrying capacity?*

Applying Science

Do you have too many crickets?

You've decided to raise crickets to sell to pet stores. A friend says you should not allow the cricket population density to go over 210 crickets/m². Use what you've learned in this section to measure the population density in your cricket tanks.

Identifying the Problem

The table on the right lists the areas and populations of your three cricket tanks. How can you determine if too many crickets are in one tank? If a tank contains too many crickets, what could you do? Explain why too many crickets in a tank might be a problem.

Cricket Population

Tank	Area (m²)	Number of Crickets
1	0.80	200
2	0.80	150
3	1.5	315

Solving the Problem

1. Do any of the tanks contain too many crickets? Could you make the population density of the three tanks equal by moving crickets from one tank to another? If so, which tank would you move crickets into?

2. Wild crickets living in a field have a population density of 2.4 crickets/m². If the field's area is 250 m², what is the approximate size of the cricket population? Why would the population density of crickets in a field be lower than the population density of crickets in a tank?

SECTION 2 Populations

Topic: Birthrates and Death Rates

Visit blue.msscience.com for Web links to information about birthrates and death rates for the human population.

Activity Find out whether the human population worldwide is increasing because of rising birthrates or declining death rates.

Biotic Potential What would happen if no limiting factors restricted the growth of a population? Think about a population that has an unlimited supply of food, water, and living space. The climate is favorable. Population growth is not limited by diseases, predators, or competition with other species. Under ideal conditions like these, the population would continue to grow.

The highest rate of reproduction under ideal conditions is a population's biotic potential. The larger the number of offspring that are produced by parent organisms, the higher the biotic potential of the species will be. Compare an avocado tree to a tangerine tree. Assume that each tree produces the same number of fruits. Each avocado fruit contains one large seed. Each tangerine fruit contains a dozen seeds or more. Because the tangerine tree produces more seeds per fruit, it has a higher biotic potential than the avocado tree.

Changes in Populations

Birthrates and death rates also influence the size of a population and its rate of growth. A population gets larger when the number of individuals born is greater than the number of individuals that die. When the number of deaths is greater than the number of births, populations get smaller. Take the squirrels living in New York City's Central Park as an example. In one year, if 900 squirrels are born and 800 die, the population increases by 100. If 400 squirrels are born and 500 die, the population decreases by 100.

The same is true for human populations. **Table 1** shows birthrates, death rates, and population changes for several countries around the world. In countries with faster population growth, birthrates are much higher than death rates. In countries with slower population growth, birthrates are only slightly higher than death rates. In Germany, where the population is getting smaller, the birthrate is lower than the death rate.

Table 1 Population Growth	Birthrate*	Death Rate*	Population Increase (percent)
Rapid-Growth Countries			
Jordan	38.8	5.5	3.3
Uganda	50.8	21.8	2.9
Zimbabwe	34.3	9.4	5.2
Slow-Growth Countries			
Germany	9.4	10.8	−1.5
Sweden	10.8	10.6	0.1
United States	14.8	8.8	0.6

*Number per 1,000 people

Figure 9 Mangrove seeds sprout while they are still attached to the parent tree. Some sprouted seeds drop into the mud below the parent tree and continue to grow. Others drop into the water and can be carried away by tides and ocean currents. When they wash ashore, they might start a new population of mangroves or add to an existing mangrove population.

Moving Around Most animals can move easily from place to place, and these movements can affect population size. For example, a male mountain sheep might wander many miles in search of a mate. After he finds a mate, their offspring might establish a completely new population of mountain sheep far from the male's original population.

Many bird species move from one place to another during their annual migrations. During the summer, populations of Baltimore orioles are found throughout eastern North America. During the winter, these populations disappear because the birds migrate to Central America. They spend the winter there, where the climate is mild and food supplies are plentiful. When summer approaches, the orioles migrate back to North America.

Even plants and microscopic organisms can move from place to place, carried by wind, water, or animals. The tiny spores of mushrooms, mosses, and ferns float through the air. The seeds of dandelions, maple trees, and other plants have feathery or winglike growths that allow them to be carried by wind. Spine-covered seeds hitch rides by clinging to animal fur or people's clothing. Many kinds of seeds can be transported by river and ocean currents. Mangrove trees growing along Florida's Gulf Coast, shown in **Figure 9,** provide an example of how water moves seeds.

Mini LAB

Comparing Biotic Potential

Procedure
1. Remove all the seeds from a **whole fruit.** Do not put fruit or seeds in your mouth.
2. Count the total number of seeds in the fruit. Wash your hands, then record these data in your **Science Journal.**
3. Compare your seed totals with those of classmates who examined other types of fruit.

Analysis
1. Which type of fruit had the most seeds? Which had the fewest seeds?
2. What is an advantage of producing many seeds? Can you think of a possible disadvantage?
3. To estimate the total number of seeds produced by a tomato plant, what would you need to know?

SECTION 2 Populations **103**

NATIONAL GEOGRAPHIC VISUALIZING POPULATION GROWTH

Figure 10

When a species enters an ecosystem that has abundant food, water, and other resources, its population can flourish. Beginning with a few organisms, the population increases until the number of organisms and available resources are in balance. At that point, population growth slows or stops. A graph of these changes over time produces an S-curve, as shown here for coyotes.

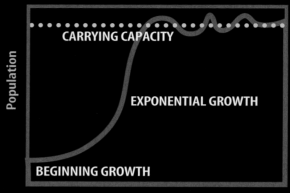

BEGINNING GROWTH During the first few years, population growth is slow, because there are few adults to produce young. As the population grows, so does the number of breeding adults.

EXPONENTIAL GROWTH As the number of adults in the population grows, so does the number of births. The coyote population undergoes exponential growth, quickly increasing in size.

CARRYING CAPACITY As resources become less plentiful, the birthrate declines and the death rate may rise. Population growth slows. The coyote population has reached the environmental carrying capacity—the maximum number of coyotes that the environment can sustain.

Exponential Growth When a species moves into a new area with plenty of food, living space, and other resources, the population grows quickly, in a pattern called exponential growth. Exponential growth means that the larger a population gets, the faster it grows. Over time, the population will reach the ecosystem's carrying capacity for that species. **Figure 10** shows each stage in this pattern of population growth.

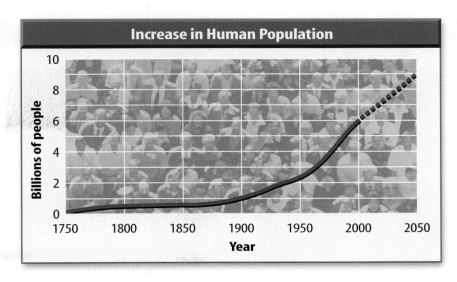

As a population approaches its ecosystem's carrying capacity, competition for living space and other resources increases. As you can see in **Figure 11,** Earth's human population shows exponential growth. By the year 2050, the population could reach 9 billion. You probably have read about or experienced some of the competition associated with human population growth, such as freeway traffic jams, crowded subways and buses, or housing shortages. As population density increases, people are forced to live closer to one another. Infectious diseases can spread easily when people are crowded together.

Figure 11 The size of the human population is increasing by about 1.6 percent per year. **Identify** *the factors that affect human population growth.*

section 2 review

Summary

Competition
- When more than one organism needs the same resource, competition occurs.
- Competition limits population size.

Population Size
- Population density is the number of individuals per unit area.
- Limiting factors are resources that restrict population size.
- An ecosystem's carrying capacity is the largest population it can support.
- Biotic potential is the highest possible rate of growth for a population.

Changes in Populations
- Birthrates, death rates, and movement from place to place affect population size.

Self Check

1. **Describe** three ways in which ecologists can estimate the size of a population.
2. **Explain** how birthrates and death rates influence the size of a population.
3. **Explain** how carrying capacity influences the number of organisms in an ecosystem.
4. **Think Critically** Why are food and water the limiting factors that usually have the greatest effect on population size?

Applying Skills

5. **Make and use a table** on changes in the size of a deer population in Arizona. Use the following data. In 1910 there were 6 deer; in 1915, 36 deer; in 1920, 143 deer; in 1925, 86 deer; and in 1935, 26 deer. Explain what might have caused these changes.

blue.msscience.com/self_check_quiz

section 3

Interactions Within Communities

as you read

What You'll Learn
- **Describe** how organisms obtain energy for life.
- **Explain** how organisms interact.
- **Recognize** that every organism occupies a niche.

Why It's Important
Obtaining food, shelter, and other needs is crucial to the survival of all living organisms, including you.

Review Vocabulary
social behavior: interactions among members of the same species

New Vocabulary
- producer
- consumer
- symbiosis
- mutualism
- commensalism
- parasitism
- niche

Obtaining Energy

Just as a car engine needs a constant supply of gasoline, living organisms need a constant supply of energy. The energy that fuels most life on Earth comes from the Sun. Some organisms use the Sun's energy to create energy-rich molecules through the process of photosynthesis. The energy-rich molecules, usually sugars, serve as food. They are made up of different combinations of carbon, hydrogen, and oxygen atoms. Energy is stored in the chemical bonds that hold the atoms of these molecules together. When the molecules break apart—for example, during digestion—the energy in the chemical bonds is released to fuel life processes.

Producers Organisms that use an outside energy source like the Sun to make energy-rich molecules are called **producers.** Most producers contain chlorophyll (KLOR uh fihl), a chemical that is required for photosynthesis. As shown in **Figure 12,** green plants are producers. Some producers do not contain chlorophyll and do not use energy from the Sun. Instead, they make energy-rich molecules through a process called chemosynthesis (kee moh SIHN thuh sus). These organisms can be found near volcanic vents on the ocean floor. Inorganic molecules in the water provide the energy source for chemosynthesis.

Figure 12 Green plants, including the grasses that surround this pond, are producers. The pond water also contains producers, including microscopic organisms like *Euglena* and algae.

Euglena
LM Magnification: 125×

Algae
LM Magnification: 25×

106 CHAPTER 4 Interactions of Life

Figure 13 Four categories of consumers are shown.
Identify *the consumer category that would apply to a bear. What about a mushroom?*

Consumers
Organisms that cannot make their own energy-rich molecules are called **consumers.** Consumers obtain energy by eating other organisms. **Figure 13** shows the four general categories of consumers. Herbivores are the vegetarians of the world. They include rabbits, deer, and other plant eaters. Carnivores are animals that eat other animals. Frogs and spiders are carnivores that eat insects. Omnivores, including pigs and humans, eat mostly plants and animals. Decomposers, including fungi, bacteria, and earthworms, consume wastes and dead organisms. Decomposers help recycle once-living matter by breaking it down into simple, energy-rich substances. These substances might serve as food for decomposers, be absorbed by plant roots, or be consumed by other organisms.

Reading Check *How are producers different from consumers?*

Food Chains
Ecology includes the study of how organisms depend on each other for food. A food chain is a simple model of the feeding relationships in an ecosystem. For example, shrubs are food for deer, and deer are food for mountain lions, as illustrated in **Figure 14.** What food chain would include you?

Glucose The nutrient molecule produced during photosynthesis is glucose. Look up the chemical structure of glucose and draw it in your Science Journal.

Figure 14 Food chains illustrate how consumers obtain energy from other organisms in an ecosystem.

Figure 15 Many examples of symbiotic relationships exist in nature.

Symbiotic Relationships

Not all relationships among organisms involve food. Many organisms live together and share resources in other ways. Any close relationship between species is called **symbiosis.**

Lichens are a result of mutualism.

Mutualism You may have noticed crusty lichens growing on fences, trees, or rocks. Lichens, like those shown in **Figure 15,** are made up of an alga or a cyanobacterium that lives within the tissues of a fungus. Through photosynthesis, the cyanobacterium or alga supplies energy to itself and the fungus. The fungus provides a protected space in which the cyanobacterium or alga can live. Both organisms benefit from this association. A symbiotic relationship in which both species benefit is called **mutualism** (MYEW chuh wuh lih zum).

Clown fish and sea anemones have a commensal relationship.

Commensalism If you've ever visited a marine aquarium, you might have seen the ocean organisms shown in **Figure 15.** The creature with gently waving, tubelike tentacles is a sea anemone. The tentacles contain a mild poison. Anemones use their tentacles to capture shrimp, fish, and other small animals to eat. The striped clown fish can swim among the tentacles without being harmed. The anemone's tentacles protect the clown fish from predators. In this relationship, the clown fish benefits but the sea anemone is not helped or hurt. A symbiotic relationship in which one organism benefits and the other is not affected is called **commensalism** (kuh MEN suh lih zum).

LM Magnification: 128×

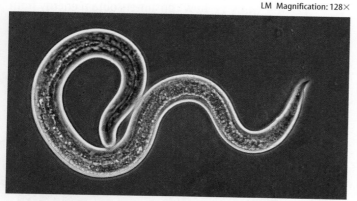

Some roundworms are parasites that rob nutrients from their hosts.

Parasitism Pet cats or dogs sometimes have to be treated for worms. Roundworms, like the one shown in **Figure 15,** are common in puppies. This roundworm attaches itself to the inside of the puppy's intestine and feeds on nutrients in the puppy's blood. The puppy may have abdominal pain, bloating, and diarrhea. If the infection is severe, the puppy might die. A symbiotic relationship in which one organism benefits but the other is harmed is called **parasitism** (PER uh suh tih zum).

Niches

One habitat might contain hundreds or even thousands of species. Look at the rotting log habitat shown in **Figure 16**. A rotting log in a forest can be home to many species of insects, including termites that eat decaying wood and ants that feed on the termites. Other species that live on or under the rotting log include millipedes, centipedes, spiders, and worms. You might think that competition for resources would make it impossible for so many species to live in the same habitat. However, each species has different requirements for its survival. As a result, each species has its own niche (NICH). An organism's **niche** is its role in its environment—how it obtains food and shelter, finds a mate, cares for its young, and avoids danger.

Reading Check *Why does each species have its own niche?*

Special adaptations that improve survival are often part of an organism's niche. Milkweed plants contain a poison that prevents many insects from feeding on them. Monarch butterfly caterpillars have an adaptation that allows them to eat milkweed. Monarchs can take advantage of a food resource that other species cannot use. Milkweed poison also helps protect monarchs from predators. When the caterpillars eat milkweed, they become slightly poisonous. Birds avoid eating monarchs because they learn that the caterpillars and adult butterflies have an awful taste and can make them sick.

Plant Poisons The poison in milkweed is similar to the drug digitalis. Small amounts of digitalis are used to treat heart ailments in humans, but it is poisonous in large doses. Research the history of digitalis as a medicine. In your Science Journal, list diseases for which it was used but is no longer used.

Figure 16 Different adaptations enable each species living in this rotting log to have its own niche. Termites eat wood. They make tunnels inside the log. Millipedes feed on plant matter and find shelter beneath the log. Wolf spiders capture insects living in and around the log.

Termites

Millipede

Wolf spider

Figure 17 The alligator is a predator. The turtle is its prey.

Predator and Prey When you think of survival in the wild, you might imagine an antelope running away from a lion. An organism's niche includes how it avoids being eaten and how it finds or captures its food. Predators, like the one shown in **Figure 17**, are consumers that capture and eat other consumers. The prey is the organism that is captured by the predator. The presence of predators usually increases the number of different species that can live in an ecosystem. Predators limit the size of prey populations. As a result, food and other resources are less likely to become scarce, and competition between species is reduced.

Cooperation Individual organisms often cooperate in ways that improve survival. For example, a white-tailed deer that detects the presence of wolves or coyotes will alert the other deer in the herd. Many insects, such as ants and honeybees, live in social groups. Different individuals perform different tasks required for the survival of the entire nest. Soldier ants protect workers that go out of the nest to gather food. Worker ants feed and care for ant larvae that hatch from eggs laid by the queen. These cooperative actions improve survival and are a part of the specie's niche.

section 3 review

Summary

Obtaining Energy
- All life requires a constant supply of energy.
- Most producers make food by photosynthesis using light energy.
- Consumers cannot make food. They obtain energy by eating producers or other consumers.
- A food chain models the feeding relationships between species.

Symbiotic Relationships
- Symbiosis is any close relationship between species.
- Mutualism, commensalism, and parasitism are types of symbiosis.
- An organism's niche describes the ways in which the organism obtains food, avoids danger, and finds shelter.

Self Check

1. **Explain** why all consumers depend on producers for food.
2. **Describe** a mutualistic relationship between two imaginary organisms. Name the organisms and explain how each benefits.
3. **Compare and contrast** the terms *habitat* and *niche*.
4. **Think Critically** A parasite can obtain food only from a host organism. Explain why most parasites weaken, but do not kill, their hosts.

Applying Skills

5. **Design an experiment** to classify the symbiotic relationship that exists between two hypothetical organisms. Animal A definitely benefits from its relationship with Plant B, but it is not clear whether Plant B benefits, is harmed, or is unaffected.

Feeding Habits of Planaria

You probably have watched minnows darting about in a stream. It is not as easy to observe organisms that live at the bottom of a stream, beneath rocks, logs, and dead leaves. Countless stream organisms, including insect larvae, worms, and microscopic organisms, live out of your view. One such organism is a type of flatworm called a planarian. In this lab, you will find out about the eating habits of planarians.

Real-World Question

What food items do planarians prefer to eat?

Goals
- **Observe** the food preference of planarians.
- **Infer** what planarians eat in the wild.

Materials
small bowl
planarians (several)
lettuce leaf
raw liver or meat
guppies (several)
pond or stream water
magnifying lens

Safety Precautions

Procedure

1. Fill the bowl with stream water.
2. Place a lettuce leaf, piece of raw liver, and several guppies in the bowl. Add the planarians. Wash your hands.
3. **Observe** what happens inside the bowl for at least 20 minutes. Do not disturb the bowl or its contents. Use a magnifying lens to look at the planarians.
4. **Record** all of your observations in your Science Journal.

Conclude and Apply

1. **Name** the food the planarians preferred.
2. **Infer** what planarians might eat when in their natural environment.
3. **Describe,** based on your observations during this lab, a planarian's niche in a stream ecosystem.
4. **Predict** where in a stream you might find planarians. Use references to find out whether your prediction is correct.

Communicating Your Data

Share your results with other students in your class. Plan an adult-supervised trip with several classmates to a local stream to search for planarians in their native habitat. **For more help, refer to the** Science Skill Handbook.

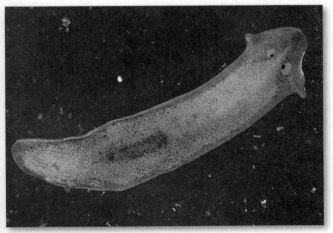

Magnification: Unknown

LAB Design Your Own

Population Growth in Fruit Flies

Real-World Question

Populations can grow at an exponential rate only if the environment provides the right amount of food, shelter, air, moisture, heat, living space, and other factors. You probably have seen fruit flies hovering near ripe bananas or other fruit. Fruit flies are fast-growing organisms often raised in science laboratories. The flies are kept in culture tubes and fed a diet of specially prepared food flakes. Can you improve on this standard growing method to achieve faster population growth? Will a change in one environmental factor affect the growth of a fruit fly population?

Goals
- **Identify** the environmental factors needed by a population of fruit flies.
- **Design** an experiment to investigate how a change in one environmental factor affects in any way the size of a fruit fly population.
- **Observe** and **measure** changes in population size.

Possible Materials
fruit flies
standard fruit fly culture kit
food items (banana, orange peel, or other fruit)
water
heating or cooling source
culture containers
cloth, plastic, or other tops for culture containers
magnifying lens

Safety Precautions

Form a Hypothesis

Based on your reading about fruit flies, state a hypothesis about how changing one environmental factor will affect the rate of growth of a fruit fly population.

Test Your Hypothesis

Make a Plan

1. As a group, decide on one environmental factor to investigate. Agree on a hypothesis about how a change in this factor will affect population growth. Decide how you will test your hypothesis, and identify the experimental results that would support your hypothesis.

2. **List** the steps you will need to take to test your hypothesis. Describe exactly what you will do. List your materials.

3. **Determine** the method you will use to measure changes in the size of your fruit fly populations.

Using Scientific Methods

4. Prepare a data table in your Science Journal to record weekly measurements of your fruit fly populations.
5. Read the entire experiment and make sure all of the steps are in a logical order.
6. **Research** the standard method used to raise fruit flies in the laboratory. Use this method as the control in your experiment.
7. **Identify** all constants, variables, and controls in your experiment.

Follow Your Plan

1. Make sure your teacher approves your plan before you start.
2. Carry out your experiment.
3. **Measure** the growth of your fruit fly populations weekly and record the data in your data table.

Analyze Your Data

1. **Identify** the constants and the variables in your experiment.
2. **Compare** changes in the size of your control population with changes in your experimental population. Which population grew faster?
3. **Make and Use Graphs** Using the information in your data table, make a line graph that shows how the sizes of your two fruit fly populations changed over time. Use a different colored pencil for each population's line on the graph.

Conclude and Apply

1. **Explain** whether or not the results support your hypothesis.
2. **Compare** the growth of your control and experimental populations. Did either population reach exponential growth? How do you know?

Communicating Your Data

Compare the results of your experiment with those of other students in your class. **For more help, refer to the** Science Skill Handbook.

LAB **113**

TIME SCIENCE AND HISTORY

SCIENCE CAN CHANGE THE COURSE OF HISTORY!

The Census measures a human population

Counting people is important to the United States and to many other countries around the world. It helps governments determine the distribution of people in the various regions of a nation. To obtain this information, the government takes a census—a count of how many people are living in their country on a particular day at a particular time, and in a particular place. A census is a snapshot of a country's population.

Counting on the Count

When the United States government was formed, its founders set up the House of Representatives based on population. Areas with more people had more government representatives, and areas with fewer people had fewer representatives. In 1787, the requirement for a census became part of the U.S. Constitution. A census must be taken every ten years so the proper number of representatives for each state can be calculated.

The Short Form

Before 1970, United States census data was collected by field workers. They went door to door to count the number of people living in each household. Since then, the census has been done mostly by mail. Census data are important in deciding how to distribute government services and funding.

The 2000 Snapshot

One of the findings of the 2000 Census is that the U.S. population is becoming more equally spread out across age groups. Census officials estimate that by 2020 the population of children, middle-aged people, and senior citizens will be about equal. It's predicted also that there will be more people who are over 100 years old than ever before. Federal, state, and local governments will be using the results of the 2000 Census for years to come as they plan our future.

Census Develop a school census. What questions will you ask? (Don't ask questions that are too personal.) Who will ask them? How will you make sure you counted everyone? Using the results, can you make any predictions about your school's future or its current students?

For more information, visit blue.msscience.com/time

Chapter 4 Study Guide

Reviewing Main Ideas

Section 1 Living Earth

1. Ecology is the study of interactions that take place in the biosphere.
2. A population is made up of all organisms of one species living in an area at the same time.
3. A community is made up of all the populations living in one ecosystem.
4. Living and nonliving factors affect an organism's ability to survive in its habitat.

Section 2 Populations

1. Population size can be estimated by counting a sample of a total population.
2. Competition for limiting factors can restrict the size of a population.
3. Population growth is affected by birthrate, death rate, and the movement of individuals into or out of a community.
4. Exponential population growth can occur in environments that provide a species with plenty of food, shelter, and other resources.

Section 3 Interactions Within Communities

1. All life requires energy.
2. Most producers use light to make food in the form of energy-rich molecules. Consumers obtain energy by eating other organisms.
3. Mutualism, commensalism, and parasitism are the three kinds of symbiosis.
4. Every species has its own niche, which includes adaptations for survival.

Visualizing Main Ideas

Copy and complete the following concept map on communities.

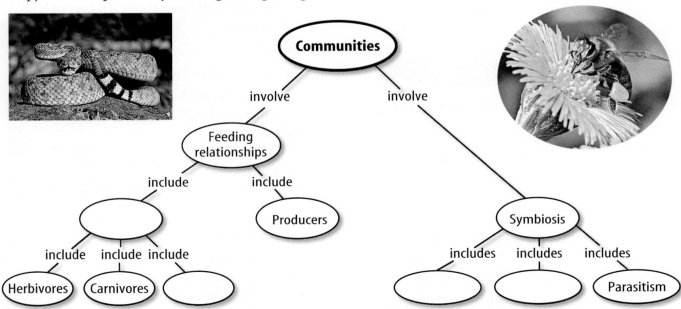

chapter 4 Review

Using Vocabulary

biosphere p. 94	limiting factor p. 100
carrying capacity p. 101	mutualism p. 108
commensalism p. 108	niche p. 109
community p. 96	parasitism p. 108
consumer p. 107	population p. 96
ecology p. 95	producer p. 106
ecosystem p. 95	symbiosis p. 108
habitat p. 97	

Explain the difference between the vocabulary words in each of the following sets.

1. niche—habitat
2. mutualism—commensalism
3. limiting factor—carrying capacity
4. biosphere—ecosystem
5. producer—consumer
6. population—ecosystem
7. community—population
8. parasitism—symbiosis
9. ecosystem—ecology
10. parasitism—commensalism

Checking Concepts

Choose the word or phrase that best answers the question.

11. Which of the following is a living factor in the environment?
    A) animals   C) sunlight
    B) air   D) soil

12. What is made up of all the populations in an area?
    A) niches   C) community
    B) habitats   D) ecosystem

13. What does the number of individuals in a population that occupies an area of a specific size describe?
    A) clumping   C) spacing
    B) size   D) density

14. Which of the following animals is an example of an herbivore?
    A) wolf   C) tree
    B) moss   D) rabbit

15. What term best describes a symbiotic relationship in which one species is helped and the other is harmed?
    A) mutualism   C) commensalism
    B) parasitism   D) consumerism

16. Which of the following conditions tends to increase the size of a population?
    A) births exceed deaths
    B) population size exceeds the carrying capacity
    C) movements out of an area exceed movements into the area
    D) severe drought

17. Which of the following is most likely to be a limiting factor in a population of fish living in the shallow water of a large lake?
    A) sunlight   C) food
    B) water   D) soil

18. In which of the following categories does the pictured organism belong?
    A) herbivore
    B) carnivore
    C) producer
    D) consumer

19. Which pair of words is incorrect?
    A) black bear—carnivore
    B) grasshopper—herbivore
    C) pig—omnivore
    D) lion—carnivore

116 CHAPTER REVIEW   blue.msscience.com/vocabulary_puzzlemaker

chapter 4 Review

Thinking Critically

20. **Infer** why a parasite has a harmful effect on the organism it infects.

21. **Explain** what factors affect carrying capacity.

22. **Describe** your own habitat and niche.

23. **Make and Use Tables** Copy and complete the following table.

Types of Symbiosis		
Organism A	Organism B	Relationship
Gains	Doesn't gain or lose	
Gains		Mutualism
Gains	Loses	

24. **Explain** how several different niches can exist in the same habitat.

25. **Make a model** of a food chain using the following organisms: grass, snake, mouse, and hawk.

26. **Predict** Dandelion seeds can float great distances on the wind with the help of white, featherlike attachments. Predict how a dandelion seed's ability to be carried on the wind helps reduce competition among dandelion plants.

27. **Classify** the following relationships as parasitism, commensalism, or mutualism: a shark and a remora fish that cleans and eats parasites from the shark's gills; head lice and a human; a spiny sea urchin and a tiny fish that hides from predators by floating among the sea urchin's spines.

28. **Compare and contrast** the diets of omnivores and herbivores. Give examples of each.

29. **List** three ways exponential growth in the human population affects people's lives.

Performance Activities

30. **Poster** Use photographs from old magazines to create a poster that shows at least three different food chains. Illustrate energy pathways from organism to organism and from organisms to the environment. Display your poster for your classmates.

Applying Math

31. **Measuring Populations** An ecologist wants to know the size of a population of wild daisy plants growing in a meadow that measures 1,000 m^2. The ecologist counts 30 daisy plants in a sample area of 100 m^2. What is the estimated population of daisies in the entire meadow?

Use the table below to answer question 32.

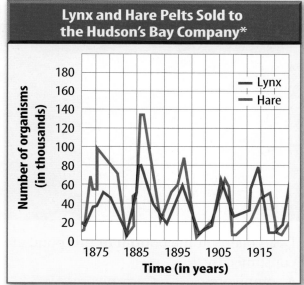

* Data from 1875 through 1904 reflects actual pelts counted. Data from 1905 through 1915 is based on answers to questionnaire.

32. **Changes in Populations** The graph above shows changes over time in the sizes of lynx and rabbit populations in an ecosystem. What does the graph tell you about the relationship between these two species? Explain how they influence each other's population size.

Chapter 4 Standardized Test Practice

Part 1 Multiple Choice

Record your answers on the answer sheet provided by your teacher or on a sheet of paper.

1. Which of the following terms is defined in part by nonliving factors?
   A. population   C. ecosystem
   B. community   D. niche

2. Which of the follow terms would include all places where organisms live on Earth?
   A. ecosystem   C. biosphere
   B. habitat   D. community

3. Which of the following is not a method of measuring populations?
   A. total count   C. sample count
   B. trap-release   D. trap-mark-release

Use the photo below to answer questions 4 and 5.

4. Dead plants at the bottom of this pond are consumed by
   A. omnivores.   C. carnivores.
   B. herbivores.   D. decomposers.

5. If the pond shrinks in size, what effect will this have on the population density of the pond's minnow species?
   A. It will increase.
   B. It will decrease.
   C. It will stay the same.
   D. No effect; it is not a limiting factor.

6. Which of the following includes organisms that can directly convert energy from the Sun into food?
   A. producers   C. omnivores
   B. decomposers   D. consumers

7. You have a symbiotic relationship with bacteria in your digestive system. These bacteria break down food you ingest, and you get vital nutrients from them. Which type of symbiosis is this?
   A. mutualism   C. commensalism
   B. barbarism   D. parasitism

Use the photo below to answer questions 8 and 9.

8. An eastern screech owl might compete with which organism most intensely for resources?
   A. mouse   C. mountain lion
   B. hawk   D. wren

9. Which of the following organisms might compete with the mouse for seeds?
   A. hawk   C. fox
   B. lion   D. sparrow

10. Which of the following is an example of a community?
    A. all the white-tailed deer in a forest
    B. all the trees, soil, and water in a forest
    C. all the plants and animals in a wetland
    D. all the cattails in a wetland

Standardized Test Practice

Part 2 Short Response/Grid In

Record your answers on the answer sheet provided by your teacher or on a sheet of paper.

Use the graph below to answer question 11.

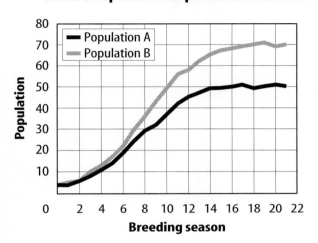

Mouse Population Exposed to Predators

11. The graph depicts the growth of two white-footed mice populations, one exposed to hawks (population A) and one without hawks (population B). Are hawks a limiting factor for either mouse population? If not, then what other factor could be a limiting factor for that population?

12. Diagram the flow of energy through an ecosystem. Include the sources of energy, producers, consumers, and decomposers in the ecosystem.

Test-Taking Tip

Understand the Question Be sure you understand the question before you read the answer choices. Make special note of words like NOT or EXCEPT. Read and consider choices before you mark your answer sheet.

Question 11 Make sure you understand which mouse population is subject to predation by hawks and which mouse population do hawks not affect.

Part 3 Open Ended

Record your answers on a sheet of paper.

13. The colors and patterns of the viceroy butterfly are similar to the monarch butterfly, however, the viceroy caterpillars don't feed on milkweed. How does the viceroy butterfly benefit from this adaptation of its appearance? Under what circumstance would this adaptation not benefit the viceroy? Why?

Use the illustration below to answer question 14.

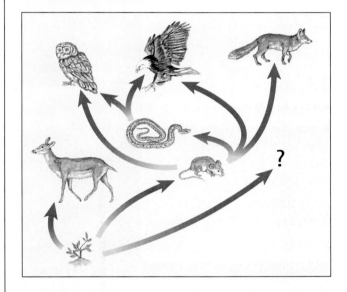

14. The illustration depicts a food web for a particular ecosystem. If the "?" is another mouse species population that is introduced into the ecosystem, explain what impact this would have on the species populations in the ecosystem.

15. Identify and explain possible limiting factors that would control the size of an ant colony.

16. How would you measure the size of a population of gray squirrels in a woodland? Explain which method you would choose and why.

chapter 5

The Nonliving Environment

The BIG Idea
Environments have both living and nonliving elements.

SECTION 1
Abiotic Factors
Main Idea Both living and nonliving parts of an environment are needed for organisms to survive.

SECTION 2
Cycles in Nature
Main Idea Many nonliving elements on Earth, such as water and oxygen, are recycled over and over.

SECTION 3
Energy Flow
Main Idea All living things use energy.

Sun, Surf, and Sand
Living things on this coast directly or indirectly depend on nonliving things, such as sunlight, water, and rocks, for energy and raw materials needed for their life processes. In this chapter, you will read how these and other nonliving things affect life on Earth.

Science Journal List all the nonliving things that you can see in this picture in order of importance. Explain your reasoning for the order you chose.

Start-Up Activities

Earth Has Many Ecosystems

Do you live in a dry, sandy region covered with cactus plants or desert scrub? Is your home in the mountains? Does snow fall during the winter? In this chapter, you'll learn why the nonliving factors in each ecosystem are different. The following lab will get you started.

1. Locate your city or town on a globe or world map. Find your latitude. Latitude shows your distance from the equator and is expressed in degrees, minutes, and seconds.
2. Locate another city with the same latitude as your city but on a different continent.
3. Locate a third city with latitude close to the equator.
4. Using references, compare average annual precipitation and average high and low temperatures for all three cities.
5. **Think Critically** Hypothesize how latitude affects average temperatures and rainfall.

Preview this chapter's content and activities at blue.msscience.com

FOLDABLES Study Organizer

Nonliving Factors Make the following Foldable to help you understand the cause and effect relationships within the nonliving environment.

STEP 1 Fold two vertical sheets of paper in half from top to bottom. Cut the papers in half along the folds.

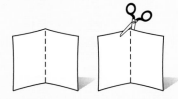

STEP 2 Discard one piece and fold the three vertical pieces in half from top to bottom.

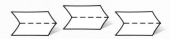

STEP 3 Turn the papers horizontally. Tape the short ends of the pieces together (overlapping the edges slightly).

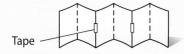

STEP 4 On one side, label the folds: *Nonliving, Water, Soil, Wind, Temperature,* and *Elevation.* Draw a picture of a familiar ecosystem on the other side.

Sequence As you read the chapter, write on the folds how each nonliving factor affects the environment that you draw.

Get Ready to Read

Make Inferences

1 Learn It! When you make inferences, you draw conclusions that are not directly stated in the text. This means you "read between the lines." You interpret clues and draw upon prior knowledge. Authors rely on a reader's ability to infer because all the details are not always given.

2 Practice It! Read the excerpt below and pay attention to highlighted words as you make inferences. Use this Think-Through chart to help you make inferences.

Water vapor that has been released into the atmosphere eventually comes into contact with colder air. The temperature of the water vapor drops. Over time, the water vapor cools enough to change back into liquid water. The process of changing from a gas to a liquid is called condensation. Water vapor condenses on particles of dust in the air, forming tiny droplets. At first, the droplets clump together to form clouds.

—*from page 131*

Text	Question	Inferences
Released into the atmosphere	How is the water vapor released back into the atmosphere?	Evaporation? Animals?
Comes into contact with colder air	Where does it come into contact with cold air?	High in the atmosphere? Cool air from cool weather?
Condenses on particles of dust	Where have the dust particles come from?	Pollution? Wind?

3 Apply It! As you read this chapter, practice your skill at making inferences by making connections and asking questions.

122 A CHAPTER 5 The Nonliving Environment

Target Your Reading

Use this to focus on the main ideas as you read the chapter.

1 Before you read the chapter, respond to the statements below on your worksheet or on a numbered sheet of paper.
- Write an **A** if you **agree** with the statement.
- Write a **D** if you **disagree** with the statement.

2 After you read the chapter, look back to this page to see if you've changed your mind about any of the statements.
- If any of your answers changed, explain why.
- Change any false statements into true statements.
- Use your revised statements as a study guide.

Reading Tip

Sometimes you make inferences by using other reading skills, such as questioning and predicting.

Science Online
Print out a worksheet of this page at blue.msscience.com

Before You Read A or D		Statement	After You Read A or D
	1	The nonliving part on an environment often determines what living organisms are found there.	
	2	Most living organisms are mainly made up of water.	
	3	Heat from the Sun is responsible for wind.	
	4	Animals do not release water vapor.	
	5	The air we breathe mostly contains nitrogen.	
	6	Photosynthesis uses oxygen to produce energy.	
	7	Energy can be both converted to other forms and recycled.	
	8	Matter can be converted to other forms, but cannot be recycled.	
	9	The majority of energy is found at the bottom on an energy pyramid.	
	10	Water is a living part of the environment.	

section 1
Abiotic Factors

as you read

What You'll Learn
- **Identify** common abiotic factors in most ecosystems.
- **List** the components of air that are needed for life.
- **Explain** how climate influences life in an ecosystem.

Why It's Important
Knowing how organisms depend on the nonliving world can help humans maintain a healthy environment.

Review Vocabulary
environment: everything, such as climate, soil, and living things, that surrounds and affects an organism

New Vocabulary
- biotic
- abiotic
- atmosphere
- soil
- climate

Environmental Factors

Living organisms depend on one another for food and shelter. The leaves of plants provide food and a home for grasshoppers, caterpillars, and other insects. Many birds depend on insects for food. Dead plants and animals decay and become part of the soil. The features of the environment that are alive, or were once alive, are called **biotic** (bi AH tihk) factors. The term *biotic* means "living."

Biotic factors are not the only things in an environment that are important to life. Most plants cannot grow without sunlight, air, water, and soil. Animals cannot survive without air, water, or the warmth that sunlight provides. The nonliving, physical features of the environment are called **abiotic** (ay bi AH tihk) factors. The prefix *a* means "not." The term *abiotic* means "not living." Abiotic factors include air, water, soil, sunlight, temperature, and climate. The abiotic factors in an environment often determine which kinds of organisms can live there. For example, water is an important abiotic factor in the environment, as shown in **Figure 1.**

Figure 1 Abiotic factors—air, water, soil, sunlight, temperature, and climate—influence all life on Earth.

Air

Air is invisible and plentiful, so it is easily overlooked as an abiotic factor of the environment. The air that surrounds Earth is called the **atmosphere**. Air contains 78 percent nitrogen, 21 percent oxygen, 0.94 percent argon, 0.03 percent carbon dioxide, and trace amounts of other gases. Some of these gases provide substances that support life.

Carbon dioxide (CO_2) is required for photosynthesis. Photosynthesis—a series of chemical reactions—uses CO_2, water, and energy from sunlight to produce sugar molecules. Organisms, like plants, that can use photosynthesis are called producers because they produce their own food. During photosynthesis, oxygen is released into the atmosphere.

When a candle burns, oxygen from the air chemically combines with the molecules of candle wax. Chemical energy stored in the wax is converted and released as heat and light energy. In a similar way, cells use oxygen to release the chemical energy stored in sugar molecules. This process is called respiration. Through respiration, cells obtain the energy needed for all life processes. Air-breathing animals aren't the only organisms that need oxygen. Plants, some bacteria, algae, fish, and other organisms need oxygen for respiration.

Water

Water is essential to life on Earth. It is a major ingredient of the fluid inside the cells of all organisms. In fact, most organisms are 50 percent to 95 percent water. Respiration, digestion, photosynthesis, and many other important life processes can take place only in the presence of water. As **Figure 2** shows, environments that have plenty of water usually support a greater diversity of and a larger number of organisms than environments that have little water.

Figure 2 Water is an important abiotic factor in deserts and rain forests.

Life in deserts is limited to species that can survive for long periods without water.

Thousands of species can live in lush rain forests where rain falls almost every day.

Mini LAB

Determining Soil Makeup

Procedure

1. Collect 2 cups of **soil**. Remove large pieces of debris and break up clods.
2. Put the soil in a **quart jar** or **similar container that has a lid**.
3. Fill the container with **water** and add 1 teaspoon of **dishwashing liquid**.
4. Put the lid on tightly and shake the container.
5. After 1 min, measure and record the depth of sand that settled on the bottom.
6. After 2 h, measure and record the depth of silt that settles on top of the sand.
7. After 24 h, measure and record the depth of the layer between the silt and the floating organic matter.

Analysis

1. Clay particles are so small that they can remain suspended in water. Where is the clay in your sample?
2. Is sand, silt, or clay the greatest part of your soil sample?

Try at Home

Soil

Soil is a mixture of mineral and rock particles, the remains of dead organisms, water, and air. It is the topmost layer of Earth's crust, and it supports plant growth. Soil is formed, in part, of rock that has been broken down into tiny particles.

Soil is considered an abiotic factor because most of it is made up of nonliving rock and mineral particles. However, soil also contains living organisms and the decaying remains of dead organisms. Soil life includes bacteria, fungi, insects, and worms. The decaying matter found in soil is called humus. Soils contain different combinations of sand, clay, and humus. The type of soil present in a region has an important influence on the kinds of plant life that grow there.

Sunlight

All life requires energy, and sunlight is the energy source for almost all life on Earth. During photosynthesis, producers convert light energy into chemical energy that is stored in sugar molecules. Consumers are organisms that cannot make their own food. Energy is passed to consumers when they eat producers or other consumers. As shown in **Figure 3,** photosynthesis cannot take place if light is never available.

Shady forest

Bottom of deep ocean

Figure 3 Photosynthesis requires light. Little sunlight reaches the shady forest floor, so plant growth beneath trees is limited. Sunlight does not reach into deep lake or ocean waters. Photosynthesis can take place only in shallow water or near the water's surface.
Infer *how fish that live at the bottom of the deep ocean obtain energy.*

Figure 4 Temperature is an abiotic factor that can affect an organism's survival.

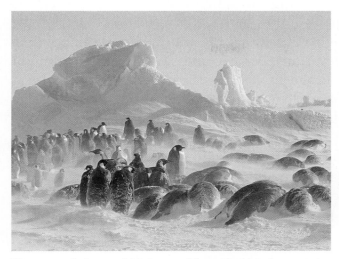

The penguin has a thick layer of fat to hold in heat and keep the bird from freezing. These emperor penguins huddle together for added warmth.

The Arabian camel stores fat only in its hump. This way, the camel loses heat from other parts of its body, which helps it stay cool in the hot desert.

Temperature

Sunlight supplies life on Earth with light energy for photosynthesis and heat energy for warmth. Most organisms can survive only if their body temperatures stay within the range of 0°C to 50°C. Water freezes at 0°C. The penguins in **Figure 4** are adapted for survival in the freezing Antarctic. Camels can survive the hot temperatures of the Arabian Desert because their bodies are adapted for staying cool. The temperature of a region depends in part on the amount of sunlight it receives. The amount of sunlight depends on the land's latitude and elevation.

Figure 5 Because Earth is curved, latitudes farther from the equator are colder than latitudes near the equator.

 *What does sunlight provide for life on Earth?*

Latitude In this chapter's Launch Lab, you discovered that temperature is affected by latitude. You found that cities located at latitudes farther from the equator tend to have colder temperatures than cities at latitudes nearer to the equator. As **Figure 5** shows, polar regions receive less of the Sun's energy than equatorial regions. Near the equator, sunlight strikes Earth directly. Near the poles, sunlight strikes Earth at an angle, which spreads the energy over a larger area.

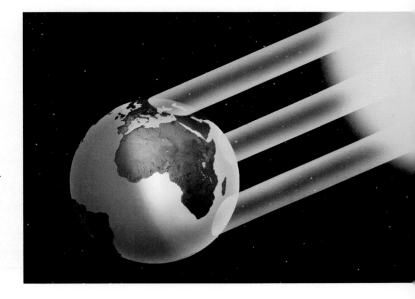

SECTION 1 Abiotic Factors **125**

Figure 6 The stunted growth of these trees is a result of abiotic factors.

Elevation If you have climbed or driven up a mountain, you probably noticed that the temperature got cooler as you went higher. A region's elevation, or distance above sea level, affects its temperature. Earth's atmosphere acts as insulation that traps the Sun's heat. At higher elevations, the atmosphere is thinner than it is at lower elevations. Air becomes warmer when sunlight heats molecules in the air. Because there are fewer molecules at higher elevations, air temperatures there tend to be cooler.

At higher elevations, trees are shorter and the ground is rocky, as shown in **Figure 6.** Above the timberline—the elevation beyond which trees do not grow—plant life is limited to low-growing plants. The tops of some mountains are so cold that no plants can survive. Some mountain peaks are covered with snow year-round.

Applying Math — Solve for an Unknown

TEMPERATURE CHANGES You climb a mountain and record the temperature every 1,000 m of elevation. The temperature is 30°C at 304.8 m, 25°C at 609.6 m, 20°C at 914.4 m, 15°C at 1,219.2 m, and 5°C at 1,828.8 m. Make a graph of the data. Use your graph to predict the temperature at an altitude of 2,133.6 m.

Solution

1. *This is what you know:* The data can be written as ordered pairs (elevation, temperature). The ordered pairs for these data are (304.8, 30), (609.6, 25), (914.4, 20), (1,219.2, 15), (1,828.8, 5).

2. *This is what you want to find:* Predict the temperature at an elevation of 2,133.6 m.

3. *This is what you need to do:* Graph the data by plotting elevation on the *x*-axis and temperature on the *y*-axis.

4. *Predict the temperature at 2,133.6 m:* Extend the graph line to predict the temperature at 2,133.6 m.

Practice Problems

1. Temperatures on another mountain are 33°C at sea level, 31°C at 125 m, 29°C at 250 m, and 26°C at 425 m. Graph the data and predict the temperature at 550 m.
2. Predict what the temperature would be at 375 m.

For more practice, visit blue.msscience.com/math_practice

Climate

In Fairbanks, Alaska, winter temperatures may be as low as −52°C, and more than a meter of snow might fall in one month. In Key West, Florida, snow never falls and winter temperatures rarely dip below 5°C. These two cities have different climates. **Climate** refers to an area's average weather conditions over time, including temperature, rainfall or other precipitation, and wind.

For the majority of living things, temperature and precipitation are the two most important components of climate. The average temperature and rainfall in an area influence the type of life found there. Suppose a region has an average temperature of 25°C and receives an average of less than 25 cm of rain every year. It is likely to be the home of cactus plants and other desert life. A region with similar temperatures that receives more than 300 cm of rain every year is probably a tropical rain forest.

Wind Heat energy from the Sun not only determines temperature, but also is responsible for the wind. The air is made up of molecules of gas. As the temperature increases, the molecules spread farther apart. As a result, warm air is lighter than cold air. Colder air sinks below warmer air and pushes it upward, as shown in **Figure 7.** These motions create air currents that are called wind.

Farmer Changes in weather have a strong influence in crop production. Farmers sometimes adapt by changing planting and harvesting dates, selecting a different crop, or changing water use. In your Science Journal, describe another profession affected by climate.

Topic: Weather Data
Visit blue.msscience.com for Web links to information about recent weather data for your area.

Activity In your Science Journal, describe how these weather conditions affect plants or animals that live in your area.

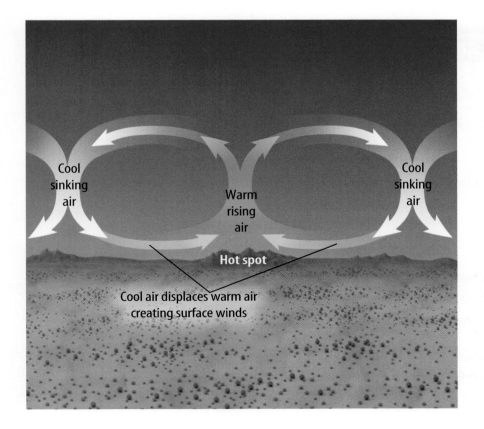

Figure 7 Winds are created when sunlight heats some portions of Earth's surface more than others. In areas that receive more heat, the air becomes warmer. Cold air sinks beneath the warm air, forcing the warm air upward.

Figure 8 In Washington State, the western side of the Cascade Mountains receives an average of 101 cm of rain each year. The eastern side of the Cascades is in a rain shadow that receives only about 25 cm of rain per year.

The Rain Shadow Effect The presence of mountains can affect rainfall patterns. As **Figure 8** shows, wind blowing toward one side of a mountain is forced upward by the mountain's shape. As the air nears the top of the mountain, it cools. When air cools, the moisture it contains falls as rain or snow. By the time the cool air crosses over the top of the mountain, it has lost most of its moisture. The other side of the mountain range receives much less precipitation. It is not uncommon to find lush forests on one side of a mountain range and desert on the other side.

section 1 review

Summary

Environmental Factors
- Organisms depend on one another as well as sunlight, air, water, and soil.

Air, Water, and Soil
- Some of the gases in air provide substances to support life.
- Water is a major component of the cells in all organisms.
- Soil supports plant growth.

Sunlight, Temperature, and Climate
- Light energy supports almost all life on Earth.
- Most organisms require temperature between 0°C and 50°C to survive.
- For most organisms, temperature and precipitation are the two most important components of climate.

Self Check

1. **Compare and contrast** biotic factors and abiotic factors in ecosystems.
2. **Explain** why soil is considered an abiotic factor and a biotic factor.
3. **Think Critically** On day 1, you hike in shade under tall trees. On day 2, the trees are shorter and farther apart. On day 3, you see small plants but no trees. On day 4, you see snow. What abiotic factors might contribute to these changes?

Applying Math

4. **Use an Electronic Spreadsheet** Obtain two months of temperature and precipitation data for two cities in your state. Enter the data in a spreadsheet and calculate average daily temperature and precipitation. Use your calculations to compare the two climates.

128 **CHAPTER 5** The Nonliving Environment

 blue.msscience.com/self_check_quiz

Humus Farm

Besides abiotic factors, such as rock particles and minerals, soil also contains biotic factors, including bacteria, molds, fungi, worms, insects, and decayed organisms. Crumbly, dark brown soil contains a high percentage of humus that is formed primarily from the decayed remains of plants, animals, and animal droppings. In this lab, you will cultivate your own humus.

Real-World Question
How does humus form?

Goals
- **Observe** the formation of humus.
- **Observe** biotic factors in the soil.
- **Infer** how humus forms naturally.

Materials
widemouthed jar
soil
grass clippings
   or green leaves
water
marker
metric ruler
graduated cylinder

Safety Precautions
Wash your hands thoroughly after handling soil, grass clippings, or leaves.

Procedure

1. Copy the data table below into your Science Journal.
2. Place 4 cm of soil in the jar. Pour 30 mL of water into the jar to moisten the soil.
3. Place 2 cm of grass clippings or green leaves on top of the soil in the jar.
4. Use a marker to mark the height of the grass clippings or green leaves in the jar.
5. Put the jar in a sunny place. Every other day, add 30 mL of water to it. In your Science Journal, write a prediction of what you think will happen in your jar.
6. **Observe** your jar every other day for four weeks. Record your observations in your data table.

Conclude and Apply

1. **Describe** what happened during your investigation.
2. **Infer** how molds and bacteria help the process of humus formation.
3. **Infer** how humus forms on forest floors or in grasslands.

Humus Formation	
Date	Observations
	Do not write in this book.

Communicating Your Data

Compare your humus farm with those of your classmates. With several classmates, write a recipe for creating the richest humus. Ask your teacher to post your recipe in the classroom. **For more help, refer to the** Science Skill Handbook.

LAB **129**

section 2

Cycles in Nature

as you read

What You'll Learn
- **Explain** the importance of Earth's water cycle.
- **Diagram** the carbon cycle.
- **Recognize** the role of nitrogen in life on Earth.

Why It's Important
The recycling of matter on Earth demonstrates natural processes.

Review Vocabulary
biosphere: the part of the world in which life can exist

New Vocabulary
- evaporation
- condensation
- water cycle
- nitrogen fixation
- nitrogen cycle
- carbon cycle

The Cycles of Matter

Imagine an aquarium containing water, fish, snails, plants, algae, and bacteria. The tank is sealed so that only light can enter. Food, water, and air cannot be added. Will the organisms in this environment survive? Through photosynthesis, plants and algae produce their own food. They also supply oxygen to the tank. Fish and snails take in oxygen and eat plants and algae. Wastes from fish and snails fertilize plants and algae. Organisms that die are decomposed by the bacteria. The organisms in this closed environment can survive because the materials are recycled. A constant supply of light energy is the only requirement. Earth's biosphere also contains a fixed amount of water, carbon, nitrogen, oxygen, and other materials required for life. These materials cycle through the environment and are reused by different organisms.

The Water Cycle

If you leave a glass of water on a sunny windowsill, the water will evaporate. **Evaporation** takes place when liquid water changes into water vapor, which is a gas, and enters the atmosphere, shown in **Figure 9.** Water evaporates from the surfaces of lakes, streams, puddles, and oceans. Water vapor enters the atmosphere from plant leaves in a process known as transpiration (trans puh RAY shun). Animals release water vapor into the air when they exhale. Water also returns to the environment from animal wastes.

Figure 9 Water vapor is a gas that is present in the atmosphere.

130 CHAPTER 5 The Nonliving Environment

Condensation Water vapor that has been released into the atmosphere eventually comes into contact with colder air. The temperature of the water vapor drops. Over time, the water vapor cools enough to change back into liquid water. The process of changing from a gas to a liquid is called **condensation.** Water vapor condenses on particles of dust in the air, forming tiny droplets. At first, the droplets clump together to form clouds. When they become large and heavy enough, they fall to the ground as rain or other precipitation. As the diagram in **Figure 10** shows, the **water cycle** is a model that describes how water moves from the surface of Earth to the atmosphere and back to the surface again.

Figure 10 The water cycle involves evaporation, condensation, and precipitation. Water molecules can follow several pathways through the water cycle.
Identify *as many water cycle pathways as you can from this diagram.*

Water Use Data about the amount of water people take from reservoirs, rivers, and lakes for use in households, businesses, agriculture, and power production is shown in **Table 1.** These actions can reduce the amount of water that evaporates into the atmosphere. They also can influence how much water returns to the atmosphere by limiting the amount of water available to plants and animals.

Table 1 U.S. Estimated Water Use in 1995		
Water Use	Millions of Gallons per Day	Percent of Total
Homes and Businesses	41,600	12.2
Industry and Mining	28,000	8.2
Farms and Ranches	139,200	40.9
Electricity Production	131,800	38.7

SECTION 2 Cycles in Nature **131**

The Nitrogen Cycle

The element nitrogen is important to all living things. Nitrogen is a necessary ingredient of proteins. Proteins are required for the life processes that take place in the cells of all organisms. Nitrogen is also an essential part of the DNA of all organisms. Although nitrogen is the most plentiful gas in the atmosphere, most organisms cannot use nitrogen directly from the air. Plants need nitrogen that has been combined with other elements to form nitrogen compounds. Through a process called **nitrogen fixation,** some types of soil bacteria can form the nitrogen compounds that plants need. Plants absorb these nitrogen compounds through their roots. Animals obtain the nitrogen they need by eating plants or other animals. When dead organisms decay, the nitrogen in their bodies returns to the soil or to the atmosphere. This transfer of nitrogen from the atmosphere to the soil, to living organisms, and back to the atmosphere is called the **nitrogen cycle,** shown in **Figure 11**.

Reading Check *What is nitrogen fixation?*

Figure 11 During the nitrogen cycle, nitrogen gas from the atmosphere is converted to a soil compound that plants can use. **State** *one source of recycled nitrogen.*

Nitrogen gas is changed into usable compounds by lightning or by nitrogen-fixing bacteria that live on the roots of certain plants.

Plants use nitrogen compounds to build cells.

Animals eat plants. Animal wastes return some nitrogen compounds to the soil.

Animals and plants die and decompose, releasing nitrogen compounds back into the soil.

132 **CHAPTER 5** The Nonliving Environment

Figure 12 The swollen nodules on the roots of soybean plants contain colonies of nitrogen-fixing bacteria that help restore nitrogen to the soil. The bacteria depend on the plant for food, while the plant depends on the bacteria to form the nitrogen compounds the plant needs.

Soybeans

Nodules on roots

Nitrogen-fixing bacteria
Stained LM Magnification: 1000×

Soil Nitrogen Human activities can affect the part of the nitrogen cycle that takes place in the soil. If a farmer grows a crop, such as corn or wheat, most of the plant material is taken away when the crop is harvested. The plants are not left in the field to decay and return their nitrogen compounds to the soil. If these nitrogen compounds are not replaced, the soil could become infertile. You might have noticed that adding fertilizer to soil can make plants grow greener, bushier, or taller. Most fertilizers contain the kinds of nitrogen compounds that plants need for growth. Fertilizers can be used to replace soil nitrogen in crop fields, lawns, and gardens. Compost and animal manure also contain nitrogen compounds that plants can use. They also can be added to soil to improve fertility.

Another method farmers use to replace soil nitrogen is to grow nitrogen-fixing crops. Most nitrogen-fixing bacteria live on or in the roots of certain plants. Some plants, such as peas, clover, and beans, including the soybeans shown in **Figure 12,** have roots with swollen nodules that contain nitrogen-fixing bacteria. These bacteria supply nitrogen compounds to the soybean plants and add nitrogen compounds to the soil.

Mini LAB

Comparing Fertilizers
Procedure
1. Examine the three numbers (e.g., 5-10-5) on the **labels of three brands of houseplant fertilizer.** The numbers indicate the percentages of nitrogen, phosphorus, and potassium, respectively, that the product contains.
2. Compare the prices of the three brands of fertilizer.
3. Compare the amount of each brand needed to fertilize a typical houseplant.

Analysis
1. **Identify** the brand with the highest percentage of nitrogen.
2. **Calculate** which brand is the most expensive source of nitrogen. The least expensive.

SECTION 2 Cycles in Nature

NATIONAL GEOGRAPHIC VISUALIZING THE CARBON CYCLE

Figure 13

Carbon—in the form of different kinds of carbon-containing molecules—moves through an endless cycle. The diagram below shows several stages of the carbon cycle. It begins when plants and algae remove carbon from the environment during photosynthesis. This carbon returns to the atmosphere via several carbon-cycle pathways.

A Air contains carbon in the form of carbon dioxide gas. Plants and algae use carbon dioxide to make sugars, which are energy-rich, carbon-containing compounds.

B Organisms break down sugar molecules made by plants and algae to obtain energy for life and growth. Carbon dioxide is released as a waste.

C Burning fossil fuels and wood releases carbon dioxide into the atmosphere.

D When organisms die, their carbon-containing molecules become part of the soil. The molecules are broken down by fungi, bacteria, and other decomposers. During this decay process, carbon dioxide is released into the air.

E Under certain conditions, the remains of some dead organisms may gradually be changed into fossil fuels such as coal, gas, and oil. These carbon compounds are energy rich.

The Carbon Cycle

Carbon atoms are found in the molecules that make up living organisms. Carbon is an important part of soil humus, which is formed when dead organisms decay, and it is found in the atmosphere as carbon dioxide gas (CO_2). The **carbon cycle** describes how carbon molecules move between the living and nonliving world, as shown in **Figure 13.**

The carbon cycle begins when producers remove CO_2 from the air during photosynthesis. They use CO_2, water, and sunlight to produce energy-rich sugar molecules. Energy is released from these molecules during respiration—the chemical process that provides energy for cells. Respiration uses oxygen and releases CO_2. Photosynthesis uses CO_2 and releases oxygen. These two processes help recycle carbon on Earth.

Reading Check *How does carbon dioxide enter the atmosphere?*

Human activities also release CO_2 into the atmosphere. Fossil fuels such as gasoline, coal, and heating oil are the remains of organisms that lived millions of years ago. These fuels are made of energy-rich, carbon-based molecules. When people burn these fuels, CO_2 is released into the atmosphere as a waste product. People also use wood for construction and for fuel. Trees that are harvested for these purposes no longer remove CO_2 from the atmosphere during photosynthesis. The amount of CO_2 in the atmosphere is increasing. Extra CO_2 could trap more heat from the Sun and cause average temperatures on Earth to rise.

Science Online

Topic: Life Processes
Visit blue.msscience.com for Web links to information about chemical equations that describe photosynthesis and respiration.

Activity Use these equations to explain how respiration is the reverse of photosynthesis.

section 2 review

Summary

The Cycles of Matter
- Earth's biosphere contains a fixed amount of water, carbon, nitrogen, oxygen, and other materials that cycle through the environment.

The Water Cycle
- Water cycles through the environment using several pathways.

The Nitrogen Cycle
- Some types of bacteria can form nitrogen compounds that plants and animals can use.

The Carbon Cycle
- Producers remove CO_2 from the air during photosynthesis and produce O_2.
- Consumers remove O_2 and produce CO_2.

Self Check

1. **Describe** the water cycle.
2. **Infer** how burning fossil fuels might affect the makeup of gases in the atmosphere.
3. **Explain** why plants, animals, and other organisms need nitrogen.
4. **Think Critically** Most chemical fertilizers contain nitrogen, phosphorous, and potassium. If they do not contain carbon, how do plants obtain carbon?

Applying Skills

5. **Identify and Manipulate Variables and Controls** Describe an experiment that would determine whether extra carbon dioxide enhances the growth of tomato plants.

section 3
Energy Flow

as you read

What You'll Learn
- **Explain** how organisms produce energy-rich compounds.
- **Describe** how energy flows through ecosystems.
- **Recognize** how much energy is available at different levels in a food chain.

Why It's Important
All living things, including people, need a constant supply of energy.

Review Vocabulary
energy: the capacity for doing work

New Vocabulary
- chemosynthesis
- food web
- energy pyramid

Converting Energy

All living things are made of matter, and all living things need energy. Matter and energy move through the natural world in different ways. Matter can be recycled over and over again. The recycling of matter requires energy. Energy is not recycled, but it is converted from one form to another. The conversion of energy is important to all life on Earth.

Photosynthesis During photosynthesis, producers convert light energy into the chemical energy in sugar molecules. Some of these sugar molecules are broken down as energy. Others are used to build complex carbohydrate molecules that become part of the producer's body. Fats and proteins also contain stored energy.

Chemosynthesis Not all producers rely on light for energy. During the 1970s, scientists exploring the ocean floor were amazed to find communities teeming with life. These communities were at a depth of almost 3.2 km and living in total darkness. They were found near powerful hydrothermal vents like the one shown in **Figure 14**.

Figure 14 Chemicals in the water that flows from hydrothermal vents provide bacteria with a source of energy. The bacterial producers use this energy to make nutrients through the process of chemosynthesis. Consumers, such as tubeworms, feed on the bacteria.

Hydrothermal Vents A hydrothermal vent is a deep crack in the ocean floor through which the heat of molten magma can escape. The water from hydrothermal vents is extremely hot from contact with molten rock that lies deep in Earth's crust.

Because no sunlight reaches these deep ocean regions, plants or algae cannot grow there. How do the organisms living in this community obtain energy? Scientists learned that the hot water contains nutrients such as sulfur molecules that bacteria use to produce their own food. The production of energy-rich nutrient molecules from chemicals is called **chemosynthesis** (kee moh SIHN thuh sus). Consumers living in the hydrothermal vent communities rely on chemosynthetic bacteria for nutrients and energy. Chemosynthesis and photosynthesis allow producers to make their own energy-rich molecules.

Reading Check *What is chemosynthesis?*

Hydrothermal Vents The first hydrothermal vent community discovered was found along the Galápagos rift zone. A rift zone forms where two plates of Earth's crust are spreading apart. In your Science Journal, describe the energy source that heats the water in the hydrothermal vents of the Galápagos rift zone.

Energy Transfer

Energy can be converted from one form to another. It also can be transferred from one organism to another. Consumers cannot make their own food. Instead, they obtain energy by eating producers or other consumers. The energy stored in the molecules of one organism is transferred to another organism. That organism can oxidize food to release energy that it can use for maintenance and growth or is transformed into heat. At the same time, the matter that makes up those molecules is transferred from one organism to another.

Food Chains A food chain is a way of showing how matter and energy pass from one organism to another. Producers—plants, algae, and other organisms that are capable of photosynthesis or chemosynthesis—are always the first step in a food chain. Animals that consume producers such as herbivores are the second step. Carnivores and omnivores—animals that eat other consumers—are the third and higher steps of food chains. One example of a food chain is shown in **Figure 15**.

Figure 15 In this food chain, grasses are producers, marmots are herbivores that eat the grasses, and grizzly bears are consumers that eat marmots. The arrows show the direction in which matter and energy flow.
Infer *what might happen if grizzly bears disappeared from this ecosystem.*

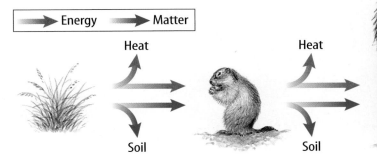

SECTION 3 Energy Flow **137**

Food Webs A forest community includes many feeding relationships. These relationships can be too complex to show with a food chain. For example, grizzly bears eat many different organisms, including berries, insects, chipmunks, and fish. Berries are eaten by bears, birds, insects, and other animals. A bear carcass might be eaten by wolves, birds, or insects. A **food web** is a model that shows all the possible feeding relationships among the organisms in a community. A food web is made up of many different food chains, as shown in **Figure 16**.

Energy Pyramids

Food chains usually have at least three links, but rarely more than five. This limit exists because the amount of available energy is reduced as you move from one level to the next in a food chain. Imagine a grass plant that absorbs energy from the Sun. The plant uses some of this energy to grow and produce seeds. Some of the energy is stored in the seeds.

Figure 16 Compared to a food chain, a food web provides a more complete model of the feeding relationships in a community.

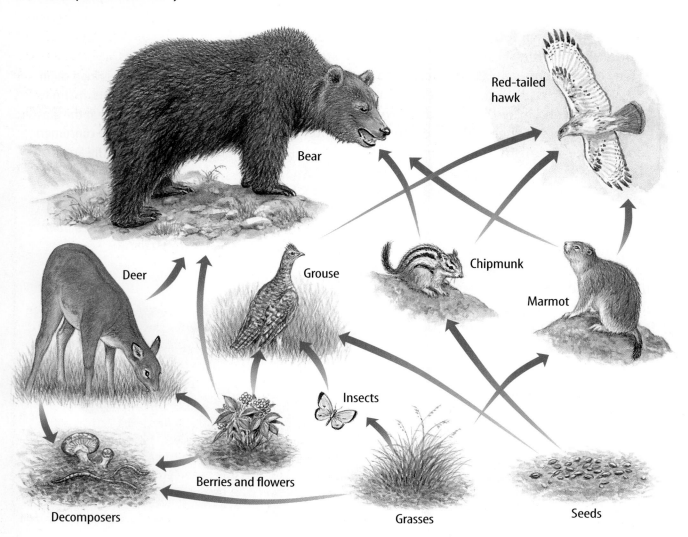

Available Energy When a mouse eats grass seeds, energy stored in the seeds is transferred to the mouse. However, most of the energy the plant absorbed from the Sun was used for the plant's growth. The mouse uses energy from the seed for its own life processes, including respiration, digestion, and growth. Some of this energy was given off as heat. A hawk that eats the mouse obtains even less energy. The amount of available energy is reduced from one feeding level of a food chain to another.

An **energy pyramid,** like the one in **Figure 17,** shows the amount of energy available at each feeding level in an ecosystem. The bottom of the pyramid, which represents all of the producers, is the first feeding level. It is the largest level because it contains the most energy and the largest number of organisms. As you move up the pyramid, the transfer of energy is less efficient and each level becomes smaller. Only about ten percent of the energy available at each feeding level of an energy pyramid is transferred to the next higher level.

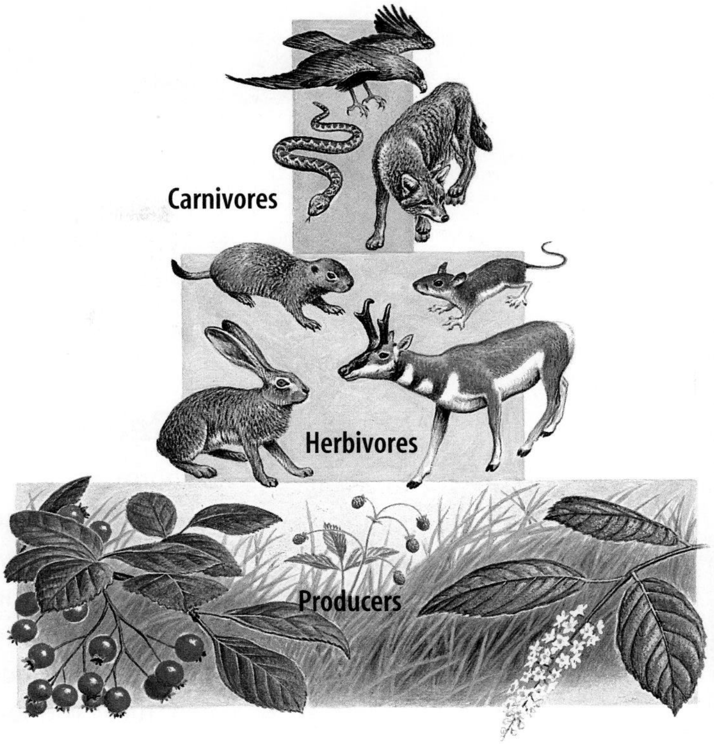

Figure 17 An energy pyramid shows that each feeding level has less energy than the one below it. **Describe** what would happen if the hawks and snakes outnumbered the rabbits and mice in this ecosystem.

Reading Check Why does the first feeding level of an energy pyramid contain the most energy?

section 3 review

Summary

Converting Energy
- Most producers convert light energy into chemical energy.
- Some producers can produce their own food using energy in chemicals such as sulfur.

Energy Transfer
- Producers convert energy into forms that other organisms can use.
- Food chains show how matter and energy pass from one organism to another.

Energy Pyramids
- Energy pyramids show the amount of energy available at each feeding level.
- The amount of available energy decreases from the base to the top of the energy pyramid.

Self Check

1. **Compare and contrast** a food web and an energy pyramid.
2. **Explain** why there is a limit to the number of links in a food chain.
3. **Think Critically** Use your knowledge of food chains and the energy pyramid to explain why the number of mice in a grassland ecosystem is greater than the number of hawks.

Applying Math

4. **Solve One-Step Equations** A forest has 24,055,000 kilocalories (kcals) of producers, 2,515,000 kcals of herbivores, and 235,000 kcals of carnivores. How much energy is lost between producers and herbivores? Between herbivores and carnivores?

blue.mssience.com/self_check_quiz

SECTION 3 Energy Flow **139**

Where does the mass of a plant come from?

Real-World Question

An enormous oak tree starts out as a tiny acorn. The acorn sprouts in dark, moist soil. Roots grow down through the soil. Its stem and leaves grow up toward the light and air. Year after year, the tree grows taller, its trunk grows thicker, and its roots grow deeper. It becomes a towering oak that produces thousands of acorns of its own. An oak tree has much more mass than an acorn. Where does this mass come from? The soil? The air? In this activity, you'll find out by conducting an experiment with radish plants. Does all of the matter in a radish plant come from the soil?

Goals
- **Measure** the mass of soil before and after radish plants have been grown in it.
- **Measure** the mass of radish plants grown in the soil.
- **Analyze** the data to determine whether the mass gained by the plants equals the mass lost by the soil.

Materials
8-oz plastic or paper cup
potting soil to fill cup
scale or balance
radish seeds (4)
water
paper towels

Safety Precautions

Using Scientific Methods

▶ Procedure

1. Copy the data table into your Science Journal.
2. Fill the cup with dry soil.
3. Find the mass of the cup of soil and record this value in your data table.
4. Moisten the soil in the cup. Plant four radish seeds 2 cm deep in the soil. Space the seeds an equal distance apart. Wash your hands.
5. Add water to keep the soil barely moist as the seeds sprout and grow.
6. When the plants have developed four to six true leaves, usually after two to three weeks, carefully remove the plants from the soil. Gently brush the soil off the roots. Make sure all the soil remains in the cup.
7. Spread the plants out on a paper towel. Place the plants and the cup of soil in a warm area to dry out.
8. When the plants are dry, measure their mass and record this value in your data table. Write this number with a plus sign in the Gain or Loss column.
9. When the soil is dry, find the mass of the cup of soil. Record this value in your data table. Subtract the End mass from the Start mass and record this number with a minus sign in the Gain or Loss column.

Mass of Soil and Radish Plants			
	Start	End	Gain (+) or Loss (−)
Mass of dry soil and cup		Do not write in this book.	
Mass of dried radish plants	0 g		

▶ Analyze Your Data

1. **Calculate** how much mass was gained or lost by the soil. By the radish plants.
2. Did the mass of the plants come completely from the soil? How do you know?

▶ Conclude and Apply

1. In the early 1600s, a Belgian scientist named J. B. van Helmont conducted this experiment with a willow tree. What is the advantage of using radishes instead of a tree?
2. **Predict** where all of the mass gained by the plants came from.

Compare your conclusions with those of other students in your class. **For more help, refer to the** Science Skill Handbook.

SCIENCE Stats

Extreme Climates

Did you know...

... The greatest snowfall in one year occurred at Mount Baker in Washington State. Approximately 2,896 cm of snow fell during the 1998–99, 12-month snowfall season. That's enough snow to bury an eight-story building.

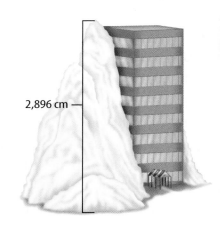

2,896 cm

Applying Math What was the average monthly snowfall at Mount Baker during the 1998–99 snowfall season?

... The hottest climate in the United States is found in Death Valley, California. In July 1913, Death Valley reached approximately 57°C. As a comparison, a comfortable room temperature is about 20°C.

... The record low temperature of a frigid −89°C was set in Antarctica in 1983. As a comparison, the temperature of a home freezer is about −15°C.

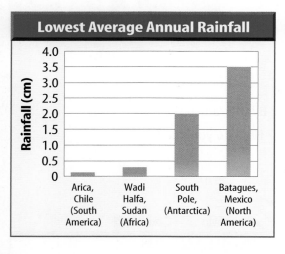

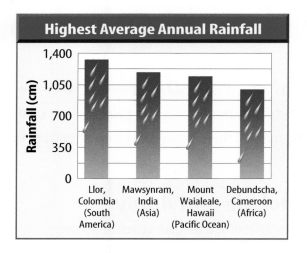

Graph It

Visit blue.msscience.com/science_stats to find the average monthly rainfall in a tropical rain forest. Make a line graph to show how the amount of precipitation changes during the 12 months of the year.

Chapter 5 Study Guide

Reviewing Main Ideas

Section 1 Abiotic Factors

1. Abiotic factors include air, water, soil, sunlight, temperature, and climate.
2. The availability of water and light influences where life exists on Earth.
3. Soil and climate have an important influence on the types of organisms that can survive in different environments.
4. High latitudes and elevations generally have lower average temperatures.

Section 2 Cycles in Nature

1. Matter is limited on Earth and is recycled through the environment.
2. The water cycle involves evaporation, condensation, and precipitation.
3. The carbon cycle involves photosynthesis and respiration.
4. Nitrogen in the form of soil compounds enters plants, which then are consumed by other organisms.

Section 3 Energy Flow

1. Producers make energy-rich molecules through photosynthesis or chemosynthesis.
2. When organisms feed on other organisms, they obtain matter and energy.
3. Matter can be recycled, but energy cannot.
4. Food webs are models of the complex feeding relationships in communities.
5. Available energy decreases as you go to higher feeding levels in an energy pyramid.

Visualizing Main Ideas

This diagram represents photosynthesis in a leaf. Match each letter with one of the following terms: light, carbon dioxide, *or* oxygen.

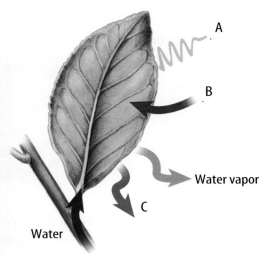

CHAPTER STUDY GUIDE 143

chapter 5 Review

Using Vocabulary

abiotic p. 122
atmosphere p. 123
biotic p. 122
carbon cycle p. 135
chemosynthesis p. 137
climate p. 127
condensation p. 131
energy pyramid p. 139
evaporation p. 130
food web p. 138
nitrogen cycle p. 132
nitrogen fixation p. 132
soil p. 124
water cycle p. 131

Which vocabulary word best corresponds to each of the following events?

1. A liquid changes to a gas.
2. Some types of bacteria form nitrogen compounds in the soil.
3. Decaying plants add nitrogen to the soil.
4. Chemical energy is used to make energy-rich molecules.
5. Decaying plants add carbon to the soil.
6. A gas changes to a liquid.
7. Water flows downhill into a stream. The stream flows into a lake, and water evaporates from the lake.
8. Burning coal and exhaust from automobiles release carbon into the air.

Checking Concepts

Choose the word or phrase that best answers the question.

9. Which of the following is an abiotic factor?
   A) penguins   C) soil bacteria
   B) rain   D) redwood trees

Use the equation below to answer question 10.

$$CO_2 + H_2O \xrightarrow{\text{light energy}} \text{sugar} + O_2$$

10. Which of the following processes is shown in the equation above?
    A) condensation   C) burning
    B) photosynthesis D) respiration

11. Which of the following applies to latitudes farther from the equator?
    A) higher elevations
    B) higher temperatures
    C) higher precipitation levels
    D) lower temperatures

12. Water vapor forming droplets that form clouds directly involves which process?
    A) condensation   C) evaporation
    B) respiration   D) transpiration

13. Which one of the following components of air is least necessary for life on Earth?
    A) argon   C) carbon dioxide
    B) nitrogen   D) oxygen

14. Which group makes up the largest level of an energy pyramid?
    A) herbivores   C) decomposers
    B) producers   D) carnivores

15. Earth receives a constant supply of which of the following items?
    A) light energy   C) nitrogen
    B) carbon   D) water

16. Which of these is an energy source for chemosynthesis?
    A) sunlight   C) sulfur molecules
    B) moonlight   D) carnivores

Use the illustration below to answer question 17.

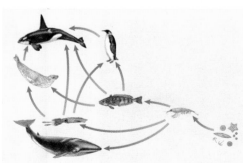

17. What is the illustration above an example of?
    A) food chain   C) energy pyramid
    B) food web   D) carbon cycle

chapter 5 Review

Thinking Critically

18. **Draw a Conclusion** A country has many starving people. Should they grow vegetables and corn to eat, or should they grow corn to feed cattle so they can eat beef? Explain.

19. **Explain** why a food web is a better model of energy flow than a food chain.

20. **Infer** Do bacteria need nitrogen? Why or why not?

21. **Describe** why it is often easier to walk through an old, mature forest of tall trees than through a young forest of small trees.

22. **Explain** why giant sequoia trees grow on the west side of California's Inyo Mountains and Death Valley, a desert, is on the east side of the mountains.

23. **Concept Map** Copy and complete this food web using the following information: *caterpillars and rabbits eat grasses, raccoons eat rabbits and mice, mice eat grass seeds,* and *birds eat caterpillars.*

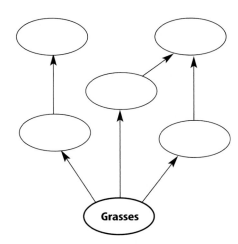

24. **Form a Hypothesis** For each hectare of land, ecologists found 10,000 kcals of producers, 10,000 kcals of herbivores, and 2,000 kcals of carnivores. Suggest a reason why producer and herbivore levels are equal.

25. **Recognize Cause and Effect** A lake in Kenya has been taken over by a floating weed. How could you determine if nitrogen fertilizer runoff from farms is causing the problem?

Performance Activities

26. **Poster** Use magazine photographs to make a visual representation of the water cycle.

Applying Math

27. **Energy Budget** Raymond Lindeman, from the University of Minnesota, was the first person to calculate the total energy budget of an entire community at Cedar Bog Lake in MN. He found the total amount of energy produced by producers was 1,114 kilocalories per meter squared per year. About 20% of the 1,114 kilocalories were used up during respiration. How many kilocalories were used during respiration?

28. **Kilocalorie Use** Of the 600 kilocalories of producers available to a caterpillar, the caterpillar consumes about 150 kilocalories. About 25% of the 150 kilocalories is used to maintain its life processes and is lost as heat, while 16% cannot be digested. How many kilocalories are lost as heat? What percentage of the 600 kilocalories is available to the next feeding level?

Use the table below to answer question 29.

Mighty Migrators	
Species	Distance (km)
Desert locust	4,800
Caribou	800
Green turtle	1,900
Arctic tern	35,000
Gray whale	19,000

29. **Make and Use Graphs** Climate can cause populations to move from place to place. Make a bar graph of migration distances shown above.

chapter 5 Standardized Test Practice

Part 1 Multiple Choice

Record your answers on the answer sheet provided by your teacher or on a sheet of paper.

1. The abiotic factor that provides energy for nearly all life on Earth is
   A. air.
   B. sunlight.
   C. water.
   D. soil.

2. Which of the following is characteristic of places at high elevations?
   A. fertile soil
   B. fewer molecules in the air
   C. tall trees
   D. warm temperatures

Use the diagram below to answer questions 3 and 4.

3. The air at point C is
   A. dry and warm.
   B. dry and cool.
   C. moist and warm.
   D. moist and cool.

4. The air at point A is
   A. dry and warm.
   B. dry and cool.
   C. moist and warm.
   D. moist and cool.

5. What process do plants use to return water vapor to the atmosphere?
   A. transpiration
   B. evaporation
   C. respiration
   D. condensation

6. Clouds form as a result of what process?
   A. evaporation
   B. transpiration
   C. respiration
   D. condensation

Use the illustration of the nitrogen cycle below to answer questions 7 and 8.

7. Which of the following items shown in the diagram contribute to the nitrogen cycle by releasing AND absorbing nitrogen?
   A. the decaying organism only
   B. the trees only
   C. the trees and the grazing cows
   D. the lightning and the decaying organism

8. Which of the following items shown in the diagram contribute to the nitrogen cycle by ONLY releasing nitrogen?
   A. the decaying organism only
   B. the trees only
   C. the trees and the grazing cows
   D. the lightning and the decaying organism

9. Where is most of the energy found in an energy pyramid?
   A. at the top level
   B. in the middle levels
   C. at the bottom level
   D. all levels are the same

10. What organisms remove carbon dioxide gas from the air during photosynthesis?
    A. consumers
    B. producers
    C. herbivores
    D. omnivores

Part 2 | Short Response/Grid In

Record your answers on the answer sheet provided by your teacher or on a sheet of paper.

11. Give two examples of abiotic factors and describe how each one is important to biotic factors.

Use the table below to answer questions 12 and 13.

U.S. Estimated Water Use in 1995		
Water Use	Millions of Gallons per Day	Percent of Total
Homes and Businesses	41,600	12.2
Industry and Mining	28,000	8.2
Farms and Ranches	139,200	40.9
Electricity Production	131,800	38.7

12. According to the table above, what accounted for the highest water use in the U.S. in 1995?

13. What percentage of the total amount of water use results from electricity production and homes and business combined?

14. Where are nitrogen-fixing bacteria found?

15. Describe two ways that carbon is released into the atmosphere.

16. How are organisms near hydrothermal vents deep in the ocean able to survive?

17. Use a diagram to represent the transfer of energy among these organisms: a weasel, a rabbit, grasses, and a coyote.

> **Test-Taking Tip**
> **Answer All Questions** Never leave any answer blank.

Part 3 | Open Ended

Record your answers on a sheet of paper.

18. Explain how a decrease in the amount of sunlight would affect producers that use photosynthesis, and producers that use chemosynthesis.

19. Describe how wind and wind currents are produced.

20. Use the water cycle to explain why beads of water form on the outside of a glass of iced water on a hot day.

21. Draw a flowchart that shows how soy beans, deer, and nitrogen-fixing bacteria help cycle nitrogen from the atmosphere, to the soil, to living organisms, and back to the atmosphere.

Use the diagram below to answer questions 22 and 23.

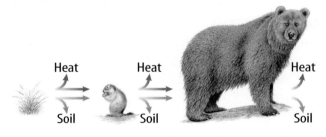

22. What term is used for the diagram above? Explain how the diagram represents energy transfer.

23. Explain how the grass and bear populations would be affected if the marmot population suddenly declined.

24. Compare and contrast an energy pyramid and a food web.

25. What happens to the energy in organisms at the top of an energy pyramid when they die?

chapter 6

Ecosystems

The BIG Idea
Earth has many diverse ecosystems on land and in water.

SECTION 1
How Ecosystems Change
Main Idea Ecosystems gradually change over time.

SECTION 2
Biomes
Main Idea Land on Earth is divided into large geographic areas that have similar climates and ecosystems.

SECTION 3
Acquatic Ecosystems
Main Idea Both Earth's salt water and freshwater are divided into a variety of ecosystems.

The Benefits of Wildfires

Ecosystems are places where organisms, including humans, interact with each other and with their physical environment. In some ecosystems, wildfires are an essential part of the physical environment. Organisms in these ecosystems are well adapted to the changes that fire brings, and can benefit from wildfires.

Science Journal What traits might plants on this burning Montana hillside have that enable them to survive?

Start-Up Activities

What environment do houseplants need?

The plants growing in your classroom or home may not look like the same types of plants that you find growing outside. Many indoor plants don't grow well outside in most North American climates. Do the lab below to determine what type of environment most houseplants thrive in.

1. Examine a healthy houseplant in your classroom or home.
2. Describe the environmental conditions found in your classroom or home. For example, is the air humid or dry? Is the room warm or cool? Does the temperature stay about the same, or change during the day?
3. Using observations from step 1 and descriptions from step 2, hypothesize about the natural environment of the plants in your classroom or home.
4. **Think Critically** In your Science Journal, record the observations that led to your hypothesis. How would you design an experiment to test your hypothesis?

FOLDABLES Study Organizer

Primary and Secondary Succession Make the following Foldable to help you illustrate the main ideas about succession.

STEP 1 Fold a vertical sheet of paper in half from top to bottom.

STEP 2 Fold in half from side to side with the fold at the top.

STEP 3 Unfold the paper once. Cut only the fold of the top flap to make two tabs.

STEP 4 Turn the paper vertically and label on the front tabs as shown.

Illustrate and Label As you read the chapter, define terms and collect information under the appropriate tabs.

Preview this chapter's content and activities at
blue.msscience.com

Get Ready to Read

Take Notes

① Learn It! The best way for you to remember information is to write it down, or take notes. Good note-taking is useful for studying and research. When you are taking notes, it is helpful to
- phrase the information in your own words;
- restate ideas in short, memorable phrases;
- stay focused on main ideas and only the most important supporting details.

② Practice It! Make note-taking easier by using a chart to help you organize information clearly. Write the main ideas in the left column. Then write at least three supporting details in the right column. Read the text from Section 3 of this chapter under the heading *Water Pollution,* page 175. Then take notes using a chart, such as the one below.

Main Idea	Supporting Details
	1. 2. 3. 4. 5.
	1. 2. 3. 4. 5.

③ Apply It! As you read this chapter, make a chart of the main ideas. Next to each main idea, list at least two supporting details.

Target Your Reading

Use this to focus on the main ideas as you read the chapter.

① Before you read the chapter, respond to the statements below on your worksheet or on a numbered sheet of paper.
- Write an **A** if you **agree** with the statement.
- Write a **D** if you **disagree** with the statement.

② After you read the chapter, look back to this page to see if you've changed your mind about any of the statements.
- If any of your answers changed, explain why.
- Change any false statements into true statements.
- Use your revised statements as a study guide.

Reading Tip

Read one or two paragraphs first and take notes after you read. You are likely to take down too much information if you take notes as you read.

Before You Read A or D		Statement	After You Read A or D
	1	Gradual changes to the types of species in an ecosystem always follow the same pattern.	
	2	A variety of organisms can live on bare rock.	
	3	Plant communities are always changing and unstable.	
	4	Deserts in different parts of the world do not have any similarities with one another.	
	5	Rainforests are all located near the equator.	
	6	Most of the soil in the tundra is frozen year round.	
	7	Aquatic ecosystems are either freshwater or saltwater.	
	8	Coral reefs are durable ecosystems that adapt quickly to stress.	
	9	The amount of sunlight available in ocean and lake waters affects the number of organisms found there.	

Science Online
Print out a worksheet of this page at blue.msscience.com

section 1

How Ecosystems Change

as you read

What You'll Learn
- **Explain** how ecosystems change over time.
- **Describe** how new communities begin in areas without life.
- **Compare** pioneer species and climax communities.

Why It's Important
Understanding ecosystems and your role in them can help you manage your impact on them and predict the changes that may happen in the future.

Review Vocabulary
ecosystem: community of living organisms interacting with each other and their physical environment

New Vocabulary
- succession
- pioneer species
- climax community

Ecological Succession

What would happen if the lawn at your home were never cut? The grass would get longer, as in **Figure 1,** and soon it would look like a meadow. Later, larger plants would grow from seeds brought to the area by animals or wind. Then, trees might sprout. In fact, in 20 years or less you wouldn't be able to tell that the land was once a mowed lawn. An ecologist can tell you what type of ecosystem your lawn would become. If it would become a forest, they can tell you how long it would take and predict the type of trees that would grow there. **Succession** refers to the normal, gradual changes that occur in the types of species that live in an area. Succession occurs differently in different places around the world.

Primary Succession As lava flows from the mouth of a volcano, it is so hot that it destroys everything in its path. When it cools, lava forms new land composed of rock. It is hard to imagine that this land eventually could become a forest or grassland someday.

The process of succession that begins in a place previously without plants is called primary succession. It starts with the arrival of living things such as lichens (LI kunz). These living things, called **pioneer species,** are the first to inhabit an area. They survive drought, extreme heat and cold, and other harsh conditions and often start the soil-building process.

Figure 1 Open areas that are not maintained will become overgrown with grasses and shrubs as succession proceeds.

Figure 2 Lichens, like these in Colorado, are fragile and take many years to grow. They often cling to bare rock where many other organisms can't survive. **Describe** how lichens form soil.

New Soil During primary succession, shown in **Figure 2,** soil begins to form as lichens and the forces of weather and erosion help break down rocks into smaller pieces. When lichens die, they decay, adding small amounts of organic matter to the rock. Plants such as mosses and ferns can grow in this new soil. Eventually, these plants die, adding more organic material. The soil layer thickens, and grasses, wildflowers, and other plants begin to take over. When these plants die, they add more nutrients to the soil. This buildup is enough to support the growth of shrubs and trees. All the while, insects, small birds, and mammals have begun to move in. What was once bare rock now supports a variety of life.

Secondary Succession What happens when a fire, such as the one in **Figure 3,** disturbs a forest or when a building is torn down in a city? After a forest fire, not much seems to be left except dead trees and ash-covered soil. After the rubble of a building is removed, all that remains is bare soil. However, these places do not remain lifeless for long. The soil already contains the seeds of weeds, grasses, and trees. More seeds are carried to the area by wind and birds. Other wildlife may move in. Succession that begins in a place that already has soil and was once the home of living organisms is called secondary succession. Because soil already is present, secondary succession occurs faster and has different pioneer species than primary succession does.

Topic: Eutrophication
Visit for blue.msscience.com Web links to information about eutrophication (yoo truh fih KAY shun)—secondary succession in an aquatic ecosystem.

Activity Using the information that you find, illustrate or describe in your Science Journal this process for a small freshwater lake.

 *Which type of succession usually starts without soil?*

SECTION 1 How Ecosystems Change **151**

NATIONAL GEOGRAPHIC VISUALIZING SECONDARY SUCCESSION

Figure 3

In the summer of 1988, wind-driven flames like those shown in the background photo swept through Yellowstone National Park, scorching nearly a million acres. The Yellowstone fire was one of the largest forest fires in United States history. The images on this page show secondary succession—the process of ecological regeneration—triggered by the fire.

▶ After the fire, burned timber and blackened soil seemed to be all that remained. However, the fire didn't destroy the seeds that were protected under the soil.

◀ Within weeks, grasses and other plants were beginning to grow in the burned areas. Ecological succession was underway.

▶ Many burned areas in the park opened new plots for stands of trees. This picture shows young lodgepole pines in August 1999. The forest habitat of America's oldest national park is being restored gradually through secondary succession.

Figure 4 This beech-maple forest is an example of a climax community.

Climax Communities A community of plants that is relatively stable and undisturbed and has reached a stage of succession is called a **climax community.** The beech-maple forest shown in **Figure 4** is an example of a community that has reached the end of succession. New trees grow when larger, older trees die. The individual trees change, but the species remain stable. There are fewer changes of species in a climax community over time, as long as the community isn't disturbed by wildfire, avalanche, or human activities.

Primary succession begins in areas with no previous vegetation. It can take hundreds or even thousands of years to develop into a climax community. Secondary succession is usually a shorter process, but it still can take a century or more.

section 1 review

Summary

Ecological Succession
- Succession is the natural, gradual changes over time of species in a community.
- Primary succession occurs in areas that previously were without soil or plants.
- Secondary succession occurs in areas where soil has been disturbed.
- Climax communities have reached an end stage of succession and are stable.
- Climax communities have less diversity than communities in mid-succession.

Self Check

1. **Compare** primary and secondary succession.
2. **Describe** adaptations of pioneer species.
3. **Infer** the kind of succession that will take place on an abandoned, unpaved country road.
4. **Think Critically** Show the sequence of events in primary succession. Include the term *climax community*.

Applying Math

5. **Solve One-Step Equations** A tombstone etched with 1802 as the date of death has a lichen on it that is 6 cm in diameter. If the lichen began growing in 1802, calculate its average yearly rate of growth.

section 2

Biomes

as you read

What You'll Learn
- **Explain** how climate influences land environments.
- **Identify** seven biomes of Earth.
- **Describe** the adaptations of organisms found in each biome.

Why It's Important
Resources that you need to survive are found in a variety of biomes.

Review Vocabulary
climate: the average weather conditions of an area over many years

New Vocabulary
- biome
- tundra
- taiga
- temperate deciduous forest
- temperate rain forest
- tropical rain forest
- desert
- grassland

Factors That Affect Biomes

Does a desert in Arizona have anything in common with a desert in Africa? Both have heat, little rain, poor soil, water-conserving plants with thorns, and lizards. Even widely separated regions of the world can have similar biomes because they have similar climates. Climate is the average weather pattern in an area over a long period of time. The two most important climatic factors that affect life in an area are temperature and precipitation.

Major Biomes

Large geographic areas that have similar climates and ecosystems are called **biomes** (BI ohmz). Seven common types of land biomes are mapped in **Figure 5.** Areas with similar climates produce similar climax communities. Tropical rain forests are climax communities found near the equator, where temperatures are warm and rainfall is plentiful. Coniferous forests grow where winter temperatures are cold and rainfall is moderate.

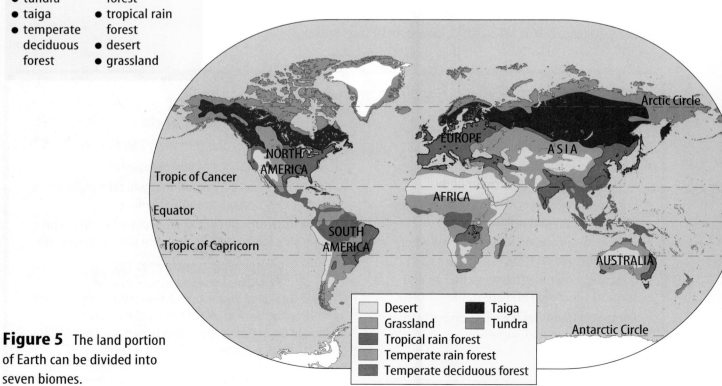

Figure 5 The land portion of Earth can be divided into seven biomes.

154 CHAPTER 6 Ecosystems

Tundra At latitudes just south of the north pole or at high elevations, a biome can be found that receives little precipitation but is covered with ice most of the year. The **tundra** is a cold, dry, treeless region, sometimes called a cold desert. Precipitation averages less than 25 cm per year. Winters in the Arctic can be six to nine months long. For some of these months, the Sun never appears above the horizon and it is dark 24 hours a day. The average daily temperature is about −12°C. For a few days during the short, cold summer, the Sun is always visible. Only the top portion of soil thaws in the summer. Below the thawed surface is a layer of permanently frozen soil called permafrost, shown in **Figure 6.** Alpine tundra, found above the treeline on high mountains, have similar climates. Tundra soil has few nutrients because the cold temperatures slow the process of decomposition.

Tundra Life Tundra plants are adapted to drought and cold. They include mosses, grasses, and small shrubs, as seen in **Figure 7.** Many lichens grow on the tundra. During the summer, mosquitoes, blackflies, and other biting insects fill the air. Migratory birds such as ducks, geese, shorebirds, and songbirds nest on the Arctic tundra during the summer. Other inhabitants include hawks, snowy owls, and willow grouse. Mice, voles, lemmings, arctic hares, caribou, reindeer, and musk oxen also are found there.

People are concerned about overgrazing by animals on the tundra. Fences, roads, and pipelines have disrupted the migratory routes of some animals and forced them to stay in a limited area. Because the growing season is so short, plants and other vegetation can take decades to recover from damage.

Figure 6 This permafrost in Alaska is covered by soil that freezes in the winter and thaws in the summer.
Infer *what types of problems this might cause for people living in this area.*

Figure 7 Lichens, mosses, grasses, and small shrubs thrive on the tundra. Ptarmigan also live on the tundra. In winter, their feathers turn white. Extra feathers on their feet keep them warm and prevent them from sinking into the snow.

Tundra

Ptarmigan

Taiga South of the tundra—between latitudes 50°N and 60°N and stretching across North America, northern Europe, and Asia—is the world's largest biome. The **taiga** (TI guh), shown in **Figure 8,** is a cold, forest region dominated by cone-bearing evergreen trees. Although the winter is long and cold, the taiga is warmer and wetter than the tundra. Precipitation is mostly snow and averages 35 cm to 100 cm each year.

Most soils of the taiga thaw completely during the summer, making it possible for trees to grow. However, permafrost is present in the extreme northern regions of the taiga. The forests of the taiga might be so dense that little sunlight penetrates the trees to reach the forest floor. However, some lichens and mosses do grow on the forest floor. Moose, lynx, shrews, bears, and foxes are some of the animals that live in the taiga.

Figure 8 The taiga is dominated by cone-bearing trees. The lynx, a mammal adapted to life in the taiga, has broad, heavily furred feet that act like snowshoes to prevent it from sinking in the snow.
Infer why "snowshoe feet" are important for a lynx.

Temperate Deciduous Forest Temperate regions usually have four distinct seasons each year. Annual precipitation ranges from about 75 cm to 150 cm and is distributed throughout the year. Temperatures range from below freezing during the winter to 30°C or more during the warmest days of summer.

Figure 9 White-tailed deer are one of many species that you can find in a deciduous forest. In autumn, the leaves on deciduous trees change color and fall to the ground.

Temperate Forest Life Many evergreen trees grow in the temperate regions of the world. However, most of the temperate forests in Europe and North America are dominated by climax communities of deciduous trees, which lose their leaves every autumn. These forests, like the one in **Figure 9,** are called **temperate deciduous forests.** In the United States, most of them are located east of the Mississippi River.

When European settlers first came to America, they cut trees to create farmland and to supply wood. As forests were cut, organisms lost their habitats. When agriculture shifted from the eastern to the midwestern and western states, secondary succession began, and trees eventually returned to some areas. Now, nearly as many trees grow in the New England states as did before the American Revolutionary War. Many trees are located in smaller patches. Yet, the recovery of large forests such as those in the Adirondack Mountains in New York State shows the result of secondary succession.

Temperate Rain Forest New Zealand, southern Chile, and the Pacific Northwest of the United States are some of the places where **temperate rain forests,** shown in **Figure 10,** are found. The average temperature of a temperate rain forest ranges from 9°C to 12°C. Precipitation ranges from 200 cm to 400 cm per year.

Trees with needlelike leaves dominate these forests, including the Douglas fir, western red cedar, and spruce. Many grow to great heights. Animals of the temperate rain forest include the black bear, cougar, bobcat, northern spotted owl, and marbled murrelet. Many species of amphibians also inhabit the temperate rain forest, including salamanders.

The logging industry in the Northwest provides jobs for many people. However, it also removes large parts of the temperate rain forest and destroys the habitat of many organisms. Many logging companies now are required to replant trees to replace the ones they cut down. Also, some rain forest areas are protected as national parks and forests.

Figure 10 In the Olympic rain forest in Washington State, mosses and lichens blanket the ground and hang from the trees. Wet areas are perfect habitats for amphibians like the Pacific giant salamander above.

SECTION 2 Biomes **157**

Figure 11 Tropical rain forests are lush environments that contain such a large variety of species that many have not been discovered.

Mini LAB

Modeling Rain Forest Leaves

Procedure
1. Draw an oval leaf about 10 cm long on a piece of **poster board**. Cut it out.
2. Draw a second leaf the same size but make one end pointed. This is called a drip tip. Cut this leaf out.
3. Hold your hands palm-side up over a **sink** and have someone lay a leaf on each one. Point the drip tip away from you. Tilt your hands down but do not allow the leaves to fall off.
4. Have someone gently spray water on the leaves and observe what happens.

Analysis
1. From which leaf does water drain faster?
2. Infer why it is an advantage for a leaf to get rid of water quickly in a rain forest.

Try at Home

Tropical Rain Forest Warm temperatures, wet weather, and lush plant growth are found in **tropical rain forests.** These forests are warm because they are near the equator. The average temperature, about 25°C, doesn't vary much between night and day. Most tropical rain forests receive at least 200 cm of rain annually. Some receive as much as 600 cm of rain each year.

Tropical rain forests, like the one in **Figure 11,** are home to an astonishing variety of organisms. They are one of the most biologically diverse places in the world. For example, one tree in a South American rain forest might contain more species of ants than exist in all of the British Isles.

Tropical Rain Forest Life Different animals and plants live in different parts of the rain forest. Scientists divide the rain forest into zones based on the types of plants and animals that live there, just as a library separates books about different topics onto separate shelves. The zones include: forest floor, understory, canopy, and emergents, as shown in **Figure 12.** These zones often blend together, but their existence provide different habitats for many diverse organisms to live in the tropical rain forest.

Reading Check *What are the four zones of a tropical rain forest?*

Although tropical rain forests support a huge variety of organisms, the soil of the rain forest contains few nutrients. Over the years, nutrients have been washed out of the soil by rain. On the forest floor, decomposers immediately break down organic matter, making nutrients available to the plants again.

Human Impact Farmers that live in tropical areas clear the land to farm and to sell the valuable wood. After a few years, the crops use up the nutrients in the soil, and the farmers must clear more land. As a result, tropical rain forest habitats are being destroyed. Through education, people are realizing the value and potential value of preserving the species of the rain forest. In some areas, logging is prohibited. In other areas, farmers are taught new methods of farming so they do not have to clear rain forest lands continually.

Figure 12 Tropical rain forests contain abundant and diverse organisms.

Emergents These giant trees are much higher than the average canopy tree. Birds, such as the macaw, and insects are found here.

Canopy The canopy includes the upper parts of the trees. It's full of life—insects, birds, reptiles, and mammals.

Understory This dark, cool environment is under the canopy leaves but above the ground. Many insects, reptiles, and amphibians live in the understory.

Forest Floor The forest floor is home to many insects and the largest mammals in the rain forest generally live here.

Desertification When vegetation is removed from soil in areas that receive little rain, the dry, unprotected surface can be blown away. If the soil remains bare, a desert might form. This process is called desertification. Look on a biome map and hypothesize about which areas of the United States are most likely to become deserts.

Figure 13 Desert plants, like these in the Sonoran Desert, are adapted for survival in the extreme conditions of the desert biome. The giant hairy scorpion found in some deserts has a venomous sting.

Desert The driest biome on Earth is the **desert**. Deserts receive less than 25 cm of rain each year and support little plant life. Some desert areas receive no rain for years. When rain does come, it quickly drains away. Any water that remains on the ground evaporates rapidly.

Most deserts, like the one in **Figure 13,** are covered with a thin, sandy, or gravelly soil that contains little organic matter. Due to the lack of water, desert plants are spaced far apart and much of the ground is bare. Barren, windblown sand dunes are characteristics of the driest deserts.

✓ **Reading Check** *Why is much of a desert bare ground?*

Desert Life Desert plants are adapted for survival in the extreme dryness and hot and cold temperatures of this biome. Most desert plants are able to store water. Cactus plants are probably the most familiar desert plants of the western hemisphere. Desert animals also have adaptations that help them survive the extreme conditions. Some, like the kangaroo rat, never need to drink water. They get all the moisture they need from the breakdown of food during digestion. Most animals are active only during the night, late afternoon, or early morning when temperatures are less extreme. Few large animals are found in the desert.

In order to provide water for desert cities, rivers and streams have been diverted. When this happens, wildlife tends to move closer to cities in their search for food and water. Education about desert environments has led to an awareness of the impact of human activities. As a result, large areas of desert have been set aside as national parks and wilderness areas to protect desert habitats.

Grasslands Temperate and tropical regions that receive between 25 cm and 75 cm of precipitation each year and are dominated by climax communities of grasses are called **grasslands.** Most grasslands have a dry season, when little or no rain falls. This lack of moisture prevents the development of forests. Grasslands are found in many places around the world, and they have a variety of names. The prairies and plains of North America, the steppes of Asia, the savannas of Africa shown in **Figure 14,** and the pampas of South America are types of grasslands.

Grasslands Life The most noticeable animals in grassland biomes are usually mammals that graze on the stems, leaves, and seeds of grass plants. Kangaroos graze in the grasslands of Australia. In Africa, communities of animals such as wildebeests, impalas, and zebras thrive in the savannas.

Grasslands are perfect for growing many crops such as wheat, rye, oats, barley, and corn. Grasslands also are used to raise cattle and sheep. However, overgrazing can result in the death of grasses and the loss of valuable topsoil from erosion. Most farmers and ranchers take precautions to prevent the loss of valuable habitats and soil.

Figure 14 Animals such as zebras and wildebeests are adapted to life on the savannas in Africa.

section 2 review

Summary

Major Biomes

- Tundra, sometimes called a cold desert, can be divided into two types: arctic and alpine.
- Taiga is the world's largest biome. It is a cold forest region with long winters.
- Temperate regions have either a deciduous forest biome or a rain forest biome.
- Tropical rain forests are one of the most biologically diverse biomes.
- Humans have a huge impact on tropical rain forests.
- The driest biome is the desert. Desert organisms are adapted for extreme dryness and temperatures.
- Grasslands provide food for wildlife, livestock, and humans.

Self Check

1. **Determine** which two biomes are the driest.
2. **Compare and contrast** tundra organisms and desert organisms.
3. **Identify** the biggest climatic difference between a temperate rain forest and a tropical rain forest.
4. **Explain** why the soil of tropical rain forests make poor farmland.
5. **Think Critically** If you climb a mountain in Arizona, you might reach an area where the trees resemble the taiga trees in northern Canada. Why would a taiga forest exist in Arizona?

Applying Skills

6. **Record Observations** Animals have adaptations that help them survive in their environments. Make a list of animals that live in your area, and record the physical or behavioral adaptations that help them survive.

Studying a Land Ecosystem

An ecological study includes observation and analysis of organisms and the physical features of the environment.

Real-World Question
How do you study an ecosystem?

Goals
- **Observe** biotic factors and abiotic factors of an ecosystem.
- **Analyze** the relationships among organisms and their environments.

Materials
graph paper
binoculars
thermometer
pencil
magnifying lens
field guides
notebook
compass
tape measure

Safety Precautions

Procedure
1. Choose a portion of an ecosystem to study. You might choose a decaying log, a pond, a garden, or even a crack in the sidewalk.
2. Determine the boundaries of your study area.
3. Using a tape measure and graph paper, make a map of your area. Determine north.
4. **Record** your observations in a table similar to the one shown on this page.
5. **Observe** the organisms in your study area. Use field guides to identify them. Use a magnifying lens to study small organisms and binoculars to study animals you can't get near. Look for evidence (such as tracks or feathers) of organisms you do not see.
6. Measure and record the air temperature in your study area.
7. Visit your study area many times and at different times of day for one week. At each visit, make the same measurements and record all observations. Note how the living and nonliving parts of the ecosystem interact.

Environmental Observations		
Date		
Time of day		
Temperature		
Organisms observed	Do not write in this book.	
Comments		

Conclude and Apply
1. **Predict** what might happen if one or more abiotic factors were changed suddenly.
2. **Infer** what might happen if one or more populations of plants or animals were removed from the area.
3. **Form a hypothesis** to explain how a new population of organisms might affect your ecosystem.

Communicating Your Data
Make a classroom display of all data recorded. **For more help, refer to the** Science Skill Handbook.

Section 3
Aquatic Ecosystems

Freshwater Ecosystems

In a land environment, temperature and precipitation are the most important factors that determine which species can survive. In aquatic environments, water temperature, the amount of sunlight present, and the amounts of dissolved oxygen and salt in the water are important. Earth's freshwater ecosystems include flowing water such as rivers and streams and standing water such as lakes, ponds, and wetlands.

Rivers and Streams Flowing freshwater environments vary from small, gurgling brooks to large, slow-moving rivers. Currents can quickly wash loose particles downstream, leaving a rocky or gravelly bottom. As the water tumbles and splashes, as shown in **Figure 15,** air from the atmosphere mixes in. Naturally fast-flowing streams usually have clearer water and higher oxygen content than slow-flowing streams.

Most nutrients that support life in flowing-water ecosystems are washed into the water from land. In areas where the water movement slows, such as in the pools of streams or in large rivers, debris settles to the bottom. These environments tend to have higher nutrient levels and more plant growth. They contain organisms that are not as well adapted to swiftly flowing water, such as freshwater mussels, minnows, and leeches.

as you read

What You'll Learn
- **Compare** flowing freshwater and standing freshwater ecosystems.
- **Identify** and describe important saltwater ecosystems.
- **Identify** problems that affect aquatic ecosystems.

Why It's Important
All of the life processes in your body depend on water.

Review Vocabulary
aquatic: growing or living in water

New Vocabulary
- wetland
- intertidal zone
- coral reef
- estuary

Figure 15 Streams like this one are high in oxygen because of the swift, tumbling water.
Determine where most nutrients in streams come from.

Mini LAB

Modeling Freshwater Environments

Procedure

1. Obtain a sample of **pond sediment or debris, plants, water, and organisms** from your teacher.
2. Cover the bottom of a **clear-plastic container** with about 2 cm of the debris.
3. Add one or two plants to the container.
4. Carefully pour pond water into the container until it is about two-thirds full.
5. Use a **net** to add several organisms to the water. Seal the container.
6. Using a **magnifying lens**, observe as many organisms as possible. Record your observations. Return your sample to its original habitat.

Analysis
Write a short paragraph describing the organisms in your sample. How did the organisms interact with each other?

Human Impact People use rivers and streams for many activities. Once regarded as a free place to dump sewage and other pollutants, many people now recognize the damage this causes. Treating sewage and restricting pollutants have led to an improvement in the water quality in some rivers.

Lakes and Ponds When a low place in the land fills with rainwater, snowmelt, or water from an overflowing stream, a lake or pond might form. Pond or lake water hardly moves. It contains more plants than flowing-water environments contain.

Lakes, such as the one shown in **Figure 16,** are larger and deeper than ponds. They have more open water because most plant growth is limited to shallow areas along the shoreline. In fact, organisms found in the warm, sunlit waters of the shorelines often are similar to those found in ponds. If you were to dive to the bottom, you would discover few, if any, plants or algae growing. Colder temperatures and lower light levels limit the types of organisms that can live in deep lake waters. Floating in the warm, sunlit waters near the surface of freshwater lakes and ponds are microscopic algae, plants, and other organisms known as plankton.

A pond is a small, shallow body of water. Because ponds are shallow, they are filled with animal and plant life. Sunlight usually penetrates to the bottom. The warm, sunlit water promotes the growth of plants and algae. In fact, many ponds are filled almost completely with plant material, so the only clear, open water is at the center. Because of the lush growth in pond environments, they tend to be high in nutrients.

Figure 16 Ponds contain more vegetation than lakes contain. The population of organisms in the shallow water of lakes is high. Fewer types of organisms live in the deeper water.

164 CHAPTER 6 Ecosystems

Water Pollution Human activities can harm freshwater environments. Fertilizer-filled runoff from farms and lawns, as well as sewage dumped into the water, can lead to excessive growth of algae and plants in lakes and ponds. The growth and decay of these organisms reduces the oxygen level in the water, which makes it difficult for some organisms to survive. To prevent problems, sewage is treated before it is released. People also are being educated about problems associated with polluting lakes and ponds. Fines and penalties are issued to people caught polluting waterways. These controls have led to the recovery of many freshwater ecosystems.

Wetlands As the name suggests, **wetlands,** shown in **Figure 17,** are regions that are wet for all or most of a year. They are found in regions that lie between landmasses and water. Other names for wetlands include swamps, bogs, and fens. Some people refer to wetlands as biological supermarkets. They are fertile ecosystems, but only plants that are adapted to water-logged soil survive there. Wetland animals include beavers, muskrats, alligators, and the endangered bog turtle. Many migratory bird populations use wetlands as breeding grounds.

Reading Check *Where are wetlands found?*

Wetlands once were considered to be useless, disease-ridden places. Many were drained and destroyed to make roads, farmland, shopping centers, and housing developments. Only recently have people begun to understand the importance of wetlands. Products that come from wetlands, including fish, shellfish, cranberries, and plants, are valuable resources. Now many developers are restoring wetlands, and in most states access to land through wetlands is prohibited.

Figure 17 Life in the Florida Everglades was threatened due to pollution, drought, and draining of the water. Conservation efforts are being made in an attempt to preserve this ecosystem.

Environmental Author Rachel Carson (1907–1964) was a scientist that turned her knowledge and love of the environment into articles and books. After 15 years as an editor for the U.S. Fish and Wildlife Service, she resigned and devoted her time to writing. She probably is known best for her book *Silent Spring,* in which she warned about the long-term effects of the misuse of pesticides. In your Science Journal, compile a list of other authors who write about environmental issues.

Saltwater Ecosystems

About 95 percent of the water on the surface of Earth contains high concentrations of various salts. The amount of dissolved salts in water is called salinity. The average ocean salinity is about 35 g of salts per 1,000 g of water. Saltwater ecosystems include oceans, seas, a few inland lakes such as the Great Salt Lake in Utah, coastal inlets, and estuaries.

Applying Math — Convert Units

TEMPERATURE Organisms that live around hydrothermal vents in the ocean deal with temperatures that range from 1.7°C to 371°C. You have probably seen temperatures measured in degrees Celsius (°C) and degrees Fahrenheit (°F). Which one are you familiar with? If you know the temperature in one system, you can convert it to the other.

You have a Fahrenheit thermometer and measure the water temperature of a pond at 59°F. What is that temperature in degrees Celsius?

Solution

1. *This is what you know:* water temperature in degrees Fahrenheit = 59°F

2. *This is what you need to find out:* The water temperature in degrees Celsius.

3. *This is the procedure you need to use:*
   - Solve the equation for degrees Celsius:
     (°C × 1.8) + 32 = °F
     °C = (°F − 32)/1.8
   - Substitute the known value:
     °C = (59°F − 32)/1.8 = 15°C
   - Water temperature that is 59°F is 15°C.

4. *Check your answer:* Substitute the Celsius temperature back into the original equation. You should get 59.

Practice Problems

1. The thermometer outside your classroom reads 78°F. What is the temperature in degrees Celsius?

2. If lake water was 12°C in October and 23°C in May, what is the difference in degrees Fahrenheit?

 For more practice, visit blue.msscience.com/math_practice

166 CHAPTER 6 Ecosystems

Open Oceans Life abounds in the open ocean. Scientists divide the ocean into different life zones, based on the depth to which sunlight penetrates the water. The lighted zone of the ocean is the upper 200 m or so. It is the home of the plankton that make up the foundation of the food chain in the open ocean. Below about 200 m is the dark zone of the ocean. Animals living in this region feed on material that floats down from the lighted zone, or they feed on each other. A few organisms are able to produce their own food.

Coral Reefs One of the most diverse ecosystems in the world is the coral reef. **Coral reefs** are formed over long periods of time from the calcium carbonate skeletons secreted by animals called corals. When corals die, their skeletons remain. Over time, the skeletal deposits form reefs such as the Great Barrier Reef off the coast of Australia, shown in **Figure 18.**

Reefs do not adapt well to long-term stress. Runoff from fields, sewage, and increased sedimentation from cleared land harm reef ecosystems. Organizations like the Environmental Protection Agency have developed management plans to protect the diversity of coral reefs. These plans treat a coral reef as a system that includes all the areas that surround the reef. Keeping the areas around reefs healthy will result in a healthy environment for the coral reef ecosystem.

Science Online

Topic: Coral Reefs
Visit blue.msscience.com for Web links to information about coral reef ecosystems.

Activity Construct a diorama of a coral reef. Include as many different kinds of organisms as you can for a coral reef ecosystem.

Figure 18 The lighter areas around this island are part of the Great Barrier Reef. It comprises about 3,000 reefs and about 900 islands. Reefs contain colorful fish and a large variety of other organisms.

Figure 19 As the tide recedes, small pools of seawater are left behind. These pools contain a variety of organisms such as sea stars and periwinkles.

Sea star

Periwinkles

Seashores All of Earth's landmasses are bordered by ocean water. The shallow waters along the world's coastlines contain a variety of saltwater ecosystems, all of which are influenced by the tides and by the action of waves. The gravitational pull of the Moon, and to a lesser extent, the Sun, on Earth causes the tides to rise and fall each day. The height of the tides varies according to the phases of the Moon, the season, and the slope of the shoreline. The **intertidal zone** is the portion of the shoreline that is covered with water at high tide and exposed to the air during low tide. Organisms that live in the intertidal zone, such as those in **Figure 19,** must be adapted to dramatic changes in temperature, moisture, and salinity and must be able to withstand the force of wave action.

Estuaries Almost every river on Earth eventually flows into an ocean. The area where a river meets an ocean and contains a mixture of freshwater and salt water is called an **estuary** (ES chuh wer ee). Other names for estuaries include bays, lagoons, harbors, inlets, and sounds. They are located near coastlines and border the land. Salinity in estuaries changes with the amount of freshwater brought in by rivers and streams, and with the amount of salt water pushed inland by the ocean tides.

Estuaries, shown in **Figure 20,** are extremely fertile, productive environments because freshwater streams bring in tons of nutrients washed from inland soils. Therefore, nutrient levels in estuaries are higher than in freshwater ecosystems or other saltwater ecosystems.

Figure 20 The Chesapeake Bay is an estuary rich in resources. Fish and shrimp are harvested by commercial fishing boats.
Describe what other resources can be found in estuaries.

Estuary Life Organisms found in estuaries include many species of algae, salt-tolerant grasses, shrimp, crabs, clams, oysters, snails, worms, and fish. Estuaries also serve as important nurseries for many species of ocean fish. Estuaries provide much of the seafood consumed by humans.

Reading Check *Why are estuaries more fertile than other aquatic ecosystems?*

section 3 review

Summary

Freshwater Ecosystems
- Temperature, light, salt, and dissolved oxygen are important factors.
- Rivers, streams, lakes, ponds, and wetlands are freshwater ecosystems.
- Human activities, such as too much lawn fertilizer, can pollute aquatic ecosystems.

Saltwater Ecosystems
- About 95 percent of Earth's water contains dissolved salts.
- Saltwater ecosystems include open oceans, coral reefs, seashores, and estuaries.
- Organisms that live on seashores have adaptations that enable them to survive dramatic changes in temperature, moisture, and salinity.
- Estuaries serve as nursery areas for many species of ocean fish.

Self Check

1. **Identify** the similarities and differences between a lake and a stream.
2. **Compare and contrast** the dark zone of the ocean with the forest floor of a tropical rain forest. What living or nonliving factors affect these areas?
3. **Explain** why fewer plants are at the bottom of deep lakes.
4. **Infer** what adaptations are necessary for organisms that live in the intertidal zone.
5. **Think Critically** Would you expect a fast moving mountain stream or the Mississippi River to have more dissolved oxygen? Explain.

Applying Skills

6. **Communicate** Wetlands trap and slowly release rain, snow, and groundwater. Describe in your Science Journal what might happen to a town located on a floodplain if nearby wetlands are destroyed.

Science Online blue.msscience.com/self_check_quiz

LAB Use the Internet

Exploring Wetlands

Real-World Question

Wetlands, such as the one shown below, are an important part of the environment. These fertile ecosystems support unique plants and animals that can survive only in wetland conditions. The more you understand the importance of wetlands, the more you can do to preserve and protect them. Why are wetlands an important part of the ecosystem?

Goals
- **Identify** wetland regions in the United States.
- **Describe** the significance of the wetland ecosystem.
- **Identify** plant and animal species native to a wetland region.
- **Identify** strategies for supporting the preservation of wetlands.

Data Source

Science Online
Visit blue.msscience.com/ for more information about wetland environments and for data collected by other students.

170 CHAPTER 6 Ecosystems

Using Scientific Methods

▶ Make a Plan

1. **Determine** where some major wetlands are located in the United States.
2. **Identify** one wetland area to study in depth. Where is it located? Is it classified as a marsh, bog, or something else?
3. **Explain** the role this ecosystem plays in the overall ecology of the area.
4. **Research information** about the plants and animals that live in the wetland environment you are researching.
5. **Investigate** what laws protect the wetland you are studying.

▶ Follow Your Plan

1. Make sure your teacher approves your plan before you start.
2. Perform the investigation.
3. Post your data at the link shown below.

▶ Analyze Your Data

1. **Describe** the wetland area you have researched. What region of the United States is it located in? What other ecological factors are found in that region?
2. **Outline** the laws protecting the wetland you are investigating. How long have the laws been in place?
3. **List** the plants and animals native to the wetland area you are researching. Are those plants and animals found in other parts of the region or the United States? What adaptations do the plants and animals have that help them survive in a wetland environment?

▶ Conclude and Apply

1. **Infer** Are all wetlands the same?
2. **Determine** what the ecological significance of the wetland area that you studied for that region of the country is.
3. **Draw Conclusions** Why should wetland environments be protected?
4. **Summarize** what people can do to support the continued preservation of wetland environments in the United States.

Communicating Your Data

Find this lab using the link below. **Post** your data in the table provided. **Review** other students' data to learn about other wetland environments in the United States.

Science Online
blue.msscience.com/internet_lab

LAB **171**

TIME SCIENCE AND Society

SCIENCE ISSUES THAT AFFECT YOU!

Creating Wetlands to Purify Wastewater

Pebbles were added to the Corrales wetlands to help with drainage.

Water irises thrived in the wetlands, less than a year after planting.

Students enjoy pure water from the Corrales wetlands after it is filtered.

When you wash your hands or flush the toilet, do you think about where the wastewater goes? In most places, it eventually ends up being processed in a traditional sewage-treatment facility. But some places are experimenting with a new method that processes wastewater by creating wetlands. Wetlands are home to filtering plants, such as cattails, and sewage-eating bacteria.

In 1996, school officials at the Corrales Elementary School in Albuquerque, New Mexico, faced a big problem. The old wastewater-treatment system had failed. Replacing it was going to cost a lot of money. Instead of constructing a new sewage-treatment plant, school officials decided to create a natural wetlands system. The wetlands system could do the job less expensively, while protecting the environment.

Today, this wetlands efficiently converts polluted water into cleaner water that's good for the environment. U.S. government officials are monitoring this alternative sewage-treatment system to see if it is successful. So far, so good!

Wetlands filter water through the actions of the plants and microorganisms that live there. When plants absorb water into their roots, some also take up pollutants. The plants convert the pollutants to forms that are not dangerous. At the same time, bacteria and other microorganisms are filtering water as they feed. Water moves slowly through wetlands, so the organisms have plenty of time to do their work. Wetlands built by people to filter small amounts of pollutants are called "constructed wetlands". In many places, constructed wetlands are better at cleaning wastewater than sewers or septic systems.

Visit and Observe Visit a wetlands and create a field journal of your observations. Draw the plants and animals you see. Use a field guide to help identify the wildlife. If you don't live near a wetlands, use resources to research wetlands environments.

For more information, visit
blue.msscience.com/time

Chapter 6 Study Guide

Reviewing Main Ideas

Section 1 — How Ecosystems Change

1. Ecological succession is the gradual change from one plant community to another.
2. Primary succession begins in a place where no plants were before.
3. Secondary succession begins in a place that has soil and was once the home of living organisms.
4. A climax community has reached a stable stage of ecological succession.

Section 2 — Biomes

1. Temperature and precipitation help determine the climate of a region.
2. Large geographic areas with similar climax communities are called biomes.
3. Earth's land biomes include tundra, taiga, temperate deciduous forest, temperate rain forest, tropical rain forest, grassland, and desert.

Section 3 — Aquatic Ecosystems

1. Freshwater ecosystems include streams, rivers, lakes, ponds, and wetlands.
2. Wetlands are areas that are covered with water most of the year. They are found in regions that lie between land-masses and water.
3. Saltwater ecosystems include estuaries, seashores, coral reefs, a few inland lakes, and the deep ocean.
4. Estuaries are fertile transitional zones between freshwater and saltwater environments.

Visualizing Main Ideas

Copy and complete this concept map about land biomes.

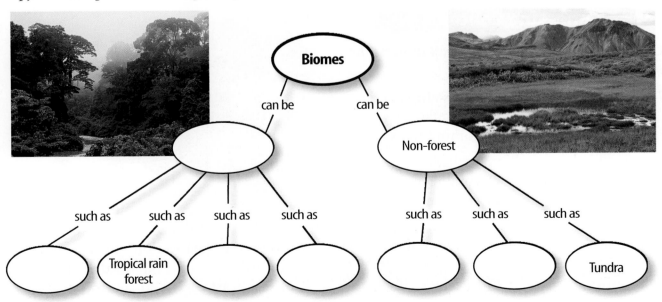

CHAPTER STUDY GUIDE 173

Chapter 6 Review

Using Vocabulary

biome p. 154
climax community p. 153
coral reef p. 167
desert p. 160
estuary p. 168
grassland p. 161
intertidal zone p. 168
pioneer species p. 150
succession p. 150
taiga p. 156
temperate deciduous forest p. 157
temperate rain forest p. 157
tropical rain forest p. 158
tundra p. 155
wetland p. 165

Fill in the blanks with the correct vocabulary word or words.

1. _____ refers to the normal changes in the types of species that live in communities.

2. A(n) _____ is a group of organisms found in a stable stage of succession.

3. Deciduous trees are dominant in the _____.

4. The average temperature in _____ is between 9°C and 12°C.

5. _____ are the most biologically diverse biomes in the world.

6. A(n) _____ is an area where freshwater meets the ocean.

Checking Concepts

Choose the word or phrase that best answers the question.

7. What are tundra and desert examples of?
   A) ecosystems
   B) biomes
   C) habitats
   D) communities

8. What is a hot, dry biome called?
   A) desert
   B) tundra
   C) coral reef
   D) grassland

9. Where would organisms that are adapted to live in slightly salty water be found?
   A) lake
   B) estuary
   C) open ocean
   D) intertidal zone

10. Which biome contains mostly frozen soil called permafrost?
    A) taiga
    B) temperate rain forest
    C) tundra
    D) temperate deciduous forest

11. A new island is formed from a volcanic eruption. Which species probably would be the first to grow and survive?
    A) palm trees
    B) lichens
    C) grasses
    D) ferns

12. What would the changes in communities that take place on a recently formed volcanic island best be described as?
    A) primary succession
    B) secondary succession
    C) tertiary succession
    D) magma

13. What is the stable end stage of succession?
    A) pioneer species
    B) climax community
    C) limiting factor
    D) permafrost

Use the illustration below to answer question 14.

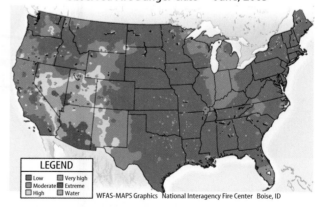

Observed Fire Danger Class—June, 2003

14. Which area of the U.S. had the highest observed fire danger on June 20, 2003?
    A) northeast
    B) southeast
    C) northwest
    D) southwest

chapter 6 Review

Thinking Critically

15. Explain In most cases, would a soil sample from a temperate deciduous forest be more or less nutrient-rich than a soil sample from a tropical rain forest?

16. Explain why some plant seeds need fire in order to germinate. How does this give these plants an advantage in secondary succession?

17. Determine A grassy meadow borders a beech-maple forest. Is one of these ecosystems undergoing succession? Why?

18. Infer why tundra plants are usually small.

19. Make and Use a Table Copy and complete the following table about aquatic ecosystems. Include these terms: *intertidal zone, lake, pond, coral reef, open ocean, river, estuary,* and *stream.*

Aquatic Ecosystems	
Saltwater	Freshwater
Do not write in this book.	

20. Recognize Cause and Effect Wildfires like the one in Yellowstone National Park in 1988, cause many changes to the land. Determine the effect of a fire on an area that has reached its climax community.

Performance Activities

21. Oral Presentation Research a biome not in this chapter. Find out about its climate and location, and which organisms live there. Present this information to your class.

Applying Math

Use the table below to answer question 22.

Rainfall Amounts	
Biome	Average Precipitation/Year (cm)
Taiga	50
Temperate rain forest	200
Tropical rain forest	400
Desert	25
Temperate deciduous forest	150
Tundra	25

22. Biome Precipitation How many times more precipitation does the tropical rain forest biome receive than the taiga or desert?

Use the graph below to answer question 23.

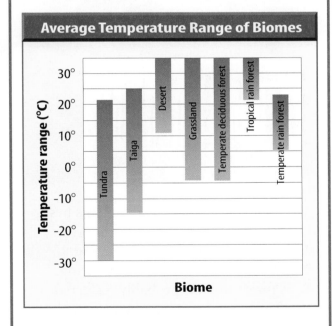

23. Biome Temperatures According to the graph, which biome has the greatest and which biome has the least variation in temperature throughout the year? Estimate the difference between the two.

Chapter 6 Standardized Test Practice

Part 1 Multiple Choice

Record your answers on the answer sheet provided by your teacher or on a sheet of paper.

1. What two factors are most responsible for limiting life in a particular area?
   A. sunlight and temperature
   B. precipitation and temperature
   C. precipitation and sunlight
   D. soil conditions and precipitation

2. Which of the following forms during primary succession?
   A. trees   C. wildlife
   B. soil   D. grasses

Use the illustrations below to answer questions 3 and 4.

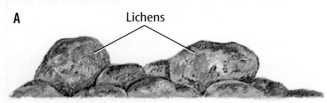

3. Which of the following statements best describes what is represented by A?
   A. Primary succession is occurring.
   B. Secondary succession is occurring.
   C. A forest fire has probably occurred.
   D. The climax stage has been reached.

4. Which of the following statements best describes what is represented by B?
   A. The climax stage has been reached.
   B. Pioneer species are forming soil.
   C. Bare rock covers most of the area.
   D. Secondary succession is occurring.

Use the map below to answer questions 5 and 6.

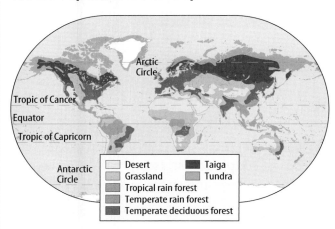

5. What biome is located in the latitudes just south of the north pole?
   A. taiga
   B. temperate deciduous rain forest
   C. tundra
   D. temperate rain forest

6. The tropical rainforest biome is found primarily near the
   A. Arctic Circle.
   B. Tropic of Cancer.
   C. equator.
   D. Tropic of Capricorn.

7. Which of the following is composed of a mix of salt water and freshwater?
   A. an intertidal zone
   B. an estuary
   C. a seashore
   D. a coral reef

Test-Taking Tip

Come Prepared Bring at least two sharpened No. 2 pencils and a good eraser to the test. Check to make sure that your eraser completely removes all pencil marks.

176 STANDARDIZED TEST PRACTICE

Part 2 Short Response/Grid In

Record your answers on the answer sheet provided by your teacher or on a sheet of paper.

8. Name two products that come from wetlands.

9. Which takes longer, primary succession or secondary succession? Why?

Use the photos below to answer questions 10 and 11.

10. In what biome would you most likely find A? How is this animal adapted to survive in its biome?

11. In what biome would you most likely find B? How is this animal adapted to survive in its biome?

12. Which biome receives the most rainfall per year? Which receives the least rainfall?

13. Why are forests unlikely to develop in grasslands?

14. What are two kinds of wetlands? What kinds of animals are found in wetlands?

15. What organisms inhabit the upper zone of the open ocean and why are they so important?

Part 3 Open Ended

Record your answers on a sheet of paper.

16. Explain how lichens contribute to the process of soil formation.

17. Compare and contrast a freshwater lake ecosystem with a freshwater pond ecosystem.

18. What special adaptations must all of the organisms that live in the intertidal zone have?

19. What are the differences between the temperate rain forest biome and the tropical rain forest biome?

Use the illustration below to answer questions 20 and 21.

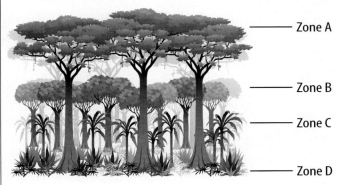

20. Identify and describe zone C in the diagram. What kinds of wildlife are found there?

21. Identify zone D and zone A. Describe the environment in each zone. Why might an organism that lives in zone A not be able to survive in zone D?

22. Discuss the effects of human impact on freshwater environments like lakes and ponds.

unit 3
Earth's Changes over Time

How Are Volcanoes & Fish Connected?

It's hard to know exactly what happened four and a half billion years ago, when Earth was very young. But it's likely that Earth was much more volcanically active than it is today. Along with lava and ash, volcanoes emit gases—including water vapor. Some scientists think that ancient volcanoes spewed tremendous amounts of water vapor into the early atmosphere. When the water vapor cooled, it would have condensed to form liquid water. Then the water would have fallen to the surface and collected in low areas, creating the oceans. Scientists hypothesize that roughly three and a half billion to four billion years ago, the first living things developed in the oceans. According to this hypothesis, these early life-forms gradually gave rise to more and more complex organisms—including the multitudes of fish that swim through the world's waters.

NATIONAL GEOGRAPHIC

unit projects

Visit blue.msscience.com/unit_project to find project ideas and resources. Projects include:

- **History** Create a time line of volcano trivia with facts such as location, greatest magnitude, most destructive, and first volcano recorded. Can volcanoes be predicted?
- **Careers** Study the specialized skills of various careers as you design and prepare a city for a natural disaster.
- **Model** Research, design, construct, test, evaluate, and present your home seismograph in a 5-minute infomercial.

WebQuest *Volcanoes and the Ring of Fire* is an online study of plate tectonics. Design a chart of recent volcano activity, and use it to produce a map of the Ring of Fire with the names and ages of each volcano.

chapter 7

Plate Tectonics

The BIG Idea

The combination of ideas from continental drift, seafloor spreading, and many other discoveries led to the theory of plate tectonics.

SECTION 1
Continental Drift
Main Idea The continental drift hypothesis states that continents have moved slowly to their current locations.

SECTION 2
Seafloor Spreading
Main Idea New discoveries led to the theory of seafloor spreading as an explanation for continental drift.

SECTION 3
Theory of Plate Tectonics
Main Idea The theory of plate tectonics explains the formation of many of Earth's features and geologic events.

Will this continent split?

Ol Doinyo Lengai is an active volcano in the East African Rift Valley, a place where Earth's crust is being pulled apart. If the pulling continues over millions of years, Africa will separate into two landmasses. In this chapter, you'll learn about rift valleys and other clues that the continents move over time.

Science Journal Pretend you're a journalist with an audience that assumes the continents have never moved. Write about the kinds of evidence you'll need to convince people otherwise.

Start-Up Activities

Reassemble an Image

Can you imagine a giant landmass that broke into many separate continents and Earth scientists working to reconstruct Earth's past? Do this lab to learn about clues that can be used to reassemble a supercontinent.

1. Collect interesting photographs from an old magazine.
2. You and a partner each select one photo, but don't show them to each other. Then each of you cut your photos into pieces no smaller than about 5 cm or 6 cm.
3. Trade your cut-up photo for your partner's.
4. Observe the pieces, and reassemble the photograph your partner has cut up.
5. **Think Critically** Write a paragraph describing the characteristics of the cut-up photograph that helped you put the image back together. Think of other examples in which characteristics of objects are used to match them up with other objects.

Preview this chapter's content and activities at
blue.msscience.com

Plate Tectonics Make the following Foldable to help identify what you already know, what you want to know, and what you learned about plate tectonics.

STEP 1 Fold a vertical sheet of paper from side to side. Make the front edge about 1.25 cm shorter than the back edge.

STEP 2 Turn lengthwise and fold into thirds.

STEP 3 Unfold and cut only the layer along both folds to make three tabs.

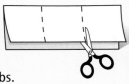

STEP 4 Label each tab.

Identify Questions Before you read the chapter, write what you already know about plate tectonics under the left tab of your Foldable, and write questions about what you'd like to know under the center tab. After you read the chapter, list what you learned under the right tab.

Get Ready to Read

Summarize

① Learn It! Summarizing helps you organize information, focus on main ideas, and reduce the amount of information to remember. To summarize, restate the important facts in a short sentence or paragraph. Be brief and do not include too many details.

② Practice It! Read the text on pages 182–184 labeled *Evidence for Continental Drift*. Then read the summary below and look at the important facts from that passage.

Important Facts

- The edges of some continents look as though they could fit together like a puzzle.
- Fossils of the freshwater reptile *Mesosaurus* have been found in South America and Africa. The fossil plant *Glossopteris* has been found in Africa, Australia, India, South America, and Antarctica.
- Fossils of warm-weather plants were found on an island in the Arctic Ocean. Evidence of glaciers was found in temperate and tropical areas.
- Rocks and rock structures found in parts of the Appalachian Mountains of the eastern United States are similar to those found in Greenland and western Europe.

Summary

The puzzle-like fit of the continents, along with fossil, climate, and rock clues, were the main types of evidence supporting Wegener's hypothesis of continental drift.

③ Apply It! Practice summarizing as you read this chapter. Stop after each section and write a brief summary.

Target Your Reading

Reading Tip

Reread your summary to make sure you didn't change the author's original meaning or ideas.

Use this to focus on the main ideas as you read the chapter.

1 Before you read the chapter, respond to the statements below on your worksheet or on a numbered sheet of paper.
- Write an **A** if you **agree** with the statement.
- Write a **D** if you **disagree** with the statement.

2 After you read the chapter, look back to this page to see if you've changed your mind about any of the statements.
- If any of your answers changed, explain why.
- Change any false statements into true statements.
- Use your revised statements as a study guide.

Science Online
Print out a worksheet of this page at blue.msscience.com

Before You Read A or D		Statement	After You Read A or D
	1	Fossils of tropical plants are never found in Antarctica.	
	2	Because of all the evidence that Alfred Wegener collected, scientists initially accepted his hypothesis of continental drift.	
	3	Wegener's continental drift hypothesis explains how, when, and why the continents drifted apart.	
	4	Earthquakes and volcanic eruptions often occur underwater along mid-ocean ridges.	
	5	Seafloor spreading provided part of the explanation of how continents could move.	
	6	Earth's broken crust rides on several large plates that move on a plastic-like layer of Earth's mantle.	
	7	The San Andreas Fault is part of a plate boundary.	
	8	When two continental plates move toward each other, one continent sinks beneath the other.	
	9	Scientists have proposed several explanations of how heat moves in Earth's interior.	

182 B

section 1

Continental Drift

as you read

What You'll Learn
- **Describe** the hypothesis of continental drift.
- **Identify** evidence supporting continental drift.

Why It's Important
The hypothesis of continental drift led to plate tectonics—a theory that explains many processes in Earth.

Review Vocabulary
continent: one of the six or seven great divisions of land on the globe

New Vocabulary
- continental drift
- Pangaea

Evidence for Continental Drift

If you look at a map of Earth's surface, you can see that the edges of some continents look as though they could fit together like a puzzle. Other people also have noticed this fact. For example, Dutch mapmaker Abraham Ortelius noted the fit between the coastlines of South America and Africa more than 400 years ago.

Pangaea German meteorologist Alfred Wegener (VEG nur) thought that the fit of the continents wasn't just a coincidence. He suggested that all the continents were joined together at some time in the past. In a 1912 lecture, he proposed the hypothesis of continental drift. According to the hypothesis of **continental drift,** continents have moved slowly to their current locations. Wegener suggested that all continents once were connected as one large landmass, shown in **Figure 1,** that broke apart about 200 million years ago. He called this large landmass **Pangaea** (pan JEE uh), which means "all land."

Reading Check *Who proposed continental drift?*

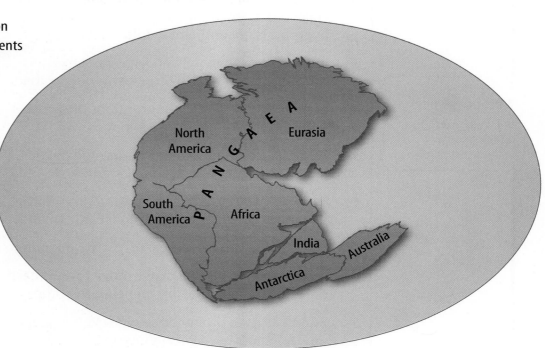

Figure 1 This illustration represents how the continents once were joined to form Pangaea. This fitting together of continents according to shape is not the only evidence supporting the past existence of Pangaea.

182 CHAPTER 7 Plate Tectonics

A Controversial Idea Wegener's ideas about continental drift were controversial. It wasn't until long after Wegener's death in 1930 that his basic hypothesis was accepted. The evidence Wegener presented hadn't been enough to convince many people during his lifetime. He was unable to explain exactly how the continents drifted apart. He proposed that the continents plowed through the ocean floor, driven by the spin of Earth. Physicists and geologists of the time strongly disagreed with Wegener's explanation. They pointed out that continental drift would not be necessary to explain many of Wegener's observations. Other important observations that came later eventually supported Wegener's earlier evidence.

Fossil Clues Besides the puzzlelike fit of the continents, fossils provided support for continental drift. Fossils of the reptile *Mesosaurus* have been found in South America and Africa, as shown in **Figure 2.** This swimming reptile lived in freshwater and on land. How could fossils of *Mesosaurus* be found on land areas separated by a large ocean of salt water? It probably couldn't swim between the continents. Wegener hypothesized that this reptile lived on both continents when they were joined.

Topic: Continental Drift
Visit blue.msscience.com for Web links to information about the continental drift hypothesis.

Activity Research and write a brief report about the initial reactions, from the public and scientific communities, toward Wegener's continental drift hypothesis.

 *How do* Mesosaurus *fossils support the past existence of Pangaea?*

Figure 2 Fossil remains of plants and animals that lived in Pangaea have been found on more than one continent.
Evaluate *How do the locations of* Glossopteris, Mesosaurus, Kannemeyerid, Labyrinthodont, *and other fossils support Wegener's hypothesis of continental drift?*

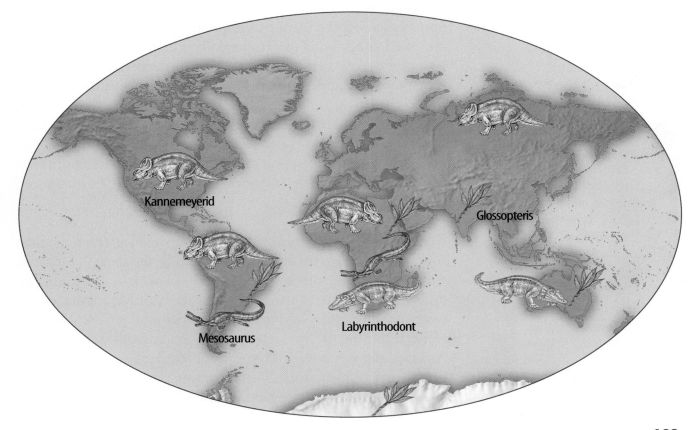

Figure 3 This fossil plant, *Glossopteris,* grew in a temperate climate.

Mini LAB

Interpreting Fossil Data

Procedure
1. Build a three-layer landmass using **clay** or **modeling dough.**
2. Mold the clay into mountain ranges.
3. Place similar "**fossils**" into the clay at various locations around the landmass.
4. Form five continents from the one landmass. Also, form two smaller landmasses out of different clay with different mountain ranges and fossils.
5. Place the five continents and two smaller landmasses around the room.
6. Have someone who did not make or place the landmasses make a model that shows how they once were positioned.
7. Return the clay to its container so it can be used again.

Analysis
What clues were useful in reconstructing the original landmass?

Try at Home

A Widespread Plant Another fossil that supports the hypothesis of continental drift is *Glossopteris* (glahs AHP tur us). **Figure 3** shows this fossil plant, which has been found in Africa, Australia, India, South America, and Antarctica. The presence of *Glossopteris* in so many areas also supported Wegener's idea that all of these regions once were connected and had similar climates.

Climate Clues Wegener used continental drift to explain evidence of changing climates. For example, fossils of warm-weather plants were found on the island of Spitsbergen in the Arctic Ocean. To explain this, Wegener hypothesized that Spitsbergen drifted from tropical regions to the arctic. Wegener also used continental drift to explain evidence of glaciers found in temperate and tropical areas. Glacial deposits and rock surfaces scoured and polished by glaciers are found in South America, Africa, India, and Australia. This shows that parts of these continents were covered with glaciers in the past. How could you explain why glacial deposits are found in areas where no glaciers exist today? Wegener thought that these continents were connected and partly covered with ice near Earth's south pole long ago.

Rock Clues If the continents were connected at one time, then rocks that make up the continents should be the same in locations where they were joined. Similar rock structures are found on different continents. Parts of the Appalachian Mountains of the eastern United States are similar to those found in Greenland and western Europe. If you were to study rocks from eastern South America and western Africa, you would find other rock structures that also are similar. Rock clues like these support the idea that the continents were connected in the past.

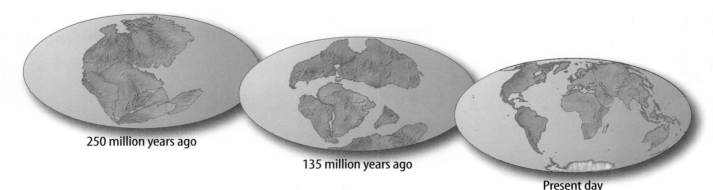

250 million years ago

135 million years ago

Present day

How could continents drift?

Although Wegener provided evidence to support his hypothesis of continental drift, he couldn't explain how, when, or why these changes, shown in **Figure 4,** took place. The idea suggested that lower-density, continental material somehow had to plow through higher-density, ocean-floor material. The force behind this plowing was thought to be the spin of Earth on its axis—a notion that was quickly rejected by physicists. Because other scientists could not provide explanations either, Wegener's idea of continental drift was initially rejected. The idea was so radically different at that time that most people closed their minds to it.

Rock, fossil, and climate clues were the main types of evidence for continental drift. After Wegener's death, more clues were found, largely because of advances in technology, and new ideas that related to continental drift were developed. You'll learn about a new idea, seafloor spreading, in the next section.

Figure 4 These computer models show the probable course the continents have taken. On the far left is their position 250 million years ago. In the middle is their position 135 million years ago. At right is their current position.

section 1 review

Summary

Evidence for Continental Drift

- Alfred Wegener proposed in his hypothesis of continental drift that all continents were once connected as one large landmass called Pangaea.
- Evidence of continental drift came from fossils, signs of climate change, and rock structures from different continents.

How could continents drift?

- During his lifetime, Wegener was unable to explain how, when, or why the continents drifted.
- After his death, advances in technology permitted new ideas to be developed to help explain his hypothesis.

Self Check

1. **Explain** how Wegener used climate clues to support his hypothesis of continental drift.
2. **Describe** how rock clues were used to support the hypothesis of continental drift.
3. **Summarize** the ways that fossils helped support the hypothesis of continental drift.
4. **Think Critically** Why would you expect to see similar rocks and rock structures on two landmasses that were connected at one time.

Applying Skills

5. **Compare and contrast** the locations of fossils of the temperate plant *Glossopteris,* as shown in **Figure 2,** with the climate that exists at each location today.

section 2

Seafloor Spreading

as you read

What You'll Learn
- **Explain** seafloor spreading.
- **Recognize** how age and magnetic clues support seafloor spreading.

Why It's Important
Seafloor spreading helps explain how continents moved apart.

Review Vocabulary
seafloor: portion of Earth's crust that lies beneath ocean waters

New Vocabulary
- seafloor spreading

Mapping the Ocean Floor

If you were to lower a rope from a boat until it reached the seafloor, you could record the depth of the ocean at that particular point. In how many different locations would you have to do this to create an accurate map of the seafloor? This is exactly how it was done until World War I, when the use of sound waves was introduced by German scientists to detect submarines. During the 1940s and 1950s, scientists began using sound waves on moving ships to map large areas of the ocean floor in detail. Sound waves echo off the ocean bottom—the longer the sound waves take to return to the ship, the deeper the water is.

Using sound waves, researchers discovered an underwater system of ridges, or mountains, and valleys like those found on the continents. In some of these underwater ridges are rather long rift valleys where volcanic eruptions and earthquakes occur from time to time. Some of these volcanoes actually are visible above the ocean surface. In the Atlantic, the Pacific, and in other oceans around the world, a system of ridges, called the mid-ocean ridges, is present. These underwater mountain ranges, shown in **Figure 5,** stretch along the center of much of Earth's ocean floor. This discovery raised the curiosity of many scientists. What formed these mid-ocean ridges?

Reading Check *How were mid-ocean ridges discovered?*

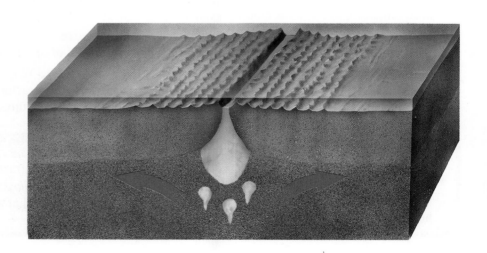

Figure 5 As the seafloor spreads apart at a mid-ocean ridge, new seafloor is created. The older seafloor moves away from the ridge in opposite directions.

186 CHAPTER 7 Plate Tectonics

The Seafloor Moves In the early 1960s, Princeton University scientist Harry Hess suggested an explanation. His now-famous theory is known as **seafloor spreading.** Hess proposed that hot, less dense material below Earth's crust rises toward the surface at the mid-ocean ridges. Then, it flows sideways, carrying the seafloor away from the ridge in both directions, as seen in **Figure 5.**

As the seafloor spreads apart, magma is forced upward and flows from the cracks. It becomes solid as it cools and forms new seafloor. As new seafloor moves away from the mid-ocean ridge, it cools, contracts, and becomes denser. This denser, colder seafloor sinks, helping to form the ridge. The theory of seafloor spreading was later supported by the following observations.

Reading Check *How does new seafloor form at mid-ocean ridges?*

Figure 6 Many new discoveries have been made on the seafloor. These giant tube worms inhabit areas near hot water vents along mid-ocean ridges.

Evidence for Spreading

In 1968, scientists aboard the research ship *Glomar Challenger* began gathering information about the rocks on the seafloor. *Glomar Challenger* was equipped with a drilling rig that allowed scientists to drill into the seafloor to obtain rock samples. Scientists found that the youngest rocks are located at the mid-ocean ridges. The ages of the rocks become increasingly older in samples obtained farther from the ridges, adding to the evidence for seafloor spreading.

Using submersibles along mid-ocean ridges, new seafloor features and life-forms also were discovered there, as shown in **Figure 6.** As molten material is forced upward along the ridges, it brings heat and chemicals that support exotic life-forms in deep, ocean water. Among these are giant clams, mussels, and tube worms.

Curie Point Find out what the Curie point is and describe in your Science Journal what happens to iron-bearing minerals when they are heated to the Curie point. Explain how this is important to studies of seafloor spreading.

 **Magnetic Clues** Earth's magnetic field has a north and a south pole. Magnetic lines, or directions, of force leave Earth near the south pole and enter Earth near the north pole. During a magnetic reversal, the lines of magnetic force run the opposite way. Scientists have determined that Earth's magnetic field has reversed itself many times in the past. These reversals occur over intervals of thousands or even millions of years. The reversals are recorded in rocks forming along mid-ocean ridges.

SECTION 2 Seafloor Spreading **187**

■ Normal magnetic polarity
■ Reverse magnetic polarity

Figure 7 Changes in Earth's magnetic field are preserved in rock that forms on both sides of mid-ocean ridges.
Explain why this is considered to be evidence of seafloor spreading.

Magnetic Time Scale Iron-bearing minerals, such as magnetite, that are found in the rocks of the seafloor can record Earth's magnetic field direction when they form. Whenever Earth's magnetic field reverses, newly forming iron minerals will record the magnetic reversal.

Using a sensing device called a magnetometer (mag nuh TAH muh tur) to detect magnetic fields, scientists found that rocks on the ocean floor show many periods of magnetic reversal. The magnetic alignment in the rocks reverses back and forth over time in strips parallel to the mid-ocean ridges, as shown in **Figure 7**. A strong magnetic reading is recorded when the polarity of a rock is the same as the polarity of Earth's magnetic field today. Because of this, normal polarities in rocks show up as large peaks. This discovery provided strong support that seafloor spreading was indeed occurring. The magnetic reversals showed that new rock was being formed at the mid-ocean ridges. This helped explain how the crust could move—something that the continental drift hypothesis could not do.

section 2 review

Summary

Mapping the Ocean Floor

- Mid-ocean ridges, along the center of the ocean floor, have been found by using sound waves, the same method once used to detect submarines during World War I.
- Harry Hess suggested, in his seafloor spreading hypothesis, that the seafloor moves.

Evidence for Spreading

- Scientists aboard *Glomar Challenger* provided evidence of spreading by discovering that the youngest rocks are located at ridges and become increasingly older farther from the ridges.
- Magnetic alignment of rocks, in alternating strips that run parallel to ridges, indicates reversals in Earth's magnetic field and provides further evidence of seafloor spreading.

Self Check

1. **Summarize** What properties of iron-bearing minerals on the seafloor support the theory of seafloor spreading?
2. **Explain** how the ages of the rocks on the ocean floor support the theory of seafloor spreading.
3. **Summarize** How did Harry Hess's hypothesis explain seafloor movement?
4. **Explain** why some partly molten material rises toward Earth's surface.
5. **Think Critically** The ideas of Hess, Wegener, and others emphasize that Earth is a dynamic planet. How is seafloor spreading different from continental drift?

Applying Skills

6. **Solve One-Step Equations** North America is moving about 1.25 cm per year away from a ridge in the middle of the Atlantic Ocean. Using this rate, how much farther apart will North America and the ridge be in 200 million years?

Seafloor Spreading Rates

How did scientists use their knowledge of seafloor spreading and magnetic field reversals to reconstruct Pangaea? Try this lab to see how you can determine where a continent may have been located in the past.

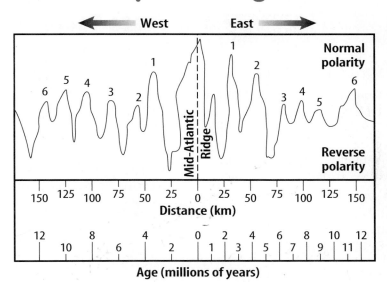

Real-World Question

Can you use clues, such as magnetic field reversals on Earth, to help reconstruct Pangaea?

Goals
- **Interpret** data about magnetic field reversals. Use these magnetic clues to reconstruct Pangaea.

Materials
metric ruler
pencil

Procedure

1. Study the magnetic field graph above. You will be working only with normal polarity readings, which are the peaks above the baseline in the top half of the graph.
2. Place the long edge of a ruler vertically on the graph. Slide the ruler so that it lines up with the center of peak 1 west of the Mid-Atlantic Ridge.
3. **Determine** and record the distance and age that line up with the center of peak 1 west. Repeat this process for peak 1 east of the ridge.
4. **Calculate** the average distance and age for this pair of peaks.
5. Repeat steps 2 through 4 for the remaining pairs of normal-polarity peaks.
6. **Calculate** the rate of movement in cm per year for the six pairs of peaks. Use the formula rate = distance/time. Convert kilometers to centimeters. For example, to calculate a rate using normal-polarity peak 5, west of the ridge:

$$\text{rate} = \frac{125 \text{ km}}{10 \text{ million years}} = \frac{12.5 \text{ km}}{\text{million years}} = \frac{1{,}250{,}000 \text{ cm}}{1{,}000{,}000 \text{ years}} = 1.25 \text{ cm/year}$$

Conclude and Apply

1. **Compare** the age of igneous rock found near the mid-ocean ridge with that of igneous rock found farther away from the ridge.
2. If the distance from a point on the coast of Africa to the Mid-Atlantic Ridge is approximately 2,400 km, calculate how long ago that point in Africa was at or near the Mid-Atlantic Ridge.
3. How could you use this method to reconstruct Pangaea?

LAB **189**

section 3
Theory of Plate Tectonics

as you read

What You'll Learn
- **Compare and contrast** different types of plate boundaries.
- **Explain** how heat inside Earth causes plate tectonics.
- **Recognize** features caused by plate tectonics.

Why It's Important
Plate tectonics explains how many of Earth's features form.

Review Vocabulary
converge: to come together
diverge: to move apart
transform: to convert or change

New Vocabulary
- plate tectonics
- plate
- lithosphere
- asthenosphere
- convection current

Plate Tectonics

The idea of seafloor spreading showed that more than just continents were moving, as Wegener had thought. It was now clear to scientists that sections of the seafloor and continents move in relation to one another.

Plate Movements In the 1960s, scientists developed a new theory that combined continental drift and seafloor spreading. According to the theory of **plate tectonics,** Earth's crust and part of the upper mantle are broken into sections. These sections, called **plates,** move on a plasticlike layer of the mantle. The plates can be thought of as rafts that float and move on this layer.

Composition of Earth's Plates Plates are made of the crust and a part of the upper mantle, as shown in **Figure 8.** These two parts combined are the **lithosphere** (LIH thuh sfihr). This rigid layer is about 100 km thick and generally is less dense than material underneath. The plasticlike layer below the lithosphere is called the **asthenosphere** (as THE nuh sfihr). The rigid plates of the lithosphere float and move around on the asthenosphere.

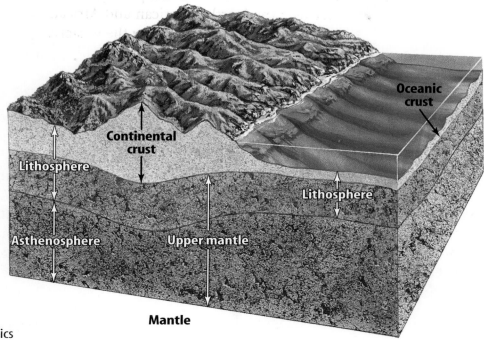

Figure 8 Plates of the lithosphere are composed of oceanic crust, continental crust, and rigid upper mantle.

190 CHAPTER 7 Plate Tectonics

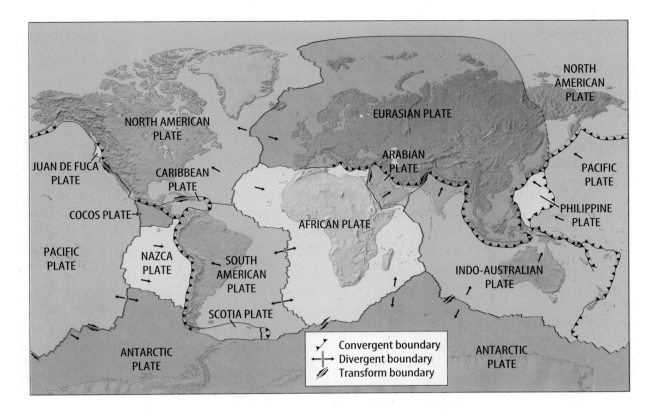

Plate Boundaries

When plates move, they can interact in several ways. They can move toward each other and converge, or collide. They also can pull apart or slide alongside one another. When the plates interact, the result of their movement is seen at the plate boundaries, as in **Figure 9.**

Reading Check *What are the general ways that plates interact?*

Movement along any plate boundary means that changes must happen at other boundaries. What is happening to the Atlantic Ocean floor between the North American and African Plates? Compare this with what is happening along the western margin of South America.

Plates Moving Apart The boundary between two plates that are moving apart is called a divergent boundary. You learned about divergent boundaries when you read about seafloor spreading. In the Atlantic Ocean, the North American Plate is moving away from the Eurasian and the African Plates, as shown in **Figure 9.** That divergent boundary is called the Mid-Atlantic Ridge. The Great Rift Valley in eastern Africa might become a divergent plate boundary. There, a valley has formed where a continental plate is being pulled apart. **Figure 10** shows a side view of what a rift valley might look like and illustrates how the hot material rises up where plates separate.

Figure 9 This diagram shows the major plates of the lithosphere, their direction of movement, and the type of boundary between them.
Analyze and Conclude *Based on what is shown in this figure, what is happening where the Nazca Plate meets the Pacific Plate?*

SECTION 3 Theory of Plate Tectonics **191**

Topic: Earthquakes and Volcanoes

Visit blue.msscience.com for Web links to recent news or magazine articles about earthquakes and volcanic activity related to plate tectonics.

Activity Prepare a group demonstration about recent volcanic and earthquake events. Divide tasks among group members. Find and copy maps, diagrams, photographs, and charts to highlight your presentation. Emphasize the locations of events and the relationship to plate tectonics.

Plates Moving Together If new crust is being added at one location, why doesn't Earth's surface keep expanding? As new crust is added in one place, it disappears below the surface at another. The disappearance of crust can occur when seafloor cools, becomes denser, and sinks. This occurs where two plates move together at a convergent boundary.

When an oceanic plate converges with a less dense continental plate, the denser oceanic plate sinks under the continental plate. The area where an oceanic plate subducts, or goes down, into the mantle is called a subduction zone. Some volcanoes form above subduction zones. **Figure 10** shows how this type of convergent boundary creates a deep-sea trench where one plate bends and sinks beneath the other. High temperatures cause rock to melt around the subducting slab as it goes under the other plate. The newly formed magma is forced upward along these plate boundaries, forming volcanoes. The Andes mountain range of South America contains many volcanoes. They were formed at the convergent boundary of the Nazca and the South American Plates.

Applying Science

How well do the continents fit together?

Recall the Launch Lab you performed at the beginning of this chapter. While you were trying to fit pieces of a cut-up photograph together, what clues did you use?

Identifying the Problem

Take a copy of a map of the world and cut out each continent. Lay them on a tabletop and try to fit them together, using techniques you used in the Launch Lab. You will find that the pieces of your Earth puzzle—the continents—do not fit together well. Yet, several of the areas on some continents fit together extremely well.

Take out another world map—one that shows the continental shelves as well as the continents. Copy it and cut out the continents, this time including the continental shelves.

Solving the Problem
1. Does including the continental shelves solve the problem of fitting the continents together?
2. Why should continental shelves be included with maps of the continents?

NATIONAL GEOGRAPHIC VISUALIZING PLATE BOUNDARIES

Figure 10

By diverging at some boundaries and converging at others, Earth's plates are continually—but gradually—reshaping the landscape around you. The Mid-Atlantic Ridge, for example, was formed when the North and South American Plates pulled apart from the Eurasian and African Plates (see globe). Some features that occur along plate boundaries—rift valleys, volcanoes, and mountain ranges—are shown on the right and below.

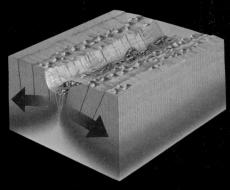

A RIFT VALLEY When continental plates pull apart, they can form rift valleys. The African continent is separating now along the East African Rift Valley.

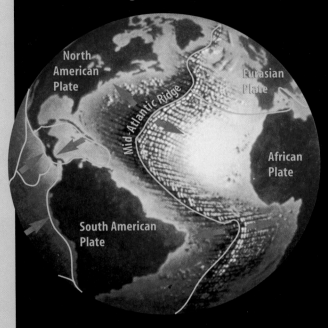

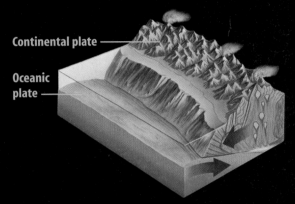

SUBDUCTION Where oceanic and continental plates collide, the oceanic plate plunges beneath the less dense continental plate. As the plate descends, molten rock (yellow) forms and rises toward the surface, creating volcanoes.

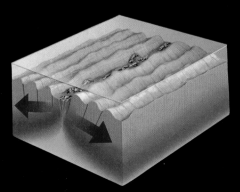

SEAFLOOR SPREADING A mid-ocean ridge, like the Mid-Atlantic Ridge, forms where oceanic plates continue to separate. As rising magma (yellow) cools, it forms new oceanic crust.

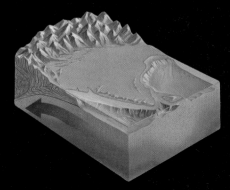

CONTINENTAL COLLISION Where two continental plates collide, they push up the crust to form mountain ranges such as the Himalaya.

SECTION 3 Theory of Plate Tectonics

Where Plates Collide A subduction zone also can form where two oceanic plates converge. In this case, the colder, older, denser oceanic plate bends and sinks down into the mantle. The Mariana Islands in the western Pacific are a chain of volcanic islands formed where two oceanic plates collide.

Usually, no subduction occurs when two continental plates collide, as shown in **Figure 10.** Because both of these plates are less dense than the material in the asthenosphere, the two plates collide and crumple up, forming mountain ranges. Earthquakes are common at these convergent boundaries. However, volcanoes do not form because there is no, or little, subduction. The Himalaya in Asia are forming where the Indo-Australian Plate collides with the Eurasian Plate.

Where Plates Slide Past Each Other The third type of plate boundary is called a transform boundary. Transform boundaries occur where two plates slide past one another. They move in opposite directions or in the same direction at different rates. When one plate slips past another suddenly, earthquakes occur. The Pacific Plate is sliding past the North American Plate, forming the famous San Andreas Fault in California, as seen in **Figure 11.** The San Andreas Fault is part of a transform plate boundary. It has been the site of many earthquakes.

Figure 11 The San Andreas Fault in California occurs along the transform plate boundary where the Pacific Plate is sliding past the North American Plate.

Overall, the two plates are moving in roughly the same direction.
Explain Why, then, do the red arrows show movement in opposite directions?

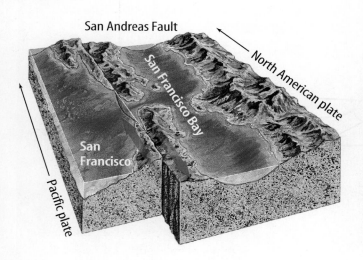

This photograph shows an aerial view of the San Andreas Fault.

Causes of Plate Tectonics

Many new discoveries have been made about Earth's crust since Wegener's day, but one question still remains. What causes the plates to move? Scientists now think they have a good idea. They think that plates move by the same basic process that occurs when you heat soup.

Convection Inside Earth Soup that is cooking in a pan on the stove contains currents caused by an unequal distribution of heat in the pan. Hot, less dense soup is forced upward by the surrounding, cooler, denser soup. As the hot soup reaches the surface, it cools and sinks back down into the pan. This entire cycle of heating, rising, cooling, and sinking is called a **convection current.** A version of this same process, occurring in the mantle, is thought to be the force behind plate tectonics. Scientists suggest that differences in density cause hot, plasticlike rock to be forced upward toward the surface.

Moving Mantle Material Wegener wasn't able to come up with an explanation for why plates move. Today, researchers who study the movement of heat in Earth's interior have proposed several possible explanations. All of the hypotheses use convection in one way or another. It is, therefore, the transfer of heat inside Earth that provides the energy to move plates and causes many of Earth's surface features. One hypothesis is shown in **Figure 12.** It relates plate motion directly to the movement of convection currents. According to this hypothesis, convection currents cause the movements of plates.

Mini LAB

Modeling Convection Currents

Procedure
1. Pour **water** into **a clear, colorless casserole dish** until it is 5 cm from the top.
2. Center the dish on a **hot plate** and heat it. WARNING: *Wear thermal mitts to protect your hands.*
3. Add a few drops of **food coloring** to the water above the center of the hot plate.
4. Looking from the side of the dish, observe what happens in the water.
5. Illustrate your observations in your **Science Journal.**

Analysis
1. Determine whether any currents form in the water.
2. Infer what causes the currents to form.

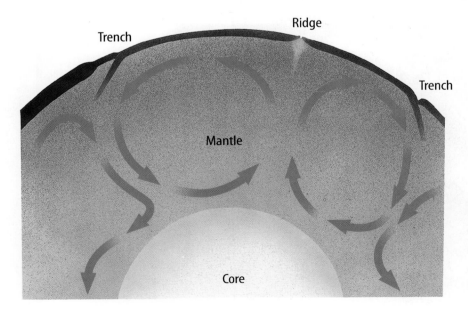

Figure 12 In one hypothesis, convection currents occur throughout the mantle. Such convection currents (see arrows) are the driving force of plate tectonics.

Features Caused by Plate Tectonics

Earth is a dynamic planet with a hot interior. This heat leads to convection, which powers the movement of plates. As the plates move, they interact. The interaction of plates produces forces that build mountains, create ocean basins, and cause volcanoes. When rocks in Earth's crust break and move, energy is released in the form of seismic waves. Humans feel this release as earthquakes. You can see some of the effects of plate tectonics in mountainous regions, where volcanoes erupt, or where landscapes have changed from past earthquake or volcanic activity.

Reading Check *What happens when seismic energy is released as rocks in Earth's crust break and move?*

Normal Faults and Rift Valleys Tension forces, which are forces that pull apart, can stretch Earth's crust. This causes large blocks of crust to break and tilt or slide down the broken surfaces of crust. When rocks break and move along surfaces, a fault forms. Faults interrupt rock layers by moving them out of place. Entire mountain ranges can form in the process, called fault-block mountains, as shown in **Figure 13**. Generally, the faults that form from pull-apart forces are normal faults—faults in which the rock layers above the fault move down when compared with rock layers below the fault.

Rift valleys and mid-ocean ridges can form where Earth's crust separates. Examples of rift valleys are the Great Rift Valley in Africa, and the valleys that occur in the middle of mid-ocean ridges. Examples of mid-ocean ridges include the Mid-Atlantic Ridge and the East Pacific Rise.

Figure 13 Fault-block mountains can form when Earth's crust is stretched by tectonic forces. The arrows indicate the directions of moving blocks.
Name *the type of force that occurs when Earth's crust is pulled in opposite directions.*

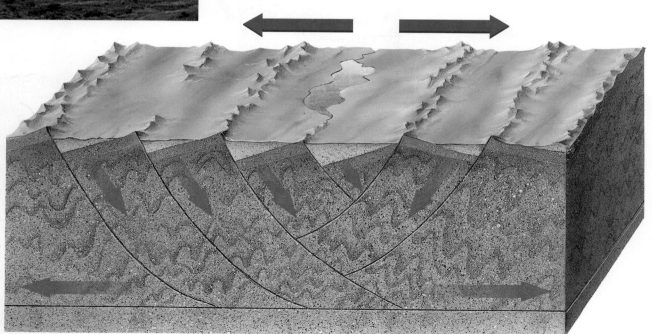

Mountains and Volcanoes Compression forces squeeze objects together. Where plates come together, compression forces produce several effects. As continental plates collide, the forces that are generated cause massive folding and faulting of rock layers into mountain ranges such as the Himalaya, shown in **Figure 14,** or the Appalachian Mountains. The type of faulting produced is generally reverse faulting. Along a reverse fault, the rock layers above the fault surface move up relative to the rock layers below the fault.

Reading Check *What features occur where plates converge?*

As you learned earlier, when two oceanic plates converge, the denser plate is forced beneath the other plate. Curved chains of volcanic islands called island arcs form above the sinking plate. If an oceanic plate converges with a continental plate, the denser oceanic plate slides under the continental plate. Folding and faulting at the continental plate margin can thicken the continental crust to produce mountain ranges. Volcanoes also typically are formed at this type of convergent boundary.

INTEGRATE Career

Volcanologist This person's job is to study volcanoes in order to predict eruptions. Early warning of volcanic eruptions gives nearby residents time to evacuate. Volcanologists also educate the public about the hazards of volcanic eruptions and tell people who live near volcanoes what they can do to be safe in the event of an eruption. Volcanologists travel all over the world to study new volcanic sites.

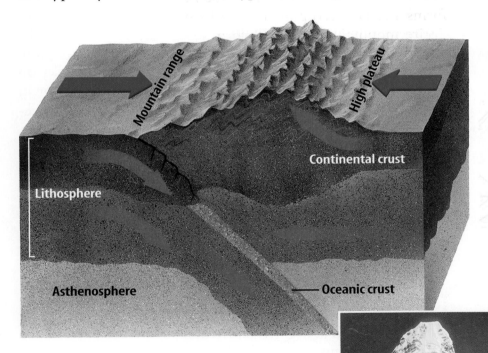

Figure 14 The Himalaya still are forming today as the Indo-Australian Plate collides with the Eurasian Plate.

Figure 15 Most of the movement along a strike-slip fault is parallel to Earth's surface. When movement occurs, human-built structures along a strike-slip fault are offset, as shown here in this road.

Strike-Slip Faults At transform boundaries, two plates slide past one another without converging or diverging. The plates stick and then slide, mostly in a horizontal direction, along large strike-slip faults. In a strike-slip fault, rocks on opposite sides of the fault move in opposite directions, or in the same direction at different rates. This type of fault movement is shown in **Figure 15.** One such example is the San Andreas Fault. When plates move suddenly, vibrations are generated inside Earth that are felt as an earthquake.

Earthquakes, volcanoes, and mountain ranges are evidence of plate motion. Plate tectonics explains how activity inside Earth can affect Earth's crust differently in different locations. You've seen how plates have moved since Pangaea separated. Is it possible to measure how far plates move each year?

Testing for Plate Tectonics

Until recently, the only tests scientists could use to check for plate movement were indirect. They could study the magnetic characteristics of rocks on the seafloor. They could study volcanoes and earthquakes. These methods supported the theory that the plates have moved and still are moving. However, they did not provide proof—only support—of the idea.

New methods had to be discovered to be able to measure the small amounts of movement of Earth's plates. One method, shown in **Figure 16,** uses lasers and a satellite. Now, scientists can measure exact movements of Earth's plates of as little as 1 cm per year.

Direction of Forces In which directions do forces act at convergent, divergent, and transform boundaries? Demonstrate these forces using wooden blocks or your hands.

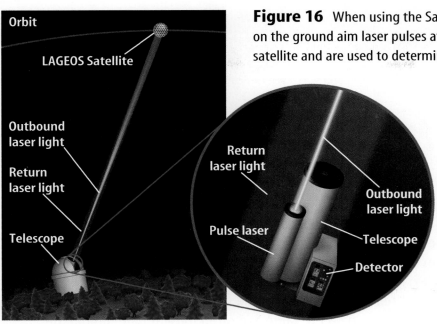

Figure 16 When using the Satellite Laser Ranging System, scientists on the ground aim laser pulses at a satellite. The pulses reflect off the satellite and are used to determine a precise location on the ground.

Current Data Satellite Laser Ranging System data show that Hawaii is moving toward Japan at a rate of about 8.3 cm per year. Maryland is moving away from England at a rate of 1.7 cm per year. Using such methods, scientists have observed that the plates move at rates ranging from about 1 cm to 12 cm per year.

section 3 review

Summary

Plate Tectonics
- The theory of plate tectonics states that sections of the seafloor and continents move as plates on a plasticlike layer of the mantle.

Plate Boundaries
- The boundary between two plates moving apart is called a divergent boundary.
- Plates move together at a convergent boundary.
- Transform boundaries occur where two plates slide past one another.

Causes of Plate Tectonics
- Convection currents are thought to cause the movement of Earth's plates.

Features Caused by Plate Tectonics
- Tension forces cause normal faults, rift valleys, and mid-ocean ridges at divergent boundaries.
- At convergent boundaries, compression forces cause folding, reverse faults, and mountains.
- At transform boundaries, two plates slide past one another along strike-slip faults.

Self Check

1. **Describe** what occurs at plate boundaries that are associated with seafloor spreading.
2. **Describe** three types of plate boundaries where volcanic eruptions can occur.
3. **Explain** how convection currents are related to plate tectonics.
4. **Think Critically** Using **Figure 9** and a world map, determine what natural disasters might occur in Iceland. Also determine what disasters might occur in Tibet. Explain why some Icelandic disasters are not expected to occur in Tibet.

Applying Skills

5. **Predict** Plate tectonic activity causes many events that can be dangerous to humans. One of these events is a seismic sea wave, or tsunami. Learn how scientists predict the arrival time of a tsunami in a coastal area.
6. **Use a Word Processor** Write three separate descriptions of the three basic types of plate boundaries—divergent boundaries, convergent boundaries, and transform boundaries. Then draw a sketch of an example of each boundary next to your description.

LAB Use the Internet

Predicting Tectonic Activity

Goals
- **Research** the locations of earthquakes and volcanic eruptions around the world.
- **Plot** earthquake epicenters and the locations of volcanic eruptions.
- **Predict** locations that are tectonically active based on a plot of the locations of earthquake epicenters and active volcanoes.

Data Source

Science Online
Visit blue.msscience.com/internet_lab for more information about earthquake and volcano sites, and data from other students.

Real-World Question

The movement of plates on Earth causes forces that build up energy in rocks. The release of this energy can produce vibrations in Earth that you know as earthquakes. Earthquakes occur every day. Many of them are too small to be felt by humans, but each event tells scientists something more about the planet. Active volcanoes can do the same and often form at plate boundaries.

Can you predict tectonically active areas by plotting locations of earthquake epicenters and volcanic eruptions?

Think about where earthquakes and volcanoes have occurred in the past. Make a hypothesis about whether the locations of earthquake epicenters and active volcanoes can be used to predict tectonically active areas.

200 CHAPTER 7 Plate Tectonics

Using Scientific Methods

▶ Make a Plan

1. Make a data table in your Science Journal like the one shown.
2. Collect data for earthquake epicenters and volcanic eruptions for at least the past two weeks. Your data should include the longitude and latitude for each location. For help, refer to the data sources given on the opposite page.

Locations of Epicenters and Eruptions		
Earthquake Epicenter/ Volcanic Eruption	Longitude	Latitude
	Do not write in this book.	

▶ Follow Your Plan

1. Make sure your teacher approves your plan before you start.
2. **Plot** the locations of earthquake epicenters and volcanic eruptions on a map of the world. Use an overlay of tissue paper or plastic.
3. After you have collected the necessary data, predict where the tectonically active areas on Earth are.
4. **Compare and contrast** the areas that you predicted to be tectonically active with the plate boundary map shown in **Figure 9**.

▶ Analyze Your Data

1. What areas on Earth do you predict to be the locations of tectonic activity?
2. How close did your prediction come to the actual location of tectonically active areas?

▶ Conclude and Apply

1. How could you make your predictions closer to the locations of actual tectonic activity?
2. Would data from a longer period of time help? Explain.
3. What types of plate boundaries were close to your locations of earthquake epicenters? Volcanic eruptions?
4. **Explain** which types of plate boundaries produce volcanic eruptions. Be specific.

Communicating Your Data

Find this lab using the link below. Post your data in the table provided. **Compare** your data to those of other students. Combine your data with those of other students and **plot** these combined data on a map to recognize the relationship between plate boundaries, volcanic eruptions, and earthquake epicenters.

Science online

blue.msscience.com/internet_lab

Science and Language Arts

Listening In
by Gordon Judge

I'm just a bit of seafloor on this mighty solid sphere.
With no mind to be broadened, I'm quite content
  down here.
The mantle churns below me, and the sea's in turmoil, too;
But nothing much disturbs me, I'm rock solid through
  and through.

I do pick up occasional low-frequency vibrations –
(I think, although I can't be sure, they're sperm whales'
  conversations).
I know I shouldn't listen in, but what else can I do?
It seems they are all studying for degrees from the OU.

They've mentioned me in passing, as their minds begin
  improving:

I think I've heard them say
"The theory says the sea-
floor's moving…".
They call it "Plate Tectonics", this
new theory in their noddle.
If they would only ask me, I
could tell them it's all
twaddle….

But, how can I be moving, when I know full well myself
That I'm quite firmly anchored to a continental shelf?
"Well, the continent is moving, too; you're *pushing* it,
  you see,"
I hear those OU whales intone, hydro-acoustically….

Well, thank you very much, OU. You've upset my
  composure.
Next time you send your student whales to look at
  my exposure
I'll tell them it's a load of tosh: it's *they* who move,
  not me,
Those arty-smarty blobs of blubber, clogging up the sea!

Understanding Literature

Point of View Point of view refers to the perspective from which an author writes. This poem begins, "I'm just a bit of seafloor…." Right away, you know that the poem, or story, is being told from the point of view of the speaker, or the "first person." What effect does the first-person narration have on the story?

Respond to the Reading

1. Who is narrating the poem?
2. Why might the narrator think he or she hasn't moved?
3. **Linking Science and Writing** Using the first-person point of view, write an account from the point of view of a living or nonliving thing.

Volcanoes can occur where two plates move toward each other. When an oceanic plate and a continental plate collide, a volcano will form. Subduction zones occur when one plate sinks under another plate. Rocks melt in the zones where these plates converge, causing magma to move upward and form volcanic mountains.

Chapter 7 Study Guide

Reviewing Main Ideas

Section 1 Continental Drift

1. Alfred Wegener suggested that the continents were joined together at some point in the past in a large landmass he called Pangaea. Wegener proposed that continents have moved slowly, over millions of years, to their current locations.

2. The puzzlelike fit of the continents, fossils, climatic evidence, and similar rock structures support Wegener's idea of continental drift. However, Wegener could not explain what process could cause the movement of the landmasses.

Section 2 Seafloor Spreading

1. Detailed mapping of the ocean floor in the 1950s showed underwater mountains and rift valleys.

2. In the 1960s, Harry Hess suggested seafloor spreading as an explanation for the formation of mid-ocean ridges.

3. The theory of seafloor spreading is supported by magnetic evidence in rocks and by the ages of rocks on the ocean floor.

Section 3 Theory of Plate Tectonics

1. In the 1960s, scientists combined the ideas of continental drift and seafloor spreading to develop the theory of plate tectonics. The theory states that the surface of Earth is broken into sections called plates that move around on the asthenosphere.

2. Currents in Earth's mantle called convection currents transfer heat in Earth's interior. It is thought that this transfer of heat energy moves plates.

3. Earth is a dynamic planet. As the plates move, they interact, resulting in many of the features of Earth's surface.

Visualizing Main Ideas

Copy and complete the concept map below about continental drift, seafloor spreading, and plate tectonics.

chapter 7 Review

Using Vocabulary

asthenosphere p. 190
continental drift p. 182
convection current p. 195
lithosphere p. 190
Pangaea p. 182
plate p. 190
plate tectonics p. 190
seafloor spreading p. 187

Each phrase below describes a vocabulary term from the list. Write the term that matches the phrase describing it.

1. plasticlike layer below the lithosphere
2. idea that continents move slowly across Earth's surface
3. large, ancient landmass that consisted of all the continents on Earth
4. composed of oceanic or continental crust and upper mantle
5. explains locations of mountains, trenches, and volcanoes
6. theory proposed by Harry Hess that includes processes along mid-ocean ridges

Checking Concepts

Choose the word or phrase that best answers the question.

7. Which layer of Earth contains the asthenosphere?
   A) crust
   B) mantle
   C) outer core
   D) inner core

8. What type of plate boundary is the San Andreas Fault part of?
   A) divergent
   B) subduction
   C) convergent
   D) transform

9. What hypothesis states that continents slowly moved to their present positions on Earth?
   A) subduction
   B) erosion
   C) continental drift
   D) seafloor spreading

Use the illustration below to answer question 10.

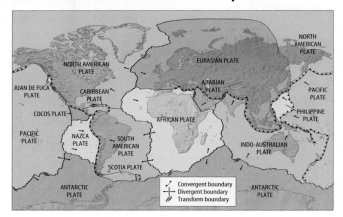

10. Which plate is subducting beneath the South American Plate?
    A) Nazca
    B) African
    C) North American
    D) Indo-Australian

11. Which of the following features are evidence that many continents were at one time near Earth's south pole?
    A) glacial deposits
    B) earthquakes
    C) volcanoes
    D) mid-ocean ridges

12. What evidence in rocks supports the theory of seafloor spreading?
    A) plate movement
    B) magnetic reversals
    C) subduction
    D) convergence

13. Which type of plate boundary is the Mid-Atlantic Ridge a part of?
    A) convergent
    B) divergent
    C) transform
    D) subduction

14. What theory states that plates move around on the asthenosphere?
    A) continental drift
    B) seafloor spreading
    C) subduction
    D) plate tectonics

204 CHAPTER REVIEW

chapter 7 Review

Thinking Critically

15. **Infer** Why do many earthquakes but few volcanic eruptions occur in the Himalaya?

16. **Explain** Glacial deposits often form at high latitudes near the poles. Explain why glacial deposits have been found in Africa.

17. **Describe** how magnetism is used to support the theory of seafloor spreading.

18. **Explain** why volcanoes do not form along the San Andreas Fault.

19. **Explain** why the fossil of an ocean fish found on two different continents would not be good evidence of continental drift.

20. **Form Hypotheses** Mount St. Helens in the Cascade Range is a volcano. Use **Figure 9** and a U.S. map to hypothesize how it might have formed.

21. **Concept Map** Make an events-chain concept map that describes seafloor spreading along a divergent plate boundary. Choose from the following phrases: *magma cools to form new seafloor, convection currents circulate hot material along divergent boundary,* and *older seafloor is forced apart.*

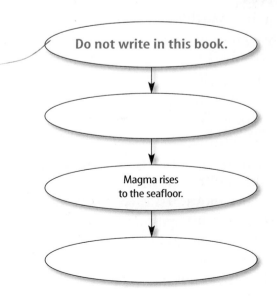

Performance Activities

22. **Observe and Infer** In the MiniLab called "Modeling Convection Currents," you observed convection currents produced in water as it was heated. Repeat the experiment, placing sequins, pieces of wood, or pieces of rubber bands into the water. How do their movements support your observations and inferences from the MiniLab?

Applying Math

23. **A Growing Rift** Movement along the African Rift Valley is about 2.1 cm per year. If plates continue to move apart at this rate, how much larger will the rift be (in meters) in 1,000 years? In 15,500 years?

Use the illustration below to answer questions 24 and 25.

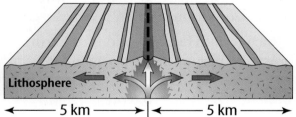

24. **New Seafloor** 10 km of new seafloor has been created in 50,000 years, with 5 km on each side of a mid-ocean ridge. What is the rate of movement, in km per year, of each plate? In cm per year?

25. **Use a Ratio** If 10 km of seafloor were created in 50,000 years, how many kilometers of seafloor were created in 10,000 years? How many years will it take to create a total of 30 km of seafloor?

chapter 7 Standardized Test Practice

Part 1 Multiple Choice

Record your answers on the answer sheet provided by your teacher or on a sheet of paper.

Use the illustration below to answer question 1.

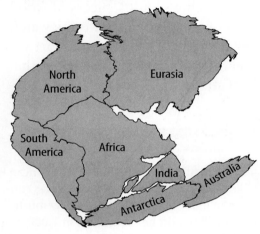

1. What is the name of the ancient supercontinent shown above?
   A. Pangaea   C. Laurasia
   B. Gondwanaland   D. North America

2. Who developed the continental drift hypothesis?
   A. Harry Hess   C. Alfred Wegener
   B. J. Tuzo Wilson D. W. Jason Morgan

3. Which term refers to sections of Earth's crust and part of the upper mantle?
   A. asthenosphere   C. lithosphere
   B. plate   D. core

4. About how fast do plates move?
   A. a few millimeters each year
   B. a few centimeters each year
   C. a few meters each year
   D. a few kilometers each year

Test-Taking Tip

Marking Answers Be sure to ask if it is okay to mark in the test booklet when taking the test, but make sure you mark all answers on your answer sheet.

5. Where do Earth's plates slide past each other?
   A. convergent boundaries
   B. divergent boundaries
   C. transform boundaries
   D. subduction zones

Study the diagram below before answering questions 6 and 7.

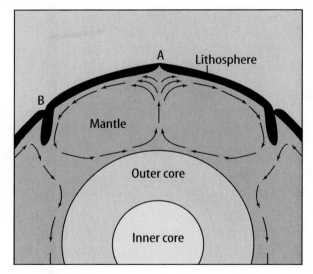

6. Suppose that the arrows in the diagram represent patterns of convection in Earth's mantle. Which type of plate boundary is most likely to occur along the region labeled "A"?
   A. transform
   B. reverse
   C. convergent
   D. divergent

7. Which statement is true of the region marked "B" on the diagram?
   A. Plates move past each other sideways.
   B. Plates move apart and volcanoes form.
   C. Plates move toward each other and volcanoes form.
   D. Plates are not moving.

206 STANDARDIZED TEST PRACTICE

Standardized Test Practice

Part 2 | Short Response/Grid In

Record your answers on the answer sheet provided by your teacher or on a sheet of paper.

8. What is an ocean trench? Where do they occur?
9. How do island arcs form?
10. Why do earthquakes occur along the San Andreas Fault?
11. Describe a mid-ocean ridge.
12. Why do plates sometimes sink into the mantle?

Use the graph below to answer questions 13–15.

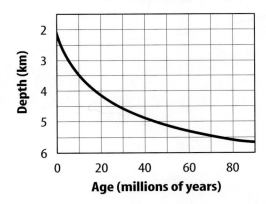

Relationship Between Depth and Age of Seafloor

13. Use the graph to estimate the average depth below the ocean of ocean crust that has just formed.
14. Estimate the average depth of ocean crust that is 60 million years old.
15. Describe how the depth of ocean crust is related to the age of ocean crust.
16. On average, about how fast do plates move?
17. What layer in Earth's mantle do plates slide over?
18. Describe how scientists make maps of the ocean floor.

Part 3 | Open Ended

Record your answers on a sheet of paper.

Use the illustration below to answer question 19.

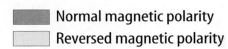
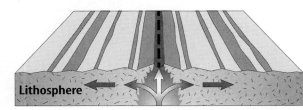

19. Examine the diagram above. Explain how the magnetic stripes form in rock that makes up the ocean crust.
20. What causes convection in Earth's mantle?
21. Explain the theory of plate tectonics.
22. What happened to the continents that made up Pangaea after it started to break up?
23. How does Earth's lithosphere differ from Earth's asthenosphere?
24. What types of life have been discovered near mid-ocean ridges?
25. What are the three types of motion that occur at plate boundaries? Describe each motion.
26. What forms when continents collide? Describe the process.
27. What occurs at the center of a mid-ocean ridge? What might you find there?
28. What evidence do we have that supports the hypothesis of continental drift?
29. Who proposed the first theories about plate tectonics? Explain why other scientists questioned these theories.

chapter 8

Earthquakes and Volcanoes

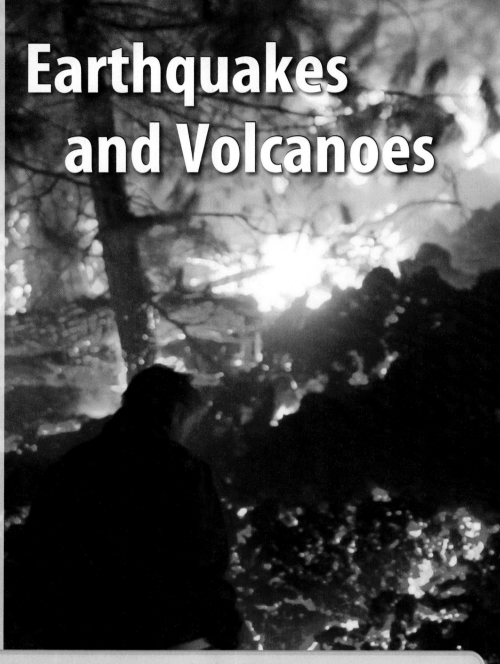

The BIG Idea

Most earthquakes and volcanic events occur along plate boundaries where Earth's plates move relative to one another.

SECTION 1
Earthquakes
Main Idea Earthquakes are vibrations, or seismic waves, often caused by a rupture and elastic rebound of rocks along a fault.

SECTION 2
Volcanoes
Main Idea Magma, solids, and gas can erupt explosively or quietly from cracks in Earth's surface, producing a variety of volcanic materials, events, and landforms.

SECTION 3
Earthquakes, Volcanoes, and Plate Tectonics
Main Idea Convection in Earth can cause plate motion, which produces conditions that cause volcanoes and earthquakes.

Earth's upset stomach?
Rivers of boiling lava poured down the mountain, engulfing small buildings and threatening a lodge after a series of earthquakes awakened this volcano. What causes Earth to behave this way? Are earthquakes and volcanoes related?

Science Journal Are earthquakes and volcanoes completely unrelated, or could there be a possible connection? Propose several ideas that might explain what causes these events.

Start-Up Activities

Construct with Strength

One of the greatest dangers associated with an earthquake occurs when people are inside buildings during the event. In the following lab, you will see how construction materials can be used to help strengthen a building.

1. Using wooden blocks, construct a building with four walls. Place a piece of cardboard over the four walls as a ceiling.
2. Gently shake the table under your building. Describe what happens.
3. Reconstruct the building. Wrap large rubber bands around each section, or wall, of blocks. Then wrap large rubber bands around the entire building.
4. Gently shake the table again.
5. **Think Critically** In your Science Journal, note any differences you observed as the two buildings were shaken. Hypothesize how the construction methods you used in this activity might be applied to the construction of real buildings.

Preview this chapter's content and activities at blue.msscience.com

FOLDABLES
Study Organizer

Earthquakes and Volcanoes Make the following Foldable to help you compare and contrast the characteristics of earthquakes and volcanoes.

STEP 1 Draw a mark at the midpoint of a vertical sheet of paper.

STEP 2 Turn the paper horizontally and **fold** the outside edges in to touch at the midpoint mark.

STEP 3 Draw a volcano on one flap and label the flap *Volcanoes*. Draw an earthquake on the other flap and label it *Earthquakes*. The inside portion should be labeled *Both* and include characteristics that both events share.

Analyze and Critique Before you read the chapter, write what you know about earthquakes and volcanoes on the back of each flap. As you read the chapter, add more information about earthquakes and volcanoes.

Get Ready to Read

Monitor

① Learn It! An important strategy to help you improve your reading is monitoring, or finding your reading strengths and weaknesses. As you read, monitor yourself to make sure the text makes sense. Discover different monitoring techniques you can use at different times, depending on the type of test and situation.

② Practice It! The paragraph below appears in Section 1. Read the passage and answer the questions that follow. Discuss your answers with other students to see how they monitor their reading.

> If enough force is applied, rocks become strained, which means they change shape. They may even break, and the ends of the broken pieces may snap back. This snapping back is called elastic rebound.
>
> —from page 210

- What questions do you still have after reading?
- Do you understand all of the words in the passage?
- Did you have to stop reading often? Is the reading level appropriate for you?

③ Apply It! Identify one paragraph that is difficult to understand. Discuss it with a partner to improve your understanding.

Target Your Reading

Reading Tip

Monitor your reading by slowing down or speeding up depending on your understanding of the text.

Use this to focus on the main ideas as you read the chapter.

① Before you read the chapter, respond to the statements below on your worksheet or on a numbered sheet of paper.
- Write an **A** if you **agree** with the statement.
- Write a **D** if you **disagree** with the statement.

② After you read the chapter, look back to this page to see if you've changed your mind about any of the statements.
- If any of your answers changed, explain why.
- Change any false statements into true statements.
- Use your revised statements as a study guide.

Science Online
Print out a worksheet of this page at blue.msscience.com

Before You Read A or D		Statement	After You Read A or D
	1	Portions of Earth's rocky crust can rebound elastically, similar to a diving board.	
	2	Primary seismic waves originate at the epicenter of an earthquake.	
	3	Tsunamis are huge tidal waves.	
	4	An earthquake with a Richter magnitude of 7.5 releases about 32 times more energy than a 6.5 magnitude earthquake.	
	5	Molten rock material that forms deep below Earth's surface is called lava.	
	6	The composition of magma affects whether a volcano erupts explosively or quietly.	
	7	Movement of Earth's plates puts most of the stress on the rocks in the middle of the plates.	
	8	Most volcanic eruptions occur at, or near, plate boundaries.	
	9	The volcanic Hawaiian Islands are located near a plate boundary.	

section 1

Earthquakes

as you read

What You'll Learn
- **Explain** how earthquakes are caused by a buildup of strain in Earth's crust.
- **Compare and contrast** primary, secondary, and surface waves.
- **Recognize** earthquake hazards and how to prepare for them.

Why It's Important
Studying earthquakes will help you learn where they might occur and how you can prepare for their hazards.

Review Vocabulary
energy: the ability to cause change

New Vocabulary
- earthquake
- fault
- seismic wave
- focus
- epicenter
- seismograph
- magnitude
- tsunami
- seismic safe

What causes earthquakes?

If you've gone for a walk in the woods lately, maybe you picked up a stick along the way. If so, did you try to bend or break it? If you've ever bent a stick slowly, you might have noticed that it changes shape but usually springs back to normal form when you stop bending it. If you continue to bend the stick, you can do it for only so long before it changes permanently. When this elastic limit is passed, the stick may break, as shown in **Figure 1.** When the stick snaps, you can feel vibrations in the stick.

Elastic Rebound As hard as they seem, rocks act in much the same way when forces push or pull on them. If enough force is applied, rocks become strained, which means they change shape. They may even break, and the ends of the broken pieces may snap back. This snapping back is called elastic rebound.

Rocks usually change shape, or deform, slowly over long periods of time. As they are strained, potential energy builds up in them. This energy is released suddenly by the action of rocks breaking and moving. Such breaking, and the movement that follows, causes vibrations that move through rock or other earth materials. If they are large enough, these vibrations are felt as **earthquakes.**

Reading Check *What is an earthquake?*

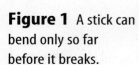

Figure 1 A stick can bend only so far before it breaks.

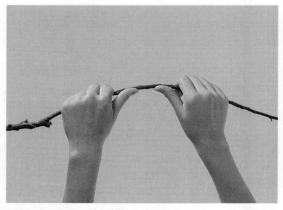

When a stick is bent, potential energy is stored in the stick.

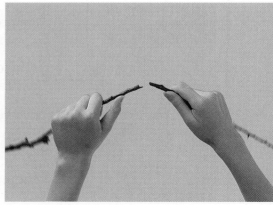

The energy is released as vibrations when the stick breaks.

210 CHAPTER 8 Earthquakes and Volcanoes

Figure 2 When rocks change shape by breaking, faults form. The type of fault formed depends on the type of stress exerted on the rock.

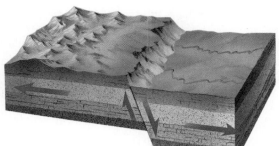

A When rocks are pulled apart, a normal fault may form.

B When rocks are compressed, a reverse fault may form.

C When rocks are sheared, a strike-slip fault may form.

Types of Faults When a section of rock breaks, rocks on either side of the break along which rocks move might move as a result of elastic rebound. The surface of such a break along which rocks move is called a **fault**. Several types of faults exist. The type that forms depends on how forces were applied to the rocks.

When rocks are pulled apart under tension forces, normal faults form, as shown in **Figure 2A**. Along a normal fault, rock above the fault moves down compared to rock below the fault. Compression forces squeeze rocks together, like an accordion. Compression might cause rock above a fault to move up compared to rock below the fault. This movement forms reverse faults, as shown in **Figure 2B**. As illustrated in **Figure 2C**, rock experiencing shear forces can break to form a strike-slip fault. Shear forces cause rock on either side of a strike-slip fault to move past one another in opposite directions along Earth's surface. You could infer the motion of a strike-slip fault while walking along and observing an offset feature, such as a displaced fence line, on Earth's surface.

Where do the forces come from that cause rocks to deform by bending or breaking? Why do faults form and why do earthquakes occur in certain areas? As you'll learn later in this chapter, forces inside Earth are caused by the constant motion of plates, or sections, of Earth's crust and upper mantle.

Mini LAB

Observing Deformation

WARNING: *Do not taste or eat any lab materials. Wash hands when finished.*

Procedure
1. Remove the wrapper from three bars of **taffy**.
2. Hold a bar of taffy lengthwise between your hands and gently push on it from opposite directions.
3. Hold another bar of taffy and pull it in opposite directions.

Analysis
1. Which of the procedures that you performed on the taffy involved applying tension? Which involved applying compression?
2. Infer how to apply a shear stress to the third bar of taffy.

SECTION 1 Earthquakes

Making Waves

Do you recall the last time you shouted for a friend to save you a seat on the bus? When you called out, energy was transmitted through the air to your friend, who interpreted the familiar sound of your voice as belonging to you. These sound waves were released by your vocal cords and were affected by your tongue and mouth. They traveled outward through the air. Earthquakes also release waves. Earthquake waves are transmitted through materials in Earth and along Earth's surface. Earthquake waves are called **seismic waves.** In the two-page activity, you'll make waves similar to seismic waves by moving a coiled spring toy.

Earthquake Focus and Epicenter Movement along a fault releases strain energy. Strain energy is potential energy that builds up in rock when it is bent. When this potential energy is released, it moves outward from the fault in the form of seismic waves. The point inside Earth where this movement first occurs and energy is released is called the **focus** of an earthquake, as shown in **Figure 3.** The point on Earth's surface located directly above the earthquake focus is called the **epicenter** of the earthquake.

Reading Check *Where is the focus of an earthquake located?*

Figure 3 During an earthquake, several types of seismic waves form. Primary and secondary waves travel in all directions from the focus and can travel through Earth's interior. Surface waves travel at shallow depths and along Earth's surface.
Infer Which seismic waves are the most destructive?

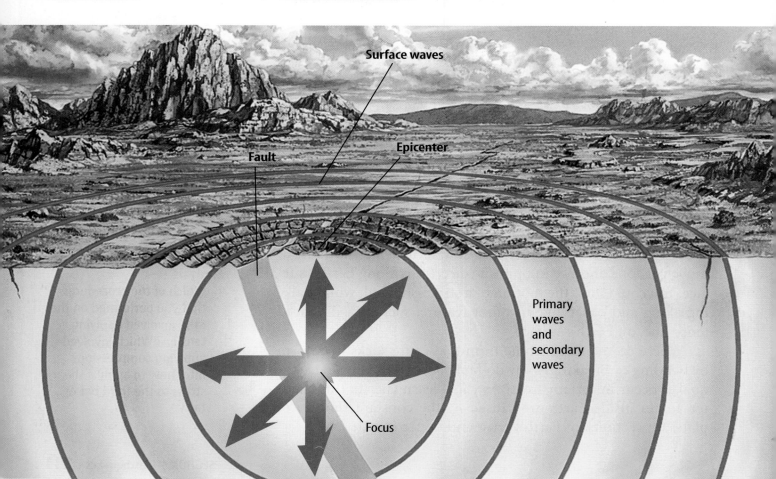

Seismic Waves After they are produced at the focus, seismic waves travel away from the focus in all directions, as illustrated in **Figure 3**. Some seismic waves travel throughout Earth's interior, and others travel along Earth's surface. The surface waves cause the most damage during an earthquake event.

Primary waves, also known as P-waves, travel the fastest through rock material by causing particles in the rock to move back and forth, or vibrate, in the same direction as the waves are moving. Secondary waves, known as S-waves, move through rock material by causing particles in the rock to vibrate at right angles to the direction in which the waves are moving. P- and S-waves travel through Earth's interior. Studying them has revealed much information about Earth's interior.

Surface waves are the slowest and largest of the seismic waves, and they cause most of the destruction during an earthquake. The movements of surface waves are complex. Some surface waves move along Earth's surface in a manner that moves rock and soil in a backward rolling motion. They have been observed moving across the land like waves of water. Some surface waves vibrate in a side-to-side, or swaying, motion parallel to Earth's surface. This motion can be particularly devastating to human-built structures.

Learning from Earthquakes

On your way to lunch tomorrow, suppose you were to walk twice as fast as your friend does. What would happen to the distance between the two of you as you walked to the lunchroom? The distance between you and your friend would become greater the farther you walked, and you would arrive first. Using this same line of reasoning, scientists use the different speeds of seismic waves and their differing arrival times to calculate the distance to an earthquake epicenter.

Earthquake Measurements Seismologists are scientists who study earthquakes and seismic waves. The instrument they use to obtain a record of seismic waves from all over the world is called a **seismograph,** shown in the top photo of **Figure 4**.

One type of seismograph has a drum holding a roll of paper on a fixed frame. A pendulum with an attached pen is suspended from the frame. When seismic waves are received at the station, the drum vibrates but the pendulum remains at rest. The pen on the pendulum traces a record of the vibrations on the paper. The height of the lines traced on the paper is a measure of the energy released by the earthquake, also known as its **magnitude.**

Figure 4 Scientists study seismic waves using seismographs located around the world.

This seismograph records incoming seismic waves using a fixed mass.

Some seismographs collect and store data on a computer.

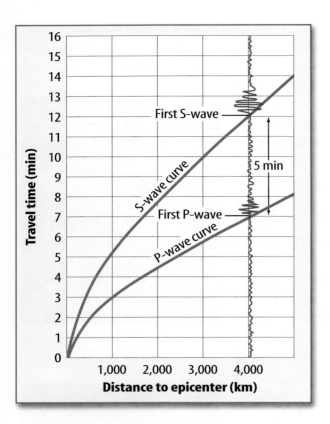

Figure 5 P-waves and S-waves travel at different speeds. These speeds are used to determine how close a seismograph station is to an earthquake.

Figure 6 After distances from at least three seismograph stations are determined, they are plotted as circles with radii equal to these distances on a map. The epicenter is the point at which the circles intersect.

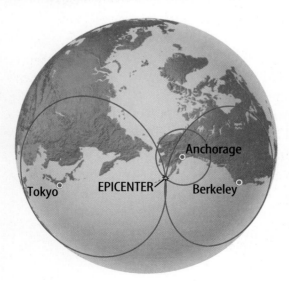

Epicenter Location When seismic-wave arrival times are recorded at a seismograph station, the distance from that station to the epicenter can be determined. The farther apart the arrival times for the different waves are, the farther away the earthquake epicenter is. This difference is shown by the graph in **Figure 5**. Using this information, scientists draw a circle with a radius equal to the distance from the earthquake for each of at least three seismograph stations, as illustrated in **Figure 6**. The point where the three circles meet is the location of the earthquake epicenter. Data from many stations normally are used to determine an epicenter location.

How strong are earthquakes?

As shown in **Table 1**, major earthquakes cause much loss of life. For example, on September 20, 1999, a major earthquake struck Taiwan, leaving more than 2,400 people dead, more than 8,700 injured, and at least 100,000 homeless. Sometimes earthquakes are felt and can cause destruction in areas hundreds of kilometers away from their epicenters. The Mexico City earthquake in 1985 is an example of this. The movement of the soft sediment underneath Mexico City caused extensive damage to this city, even though the epicenter was nearly 400 km away.

The Richter Scale Richter (RIHK tur) magnitude is based on measurements of amplitudes, or heights, of seismic waves as recorded on seismographs. Richter magnitude describes how much energy an earthquake releases. For each increase of 1.0 on the Richter scale, the amplitude of the highest recorded seismic wave increases by 10. However, about 32 times more energy is released for every increase of 1.0 on the scale. For example, an earthquake with a magnitude of 7.5 releases about 32 times more energy than one with a magnitude of 6.5, and the wave height for a 7.5-magnitude quake is ten times higher than for a quake with a magnitude of 6.5.

Earthquake Damage Another way to measure earthquakes is available. The modified Mercalli intensity scale measures the intensity of an earthquake. Intensity is a measure of the amount of structural and geologic damage done by an earthquake in a specific location. The range of intensities spans Roman numerals I through XII. The amount of damage done depends on several factors—the strength of the earthquake, the nature of the surface material, the design of structures, and the distance from the epicenter. An intensity-I earthquake would be felt only by a few people under ideal conditions. An intensity-VI earthquake would be felt by everyone. An intensity-XII earthquake would cause major destruction to human-built structures and Earth's surface. The 1994 earthquake in Northridge, California was a Richter magnitude 6.7, and its intensity was listed at IX. An intensity-IX earthquake causes considerable damage to buildings and could cause cracks in the ground.

Table 1 Strong Earthquakes			
Year	Location	Magnitude	Deaths
1989	Loma Prieta, CA	7.1	62
1990	Iran	7.7	50,000
1993	Guam	8.1	none
1993	Maharashtra, India	6.4	30,000
1994	Northridge, CA	6.7	61
1995	Kobe, Japan	6.8	5,378
1999	Taiwan	7.7	2,400
2000	Indonesia	7.9	103
2001	India	7.7	20,000
2003	Iran	6.6	30,000

Tsunamis Most damage from an earthquake is caused by surface waves. Buildings can crack or fall down. Elevated bridges and highways can collapse. However, people living near the seashore must protect themselves against another hazard from earthquakes. When an earthquake occurs on the ocean floor, the sudden movement pushes against the water and powerful water waves are produced. These waves can travel outward from the earthquake thousands of kilometers in all directions.

When these seismic sea waves, or **tsunamis,** are far from shore, their energy is spread out over large distances and great water depths. The wave heights of tsunamis are less than a meter in deep water, and large ships can ride over them and not even know it. In the open ocean, the speed of tsunamis can reach 950 km/h. However, when tsunamis approach land, the waves slow down and their wave heights increase as they encounter the bottom of the seafloor. This creates huge tsunami waves that can be as much as 30 m in height. Just before a tsunami crashes to shore, the water near a shoreline may move rapidly out toward the sea. If this should happen, there is immediate danger that a tsunami is about to strike. **Figure 7** illustrates the behavior of a tsunami as it approaches the shore.

Topic: Earthquake Magnitude
Visit blue.msscience.com for Web links to information about determining earthquake magnitudes.

Activity Create a table that compares the damage in dollars, the magnitude, and the general location of six recent earthquakes.

SECTION 1 Earthquakes **215**

NATIONAL GEOGRAPHIC VISUALIZING TSUNAMIS

Figure 7

The diagram below shows stages in the development of a tsunami. A tsunami is an ocean wave that is usually generated by an earthquake and is capable of inflicting great destruction.

▶ **TSUNAMI ALERT** The red dots on this map show the tide monitoring stations that make up part of the Tsunami Warning System for the Pacific Ocean. The map shows approximately how long it would take for tsunamis that originate at different places in the Pacific to reach Hawaii. Each ring represents two hours of travel time.

A The vibrations set off by a sudden movement along a fault in Earth's crust are transferred to the water's surface and spread across the ocean in a series of long waves.

B The waves travel across the ocean at speeds ranging from about 500 to 950 km/h

C When a tsunami wave reaches shallow water, friction slows it down and causes it to roll up into a wall of water—sometimes 30 m high—before it breaks against the shore.

Tsunami Warning System buoy

Earthquake Safety

You've just read about the destruction that earthquakes cause. Fortunately, there are ways to reduce the damage and the loss of life associated with earthquakes.

Learning the earthquake history of an area is one of the first things to do to protect yourself. If the area you are in has had earthquakes before, chances are it will again and you can prepare for that.

Is your home seismic safe? What could you do to make your home earthquake safe? As shown in **Figure 8,** it's a good idea to move all heavy objects to lower shelves so they can't fall on you. Make sure your gas hot-water heater and appliances are well secured. A new method of protecting against fire is to place sensors on your gas line that would shut off the gas when the vibrations of an earthquake are felt.

In the event of an earthquake, keep away from all windows and avoid anything that might fall on you. Watch for fallen power lines and possible fire hazards. Collapsed buildings and piles of rubble can contain many sharp edges, so keep clear of these areas.

Seismic-Safe Structures If a building is considered **seismic safe,** it will be able to stand up against the vibrations caused by most earthquakes. Residents in earthquake-prone areas are constantly improving the way structures are built. Since 1971, stricter building codes have been enforced in California. Older buildings have been reinforced. Many high-rise office buildings now stand on huge steel-and-rubber supports that could enable them to ride out the vibrations of an earthquake. Underground water and gas pipes are replaced with pipes that will bend during an earthquake. This can help prevent broken gas lines and therefore reduce damage from fires.

Seismic-safe highways have cement pillars with spiral reinforcing rods placed within them. One structure that was severely damaged in the 1989 Loma Prieta, California earthquake was Interstate Highway 880. The collapsed highway was due to be renovated to make it seismic safe. It was built in the 1950s and did not have spiral reinforcing rods in its concrete columns. When the upper highway went in one direction, the lower one went in the opposite direction. The columns collapsed and the upper highway came down onto the lower one.

Figure 8 You can minimize your risk of getting hurt by preparing for an earthquake in advance.

Placing heavy or breakable objects on lower shelves means they won't fall too far during an earthquake.

Vibration sensors on gas lines shut off the supply of gas automatically during an earthquake. **Draw Conclusions** What hazard can be prevented if the gas is turned off?

SECTION 1 Earthquakes **217**

Figure 9 One way to monitor changes along a fault is to detect any movement that occurs.

Predicting Earthquakes Imagine how many lives could be saved if only the time and location of a major earthquake could be predicted. Because most injuries from earthquakes occur when structures fall on top of people, it would help if people could be warned to move outside of buildings.

Researchers try to predict earthquakes by noting changes that precede them. That way, if such changes are observed again, an earthquake warning may be issued.

For example, movement along faults is monitored using laser-equipped, distance-measuring devices, such as the one shown in **Figure 9**. Changes in groundwater level or in electrical properties of rocks under stress have been measured by some scientists. Some people even study rock layers that have been affected by ancient earthquakes. Whether any of these studies will lead to the accurate and reliable prediction of earthquakes, no one knows. A major problem is that no single change in Earth occurs for all earthquakes. Each earthquake is unique.

Long-range forecasts predict whether an earthquake of a certain magnitude is likely to occur in a given area within 30 to 100 years. Forecasts of this nature are used to update building codes to make a given area more seismic safe.

section 1 review

Summary

What causes earthquakes?
- The sudden release of energy in rock and the resulting movement causes an earthquake.
- Faults are breaks in rocks along which movement occurs.

Making Waves
- The focus is where an earthquake occurs. The epicenter is directly above it.
- Earthquakes generate seismic waves.

How strong are earthquakes?
- The Richter Scale measures magnitude.
- The modified Mercalli scale measures intensity.

Earthquake Safety
- Structures can be made seismic safe.

Self Check

1. **Explain** what happens to rocks after their elastic limit is passed.
2. **Identify** Which seismic waves cause most of the damage during an earthquake?
3. **Apply** What has been done to make structures more seismic safe?
4. **Summarize** How can seismic waves be used to determine an earthquake's epicenter?
5. **Think Critically** Explain how a magnitude-8.0 earthquake could be classified as a low-intensity earthquake.

Applying Skills

6. **Make and Use Tables** Use **Table 1** to research the earthquakes that struck Indonesia in 2000, Loma Prieta, California in 1989, and Iran in 1990. Why was there such a great difference in the number of deaths?

section 2 Volcanoes

How do volcanoes form?

Much like air bubbles that are forced upward toward the bottom of an overturned bottle of denser syrup, molten rock material, or magma, is forced upward toward Earth's surface by denser surrounding rock. Rising magma eventually can lead to an eruption, where magma, solids, and gas are spewed out to form cone-shaped mountains called **volcanoes**. As magma flows onto Earth's surface through a vent, or opening, it is called **lava**. Volcanoes have circular holes near their summits called craters. Lava and other volcanic materials can be expelled through a volcano's crater.

Some explosive eruptions throw lava and rock thousands of meters into the air. Bits of rock or solidified lava dropped from the air are called tephra. Tephra varies in size from volcanic ash to cinders to larger rocks called bombs or blocks.

Where Plates Collide Some volcanoes form because of collision of large plates of Earth's crust and upper mantle. This process has produced a string of volcanic islands, much like those illustrated in **Figure 10,** which includes Montserrat. These islands are forming as plates made up of oceanic crust and mantle collide. The older and denser oceanic plate subducts, or sinks beneath, the less dense plate, as shown in **Figure 10.** When one plate sinks under another plate, rock in and above the sinking plate melts, forming chambers of magma. This magma is the source for volcanic eruptions that have formed the Caribbean Islands.

as you read

What You'll Learn
- **Explain** how volcanoes can affect people.
- **Describe** how types of materials are produced by volcanoes.
- **Compare** how three different volcano forms develop.

Why It's Important
Volcanic eruptions can cause serious consequences for humans and other organisms.

Review Vocabulary
plate: a large section of Earth's crust and rigid upper mantle that moves around on the asthenosphere

New Vocabulary
- volcano
- lava
- shield volcano
- cinder cone volcano
- composite volcano

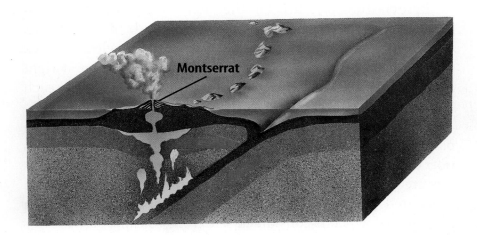

Figure 10 A string of Caribbean Islands known as the Lesser Antilles formed because of subduction. The island of Montserrat is among these.

Figure 11 Several volcanic hazards are associated with explosive activity.

Volcanic ash blanketing an area can cause collapse of structures or—when mixed with precipitation—mudflows.

Objects in the path of a pyroclastic flow are subject to complete destruction.

Mini LAB

Modeling an Eruption

Procedure

1. Place **red-colored gelatin** into a **self-sealing plastic bag** until the bag is half full.
2. Seal the bag and press the gelatin to the bottom of the bag.
3. Put a hole in the bottom of the bag with a **pin**.

Analysis

1. What parts of a volcano do the gelatin, the plastic bag, and the hole represent?
2. What force in nature did you mimic as you moved the gelatin to the bottom of the bag?
3. What factors in nature cause this force to increase and lead to an eruption?

Try at Home

Eruptions on a Caribbean Island Soufrière (soo free UR) Hills volcano on the island of Montserrat was considered dormant until recently. However, in 1995, Soufrière Hills volcano surprised its inhabitants with explosive activity. In July 1995, plumes of ash soared to heights of more than 10,000 m. This ash covered the capital city of Plymouth and many other villages, as shown at left in **Figure 11.**

Every aspect of a once-calm tropical life changed when the volcano erupted. Glowing avalanches and hot, boiling mudflows destroyed villages and shut down the main harbor of the island and its airport. During activity on July 3, 1998, volcanic ash reached heights of more than 14,000 m. This ash settled over the entire island and was followed by mudflows brought on by heavy rains.

Pyroclastic flows are another hazard for inhabitants of Montserrat. They can occur anytime on any side of the volcano. Pyroclastic flows are massive avalanches of hot, glowing rock flowing on a cushion of intensely hot gases, as shown at right in **Figure 11.** Speeds at which these flows travel can reach 200 km/h.

More than one half of Montserrat has been converted to a barren wasteland by the volcano. Virtually all of the farmland is now unusable, and most of the island's business and leisure centers are gone. Many of the inhabitants of the island have been evacuated to England, surrounding islands, or northern Montserrat, which is considered safe from volcanic activity.

Volcanic Risks According to the volcanic-risk map shown in **Figure 12,** inactive volcanic centers exist at Silver Hill, Centre Hill, and South Soufrière Hills. The active volcano, Soufrière Hills volcano, is located just north of South Soufrière Hills. The risk map shows different zones of the island where inhabitants still are able to stay and locations from which they have been evacuated. Twenty people who had ignored evacuation orders were killed by pyroclastic flows from the June 25, 1997, event. These are the first and only deaths that have occurred since July 1995.

Forms of Volcanoes

As you have learned, volcanoes can cause great destruction. However, volcanoes also add new rock to Earth's crust with each eruption. The way volcanoes add this new material to Earth's surface varies greatly. Different types of eruptions produce different types of volcanoes.

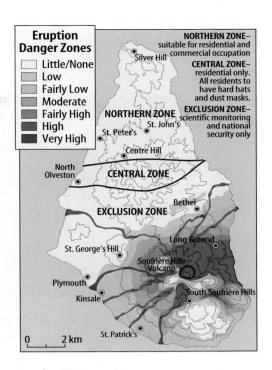

Figure 12 A volcanic risk map for Montserrat was prepared to warn inhabitants and visitors about unsafe areas on the island.

 **What determines how a volcano erupts?** Some volcanic eruptions are violent, while during others lava flows out quietly around a vent. The composition of the magma plays a big part in determining the manner in which energy is released during a volcanic eruption. Lava that contains more silica, which is a compound consisting of silicon and oxygen, tends to be thicker and is more resistant to flow. Lava containing more iron and magnesium and less silica tends to flow easily. The amount of water vapor and other gases trapped in the lava also influences how lava erupts.

When you shake a bottle of carbonated soft drink before opening it, the pressure from the gas in the drink builds up and is released suddenly when the container is opened. Similarly, steam builds pressure in magma. This pressure is released as magma rises toward Earth's surface and eventually erupts. Sticky, silica-rich lava tends to trap water vapor and other gases.

Water is carried down from the surface of Earth into the mantle when one plate subducts beneath another, as in the case of the Lesser Antilles volcanoes. In hotter regions of Earth's interior, part of a descending plate and nearby rock will melt to form magma. The magma produced is more silica rich than the rock that melts to form the magma. Superheated steam produces tremendous pressure in such thick, silica-rich magmas. After enough pressure builds up, an eruption occurs. The type of lava and the gases contained in that lava determine the type of eruption that occurs.

Topic: Montserrat Volcano
Visit blue.msscience.com for Web links to an update of data on Soufrière Hills volcano on Montserrat.

Activity Compare the recent activity of the Soufrière Hills volcano to another recently active volcano. Gather your findings into a table and include dates as well as amount of area destroyed in your report.

Figure 13 Volcanic landforms vary greatly in size and shape.

A The fluid nature of basaltic lava has produced extensive flows at Mauna Loa, Hawaii—the largest active volcano on Earth.

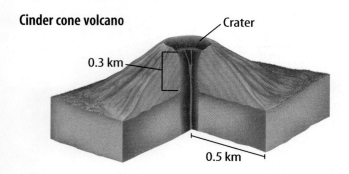

B Sunset Crater is small and steep along its flanks—typical of a cinder cone. Compare the scale given for Sunset Crater with that shown in Figure 13A.

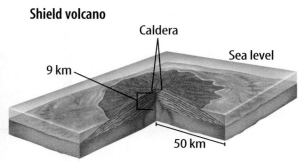

Shield Volcanoes Basaltic lava, which is high in iron and magnesium and low in silica, flows in broad, flat layers. The buildup of basaltic layers forms a broad volcano with gently sloping sides called a **shield volcano.** Shield volcanoes, shown in **Figure 13A,** are the largest type of volcano. They form where magma is being forced up from extreme depths within Earth, or in areas where Earth's plates are moving apart. The separation of plates enables magma to be forced upward to Earth's surface.

Reading Check *What materials are shield volcanoes composed of?*

Cinder Cone Volcanoes Rising magma accumulates gases on its way to the surface. When the gas builds up enough pressure, it erupts. Moderate to violent eruptions throw volcanic ash, cinders, and lava high into the air. The lava cools quickly in midair and the particles of solidified lava, ash, and cinders fall back to Earth. This tephra forms a relatively small cone of volcanic material called a **cinder cone volcano.** Cinder cones are usually less than 300 m in height and often form in groups near other larger volcanoes. Because the eruption is powered by the high gas content, it usually doesn't last long. After the gas is released, the force behind the eruption is gone. Sunset Crater, an example of a cinder cone near Flagstaff, Arizona, is shown in **Figure 13B.**

Composite Volcanoes Steep-sided mountains composed of alternating layers of lava and tephra are **composite volcanoes**. They sometimes erupt violently, releasing large quantities of ash and gas. This forms a tephra layer of solid materials. Then a quieter eruption forms a lava layer.

Composite volcanoes form where one plate sinks beneath another. Soufrière Hills volcano is an example of a composite volcano. Another volcanic eruption from a composite volcano was the May 1980 eruption of Mount St. Helens in the state of Washington. It erupted explosively, spewing ash that fell on regions hundreds of kilometers away from the volcano. A composite volcano is shown in **Figure 13C.**

C Composite cones are intermediate in size and shape compared to shield volcanoes and cinder cone volcanoes.

Mount Rainier, Washington

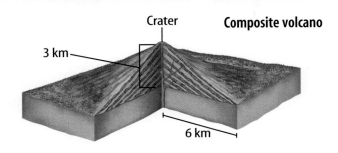

Fissure Eruptions Magma that is highly fluid can ooze from cracks or fissures in Earth's surface. This is the type of magma that usually is associated with fissure eruptions. The lava that erupts has a low viscosity, which means it can flow freely across the land to form flood basalts. Flood basalts that have been exposed to erosion for millions of years can become large, relatively flat landforms known as lava plateaus, as shown in **Figure 13D.** The Columbia River Plateau in the northwestern United States was formed about 15 million years ago when several fissures erupted and the flows built up layer upon layer.

D No modern example compares with the extensive flood basalts making up the Columbia River Plateau.

SECTION 2 Volcanoes **223**

Table 2 Seven Selected Eruptions in History

Volcano (Year)	Type	Eruptive Force	Silica Content	Gas Content	Eruption Products
Krakatau, Indonesia (1883)	composite	high	high	high	gas, cinders, ash
Katmai, Alaska (1912)	composite	high	high	high	lava, ash, gas
Paricutín, Mexico (1943)	cinder cone	moderate	high	low	gas, cinders, ash
Helgafell, Iceland (1973)	cinder cone	moderate	low	high	gas, ash
Mount St. Helens, Washington (1980)	composite	high	high	high	gas, ash
Kilauea Iki, Hawaii (1989)	shield	low	low	low	gas, lava
Soufrière Hills, Montserrat (1995–)	composite	high	high	high	gas, ash, rocks

You have read about some variables that control the type of volcanic eruption that will occur. Examine **Table 2** for a summary of these important factors. In the next section, you'll learn that the type of magma produced is associated with properties of Earth's plates and how these plates interact.

section 2 review

Summary

How do volcanoes form?
- Some volcanoes form as two or more large plates collide.
- The Caribbean Islands formed from volcanic eruptions as one plate sinks under another plate.

Forms of volcanoes
- Lava high in silica produces explosive eruptions, lava low in silica but high in iron and magnesium produces more fluid eruptions.
- The amount of water vapor and gases impacts how volcanoes erupt.
- The types of volcanoes include shield, cinder cone and composite volcanoes, and fissure eruptions.

Self Check

1. **Identify** Which types of lava eruptions cover the largest area on Earth's surface?
2. **Describe** the processes that have led to the formation of the Soufrière Hills volcano.
3. **Explain** why a cinder cone has steep sides?
4. **List** What types of materials are volcanoes like Mount St. Helens made of?
5. **Think Critically** Why is silica-rich magma explosive?

Applying Math

6. **Solve One-Step Equations** Mauna Loa in Hawaii rises 9 km above the seafloor. Sunset Crater in Arizona rises to an elevation of 300 m. How many times higher is Mauna Loa than Sunset Crater?

Disruptive Eruptions

A volcano's structure can influence how it erupts. Some volcanoes have only one central vent, while others have numerous fissures that allow lava to escape. Materials in magma influence its viscosity, or how it flows. If magma is a thin fluid—not viscous—gases can escape easily. But if magma is thick—viscous—gases cannot escape as easily. This builds up pressure within a volcano.

▶ Real-World Question

What determines the explosiveness of a volcanic eruption?

Goals
- **Infer** how a volcano's opening contributes to how explosive an eruption might be.
- **Hypothesize** how the viscosity of magma can influence an eruption.

Materials
plastic film canisters
baking soda ($NaHCO_3$)
vinegar (CH_3COOH)
50-mL graduated cylinder
teaspoon

Safety Precautions

This lab should be done outdoors. Goggles must be worn at all times. The caps of the film canisters fly off due to the chemical reaction that occurs inside them. Never put anything in your mouth while doing the experiment.

▶ Procedure

1. Watch your teacher demonstrate this lab before attempting to do it yourself.
2. Add 15 mL of vinegar to a film canister.
3. Place 1 teaspoon of baking soda in the film canister's lid, using it as a type of plate.
4. Place the lid on top of the film canister, but do not cap it. The baking soda will fall into the vinegar. Move a safe distance away. Record your observations in your Science Journal.
5. Clean out your film canister and repeat the lab, but this time cap the canister quickly and tightly. Record your observations.

▶ Conclude and Apply

1. **Identify** Which of the two labs models a more explosive eruption?
2. **Explain** Was the pressure greater inside the canister during the first or second lab? Why?
3. **Explain** What do the bubbles have to do with the explosion? How do they influence the pressure in the container?
4. **Infer** If the vinegar were a more viscous substance, how would the eruption be affected?

Communicating Your Data

Research three volcanic eruptions that have occurred in the past five years. Compare each eruption to one of the eruption styles you modeled in this lab. Communicate to your class what you learn.

LAB **225**

Section 3

Earthquakes, Volcanoes, and Plate Tectonics

as you read

What You'll Learn
- **Explain** how the locations of volcanoes and earthquake epicenters are related to tectonic plate boundaries.
- **Explain** how heat within Earth causes Earth's plates to move.

Why It's Important
Most volcanoes and earthquakes are caused by the motion and interaction of Earth's plates.

Review Vocabulary
asthenosphere: plasticlike layer of mantle under the lithosphere

New Vocabulary
- rift
- hot spot

Earth's Moving Plates

At the beginning of class, your teacher asks for volunteers to help set up the cafeteria for a special assembly. You and your classmates begin to move the tables carefully, like the students shown in **Figure 14.** As you move the tables, two or three of them crash into each other. Think about what could happen if the students moving those tables kept pushing on them. For a while one or two of the tables might keep another from moving. However, if enough force were used, the tables would slide past one another. One table might even slide up on top of another. It is because of this possibility that your teacher has asked that you move the tables carefully.

The movement of the tables and the possible collisions among them is like the movement of Earth's crust and uppermost mantle, called the lithosphere. Earth's lithosphere is broken into separate sections, or plates. When these plates move around, they collide, move apart, or slide past each other. The movement of these plates can cause vibrations known as earthquakes and can create conditions that cause volcanoes to form.

Figure 14 Like the tables pictured here, Earth's plates are in contact with one another and can slide beneath each other. The way Earth's plates interact at boundaries is an important factor in the locations of earthquakes and volcanoes.

226 CHAPTER 8 Earthquakes and Volcanoes

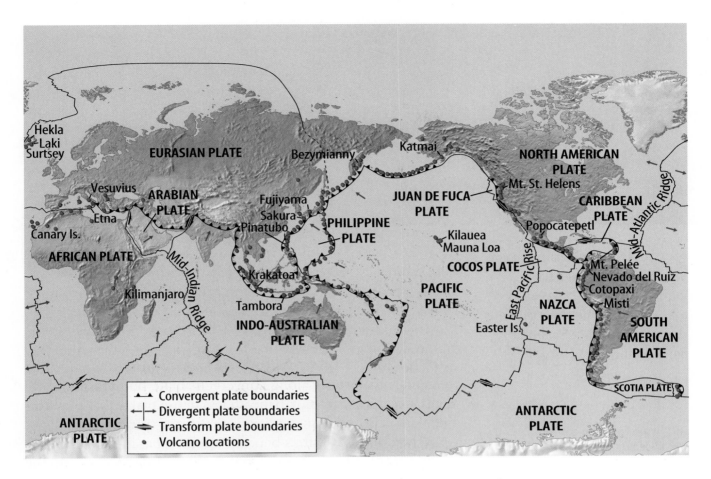

Figure 15 Earth's lithosphere is divided into about 13 major plates. Where plates collide, separate, and slip past one another at plate boundaries, interesting geological activity results.

Where Volcanoes Form

A plot of the location of plate boundaries and volcanoes on Earth shows that most volcanoes form along plate boundaries. Examine the map in **Figure 15.** Can you see how this indicates that plate tectonics and volcanic activity are related? Perhaps the energy involved in plate tectonics is causing magma to form deep under Earth's surface. You'll recall that the Soufrière Hills volcano formed where plates converge. Plate movement often explains why volcanoes form in certain areas.

Divergent Plate Boundaries Tectonic plates move apart at divergent plate boundaries. As the plates separate, long cracks called **rifts** form between them. Rifts contain fractures that serve as passageways for magma originating in the mantle. Rift zones account for most of the places where lava flows onto Earth's surface. Fissure eruptions often occur along rift zones. These eruptions form lava that cools and solidifies into basalt, the most abundant type of rock in Earth's crust.

 *Where does magma along divergent boundaries originate?*

SECTION 3 Earthquakes, Volcanoes, and Plate Tectonics

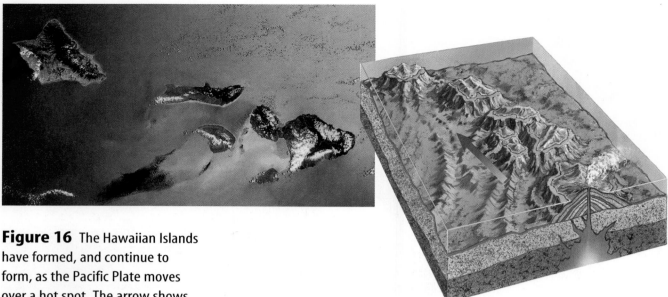

Figure 16 The Hawaiian Islands have formed, and continue to form, as the Pacific Plate moves over a hot spot. The arrow shows that the Pacific Plate is moving north-northwest.

Melting Points The melting point of a substance is the temperature at which a solid changes to a liquid. Depending on the substance, a change in pressure can raise or lower the melting point. Do research to find out how pressure affects the formation of magma in a mantle plume in a process called decompression melting.

Convergent Plate Boundaries A common location for volcanoes to form is along convergent plate boundaries. More dense oceanic plates sink beneath less dense plates that they collide with. This sets up conditions that form volcanoes.

When one plate sinks beneath another, basalt and sediment on an oceanic plate move down into the mantle. Water from the sediment and altered basalt lowers the melting point of the surrounding rock. Heat in the mantle causes part of the sinking plate and overlying mantle to melt. This melted material then is forced upward. Volcanoes have formed in this way all around the Pacific Ocean, where the Pacific Plate, among others, collides with several other plates. This belt of volcanoes surrounding the Pacific Ocean is called the Pacific Ring of Fire.

Hot Spots The Hawaiian Islands are volcanic islands that have not formed along a plate boundary. In fact, they are located well within the Pacific Plate. What process causes them to form? Large bodies of magma, called **hot spots,** are forced upward through Earth's mantle and crust, as shown in **Figure 16**. Scientists suggest that this is what is occurring at a hot spot that exists under the present location of Hawaii.

Reading Check *What is a hot spot?*

Volcanoes on Earth usually form along rift zones, subduction zones (where one plate sinks beneath another), or over hot spots. At each of these locations, magma from deep within Earth is forced upward toward the surface. Lava breaks through and flows out, where it piles up into layers or forms a volcanic cone.

Moving Plates Cause Earthquakes

Place two notebooks on your desk with the page edges facing each other. Then push them together slowly. The individual sheets of paper gradually will bend upward from the stress. If you continue to push on the notebooks, one will slip past the other suddenly. This sudden movement is like an earthquake.

Now imagine what would happen if tectonic plates were moving like the notebooks. What would happen if the plates collided and stopped moving? Forces generated by the locked-up plates would cause strain to build up. Both plates would begin to deform until the elastic limit was passed. The breaking and elastic rebound of the deformed material would produce vibrations felt as earthquakes.

Earthquakes often occur where tectonic plates come together at a convergent boundary, where tectonic plates move apart at a divergent boundary, and where tectonic plates grind past each other, called a transform boundary.

Earthquake Locations If you look at a map of earthquakes, you'll see that most occur in well-known belts. About 80 percent of them occur in the Pacific Ring of Fire—the same belt in which many of Earth's volcanoes occur. If you compare **Figure 17** with **Figure 15,** you will notice a definite relationship between earthquake epicenters and tectonic plate boundaries. Movement of the plates produces forces that generate the energy to cause earthquakes.

Friction Friction is a force that opposes the motion of two objects in contact. Do research to find out different types of friction in a literary and figurative sense.

Figure 17 Locations of earthquakes that have occurred between 1990 and 2000 are plotted below.

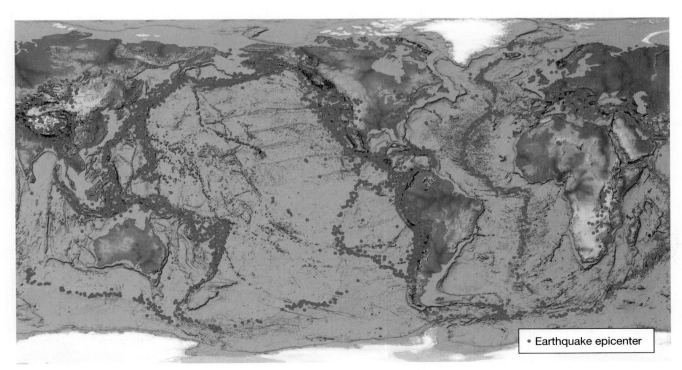

• Earthquake epicenter

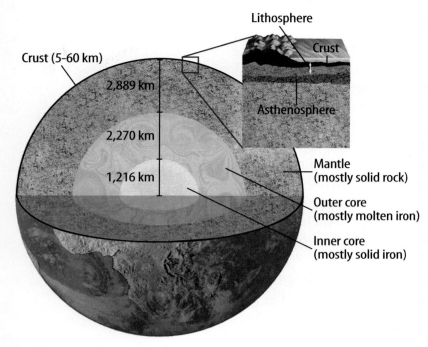

Figure 18 Seismic waves generated by earthquakes allow researchers to figure out the structure and composition of Earth's layers.

Earth's Plates and Interior

Researchers have learned much about Earth's interior and plate tectonics by studying seismic waves. The way in which seismic waves pass through a material depends on the properties of that material. Seismic wave speeds, and how they travel through different levels in the interior, have allowed scientists to map out the major layers of Earth, as shown in **Figure 18**.

For example, the asthenosphere was discovered when seismologists noted that seismic waves slowed when they reached the base of the lithosphere of the Earth. This partially molten layer forms a warmer, softer layer over which the colder, brittle, rocky plates move.

Applying Math — Calculate

P-WAVE TRAVEL TIME There is a relationship between the density of a region in Earth and the velocity of P-waves. How can you calculate the time it would take P-waves to travel 100 km in the crust of Earth?

Density and Wave Velocity

Region	Density	P-Wave Velocity
Crust	2.8 g/cm^3	6 km/s
Upper mantle	3.3 g/cm^3	8 km/s

Solution

1. *This is what you know:*
   - velocity: $v = 6$ km/s
   - distance: $d = 100$ km

2. *This is what you need to find:* How long would it take a P wave to travel?

3. *This is the procedure you need to use:*
   - $t = d/v$
   - $t = (100 \text{ km})/(6 \text{ km/s}) = 16.7$ s

4. *Check your answer:* Solve $v = d/t = (100 \text{ km})/(16.7 \text{ s}) = 6$ km/s

Practice Problems

1. Calculate the time it takes P-waves to travel 300 km in the upper mantle.
2. How long will it take a P-wave to travel 500 km in the crust?

For more practice, visit blue.msscience.com/math_practice

What is driving Earth's plates? There are several hypotheses about where all the energy comes from to power the movement of Earth's plates.

In one case, mantle material deep inside Earth is heated by Earth's core. This hot, less dense rock material is forced toward the surface. The hotter, rising mantle material eventually cools. The cooler material then sinks into the mantle toward Earth's core, completing the convection current. Convection currents inside Earth, shown in **Figure 19,** provide the mechanism for plate motion, which then produces the conditions that cause volcanoes and earthquakes. Sometimes magma is forced up directly within a plate. Volcanic activity in Yellowstone National Park is caused by a hot spot beneath the North American Plate. Such hot spots might be related to larger-scale convection in Earth's mantle.

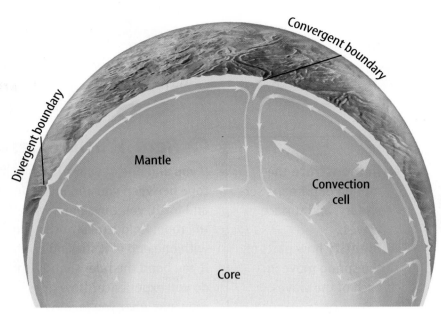

Figure 19 Convection of material in Earth's interior drives the motion of tectonic plates.

section 3 review

Summary

Earth's Moving Plates
- Earth's lithosphere is broken into plates that move around the planet.

Where Volcanoes Form
- Plates move apart at divergent plate boundaries, creating fissure eruptions.
- Plates collide at convergent plate boundaries.
- Many volcanoes form at convergent plate boundaries.
- Volcanoes may also form along rift zones, subduction zones, or over hot spots.

Moving Plates Cause Earthquakes
- Earthquakes often form at plate boundaries.
- Seismic waves have been used to determine the characteristics of Earth's interior.
- Convection currently may drive tectonic plate movement.

Self Check

1. **Identify** Along which type of plate boundary has the Soufrière Hills volcano formed?
2. **Predict** At which type of plate boundary does rift-volcanism occur?
3. **Explain** how volcanoes in Hawaii form.
4. **Recognize Cause and Effect** Why do most deep earthquakes occur at convergent boundaries?
5. **Think Critically** Subduction occurs where plates converge. This causes water-rich sediment and altered rock to be forced down to great depths. Explain how water can help form a volcano.

Applying Skills

6. **Form Hypotheses** Write a hypothesis concerning the type of lava that will form a hot spot volcano. Consider that magma in a hot spot comes from deep inside Earth's mantle.

Seismic Waves

Goals
- **Demonstrate** the motion of primary, secondary, and surface waves.
- **Identify** how parts of the spring move in each of the waves.

Materials
coiled spring toy
yarn or string
metric ruler

Safety Precautions

Real-World Question

If you and one of your friends hold a long piece of rope between you and move one end of the rope back and forth, you can send a wave through the length of the rope. Hold a ruler at the edge of a table securely with one end of it sticking out from the table's edge. If you bend the ruler slightly and then release it, what do you experience? How does what you see in the rope and what you feel in the ruler relate to seismic waves? How do seismic waves differ?

Procedure

1. Copy the following data table in your Science Journal.
2. Tie a small piece of yarn or string to every tenth coil of the spring.
3. Place the spring on a smooth, flat surface. Stretch it so it is about 2 m long (1 m for shorter springs).
4. Hold your end of the spring firmly. Make a wave by having your partner snap the spring from side to side quickly.
5. **Record** your observations in your Science Journal and draw the wave you and your partner made in the data table.
6. Have your lab partner hold his or her end of the spring firmly. Make a wave by quickly pushing your end of the spring toward your partner and bringing it back to its original position.

Comparing Seismic Waves			
Observation of Wave	Observation of Yarn or String	Drawing	Wave Type
	Do not write in this book.		

232 CHAPTER 8 Earthquakes and Volcanoes

Using Scientific Methods

7. **Record** your observations of the wave and of the yarn or string and draw the wave in the data table.

8. Have your lab partner hold his or her end of the spring firmly. Move the spring off of the table. Gently move your end of the spring side to side while at the same time moving it in a rolling motion, first up and away and then down and toward your partner.

9. **Record** your observations and draw the wave in the data table.

▶ Conclude and Apply

1. Based on your observations, determine which of the waves that you and your partner have generated demonstrates a primary, or pressure, wave. Record in your data table and explain why you chose the wave you did.

2. Do the same for the secondary, or shear wave, and for the surface wave. Explain why you chose the wave you did.

3. **Explain** Based on your observations of wave motion, which of the waves that you and your partner generated probably would cause the most damage during an earthquake?

4. **Observe** What was the purpose of the yarn or string?

5. **Compare and contrast** the motion of the yarn or string when primary and secondary waves travel through the spring. Which of these waves is a compression wave? Explain your answer.

6. **Compare and Contrast** Which wave most closely resembled wave motion in a body of water? How was it different? Explain.

Communicating Your Data

Compare your conclusions with those of other students in your class. **For more help, refer to the** Science Skill Handbook.

TIME SCIENCE AND HISTORY

SCIENCE CAN CHANGE THE COURSE OF HISTORY!

Quake

The 1906 San Francisco earthquake taught people valuable lessons

It struck without warning. "We found ourselves staggering and reeling. It was as if the earth was slipping gently from under our feet. Then came the sickening swaying of the earth that threw us flat upon our faces. We struggled in the street. We could not get on our feet. Then it seemed as though my head were split with the roar that crashed into my ears. Big buildings were crumbling as one might crush a biscuit in one's hand."

That's how survivor P. Barrett described the San Francisco earthquake of 1906. Duration of the quake on the morning of April 18—one minute. Yet, in that short time, Earth opened a gaping hole stretching more than 430 km. The tragic result was one of the worst natural disasters in U.S. history.

Fires caused by falling chimneys and fed by broken gas mains raged for three days. Despite the estimated 3,000 deaths and enormous devastation to San Francisco, the earthquake did have a positive effect. It led to major building changes that would help protect people and property from future quakes.

Computers analyze information from seismographs that have helped to map the San Andreas Fault—the area along which many California earthquakes take place. This information is helping scientists better understand how and when earthquakes might strike.

The 1906 quake also has led to building codes that require stronger construction materials for homes, offices, and bridges. Laws have been passed saying where hospitals, homes, and nuclear power plants can be built—away from soft ground and away from the San Andreas Fault.

Even today, scientists can't predict an earthquake. But thanks to what they learned from the 1906 quake—and others—people are safer today than ever before.

Write Prepare a diary entry pretending to be a person who experienced the 1906 San Francisco earthquake. Possible events to include in your entry: What were you doing at 5:15 A.M.? What began to happen around you? What did you see and hear?

For more information, visit
blue.msscience.com/time

chapter 8 Study Guide

Reviewing Main Ideas

Section 1 Earthquakes

1. Earthquakes occur whenever rocks inside Earth pass their elastic limit, break, and experience elastic rebound.

2. Seismic waves are vibrations inside Earth. P- and S-waves travel in all directions away from the earthquake focus. Surface waves travel along the surface.

3. Earthquakes are measured by their magnitudes—the amount of energy they release—and by their intensity—the amount of damage they produce.

Section 2 Volcanoes

1. The Soufrière Hills volcano is a composite volcano formed by converging plates.

2. The way a volcano erupts is determined by the composition of the lava and the amount of water vapor and other gases in the lava.

3. Three different forms of volcanoes are shield volcanoes, cinder cone volcanoes, and composite volcanoes.

Section 3 Earthquakes, Volcanoes, and Plate Tectonics

1. The locations of volcanoes and earthquake epicenters are related to the locations of plate boundaries.

2. Volcanoes occur along rift zones, subduction zones, and at hot spots.

3. Most earthquakes occur at convergent, divergent, and transform plate boundaries.

Visualizing Main Ideas

Copy and complete the following table comparing characteristics of shield, composite, and cinder cone volcanoes.

Volcanoes

Characteristic	Shield Volcano	Cinder Cone Volcano	Composite Volcano
Relative size	large		
Nature of eruption			moderate to high eruptive force
Materials extruded	lava, gas	cinders, gas	
Composition of lava			high silica
Ability of lava to flow		low	variable

chapter 8 Review

Using Vocabulary

cinder cone volcano p. 222
composite volcano p. 223
earthquake p. 210
epicenter p. 212
fault p. 211
focus p. 212
hot spot p. 228
lava p. 219
magnitude p. 213
rift p. 227
seismic safe p. 217
seismic wave p. 212
seismograph p. 213
shield volcano p. 222
tsunami p. 215
volcano p. 219

Explain the differences between the vocabulary words in each of the following sets.

1. fault—earthquake
2. shield volcano—composite volcano
3. focus—epicenter
4. seismic wave—seismograph
5. tsunami—seismic wave
6. epicenter—earthquake
7. cinder cone volcano—shield volcano

Checking Concepts

Choose the word or phrase that best answers the question.

8. Which type of plate boundary caused the formation of the Soufrière Hills volcano?
    A) divergent   C) rift
    B) transform   D) convergent

9. What is a cone-shaped mountain that is built from layers of lava?
    A) volcano   C) vent
    B) lava flow   D) crater

10. What is the cause of the volcanoes on Hawaii?
    A) rift zone
    B) hot spot
    C) divergent plate boundary
    D) convergent plate boundary

11. Which type of lava flows easily?
    A) silica-rich   C) basaltic
    B) composite   D) smooth

12. Which type of volcano is built from alternating layers of lava and tephra?
    A) shield   C) lava dome
    B) cinder cone   D) composite

13. Which type of volcano is relatively small with steep sides?
    A) shield   C) lava dome
    B) cinder cone   D) composite

14. Which seismic wave moves through Earth at the fastest speed?
    A) primary wave
    B) secondary wave
    C) surface wave
    D) tsunami

15. Which of the following is a wave of water caused by an earthquake under the ocean?
    A) primary wave
    B) secondary wave
    C) surface wave
    D) tsunami

Use the illustration below to answer question 16.

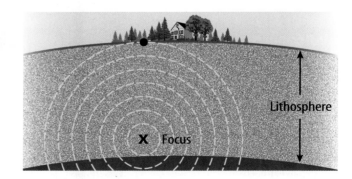

16. What is the point on Earth's surface directly above an earthquake's focus?
    A) earthquake center
    B) epicenter
    C) fault
    D) focus

chapter 8 Review

Thinking Critically

17. **Infer** Why does the Soufrière Hills volcano erupt so explosively?

18. **Compare and contrast** composite and cinder cone volcanoes.

19. **Explain** how the composition of magma can affect the way a volcano erupts.

20. **Evaluate** What factors determine an earthquake's intensity on the modified Mercalli scale?

21. **Compare and contrast** magnitude and intensity.

22. **Make Models** Select one of the three forms of volcanoes and make a model, using appropriate materials.

23. **Draw Conclusions** You are flying over an area that has just experienced an earthquake. You see that most of the buildings are damaged or destroyed and much of the surrounding countryside is disrupted. What level of intensity would you conclude for this earthquake?

24. **Concept Map** Copy and complete this concept map on examples of features produced along plate boundaries. Use the following terms: *Mid-Atlantic Ridge, Soufrière Hills volcano, divergent, San Andreas Fault, convergent,* and *transform.*

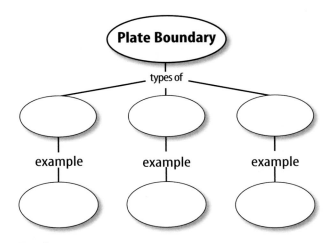

Performance Activities

25. **Oral Presentation** Research the earthquake or volcano history of your state or community. Find out how long ago your area experienced earthquake- or volcano-related problems. Present your findings in a speech to your class.

Applying Math

Use the graph below to answer questions 26 and 27.

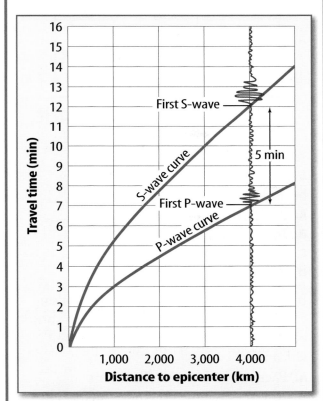

26. **Earthquake Epicenter** If a P-wave arrives at a seismograph station at 9:07 AM and the S-wave arrives at the same seismograph station at 9:09 AM, about how far is that station from the epicenter of the earthquake?

27. **Arrival Time** A seismograph station is 2,500 km from the epicenter of an earthquake. What is the difference in time between the P-wave arrival time and the S-wave arrival time?

CHAPTER REVIEW **237**

Chapter 8 Standardized Test Practice

Part 1 Multiple Choice

Record your answers on the answer sheet provided by your teacher or on a sheet of paper.

Use the table below to answer questions 1 and 2.

Plate Boundaries		
Plate	Number of convergent boundaries	Number of divergent boundaries
African	1	4
Antarctic	1	2
Indo-Australian	4	2
Eurasian	4	1
North American	2	1
Pacific	6	2
South American	2	1

1. Which plate has the most spreading boundaries?
   A. African
   B. Indo-Australian
   C. Pacific
   D. Antarctic

2. If composite volcanoes often form along convergent boundaries, which plate should be surrounded by the most composite volcanoes?
   A. Pacific   C. Eurasian
   B. Antarctic   D. Indo-Australian

3. Which of the following best describes a fault?
   A. the point on Earth's surface located directly above the earthquake focus
   B. the point inside Earth where movement first occurs during an earthquake
   C. the surface of a break in a rock along which there is movement
   D. the snapping back of a rock that has been strained by force

4. Waves created by earthquakes that travel through Earth's interior and along Earth's surface are called
   A. sound waves.   C. light waves.
   B. energy waves.   D. seismic waves.

5. Volcanoes are associated with all of the following areas EXCEPT
   A. rift zones.   C. subduction zones.
   B. epicenters.   D. hot spots.

Use the figure below to answer question 6 and 7.

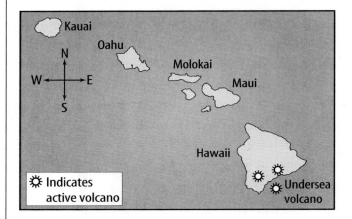

6. In which direction is the Pacific Plate moving?
   A. north-northwest
   B. north-northeast
   C. south-southwest
   D. south-southeast

7. Which of the following islands is the oldest?
   A. Kauai   C. Maui
   B. Molokai   D. Hawaii

Test-Taking Tip

Watch the Time If you are taking a timed test, keep track of time during the test. If you find you're spending too much time on a multiple-choice question, mark your best guess and move on.

Part 2 | Short Response/Grid In

Record your answers on the answer sheet provided by your teacher or on a sheet of paper.

8. What is an earthquake?

Use the figure below to answer questions 9 and 10.

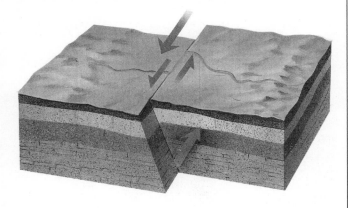

9. Identify the type of fault shown here.

10. Explain how this type of fault is formed.

11. What is a tsunami? What happens when a tsunami enters shallow water?

12. The lava of a particular volcano is high in silica and water vapor and other gases. What kind of eruptive force will likely result from this volcano?

13. Suppose a tsunami begins near the Aleutian Islands in Alaska. The wave reaches the Hawaiian Islands, a distance of 3800 km, 5 hours later. At what speed is the wave traveling?

14. What is elastic rebound? How does it relate to strain energy and earthquakes?

15. Describe a volcanic crater. Where is it located? What is its shape?

16. What is a seismograph? How does it work?

Part 1 | Multiple Choice

Record your answers on a sheet of paper.

17. Identify the characteristics associated with a shield volcano. How does a shield volcano form?

18. Compare and contrast divergent plate boundaries and convergent plate boundaries.

19. Explain how convection currents may be related to plate tectonics.

20. Describe each of the three types of seismic waves.

Use the figure below to answer question 21.

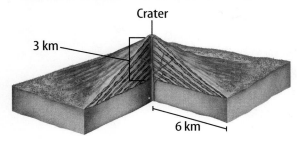

21. Which type of volcano is shown in the figure? Explain how you know this. Where does this type of volcano form?

22. Explain the relationship between faults and earthquakes.

23. Some surface waves vibrate in a side-to-side, or swaying, motion parallel to Earth's surface. Why is this type of motion so devastating to human-built structures?

24. What type of magma is associated with fissure eruptions? What type of features can these eruptions form?

chapter 9

Clues to Earth's Past

The BIG Idea

Fossils, along with the relative ages and absolute ages of rocks, provide evidence of past life, climates, and environments on Earth.

SECTION 1
Fossils
Main Idea Fossils are used as evidence of past life on Earth.

SECTION 2
Relative Ages of Rocks
Main Idea Relative ages of rocks can be determined by examining their locations within a sequence of rock layers.

SECTION 3
Absolute Ages of Rocks
Main Idea Absolute ages of rocks can be determined by using properties of the atoms that make up materials.

Reading the Past

The pages of Earth's history, much like the pages of human history, can be read if you look in the right place. Unlike the pages of a book, the pages of Earth's past are written in stone. In this chapter you will learn how to read the pages of Earth's history to understand what the planet was like in the distant past.

Science Journal   List three fossils that you would expect to find a million years from now in the place you live today.

Start-Up Activities

Clues to Life's Past

Fossil formation begins when dead plants or animals are buried in sediment. In time, if conditions are right, the sediment hardens into sedimentary rock. Parts of the organism are preserved along with the impressions of parts that don't survive. Any evidence of once-living things contained in the rock record is a fossil.

1. Fill a small jar (about 500 mL) one-third full of plaster of paris. Add water until the jar is half full.
2. Drop in a few small shells.
3. Cover the jar and shake it to model a swift, muddy stream.
4. Now model the stream flowing into a lake by uncovering the jar and pouring the contents into a paper or plastic bowl. Let the mixture sit for an hour.
5. Crack open the hardened plaster to locate the model fossils.
6. **Think Critically** Remove the shells from the plaster and study the impressions they made. In your Science Journal, list what the impressions would tell you if found in a rock.

Age of Rocks Make the following Foldable to help you understand how scientists determine the age of a rock.

STEP 1 Fold a sheet of paper in half lengthwise.

STEP 2 Fold paper down 2.5 cm from the top. (Hint: From the tip of your index finger to your middle knuckle is about 2.5 cm.)

STEP 3 Open and draw lines along the 2.5-cm fold. Label as shown.

Summarize in a Table As you read the chapter, in the left column, list four different ways in which one could determine the age of a rock. In the right column, note whether each method gives an absolute or a relative age.

Preview this chapter's content and activities at
blue.msscience.com

Get Ready to Read

Take Notes

1 Learn It! The best way for you to remember information is to write it down, or take notes. Good note-taking is useful for studying and research. When you are taking notes, it is helpful to
- phrase the information in your own words;
- restate ideas in short, memorable phrases;
- stay focused on main ideas and only the most important supporting details.

2 Practice It! Make note-taking easier by using a chart to help you organize information clearly. Write the main ideas in the left column. Then write at least three supporting details in the right column. Read the text from Section 1 of this chapter under the heading *Conditions Needed for Fossil Formation,* page 243. Then take notes using a chart, such as the one below.

Main Idea	Supporting Details
	1. 2. 3. 4. 5.
	1. 2. 3. 4. 5.

3 Apply It! As you read this chapter, make a chart of the main ideas. Next to each main idea, list at least three supporting details.

Target Your Reading

Use this to focus on the main ideas as you read the chapter.

Reading Tip

Read one or two paragraphs first and take notes after you read. You are likely to take down too much information if you take notes as you read.

① Before you read the chapter, respond to the statements below on your worksheet or on a numbered sheet of paper.
- Write an **A** if you **agree** with the statement.
- Write a **D** if you **disagree** with the statement.

② After you read the chapter, look back to this page to see if you've changed your mind about any of the statements.
- If any of your answers changed, explain why.
- Change any false statements into true statements.
- Use your revised statements as a study guide.

Science Online
Print out a worksheet of this page at blue.msscience.com

Before You Read A or D		Statement	After You Read A or D
	1	All fossils are made from the hard parts of animals.	
	2	Fossils can be used as evidence to show that past climates and environments have changed.	
	3	A trace fossil is the outline, or copy, of a fossil.	
	4	Sediment typically accumulates in horizontal beds, which can later form layers of sedimentary rock.	
	5	The relative age of a rock layer indicates whether the layer is older or younger when compared to other rock layers.	
	6	The principle of superposition refers to a high concentration of fossils within a small area.	
	7	Most sequences of rock layers are complete.	
	8	Geologists often can match up, or correlate, layers of rock over great distances.	
	9	The absolute age of a material refers to the actual age, in years, of the material.	

242 B

section 1
Fossils

as you read

What You'll Learn
- **List** the conditions necessary for fossils to form.
- **Describe** several processes of fossil formation.
- **Explain** how fossil correlation is used to determine rock ages.
- **Determine** how fossils can be used to explain changes in Earth's surface, life forms, and environments.

Why It's Important
Fossils help scientists find oil and other sources of energy necessary for society.

Review Vocabulary
paleontologist: a scientist who studies fossils

New Vocabulary
- fossil
- permineralized remains
- carbon film
- mold
- cast
- index fossil

Traces of the Distant Past

A giant crocodile lurks in the shallow water of a river. A herd of *Triceratops* emerges from the edge of the forest and cautiously moves toward the river. The dinosaurs are thirsty, but danger waits for them in the water. A large bull *Triceratops* moves into the river. The others follow.

Does this scene sound familiar to you? It's likely that you've read about dinosaurs and other past inhabitants of Earth. But how do you know that they really existed or what they were like? What evidence do humans have of past life on Earth? The answer is fossils. Paleontologists, scientists who study fossils, can learn about extinct animals from their fossil remains, as shown in **Figure 1**.

Figure 1 Scientists can learn how dinosaurs looked and moved using fossil remains. A skeleton can then be reassembled and displayed in a museum.

Formation of Fossils

Fossils are the remains, imprints, or traces of prehistoric organisms. Fossils have helped scientists determine approximately when life first appeared, when plants and animals first lived on land, and when organisms became extinct. Fossils are evidence of not only when and where organisms once lived, but also how they lived.

For the most part, the remains of dead plants and animals disappear quickly. Scavengers eat and scatter the remains of dead organisms. Fungi and bacteria invade, causing the remains to rot and disappear. If you've ever left a banana on the counter too long, you've seen this process begin. In time, compounds within the banana cause it to break down chemically and soften. Microorganisms, such as bacteria, cause it to decay. What keeps some plants and animals from disappearing before they become fossils? Which organisms are more likely to become fossils?

Figure 2 These fossil shark teeth are hard parts. Soft parts of animals do not become fossilized as easily.

Conditions Needed for Fossil Formation Whether or not a dead organism becomes a fossil depends upon how well it is protected from scavengers and agents of physical destruction, such as waves and currents. One way a dead organism can be protected is for sediment to bury the body quickly. If a fish dies and sinks to the bottom of a lake, sediment carried into the lake by a stream can cover the fish rapidly. As a result, no waves or scavengers can get to it and tear it apart. The body parts then might be fossilized and included in a sedimentary rock like shale. However, quick burial alone isn't always enough to make a fossil.

Organisms have a better chance of becoming fossils if they have hard parts such as bones, shells, or teeth. One reason is that scavengers are less likely to eat these hard parts. Hard parts also decay more slowly than soft parts do. Most fossils are the hard parts of organisms, such as the fossil teeth in **Figure 2.**

Types of Preservation

Perhaps you've seen skeletal remains of *Tyrannosaurus rex* towering above you in a museum. You also have some idea of what this dinosaur looked like because you've seen illustrations. Artists who draw *Tyrannosaurus rex* and other dinosaurs base their illustrations on fossil bones. What preserves fossil bones?

Mini LAB

Predicting Fossil Preservation

Procedure
1. Take a brief walk outside and observe your neighborhood.
2. Look around and notice what kinds of plants and animals live nearby.

Analysis
1. Predict what remains from your time might be preserved far into the future.
2. Explain what conditions would need to exist for these remains to be fossilized.

Figure 3 Opal and various minerals have replaced original materials and filled the hollow spaces in this permineralized dinosaur bone. **Explain** why this fossil retained the shape of the original bone.

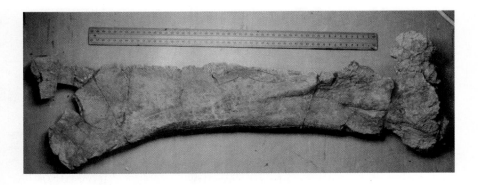

Mineral Replacement Most hard parts of organisms such as bones, teeth, and shells have tiny spaces within them. In life, these spaces can be filled with cells, blood vessels, nerves, or air. When the organism dies and the soft materials inside the hard parts decay, the tiny spaces become empty. If the hard part is buried, groundwater can seep in and deposit minerals in the spaces. **Permineralized remains** are fossils in which the spaces inside are filled with minerals from groundwater. In permineralized remains, some original material from the fossil organism's body might be preserved—encased within the minerals from groundwater. It is from these original materials that DNA, the chemical that contains an organism's genetic code, can sometimes be recovered.

Sometimes minerals replace the hard parts of fossil organisms. For example, a solution of water and dissolved silica (the compound SiO_2) might flow into and through the shell of a dead organism. If the water dissolves the shell and leaves silica in its place, the original shell is replaced.

Often people learn about past forms of life from bones, wood, and other remains that became permineralized or replaced with minerals from groundwater, as shown in **Figure 3,** but many other types of fossils can be found.

Figure 4 Graptolites lived hundreds of millions of years ago and drifted on currents in the oceans. These organisms often are preserved as carbon films.

Carbon Films The tissues of organisms are made of compounds that contain carbon. Sometimes fossils contain only carbon. Fossils usually form when sediments bury a dead organism. As sediment piles up, the organism's remains are subjected to pressure and heat. These conditions force gases and liquids from the body. A thin film of carbon residue is left, forming a silhouette of the original organism called a **carbon film. Figure 4** shows the carbonized remains of graptolites, which were small marine animals. Graptolites have been found in rocks as old as 500 million years.

244 **CHAPTER 9** Clues to Earth's Past

Coal In swampy regions, large volumes of plant matter accumulate. Over millions of years, these deposits become completely carbonized, forming coal. Coal is an important fuel source, but since the structure of the original plant is usually lost, it cannot reveal as much about the past as other kinds of fossils.

Reading Check *In what sort of environment does coal form?*

Molds and Casts In nature, impressions form when seashells or other hard parts of organisms fall into a soft sediment such as mud. The object and sediment are then buried by more sediment. Compaction, together with cementation, which is the deposition of minerals from water into the pore spaces between sediment particles, turns the sediment into rock. Other open pores in the rock then let water and air reach the shell or hard part. The hard part might decay or dissolve, leaving behind a cavity in the rock called a **mold**. Later, mineral-rich water or other sediment might enter the cavity, form new rock, and produce a copy or **cast** of the original object, as shown in **Figure 5.**

INTEGRATE Social Studies

Coal Mining Many of the first coal mines in the United States were located in eastern states like Pennsylvania and West Virginia. In your Science Journal, discuss how the environments of the past relate to people's lives today.

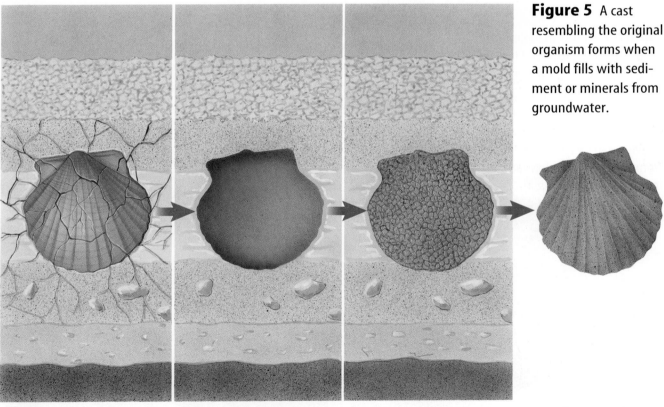

Figure 5 A cast resembling the original organism forms when a mold fills with sediment or minerals from groundwater.

The fossil begins to dissolve as water moves through spaces in the rock layers.

The fossil has been dissolved away. The harder rock once surrounding it forms a mold.

Sediment washes into the mold and is deposited, or mineral crystals form.

A cast results.

SECTION 1 Fossils **245**

Figure 6 The original soft parts of this mosquito have been preserved in amber for millions of years.

Original Remains Sometimes conditions allow original soft parts of organisms to be preserved for thousands or millions of years. For example, insects can be trapped in amber, a hardened form of sticky tree resin. The amber surrounds and protects the original material of the insect's exoskeleton from destruction, as shown in **Figure 6.** Some organisms, such as the mammoth, have been found preserved in frozen ground in Siberia. Original remains also have been found in natural tar deposits, such as the La Brea tar pits in California.

Trace Fossils Do you have a handprint in plaster that you made when you were in kindergarten? If so, it's a record that tells something about you. From it, others can guess your size and maybe your weight at that age. Animals walking on Earth long ago left similar tracks, such as those in **Figure 7.** Trace fossils are fossilized tracks and other evidence of the activity of organisms. In some cases, tracks can tell you more about how an organism lived than any other type of fossil. For example, from a set of tracks at Davenport Ranch, Texas, you might be able to learn something about the social life of sauropods, which were large, plant-eating dinosaurs. The largest tracks of the herd are on the outer edges and the smallest are on the inside. These tracks led some scientists to hypothesize that adult sauropods surrounded their young as they traveled—perhaps to protect them from predators. A nearby set of tracks might mean that another type of dinosaur, an allosaur, was stalking the herd.

Figure 7 Tracks made in soft mud, and now preserved in solid rock, can provide information about animal size, speed, and behavior.

The dinosaur track below is from the Glen Rose Formation in north-central Texas.

The tracks to the right are located on a Navajo reservation in Arizona.

Trails and Burrows Other trace fossils include trails and burrows made by worms and other animals. These, too, tell something about how these animals lived. For example, by examining fossil burrows you can sometimes tell how firm the sediment the animals lived in was. As you can see, fossils can tell a great deal about the organisms that have inhabited Earth.

 *How are trace fossils different from fossils that are the remains of an organism's body?*

Index Fossils

One thing you can learn by studying fossils is that species of organisms have changed over time. Some species of organisms inhabited Earth for long periods of time without changing. Other species changed a lot in comparatively short amounts of time. It is these organisms that scientists use as index fossils.

Index fossils are the remains of species that existed on Earth for relatively short periods of time, were abundant, and were widespread geographically. Because the organisms that became index fossils lived only during specific intervals of geologic time, geologists can estimate the ages of rock layers based on the particular index fossils they contain. However, not all rocks contain index fossils. Another way to approximate the age of a rock layer is to compare the spans of time, or ranges, over which more than one fossil appears. The estimated age is the time interval where fossil ranges overlap, as shown in **Figure 8.**

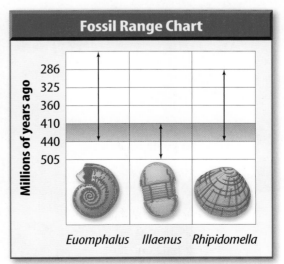

Figure 8 The fossils in a sequence of sedimentary rock can be used to estimate the ages of each layer. The chart shows when each organism inhabited Earth.
Explain *why it is possible to say that the middle layer of rock was deposited between 440 million and 410 million years ago.*

Ancient Ecology
Ecology is the study of how organisms interact with each other and with their environment. Some paleontologists study the ecology of ancient organisms. Discuss the kinds of information you could use to determine how ancient organisms interacted with their environment.

Fossils and Ancient Environments

Scientists can use fossils to determine what the environment of an area was like long ago. Using fossils, you might be able to find out whether an area was land or whether it was covered by an ocean at a particular time. If the region was covered by ocean, it might even be possible to learn the depth of the water. What clues about the depth of water do you think fossils could provide?

Fossils also are used to determine the past climate of a region. For example, rocks in parts of the eastern United States contain fossils of tropical plants. The environment of this part of the United States today isn't tropical. However, because of the fossils, scientists know that it was tropical when these plants were living. **Figure 9** shows that North America was located near the equator when these fossils formed.

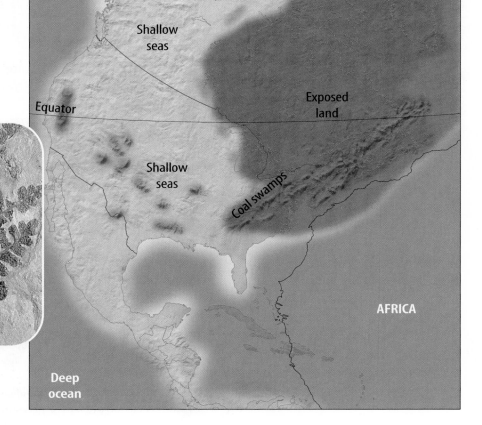

Figure 9 The equator passed through North America 310 million years ago. At this time, warm, shallow seas and coal swamps covered much of the continent, and ferns like the *Neuropteris,* below, were common.

248 CHAPTER 9 Clues to Earth's Past

Shallow Seas How would you explain the presence of fossilized crinoids—animals that lived in shallow seas—in rocks found in what is today a desert? **Figure 10** shows a fossil crinoid and a living crinoid. When the fossil crinoids were alive, a shallow sea covered much of western and central North America. The crinoid hard parts were included in rocks that formed from the sediments at the bottom of this sea. Fossils provide information about past life on Earth and also about the history of the rock layers that contain them. Fossils can provide information about the ages of rocks and the climate and type of environment that existed when the rocks formed.

Figure 10 The crinoid on the left lived in warm, shallow seas that once covered part of North America. Crinoids like the one on the right typically live in warm, shallow waters in the Pacific Ocean.

section 1 review

Summary

Formation of Fossils
- Fossils are the remains, imprints, or traces of past organisms.
- Fossilization is most likely if the organism had hard parts and was buried quickly.

Fossil Preservation
- Permineralized remains have open spaces filled with minerals from groundwater.
- Thin carbon films remain in the shapes of dead organisms.
- Hard parts dissolve to leave molds.
- Trace fossils are evidence of past activity.

Index Fossils
- Index fossils are from species that were abundant briefly, but over wide areas.
- Scientists can estimate the ages of rocks containing index fossils.

Fossils and Ancient Environments
- Fossils tell us about the environment in which the organisms lived.

Self Check

1. **Describe** the typical conditions necessary for fossil formation.
2. **Explain** how a fossil mold is different from a fossil cast.
3. **Discuss** how the characteristics of an index fossil are useful to geologists.
4. **Describe** how carbon films form.
5. **Think Critically** What can you say about the ages of two widely separated layers of rock that contain the same type of fossil?

Applying Skills

6. **Communicate** what you learn about fossils. Visit a museum that has fossils on display. Make an illustration of each fossil in your Science Journal. Write a brief description, noting key facts about each fossil and how each fossil might have formed.
7. **Compare and contrast** original remains with other kinds of fossils. What kinds of information would only be available from original remains? Are there any limitations to the use of original remains?

blue.msscience.com/self_check_quiz

section 2
Relative Ages of Rocks

as you read

What You'll Learn
- **Describe** methods used to assign relative ages to rock layers.
- **Interpret** gaps in the rock record.
- **Give** an example of how rock layers can be correlated with other rock layers.

Why It's Important
Being able to determine the age of rock layers is important in trying to understand a history of Earth.

Review Vocabulary
sedimentary rock: rock formed when sediments are cemented and compacted or when minerals are precipitated from solution

New Vocabulary
- principle of superposition
- relative age
- unconformity

Superposition

Imagine that you are walking to your favorite store and you happen to notice an interesting car go by. You're not sure what kind it is, but you remember that you read an article about it. You decide to look it up. At home you have a stack of magazines from the past year, as seen in **Figure 11.**

You know that the article you're thinking of came out in the January edition, so it must be near the bottom of the pile. As you dig downward, you find magazines from March, then February. January must be next. How did you know that the January issue of the magazine would be on the bottom? To find the older edition under newer ones, you applied the principle of superposition.

Oldest Rocks on the Bottom According to the **principle of superposition,** in undisturbed layers of rock, the oldest rocks are on the bottom and the rocks become progressively younger toward the top. Why is this the case?

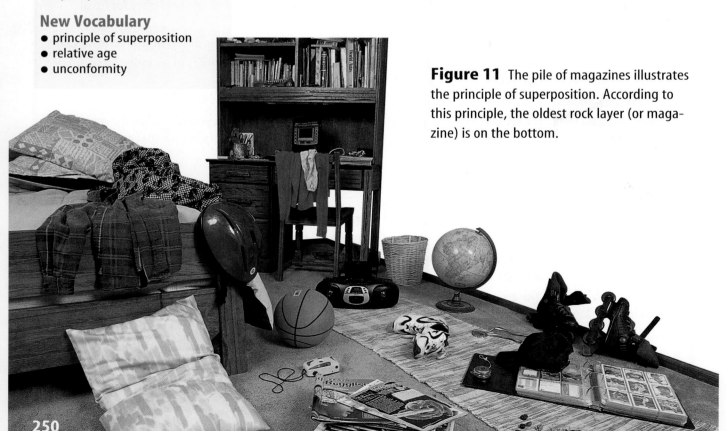

Figure 11 The pile of magazines illustrates the principle of superposition. According to this principle, the oldest rock layer (or magazine) is on the bottom.

Rock Layers Sediment accumulates in horizontal beds, forming layers of sedimentary rock. The first layer to form is on the bottom. The next layer forms on top of the previous one. Because of this, the oldest rocks are at the bottom. However, forces generated by mountain formation sometimes can turn layers over. When layers have been turned upside down, it's necessary to use other clues in the rock layers to determine their original positions and relative ages.

Relative Ages

Now you want to look for another magazine. You're not sure how old it is, but you know it arrived after the January issue. You can find it in the stack by using the principle of relative age.

The **relative age** of something is its age in comparison to the ages of other things. Geologists determine the relative ages of rocks and other structures by examining their places in a sequence. For example, if layers of sedimentary rock are offset by a fault, which is a break in Earth's surface, you know that the layers had to be there before a fault could cut through them. The relative age of the rocks is older than the relative age of the fault. Relative age determination doesn't tell you anything about the age of rock layers in actual years. You don't know if a layer is 100 million or 10,000 years old. You only know that it's younger than the layers below it and older than the fault cutting through it.

Other Clues Help Determination of relative age is easy if the rocks haven't been faulted or turned upside down. For example, look at **Figure 12**. Which layer is the oldest? In cases where rock layers have been disturbed you might have to look for fossils and other clues to date the rocks. If you find a fossil in the top layer that's older than a fossil in a lower layer, you can hypothesize that layers have been turned upside down by folding during mountain building.

Science Online

Topic: Relative Dating
Visit blue.msscience.com for Web links to information about relative dating of rocks and other materials.

Activity Imagine yourself at an archaeological dig. You have found a rare artifact and want to know its age. Make a list of clues you might look for to provide a relative date and explain how each would allow you to approximate the artifact's age.

Figure 12 In a stack of undisturbed sedimentary rocks, the oldest rocks are at the bottom. This stack of rocks can be folded by forces within Earth.
Explain *how you can tell if an older rock is above a younger one.*

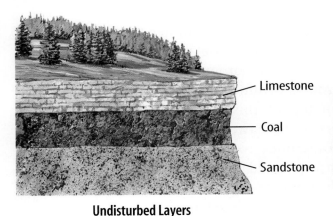

Undisturbed Layers

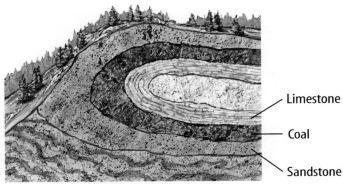

Folded Layers

SECTION 2 Relative Ages of Rocks **251**

Figure 13 An angular unconformity results when horizontal layers cover tilted, eroded layers.

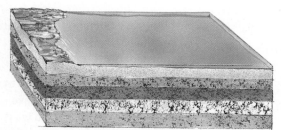

A Sedimentary rocks are deposited originally as horizontal layers.

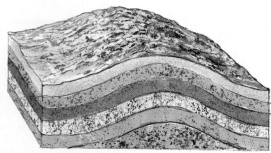

B The horizontal rock layers are tilted as forces within Earth deform them.

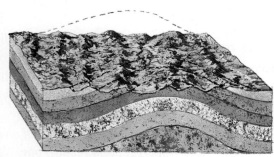

C The tilted layers erode.

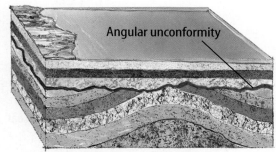

D An angular unconformity results when new layers form on the tilted layers as deposition resumes.

Unconformities

A sequence of rock is a record of past events. But most rock sequences are incomplete—layers are missing. These gaps in rock sequences are called **unconformities** (un kun FOR muh teez). Unconformities develop when agents of erosion such as running water or glaciers remove rock layers by washing or scraping them away.

Reading Check *How do unconformities form?*

Angular Unconformities Horizontal layers of sedimentary rock often are tilted and uplifted. Erosion and weathering then wear down these tilted rock layers. Eventually, younger sediment layers are deposited horizontally on top of the tilted and eroded layers. Geologists call such an unconformity an angular unconformity. **Figure 13** shows how angular unconformities develop.

Disconformity Suppose you're looking at a stack of sedimentary rock layers. They look complete, but layers are missing. If you look closely, you might find an old surface of erosion. This records a time when the rocks were exposed and eroded. Later, younger rocks formed above the erosion surface when deposition of sediment began again. Even though all the layers are parallel, the rock record still has a gap. This type of unconformity is called a disconformity. A disconformity also forms when a period of time passes without any new deposition occurring to form new layers of rock.

Nonconformity Another type of unconformity, called a nonconformity, occurs when metamorphic or igneous rocks are uplifted and eroded. Sedimentary rocks are then deposited on top of this erosion surface. The surface between the two rock types is a nonconformity. Sometimes rock fragments from below are incorporated into sediments deposited above the nonconformity. All types of unconformities are shown in **Figure 14.**

NATIONAL GEOGRAPHIC VISUALIZING UNCONFORMITIES

Figure 14

An unconformity is a gap in the rock record caused by erosion or a pause in deposition. There are three major kinds of unconformities—nonconformity, angular unconformity, and disconformity.

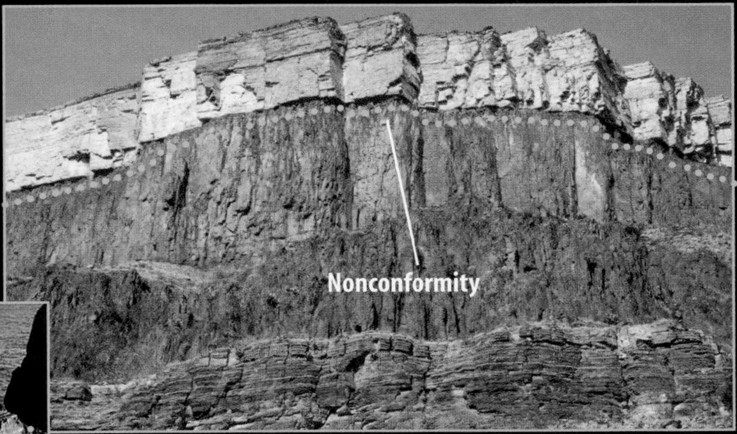

▲ In a nonconformity, horizontal layers of sedimentary rock overlie older igneous or metamorphic rocks. A nonconformity in Big Bend National Park, Texas, is shown above.

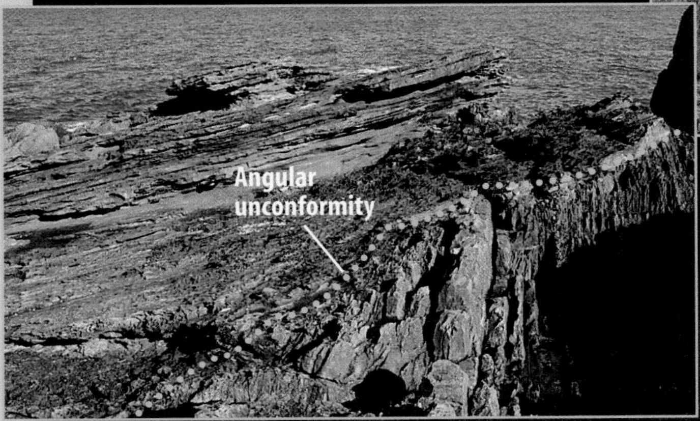

▲ An angular unconformity develops when new horizontal layers of sedimentary rock form on top of older sedimentary rock layers that have been folded by compression. An example of an angular unconformity at Siccar Point in southeastern Scotland is shown above.

▼ A disconformity develops when horizontal rock layers are exposed and eroded, and new horizontal layers of rock are deposited on the eroded surface. The disconformity shown below is in the Grand Canyon.

SECTION 2 Relative Ages of Rocks

Science Online

Topic: Correlating with Index Fossils
Visit blue.msscience.com for Web links to information about using index fossils to match up layers of rock.

Activity Make a chart that shows the rock layers of both the Grand Canyon and Capitol Reef National Park in Utah. For each layer that appears in both parks, list an index fossil you could find to correlate the layers.

Matching Up Rock Layers

Suppose you're studying a layer of sandstone in Bryce Canyon in Utah. Later, when you visit Canyonlands National Park, Utah, you notice that a layer of sandstone there looks just like the sandstone in Bryce Canyon, 250 km away. Above the sandstone in the Canyonlands is a layer of limestone and then another sandstone layer. You return to Bryce Canyon and find the same sequence—sandstone, limestone, and sandstone. What do you infer? It's likely that you're looking at the same layers of rocks in two different locations. **Figure 15** shows that these rocks are parts of huge deposits that covered this whole area of the western United States. Geologists often can match up, or correlate, layers of rocks over great distances.

Evidence Used for Correlation It's not always easy to say that a rock layer exposed in one area is the same as a rock layer exposed in another area. Sometimes it's possible to walk along the layer for kilometers and prove that it's continuous. In other cases, such as at the Canyonlands area and Bryce Canyon as seen in **Figure 16,** the rock layers are exposed only where rivers have cut through overlying layers of rock and sediment. How can you show that the limestone sandwiched between the two layers of sandstone in Canyonlands is likely the same limestone as at Bryce Canyon? One way is to use fossil evidence. If the same types of fossils were found in the limestone layer in both places, it's a good indication that the limestone at each location is the same age, and, therefore, one continuous deposit.

Reading Check *How do fossils help show that rocks at different locations belong to the same rock layer?*

Figure 15 These rock layers, exposed at Hopi Point in Grand Canyon National Park, Arizona, can be correlated, or matched up, with rocks from across large areas of the western United States.

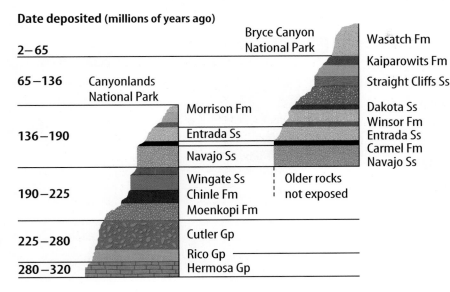

Figure 16 Geologists have named the many rock layers, or formations, in Canyonlands and in Bryce Canyon, Utah. They also have correlated some formations between the two canyons.
List *the labeled layers present at both canyons.*

Can layers of rock be correlated in other ways? Sometimes determining relative ages isn't enough and other dating methods must be used. In Section 3, you'll see how the numerical ages of rocks can be determined and how geologists have used this information to estimate the age of Earth.

section 2 review

Summary

Superposition
- Superposition states that in undisturbed rock, the oldest layers are on the bottom.

Relative Ages
- Rock layers can be ranked by relative age.

Unconformities
- Angular unconformities are new layers deposited over tilted and eroded rock layers.
- Disconformities are gaps in the rock record.
- Nonconformities divide uplifted igneous or metamorphic rock from new sedimentary rock.

Matching Up Rock Layers
- Rocks from different areas may be correlated if they are part of the same layer.

Self Check

1. **Discuss** how to find the oldest paper in a stack of papers.
2. **Explain** the concept of relative age.
3. **Illustrate** a disconformity.
4. **Describe** one way to correlate similar rock layers.
5. **Think Critically** Explain the relationship between the concept of relative age and the principle of superposition.

Applying Skills

6. **Interpret data** to determine the oldest rock bed. A sandstone contains a 400-million-year-old fossil. A shale has fossils that are over 500 million years old. A limestone, below the sandstone, contains fossils between 400 million and 500 million years old. Which rock bed is oldest? Explain.

Science online   blue.msscience.com/self_check_quiz

SECTION 2 Relative Ages of Rocks **255**

Relative Ages

Which of your two friends is older? To answer this question, you'd need to know their relative ages. You wouldn't need to know the exact age of either of your friends—just who was born first. The same is sometimes true for rock layers.

Real-World Question
Can you determine the relative ages of rock layers?

Goals
- **Interpret** illustrations of rock layers and other geological structures and determine the relative order of events.

Materials
paper   pencil

Procedure
1. **Analyze** Figures **A** and **B**.
2. Make a sketch of **Figure A**. On it, identify the relative age of each rock layer, igneous intrusion, fault, and unconformity. For example, the shale layer is the oldest, so mark it with a 1. Mark the next-oldest feature with a 2, and so on.
3. Repeat step 2 for **Figure B**.

Conclude and Apply
Figure A
1. **Identify** the type of unconformity shown. Is it possible that there were originally more layers of rock than are shown?
2. **Describe** how the rocks above the fault moved in relation to rocks below the fault.
3. **Hypothesize** how the hill on the left side of the figure formed.

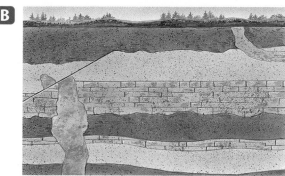

Granite   Limestone
Sandstone   Shale

Figure B
4. Is it possible to conclude if the igneous intrusion on the left is older or younger than the unconformity nearest the surface?
5. **Describe** the relative ages of the two igneous intrusions. How did you know?
6. **Hypothesize** which two layers of rock might have been much thicker in the past.

Communicating Your Data
Compare your results with other students' results. **For more help, refer to the** Science Skill Handbook.

256 **CHAPTER 9** Clues to Earth's Past

section 3
Absolute Ages of Rocks

Absolute Ages

As you sort through your stack of magazines looking for that article about the car you saw, you decide that you need to restack them into a neat pile. By now, they're in a jumble and no longer in order of their relative age, as shown in **Figure 17.** How can you stack them so the oldest are on the bottom and the newest are on top? Fortunately, magazine dates are printed on the cover. Thus, stacking magazines in order is a simple process. Unfortunately, rocks don't have their ages stamped on them. Or do they? **Absolute age** is the age, in years, of a rock or other object. Geologists determine absolute ages by using properties of the atoms that make up materials.

Radioactive Decay

INTEGRATE Physics Atoms consist of a dense central region called the nucleus, which is surrounded by a cloud of negatively charged particles called electrons. The nucleus is made up of protons, which have a positive charge, and neutrons, which have no electric charge. The number of protons determines the identity of the element, and the number of neutrons determines the form of the element, or isotope. For example, every atom with a single proton is a hydrogen atom. Hydrogen atoms can have no neutrons, a single neutron, or two neutrons. This means that there are three isotopes of hydrogen.

Reading Check What particles make up an atom's nucleus?

Some isotopes are unstable and break down into other isotopes and particles. Sometimes a lot of energy is given off during this process. The process of breaking down is called **radioactive decay.** In the case of hydrogen, atoms with one proton and two neutrons are unstable and tend to break down. Many other elements have stable and unstable isotopes.

as you read

What **You'll Learn**
- **Identify** how absolute age differs from relative age.
- **Describe** how the half-lives of isotopes are used to determine a rock's age.

Why **It's Important**

Events in Earth's history can be better understood if their absolute ages are known.

Review Vocabulary
isotopes: atoms of the same element that have different numbers of neutrons

New Vocabulary
- absolute age
- radioactive decay
- half-life
- radiometric dating
- uniformitarianism

Figure 17 The magazines that have been shuffled through no longer illustrate the principle of superposition.

SECTION 3 Absolute Ages of Rocks **257**

Mini LAB

Modeling Carbon-14 Dating

Procedure
1. Count out 80 **red jelly beans**.
2. Remove half the red jelly beans and replace them with **green jelly beans**.
3. Continue replacing half the red jelly beans with green jelly beans until only 5 red jelly beans remain. Count the number of times you replace half the red jelly beans.

Analysis
1. How did this activity model the decay of carbon-14 atoms?
2. How many half lives of carbon-14 did you model during this activity?
3. If the atoms in a bone experienced the same number of half lives as your jelly beans, how old would the bone be?

Alpha and Beta Decay In some isotopes, a neutron breaks down into a proton and an electron. This type of radioactive decay is called beta decay because the electron leaves the atom as a beta particle. The nucleus loses a neutron but gains a proton. When the number of protons in an atom is changed, a new element forms. Other isotopes give off two protons and two neutrons in the form of an alpha particle. Alpha and beta decay are shown in **Figure 18**.

Half-Life In radioactive decay reactions, the parent isotope undergoes radioactive decay. The daughter product is produced by radioactive decay. Each radioactive parent isotope decays to its daughter product at a certain rate. Based on this decay rate, it takes a certain period of time for one half of the parent isotope to decay to its daughter product. The **half-life** of an isotope is the time it takes for half of the atoms in the isotope to decay. For example, the half-life of carbon-14 is 5,730 years. So it will take 5,730 years for half of the carbon-14 atoms in an object to change into nitrogen-14 atoms. You might guess that in another 5,730 years, all of the remaining carbon-14 atoms will decay to nitrogen-14. However, this is not the case. Only half of the atoms of carbon-14 remaining after the first 5,730 years will decay during the second 5,730 years. So, after two half-lives, one fourth of the original carbon-14 atoms still remain. Half of them will decay during another 5,730 years. After three half-lives, one eighth of the original carbon-14 atoms still remain. After many half-lives, such a small amount of the parent isotope remains that it might not be measurable.

Figure 18 In beta decay, a neutron changes into a proton by giving off an electron. This electron has a lot of energy and is called a beta particle.

In the process of alpha decay, an unstable parent isotope nucleus gives off an alpha particle and changes into a new daughter product. Alpha particles contain two neutrons and two protons.

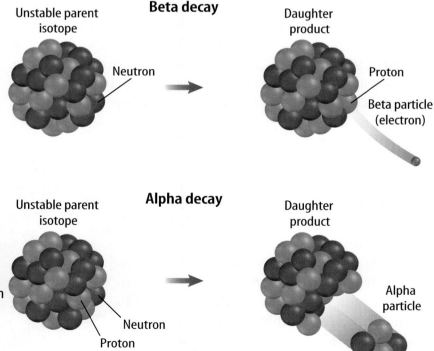

258 CHAPTER 9 Clues to Earth's Past

Radiometric Ages

Decay of radioactive isotopes is like a clock keeping track of time that has passed since rocks have formed. As time passes, the amount of parent isotope in a rock decreases as the amount of daughter product increases, as in **Figure 19**. By measuring the ratio of parent isotope to daughter product in a mineral and by knowing the half-life of the parent, in many cases you can calculate the absolute age of a rock. This process is called **radiometric dating**.

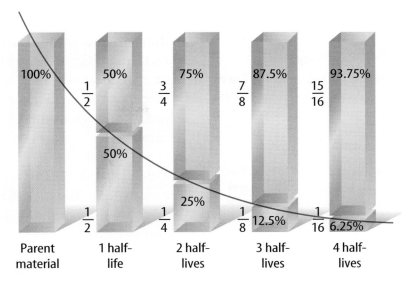

Figure 19 During each half-life, one half of the parent material decays to the daughter product. **Explain** *how one uses both parent and daughter material to estimate age.*

A scientist must decide which parent isotope to use when measuring the age of a rock. If the object to be dated seems old, then the geologist will use an isotope with a long half-life. The half-life for the decay of potassium-40 to argon-40 is 1.25 billion years. As a result, this isotope can be used to date rocks that are many millions of years old. To avoid error, conditions must be met for the ratios to give a correct indication of age. For example, the rock being studied must still retain all of the argon-40 that was produced by the decay of potassium-40. Also, it cannot contain any contamination of daughter product from other sources. Potassium-argon dating is good for rocks containing potassium, but what about other things?

Radiocarbon Dating Carbon-14 is useful for dating bones, wood, and charcoal up to 75,000 years old. Living things take in carbon from the environment to build their bodies. Most of that carbon is carbon-12, but some is carbon-14, and the ratio of these two isotopes in the environment is always the same. After the organism dies, the carbon-14 slowly decays. By determining the amounts of the isotopes in a sample, scientists can evaluate how much the isotope ratio in the sample differs from that in the environment. For example, during much of human history, people built campfires. The wood from these fires often is preserved as charcoal. Scientists can determine the amount of carbon-14 remaining in a sample of charcoal by measuring the amount of radiation emitted by the carbon-14 isotope in labs like the one in **Figure 20**. Once they know the amount of carbon-14 in a charcoal sample, scientists can determine the age of the wood used to make the fire.

Figure 20 Radiometric ages are determined in labs like this one.

Topic: Isotopes in Ice Cores
Visit blue.msscience.com for Web links to information about ice cores and how isotopes in ice are used to learn about Earth's past.

Activity Prepare a report that shows how isotopes in ice cores can tell us about past Earth environments. Include how these findings can help us understand today's climate.

Age Determinations Aside from carbon-14 dating, rocks that can be radiometrically dated are mostly igneous and metamorphic rocks. Most sedimentary rocks cannot be dated by this method. This is because many sedimentary rocks are made up of particles eroded from older rocks. Dating these pieces only gives the age of the preexisting rock from which it came.

The Oldest Known Rocks Radiometric dating has been used to date the oldest rocks on Earth. These rocks are about 3.96 billion years old. By determining the age of meteorites, and using other evidence, scientists have estimated the age of Earth to be about 4.5 billion years. Earth rocks greater than 3.96 billion years old probably were eroded or changed by heat and pressure.

Reading Check *Why can't most sedimentary rocks be dated radiometrically?*

Applying Science

When did the Iceman die?

Carbon-14 dating has been used to date charcoal, wood, bones, mummies from Egypt and Peru, the Dead Sea Scrolls, and the Italian Iceman. The Iceman was found in 1991 in the Italian Alps, near the Austrian border. Based on carbon-14 analysis, scientists determined that the Iceman is 5,300 years old. Determine approximately in what year the Iceman died.

Half-Life of Carbon-14	
Percent Carbon-14	Years Passed
100	0
50	5,730
25	11,460
12.5	17,190
6.25	22,920
3.125	

Reconstruction of Iceman

Identifying the Problem

The half-life chart shows the decay of carbon-14 over time. Half-life is the time it takes for half of a sample to decay. Fill in the years passed when only 3.125 percent of carbon-14 remain. Is there a point at which no carbon-14 would be present? Explain.

Solving the Problem

1. Estimate, using the data table, how much carbon-14 still was present in the Iceman's body that allowed scientists to determine his age.
2. If you had an artifact that originally contained 10.0 g of carbon-14, how many grams would remain after 17,190 years?

260 CHAPTER 9 Clues to Earth's Past

Uniformitarianism

Can you imagine trying to determine the age of Earth without some of the information you know today? Before the discovery of radiometric dating, many people estimated that Earth is only a few thousand years old. But in the 1700s, Scottish scientist James Hutton estimated that Earth is much older. He used the principle of **uniformitarianism.** This principle states that Earth processes occurring today are similar to those that occurred in the past. Hutton's principle is often paraphrased as "the present is the key to the past."

Hutton observed that the processes that changed the landscape around him were slow, and he inferred that they were just as slow throughout Earth's history. Hutton hypothesized that it took much longer than a few thousand years to form the layers of rock around him and to erode mountains that once stood kilometers high. **Figure 21** shows Hutton's native Scotland, a region shaped by millions of years of geologic processes.

Today, scientists recognize that Earth has been shaped by two types of change: slow, everyday processes that take place over millions of years, and violent, unusual events such as the collision of a comet or asteroid about 65 million years ago that might have caused the extinction of the dinosaurs.

Figure 21 The rugged highlands of Scotland were shaped by erosion and uplift.

section 3 review

Summary

Absolute Ages
- The absolute age is the actual age of an object.

Radioactive Decay
- Some isotopes are unstable and decay into other isotopes and particles.
- Decay is measured in half-lives, the time it takes for half of a given isotope to decay.

Radiometric Ages
- By measuring the ratio of parent isotope to daughter product, one can determine the absolute age of a rock.
- Living organisms less than 75,000 years old can be dated using carbon-14.

Uniformitarianism
- Processes observable today are the same as the processes that took place in the past.

Self Check

1. **Evaluate** the age of rocks. You find three undisturbed rock layers. The middle layer is 120 million years old. What can you say about the ages of the layers above and below it?
2. **Determine** the age of a fossil if it had only one eighth of its original carbon-14 content remaining.
3. **Explain** the concept of uniformitarianism.
4. **Describe** how radioactive isotopes decay.
5. **Think Critically** Why can't scientists use carbon-14 to determine the age of an igneous rock?

Applying Math

6. **Make and use a table** that shows the amount of parent material of a radioactive element that is left after four half-lives if the original parent material had a mass of 100 g.

LAB Model and Invent

Trace Fossils

Goals
- **Construct** a model of trace fossils.
- **Describe** the information that you can learn from looking at your model.

Possible Materials
construction paper
wire
plastic (a fairly rigid type)
scissors
plaster of paris
toothpicks
sturdy cardboard
clay
pipe cleaners
glue

Safety Precautions

Real-World Question
Trace fossils can tell you a lot about the activities of organisms that left them. They can tell you how an organism fed or what kind of home it had. How can you model trace fossils that can provide information about the behavior of organisms? What materials can you use to model trace fossils? What types of behavior could you show with your trace fossil model?

Make a Model
1. **Decide** how you are going to make your model. What materials will you need?
2. **Decide** what types of activities you will demonstrate with your model. Were the organisms feeding? Resting? Traveling? Were they predators? Prey? How will your model indicate the activities you chose?
3. What is the setting of your model? Are you modeling the organism's home? Feeding areas? Is your model on land or water? How can the setting affect the way you build your model?
4. Will you only show trace fossils from a single species or multiple species? If you include more than one species, how will you provide evidence of any interaction between the species?

Check the Model Plans
1. Compare your plans with those of others in your class. Did other groups mention details that you had forgotten to think about? Are there any changes you would like to make to your plan before you continue?
2. Make sure your teacher approves your plan before you continue.

262 CHAPTER 9 Clues to Earth's Past

Using Scientific Methods

▶ Test Your Model

1. Following your plan, construct your model of trace fossils.
2. Have you included evidence of all the behaviors you intended to model?

▶ Analyze Your Data

1. **Evaluate** Now that your model is complete, do you think that it adequately shows the behaviors you planned to demonstrate? Is there anything that you think you might want to do differently if you were going to make the model again?
2. **Describe** how using different kinds of materials might have affected your model. Can you think of other materials that would have allowed you to show more detail than you did?

▶ Conclude and Apply

1. **Compare and contrast** your model of trace fossils with trace fossils left by real organisms. Is one more easily interpreted than the other? Explain.
2. **List** behaviors that might not leave any trace fossils. Explain.

Communicating Your Data

Ask other students in your class or another class to look at your model and describe what information they can learn from the trace fossils. Did their interpretations agree with what you intended to show?

LAB 263

Oops! Accidents in SCIENCE

SOMETIMES GREAT DISCOVERIES HAPPEN BY ACCIDENT!

The World's Oldest Fish Story

A catch-of-the-day set science on its ears

Camouflage marks • First dorsal fin • Second dorsal fin • Pectoral fin • Anal fin • Pelvic fin

Some scientists call the coelacanth "Old Four Legs." It got its nickname because the fish has paired fins that look something like legs.

On a December day in 1938, just before Christmas, Marjorie Courtenay-Latimer went to say hello to her friends on board a fishing boat that had just returned to port in South Africa. Courtenay-Latimer, who worked at a museum, often went aboard her friends' ship to check out the catch. On this visit, she received a surprise Christmas present—an odd-looking fish. As soon as the woman spotted its strange blue fins among the piles of sharks and rays, she knew it was special.

Courtenay-Latimer took the fish back to her museum to study it. "It was the most beautiful fish I had ever seen, five feet long, and a pale mauve blue with iridescent silver markings," she later wrote. Courtenay-Latimer sketched it and sent the drawing to a friend of hers, J. L. B. Smith.

Smith was a chemistry teacher who was passionate about fish. After a time, he realized it was a coelacanth (SEE luh kanth). Fish experts knew that coelacanths had first appeared on Earth 400 million years ago. But the experts thought the fish were extinct. People had found fossils of coelacanths, but no one had seen one alive. It was assumed that the last coelacanth species had died out 65 million years ago. They were wrong. The ship's crew had caught one by accident.

Smith figured there might be more living coelacanths. So he decided to offer a reward for anyone who could find a living specimen. After 14 years of silence, a report came in that a coelacanth had been caught off the east coast of Africa.

Today, scientists know that there are at least several hundred coelacanths living in the Indian Ocean, just east of central Africa. Many of these fish live near the Comoros Islands. The coelacanths live in underwater caves during the day but move out at night to feed. The rare fish are now a protected species. With any luck, they will survive for another hundred million years.

Write a short essay describing the discovery of the coelacanths and describe the reaction of scientists to this discovery.

Science online
For more information, visit blue.msscience.com/oops

chapter 9 Study Guide

Reviewing Main Ideas

Section 1 Fossils

1. Fossils are more likely to form if hard parts of the dead organisms are buried quickly.

2. Some fossils form when original materials that made up the organisms are replaced with minerals. Other fossils form when remains are subjected to heat and pressure, leaving only a carbon film behind. Some fossils are the tracks or traces left by ancient organisms.

Section 2 Relative Ages of Rocks

1. The principle of superposition states that, in undisturbed layers, older rocks lie underneath younger rocks.

2. Unconformities, or gaps in the rock record, are due to erosion or periods of time during which no deposition occurred.

3. Rock layers can be correlated using rock types and fossils.

Section 3 Absolute Ages of Rocks

1. Absolute dating provides an age in years for the rocks.

2. The half-life of a radioactive isotope is the time it takes for half of the atoms of the isotope to decay into another isotope.

Visualizing Main Ideas

Copy and complete the following concept map on fossils.

Chapter 9 Review

Using Vocabulary

absolute age p. 257
carbon film p. 244
cast p. 245
fossil p. 243
half-life p. 258
index fossil p. 247
mold p. 245
permineralized remains p. 244
principle of superposition p. 250
radioactive decay p. 257
radiometric dating p. 259
relative age p. 251
unconformity p. 252
uniformitarianism p. 261

Write an original sentence using the vocabulary word to which each phrase refers.

1. thin film of carbon preserved as a fossil
2. older rocks lie under younger rocks
3. processes occur today as they did in the past
4. gap in the rock record
5. time needed for half the atoms to decay
6. fossil organism that lived for a short time
7. gives the age of rocks in years
8. minerals fill spaces inside fossil
9. a copy of a fossil produced by filling a mold with sediment or crystals

Checking Concepts

Choose the word or phrase that best answers the question.

10. What is any evidence of ancient life called?
    A) half-life
    B) fossil
    C) unconformity
    D) disconformity

11. Which of the following conditions makes fossil formation more likely?
    A) buried slowly
    B) attacked by scavengers
    C) made of hard parts
    D) composed of soft parts

12. What are cavities left in rocks when a shell or bone dissolves called?
    A) casts
    B) molds
    C) original remains
    D) carbon films

13. To say "the present is the key to the past" is a way to describe which of the following principles?
    A) superposition
    B) succession
    C) radioactivity
    D) uniformitarianism

14. A fault can be useful in determining which of the following for a group of rocks?
    A) absolute age
    B) index age
    C) radiometric age
    D) relative age

15. Which of the following is an unconformity between parallel rock layers?
    A) angular unconformity
    B) fault
    C) disconformity
    D) nonconformity

Use the illustration below to answer question 16.

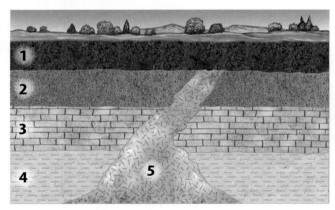

16. Which of the following puts the layers in order from oldest to youngest?
    A) 5-4-3-2-1
    B) 1-2-3-4-5
    C) 2-3-4-5-1
    D) 4-3-2-5-1

17. Which process forms new elements?
    A) superposition
    B) uniformitarianism
    C) permineralization
    D) radioactive decay

266 CHAPTER REVIEW

blue.msscience.com/vocabulary_puzzlemaker

Chapter 9 Review

Thinking Critically

18. **Explain** why the fossil record of life on Earth is incomplete. Give some reasons why.

19. **Infer** Suppose a lava flow was found between two sedimentary rock layers. How could you use the lava flow to learn about the ages of the sedimentary rock layers? *(Hint: Most lava contains radioactive isotopes.)*

20. **Infer** Suppose you're correlating rock layers in the western United States. You find a layer of volcanic ash deposits. How can this layer help you in your correlation over a large area?

21. **Recognize Cause and Effect** Explain how some woolly mammoths could have been preserved intact in frozen ground. What conditions must have persisted since the deaths of these animals?

22. **Classify** each of the following fossils in the correct category in the table below: *dinosaur footprint, worm burrow, dinosaur skull, insect in amber, fossil woodpecker hole,* and *fish tooth.*

Types of Fossils	
Trace Fossils	Body Fossils
Do not write in this book.	

23. **Compare and contrast** the three different kinds of unconformities. Draw sketches of each that illustrate the features that identify them.

24. **Describe** how relative and absolute ages differ. How might both be used to establish ages in a series of rock layers?

25. **Discuss** uniformitarianism in the following scenario. You find a shell on the beach, and a friend remembers seeing a similar fossil while hiking in the mountains. What does this suggest about the past environment of the mountain?

Performance Activities

26. **Illustrate** Create a model that allows you to explain how to establish the relative ages of rock layers.

27. **Use a Classification System** Start your own fossil collection. Label each find as to type, approximate age, and the place where it was found. Most state geological surveys can provide you with reference materials on local fossils.

Applying Math

28. **Calculate** how many half-lives have passed in a rock containing one-eighth the original radioactive material and seven-eighths of the daughter product.

Use the graphs below to answer question 29.

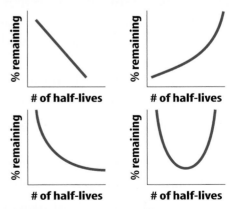

29. **Interpret Data** Which of the above curves best illustrates radioactive decay?

Chapter 9 Standardized Test Practice

Part 1 Multiple Choice

Record your answers on the answer sheet provided by your teacher or on a sheet of paper.

Use the photo below to answer question 1.

1. Which type of fossil preservation is shown above?
   A. trace fossil
   B. original remains
   C. carbon film
   D. permineralized remains

2. Which principle states that the oldest rock layer is found at the bottom in an undisturbed stack of rock layers?
   A. half-life
   B. absolute dating
   C. superposition
   D. uniformitarianism

3. Which type of scientist studies fossils?
   A. meteorologist
   B. chemist
   C. astronomer
   D. paleontologist

4. Which are the remains of species that existed on Earth for relatively short periods of time, were abundant, and were widespread geographically?
   A. trace fossils
   B. index fossils
   C. carbon films
   D. body fossils

5. Which term means matching up rock layers in different places?
   A. superposition
   B. correlation
   C. uniformitarianism
   D. absolute dating

6. Which of the following is least likely to be found as a fossil?
   A. clam shell
   B. shark tooth
   C. snail shell
   D. jellyfish imprint

7. Which type of fossil preservation is a thin carbon silhouette of the original organism?
   A. cast
   B. carbon film
   C. mold
   D. permineralized remains

8. Which isotope is useful for dating wood and charcoal that is less than about 75,000 years old?
   A. carbon-14
   B. potassium-40
   C. uranium-238
   D. argon-40

Use the diagram below to answer questions 9–11.

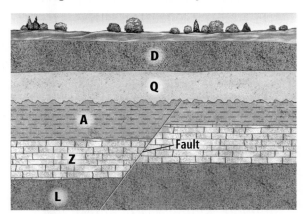

9. Which sequence of letters describes the rock layers in the diagram from oldest to youngest?
   A. D, Q, A, Z, L
   B. L, Z, A, Q, D
   C. Z, L, A, D, Q
   D. Q, D, L, Z, A

10. What does the wavy line between layers A and Q represent?
    A. a disconformity
    B. a fault
    C. a nonconformity
    D. an angular unconformity

11. Which of the following correctly describes the relative age of the fault?
    A. younger than A, but older than Q
    B. younger than Z, but older than L
    C. younger than Q, but older than A
    D. younger than D, but older than Q

Standardized Test Practice

Part 2 | Short Response/Grid In

Record your answers on the answer sheet provided by your teacher or on a sheet of paper.

12. What is a fossil?

13. How is a fossil cast different from a fossil mold?

14. Describe the principle of uniformitarianism.

15. Explain how the original remains of an insect can be preserved as a fossil in amber.

16. Why do scientists hypothesize that Earth is about 4.5 billion years old?

17. Describe the process of radioactive decay. Use the terms *isotope, nucleus,* and *half-life* in your answer.

Use the table below to answer questions 18–20.

Number of Half-lives	Parent Isotope Remaining (%)
1	100
2	X
3	25
4	12.5
5	Y

18. What value should replace the letter X in the table above?

19. What value should replace the letter Y in the table above?

20. Explain the relationship between the number of half-lives that have elapsed and the amount of parent isotope remaining.

21. Compare and contrast the three types of unconformities.

22. Why are index fossils useful for estimating the age of rock layers?

Part 3 | Open Ended

Record your answers on a sheet of paper.

23. Why are fossils important? What information do they provide?

24. List three different types of trace fossils. Explain how each type forms.

Examine the graph below and answer questions 25–27.

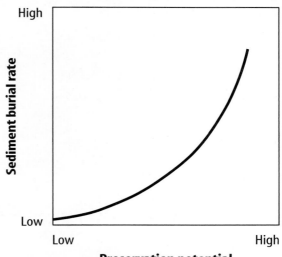

Relationship Between Sediment Burial Rate and Potential for Remains to Become Fossils

25. How does the potential for remains to be preserved change as the rate of burial by sediment increases?

26. Why do you think this relationship exists?

27. What other factors affect the potential for the remains of organisms to become fossils?

28. How could a fossil of an organism that lived in ocean water millions of years ago be found in the middle of North America?

Test-Taking Tip

Check It Again Double check your answers before turning in the test.

chapter 10

Geologic Time

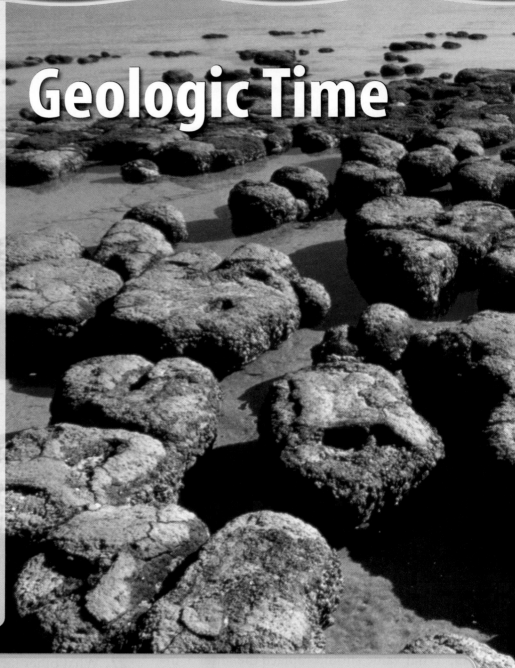

The BIG Idea
Scientists use units of geologic time to interpret the history of life on Earth.

SECTION 1
Life and Geologic Time
Main Idea Fossils provide evidence that life on Earth has evolved throughout geologic time.

SECTION 2
Early Earth History
Main Idea Primitive life forms existed on Earth during Precambrian time and the Paleozoic Era.

SECTION 3
Middle and Recent Earth History
Main Idea Life forms continued to evolve through the Mesozoic Era and the current Cenozoic Era.

Looking at the Past

The stromatolites in the picture hardly have changed since they first appeared 3.5 billion years ago. Looking at these organisms today allows us to imagine what the early Earth might have looked like. In this chapter, you will see how much Earth has changed over time, even as some parts remain the same.

Science Journal Describe how an animal or plant might change if Earth becomes hotter in the next million years.

Start-Up Activities

Survival Through Time

Environments include the living and nonliving things that surround and affect organisms. Whether or not an organism survives in its environment depends upon its characteristics. Only if an organism survives until adulthood can it reproduce and pass on its characteristics to its offspring. In this lab, you will use a model to find out how one characteristic can determine whether individuals can survive in an environment.

1. Cut 15 pieces each of green, orange, and blue yarn into 3-cm lengths.
2. Scatter them on a sheet of green construction paper.
3. Have your partner use a pair of tweezers to pick up as many pieces as possible in 15 s.
4. **Think Critically** In your Science Journal, discuss which colors your partner selected. Which color was least selected? Suppose that the construction paper represents grass, the yarn pieces represent insects, and the tweezers represent an insect-eating bird. Which color of insect do you predict would survive to adulthood?

Geological Time Make the following Foldable to help you identify the major events in each era of geologic time.

STEP 1 Fold the top of a vertical piece of paper down and the bottom up to divide the paper into thirds.

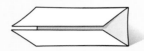

STEP 2 Turn the paper horizontally; unfold and label the three columns as shown.

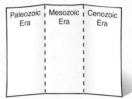

Read for Main Ideas As you read the chapter, list at least three major events that occurred in each era. Keep the events in chronological order. For each event, note the period in which it took place.

Preview this chapter's content and activities at
blue.msscience.com

Get Ready to Read

Questions and Answers

① Learn It! Knowing how to find answers to questions will help you on reviews and tests. Some answers can be found in the textbook, while other answers require you to go beyond the textbook. These answers might be based on knowledge you already have or things you have experienced.

② Practice It! Read the excerpt below. Answer the following questions and then discuss them with a partner.

> The Paleozoic Era, or era of ancient life, began about 544 million years ago and ended about 248 million years ago. Traces of life are much easier to find in Paleozoic rocks than in Precambrian rocks.
>
> —*from page 282*

- What does the term *Paleozoic* mean?
- What does the term *Precambrian* mean?
- Why are traces of life easier to find in Paleozoic rocks than in Precambrian rocks?

③ Apply It! Look at some questions in the chapter review. Which questions can be answered directly from the text? Which require you to go beyond the text?

Target Your Reading

Reading Tip: As you read, keep track of questions you answer in the chapter. This will help you remember what you read.

Use this to focus on the main ideas as you read the chapter.

1 Before you read the chapter, respond to the statements below on your worksheet or on a numbered sheet of paper.
- Write an **A** if you **agree** with the statement.
- Write a **D** if you **disagree** with the statement.

2 After you read the chapter, look back to this page to see if you've changed your mind about any of the statements.
- If any of your answers changed, explain why.
- Change any false statements into true statements.
- Use your revised statements as a study guide.

Science Online
Print out a worksheet of this page at blue.msscience.com

Before You Read A or D		Statement	After You Read A or D
	1	The fossil record shows that species have changed over geologic time.	
	2	Geologic time units are based on life-forms that lived only during certain periods of time.	
	3	Eras are longer than eons.	
	4	Precambrian time is the shortest part of Earth's history.	
	5	No life-forms existed on Earth during Precambrian time.	
	6	Oxygen gas has always been a major component of Earth's atmosphere throughout geologic time.	
	7	All dinosaurs were large, slow-moving, cold-blooded reptiles.	
	8	Dinosaurs lived during the Mesozoic Era.	
	9	Many scientists hypothesize that a comet or asteroid collision with Earth, ended Mesozoic Era.	

272 B

section 1

Life and Geologic Time

as you read

What You'll Learn
- **Explain** how geologic time can be divided into units.
- **Relate** changes of Earth's organisms to divisions on the geologic time scale.
- **Describe** how plate tectonics affects species.

Why It's Important
The life and landscape around you are the products of change through geologic time.

Review Vocabulary
fossils: remains, traces, or imprints of prehistoric organisms

New Vocabulary
- geologic time scale
- eon
- era
- period
- epoch
- organic evolution
- species
- natural selection
- trilobite
- Pangaea

Geologic Time

A group of students is searching for fossils. By looking in rocks that are hundreds of millions of years old, they hope to find many examples of trilobites (TRI loh bites) so that they can help piece together a puzzle. That puzzle is to find out what caused the extinction of these organisms. **Figure 1** shows some examples of what they are finding. The fossils are small, and their bodies are divided into segments. Some of them seem to have eyes. Could these interesting fossils be trilobites?

Trilobites are small, hard-shelled organisms that crawled on the seafloor and sometimes swam through the water. Most ranged in size from 2 cm to 7 cm in length and from 1 cm to 3 cm in width. They are considered to be index fossils because they lived over vast regions of the world during specific periods of geologic time.

The Geologic Time Scale The appearance or disappearance of types of organisms throughout Earth's history marks important occurrences in geologic time. Paleontologists have been able to divide Earth's history into time units based on the life-forms that lived only during certain periods. This division of Earth's history makes up the **geologic time scale.** However, sometimes fossils are not present, so certain divisions of the geologic time scale are based on other criteria.

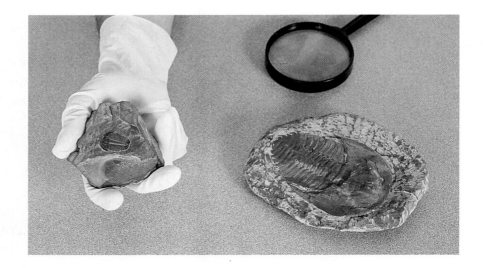

Figure 1 Many sedimentary rocks in the United States are rich in invertebrate fossils such as these trilobites.

272 CHAPTER 10 Geologic Time

Major Subdivisions of Geologic Time The oldest rocks on Earth contain no fossils. Then, for many millions of years after the first appearance of fossils, the fossil record remained sparse. Later in Earth's history came an explosion in the abundance and diversity of organisms. These organisms left a rich fossil record. As shown in **Figure 2,** four major subdivisions of geologic time are used—eons, eras, periods, and epochs. The longest subdivisions—**eons**—are based upon the abundance of certain fossils.

Reading Check *What are the major subdivisions of geologic time?*

Next to eons, the longest subdivisions are the **eras,** which are marked by major, striking, and worldwide changes in the types of fossils present. For example, at the end of the Mesozoic Era, many kinds of invertebrates, birds, mammals, and reptiles became extinct.

Eras are subdivided into periods. **Periods** are units of geologic time characterized by the types of life existing worldwide at the time. Periods can be divided into smaller units of time called **epochs.** Epochs also are characterized by differences in life-forms, but some of these differences can vary from continent to continent. Epochs of periods in the Cenozoic Era have been given specific names. Epochs of other periods usually are referred to simply as early, middle, or late. Epochs are further subdivided into units of shorter duration.

Dividing Geologic Time There is a limit to how finely geologic time can be subdivided. It depends upon the kind of rock record that is being studied. Sometimes it is possible to distinguish layers of rock that formed during a single year or season. In other cases, thick stacks of rock that have no fossils provide little information that could help in subdividing geologic time.

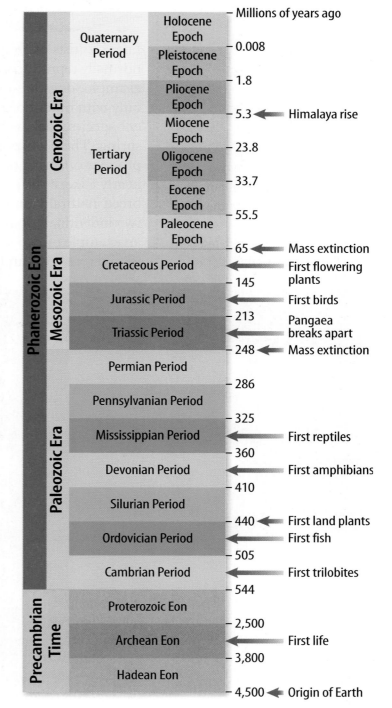

Figure 2 Scientists have divided the geologic time scale into subunits based upon the appearance and disappearance of types of organisms.
Explain *how the even blocks in this chart can be misleading.*

Organic Evolution

The fossil record shows that species have changed over geologic time. This change through time is known as **organic evolution.** According to most theories about organic evolution, environmental changes can affect an organism's survival. Those organisms that are not adapted to changes are less likely to survive or reproduce. Over time, the elimination of individuals that are not adapted can cause changes to species of organisms.

INTEGRATE Life Science

Species Many ways of defining the term species (SPEE sheez) have been proposed. Life scientists often define a **species** as a group of organisms that normally reproduces only with other members of their group. For example, dogs are a species because dogs mate and reproduce only with other dogs. In some rare cases, members of two different species, such as lions and tigers, can mate and produce offspring. These offspring, however, are usually sterile and cannot produce offspring of their own. Even though two organisms look nearly alike, if the populations they each come from do not interbreed naturally and produce offspring that can reproduce, the two individuals do not belong to the same species. **Figure 3** shows an example of two species that look similar to each other but live in different areas and do not mate naturally with each other.

Figure 3 Just because two organisms look alike does not mean that they belong to the same species.
Describe *an experiment to test if these lizards are separate species.*

The coast horned lizard lives along the coast of central and southern California.

The desert horned lizard lives in arid regions of the southwestern United States.

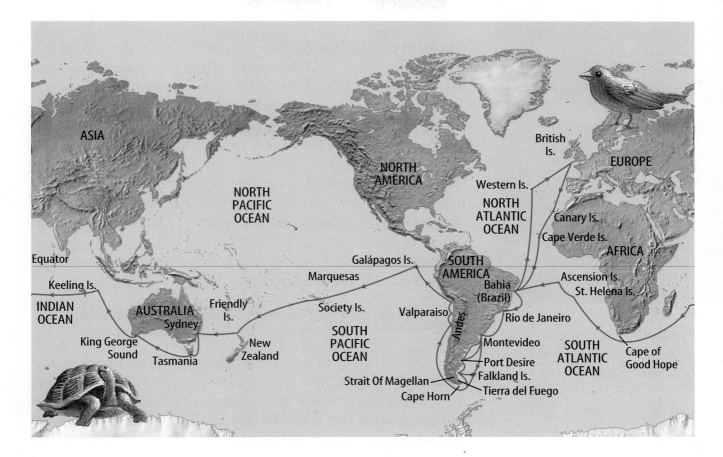

Figure 4 Charles Darwin sailed around the world between 1831 and 1836 aboard the HMS *Beagle* as a naturalist. On his journey he saw an abundance of evidence for natural selection, especially on the Galápagos Islands off the western coast of South America.

Natural Selection Charles Darwin was a naturalist who sailed around the world from 1831 to 1836 to study biology and geology. **Figure 4** shows a map of his journey. With some of the information about the plants and animals he observed on this trip in mind, he later published a book about the theory of evolution by natural selection.

In his book, he proposed that **natural selection** is a process by which organisms with characteristics that are suited to a certain environment have a better chance of surviving and reproducing than organisms that do not have these characteristics. Darwin knew that many organisms are capable of producing more offspring than can survive. This means that organisms compete with each other for resources necessary for life, such as food and living space. He also knew that individual organisms within the same species could be different, or show variations, and that these differences could help or hurt the individual organism's chance of surviving.

Some organisms that were well suited to their environment lived longer and had a better chance of producing offspring. Organisms that were poorly adapted to their environment produced few or no offspring. Because many characteristics are inherited, the characteristics of organisms that are better adapted to the environment get passed on to offspring more often. According to Darwin, this can cause a species to change over time.

SECTION 1 Life and Geologic Time

Figure 5 Giraffes can eat leaves off the branches of tall trees because of their long necks.

Figure 6 Cat breeders have succeeded in producing a great variety of cats by using the principle of artificial selection.

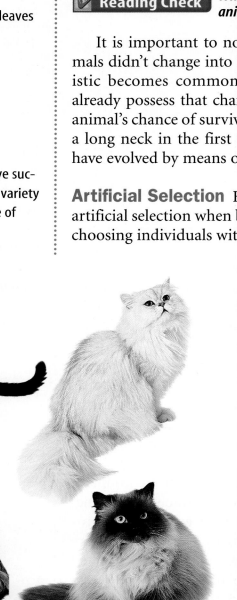

Natural Selection Within a Species Suppose that an animal species exists in which a few of the individuals have long necks, but most have short necks. The main food for the animal is the leafy foliage on trees in the area. What happens if the climate changes and the area becomes dry? The lower branches of the trees might not have any leaves. Now which of the animals will be better suited to survive? Clearly, the long-necked animals have a better chance of surviving and reproducing. Their offspring will have a greater chance of inheriting the important characteristic. Gradually, as the number of long-necked animals becomes greater, the number of short-necked animals decreases. The species might change so that nearly all of its members have long necks, as the giraffe in **Figure 5** has.

Reading Check *What might happen to the population of animals if the climate became wet again?*

It is important to notice that individual, short-necked animals didn't change into long-necked animals. A new characteristic becomes common in a species only if some members already possess that characteristic and if the trait increases the animal's chance of survival. If no animal in the species possessed a long neck in the first place, a long-necked species could not have evolved by means of natural selection.

Artificial Selection Humans have long used the principle of artificial selection when breeding domestic animals. By carefully choosing individuals with desired characteristics, animal breeders have created many breeds of cats, dogs, cattle, and chickens. **Figure 6** shows the great variety of cats produced by artificial selection.

The Evolution of New Species Natural selection explains how characteristics change and how new species arise. For example, if the short-necked animals migrated to a different location, they might have survived. They could have continued to reproduce in the new location, eventually developing enough different characteristics from the long-necked animals that they might not be able to breed with each other. At this point, at least one new species would have evolved.

Trilobites

Remember the trilobites? The term *trilobite* comes from the structure of the hard outer skeleton or exoskeleton. The exoskeleton of a **trilobite** consists of three lobes that run the length of the body. As shown in **Figure 7,** the trilobite's body also has a head (cephalon), a segmented middle section (thorax), and a tail (pygidium).

Changing Characteristics of Trilobites Trilobites inhabited Earth's oceans for more than 200 million years. Throughout the Paleozoic Era, some species of trilobites became extinct and other new species evolved. Species of trilobites that lived during one period of the Paleozoic Era showed different characteristics than species from other periods of this era. As **Figure 8** shows, paleontologists can use these different characteristics to demonstrate changes in trilobites through geologic time. These changes can tell you about how different trilobites from different periods lived and responded to changes in their environments.

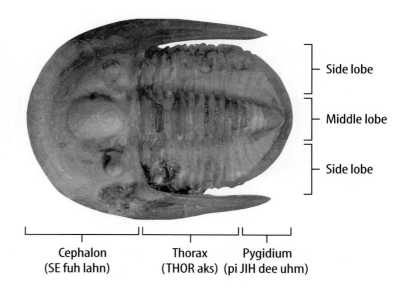

Figure 7 The trilobite's body was divided into three lobes that run the length of the body—two side lobes and one middle lobe.

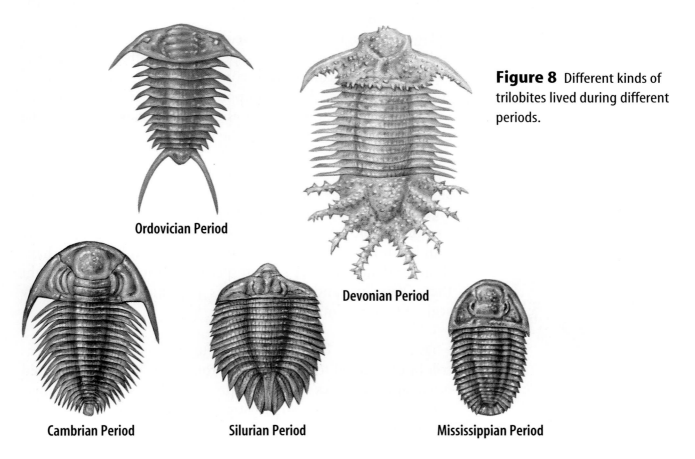

Figure 8 Different kinds of trilobites lived during different periods.

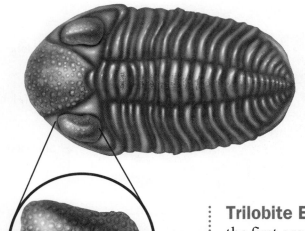
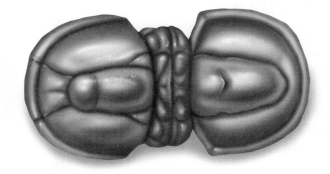

Figure 9 Trilobites had many different types of eyes. Some had eyes that contained hundreds of small circular lenses, somewhat like an insect. The blind trilobite (right) had no eyes.

Figure 10 *Olenellus* is one of the most primitive trilobite species.

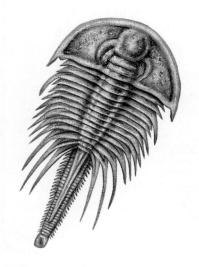

Trilobite Eyes Trilobites, shown in **Figure 9,** might have been the first organisms that could view the world with complex eyes. Trilobite eyes show the result of natural selection. The position of the eyes on an organism gives clues about where it must have lived. Eyes that are located toward the front of the head indicate an organism that was adapted for active swimming. If the eyes are located toward the back of the head, the organism could have been a bottom dweller. In most species of trilobites, the eyes were located midway on the head—a compromise for an organism that was adapted for crawling on the seafloor and swimming in the water.

Over time, the eyes in trilobites changed. In many trilobite species, the eyes became progressively smaller until they completely disappeared. Blind trilobites, such as the one on the right in **Figure 9,** might have burrowed into sediments on the seafloor or lived deeper than light could penetrate. In other species, however, the eyes became more complex. One kind of trilobite, *Aeglina,* developed large compound eyes that had numerous individual lenses. Some trilobites developed stalks that held the eyes upward. Where would this be useful?

Trilobite Bodies The trilobite body and tail also underwent significant changes in form through time, as you can see in **Figure 8** on the previous page. A special case is *Olenellus,* shown in **Figure 10.** This trilobite, which lived during the Early Cambrian Period, had an extremely segmented body—perhaps more so than any other known species of trilobite. It is thought that *Olenellus,* and other species that have so many body segments, are primitive trilobites.

Fossils Show Changes Trilobite exoskeletons changed as trilobites adapted to changing environments. Species that could not adapt became extinct. What processes on Earth caused environments to change so drastically that species adapted or became extinct?

Plate Tectonics and Earth History

Plate tectonics is one possible answer to the riddle of trilobite extinction. Earth's moving plates caused continents to collide and separate many times. Continental collisions formed mountains and closed seas caught between continents. Continental separations created wider, deeper seas between continents. By the end of the Paleozoic Era, sea levels had dropped and the continents had come together to form one giant landmass, the supercontinent **Pangaea** (pan JEE uh). Because trilobites lived in the oceans, their environment was changed or destroyed. **Figure 11** shows the arrangement of continents at the end of the Paleozoic Era. What effect might these changes have had on the trilobite populations?

Not all scientists accept the above explanation for the extinctions at the end of the Paleozoic Era, and other possibilities—such as climate change—have been proposed. As in all scientific debates, you must consider the evidence carefully and come to conclusions based on the evidence.

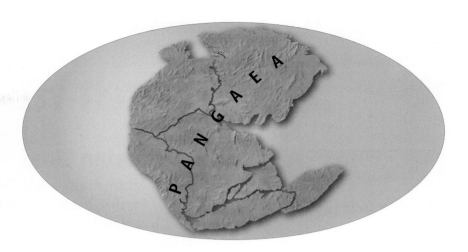

Figure 11 The amount of shallow water environment was reduced when Pangaea formed. **Describe** how this change affected organisms that lived along the coasts of continents.

section 1 review

Summary

Geologic Time
- Earth's history is divided into eons, eras, periods, and epochs, based on fossils.

Organic Evolution
- The fossil record indicates that species have changed over time.
- Charles Darwin proposed natural selection to explain change in species.
- In natural selection, organisms best suited to their environments survive and produce the most offspring.

Trilobites
- Trilobites were abundant in the Paleozoic fossil record and can be used as index fossils.

Plate Tectonics and Earth History
- Continents moving through time have influenced the environments of past organisms.

Self Check

1. **Discuss** how fossils relate to the geologic time scale.
2. **Infer** how plate tectonics might lead to extinction.
3. **Infer** how the eyes of a trilobite show how it lived.
4. **Explain** how paleontologists use trilobite fossils as index fossils for various geologic time periods.
5. **Think Critically** Aside from moving plates, what other factors could cause an organism's environment to change? How would this affect species?

Applying Skills

6. **Recognize Cause and Effect** Answer the questions below.
   a. How does natural selection cause evolutionary change to take place?
   b. How could the evolution of a characteristic within one species affect the evolution of a characteristic within another species? Give an example.

section 2

Early Earth History

As you read

What You'll Learn
- **Identify** characteristic Precambrian and Paleozoic life-forms.
- **Draw conclusions** about how species adapted to changing environments in Precambrian time and the Paleozoic Era.
- **Describe** changes in Earth and its life-forms at the end of the Paleozoic Era.

Why It's Important
The Precambrian includes most of Earth's history.

Review Vocabulary
life: state of being in which one grows, reproduces, and maintains a constant internal environment

New Vocabulary
- Precambrian time
- cyanobacteria
- Paleozoic Era

Precambrian Time

It may seem strange, but **Figure 12** is probably an accurate picture of Earth's first billion years. Over the next 3 billion years, simple life-forms began to colonize the oceans.

Look again at the geologic time scale shown in **Figure 2**. **Precambrian** (pree KAM bree un) **time** is the longest part of Earth's history and includes the Hadean, Archean, and Proterozoic Eons. Precambrian time lasted from about 4.5 billion years ago to about 544 million years ago. The oldest rocks that have been found on Earth are about 4 billion years old. However, rocks older than about 3.5 billion years are rare. This probably is due to remelting and erosion.

Although the Precambrian was the longest interval of geologic time, relatively little is known about the organisms that lived during this time. One reason is that many Precambrian rocks have been so deeply buried that they have been changed by heat and pressure. Many fossils can't withstand these conditions. In addition, most Precambrian organisms didn't have hard parts that otherwise would have increased their chances to be preserved as fossils.

Figure 12 During the early Precambrian, Earth was a lifeless planet with many volcanoes.

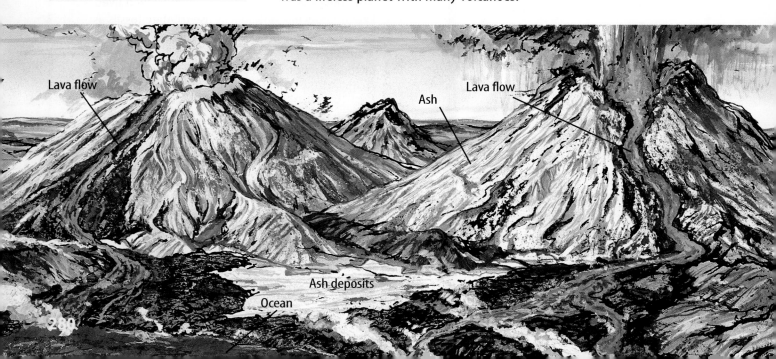

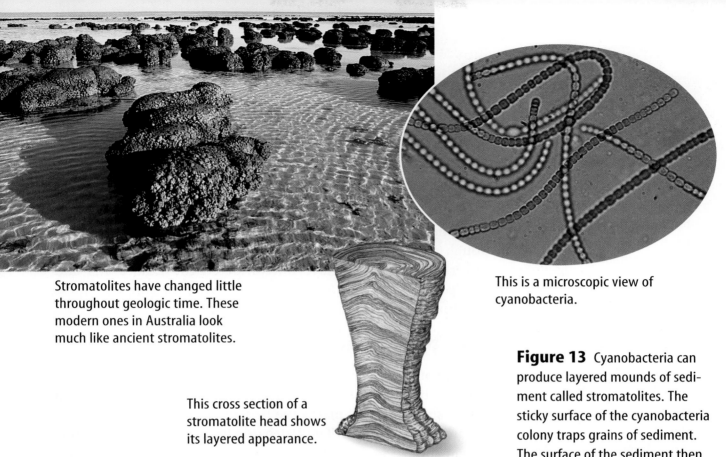

Stromatolites have changed little throughout geologic time. These modern ones in Australia look much like ancient stromatolites.

This cross section of a stromatolite head shows its layered appearance.

This is a microscopic view of cyanobacteria.

Figure 13 Cyanobacteria can produce layered mounds of sediment called stromatolites. The sticky surface of the cyanobacteria colony traps grains of sediment. The surface of the sediment then becomes colonized with cyanobacteria again, and the cycle repeats, producing the layers inside the stromatolite.

Early Life Many studies of the early history of life involve ancient stromatolites (stroh MA tuh lites). **Figure 13** shows stromatolites, which are layered mats formed by cyanobacteria colonies. **Cyanobacteria** are blue-green algae thought to be one of the earliest forms of life on Earth. Cyanobacteria first appeared about 3.5 billion years ago. They contained chlorophyll and used photosynthesis. This is important because during photosynthesis, they produced oxygen, which helped change Earth's atmosphere. Following the appearance of cyanobacteria, oxygen became a major atmospheric gas. Also of importance was that the ozone layer in the atmosphere began to develop, shielding Earth from ultraviolet rays. It is hypothesized that these changes allowed species of single-celled organisms to evolve into more complex organisms.

 **Reading Check** *What atmospheric gas is produced by photosynthesis?*

Animals without backbones, called invertebrates (ihn VUR tuh brayts), appeared toward the end of Precambrian time. Imprints of invertebrates have been found in late Precambrian rocks, but because these early invertebrates were soft bodied, they weren't often preserved as fossils. Because of this, many Precambrian fossils are trace fossils.

INTEGRATE Chemistry

Earth's First Air Cyanobacteria are thought to have been one of the mechanisms by which Earth's early atmosphere became richer in oxygen. Research the composition of Earth's early atmosphere and where these gases probably came from. Record your findings in your Science Journal.

Dating Rock Layers with Fossils

Procedure
1. Draw three rock layers.
2. Number the layers 1 to 3, bottom to top.
3. Layer 1 contains fossil A. Layer 2 contains fossils A and B. Layer 3 contains fossil C.
4. Fossil A lived from the Cambrian through the Ordovician. Fossil B lived from the Ordovician through the Silurian. Fossil C lived in the Silurian and Devonian.

Analysis
1. Which layers were you able to date to a specific period?
2. Why isn't it possible to determine during which specific period the other layers formed?

Figure 14 This giant predatory fish lived in seas that were present in North America during the Devonian Period. It grew to about 6 m in length.

Unusual Life-Forms A group of animals with shapes similar to modern jellyfish, worms, and soft corals was living late in Precambrian time. Fossils of these organisms were first found in the Ediacara Hills in southern Australia. This group of organisms has become known as the Ediacaran (ee dee uh KAR un) fauna. **Figure 15** shows some of these fossils.

 *What modern organisms do some Ediacaran organisms resemble?*

Ediacaran animals were bottom dwellers and might have had tough outer coverings like air mattresses. Trilobites and other invertebrates might have outcompeted the Ediacarans and caused their extinction, but nobody knows for sure why these creatures disappeared.

The Paleozoic Era

As you have learned, fossils are unlikely to form if organisms have only soft parts. An abundance of organisms with hard parts, such as shells, marks the beginning of the Paleozoic (pay lee uh ZOH ihk) Era. The **Paleozoic Era,** or era of ancient life, began about 544 million years ago and ended about 248 million years ago. Traces of life are much easier to find in Paleozoic rocks than in Precambrian rocks.

Paleozoic Life Because warm, shallow seas covered large parts of the continents during much of the Paleozoic Era, many of the life-forms scientists know about were marine, meaning they lived in the ocean. Trilobites were common, especially early in the Paleozoic. Other organisms developed shells that were easily preserved as fossils. Therefore, the fossil record of this era contains abundant shells. However, invertebrates were not the only animals to live in the shallow, Paleozoic seas.

Vertebrates, or animals with backbones, also evolved during this era. The first vertebrates were fishlike creatures without jaws. Armoured fish with jaws such as the one shown in **Figure 14** lived during the Devonian Period. Some of these fish were so huge that they could eat large sharks with their powerful jaws. By the Devonian Period, forests had appeared and vertebrates began to adapt to land environments, as well.

NATIONAL GEOGRAPHIC VISUALIZING UNUSUAL LIFE-FORMS

Figure 15

A variety of 600-million-year-old fossils—known as Ediacaran (eed ee uh KAR un) fauna—have been found on every continent except Antarctica. These unusual organisms were originally thought to be descendants of early animals such as jellyfish, worms, and coral. Today, paleontologists debate whether these organisms were part of the animal kingdom or belonged to an entirely new kingdom whose members became extinct about 545 million years ago.

DICKENSONIA (dihk un suh NEE uh) Impressions of *Dickensonia,* a bottom-dwelling wormlike creature, have been discovered. Some are nearly one meter long.

RANGEA (rayn JEE uh) As it lay rooted in sea-bottom sediments, *Rangea* may have snagged tiny bits of food by filtering water through its body.

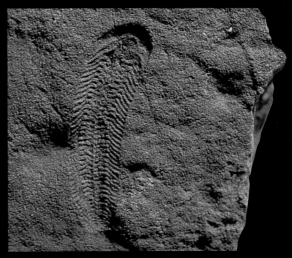

SPRIGGINA (sprih GIHN uh) Some scientists hypothesize that the four-centimeter-long *Spriggina* was a type of crawling, segmented organism. Others suggest that it sat upright while attached to the seafloor.

CYCLOMEDUSA (si kloh muh DEW suh) Although it looks a lot like a jellyfish, *Cyclomedusa* may have had more in common with modern sea anemones. Some paleontologists, however, hypothesize that it is unrelated to any living organism.

SECTION 2 Early Earth History **283**

Figure 16 Amphibians probably evolved from fish like *Panderichthys* (pan dur IHK theez), which had leglike fins and lungs.

Life on Land Based on their structure, paleontologists know that many ancient fish had lungs as well as gills. Lungs enabled these fish to live in water with low oxygen levels—when needed they could swim to the surface and breathe air. Today's lungfish also can get oxygen from the water through gills and from the air through lungs.

One kind of ancient fish had lungs and leglike fins, which were used to swim and crawl around on the ocean bottom. Paleontologists hypothesize that amphibians might have evolved from this kind of fish, shown in **Figure 16**. The characteristics that helped animals survive in oxygen-poor waters also made living on land possible. Today, amphibians live in a variety of habitats in water and on land. They all have at least one thing in common, though. They must lay their eggs in water or moist places.

 *What are some characteristics of the fish from which amphibians might have evolved?*

By the Pennsylvanian Period, some amphibians evolved an egg with a membrane that protected it from drying out. Because of this, these animals, called reptiles, no longer needed to lay eggs in water. Reptiles also have skin with hard scales that prevent loss of body fluids. This adaptation enables them to survive farther from water and in relatively dry climates, as shown in **Figure 17,** where many amphibians cannot live.

Science online

Topic: Paleozoic Life
Visit blue.msscience.com for Web links to information about Paleozoic life.

Activity Prepare a presentation on the organisms of one period of the Paleozoic Era. Describe a few animals from different groups, including how and where they lived. Are any of these creatures alive today, and if not, when did they become extinct?

Figure 17 Reptiles have scaly skins that allow them to live in dry places.

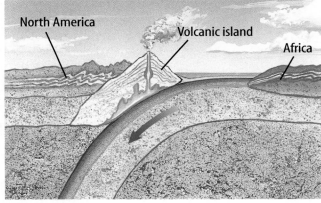

More than 375 million years ago, volcanic island chains formed in the ocean and were pushed against the coast as Africa moved toward North America.

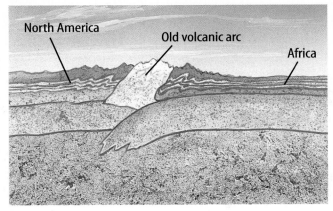

About 375 million years ago, the African plate collided with the North American plate, forming mountains on both continents.

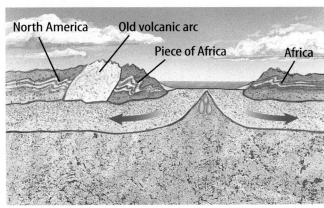

About 200 million years ago, the Atlantic Ocean opened up, separating the two continents.

Mountain Building Several mountain-building episodes occurred during the Paleozoic Era. The Appalachian Mountains, for example, formed during this time. This happened in several stages, as shown in **Figure 18.** The first mountain-building episode occurred as the ocean separating North America from Europe and Africa closed. Several volcanic island chains that had formed in the ocean collided with the North American Plate, as shown in the top picture of **Figure 18.** The collision of the island chains generated high mountains.

The next mountain-building episode was a result of the African Plate colliding with the North American Plate, as shown in the left picture of **Figure 18.** When Africa and North America collided, rock layers were folded and faulted. Some rocks originally deposited near the eastern coast of the North American Plate were pushed along faults as much as 65 km westward by the collision. Sediments were uplifted to form an immense mountain belt, part of which still remains today.

Figure 18 The Appalachian Mountains formed in several stages.
Infer how these movements affected species in the Appalachians.

Figure 19 Hyoliths were organisms that became extinct at the end of the Paleozoic Era.

End of an Era At the end of the Paleozoic Era, more than 90 percent of all marine species and 70 percent of all land species died off. **Figure 19** shows one such animal. The cause of these extinctions might have been changes in climate and a lowering of sea level.

Near the end of the Permian Period, the continental plates came together and formed the supercontinent Pangaea. Glaciers formed over most of its southern part. The slow, gradual collision of continental plates caused mountain building. Mountain-building processes caused seas to close and deserts to spread over North America and Europe. Many species, especially marine organisms, couldn't adapt to these changes, and became extinct.

Other Hypotheses Other explanations also have been proposed for this mass extinction. During the late Paleozoic Era, volcanoes were extremely active. If the volcanic activity was great enough, it could have affected the entire globe. Another recent theory is similar to the one proposed to explain the extinction of dinosaurs. Perhaps a large asteroid or comet collided with Earth some 248 million years ago. This event could have caused widespread extinctions just as many paleontologists suggest happened at the end of the Mesozoic Era, 65 million years ago. Perhaps the extinction at the end of the Paleozoic Era was caused by several or all of these events happening at about the same time.

section 2 review

Summary

Precambrian Time

- Precambrian time covers almost 4 billion years of Earth history, but little is known about the organisms of this time.
- Cyanobacteria were among the earliest life-forms.

The Paleozoic Era

- Invertebrates developed shells and other hard parts, leaving a rich fossil record.
- Vertebrates—animals with backbones—appeared during this era.
- Plants and amphibians first moved to land during the Paleozoic Era.
- Adaptations in reptiles allow them to move away from water for reproduction.
- Geologic events at the end of the Paleozoic Era led to a mass extinction.

Self Check

1. **List** the geologic events that ended the Paleozoic Era.
2. **Infer** how geologic events at the end of the Paleozoic Era might have caused extinctions.
3. **Discuss** the advance that allowed reptiles to reproduce away from water. Why was this an advantage?
4. **Identify** the major change in life-forms that occurred at the end of Precambrian time.
5. **Think Critically** How did cyanobacteria aid the evolution of complex life on land? Do you think cyanobacteria are as significant to this process today as they were during Precambrian time?

Applying Skills

6. **Use a Database** Research trilobites and describe these organisms and their habitats in your Science Journal. Include hand-drawn illustrations and compare them with the illustrations in your references.

blue.msscience.com/self_check_quiz

Changing Species

In this lab, you will observe how adaptation within a species might cause the evolution of a particular trait, leading to the development of a new species.

● Real-World Question

How might adaptation within a species cause the evolution of a particular trait?

Goals
- **Model** adaptation within a species.

Materials
deck of playing cards

● Procedure

1. **Remove** all of the kings, queens, jacks, and aces from a deck of playing cards.
2. Each remaining card represents an individual in a population of animals called "varimals." The number on each card represents the height of the individual.
3. **Calculate** the average height of the population of varimals represented by your cards.
4. Suppose varimals eat grass, shrubs, and leaves from trees. A drought causes many of these plants to die. All that's left are a few tall trees. Only varimals at least 6 units tall can reach the leaves on these trees.
5. All the varimals under 6 units leave the area or die from starvation. Discard all of the cards with a number less than 6. Calculate the new average height of the varimals.
6. Shuffle the deck of remaining cards.
7. **Draw** two cards at a time. Each pair represents a pair of varimals that will mate.
8. The offspring of each pair reaches the average height of its parents. Calculate and record the height of each offspring.
9. Discard all parents and offspring under 8 units tall and repeat steps 6–8. Now calculate the new average height of varimals. Include both the parents and offspring in your calculation.

● Conclude and Apply

1. **Describe** how the height of the population changed.
2. **Explain** If you hadn't discarded the shortest varimals, would the average height of the population have changed as much?
3. Suppose the offspring grew to the height of one of its parents. How would the results change in each of the following scenarios?
   a. The height value for the offspring is chosen by coin toss.
   b. The height value for the offspring is whichever parent is tallest.
4. **Explain** If there had been no variation in height before the droughts occurred, would the species have been able to evolve?

section 3

Middle and Recent Earth History

as you read

What You'll Learn
- **Compare and contrast** characteristic life-forms in the Mesozoic and Cenozoic Eras.
- **Explain** how changes caused by plate tectonics affected organisms during the Mesozoic Era.
- **Identify** when humans first appeared on Earth.

Why It's Important
Many important groups of animals, like birds and mammals, appeared during the Mesozoic Era.

Review Vocabulary
dinosaur: a reptile from one of two orders that dominated the Mesozoic Era

New Vocabulary
- Mesozoic Era
- Cenozoic Era

The Mesozoic Era

Dinosaurs have captured people's imaginations since their bones first were unearthed more than 150 years ago. Dinosaurs and other interesting animals lived during the Mesozoic Era, which was between 248 and 65 million years ago. The Mesozoic Era also was marked by rapid movement of Earth's plates.

The Breakup of Pangaea The **Mesozoic** (meh zuh ZOH ihk) **Era,** or era of middle life, was a time of many changes on Earth. At the beginning of the Mesozoic Era, all continents were joined as a single landmass called Pangaea, as shown in **Figure 11.**

Pangaea separated into two large landmasses during the Triassic Period, as shown in **Figure 20.** The northern mass was Laurasia (law RAY zhuh), and Gondwanaland (gahn DWAH nuh land) was the southern landmass. As the Mesozoic Era continued, Laurasia and Gondwanaland broke apart and eventually formed the present-day continents.

Species that had adapted to the new environments survived the mass extinction at the end of the Paleozoic Era. Recall that a reptile's skin helps it retain bodily fluids. This characteristic, along with their shelled eggs, enabled reptiles to adapt readily to the drier climate of the Mesozoic Era. Reptiles became the most conspicuous animals on land by the Triassic Period.

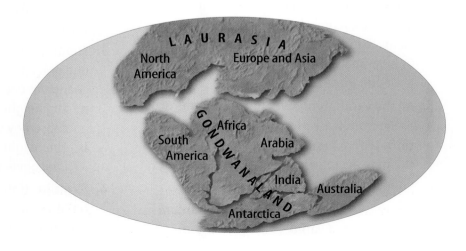

Figure 20 At the end of the Triassic Period, Pangaea began to break up into the northern supercontinent, Laurasia, and the southern supercontinent, Gondwanaland.

Dinosaurs What were the dinosaurs like? Dinosaurs ranged in height from less than 1 m to enormous creatures like *Apatosaurus* and *Tyrannosaurus*. The first small dinosaurs appeared during the Triassic Period. Larger species appeared during the Jurassic and Cretaceous Periods. Throughout the Mesozoic Era, new species of dinosaurs evolved and other species became extinct.

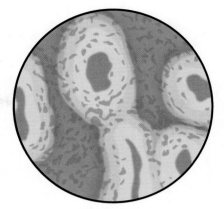
Dinosaur bone

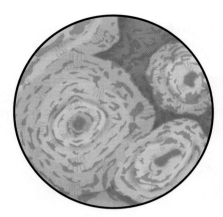
Mammal bone

Figure 21 Some dinosaur bones show structural features that are like mammal bones, leading some paleontologists to think that dinosaurs were warm blooded like mammals.

Dinosaurs Were Active Studying fossil footprints sometimes allows paleontologists to calculate how fast animals walked or ran. Some dinosaur tracks indicate that these animals were much faster runners than you might think. *Gallimimus* was 4 m long and could reach speeds of 65 km/h—as fast as a modern racehorse.

Some studies also indicate that dinosaurs might have been warm blooded, not cold blooded like present-day reptiles. The evidence that leads to this conclusion has to do with their bone structure. Slices through some cold-blooded animal bones show rings similar to growth rings in trees. The bones of some dinosaurs don't show this ring structure. Instead, they are similar to bones found in modern mammals, as you can see in **Figure 21**.

 *Why do some paleontologists think that dinosaurs were warm blooded?*

These observations indicate that some dinosaurs might have been warm-blooded, fast-moving animals somewhat like present-day mammals and birds. They might have been quite different from present-day reptiles.

Good Mother Dinosaurs The fossil record also indicates that some dinosaurs nurtured their young and traveled in herds in which the adults surrounded their young.

One such dinosaur is *Maiasaura*. This dinosaur built nests in which it laid its eggs and raised its offspring. Nests have been found in relatively close clusters, indicating that more than one family of dinosaurs built in the same area. Some fossils of hatchlings have been found near adult animals, leading paleontologists to think that some dinosaurs nurtured their young. In fact, *Maiasaura* hatchlings might have stayed in the nest while they grew in length from about 35 cm to more than 1 m.

Topic: Warm Versus Cold
Visit blue.msscience.com for Web links to information about dinosaurs.

Activity Work with a partner to research the debate on warm-blooded versus cold-blooded dinosaurs. Present your finding to the class in the form of a debate. Be sure to cover the main points of disagreement between the two sides.

Figure 22 Birds might have evolved from dinosaurs.

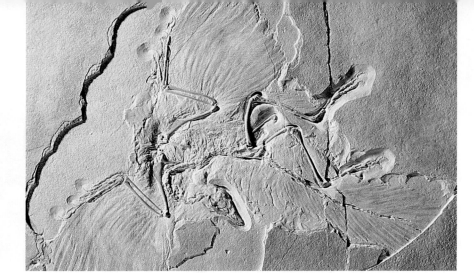

B Considered one of the world's most priceless fossils, *Archaeopteryx*, above, was first found in a limestone quarry in Germany in 1861.

A *Bambiraptor feinberger*, above, is a 75-million-year-old member of a family of meat-eating dinosaurs thought by some paleontologists to be closely related to birds.

Birds Birds appeared during the Jurassic Period. Some paleontologists think that birds evolved from small, meat-eating dinosaurs much like *Bambiraptor feinberger* in **Figure 22A**. The earliest bird, *Archaeopteryx*, shown in **Figure 22B**, had wings and feathers. However, because *Archaeopteryx* had features not shared with modern birds, scientists know it was not a direct ancestor of today's birds.

Mammals Mammals first appeared in the Triassic Period. The earliest mammals were small, mouselike creatures, as shown in **Figure 23**. Mammals are warm-blooded vertebrates that have hair covering their bodies. The females produce milk to feed their young. These two characteristics have enabled mammals to survive in many changing environments.

Gymnosperms During most of the Mesozoic Era, gymnosperms (JIHM nuh spurmz), which first appeared in the Paleozoic Era, dominated the land. Gymnosperms are plants that produce seeds but not flowers. Many gymnosperms are still around today. These include pines and ginkgo trees.

Figure 23 The earliest mammals were small creatures that resembled today's mice and shrews.

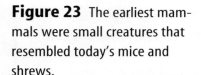

Angiosperms Angiosperms (AN jee uh spurmz), or flowering plants, first evolved during the Cretaceous Period. Angiosperms produce seeds with hard outer coverings.

Because their seeds are enclosed and protected, angiosperms can live in many environments. Angiosperms are the most diverse and abundant land plants today. Present-day angiosperms that evolved during the Mesozoic Era include magnolia and oak trees.

End of an Era The Mesozoic Era ended about 65 million years ago with a major extinction of land and marine species. Many groups of animals, including the dinosaurs, disappeared suddenly at this time. Many paleontologists hypothesize that a comet or asteroid collided with Earth, causing a huge cloud of dust and smoke to rise into the atmosphere, blocking out the Sun. Without sunlight the plants died, and all the animals that depended on these plants also died. Not everything died, however. All the organisms that you see around you today are descendants of the survivors of the great extinction at the end of the Mesozoic Era.

Applying Math — Calculate Percentages

CALCULATING EXTINCTION BY USING PERCENTAGES At the end of the Cretaceous Period, large numbers of plants and animals became extinct. Scientists still are trying to understand why some types of plants and animals survived while others died out. Looking at data about amphibians, reptiles, and mammals that lived during the Cretaceous Period, can you determine what percentage of amphibians survived this mass extinction?

Solution

1 This is what you know:

Animal Extinctions

Animal Type	Groups Living Before Extinction Event (n)	Groups Left After Extinction Event (t)
Amphibians	12	4
Reptiles	63	30
Mammals	24	8

2 This is what you need to find out: p = the percentage of amphibian groups that survived the Cretaceous extinction

3 This is the equation you need to use:
- $p = t/n \times 100$
- Both t and n are shown on the above chart.

4 Substitute the known values: $p = 4/12 \times 100 = 33.3\%$

Practice Problems

1. Using the same equation as demonstrated above, calculate the percentage of reptiles and then the percentage of mammals that survived. Which type of animal was least affected by the extinction?
2. What percentage of all groups survived?

For more practice, visit blue.msscience.com/math_practice

Mini LAB

Calculating the Age of the Atlantic Ocean

Procedure
1. On a **world map** or **globe,** measure the distance in kilometers between a point on the east coast of South America and a point on the west coast of Africa.
2. Measure in SI several times and take the average of your results.
3. Assuming that Africa has been moving away from South America at a rate of 3.5 cm per year, calculate how many years it took to create the Atlantic Ocean.

Analysis
1. Did the values used to obtain your average value vary much?
2. How close did your age come to the accepted estimate for the beginning of the breakup of Pangaea in the Triassic Period?

Try at Home

The Cenozoic Era

The **Cenozoic** (se nuh ZOH ihk) **Era,** or era of recent life, began about 65 million years ago and continues today. Many mountain ranges in North and South America and Europe began to form in the Cenozoic Era. In the late Cenozoic, the climate became much cooler and ice ages occurred. The Cenozoic Era is subdivided into two periods. The first of these is the Tertiary Period. The present-day period is the Quaternary Period. It began about 1.8 million years ago.

Reading Check *What happened to the climate during the late Cenozoic Era?*

Times of Mountain Building Many mountain ranges formed during the Cenozoic Era. These include the Alps in Europe and the Andes in South America. The Himalaya, shown in **Figure 24,** formed as India moved northward and collided with Asia. The collision crumpled and thickened Earth's crust, raising the highest mountains presently on Earth. Many people think the growth of these mountains has helped create cooler climates worldwide.

Figure 24 The Himalaya extend along the India-Tibet border and contain some of the world's tallest mountains. India drifted north and finally collided with Asia, forming the Himalaya.

292 CHAPTER 10 Geologic Time

Further Evolution of Mammals

Throughout much of the Cenozoic Era, expanding grasslands favored grazing plant eaters like horses, camels, deer, and some elephants. Many kinds of mammals became larger. Horses evolved from small, multi-toed animals into the large, hoofed animals of today. However, not all mammals remained on land. Ancestors of the present-day whales and dolphins evolved to live in the sea.

As Australia and South America separated from Antarctica during the continuing breakup of the continents, many species became isolated. They evolved separately from life-forms in other parts of the world. Evidence of this can be seen today in Australia's marsupials. Marsupials are mammals such as kangaroos, koalas, and wombats (shown in **Figure 25**) that carry their young in a pouch.

Your species, *Homo sapiens*, probably appeared about 140,000 years ago. Some people suggest that the appearance of humans could have led to the extinction of many other mammals. As their numbers grew, humans competed for food that other animals relied upon. Also, fossil bones and other evidence indicate that early humans were hunters.

Figure 25 The wombat is one of many Australian marsupials. As a result of human activities, the number and range of wombats have diminished.

section 3 review

Summary

The Mesozoic Era
- During the Triassic Period, Pangaea split into two continents.
- Dinosaurs were the dominant land animals of the Mesozoic Era.
- Birds, mammals, and flowering plants all appeared during this era.
- The Mesozoic Era ended 65 million years ago with a mass extinction.

The Cenozoic Era
- The Cenozoic Era has been a mountain-building period with cooler climates.
- Mammals became dominant with many new life-forms appearing after the dinosaurs disappeared.
- Humans also appeared in the Cenozoic Era, probably about 140,000 years ago.

Self Check

1. **List** the era, period, and epoch in which *Homo sapiens* first appeared.
2. **Discuss** whether mammals became more or less abundant after the extinction of the dinosaurs, and explain why.
3. **Infer** how seeds with a hard outer covering enabled angiosperms to survive in a wide variety of climates.
4. **Explain** why some paleontologists hypothesize that dinosaurs were warm-blooded animals.
5. **Think Critically** How could two species that evolved on separate continents have many similarities?

Applying Math

6. **Convert Units** A fossil mosasaur, a giant marine reptile, measured 9 m in length and had a skull that measured 45 cm in length. What fraction of the mosasaur's total length did the skull account for? Compare your length with the mosasaur's length.

LAB Use the Internet

Discovering the Past

Goals
- **Gather** information about fossils found in your area.
- **Communicate** details about fossils found in your area.
- **Synthesize** information from sources about the fossil record and the changes in your area over time.

Data Source

Science Online
Visit **blue.msscience.com/ internet_lab** for more information about fossils and changes over geologic time and for data collected by other students.

Real-World Question

Imagine what your state was like millions of years ago. What animals might have been roaming around the spot where you now sit? Can you picture a *Tyrannosaurus rex* roaming the area that is now your school? The animals and plants that once inhabited your region might have left some clues to their identity—fossils. Scientists use fossils to piece together what Earth looked like in the geologic past. Fossils can help determine whether an area used to be dry land or underwater. Fossils can help uncover clues about how plants and animals have evolved over the course of time. Using the resources of the Internet and by sharing data with your peers, you can start to discover how North America has changed through time. How has your area changed over geologic time? How might the area where you are now living have looked thousands or millions of years ago? Do you think that the types of animals and plants have changed much over time? Form a hypothesis concerning the change in organisms and geography from long ago to the present day in your area.

Fossils in Your Area

Fossil Name	Plant or Animal Fossil	Age of Fossils	Details About Plant or Animal Fossil	Location of Fossil	Additional Information
		Do not write in this book.			

294 **CHAPTER 10** Geologic Time

Using Scientific Methods

▶ Make a Plan

1. **Determine** the age of the rocks that make up your area. Were they formed during Precambrian time, the Paleozoic Era, the Mesozoic Era, or the Cenozoic Era?
2. Gather information about the plants and animals found in your area during one of the above geologic time intervals. Find specific information on when, where, and how the fossil organisms lived. If no fossils are known from your area, find out information about the fossils found nearest your area.

▶ Follow Your Plan

1. Make sure your teacher approves your plan before you start.
2. Go to the link below to post your data in the table. Add any additional information you think is important to understanding the fossils found in your area.

▶ Analyze Your Data

1. What present-day relatives of prehistoric animals or plants exist in your area?
2. How have the organisms in your area changed over time? Is your hypothesis supported? Why or why not?
3. What other information did you discover about your area's climate or environment from the geologic time period you investigated?

▶ Conclude and Apply

1. **Describe** the plant and animal fossils that have been discovered in your area. What clues did you discover about the environment in which these organisms lived? How do these compare to the environment of your area today?
2. **Infer** from the fossil organisms found in your area what the geography and climate were like during the geologic time period you chose.

Find this lab using the link below.

blue.msscience.com/internet_lab

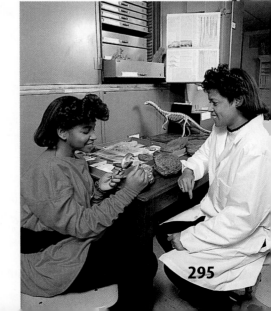

295

SCIENCE Stats

Extinct!

Did you know...

Saber-toothed cat

...The saber-toothed cat lived in the Americas from about 1.6 million to 8,000 years ago. *Smilodon,* the best-known saber-toothed cat, was among the most ferocious carnivores. It had large canine teeth, about 15 cm long, which it used to pierce the flesh of its prey.

Applying Math How many years did *Smilodon* live in the Americas before it became extinct?

Woolly mammoth

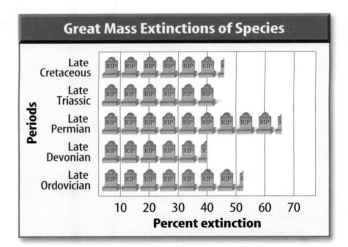

...The woolly mammoth lived in the cold tundra regions during the Ice Age. It looked rather like an elephant with long hair, had a mass between 5,300 kg and 7,300 kg, and was between 3 m and 4 m tall.

Write About It

Visit blue.msscience.com/science_stats to research extinct animals. Trace the origins of each of the species and learn how long its kind existed on Earth.

296 CHAPTER 10 Geologic Time

Chapter 10 Study Guide

Reviewing Main Ideas

Section 1 — Life and Geologic Time

1. Geologic time is divided into eons, eras, periods, and epochs.
2. Divisions within the geologic time scale are based largely on major evolutionary changes in organisms.
3. Plate movements affect organic evolution.

Section 2 — Early Earth History

1. Cyanobacteria evolved during Precambrian time. Trilobites, fish, and corals were abundant during the Paleozoic Era.
2. Plants and animals began to move onto land during the middle of the Paleozoic Era.

3. The Paleozoic Era was a time of mountain building. The Appalachian Mountains formed when several islands and finally Africa collided with North America.
4. At the end of the Paleozoic Era, many marine invertebrates became extinct.

Section 3 — Middle and Recent Earth History

1. Reptiles and gymnosperms were dominant land life-forms in the Mesozoic Era. Mammals and angiosperms began to dominate the land in the Cenozoic Era.
2. Pangaea broke apart during the Mesozoic Era. Many mountain ranges formed during the Cenozoic Era.

Visualizing Main Ideas

Copy and complete the concept map on geologic time using the following choices: Cenozoic, Trilobites in oceans, Mammals common, Paleozoic, Dinosaurs roam Earth, *and* Abundant gymnosperms.

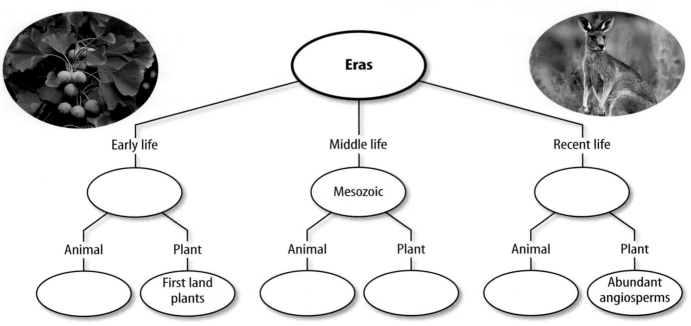

Chapter 10 Review

Using Vocabulary

Cenozoic Era p. 292
cyanobacteria p. 281
eon p. 273
epoch p. 273
era p. 273
geologic time scale p. 272
Mesozoic Era p. 288
natural selection p. 275
organic evolution p. 274
Paleozoic Era p. 282
Pangaea p. 279
period p. 273
Precambrian time p. 280
species p. 274
trilobite p. 277

Fill in the blank with the correct word or words.

1. A change in the hereditary features of a species over a long period is _____.
2. A record of events in Earth history is the _____.
3. The largest subdivision of geologic time is the _____.
4. The process by which the best-suited individuals survive in their environment is _____.
5. A group of individuals that normally breed only among themselves is a(n) _____.

Checking Concepts

Choose the word or phrase that best completes the sentence.

6. How many millions of years ago did the era in which you live begin?
   A) 650 C) 1.6
   B) 245 D) 65

7. What is the process by which better-suited organisms survive and reproduce?
   A) endangerment C) gymnosperm
   B) extinction D) natural selection

8. During what period did the most recent ice age occur?
   A) Pennsylvanian C) Tertiary
   B) Triassic D) Quaternary

9. What is the next smaller division of geologic time after the era?
   A) period C) epoch
   B) stage D) eon

10. What was one of the earliest forms of life?
    A) gymnosperm C) angiosperm
    B) cyanobacterium D) dinosaur

Use the illustration below to answer question 11.

11. Consider the undisturbed rock layers in the figure above. If fossil X were a *Tyrannosaurus rex* bone, and fossil Y were a trilobite; then fossil Z could be which of the following?
    A) stromatolite C) angiosperm
    B) sabre-tooth cat D) *Homo sapiens*

12. During which era did the dinosaurs live?
    A) Mesozoic C) Miocene
    B) Paleozoic D) Cenozoic

13. Which type of plant has seeds without protective coverings?
    A) angiosperms C) gymnosperms
    B) apples D) magnolias

14. Which group of plants evolved during the Mesozoic Era and is dominant today?
    A) gymnosperms C) ginkgoes
    B) angiosperms D) algae

15. In which era did the Ediacaran fauna live?
    A) Precambrian C) Mesozoic
    B) Paleozoic D) Cenozoic

chapter 10 Review

Thinking Critically

16. **Infer** why plants couldn't move onto land until an ozone layer formed.

17. **Discuss** why trilobites are classified as index fossils.

18. **Compare and contrast** the most significant difference between Precambrian life-forms and Paleozoic life-forms.

19. **Describe** how natural selection is related to organic evolution.

20. **Explain** In the early 1800s, a naturalist proposed that the giraffe species has a long neck as a result of years of stretching their necks to reach leaves in tall trees. Why isn't this true?

21. **Infer** Use the outlines of the present-day continents to make a sketch of Pangaea.

22. **Form Hypotheses** Suggest some reasons why trilobites might have become extinct at the end of the Paleozoic Era.

23. **Interpret Data** A student found what she thought was a piece of dinosaur bone in Pleistocene sediment. How likely is it that she is right? Explain.

24. **Infer** why mammals didn't become dominant until after the dinosaurs disappeared.

Performance Activities

25. **Make a Model** In the Section 2 Lab, you learned how a particular characteristic might evolve within a species. Modify the experimental model by using color instead of height as a characteristic. Design your activity with the understanding that varimals live in a dark-colored forest environment.

26. **Make a Display** Certain groups of animals have dominated the land throughout geologic time. Use your textbook and other references to discover some of the dominant species of each era. Make a display that illustrates some animals from each era. Be sure to include appropriate habitats.

Applying Math

Use the graph below to answer questions 27 and 28.

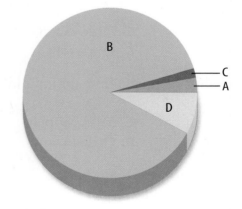

27. **Modeling Geologic Time** The circle graph above represents geologic time. Determine which interval of geologic time is represented by each portion of the graph. Which interval was longest? Which do we know the least about? Which of these intervals is getting larger?

28. **Interpret Data** The Cenozoic Era has lasted 65 million years. What percentage of Earth's 4.5-billion-year history is that?

Chapter 10 Standardized Test Practice

Part 1 Multiple Choice

Record your answers on the answer sheet provided by your teacher or on a sheet of paper.

Examine the diagram below. Then answer questions 1–3.

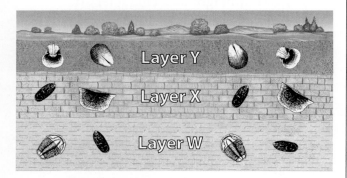

1. During which geologic time period did layer W form?
   A. Cambrian C. Devonian
   B. Ordovician D. Silurian

2. During which geologic time period did layer X form?
   A. Devonian C. Ordovician
   B. Silurian D. Cambrian

3. During which geologic time period did layer Y form?
   A. Cambrian C. Mississippian
   B. Silurian D. Ordovician

4. When did dinosaurs roam Earth?
   A. Precambrian time
   B. Paleozoic Era
   C. Mesozoic Era
   D. Cenozoic Era

5. What is the name of the supercontinent that formed at the end of the Paleozoic Era?
   A. Gondwanaland
   B. Eurasia
   C. Laurasia
   D. Pangaea

6. During which geologic period did modern humans evolve?
   A. Quaternary
   B. Triassic
   C. Ordovician
   D. Tertiary

7. How many body lobes did trilobites have?
   A. one C. three
   B. two D. four

8. Which mountain range formed because India collided with Asia?
   A. Alps C. Ural
   B. Andes D. Himalaya

Use the diagram below to answer questions 9–11.

	Quaternary Period	Holocene Epoch
		Pleistocene Epoch
Cenozoic Era	Tertiary Period	Pliocene Epoch
		Miocene Epoch
		Oligocene Epoch
		Eocene Epoch
		Paleocene Epoch

9. What is the oldest epoch in the Cenozoic Era?
   A. Pleistocene C. Miocene
   B. Paleocene D. Holocene

10. What is the youngest epoch in the Cenozoic Era?
    A. Miocene C. Paleocene
    B. Holocene D. Eocene

11. Which epoch is part of the Quaternary Period?
    A. Oligocene C. Pleistocene
    B. Eocene D. Pliocene

Standardized Test Practice

Part 2 | Short Response/Grid In

Record your answers on the answer sheet provided by your teacher or on a sheet of paper.

12. Who was Charles Darwin? How did he contribute to science?

13. Explain one hypothesis about why dinosaurs might have become extinct.

14. Describe *Archaeopteryx*. Why is this an important fossil?

15. Why do many scientists think that dinosaurs were warm-blooded?

16. What are stromatolites? How do they form?

17. Define the term *species*.

Select one of the equations below to help you answer questions 18–20.

$$\text{time} = \text{distance} \div \text{speed}$$

or

$$\text{speed} = \text{distance} \div \text{time}$$

18. It recently was estimated that *T. rex* could run no faster than about 11 m/s. At this speed, how long would it take *T. rex* to run 200 m?

19. A typical ornithopod (plant-eating dinosaur that walked on two legs) probably moved at a speed of about 2 m/s. How long would it take this dinosaur to run 200 m?

20. In 1996, Michael Johnson ran 200 m in 19.32 s. What was his average speed? How does this compare with *T. rex*?

Test-Taking Tip

Show Your Work For constructed response questions, show all of your work and any calculations on your answer sheet.

Part 3 | Open Ended

Record your answers on a sheet of paper.

Use the diagram below to answer questions 21 and 22. It shows the time ranges of various types of organisms on Earth. When a bar is wider, there were more species of that type of organism (higher diversity).

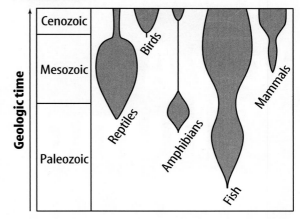

21. Describe how the diversity of reptiles changed through time.

22. How has the diversity of mammals changed through time? Do you see any relationship with how reptile diversity has changed?

23. Why might mammals in Australia be so much different than mammals on other continents?

24. Describe how natural selection might cause a species to change through time.

25. How did early photosynthetic organisms change the conditions on Earth to allow more advanced organisms to flourish?

26. What are mass extinctions? How have they affected life on Earth?

27. Write a description of what Earth was like during Precambrian time. Summarize how Earth was different than it is now.

unit 4

Earth's Place in the Universe

How Are Thunderstorms & Neutron Stars Connected?

In 1931, an engineer built an antenna to study thunderstorm static that was interfering with radio communication. The antenna did detect static from storms, but it also picked up something else: radio signals coming from beyond our solar system. That discovery marked the birth of radio astronomy. By the 1960s, radio astronomy was thriving. In 1967, astronomer Jocelyn Bell Burnell detected a peculiar series of radio pulses coming from far out in space. At first, she and her colleagues theorized that the signals might be a message from a distant civilization. Soon, however, scientists determined that the signals must be coming from something called a neutron star (below)—a rapidly spinning star that gives off a radio beam from its magnetic pole, similar to the rotating beam of a lighthouse.

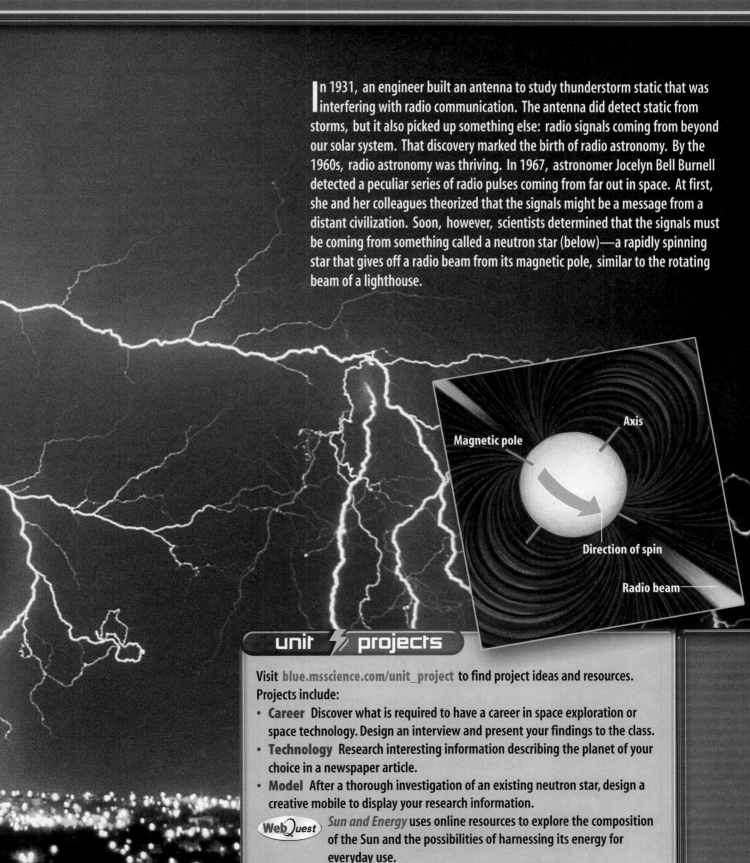

unit projects

Visit blue.msscience.com/unit_project to find project ideas and resources.
Projects include:
- **Career** Discover what is required to have a career in space exploration or space technology. Design an interview and present your findings to the class.
- **Technology** Research interesting information describing the planet of your choice in a newspaper article.
- **Model** After a thorough investigation of an existing neutron star, design a creative mobile to display your research information.

WebQuest *Sun and Energy* uses online resources to explore the composition of the Sun and the possibilities of harnessing its energy for everyday use.

chapter 11

The Sun-Earth-Moon System

The BIG Idea
Many common observations, such as seasons, eclipses, and lunar phases, are caused by interactions between the Sun, Earth, and the Moon.

SECTION 1
Earth
Main Idea Earth is a sphere that rotates on a tilted axis and revolves around the Sun.

SECTION 2
The Moon—Earth's Satellite
Main Idea Eclipses and phases of the Moon occur as the Moon moves in relation to the Sun and Earth.

SECTION 3
Exploring Earth's Moon
Main Idea Knowledge of the Moon's structure and composition has been increased by many spacecraft missions to the Moon.

Full Moon Rising—The Real Story

Why does the Moon's appearance change throughout the month? Do the Sun and Moon really rise? You will find the answers to these questions and also learn why we have summer and winter.

Science Journal Rotation or revolution—which motion of Earth brings morning and which brings summer?

Start-Up Activities

Model Rotation and Revolution

The Sun rises in the morning; at least, it seems to. Instead, it is Earth that moves. The movements of Earth cause day and night, as well as the seasons. In this lab, you will explore Earth's movements.

1. Hold a basketball with one finger at the top and one at the bottom. Have a classmate gently spin the ball.
2. Explain how this models Earth's rotation.
3. Continue to hold the basketball and walk one complete circle around another student in your class.
4. How does this model Earth's revolution?
5. **Think Critically** Write a paragraph in your Science Journal describing how these movements of the basketball model Earth's rotation and revolution.

Preview this chapter's content and activities at blue.msscience.com

Earth and the Moon All on Earth can see and feel the movements of Earth and the Moon as they circle the Sun. Make the following Foldable to organize what you learn about these movements and their effects.

STEP 1 Fold a sheet of paper in half lengthwise.

STEP 2 Fold paper down 2.5 cm from the top. (Hint: From the tip of your index finger to your middle knuckle is about 2.5 cm.)

STEP 3 Open and draw lines along the 2.5-cm fold. Label as shown.

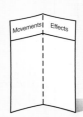

Summarize in a Table As you read the chapter, summarize the movements of Earth and the Moon in the left column and the effects of these movements in the right column.

Get Ready to Read

Summarize

① Learn It! Summarizing helps you organize information, focus on main ideas, and reduce the amount of information to remember. To summarize, restate the important facts in a short sentence or paragraph. Be brief and do not include too many details.

② Practice It! Read the text on page 310 labeled *Solstices*. Then read the summary below and look at the important facts from that passage.

Important Facts

- In the northern hemisphere, the summer solstice occurs in June, and the winter solstice occurs in December.
- In the southern hemisphere, the winter solstice occurs in June, and the summer solstice occurs in December.
- Summer solstice is about the longest period of daylight of the year.
- Winter solstice is about the shortest period of daylight of the year.

Summary

The solstice is the day when the Sun reaches its greatest distance north or south of the equator.

③ Apply It! Practice summarizing as you read this chapter. Stop after each section and write a brief summary.

Target Your Reading

Reading Tip
Reread your summary to make sure you didn't change the author's original meaning or ideas.

Use this to focus on the main ideas as you read the chapter.

1 Before you read the chapter, respond to the statements below on your worksheet or on a numbered sheet of paper.
- Write an **A** if you **agree** with the statement.
- Write a **D** if you **disagree** with the statement.

2 After you read the chapter, look back to this page to see if you've changed your mind about any of the statements.
- If any of your answers changed, explain why.
- Change any false statements into true statements.
- Use your revised statements as a study guide.

Science online
Print out a worksheet of this page at blue.msscience.com

Before You Read A or D		Statement	After You Read A or D
	1	Earth's revolution around the Sun causes day and night to occur.	
	2	Earth's magnetic poles are aligned on Earth's rotational axis.	
	3	Summer occurs in the northern hemisphere when Earth is closest to the Sun.	
	4	During an equinox, the number of daylight hours is nearly equal with the number of nighttime hours all over the world.	
	5	When observing the phases of the Moon, the Moon's lighted surface area is daylight on the Moon and the dark portion is nighttime on the Moon.	
	6	The length of one Moon day is about the same amount of time as the length of one Earth day.	
	7	A lunar eclipse occurs when the Moon comes between Earth and the Sun.	
	8	Humans first walked on the Moon during the *Apollo* spacecraft missions.	

306 B

section 1

Earth

as you read

What You'll Learn
- **Examine** Earth's physical characteristics.
- **Differentiate** between rotation and revolution.
- **Discuss** what causes seasons to change.

Why It's Important
Your life follows the rhythm of Earth's movements.

Review Vocabulary
orbit: the path taken by an object revolving around another

New Vocabulary
- sphere
- axis
- rotation
- revolution
- ellipse
- solstice
- equinox

Figure 1 For many years, sailors have observed that the tops of ships coming across the horizon appear first. This suggests that Earth is spherical, not flat, as was once widely believed.

Properties of Earth

You awaken at daybreak to catch the Sun "rising" from the dark horizon. Then it begins its daily "journey" from east to west across the sky. Finally the Sun "sinks" out of view as night falls. Is the Sun moving—or are you?

It wasn't long ago that people thought Earth was the center of the universe. It was widely believed that the Sun revolved around Earth, which stood still. It is now common knowledge that the Sun only appears to be moving around Earth. Because Earth spins as it revolves around the Sun, it creates the illusion that the Sun is moving across the sky.

Another mistaken idea about Earth concerned its shape. Even as recently as the days of Christopher Columbus, many people believed Earth to be flat. Because of this, they were afraid that if they sailed far enough out to sea, they would fall off the edge of the world. How do you know this isn't true? How have scientists determined the true shape of Earth?

Spherical Shape A round, three-dimensional object is called a **sphere** (SFIHR). Its surface is the same distance from its center at all points. Some common examples of spheres are basketballs and tennis balls.

In the late twentieth century, artificial satellites and space probes sent back pictures showing that Earth is spherical. Much earlier, Aristotle, a Greek astronomer and philosopher who lived around 350 B.C., suspected that Earth was spherical. He observed that Earth cast a curved shadow on the Moon during an eclipse.

In addition to Aristotle, other individuals made observations that indicated Earth's spherical shape. Early sailors, for example, noticed that the tops of approaching ships appeared first on the horizon and the rest appeared gradually, as if they were coming over the crest of a hill, as shown in **Figure 1.**

Additional Evidence Sailors also noticed changes in how the night sky looked. As they sailed north or south, the North Star moved higher or lower in the sky. The best explanation was a spherical Earth.

Today, most people know that Earth is spherical. They also know all objects are attracted by gravity to the center of a spherical Earth. Astronauts have clearly seen the spherical shape of Earth. However, it bulges slightly at the equator and is somewhat flattened at the poles, so it is not a perfect sphere.

Rotation Earth's **axis** is the imaginary vertical line around which Earth spins. This line cuts directly through the center of Earth, as shown in the illustration accompanying **Table 1.** The poles are located at the north and south ends of Earth's axis. The spinning of Earth on its axis, called **rotation,** causes day and night to occur. Here is how it works. As Earth rotates, you can see the Sun come into view at daybreak. Earth continues to spin, making it seem as if the Sun moves across the sky until it sets at night. During night, your area of Earth has rotated so that it is facing away from the Sun. Because of this, the Sun is no longer visible to you. Earth continues to rotate steadily, and eventually the Sun comes into view again the next morning. One complete rotation takes about 24 h, or one day. How many rotations does Earth complete during one year? As you can infer from **Table 1,** it completes about 365 rotations during its one-year journey around the Sun.

Reading Check *Why does the Sun seem to rise and set?*

Earth's Rotation
Suppose that Earth's rotation took twice as long as it does now. In your Science Journal, predict how conditions such as global temperatures, work schedules, plant growth, and other factors might change under these circumstances.

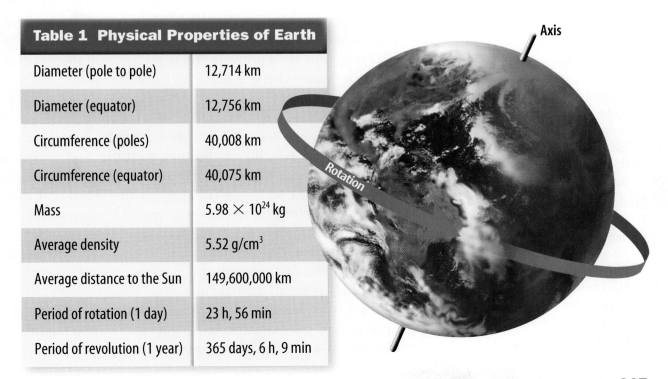

Table 1 Physical Properties of Earth	
Diameter (pole to pole)	12,714 km
Diameter (equator)	12,756 km
Circumference (poles)	40,008 km
Circumference (equator)	40,075 km
Mass	5.98×10^{24} kg
Average density	5.52 g/cm^3
Average distance to the Sun	149,600,000 km
Period of rotation (1 day)	23 h, 56 min
Period of revolution (1 year)	365 days, 6 h, 9 min

Figure 2 Earth's magnetic field is similar to that of a bar magnet, almost as if Earth contained a giant magnet. Earth's magnetic axis is angled 11.5 degrees from its rotational axis.

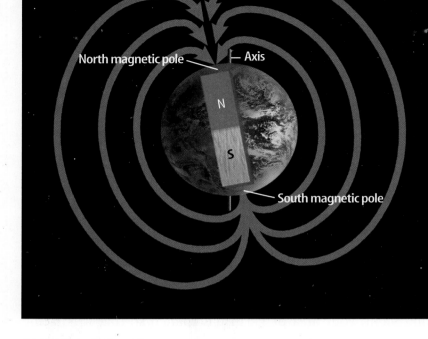

Mini LAB

Making Your Own Compass

Procedure

WARNING: *Use care when handling sharp objects.*

1. Cut off the bottom of a **plastic foam cup** to make a polystyrene disk.
2. Magnetize a **sewing needle** by continuously stroking the needle in the same direction with a **magnet** for 1 min.
3. **Tape** the needle to the center of the foam disk.
4. Fill a **plate** with **water** and float the disk, needle side up, in the water.

Analysis

1. What happened to the needle and disk when you placed them in the water? Why did this happen?
2. Infer how ancient sailors might have used magnets to help them navigate on the open seas.

Magnetic Field

INTEGRATE Physics

Scientists hypothesize that the movement of material inside Earth's core, along with Earth's rotation, generates a magnetic field. This magnetic field is much like that of a bar magnet. Earth has a north and a south magnetic pole, just as a bar magnet has opposite magnetic poles at each of its ends. When you sprinkle iron shavings over a bar magnet, the shavings align with the magnetic field of the magnet. As you can see in **Figure 2,** Earth's magnetic field is similar—almost as if Earth contained a giant bar magnet. Earth's magnetic field protects you from harmful solar radiation by trapping many charged particles from the Sun.

Magnetic Axis When you observe a compass needle pointing north, you are seeing evidence of Earth's magnetic field. Earth's magnetic axis, the line joining its north and south magnetic poles, does not align with its rotational axis. The magnetic axis is inclined at an angle of 11.5° to the rotational axis. If you followed a compass needle, you would end up at the magnetic north pole rather than the rotational north pole.

The location of the magnetic poles has been shown to change slowly over time. The magnetic poles move around the rotational (geographic) poles in an irregular way. This movement can be significant over decades. Many maps include information about the position of the magnetic north pole at the time the map was made. Why would this information be important?

308 CHAPTER 11 The Sun-Earth-Moon System

What causes changing seasons?

Flowers bloom as the days get warmer. The Sun appears higher in the sky, and daylight lasts longer. Spring seems like a fresh, new beginning. What causes these wonderful changes?

Orbiting the Sun You learned earlier that Earth's rotation causes day and night. Another important motion is **revolution,** which is Earth's yearly orbit around the Sun. Just as the Moon is Earth's satellite, Earth is a satellite of the Sun. If Earth's orbit were a circle with the Sun at the center, Earth would maintain a constant distance from the Sun. However, this is not the case. Earth's orbit is an **ellipse** (ee LIHPS)—an elongated, closed curve. The Sun is not at the center of the ellipse but is a little toward one end. Because of this, the distance between Earth and the Sun changes during Earth's yearlong orbit. Earth gets closest to the Sun—about 147 million km away—around January 3. The farthest Earth gets from the Sun is about 152 million km away. This happens around July 4 each year.

Reading Check *What is an ellipse?*

Does this elliptical orbit cause seasonal temperatures on Earth? If it did, you would expect the warmest days to be in January. You know this isn't the case in the northern hemisphere, something else must cause the change.

Even though Earth is closest to the Sun in January, the change in distance is small. Earth is exposed to almost the same amount of Sun all year. But the amount of solar energy any one place on Earth receives varies greatly during the year. Next, you will learn why.

A Tilted Axis Earth's axis is tilted 23.5° from a line drawn perpendicular to the plane of its orbit. It is this tilt that causes seasons. The number of daylight hours is greater for the hemisphere, or half of Earth, that is tilted toward the Sun. Think of how early it gets dark in the winter compared to the summer. As shown in **Figure 3,** the hemisphere that is tilted toward the Sun receives more hours of sunlight each day than the hemisphere that is tilted away from the Sun. The longer period of sunlight is one reason summer is warmer than winter, but it is not the only reason.

Science Online

Topic: Ellipses
Visit blue.msscience.com for Web links to information about orbits and ellipses.

Activity Scientists compare orbits by how close they come to being circular. To do this, they use a measurement called eccentricity. A circle has an eccentricity of zero. Ellipses have eccentricities that are greater than zero, but less than one. The closer the eccentricity is to one, the more elliptical the orbit. Compare the orbits of the four inner planets. List them in order of increasing eccentricity.

Figure 3 In summer, the northern hemisphere is tilted toward the Sun. Notice that the north pole is always lit during the summer.
Observe *Why is there a greater number of daylight hours in the summer than in the winter?*

Radiation from the Sun Earth's tilt also causes the Sun's radiation to strike the hemispheres at different angles. Sunlight strikes the hemisphere tilted towards the Sun at a higher angle, that is, closer to 90 degrees, than the hemisphere tilted away. Thus it receives more total solar radiation than the hemisphere tilted away from the Sun, where sunlight strikes at a lower angle.

Summer occurs in the hemisphere tilted toward the Sun, when its radiation strikes Earth at a higher angle and for longer periods of time. The hemisphere receiving less radiation experiences winter.

Solstices

The **solstice** is the day when the Sun reaches its greatest distance north or south of the equator. In the northern hemisphere, the summer solstice occurs on June 21 or 22, and the winter solstice occurs on December 21 or 22. Both solstices are illustrated in **Figure 4**. In the southern hemisphere, the winter solstice is in June and the summer solstice is in December. Summer solstice is about the longest period of daylight of the year. After this, the number of daylight hours become less and less, until the winter solstice, about the shortest period of daylight of the year. Then the hours of daylight start to increase again.

Figure 4 During the summer solstice in the northern hemisphere, the Sun is directly over the tropic of Cancer, the latitude line at 23.5° N latitude. During the winter solstice, the Sun is directly over the tropic of Capricorn, the latitude line at 23.5° S latitude. At fall and spring equinoxes, the Sun is directly over the equator.

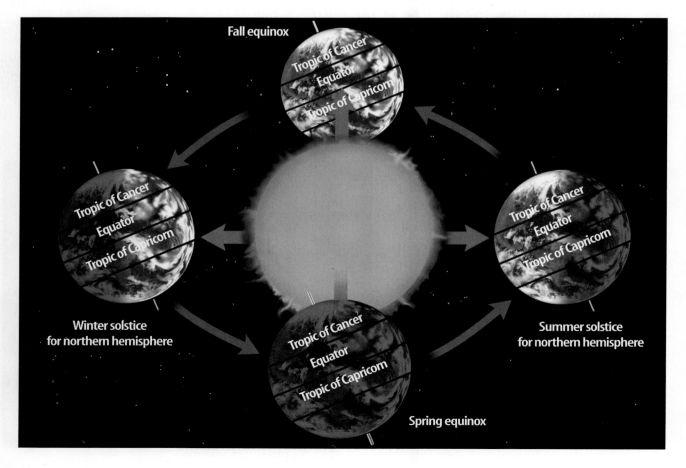

Equinoxes

An **equinox** (EE kwuh nahks) occurs when the Sun is directly above Earth's equator. Because of the tilt of Earth's axis, the Sun's position relative to the equator changes constantly. Most of the time, the Sun is either north or south of the equator, but two times each year it is directly over it, resulting in the spring and fall equinoxes. As you can see in **Figure 4,** at an equinox the Sun strikes the equator at the highest possible angle, 90°.

During an equinox, the number of daylight hours and nighttime hours is nearly equal all over the world. Also at this time, neither the northern hemisphere nor the southern hemisphere is tilted toward the Sun.

In the northern hemisphere, the Sun reaches the spring equinox on March 20 or 21, and the fall equinox occurs on September 22 or 23. In the southern hemisphere, the equinoxes are reversed. Spring occurs in September and fall occurs in March.

Earth Data Review As you have learned, Earth is a sphere that rotates on a tilted axis. This rotation causes day and night. Earth's tilted axis and its revolution around the Sun cause the seasons. One Earth revolution takes one year. In the next section, you will read how the Moon rotates on its axis and revolves around Earth.

Science Online

Topic: Seasons
Visit blue.msscience.com for Web links to information about the seasons.

Activity Make a poster describing how the seasons differ in other parts of the world. Show how holidays might be celebrated differently and how farming might vary between hemispheres.

section 1 review

Summary

Properties of Earth
- Earth is a slightly flattened sphere that rotates around an imaginary line called an axis.
- Earth has a magnetic field, much like a bar magnet.
- The magnetic axis of Earth differs from its rotational axis.

Seasons
- Earth revolves around the Sun in an elliptical orbit.
- The tilt of Earth's axis and its revolution cause the seasons.
- Solstices are days when the Sun reaches its farthest points north or south of the equator.
- Equinoxes are the points when the Sun is directly over the equator.

Self Check

1. **Explain** why Aristotle thought Earth was spherical.
2. **Compare and contrast** rotation and revolution.
3. **Describe** how Earth's distance from the Sun changes throughout the year. When is Earth closest to the Sun?
4. **Explain** why it is summer in Earth's northern hemisphere at the same time it is winter in the southern hemisphere.
5. **Think Critically Table 1** lists Earth's distance from the Sun as an average. Why isn't an exact measurement available for this distance?

Applying Skills

6. **Classify** The terms *clockwise* and *counterclockwise* are used to indicate the direction of circular motion. How would you classify the motion of the Moon around Earth as you view it from above Earth's north pole? Now try to classify Earth's movement around the Sun.

Science Online blue.msscience.com/self_check_quiz

section 2

The Moon— Earth's Satellite

as you read

What You'll Learn
- **Identify** phases of the Moon and their cause.
- **Explain** why solar and lunar eclipses occur.
- **Infer** what the Moon's surface features may reveal about its history.

Why It's Important
Learning about the Moon can teach you about Earth.

Review Vocabulary
mantle: portion of the interior of a planet or moon lying between the crust and core

New Vocabulary
- moon phase
- new moon
- waxing
- full moon
- waning
- solar eclipse
- lunar eclipse
- maria

Motions of the Moon

Just as Earth rotates on its axis and revolves around the Sun, the Moon rotates on its axis and revolves around Earth. The Moon's revolution around Earth is responsible for the changes in its appearance. If the Moon rotates on its axis, why can't you see it spin around in space? The reason is that the Moon's rotation takes 27.3 days—the same amount of time it takes to revolve once around Earth. Because these two motions take the same amount of time, the same side of the Moon always faces Earth, as shown in **Figure 5.**

You can demonstrate this by having a friend hold a ball in front of you. Direct your friend to move the ball in a circle around you while keeping the same side of it facing you. Everyone else in the room will see all sides of the ball. You will see only one side. If the moon didn't rotate, we would see all of its surface during one month.

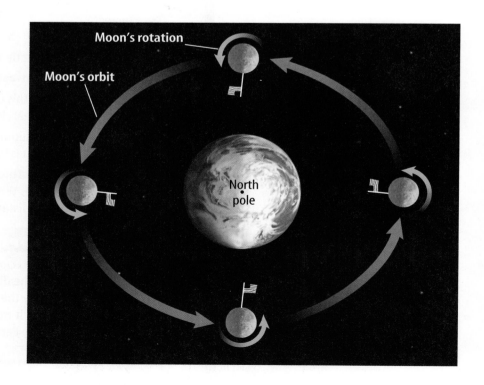

Figure 5 In about 27.3 days, the Moon orbits Earth. It also completes one rotation on its axis during the same period. **Think Critically** Explain how this affects which side of the Moon faces Earth.

312 CHAPTER 11 The Earth-Sun-Moon System

Reflection of the Sun The Moon seems to shine because its surface reflects sunlight. Just as half of Earth experiences day as the other half experiences night, half of the Moon is lighted while the other half is dark. As the Moon revolves around Earth, you see different portions of its lighted side, causing the Moon's appearance to change.

Phases of the Moon

Moon phases are the different forms that the Moon takes in its appearance from Earth. The phase depends on the relative positions of the Moon, Earth, and the Sun, as seen in **Figure 6** on the next page. A **new moon** occurs when the Moon is between Earth and the Sun. During a new moon, the lighted half of the Moon is facing the Sun and the dark side faces Earth. The Moon is in the sky, but it cannot be seen. The new moon rises and sets with the Sun.

Reading Check *Why can't you see a new moon?*

Waxing Phases After a new moon, the phases begin waxing. **Waxing** means that more of the illuminated half of the Moon can be seen each night. About 24 h after a new moon, you can see a thin slice of the Moon. This phase is called the waxing crescent. About a week after a new moon, you can see half of the lighted side of the Moon, or one quarter of the Moon's surface. This is the first quarter phase.

The phases continue to wax. When more than one quarter is visible, it is called waxing *gibbous* after the Latin word for "humpbacked." A **full moon** occurs when all of the Moon's surface facing Earth reflects light.

Waning Phases After a full moon, the phases are said to be waning. When the Moon's phases are **waning,** you see less of its illuminated half each night. Waning gibbous begins just after a full moon. When you can see only half of the lighted side, it is the third-quarter phase. The Moon continues to appear to shrink. Waning crescent occurs just before another new moon. Once again, you can see only a small slice of the Moon.

It takes about 29.5 days for the Moon to complete its cycle of phases. Recall that it takes about 27.3 days for the Moon to revolve around Earth. The discrepancy between these two numbers is due to Earth's revolution. The roughly two extra days are what it takes for the Sun, Earth, and Moon to return to their same relative positions.

Mini LAB

Comparing the Sun and the Moon

Procedure
1. Find an area where you can make a chalk mark on **pavement or similar surface.**
2. Tie a piece of **chalk** to one end of a 200-cm-long **string.**
3. Hold the other end of the string to the pavement.
4. Have a friend pull the string tight and walk around you, drawing a circle (the Sun) on the pavement.
5. Draw a 1-cm-diameter circle in the middle of the larger circle (the Moon).

Analysis
1. How big is the Sun compared to the Moon?
2. The diameter of the Sun is 1.39 million km. The diameter of Earth is 12,756 km. Draw two new circles modeling the sizes of the Sun and Earth. What scale did you use?

Try at Home

Figure 6 The phases of the Moon change during a cycle that lasts about 29.5 days.

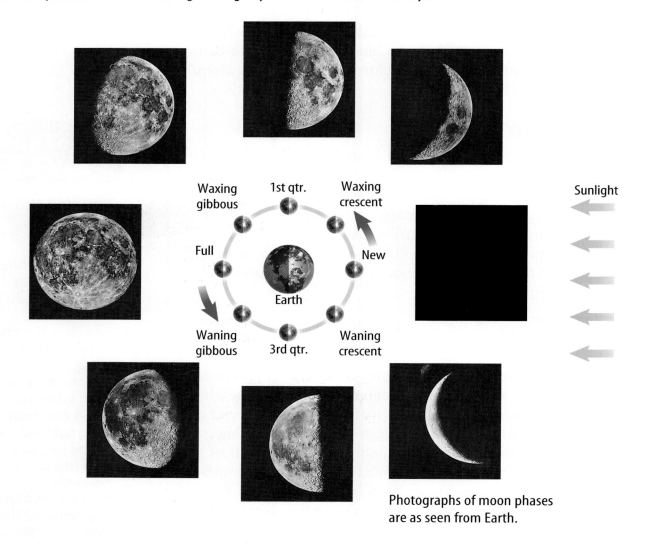

Photographs of moon phases are as seen from Earth.

Figure 7 The outer portion of the Sun's atmosphere is visible during a total solar eclipse. It looks like a halo around the Moon.

Eclipses

Imagine living 10,000 years ago. You are foraging for nuts and fruit when unexpectedly the Sun disappears from the sky. The darkness lasts only a short time, and the Sun soon returns to full brightness. You know something strange has happened, but you don't know why. It will be almost 8,000 years before anyone can explain what you just experienced.

The event just described was a total solar eclipse (ih KLIPS), shown in **Figure 7.** Today, most people know what causes such eclipses, but without this knowledge, they would have been terrifying events. During a solar eclipse, many animals act as if it is nighttime. Cows return to their barns and chickens go to sleep. What causes the day to become night and then change back into day?

Reading Check *What happens during a total solar eclipse?*

What causes an eclipse? The revolution of the Moon causes eclipses. Eclipses occur when Earth or the Moon temporarily blocks the sunlight from reaching the other. Sometimes, during a new moon, the Moon's shadow falls on Earth and causes a solar eclipse. During a full moon, Earth's shadow can be cast on the Moon, resulting in a lunar eclipse.

An eclipse can occur only when the Sun, the Moon, and Earth are lined up perfectly. Because the Moon's orbit is not in the same plane as Earth's orbit around the Sun, lunar eclipses occur only a few times each year.

Eclipses of the Sun A **solar eclipse** occurs when the Moon moves directly between the Sun and Earth and casts its shadow over part of Earth, as seen in **Figure 8.** Depending on where you are on Earth, you may experience a total eclipse or a partial eclipse. The darkest portion of the Moon's shadow is called the umbra (UM bruh). A person standing within the umbra experiences a total solar eclipse. During a total solar eclipse, the only visible portion of the Sun is a pearly white glow around the edge of the eclipsing Moon.

Surrounding the umbra is a lighter shadow on Earth's surface called the penumbra (puh NUM bruh). Persons standing in the penumbra experience a partial solar eclipse. **WARNING:** *Regardless of which eclipse you view, never look directly at the Sun. The light can permanently damage your eyes.*

Topic: Eclipses
Visit blue.msscience.com for Web links to information about solar and lunar eclipses.

Activity Make a chart showing the dates when lunar and solar eclipses will be visible in your area. Include whether the eclipses will be total or partial.

Figure 8 Only a small area of Earth experiences a total solar eclipse during the eclipse event.

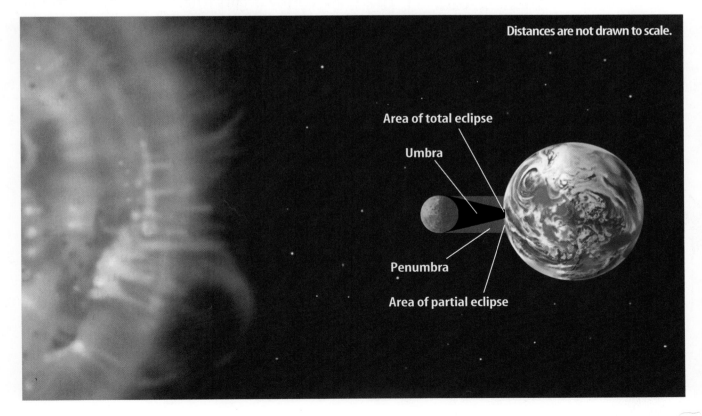

Figure 9 These photographs show the Moon moving from right to left into Earth's umbra, then out again.

Eclipses of the Moon When Earth's shadow falls on the Moon, a **lunar eclipse** occurs. A lunar eclipse begins when the Moon moves into Earth's penumbra. As the Moon continues to move, it enters Earth's umbra and you see a curved shadow on the Moon's surface, as in **Figure 9.** Upon moving completely into Earth's umbra, as shown in **Figure 10,** the Moon goes dark, signaling that a total lunar eclipse has occurred. Sometimes sunlight bent through Earth's atmosphere causes the eclipsed Moon to appear red.

A partial lunar eclipse occurs when only a portion of the Moon moves into Earth's umbra. The remainder of the Moon is in Earth's penumbra and, therefore, receives some direct sunlight. A penumbral lunar eclipse occurs when the Moon is totally within Earth's penumbra. However, it is difficult to tell when a penumbral lunar eclipse occurs because some sunlight continues to fall on the side of the Moon facing Earth.

A total lunar eclipse can be seen by anyone on the nighttime side of Earth where the Moon is not hidden by clouds. In contrast, only a lucky few people get to witness a total solar eclipse. Only those people in the small region where the Moon's umbra strikes Earth can witness one.

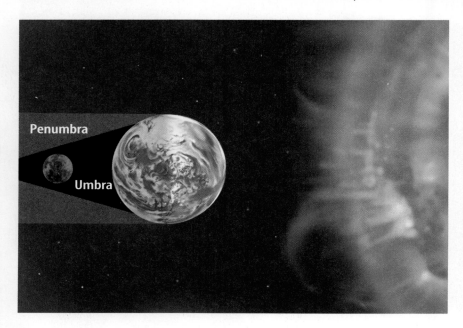

Figure 10 During a total lunar eclipse, Earth's shadow blocks light coming from the Sun.

The Moon's Surface

When you look at the Moon, as shown in **Figure 12** on the next page, you can see many depressions called craters. Meteorites, asteroids, and comets striking the Moon's surface created most of these craters, which formed early in the Moon's history. Upon impact, cracks may have formed in the Moon's crust, allowing lava to reach the surface and fill up the large craters. The resulting dark, flat regions are called **maria** (MAHR ee uh). The igneous rocks of the maria are 3 billion to 4 billion years old. So far, they are the youngest rocks to be found on the Moon. This indicates that craters formed after the Moon's surface originally cooled. The maria formed early enough in the Moon's history that molten material still remained in the Moon's interior. The Moon once must have been as geologically active as Earth is today. Before the Moon cooled to the current condition, the interior separated into distinct layers.

Inside the Moon

Earthquakes allow scientists to learn about Earth's interior. In a similar way, scientists use instruments such as the one in **Figure 11** to study moonquakes. The data they have received have led to the construction of several models of the Moon's interior. One such model, shown in **Figure 11,** suggests that the Moon's crust is about 60 km thick on the side facing Earth. On the far side, it is thought to be about 150 km thick. Under the crust, a solid mantle may extend to a depth of 1,000 km. A partly molten zone of the mantle may extend even farther down. Below this mantle may lie a solid, iron-rich core.

Seismology A seismologist is an Earth scientist who studies the propagation of seismic waves in geological materials. Usually this means studying earthquakes, but some seismologists apply their knowledge to studies of the Moon and planets. Seismologists usually study geology, physics, and applied mathematics in college and later specialize in seismology for an advanced degree.

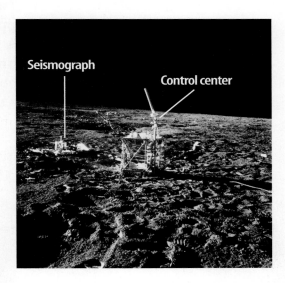

Figure 11 Equipment, such as the seismograph left on the Moon by the *Apollo 12* mission, helps scientists study moonquakes.

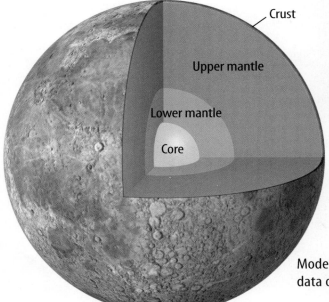

Models of the Moon's interior were created from data obtained by scientists studying moonquakes.

SECTION 2 The Moon—Earth's Satellite **317**

NATIONAL GEOGRAPHIC VISUALIZING THE MOON'S SURFACE

Figure 12

By looking through binoculars, you can see many of the features on the surface of the Moon. These include craters that are hundreds of kilometers wide, light-colored mountains, and darker patches that early astronomers called maria (Latin for "seas"). However, as the NASA Apollo missions discovered, these so-called seas do not contain water. In fact, maria (singular, mare) are flat, dry areas formed by ancient lava flows. Some of the Moon's geographic features are shown below, along with the landing sites of Apollo missions sent to investigate Earth's closest neighbor in space.

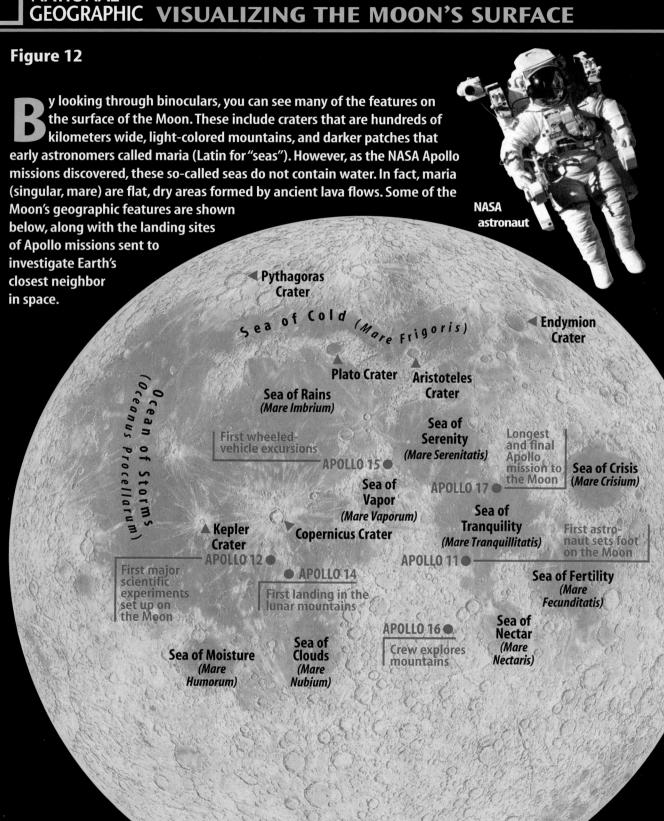

NASA astronaut

The Moon's Origin

Before the *Apollo* space missions in the 1960s and 1970s, there were three leading theories about the Moon's origin. According to one theory, the Moon was captured by Earth's gravity. Another held that the Moon and Earth condensed from the same cloud of dust and gas. An alternative theory proposed that Earth ejected molten material that became the Moon.

The Impact Theory The data gathered by the *Apollo* missions have led many scientists to support a new theory, known as the impact theory. It states that the Moon formed billions of years ago from condensing gas and debris thrown off when Earth collided with a Mars-sized object as shown in **Figure 13.**

Figure 13 According to the impact theory, a Mars-sized object collided with Earth around 4.6 billion years ago. Vaporized materials ejected by the collision began orbiting Earth and quickly consolidated into the Moon.

Applying Science

What will you use to survive on the Moon?

You have crash-landed on the Moon. It will take one day to reach a moon colony on foot. The side of the Moon that you are on will be facing away from the Sun during your entire trip. You manage to salvage the following items from your wrecked ship: food, rope, solar-powered heating unit, battery-operated heating unit, oxygen tanks, map of the constellations, compass, matches, water, solar-powered radio transmitter, three flashlights, signal mirror, and binoculars.

Identifying the Problem

The Moon lacks a magnetic field and has no atmosphere. How do the Moon's physical properties and the lack of sunlight affect your decisions?

Solving the Problem
1. Which items will be of no use to you? Which items will you take with you?
2. Describe why each of the salvaged items is useful or not useful.

SECTION 2 The Moon—Earth's Satellite

Figure 14 Moon rocks collected by astronauts provide scientists with information about the Moon and Earth.

The Moon in History Studying the Moon's phases and eclipses led to the conclusion that both Earth and the Moon were in motion around the Sun. The curved shadow Earth casts on the Moon indicated to early scientists that Earth was spherical. When Galileo first turned his telescope toward the Moon, he found a surface scarred by craters and maria. Before that time, many people believed that all planetary bodies were perfectly smooth and lacking surface features. Now, actual moon rocks are available for scientists to study, as seen in **Figure 14.** By doing so, they hope to learn more about Earth.

Reading Check *How has observing the Moon been important to science?*

section 2 review

Summary

Motions of the Moon
- The Moon rotates on its axis about once each month.
- The Moon also revolves around Earth about once every 27.3 days.
- The Moon shines because it reflects sunlight.

Phases of the Moon
- During the waxing phases, the illuminated portion of the Moon grows larger.
- During waning phases, the illuminated portion of the Moon grows smaller.
- Earth passing directly between the Sun and the Moon causes a lunar eclipse.
- The Moon passing between Earth and the Sun causes a solar eclipse.

Structure and Origin of the Moon
- The Moon's surface is covered with depressions called impact craters.
- Flat, dark regions within craters are called maria.
- The Moon may have formed as the result of a collision between Earth and a Mars-sized object.

Self Check

1. **Explain** how the Sun, Moon, and Earth are positioned relative to each other during a new moon and how this alignment changes to produce a full moon.
2. **Describe** what phase the Moon must be in to have a lunar eclipse. A solar eclipse?
3. **Define** the terms *umbra* and *penumbra* and explain how they relate to eclipses.
4. **Explain** why lunar eclipses are more common than solar eclipses and why so few people ever have a chance to view a total solar eclipse.
5. **Think Critically** What do the surface features and their distribution on the Moon's surface tell you about its history?

Applying Math

6. **Solve Simple Equations** The Moon travels in its orbit at about 3,400 km/h. Therefore, during a solar eclipse, its shadow sweeps at this speed from west to east. However, Earth rotates from west to east at about 1,670 km/h near the equator. At what speed does the shadow really move across this part of Earth's surface?

Moon Phases and Eclipses

In this lab, you will demonstrate the positions of the Sun, the Moon, and Earth during certain phases and eclipses. You also will see why only a small portion of the people on Earth witness a total solar eclipse during a particular eclipse event.

Real-World Question

Can a model be devised to show the positions of the Sun, the Moon, and Earth during various phases and eclipses?

Goals
- **Model** moon phases.
- **Model** solar and lunar eclipses.

Materials
light source (unshaded)   globe
polystyrene ball   pencil

Safety Precautions

Moon Phase Observations

Moon Phase	Observations
First quarter	
Full moon	Do not write in this book.
Third quarter	
New moon	

Procedure

1. Review the illustrations of moon phases and eclipses shown in Section 2.
2. Use the light source as a Sun model and a polystyrene ball on a pencil as a Moon model. Move the Moon around the globe to duplicate the exact position that would have to occur for a lunar eclipse to take place.
3. Move the Moon to the position that would cause a solar eclipse.
4. Place the Moon at each of the following phases: first quarter, full moon, third quarter, and new moon. Identify which, if any, type of eclipse could occur during each phase. Record your data.
5. Place the Moon at the location where a lunar eclipse could occur. Move it slightly toward Earth, then away from Earth. Note the amount of change in the size of the shadow.
6. Repeat step 5 with the Moon in a position where a solar eclipse could occur.

Conclude and Apply

1. **Identify** which phase(s) of the Moon make(s) it possible for an eclipse to occur.
2. **Describe** the effect of a small change in distance between Earth and the Moon on the size of the umbra and penumbra.
3. **Infer** why a lunar and a solar eclipse do not occur every month.
4. **Explain** why only a few people have experienced a total solar eclipse.
5. **Diagram** the positions of the Sun, Earth, and the Moon during a first-quarter moon.
6. **Infer** why it might be better to call a full moon a half moon.

Communicate your answers to other students.

LAB **321**

section 3
Exploring Earth's Moon

as you read

What You'll Learn
- **Describe** recent discoveries about the Moon.
- **Examine** facts about the Moon that might influence future space travel.

Why It's Important
Continuing moon missions may result in discoveries about Earth's origin.

Review Vocabulary
comet: space object orbiting the Sun formed from dust and rock particles mixed with frozen water, methane, and ammonia

New Vocabulary
- impact basin

Missions to the Moon

The Moon has always fascinated humanity. People have made up stories about how it formed. Children's stories even suggested it was made of cheese. Of course, for centuries astronomers also have studied the Moon for clues to its makeup and origin. In 1959, the former Soviet Union launched the first *Luna* spacecraft, enabling up-close study of the Moon. Two years later, the United States began a similar program with the first *Ranger* spacecraft and a series of *Lunar Orbiters*. The spacecraft in these early missions took detailed photographs of the Moon.

The next step was the *Surveyor* spacecraft designed to take more detailed photographs and actually land on the Moon. Five of these spacecraft successfully touched down on the lunar surface and performed the first analysis of lunar soil. The goal of the *Surveyor* program was to prepare for landing astronauts on the Moon. This goal was achieved in 1969 by the astronauts of *Apollo 11*. By 1972, when the *Apollo* missions ended, 12 U.S. astronauts had walked on the Moon. A time line of these important moon missions can be seen in **Figure 15.**

Figure 15 This time line illustrates some of the most important events in the history of moon exploration.

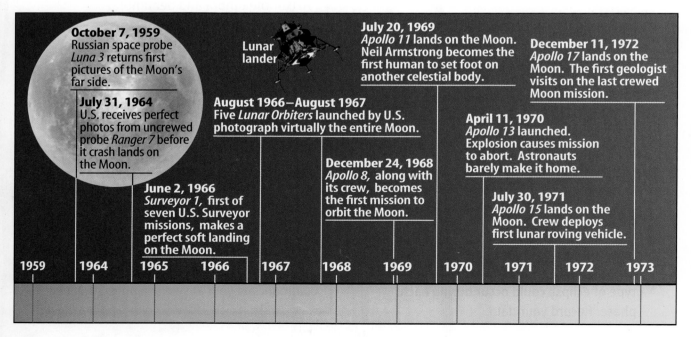

322 CHAPTER 11 The Sun-Earth-Moon System

Surveying the Moon There is still much to learn about the Moon and, for this reason, the United States resumed its studies. In 1994, the *Clementine* was placed into lunar orbit. Its goal was to conduct a two-month survey of the Moon's surface. An important aspect of this study was collecting data on the mineral content of Moon rocks. In fact, this part of its mission was instrumental in naming the spacecraft. Clementine was the daughter of a miner in the ballad *My Darlin' Clementine*. While in orbit, *Clementine* also mapped features on the Moon's surface, including huge impact basins.

Reading Check *Why was* Clementine *placed in lunar orbit?*

Impact Basins When meteorites and other objects strike the Moon, they leave behind depressions in the Moon's surface. The depression left behind by an object striking the Moon is known as an **impact basin,** or impact crater. The South Pole-Aitken Basin is the oldest identifiable impact feature on the Moon's surface. At 12 km in depth and 2,500 km in diameter, it is also the largest and deepest impact basin in the solar system.

Impact basins at the poles were of special interest to scientists. Because the Sun's rays never strike directly, the crater bottoms remain always in shadow. Temperatures in shadowed areas, as shown in **Figure 16,** would be extremely low, probably never more than −173°C. Scientists hypothesize that any ice deposited by comets impacting the Moon throughout its history would remain in these shadowed areas. Indeed, early signals from *Clementine* indicated the presence of water. This was intriguing, because it could be a source of water for future moon colonies.

Science Online

Topic: The Far Side
Visit blue.msscience.com for Web links to information about the far side of the Moon.

Activity Compare the image of the far side of the Moon with that of the near side shown in **Figure 12.** Make a list of all the differences you note and then compare them with lists made by other students.

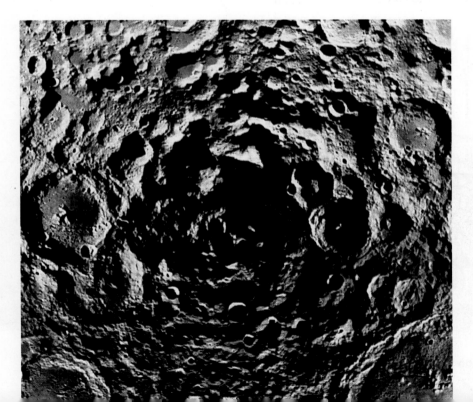

Figure 16 The South Pole-Aitken Basin is the largest of its kind found anywhere in the solar system. The deepest craters in the basin stay in shadow throughout the Moon's rotation. Ice deposits from impacting comets are thought to have collected at the bottom of these craters.

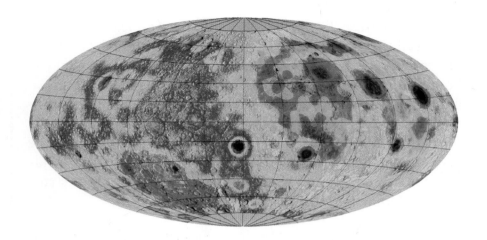

Figure 17 This computer-enhanced map based on *Clementine* data indicates the thickness of the Moon's crust. The crust of the side of the Moon facing Earth, shown mostly in red, is thinner than the crust on the far side of the Moon.

Mapping the Moon

A large part of *Clementine's* mission included taking high-resolution photographs so a detailed map of the Moon's surface could be compiled. *Clementine* carried cameras and other instruments to collect data at wavelengths ranging from infrared to ultraviolet. One camera could resolve features as small as 20 m across. One image resulting from *Clementine* data is shown in **Figure 17.** It shows that the crust on the side of the Moon that faces Earth is much thinner than the crust on the far side. Additional information shows that the Moon's crust is thinnest under impact basins. Based on analysis of the light data received from *Clementine*, a global map of the Moon also was created that shows its composition, as seen in **Figure 18.**

Reading Check *What information about the Moon did scientists learn from* Clementine?

The Lunar Prospector The success of Clementine opened the door for further moon missions. In 1998, NASA launched the desk-sized *Lunar Prospector*, shown in **Figure 18,** into lunar orbit. The spacecraft spent a year orbiting the Moon from pole to pole, once every two hours. The resulting maps confirmed the *Clementine* data. Also, data from *Lunar Prospector* confirmed that the Moon has a small, iron-rich core about 600 km in diameter. A small core supports the impact theory of how the Moon formed—only a small amount of iron could be blasted away from Earth.

Figure 18 *Lunar Prospector* performed high-resolution mapping of the lunar surface and had instruments that detected water ice at the lunar poles.

Icy Poles In addition to photographing the surface, *Lunar Prospector* carried instruments designed to map the Moon's gravity, magnetic field, and the abundances of 11 elements in the lunar crust. This provided scientists with data from the entire lunar surface rather than just the areas around the Moon's equator, which had been gathered earlier. Also, *Lunar Prospector* confirmed the findings of *Clementine* that water ice was present in deep craters at both lunar poles.

Later estimates concluded that as much as 3 billion metric tons of water ice was present at the poles, with a bit more at the north pole. Using data from *Lunar Prospector*, scientists prepared maps showing the location of water ice at each pole. **Figure 19** shows how water may be distributed at the Moon's north pole. At first it was thought that ice crystals were mixed with lunar soil, but most recent results suggest that the ice may be in the form of more compact deposits.

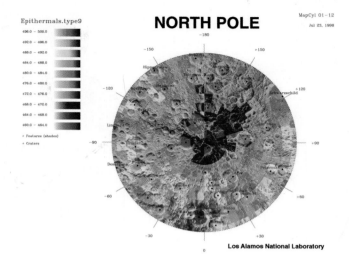

Figure 19 The *Lunar Prospector* data indicates that ice exists in crater shadows at the Moon's poles.

section 3 review

Summary

Missions to the Moon
- The first lunar surveys were done by *Luna*, launched by the former Soviet Union, and U.S.-launched *Ranger* and *Lunar Orbiters*.
- Five *Surveyor* probes landed on the Moon.
- U.S. Astronauts landed on and explored the Moon in the *Apollo* program.
- *Clementine*, a lunar orbiter, mapped the lunar surface and collected data on rocks.
- *Clementine* found that the lunar crust is thinner on the side facing Earth.
- Data from *Clementine* indicated that water ice could exist in shaded areas of impact basins.

Mapping the Moon
- *Lunar Prospector* orbited the Moon from pole to pole, collecting data that confirm *Clementine* results and that the Moon has a small iron-rich core.
- Data from *Lunar Prospector* indicate the presence of large quantities of water ice in craters at the lunar poles.

Self Check

1. **Name** the first U.S. spacecraft to successfully land on the Moon. What was the major purpose of this program?
2. **Explain** why scientists continue to study the Moon long after the *Apollo* program ended and list some of the types of data that have been collected.
3. **Explain** how water ice might be preserved in portions of deep impact craters.
4. **Describe** how the detection of a small iron-rich core supports the theory that the Moon was formed from a collision between Earth and a Mars-sized object.
5. **Think Critically** Why might the discovery of ice in impact basins at the Moon's poles be important to future space flights?

Applying Skills

6. **Infer** why it might be better to build a future moon base on a brightly lit plateau near a lunar pole in the vicinity of a deep crater. Why not build a base in the crater itself?

TILT AND TEMPERATURE

Goals
- **Measure** the temperature change in a surface after light strikes it at different angles.
- **Describe** how the angle of light relates to seasons on Earth.

Materials
tape
black construction paper (one sheet)
gooseneck lamp with 75-watt bulb
Celsius thermometer
watch
protractor

Safety Precautions

WARNING: *Do not touch the lamp without safety gloves. The lightbulb and shade can be hot even when the lamp has been turned off. Handle the thermometer carefully. If it breaks, do not touch anything. Inform your teacher immediately.*

If you walk on blacktop pavement at noon, you can feel the effect of solar energy. The Sun's rays hit at the highest angle at midday. Now consider the fact that Earth is tilted on its axis. How does this tilt affect the angle at which light rays strike an area on Earth? How is the angle of the light rays related to the amount of heat energy and the changing seasons?

Real-World Question
How does the angle at which light strikes Earth affect the amount of heat energy received by any area on Earth?

Procedure
1. Choose three angles that you will use to aim the light at the paper.
2. **Determine** how long you will shine the light at each angle before you measure the temperature. You will measure the temperature at two times for each angle. Use the same time periods for each angle.
3. Copy the following data table into your Science Journal to record the temperature the paper reaches at each angle and time.

Temperature Data			
Angle of Lamp	Initial Temperature	Temperature at ____ Minutes/Seconds	Temperature at ____ Minutes/Seconds
First angle			
Second angle		Do not write in this book.	
Third angle			

4. Form a pocket out of a sheet of black construction paper and tape it to a desk or the floor.
5. Using the protractor, set the gooseneck lamp so that it will shine on the paper at one of the angles you chose.

326 CHAPTER 11 The Sun-Earth-Moon System

Using Scientific Methods

6. Place the thermometer in the paper pocket. Turn on the lamp. Use the thermometer to measure the temperature of the paper at the end of the first time period. Continue shining the lamp on the paper until the second time period has passed. Measure the temperature again. Record your data in your data table.

7. Turn off the lamp until the paper cools to room temperature. Repeat steps 5 and 6 using your other two angles.

Conclude and Apply

1. **Describe** your experiment. Identify the variables in your experiment. Which were your independent and dependent variables?
2. **Graph** your data using a line graph. Describe what your graph tells you about the data.
3. **Describe** what happened to the temperature of the paper as you changed the angle of light.
4. **Predict** how your results might have been different if you used white paper. Explain why.
5. **Describe** how the results of this experiment apply to seasons on Earth.

Communicating Your Data

Compare your results with those of other students in your class. **Discuss** how the different angles and time periods affected the temperatures.

TIME SCIENCE AND HISTORY
SCIENCE CAN CHANGE THE COURSE OF HISTORY!

THE Mayan Calendar

Most people take for granted that a week is seven days, and that a year is 12 months. However there are other ways to divide time into useful units. Roughly 1,750 years ago, in what is now south Mexico and Central America, the Mayan people invented a calendar system based on careful observations of sun and moon cycles.

These glyphs represent four different days of the Tzolkin calendar.

In fact, the Maya had several calendars that they used at the same time.

Two calendars were most important—one was based on 260 days and the other on 365 days. The calendars were so accurate and useful that later civilizations, including the Aztecs, adopted them.

The 260-Day Calendar

This Mayan calendar, called the *Tzolkin* (tz uhl KIN), was used primarily to time planting, harvesting, drying, and storing of corn—their main crop. Each day of the *Tzolkin* had one of 20 names, as well as a number from 1 to 13 and a Mayan god associated with it.

The 365-Day Calendar

Another Mayan calendar, called the *Haab* (HAHB), was based on the orbit of Earth around the Sun. It was divided into 18 months with 20 days each, plus five extra days at the end of each year.

These calendars were used together making the Maya the most accurate reckoners of time before the modern period. In fact, they were only one day off every 6,000 years.

The Kukulkan, built around the year 1050 A.D., in what is now Chichén Itzá, Mexico, was used by the Maya as a calendar. It had four stairways, each with 91 steps, a total of 365 including the platform on top.

Drawing Symbols The Maya created picture symbols for each day of their week. Historians call these symbols glyphs. Collaborate with another student to invent seven glyphs—one for each weekday. Compare them with other glyphs at msscience.com/time.

For more information, visit blue.msscience.com/time

Chapter 11 Study Guide

Reviewing Main Ideas

Section 1 Earth

1. Earth is spherical and bulges slightly at its equator.
2. Earth rotates once per day and orbits the Sun in a little more than 365 days.
3. Earth has a magnetic field.
4. Seasons on Earth are caused by the tilt of Earth's axis as it orbits the Sun.

Section 2 The Moon—Earth's Satellite

1. Earth's Moon goes through phases that depend on the relative positions of the Sun, the Moon, and Earth.
2. Eclipses occur when Earth or the Moon temporarily blocks sunlight from reaching the other.
3. The Moon's maria are the result of ancient lava flows. Craters on the Moon's surface formed from impacts with meteorites, asteroids, and comets.

Section 3 Exploring Earth's Moon

1. The *Clementine* spacecraft took detailed photographs of the Moon's surface and collected data indicating the presence of water in deep craters.
2. NASA's *Lunar Prospector* spacecraft found additional evidence of ice.

Visualizing Main Ideas

Copy and complete the following concept map on the impact theory of the Moon's formation.

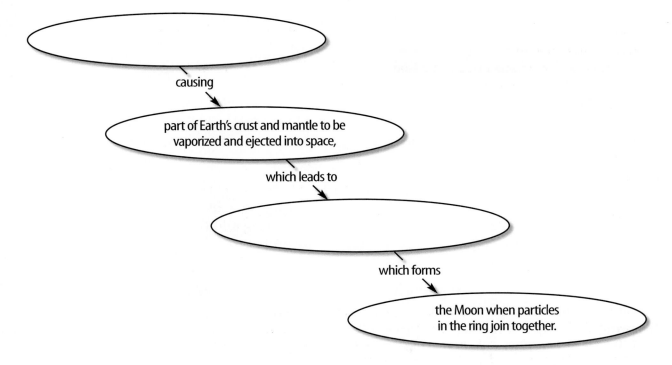

chapter 11 Review

Using Vocabulary

axis p. 307
ellipse p. 309
equinox p. 311
full moon p. 313
impact basin p. 323
lunar eclipse p. 316
maria p. 317
moon phase p. 313
new moon p. 313
revolution p. 309
rotation p. 307
solar eclipse p. 315
solstice p. 310
sphere p. 306
waning p. 313
waxing p. 313

Fill in the blanks with the correct vocabulary word or words.

1. The spinning of Earth around its axis is called _____.
2. The _____ is the point at which the Sun reaches its greatest distance north or south of the equator.
3. The Moon is said to be _____ when less and less of the side facing Earth is lighted.
4. The depression left behind by an object striking the Moon is called a(n) _____.
5. Earth's orbit is a(n) _____.

Checking Concepts

Choose the word or phrase that best answers the question.

6. How long does it take for the Moon to rotate once?
   A) 24 hours   C) 27.3 hours
   B) 365 days   D) 27.3 days

7. Where is Earth's circumference greatest?
   A) equator   C) poles
   B) mantle   D) axis

8. Earth is closest to the Sun during which season in the northern hemisphere?
   A) spring   C) winter
   B) summer   D) fall

9. What causes the Sun to appear to rise and set?
   A) Earth's revolution
   B) the Sun's revolution
   C) Earth's rotation
   D) the Sun's rotation

Use the photo below to answer question 10.

10. What phase of the Moon is shown in the photo above?
    A) waning crescent   C) third quarter
    B) waxing gibbous   D) waning gibbous

11. How long does it take for the Moon to revolve once around Earth?
    A) 24 hours   C) 27.3 hours
    B) 365 days   D) 27.3 days

12. What is it called when the phases of the Moon appear to get larger?
    A) waning   C) rotating
    B) waxing   D) revolving

13. What kind of eclipse occurs when the Moon blocks sunlight from reaching Earth?
    A) solar   C) full
    B) new   D) lunar

14. What is the darkest part of the shadow during an eclipse?
    A) waxing gibbous   C) waning gibbous
    B) umbra   D) penumbra

15. What is the name for a depression on the Moon caused by an object striking its surface?
    A) eclipse   C) phase
    B) moonquake   D) impact basin

chapter 11 Review

Thinking Critically

16. **Predict** how the Moon would appear to an observer in space during its revolution. Would phases be observable? Explain.

17. **Predict** what the effect would be on Earth's seasons if the axis were tilted at 28.5° instead of 23.5°.

18. **Infer** Seasons in the two hemispheres are opposite. Explain how this supports the statement that seasons are NOT caused by Earth's changing distance from the Sun.

19. **Draw Conclusions** How would solar eclipses be different if the Moon were twice as far from Earth? Explain.

20. **Predict** how the information gathered by moon missions could be helpful in the future for people wanting to establish a colony on the Moon.

21. **Use Variables, Constants, and Controls** Describe a simple activity to show how the Moon's rotation and revolution work to keep the same side facing Earth at all times.

22. **Draw Conclusions** Gravity is weaker on the Moon than it is on Earth. Why might more craters be present on the far side of the Moon than on the side of the Moon facing Earth?

23. **Recognize Cause and Effect** During a new phase of the Moon, we cannot see it because no sunlight reaches the side facing Earth. Yet sometimes when there is a thin crescent visible, we do see a faint image of the rest of the Moon. Explain what might cause this to happen.

24. **Describe** Earth's magnetic field. Include an explanation of how scientists believe it is generated and two ways in which it helps people on Earth.

Performance Activities

25. **Display** Draw a cross section of the Moon. Include the crust, outer and inner mantles, and possible core based on the information in this chapter. Indicate the presence of impact craters and show how the thickness of the crust varies from one side of the Moon to the other.

26. **Poem** Write a poem in which you describe the various surface features of the Moon. Be sure to include information on how these features formed.

Applying Math

27. **Orbital Tilt** The Moon's orbit is tilted at an angle of 5° to Earth's orbit around the sun. Using a protractor, draw the Moon's orbit around Earth. What fraction of a full circle (360°) is 5°?

Use the illustration below to answer question 28.

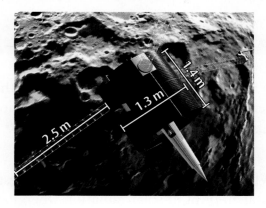

28. **Model to Scale** You are planning to make a scale model of the *Lunar Prospector* spacecraft, shown above. Assuming that the three instrument masts are of equal length, draw a labeled diagram of your model using a scale of 1 cm equals 30 cm.

29. **Spacecraft Velocity** The *Lunar Prospector* spacecraft shown above took 105 hours to reach the Moon. Assuming that the average distance from Earth to Moon is 384,000 km, calculate its average velocity on the trip.

Chapter 11 Standardized Test Practice

Part 1 Multiple Choice

Record your answers on the answer sheet provided by your teacher or on a sheet of paper.

1. Which of the following terms would you use to describe the spinning of Earth on its axis?
   A. revolution   C. rotation
   B. ellipse   D. solstice

Use the illustration below to answer questions 2 and 3.

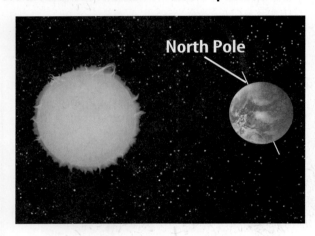

2. Which season is beginning for the southern hemisphere when Earth is in this position?
   A. spring   C. fall
   B. summer   D. winter

3. Which part of Earth receives the greatest total amount of solar radiation when Earth is in this position?
   A. northern hemisphere
   B. South Pole
   C. southern hemisphere
   D. equator

4. Which term describes the dark, flat areas on the Moon's surface which are made of cooled, hardened lava?
   A. spheres   C. highlands
   B. moonquakes D. maria

Use the illustration below to answer questions 5 and 6.

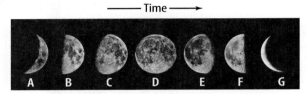

5. Which letter corresponds to the moon phase waning gibbous?
   A. G   C. E
   B. C   D. A

6. The Moon phase cycle lasts about 29.5 days. Given this information, about how long does it take the Moon to wax from new moon to full moon?
   A. about 3 days   C. about 2 weeks
   B. about 1 week   D. about 4 weeks

7. Where have large amounts of water been detected on the Moon?
   A. highlands   C. maria
   B. lunar equator D. lunar poles

8. In what month is Earth closest to the Sun?
   A. March   C. July
   B. September D. January

9. So far, where on the Moon have the youngest rocks been found?
   A. lunar highlands C. lunar poles
   B. maria   D. lunar equator

Test-Taking Tip

Eliminate Choices If you don't know the answer to a multiple-choice question, eliminate as many incorrect choices as possible. Mark your best guess from the remaining answers before moving to the next question.

Question 5 Eliminate those phases that you know are not gibbous.

Standardized Test Practice

Part 2 — Short Response/Grid In

Record your answers on the answer sheet provided by your teacher or on a sheet of paper.

10. Explain why the North Pole is always in sunlight during summer in the northern hemisphere.

11. Describe one positive effect of Earth's magnetic field.

12. Explain the difference between a solstice and an equinox. Give the dates of these events on Earth.

13. Explain how scientists know that the Moon was once geologically active.

Use the illustration below to answer questions 14 and 15.

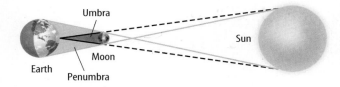

14. What type of eclipse is shown above?

15. Describe what a person standing in the Moon's umbra would see if he or she looked at the sky wearing protective eyewear.

16. The tilt of Earth on its axis causes seasons. Give two reasons why this tilt causes summer to be warmer than winter.

17. When the Apollo missions ended in 1972, 12 astronauts had visited the Moon and brought back samples of moon soil and rock. Explain why we continue to send orbiting spacecraft to study the Moon.

18. Define the term *impact basin*, and name the largest one known in the solar system.

Part 3 — Open Ended

Record your answers on a sheet of paper.

Use the illustrations below to answer questions 19 and 20.

19. As a ship comes into view over the horizon, the top appears before the rest of the ship. How does this demonstrate that Earth is spherical?

20. If Earth were flat, how would an approaching ship appear differently?

21. Explain why eclipses of the Sun occur only occasionally despite the fact that the Moon's rotation causes it to pass between Earth and the Sun every month.

22. Recent data from the spacecraft *Lunar Prospector* indicate the presence of large quantities of water in shadowed areas of lunar impact basins. Describe the hypothesis that scientists have developed to explain how this water reached the moon and how it might be preserved.

23. Compare the impact theory of lunar formation with one of the older theories proposed before the *Apollo* mission.

24. Describe how scientists study the interior of the Moon and what they have learned so far.

25. Explain why Earth's magnetic north poles must be mapped and why these maps must be kept up-to-date.

chapter 12

The Solar System

The BIG Idea
The solar system consists of planets and their moons, comets, meteoroids, and asteroids that all orbit the Sun.

SECTION 1
The Solar System
Main Idea Gravity assisted in the formation of the solar system and continues to hold the planets in their places as they orbit the Sun.

SECTION 2
The Inner Planets
Main Idea The inner planets—Mercury, Venus, Earth, and Mars—are the closest planets to the Sun.

SECTION 3
The Outer Planets
Main Idea The outer planets—Jupiter, Saturn, Uranus, and Neptune—are the farthest planets from the Sun.

SECTION 4
Other Objects in the Solar System
Main Idea Comets, meteoroids, and asteroids are smaller than planets but also orbit the Sun and are part of the solar system.

How is space explored?
You've seen the Sun and the Moon. You also might have observed some of the planets. But to get a really good look at the solar system from Earth, telescopes are needed. The optical telescope at the Keck Observatory in Hawaii allows scientists a close-up view.

Science Journal If you could command the Keck telescope, what would you view? Describe what you would see.

Start-Up Activities

Model Crater Formation

Some objects in the solar system have many craters. The Moon is covered with them. The planet Mercury also has a cratered landscape. Even Earth has some craters. All of these craters formed when rocks from space hit the surface of the planet or moon. In this lab, you'll explore crater formation.

1. Place white flour into a metal cake pan to a depth of 3 cm.
2. Cover the flour with 1 cm of colored powdered drink mix or different colors of gelatin powder.
3. From different heights, ranging from 10 cm to 25 cm, drop various-sized marbles into the pan.
4. **Think Critically** Make drawings in your Science Journal that show what happened to the surface of the powder when marbles were dropped from different heights.

 **Preview this chapter's content and activities at blue.msscience.com**

 **The Solar System** Make the following Foldable to help you identify what you already know, what you want to know, and what you learned about the solar system.

STEP 1 Fold a vertical sheet of paper from side to side. Make the front edge about 1.25 cm shorter than the back edge.

STEP 2 Turn lengthwise and fold into thirds.

STEP 3 Unfold and cut only the top layer along both folds to make three tabs.

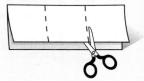

STEP 4 Label each tab.

Identify Questions Before you read the chapter, write what you already know about the solar system under the left tab of your Foldable. Write questions about what you'd like to know under the center tab. After you read the chapter, list what you learned under the right tab.

335

Get Ready to Read

Compare and Contrast

① Learn It! Good readers compare and contrast information as they read. This means they look for similarities and differences to help them to remember important ideas. Look for signal words in the text to let you know when the author is comparing or contrasting.

Compare and Contrast Signal Words	
Compare	**Contrast**
as	but
like	or
likewise	unlike
similarly	however
at the same time	although
in a similar way	on the other hand

② Practice It! Read the excerpt below and notice how the author uses contrast signal words to describe the differences between types of planets.

> You can see that the planets **closer** to the Sun travel **faster** than planets **farther** away from the Sun. Because of their **slower** speeds and the **longer** distances they must travel, the outer planets take much **longer** to orbit the Sun than the inner planets do.
>
> —*from page 340*

③ Apply It! Compare and contrast Earth's characteristics on page 344 to the other planets.

Target Your Reading

Reading Tip: As you read, use other skills, such as summarizing and connecting, to help you understand comparisons and contrasts.

Use this to focus on the main ideas as you read the chapter.

1) Before you read the chapter, respond to the statements below on your worksheet or on a numbered sheet of paper.
- Write an **A** if you **agree** with the statement.
- Write a **D** if you **disagree** with the statement.

2) After you read the chapter, look back to this page to see if you've changed your mind about any of the statements.
- If any of your answers changed, explain why.
- Change any false statements into true statements.
- Use your revised statements as a study guide.

Science Online
Print out a worksheet of this page at blue.msscience.com

Before You Read A or D		Statement	After You Read A or D
	1	The Sun contains more than 99 percent of the mass of the entire solar system.	
	2	Venus is sometimes called Earth's twin because its size and mass are similar to Earth's.	
	3	Mars is the third planet from the Sun and is nicknamed the blue planet.	
	4	Earth is the most volcanically active object in the solar system.	
	5	Jupiter, Saturn, Uranus, and Neptune all have rings that orbit these planets.	
	6	Most asteroids are located in an area called the asteroid belt, which is located between the orbits of Jupiter and Saturn.	
	7	Asteroids, meteoroids, and comets do not contain water.	

section 1

The Solar System

as you read

What You'll Learn
- **Compare** models of the solar system.
- **Explain** that gravity holds planets in orbits around the Sun.

Why It's Important
New technology has come from exploring the solar system.

Review Vocabulary
system: a portion of the universe and all of its components, processes, and interactions

New Vocabulary
- solar system

Ideas About the Solar System

People have been looking at the night sky for thousands of years. Early observers noted the changing positions of the planets and developed ideas about the solar system based on their observations and beliefs. Today, people know that objects in the solar system orbit the Sun. People also know that the Sun's gravity holds the solar system together, just as Earth's gravity holds the Moon in its orbit around Earth. However, our understanding of the solar system changes as scientists make new observations.

Earth-Centered Model Many early Greek scientists thought the planets, the Sun, and the Moon were fixed in separate spheres that rotated around Earth. The stars were thought to be in another sphere that also rotated around Earth. This is called the Earth-centered model of the solar system. It included Earth, the Moon, the Sun, five planets—Mercury, Venus, Mars, Jupiter, and Saturn—and the sphere of stars.

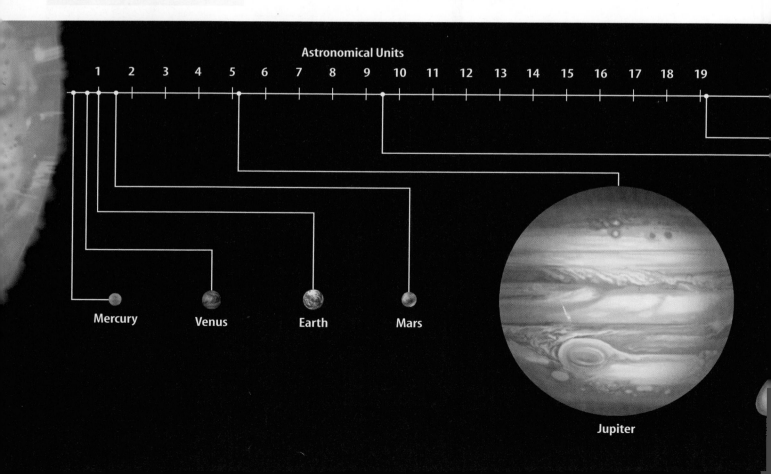

Sun-Centered Model People believed the idea of an Earth-centered solar system for centuries. Then in 1543, Nicholas Copernicus published a different view. Copernicus stated that the Moon revolved around Earth and that Earth and the other planets revolved around the Sun. He also stated that the daily movement of the planets and the stars was caused by Earth's rotation. This is the Sun-centered model of the solar system.

Using his telescope, Galileo Galilei observed that Venus went through a full cycle of phases like the Moon's. He also observed that the apparent diameter of Venus was smallest when the phase was near full. This only could be explained if Venus were orbiting the Sun. Galileo concluded that the Sun is the center of the solar system.

Modern View of the Solar System As of 2006, the **solar system** is made up of eight planets, including Earth, and many smaller objects that orbit the Sun. The eight planets and the Sun are shown in **Figure 1**. Notice how small Earth is compared with some of the other planets and the Sun.

The solar system includes a huge volume of space that stretches in all directions from the Sun. Because the Sun contains 99.86 percent of the mass of the solar system, its gravity is immense. The Sun's gravity holds the planets and other objects in the solar system in their orbits.

Science Online

Topic: Solar System
Visit blue.msscience.com for Web links to information about planets.

Figure 1 Each of the eight planets in the solar system is unique. The distances between the planets and the Sun are shown on the scale. One astronomical unit (AU) is the average distance between Earth and the Sun.

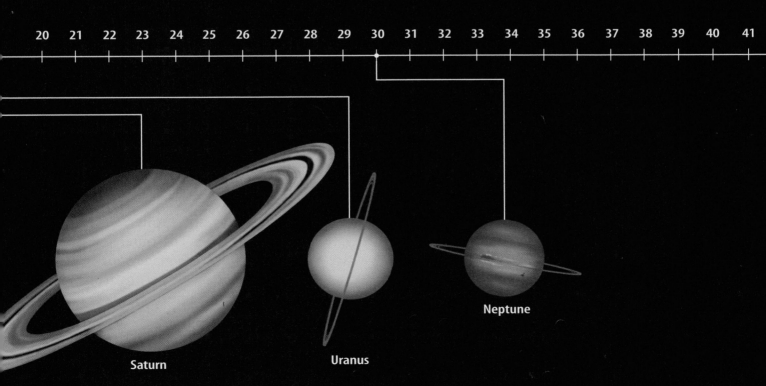

SECTION 1 The Solar System

Rotational Motion You might have noticed that when a twirling ice skater pulls in her arms, she spins faster. The same thing occurs when a cloud of gas, ice, and dust in a nebula contracts. As mass moves toward the center of the cloud, the cloud rotates faster.

How the Solar System Formed

Scientists hypothesize that the solar system formed from part of a nebula of gas, ice, and dust, like the one shown in **Figure 2.** Follow the steps shown in **Figures 3A** through **3D** to learn how this might have happened. A nearby star might have exploded and the shock waves produced by these events could have caused the cloud to start contracting. As it contracted, the nebula likely fragmented into smaller and smaller pieces. The density in the cloud fragments became greater, and the attraction of gravity pulled more gas and dust toward several centers of contraction. This in turn caused them to flatten into disks with dense centers. As the cloud fragments continued to contract, they began to rotate faster and faster.

As each cloud fragment contracted, its temperature increased. Eventually, the temperature in the core of one of these cloud fragments reached about 10 million degrees Celsius. Nuclear fusion began when hydrogen atoms started to fuse and release energy. A star was born—the beginning of the Sun.

It is unlikely that the Sun formed alone. A cluster of stars like the Sun likely formed from parts of the original cloud. The Sun, which is one of many stars in our galaxy, probably escaped from this cluster and has since revolved around the galaxy many times.

Reading Check *What is nuclear fusion?*

Planet Formation Not all of the nearby gas, ice, and dust was drawn into the core of the cloud fragment. The matter that did not get pulled into the center collided and stuck together to form the planets and asteroids. Close to the Sun, the temperature was hot, and the easily vaporized elements could not condense into solids. This is why lighter elements are scarcer in the planets near the Sun than in planets farther out in the solar system.

The inner planets of the solar system—Mercury, Venus, Earth, and Mars—are small, rocky planets with iron cores. The outer planets are Jupiter, Saturn, Uranus, and Neptune. The outer planets are much larger and are made mostly of lighter substances such as hydrogen, helium, methane, and ammonia.

Figure 2 Systems of planets, such as the solar system, form in areas of space like this, called a nebula.

NATIONAL GEOGRAPHIC VISUALIZING THE SOLAR SYSTEM'S FORMATION

Figure 3

Through careful observations, astronomers have found clues that help explain how the solar system may have formed. **A** More than 4.6 billion years ago, the solar system was a cloud fragment of gas, ice, and dust. **B** Gradually, this cloud fragment contracted into a large, tightly packed, spinning disk. The disk's center was so hot and dense that nuclear fusion reactions began to occur, and the Sun was born. **C** Eventually, the rest of the material in the disk cooled enough to clump into scattered solids. **D** Finally, these clumps collided and combined to become the eight planets that make up the solar system today.

SECTION 1 The Solar System

Table 1 Average Orbital Speed

Planet	Average Orbital Speed (km/s)
Mercury	48
Venus	35
Earth	30
Mars	24
Jupiter	13
Saturn	9.7
Uranus	6.8
Neptune	5.4

Johannes Kepler

Motions of the Planets

When Nicholas Copernicus developed his Sun-centered model of the solar system, he thought that the planets orbited the Sun in circles. In the early 1600s, German mathematician Johannes Kepler began studying the orbits of the planets. He discovered that the shapes of the orbits are not circular. They are oval shaped, or elliptical. His calculations further showed that the Sun is not at the center of the orbits but is slightly offset.

Kepler also discovered that the planets travel at different speeds in their orbits around the Sun, as shown in **Table 1.** You can see that the planets closer to the Sun travel faster than planets farther away from the Sun. Because of their slower speeds and the longer distances they must travel, the outer planets take much longer to orbit the Sun than the inner planets do.

Copernicus's ideas, considered radical at the time, led to the birth of modern astronomy. Early scientists didn't have technology such as space probes to learn about the planets. Nevertheless, they developed theories about the solar system that still are used today.

section 1 review

Summary

Ideas About the Solar System
- The planets in the solar system revolve around the Sun.
- The Sun's immense gravity holds the planets in their orbits.

How the Solar System Formed
- The solar system formed from a piece of a nebula of gas, ice, and dust.
- As the piece of nebula contracted, nuclear fusion began at its center and the Sun was born.

Motion of the Planets
- The planets' orbits are elliptical.
- Planets that are closer to the Sun revolve faster than those that are farther away from the Sun.

Self Check

1. **Describe** the Sun-centered model of the solar system. What holds the solar system together?
2. **Explain** how the planets in the solar system formed.
3. **Infer** why life is unlikely on the outer planets in spite of the presence of water, methane, and ammonia—materials needed for life to develop.
4. **List** two reasons why the outer planets take longer to orbit the Sun than the inner planets do.
5. **Think Critically** Would a year on the planet Neptune be longer or shorter than an earth year? Explain.

Applying Skills

6. **Concept Map** Make a concept map that compares and contrasts the Earth-centered model with the Sun-centered model of the solar system.

 blue.mssscience.com/self_check_quiz

Planetary Orbits

Planets travel around the Sun along paths called orbits. As you construct a model of a planetary orbit, you will observe that the shape of planetary orbits is an ellipse.

Real-World Question

How can you model planetary orbits?

Goals
- **Model** planetary orbits.
- **Calculate** the eccentricity of ellipses.

Materials
thumbtacks or pins (2)
cardboard (23 cm × 30 cm)
paper (21.5 cm × 28 cm)
metric ruler
string (25 cm)
pencil

Safety Precautions

Procedure

1. Place a blank sheet of paper on top of the cardboard and insert two thumbtacks or pins about 3 cm apart.
2. Tie the string into a circle with a circumference of 15 cm to 20 cm. Loop the string around the thumbtacks. With someone holding the tacks or pins, place your pencil inside the loop and pull it tight.
3. Moving the pencil around the tacks and keeping the string tight, mark a line until you have completed a smooth, closed curve.
4. Repeat steps 1 through 3 several times. First, vary the distance between the tacks, then vary the length of the string. However, change only one of these each time. Make a data table to record the changes in the sizes and shapes of the ellipses.
5. Eccentricity is a measure of how an orbit varies from a perfect circle. Eccentricity, e, is determined by dividing the distance, d, between the foci (fixed points—here, the tacks) by the length, l, of the major axis.

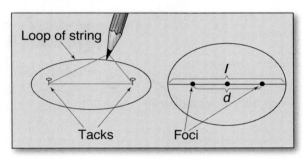

6. **Calculate** and record the eccentricity of the ellipses that you constructed.
7. **Research** the eccentricities of planetary orbits. Construct an ellipse with the same eccentricity as Earth's orbit.

Conclude and Apply

1. **Analyze** the effect that a change in the length of the string or the distance between the tacks has on the shape of the ellipse.
2. **Hypothesize** what must be done to the string or placement of tacks to decrease the eccentricity of a constructed ellipse.
3. **Describe** the shape of Earth's orbit. Where is the Sun located within the orbit?

Compare your results with those of other students. **For more help, refer to the Science Skill Handbook.**

section 2

The Inner Planets

as you read

What You'll Learn
- **List** the inner planets in order from the Sun.
- **Describe** each inner planet.
- **Compare and contrast** Venus and Earth.

Why It's Important
The planet that you live on is uniquely capable of sustaining life.

Review Vocabulary
space probe: an instrument that is sent to space to gather information and send it back to Earth

New Vocabulary
- Mercury
- Venus
- Earth
- Mars

Figure 4 Large cliffs on Mercury might have formed when the crust of the planet broke as the planet contracted.

Inner Planets

Today, people know more about the solar system than ever before. Better telescopes allow astronomers to observe the planets from Earth and space. In addition, space probes have explored much of the solar system. Prepare to take a tour of the solar system through the eyes of some space probes.

Mercury

The closest planet to the Sun is **Mercury.** The first American spacecraft mission to Mercury was in 1974–1975 by *Mariner 10*. The spacecraft flew by the planet and sent pictures back to Earth. *Mariner 10* photographed only 45 percent of Mercury's surface, so scientists don't know what the other 55 percent looks like. What they do know is that the surface of Mercury has many craters and looks much like Earth's Moon. It also has cliffs as high as 3 km on its surface. These cliffs might have formed at a time when Mercury shrank in diameter, as seen in **Figure 4.**

Why would Mercury have shrunk? *Mariner 10* detected a weak magnetic field around Mercury. This indicates that the planet has an iron core. Some scientists hypothesize that Mercury's crust solidified while the iron core was still hot and molten. As the core started to solidify, it contracted. The cliffs resulted from breaks in the crust caused by this contraction.

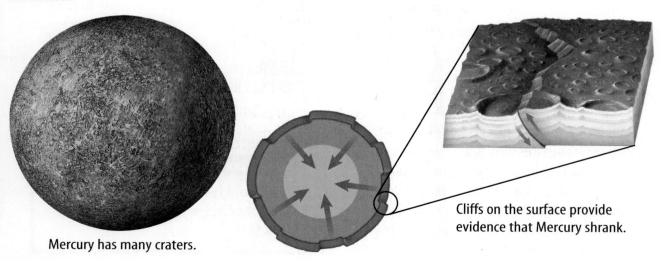

Mercury has many craters.

Cliffs on the surface provide evidence that Mercury shrank.

342 CHAPTER 12 The Solar System

Does Mercury have an atmosphere? Because of Mercury's low gravitational pull and high daytime temperatures, most gases that could form an atmosphere escape into space. *Mariner 10* found traces of hydrogen and helium gas that were first thought to be an atmosphere. However, these gases are now known to be temporarily taken from the solar wind.

The lack of atmosphere and its nearness to the Sun cause Mercury to have great extremes in temperature. Mercury's temperature can reach 425°C during the day, and it can drop to −170°C at night.

Future Mission Launched in 2004, *Messenger* is the next mission to Mercury. This space probe will fly by the planet in 2008 and orbit it in 2011. The probe will photograph and map the entire surface.

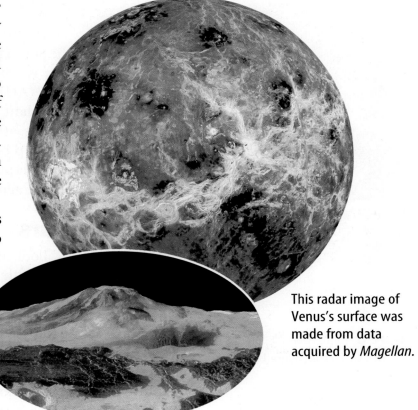

This radar image of Venus's surface was made from data acquired by *Magellan*.

Maat Mons is the highest volcano on Venus. Lava flows extend for hundreds of kilometers across the plains.

Figure 5 Venus is the second planet from the Sun.

Venus

The second planet from the Sun is **Venus,** shown in **Figure 5.** Venus is sometimes called Earth's twin because its size and mass are similar to Earth's. In 1962, *Mariner 2* flew past Venus and sent back information about Venus's atmosphere and rotation. The former Soviet Union landed the first probe on the surface of Venus in 1970. *Venera 7,* however, stopped working in less than an hour because of the high temperature and pressure. Additional *Venera* probes photographed and mapped the surface of Venus. Between 1990 and 1994, the U.S. *Magellan* probe used its radar to make the most detailed maps yet of Venus's surface. It collected radar images of 98 percent of Venus's surface. Notice the huge volcano in **Figure 5.**

Clouds on Venus are so dense that only a small percentage of the sunlight that strikes the top of the clouds reaches the planet's surface. The sunlight that does get through warms Venus's surface, which then gives off heat to the atmosphere. Much of this heat is absorbed by carbon dioxide gas in Venus's atmosphere. This causes a greenhouse effect similar to, but more intense than, Earth's greenhouse effect. Due to this intense greenhouse effect, the temperature on the surface of Venus is between 450°C and 475°C.

SECTION 2 The Inner Planets **343**

Figure 6 More than 70 percent of Earth's surface is covered by liquid water.
Explain *how Earth is unique.*

Figure 7 Many features on Mars are similar to those on Earth.

Earth

Figure 6 shows **Earth,** the third planet from the Sun. The average distance from Earth to the Sun is 150 million km, or one astronomical unit (AU). Unlike other planets, Earth has abundant liquid water and supports life. Earth's atmosphere causes most meteors to burn up before they reach the surface, and its ozone layer protects life from the effects of the Sun's intense radiation.

Mars

Look at **Figure 7.** Can you guess why **Mars,** the fourth planet from the Sun, is called the red planet? Iron oxide in soil on its surface gives it a reddish color. Other features visible from Earth are Mars's polar ice caps and changes in the coloring of the planet's surface. The ice caps are made of frozen water covered by a layer of frozen carbon dioxide.

Most of the information scientists have about Mars came from *Mariner 9,* the *Viking* probes, *Mars Pathfinder, Mars Global Surveyor, Mars Odyssey,* and the Mars Exploration Rovers. *Mariner 9* orbited Mars in 1971 and 1972. It revealed long channels on the planet that might have been carved by flowing water. *Mariner 9* also discovered the largest volcano in the solar system, Olympus Mons, shown in **Figure 7.** Olympus Mons is probably extinct. Large rift valleys in the Martian crust also were discovered. One such valley, Valles Marineris, is shown in **Figure 7.**

Mars is often called the "red planet."

Olympus Mons is the largest volcano in the solar system.

Valles Marineris is more than 4,000 km long, up to 200 km wide, and more than 7 km deep.

The *Viking* Probes The *Viking 1* and *2* probes arrived at Mars in 1976. Each probe consisted of an orbiter and a lander. The orbiters photographed the entire planet from their orbits, while the landers touched down on the surface. Instruments on the landers attempted to detect possible life by analyzing gases in the Martian soil. The tests found no conclusive evidence of life.

Pathfinder* and *Global Surveyor The *Mars Pathfinder* carried a robot rover named *Sojourner* to test samples of Martian rocks and soil. The data showed that iron in the crust might have been leached out by groundwater. Cameras onboard *Global Surveyor* showed features that looked like sediment gullies and deposits formed by flowing water. These features, shown in **Figure 8,** seem to indicate that groundwater might exist on Mars and that it reached the surface. The features are similar to those formed by flash floods on Earth, such as on Mount St. Helens.

***Odyssey* and Mars Exploration Rovers** In 2002, *Mars Odyssey* began orbiting Mars. It measured elements in Mars's crust and searched for signs of water. Instruments on *Odyssey* detected high levels of hematite, a mineral that forms in water, and subsurface ice near the poles.

Odyssey also relayed data to Earth from the Mars Exploration Rovers *Spirit* and *Opportunity* in 2004. These robot rovers analyzed Martian geology. Data from *Opportunity* confirmed that there were once bodies of water on Mars's surface.

Reading Check *What evidence indicates that Mars has water?*

Mini LAB

Inferring Effects of Gravity

Procedure
1. Suppose you are a crane operator who is sent to Mars to help build a colony.
2. Your crane can lift 4,500 kg on Earth, but the force due to gravity on Mars is only 40 percent as large as that on Earth.
3. Determine how much mass your crane could lift on Earth and Mars.

Analysis
1. How can what you have discovered be an advantage over construction on Earth?
2. How might construction advantages change the overall design of the Mars colony?

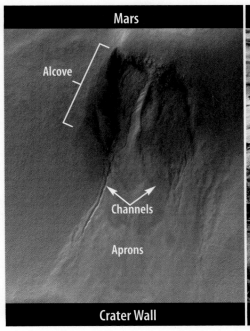
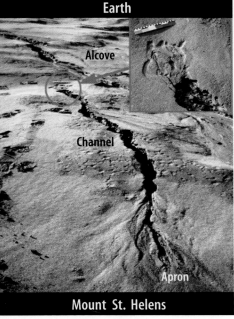

Figure 8 Compare the features found on Mars with those found on an area of Mount St. Helens in Washington state that experienced a flash flood.

SECTION 2 The Inner Planets

Science Online

Topic: Mars Exploration
Visit blue.msscience.com for Web links to information about future missions to Mars.

Activity Make a timeline that shows when each probe is scheduled to reach Mars. Include the mission objectives for each probe on your timeline.

Mars's Atmosphere The *Viking* and *Global Surveyor* probes analyzed gases in the Martian atmosphere and determined atmospheric pressure and temperature. They found that Mars's atmosphere is much thinner than Earth's. It is composed mostly of carbon dioxide, with some nitrogen and argon. Surface temperatures range from −125°C to 35°C. The temperature difference between day and night results in strong winds on the planet, which can cause global dust storms during certain seasons. This information will help in planning possible human exploration of Mars in the future.

Martian Seasons Mars's axis of rotation is tilted 25°, which is close to Earth's tilt of 23.5°. Because of this, Mars goes through seasons as it orbits the Sun, just like Earth does. The polar ice caps on Mars change with the season. During winter, carbon dioxide ice accumulates and makes the ice cap larger. During summer, carbon dioxide ice changes to carbon dioxide gas and the ice cap shrinks. As one ice cap gets larger, the other ice cap gets smaller. The color of the ice caps and other areas on Mars also changes with the season. The movement of dust and sand during dust storms causes the changing colors.

Applying Math — Use Percentages

DIAMETER OF MARS The diameter of Earth is 12,756 km. The diameter of Mars is 53.3 percent of the diameter of Earth. Calculate the diameter of Mars.

Solution

1 *This is what you know:*
- diameter of Earth: 12,756 km
- percent of Earth's diameter: 53.3%
- decimal equivalent: 0.533 (53.3% ÷ 100)

2 *This is what you need to find:* diameter of Mars

3 *This is the procedure you need to use:* Multiply the diameter of Earth by the decimal equivalent.
(12,756 km) × (0.533) = 6,799 km

Practice Problems

1. Use the same procedure to calculate the diameter of Venus. Its diameter is 94.9 percent of the diameter of Earth.
2. Calculate the diameter of Mercury. Its diameter is 38.2 percent of the diameter of Earth.

For more practice, visit blue.msscience.com/math_practice

346 CHAPTER 12 The Solar System

Martian Moons Mars has two small, irregularly shaped moons that are heavily cratered. Phobos, shown in **Figure 9,** is about 25 km in length, and Deimos is about 13 km in length. Deimos orbits Mars once every 31 h, while Phobos speeds around Mars once every 7 h.

Phobos has many interesting surface features. Grooves and chains of smaller craters seem to radiate out from the large Stickney Crater. Some of the grooves are 700 m across and 90 m deep. These features probably are the result of the large impact that formed the Stickney Crater.

Deimos is the outer of Mars's two moons. It is among the smallest known moons in the solar system. Its surface is smoother in appearance than that of Phobos because some of its craters have partially filled with soil and rock.

As you toured the inner planets through the eyes of the space probes, you saw how each planet is unique. Refer to **Table 3** following Section 3 for a summary of the planets. Mercury, Venus, Earth, and Mars are different from the outer planets, which you'll explore in the next section.

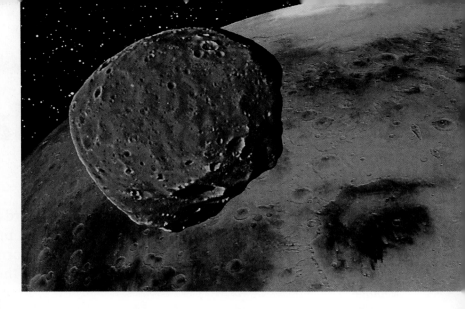

Figure 9 Phobos orbits Mars once every 7 h.
Infer *why Phobos has so many craters.*

section 2 review

Summary

Mercury
- Mercury is extremely hot during the day and extremely cold at night.
- Its surface has many craters.

Venus
- Venus's size and mass are similar to Earth's.
- Temperatures on Venus are between 450°C and 475°C.

Earth
- Earth is the only planet known to support life.

Mars
- Mars has polar ice caps, channels that might have been carved by water, and the largest volcano in the solar system, Olympus Mons.

Self Check

1. **Explain** why Mercury's surface temperature varies so much from day to night.
2. **List** important characteristics for each inner planet.
3. **Infer** why life is unlikely on Venus.
4. **Identify** the inner planet that is farthest from the Sun. Identify the one that is closest to the Sun.
5. **Think Critically** Aside from Earth, which inner planet could humans visit most easily? Explain.

Applying Math

6. **Use Statistics** The inner planets have the following average densities: Mercury, 5.43 g/cm^3; Venus, 5.24 g/cm^3; Earth, 5.52 g/cm^3; and Mars, 3.94 g/cm^3. Which planet has the highest density? Which has the lowest? Calculate the range of these data.

section 3

The Outer Planets

as you read

What You'll Learn
- **Describe** the characteristics of Jupiter, Saturn, Uranus, and Neptune.
- **Describe** the largest moons of each of the outer planets.

Why It's Important
Studying the outer planets will help scientists understand Earth.

Review Vocabulary
moon: a natural satellite of a planet that is held in its orbit around the planet by the planet's gravitational pull

New Vocabulary
- Jupiter
- Great Red Spot
- Saturn
- Uranus
- Neptune
- Pluto

Outer Planets

You might have heard about *Voyager, Galileo,* and *Cassini*. They were not the first probes to the outer planets, but they gathered a lot of new information about them. Follow the spacecrafts as you read about their journeys to the outer planets.

Jupiter

In 1979, *Voyager 1* and *Voyager 2* flew past **Jupiter,** the fifth planet from the Sun. *Galileo* reached Jupiter in 1995, and *Cassini* flew past Jupiter on its way to Saturn in 2000. The spacecrafts gathered new information about Jupiter. The *Voyager* probes revealed that Jupiter has faint dust rings around it and that one of its moons has active volcanoes on it.

Jupiter's Atmosphere Jupiter is composed mostly of hydrogen and helium, with some ammonia, methane, and water vapor. Scientists hypothesize that the atmosphere of hydrogen and helium changes to an ocean of liquid hydrogen and helium toward the middle of the planet. Below this liquid layer might be a rocky core. The extreme pressure and temperature, however, would make the core different from any rock on Earth.

You've probably seen pictures from the probes of Jupiter's colorful clouds. In **Figure 10,** you can see bands of white, red, tan, and brown clouds in its atmosphere. Continuous storms of swirling, high-pressure gas have been observed on Jupiter. The **Great Red Spot** is the most spectacular of these storms.

Figure 10 Jupiter is the largest planet in the solar system.

Notice the colorful bands of clouds in Jupiter's atmosphere.

The Great Red Spot is a giant storm about 25,000 km in size from east to west.

348 CHAPTER 12 The Solar System

Table 2 Large Moons of Jupiter

Io The most volcanically active object in the solar system; sulfurous compounds give it its distinctive reddish and orange colors; has a thin atmosphere of sulfur dioxide.	
Europa Rocky interior is covered by a smooth 5-km-thick crust of ice, which has a network of cracks; a 50-km-deep ocean might exist under the ice crust; has a thin oxygen atmosphere.	
Ganymede Has a heavily cratered crust of ice covered with grooves; has a rocky interior surrounding a molten iron core and a thin oxygen atmosphere.	
Callisto Has a heavily cratered crust with a mixture of ice and rock throughout the interior; has a rock core and a thin atmosphere of carbon dioxide.	

Moons of Jupiter At least 63 moons orbit Jupiter. In 1610, the astronomer Galileo Galilei was the first person to see Jupiter's four largest moons, shown in **Table 2.** Io (I oh) is the closest large moon to Jupiter. Jupiter's tremendous gravitational force and the gravity of Europa, Jupiter's next large moon, pull on Io. This force heats up Io, causing it to be the most volcanically active object in the solar system. You can see a volcano erupting on Io in **Figure 11.** Europa is composed mostly of rock with a thick, smooth crust of ice. Under the ice might be an ocean as deep as 50 km. If this ocean of water exists, it will be the only place in the solar system, other than Earth, where liquid water exists in large quantities. Next is Ganymede, the largest moon in the solar system—larger even than the planet Mercury. Callisto, the last of Jupiter's large moons, is composed mostly of ice and rock. Studying these moons adds to knowledge about the origin of Earth and the rest of the solar system.

Figure 11 *Voyager 2* photographed the eruption of this volcano on Io in July 1979.

Figure 12 Saturn's rings are composed of pieces of rock and ice.

Modeling Planets

Procedure
1. Research the planets to determine how the sizes of the planets in the solar system compare with Earth's size.
2. Select a scale for the diameter of Earth.
3. Make a model by drawing a circle with this diameter on **paper**.
4. Using Earth's diameter as 1.0 unit, draw each of the other planets to scale.

Analysis
1. Which planet is largest? Which is smallest?
2. Which scale diameter did you select for Earth? Was this a good choice? Why or why not?

Saturn

The *Voyager* probes next surveyed Saturn in 1980 and 1981. *Cassini* reached Saturn on July 1, 2004. **Saturn** is the sixth planet from the Sun. It is the second-largest planet in the solar system, but it has the lowest density.

Saturn's Atmosphere Similar to Jupiter, Saturn is a large, gaseous planet. It has a thick outer atmosphere composed mostly of hydrogen and helium. Saturn's atmosphere also contains ammonia, methane, and water vapor. As you go deeper into Saturn's atmosphere, the gases gradually change to liquid hydrogen and helium. Below its atmosphere and liquid layer, Saturn might have a small, rocky core.

Rings and Moons The *Voyager* and *Cassini* probes gathered information about Saturn's ring system. The probes showed that there are several broad rings. Each large ring is composed of thousands of thin ringlets. **Figure 12** shows that Saturn's rings are composed of countless ice and rock particles. These particles range in size from a speck of dust to tens of meters across. Saturn's ring system is the most complex one in the solar system.

At least 47 moons orbit Saturn. Saturn's gravity holds these moons in their orbits around Saturn, just like the Sun's gravity holds the planets in their orbits around the Sun. The largest moon, Titan, is larger than the planet Mercury. It has a thick atmosphere of nitrogen, argon, and methane. *Cassini* delivered the *Huygens* probe to analyze Titan's atmosphere in 2005.

Uranus

Beyond Saturn, *Voyager 2* flew by Uranus in 1986. **Uranus** (YOOR uh nus) is the seventh planet from the Sun and was discovered in 1781. It is a large, gaseous planet with at least 27 moons and a system of thin, dark rings. Uranus's largest moon, Titania, has many craters and deep valleys. The valleys on this moon indicate that some process reshaped its surface after it formed. Uranus's 11 rings surround the planet's equator.

Uranus's Characteristics The atmosphere of Uranus is composed of hydrogen, helium, and some methane. Methane gives the planet the bluish-green color that you see in **Figure 13**. Methane absorbs the red and yellow light, and the clouds reflect the green and blue. Few cloud bands and storm systems can be seen on Uranus. Evidence suggests that under its atmosphere, Uranus is composed primarily of rock and various ices. There is no separate core.

Figure 14 shows one of the most unusual features of Uranus. Its axis of rotation is tilted on its side compared with the other planets. The axes of rotation of the other planets are nearly perpendicular to the planes of their orbits. However, Uranus's axis of rotation is nearly parallel to the plane of its orbit. Some scientists believe a collision with another object tipped Uranus on its side.

Figure 13 The atmosphere of Uranus gives the planet its distinct bluish-green color.

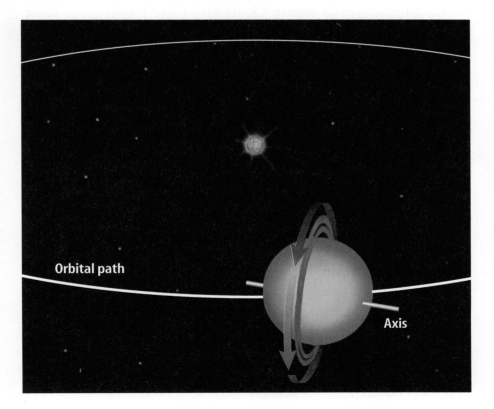

Figure 14 Uranus's axis of rotation is nearly parallel to the plane of its orbit. During its revolution around the Sun, each pole, at different times, points almost directly at the Sun.

Neptune has a distinctive bluish-green color.

The pinkish hue of Neptune's largest moon, Triton, is thought to come from an evaporating layer of nitrogen and methane ice.

Figure 15 Neptune is the eighth planet from the Sun.

Neptune

Passing Uranus, *Voyager 2* traveled to Neptune, another large, gaseous planet. Discovered in 1846, **Neptune** is the eighth planet from the Sun.

Neptune's Characteristics Like Uranus's atmosphere, Neptune's atmosphere is made up of hydrogen and helium, with smaller amounts of methane. The methane content gives Neptune, shown in **Figure 15,** its distinctive bluish-green color, just as it does for Uranus.

 *What gives Neptune its bluish-green color?*

Neptune has dark-colored storms in its atmosphere that are similar to the Great Red Spot on Jupiter. One discovered by *Voyager 2* in 1989 was called the Great Dark Spot. It was about the size of Earth with windspeeds higher than any other planet. Observations by the *Hubble Space Telescope* in 1994 showed that the Great Dark Spot disappeared and then reappeared. Bright clouds also form and then disappear. Scientists don't know what causes these changes, but they show that Neptune's atmosphere is active and changes rapidly.

Under its atmosphere, Neptune has a mixture of rock and various types of ices made from methane and ammonia. Neptune probably has a rocky core.

Neptune has at least 13 moons and several rings. Triton, shown in **Figure 15,** is Neptune's largest moon. It has a thin atmosphere composed mostly of nitrogen. Neptune's dark rings are young and probably won't last very long.

INTEGRATE Language Arts

Names of Planets The names of most of the planets in the solar system come from Roman or Greek mythology. For example, Neptune was the Roman god of the sea. Research to learn about the names of the other planets. Write a paragraph in your Science Journal that summarizes what you learn.

Dwarf Planets

From the time of its discovery in 1930 until 2006 Pluto was considered the ninth planet in the solar system. But with the discovery of Eris (EE rihs), which is larger than Pluto, the International Astronomical Union decided to define the term *planet*. Now, scientists call Pluto a dwarf planet.

Ceres Ceres was discovered in 1801. It has an average diameter of about 940 km and is located within the asteroid belt at an average distance of about 2.7 AU from the Sun. Ceres orbits the Sun about once every 4.6 years.

Pluto 1930 Pluto has a diameter of 2,300 km. It is an average distance of 39.2 AU from the Sun and takes 248 years to complete one orbit. It is surrounded by only a thin atmosphere and it has a solid, icy-rock surface. Pluto has three moons: Charon, Hydra, and Nix. The largest moon, Charon, has a diameter of about 1,200 km and orbits Pluto at a distance of about 19,500 km.

Eris Astronomers discovered Eris in 2005, originally calling it UB313. With a diameter of about 2,400 km, Eris is slightly larger than Pluto. Eris has an elliptical orbit that varies from between about 38 AU to 98 AU from the Sun. Eris orbits the Sun once every 557 years and has one moon, named Dysnomia (dihs NOH mee uh).

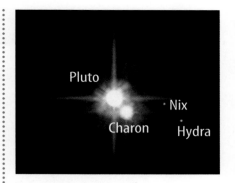

Figure 16 Hydra and Nix are about three times farther from Pluto than Charon is.

section 3 review

Summary

Jupiter
- Jupiter is the largest planet in the solar system.
- The Great Red Spot is a huge storm on Jupiter.

Saturn
- Saturn has a complex system of rings.

Uranus
- Uranus has a bluish-green color caused by methane in its atmosphere.

Neptune
- Like Uranus, Neptune has a bluish-green color.
- Neptune's atmosphere can change rapidly.

Dwarf Planets
- Pluto is made of ice and rock.
- Ceres is a dwarf planet within the asteroid belt.

Self Check

1. **Describe** the differences between the outer planets and the inner planets.
2. **Describe** what Saturn's rings are made from.
3. **Compare** Pluto to the eight planets.
4. **Explain** how Uranus's axis of rotation differs from those of most other planets.
5. **Think Critically** What would seasons be like on Uranus? Explain.

Applying Skills

6. **Identify a Question** When a probe lands on Pluto, so many questions will be answered. Think of a question about Pluto that you'd like to have answered. Then, explain why the answer is important to you.

Table 3 Planets

Mercury

- closest to the Sun
- second-smallest planet
- surface has many craters and high cliffs
- no atmosphere
- temperatures range from 425°C during the day to −170°C at night
- has no moons

Venus

- similar to Earth in size and mass
- thick atmosphere made mostly of carbon dioxide
- droplets of sulfuric acid in atmosphere give clouds a yellowish color
- surface has craters, faultlike cracks, and volcanoes
- greenhouse effect causes surface temperatures of 450°C to 475°C
- has no moons

Earth

- atmosphere protects life
- surface temperatures allow water to exist as solid, liquid, and gas
- only planet where life is known to exist
- has one large moon

Mars

- surface appears reddish-yellow because of iron oxide in soil
- ice caps are made of frozen carbon dioxide and water
- channels indicate that water had flowed on the surface; has large volcanoes and valleys
- has a thin atmosphere composed mostly of carbon dioxide
- surface temperatures range from −125°C to 35°C
- huge dust storms often blanket the planet
- has two small moons

Table 3 Planets

Jupiter

- largest planet
- has faint rings
- atmosphere is mostly hydrogen and helium; continuous storms swirl on the planet—the largest is the Great Red Spot
- has four large moons and at least 59 smaller moons; one of its moons, Io, has active volcanoes

Saturn

- second-largest planet
- thick atmosphere is mostly hydrogen and helium
- has a complex ring system
- has at least 47 moons—the largest, Titan, is larger than Mercury

Uranus

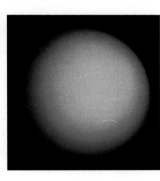

- large, gaseous planet with thin, dark rings
- atmosphere is hydrogen, helium, and methane
- axis of rotation is nearly parallel to plane of orbit
- has at least 27 moons

Neptune

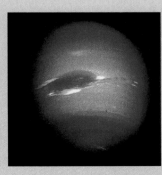

- large, gaseous planet with rings that vary in thickness
- is sometimes farther from the Sun than Pluto is
- methane atmosphere causes its bluish-green color
- has dark-colored storms in atmosphere
- has at least 13 moons

section 4

Other Objects in the Solar System

as you read

What You'll Learn
- **Describe** how comets change when they approach the Sun.
- **Distinguish** among comets, meteoroids, and asteroids.
- **Explain** that objects from space sometimes impact Earth.

Why It's Important
Comets, asteroids, and most meteorites are very old. Scientists can learn about the early solar system by studying them.

Review Vocabulary
crater: a nearly circular depression in a planet, moon, or asteroid that formed when an object from space hit its surface

New Vocabulary
- comet
- meteor
- meteorite
- asteroid

Comets

The planets and their moons are the most noticeable members of the Sun's family, but many other objects also orbit the Sun. Comets, meteoroids, and asteroids are other important objects in the solar system.

You might have heard of Halley's comet. A **comet** is composed of dust and rock particles mixed with frozen water, methane, and ammonia. Halley's comet was last seen from Earth in 1986. English astronomer Edmund Halley realized that comet sightings that had taken place about every 76 years were really sightings of the same comet. This comet, which takes about 76 years to orbit the Sun, was named after him.

Oort Cloud Astronomer Jan Oort proposed the idea that billions of comets surround the solar system. This cloud of comets, called the Oort Cloud, is located beyond the orbit of Pluto. Oort suggested that the gravities of the Sun and nearby stars interact with comets in the Oort Cloud. Comets either escape from the solar system or get captured into smaller orbits.

Comet Hale-Bopp On July 23, 1995, two amateur astronomers made an exciting discovery. A new comet, Comet Hale-Bopp, was headed toward the Sun. Larger than most that approach the Sun, it was the brightest comet visible from Earth in 20 years. Shown in **Figure 17,** Comet Hale-Bopp was at its brightest in March and April 1997.

Figure 17 Comet Hale-Bopp was most visible in March and April 1997.

356 CHAPTER 12 The Solar System

Structure of Comets The *Hubble Space Telescope* and spacecrafts such as the *International Cometary Explorer* have gathered information about comets. In 2006, a spacecraft called *Stardust* will return a capsule to Earth containing samples of dust from a comet's tail. Notice the structure of a comet shown in **Figure 18.** It is a mass of frozen ice and rock.

As a comet approaches the Sun, it changes. Ices of water, methane, and ammonia vaporize because of the heat from the Sun. This releases dust and jets of gas. The gases and released dust form a bright cloud called a coma around the nucleus, or solid part, of the comet. The solar wind pushes on the gases and dust in the coma, causing the particles to form separate tails that point away from the Sun.

After many trips around the Sun, most of the ice in a comet's nucleus has vaporized. All that's left are dust and rock, which are spread throughout the orbit of the original comet.

Meteoroids, Meteors, and Meteorites

You learned that comets vaporize and break up after they have passed close to the Sun many times. The small pieces from the comet's nucleus spread out into a loose group within the original orbit of the comet. These pieces of dust and rock, along with those derived from other sources, are called meteoroids.

Sometimes the path of a meteoroid crosses the position of Earth, and it enters Earth's atmosphere at speeds of 15 km/s to 70 km/s. Most meteoroids are so small that they completely burn up in Earth's atmosphere. A meteoroid that burns up in Earth's atmosphere is called a **meteor,** shown in **Figure 19.**

Figure 18 A comet consists of a nucleus, a coma, a dust tail, and an ion tail. Pictures of the comet Wild 2 from *Stardust* show that the comet has a rocky, cratered surface.

Figure 19 A meteoroid that burns up in Earth's atmosphere is called a meteor.

SECTION 24 Other Objects in the Solar System **357**

Figure 20 Meteorites occasionally strike Earth's surface. A large meteorite struck Arizona, forming a crater about 1.2 km in diameter and about 200 m deep.

Meteor Showers Each time Earth passes through the loose group of particles within the old orbit of a comet, many small particles of rock and dust enter the atmosphere. Because more meteors than usual are seen, the event is called a meteor shower.

When a meteoroid is large enough, it might not burn up completely in the atmosphere. If it strikes Earth, it is called a **meteorite.** Barringer Crater in Arizona, shown in **Figure 20,** was formed when a large meteorite struck Earth about 50,000 years ago. Most meteorites are probably debris from asteroid collisions or broken-up comets, but some originate from the Moon and Mars.

Reading Check *What is a meteorite?*

Asteroids

An **asteroid** is a piece of rock similar to the material that formed into the planets. Most asteroids are located in an area between the orbits of Mars and Jupiter called the asteroid belt. Find the asteroid belt in **Figure 21.** Why are they located there? The gravity of Jupiter might have kept a planet from forming in the area where the asteroid belt is located now.

Other asteroids are scattered throughout the solar system. They might have been thrown out of the belt by Jupiter's gravity. Some of these asteroids have orbits that cross Earth's orbit. Scientists monitor the positions of these asteroids. However, it is unlikely that an asteroid will hit Earth in the near future.

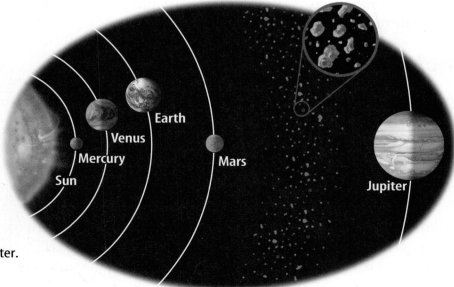

Figure 21 The asteroid belt lies between the orbits of Mars and Jupiter.

Asteroid Sizes The sizes of the asteroids in the asteroid belt range from tiny particles to objects 940 km in diameter. Ceres is the largest and the first one discovered. The next three in order of size are Vesta (530 km), Pallas (522 km), and 10 Hygiea (430 km). The asteroid Gaspra, shown in **Figure 22,** was photographed by *Galileo* on its way to Jupiter.

Exploring Asteroids On February 14, 2000, the *Near Earth Asteroid Rendezvous (NEAR)* spacecraft went into orbit around the asteroid 433 Eros and later completed its one-year mission of gathering data. Data from the probe show that Eros has many craters and is similar to meteorites on Earth. The Japanese space probe *Hayabusa* arrived at the asteroid Itokawa in November 2005. Its mission is to collect samples and return them to Earth in a capsule in June 2010.

Comets, asteroids, and most meteorites formed early in the history of the solar system. Scientists study these space objects to learn what the solar system might have been like long ago. Understanding this could help scientists better understand how Earth formed.

Figure 22 The asteroid Gaspra is about 20 km long.

section 4 review

Summary

Comets
- Comets consist of dust, rock, and different types of ice.
- The Oort Cloud was proposed as a source of comets in the solar system.

Meteoroids, Meteors, Meteorites
- When meteoroids burn up in the atmosphere, they are called meteors.
- Meteor showers occur when Earth crosses the orbital path of a comet.

Asteroids
- Many asteroids occur between the orbits of Mars and Jupiter. This region is called the asteroid belt.

Self Check

1. **Describe** how a comet changes when it comes close to the Sun.
2. **Explain** how craters form.
3. **Summarize** the differences between comets and asteroids.
4. **Describe** the mission of the *NEAR* space probe.
5. **Think Critically** A meteorite found in Antarctica is thought to have come from Mars. How could a rock from Mars get to Earth?

Applying Math

6. **Use Proportions** During the 2001 Leonid Meteor Shower, some people saw 20 meteors each minute. Assuming a constant rate, how many meteors did these people see in one hour?

LAB Model and Invent

Solar System Distance Model

Goals
- **Design** a table of scale distances and model the distances between and among the Sun and the planets.

Possible Materials
meterstick
scissors
pencil
colored markers
string (several meters)
notebook paper (several sheets)

Safety Precautions

Use care when handling scissors.

Real-World Question

Distances between the Sun and the planets of the solar system are large. These large distances can be difficult to visualize. Can you design and create a model that will demonstrate the distances in the solar system?

Make a Model

1. **List** the steps that you need to take to make your model. Describe exactly what you will do at each step.
2. **List** the materials that you will need to complete your model.
3. **Describe** the calculations that you will use to get scale distances from the Sun for all nine planets.
4. **Make** a table of scale distances that you will use in your model. Show your calculations in your table.
5. **Write** a description of how you will build your model. Explain how it will demonstrate relative distances between and among the Sun and planets of the solar system.

Planetary Distances

Planet	Distance to Sun (km)	Distance to Sun (AU)	Scale Distance (1 AU = 10 cm)	Scale Distance (1 AU = 2 cm)
Mercury	5.97×10^7	0.39		
Venus	1.08×10^8	0.72		
Earth	1.50×10^8	1.00		
Mars	2.28×10^8	1.52		
Jupiter	7.78×10^8	5.20		Do not write in this book.
Saturn	1.43×10^9	9.54		
Uranus	2.87×10^9	19.19		
Neptune	4.50×10^9	30.07		

Using Scientific Methods

Test Your Model

1. **Compare** your scale distances with those of other students. Discuss why each of you chose the scale that you did.
2. Make sure your teacher approves your plan before you start.
3. **Construct** the model using your scale distances.
4. While constructing the model, write any observations that you or other members of your group make, and complete the data table in your Science Journal. Calculate the scale distances that would be used in your model if 1 AU = 2 m.

Analyze Your Data

1. **Explain** how a scale distance is determined.
2. Was it possible to work with your scale? Explain why or why not.
3. How much string would be required to construct a model with a scale distance of 1 AU = 2 m?
4. Proxima Centauri, the closest star to the Sun, is about 270,000 AU from the Sun. Based on your scale, how much string would you need to place this star on your model?

Conclude and Apply

1. **Summarize** your observations about distances in the solar system. How are distances between the inner planets different from distances between the outer planets?
2. Using your scale distances, determine which planet orbits closest to Earth. Which planet's orbit is second closest?

Communicating Your Data

Compare your scale model with those of other students. Discuss any differences. **For more help, refer to the** Science Skill Handbook.

LAB **361**

Oops! Accidents in SCIENCE

SOMETIMES GREAT DISCOVERIES HAPPEN BY ACCIDENT!

IT CAME FROM OUTER SPACE!

On September 4, 1990, Frances Pegg was unloading bags of groceries in her kitchen in Burnwell, Kentucky. Suddenly, she heard a loud crashing sound. Her husband Arthur heard the same sound. The sound frightened the couple's goat and horse. The noise had come from an object that had crashed through the Pegg's roof, their ceiling, and the floor of their porch. They couldn't see what the object was, but the noise sounded like a gunshot, and pieces of wood from their home flew everywhere. The next day the couple looked under their front porch and found the culprit—a chunk of rock from outer space. It was a meteorite.

For seven years, the Peggs kept their "space rock" at home, making them local celebrities. The rock appeared on TV, and the couple was interviewed by newspaper reporters. In 1997, the Peggs sold the meteorite to the National Museum of Natural History in Washington, D.C., which has a collection of more than 9,000 meteorites. Scientists there study meteorites to learn more about the solar system. One astronomer explained, "Meteorites formed at about the same time as the solar system, about 4.6 billion years ago, though some are younger."

Scientists especially are interested in the Burnwell meteorite because its chemical make up is different from other meteorites previously studied. The Burnwell meteorite is richer in metallic iron and nickel than other known meteorites and is less rich in some metals such as cobalt. Scientists are comparing the rare Burnwell rock with other data to find out if there are more meteorites like the one that fell on the Peggs' roof. But so far, it seems the Peggs' visitor from outer space is one-of-a-kind.

The Burnwell meteorite crashed into the Peggs' home and landed in their basement on the right.

Research Do research to learn more about meteorites. How do they give clues to how our solar system formed? Report to the class.

Science Online

For more information, visit blue.msscience.com/oops

chapter 12 Study Guide

Reviewing Main Ideas

Section 1 The Solar System

1. Early Greek scientists thought that Earth was at the center of the solar system. They thought that the planets and stars circled Earth.
2. Today, people know that objects in the solar system revolve around the Sun.

Section 2 The Inner Planets

1. The inner planets are Mercury, Venus, Earth, and Mars.
2. The inner planets are small, rocky planets.

Section 3 The Outer Planets

1. The outer planets are Jupiter, Saturn, Uranus, and Neptune.
2. Pluto is a small icy dwarf planet. Other dwarf planets include Ceres and Eris.

Section 4 Other Objects in the Solar System

1. Comets are masses of ice and rock. When a comet approaches the Sun, some ice turns to gas and the comet glows brightly.
2. Meteors occur when small pieces of rock enter Earth's atmosphere and burn up.

Visualizing Main Ideas

Copy and complete the following concept map about the solar system.

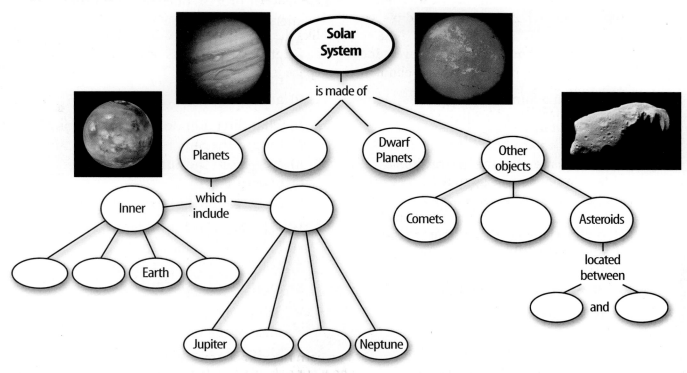

CHAPTER STUDY GUIDE 363

chapter 12 Review

Using Vocabulary

asteroid p. 358
comet p. 356
Earth p. 344
Great Red Spot p. 348
Jupiter p. 348
Mars p. 344
Mercury p. 342
meteor p. 357
meteorite p. 358
Neptune p. 352
Pluto p. 353
Saturn p. 350
solar system p. 337
Uranus p. 351
Venus p. 343

Fill in the blanks with the correct words.

1. A meteoroid that burns up in Earth's atmosphere is called a(n) _____.
2. The Great Red Spot is a giant storm on _____.
3. _____ is the second largest planet.
4. The *Viking* landers tested for life on _____.
5. The _____ includes the Sun, planets, moons, and other objects.

Checking Concepts

Choose the word or phrase that best answers the question.

6. Who proposed a Sun-centered solar system?
   A) Ptolemy   C) Galileo
   B) Copernicus   D) Oort

7. What is the shape of planetary orbits?
   A) circles   C) squares
   B) ellipses   D) rectangles

8. Which planet has extreme temperatures because it has no atmosphere?
   A) Earth   C) Saturn
   B) Jupiter   D) Mercury

9. Where is the largest volcano in the solar system?
   A) Earth   C) Mars
   B) Jupiter   D) Uranus

Use the photo below to answer question 10.

10. Which planet has a complex ring system consisting of thousands of ringlets?
    A) Jupiter   C) Uranus
    B) Saturn   D) Mars

11. What is a rock from space that strikes Earth's surface?
    A) asteroid   C) meteorite
    B) meteoroid   D) meteor

12. By what process does the Sun produce energy?
    A) magnetism
    B) nuclear fission
    C) nuclear fusion
    D) gravity

13. In what direction do comet tails point?
    A) toward the Sun
    B) away from the Sun
    C) toward Earth
    D) away from the Oort Cloud

14. Which planet has abundant surface water and is known to have life?
    A) Mars   C) Earth
    B) Jupiter   D) Venus

15. Which planet has the highest temperatures because of the greenhouse effect?
    A) Mercury   C) Saturn
    B) Venus   D) Earth

chapter 12 Review

Thinking Critically

16. Infer Why are probe landings on Jupiter not possible?

17. Concept Map Copy and complete the concept map on this page to show how a comet changes as it travels through space.

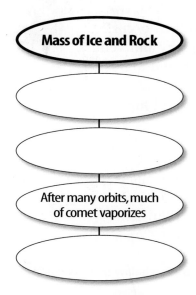

18. Recognize Cause and Effect What evidence suggests that liquid water is or once was present on Mars?

19. Venn Diagram Create a Venn diagram for Earth and Venus. Create a second Venn diagram for Uranus and Neptune. Which two planets do you think are more similar?

20. Recognize Cause and Effect Mercury is closer to the Sun than Venus, yet Venus has higher temperatures. Explain.

21. Make Models Make a model that includes the Sun, Earth, and the Moon. Use your model to demonstrate how the Moon revolves around Earth and how Earth and the Moon revolve around the Sun.

22. Form Hypotheses Why do Mars's two moons look like asteroids?

Performance Activities

23. Display Mercury, Venus, Mars, Jupiter, and Saturn can be observed with the unaided eye. Research when and where in the sky these planets can be observed during the next year. Make a display illustrating your findings. Take some time to observe some of these planets.

24. Short Story Select one of the planets or a moon in the solar system. Write a short story from the planet's or moon's perspective. Include scientifically correct facts and concepts in your story.

Applying Math

25. Saturn's Atmosphere Saturn's atmosphere consists of 96.3% hydrogen and 3.25% helium. What percentage of Saturn's atmosphere is made up of other gases?

26. Length of Day on Pluto A day on Pluto lasts 6.39 times longer than a day on Earth. If an Earth day lasts 24 h, how many hours is a day on Pluto?

Use the graph below to answer question 27.

Weight on Several Planets		
Planet	Proportion of Earth's Gravity	Melissa's Weight (kg)
Mercury	0.378	
Venus	0.903	
Earth	1.000	31.8
Mars	0.379	
Jupiter	2.54	

27. Gravity and Weight Melissa weighs 31.8 kg on Earth. Multiply Melissa's weight by the proportion of Earth's gravity for each planet to find out how much Melissa would weigh on each.

chapter 12 Standardized Test Practice

Part 1 Multiple Choice

Record your answers on the answer sheet provided by your teacher or on a sheet of paper.

Use the photo below to answer question 1.

1. What is shown in the photo above?
   A. asteroids   C. meteors
   B. comets   D. meteorites

2. Which is the eighth planet from the Sun?
   A. Earth   C. Jupiter
   B. Mars   D. Neptune

3. Which is Pluto's largest moon?
   A. Hydra   C. Charon
   B. Nix   D. Triton

4. Which object's gravity holds the planets in their orbits?
   A. Gaspra   C. Mercury
   B. Earth   D. the Sun

5. Which of the following occurs in a cycle?
   A. appearance of Halley's comet
   B. condensation of a nebula
   C. formation of a crater
   D. formation of a planet

Test-Taking Tip

No Peeking During the test, keep your eyes on your own paper. If you need to rest them, close them or look up at the ceiling.

6. Which planet likely will be visited by humans in the future?
   A. Jupiter   C. Mars
   B. Venus   D. Neptune

7. Between which two planets' orbits does the asteroid belt occur?
   A. Mercury and Venus
   B. Earth and Mars
   C. Uranus and Neptune
   D. Mars and Jupiter

8. Who discovered that planets have elliptical orbits?
   A. Galileo Galilei
   B. Johannes Kepler
   C. Albert Einstein
   D. Nicholas Copernicus

Use the illustration below to answer question 9.

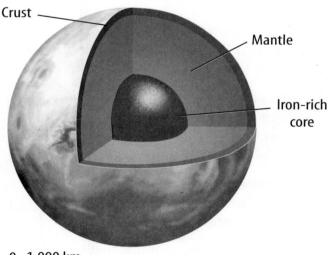

9. Which of the following answers is a good estimate for the diameter of Mars?
   A. 23,122 km   C. 1,348 km
   B. 6,794 km   D. 12,583 km

Standardized Test Practice

Part 2 Short Response/Grid In

Record your answers on the answer sheet provided by your teacher or on a sheet of paper.

10. Why does a moon remain in orbit around a planet?

11. Compare and contrast the inner planets and the outer planets.

12. Describe the difference between Pluto and the eight planets.

13. Describe Saturn's rings. What are they made of?

14. What is the Great Red Spot?

15. How is Earth different from the other planets in the solar system?

Use the graph below to answfer questions 16–19.

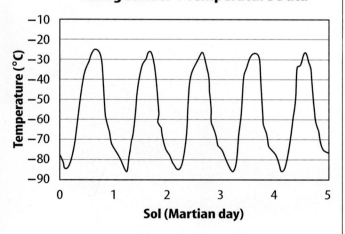

16. Why do the temperatures in the graph vary in a pattern?

17. Approximate the typical high temperature value measured by *Viking I*.

18. Approximate the typical low temperature value measured by *Viking I*.

19. What is the range of these temperature values?

Part 3 Open Ended

Record your answers on a sheet of paper.

20. How might near-Earth-asteroids affect life on Earth? Why do astronomers search for them and monitor their positions?

Use the illustration below to answer question 21.

21. Explain how scientists hypothesize that the large cliffs on Mercury formed.

22. Describe the Sun-centered model of the solar system. How is it different from the Earth-centered model?

23. What is an astronomical unit? Why is it useful?

24. Compare and contrast the distances between the planets in the solar system. Which planets are relatively close together? Which planets are relatively far apart?

25. Summarize the current hypothesis about how the solar system formed.

26. Explain how Earth's gravity affects objects that are on or near Earth.

27. Describe the shape of planets' orbits. What is the name of this shape? Where is the Sun located?

28. Describe Jupiter's atmosphere. What characteristics can be observed in images acquired by space probes?

chapter 13

Stars and Galaxies

The BIG Idea

The universe is made up of stars and galaxies.

**SECTION 1
Stars**
Main Idea For many years, people have been learning about stars by observing their locations in the sky and by studying their light.

**SECTION 2
The Sun**
Main Idea The Sun is an enormous ball of gas which produces energy by fusing hydrogen into helium.

**SECTION 3
Evolution of Stars**
Main Idea Stars pass through several stages as they evolve.

**SECTION 4
Galaxies and the Universe**
Main Idea By studying galaxies, scientists have observed that the universe is expanding.

What's your address?

You know your address at home. You also know your address at school. But do you know your address in space? You live on a planet called Earth that revolves around a star called the Sun. Earth and the Sun are part of a galaxy called the Milky Way. It looks similar to galaxy M83, shown in the photo.

Science Journal Write a description in your Science Journal of the galaxy shown on this page.

Start-Up Activities

Why do clusters of galaxies move apart?

Astronomers know that most galaxies occur in groups of galaxies called clusters. These clusters are moving away from each other in space. The fabric of space is stretching like an inflating balloon.

1. Partially inflate a balloon. Use a piece of string to seal the neck.
2. Draw six evenly spaced dots on the balloon with a felt-tipped marker. Label the dots A through F.
3. Use string and a ruler to measure the distance, in millimeters, from dot A to each of the other dots.
4. Inflate the balloon more.
5. Measure the distances from dot A again.
6. Inflate the balloon again and make new measurements.
7. **Think Critically** Imagine that each dot represents a cluster of galaxies and that the balloon represents the universe. Describe the motion of the clusters in your Science Journal.

Preview this chapter's content and activities at blue.msscience.com

Stars, Galaxies, and the Universe Make the following Foldable to show what you know about stars, galaxies, and the universe.

STEP 1 Fold a sheet of paper from side to side. Make the front edge about 1.25 cm shorter than the back edge.

STEP 2 Turn lengthwise and fold into thirds.

STEP 3 Unfold and cut only the top layer along both folds to make three tabs.

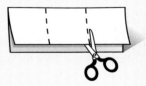

STEP 4 Label the tabs *Stars, Galaxies,* and *Universe.*

Read and Write Before you read the chapter, write what you already know about stars, galaxies, and the universe. As you read the chapter, add to or correct what you have written under the tabs.

Get Ready to Read

Make Inferences

① Learn It! When you make inferences, you draw conclusions that are not directly stated in the text. This means you "read between the lines." You interpret clues and draw upon prior knowledge. Authors rely on a reader's ability to infer because all the details are not always given.

② Practice It! Read the excerpt below and pay attention to highlighted words as you make inferences. Use this Think-Through chart to help you make inferences.

Notice how the front two stars of the Big Dipper point almost directly at Polaris, which often is called the North Star. Polaris is located at the end of the Little Dipper in the constellation Ursa Minor.

—*from page 371*

Text	Question	Inferences
the front two stars of the Big Dipper	Which are the "front" two stars?	The two "bowl" stars which are farthest from the "handle?"
point almost directly at Polaris	How do the two stars "point?"	Visualize a straight line through the two stars toward the Little Dipper?
located at the end of the Little Dipper	Which is the "end" of the Little Dipper?	The last star in the handle away from the bowl?

③ Apply It! As you read this chapter, practice your skill at making inferences by making connections and asking questions.

Target Your Reading

Reading Tip: Sometimes you make inferences by using other reading skills, such as questioning and predicting.

Use this to focus on the main ideas as you read the chapter.

① Before you read the chapter, respond to the statements below on your worksheet or on a numbered sheet of paper.
- Write an **A** if you **agree** with the statement.
- Write a **D** if you **disagree** with the statement.

② After you read the chapter, look back to this page to see if you've changed your mind about any of the statements.
- If any of your answers changed, explain why.
- Change any false statements into true statements.
- Use your revised statements as a study guide.

Science Online
Print out a worksheet of this page at blue.msscience.com

Before You Read A or D		Statement	After You Read A or D
	1	A constellation is a group of stars which are close together in space.	
	2	A light-year is a measurement of time.	
	3	The color of a star indicates its temperature.	
	4	The Sun is the closest star to Earth.	
	5	Light from the Sun reaches Earth in about eight minutes.	
	6	Most of the heat energy produced by the Sun is caused by the fission, or radioactive decay, of helium into hydrogen in the Sun's core.	
	7	A black hole is a location in space where there is no mass, gravity, or light.	
	8	The Milky Way Galaxy is located within the solar system.	
	9	A red shift in the light spectrum coming from a star means that the star is becoming hotter.	
	10	Most scientific evidence currently suggests that the universe is expanding.	

section 1

Stars

as you read

What You'll Learn
- **Explain** why some constellations are visible only during certain seasons.
- **Distinguish** between absolute magnitude and apparent magnitude.

Why It's Important
The Sun is a typical star.

Review Vocabulary
star: a large, spherical mass of gas that gives off light and other types of radiation

New Vocabulary
- constellation
- absolute magnitude
- apparent magnitude
- light-year

Constellations

It's fun to look at clouds and find ones that remind you of animals, people, or objects that you recognize. It takes more imagination to play this game with stars. Ancient Greeks, Romans, and other early cultures observed patterns of stars in the night sky called **constellations.** They imagined that the constellations represented mythological characters, animals, or familiar objects.

From Earth, a constellation looks like spots of light arranged in a particular shape against the dark night sky. **Figure 1** shows how the constellation of the mythological Greek hunter Orion appears from Earth. It also shows that the stars in a constellation often have no relationship to each other in space.

Stars in the sky can be found at specific locations within a constellation. For example, you can find the star Betelgeuse (BEE tul jooz) in the shoulder of the mighty hunter Orion. Orion's faithful companion is his dog, Canis Major. Sirius, the brightest star that is visible from the northern hemisphere, is in Canis Major.

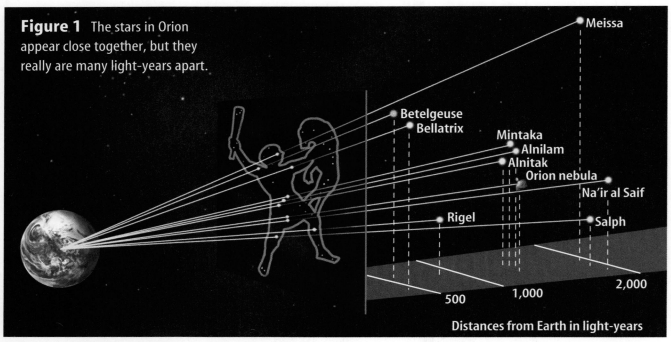

Figure 1 The stars in Orion appear close together, but they really are many light-years apart.

Distances from Earth in light-years

370 CHAPTER 13 Stars and Galaxies

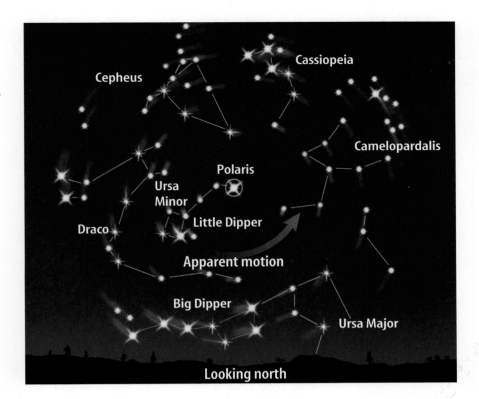

Figure 2 The Big Dipper, in red, is part of the constellation Ursa Major. It is visible year-round in the northern hemisphere. Constellations close to Polaris rotate around Polaris, which is almost directly over the north pole.

Modern Constellations Modern astronomy divides the sky into 88 constellations, many of which were named by early astronomers. You probably know some of them. Can you recognize the Big Dipper? It's part of the constellation Ursa Major, shown in **Figure 2**. Notice how the front two stars of the Big Dipper point almost directly at Polaris, which often is called the North Star. Polaris is located at the end of the Little Dipper in the constellation Ursa Minor. It is positioned almost directly over Earth's north pole.

Circumpolar Constellations As Earth rotates, Ursa Major, Ursa Minor, and other constellations in the northern sky circle around Polaris. Because of this, they are called circumpolar constellations. The constellations appear to move, as shown in **Figure 2,** because Earth is in motion. The stars appear to complete one full circle in the sky in about 24 h as Earth rotates on its axis. One circumpolar constellation that's easy to find is Cassiopeia (ka see uh PEE uh). You can look for five bright stars that form a big W or a big M in the northern sky, depending on the season.

As Earth orbits the Sun, different constellations come into view while others disappear. Because of their unique position, circumpolar constellations are visible all year long. Other constellations are not. Orion, which is visible in the winter in the northern hemisphere, can't be seen there in the summer because the daytime side of Earth is facing it.

Observing Star Patterns

Procedure
1. On a clear night, go outside after dark and study the stars. Take an adult with you.
2. Let your imagination flow to find patterns of stars that look like something familiar.
3. Draw the stars you see, note their positions, and include a drawing of what you think each star pattern resembles.

Analysis
1. Which of your constellations match those observed by your classmates?
2. How can recognizing star patterns be useful?

Try at Home

SECTION 1 Stars **371**

Absolute and Apparent Magnitudes

When you look at constellations, you'll notice that some stars are brighter than others. For example, Sirius looks much brighter than Rigel. Is Sirius a brighter star, or is it just closer to Earth, making it appear to be brighter? As it turns out, Sirius is 100 times closer to Earth than Rigel is. If Sirius and Rigel were the same distance from Earth, Rigel would appear much brighter in the night sky than Sirius would.

When you refer to the brightness of a star, you can refer to its absolute magnitude or its apparent magnitude. The **absolute magnitude** of a star is a measure of the amount of light it gives off. A measure of the amount of light received on Earth is the **apparent magnitude**. A star that's dim can appear bright in the sky if it's close to Earth, and a star that's bright can appear dim if it's far away. If two stars are the same distance away, what might cause one of them to be brighter than the other?

Reading Check *What is the difference between absolute and apparent magnitude?*

Applying Science

Are distance and brightness related?

The apparent magnitude of a star is affected by its distance from Earth. This activity will help you determine the relationship between distance and brightness.

Identifying the Problem

Luisa conducted an experiment to determine the relationship between distance and the brightness of stars. She used a meterstick, a light meter, and a lightbulb. She placed the bulb at the zero end of the meterstick, then placed the light meter at the 20-cm mark and recorded the distance and the light-meter reading in her data table. Readings are in luxes, which are units for measuring light intensity. Luisa then increased the distance from the bulb to the light meter and took more readings. By examining the data in the table, can you see a relationship between the two variables?

Effect of Distance on Light

Distance (cm)	Meter Reading (luxes)
20	4150.0
40	1037.5
60	461.1
80	259.4

Solving the Problem

1. What happened to the amount of light recorded when the distance was increased from 20 cm to 40 cm? When the distance was increased from 20 cm to 60 cm?
2. What does this indicate about the relationship between light intensity and distance? What would the light intensity be at 100 cm? Would making a graph help you visualize the relationship?

Measurement in Space

How do scientists determine the distance from Earth to nearby stars? One way is to measure parallax—the apparent shift in the position of an object when viewed from two different positions. Extend your arm and look at your thumb first with your left eye closed and then with your right eye closed, as the girl in **Figure 3A** is doing. Your thumb appears to change position with respect to the background. Now do the same experiment with your thumb closer to your face, as shown in **Figure 3B**. What do you observe? The nearer an object is to the observer, the greater its parallax is.

Astronomers can measure the parallax of relatively close stars to determine their distances from Earth. **Figure 4** shows how a close star's position appears to change. Knowing the angle that the star's position changes and the size of Earth's orbit, astronomers can calculate the distance of the star from Earth.

Because space is so vast, a special unit of measure is needed to record distances. Distances between stars and galaxies are measured in light-years. A **light-year** is the distance that light travels in one year. Light travels at 300,000 km/s, or about 9.5 trillion km in one year. The nearest star to Earth, other than the Sun, is Proxima Centauri. Proxima Centauri is a mere 4.3 light-years away, or about 40 trillion km.

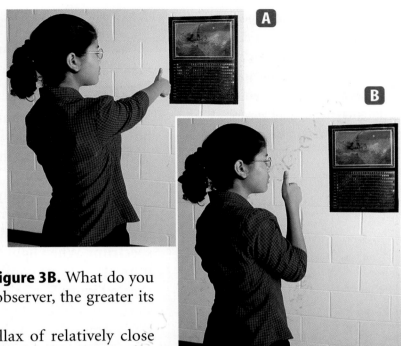

Figure 3 **A** Your thumb appears to move less against the background when it is farther away from your eyes. **B** It appears to move more when it is closer to your eyes.

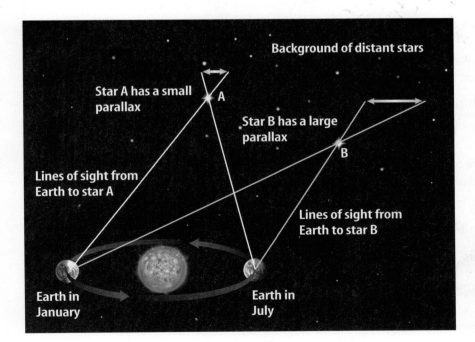

Figure 4 Parallax is determined by observing the same star when Earth is at two different points in its orbit around the Sun. The star's position relative to more distant background stars will appear to change.
Infer whether star **A** or **B** is farther from Earth.

SECTION 1 Stars **373**

Figure 5 This star spectrum was made by placing a diffraction grating over a telescope's objective lens. A diffraction grating produces a spectrum by causing interference of light waves. **Explain** *what causes the lines in spectra.*

Properties of Stars

The color of a star indicates its temperature. For example, hot stars are a blue-white color. A relatively cool star looks orange or red. Stars that have the same temperature as the Sun have a yellow color.

Astronomers study the composition of stars by observing their spectra. When fitted into a telescope, a spectroscope acts like a prism. It spreads light out in the rainbow band called a spectrum. When light from a star passes through a spectroscope, it breaks into its component colors. Look at the spectrum of a star in **Figure 5.** Notice the dark lines caused by elements in the star's atmosphere. Light radiated from a star passes through the star's atmosphere. As it does, elements in the atmosphere absorb some of this light. The wavelengths of visible light that are absorbed appear as dark lines in the spectrum. Each element absorbs certain wavelengths, producing a unique pattern of dark lines. Like a fingerprint, the patterns of lines can be used to identify the elements in a star's atmosphere.

section 1 review

Summary

Constellations
- Constellations are patterns of stars in the night sky.
- The stars in a constellation often have no relationship to each other in space.

Absolute and Apparent Magnitudes
- Absolute magnitude is a measure of how much light is given off by a star.
- Apparent magnitude is a measure of how much light from a star is received on Earth.

Measurement in Space
- Distances between stars are measured in light-years.

Properties of Stars
- Astronomers study the composition of stars by observing their spectra.

Self Check

1. **Describe** circumpolar constellations.
2. **Explain** why some constellations are visible only during certain seasons.
3. **Infer** how two stars could have the same apparent magnitude but different absolute magnitudes.
4. **Explain** how a star is similar to the Sun if it has the same absorption lines in its spectrum that occur in the Sun's spectrum.
5. **Think Critically** If a star's parallax angle is too small to measure, what can you conclude about the star's distance from Earth?

Applying Skills

6. **Recognize Cause and Effect** Suppose you viewed Proxima Centauri, which is 4.3 light-years from Earth, through a telescope. How old were you when the light that you see left this star?

 blue.msscience.com/self_check_quiz

section 2
The Sun

The Sun's Layers

The Sun is an ordinary star, but it's important to you. The Sun is the center of the solar system, and the closest star to Earth. Almost all of the life on Earth depends on energy from the Sun.

Notice the different layers of the Sun, shown in **Figure 6,** as you read about them. Like other stars, the Sun is an enormous ball of gas that produces energy by fusing hydrogen into helium in its core. This energy travels outward through the radiation zone and the convection zone. In the convection zone, gases circulate in giant swirls. Finally, energy passes into the Sun's atmosphere.

The Sun's Atmosphere

The lowest layer of the Sun's atmosphere and the layer from which light is given off is the **photosphere.** The photosphere often is called the surface of the Sun, although the surface is not a smooth feature. Temperatures there are about 6,000 K. Above the photosphere is the **chromosphere.** This layer extends upward about 2,000 km above the photosphere. A transition zone occurs between 2,000 km and 10,000 km above the photosphere. Above the transition zone is the **corona.** This is the largest layer of the Sun's atmosphere and extends millions of kilometers into space. Temperatures in the corona are as high as 2 million K. Charged particles continually escape from the corona and move through space as solar wind.

as you read

What You'll Learn
- **Explain** that the Sun is the closest star to Earth.
- **Describe** the structure of the Sun.
- **Describe** sunspots, prominences, and solar flares.

Why It's Important
The Sun is the source of most energy on Earth.

Review Vocabulary
cycle: a repeating sequence of events, such as the sunspot cycle

New Vocabulary
- photosphere
- corona
- chromosphere
- sunspot

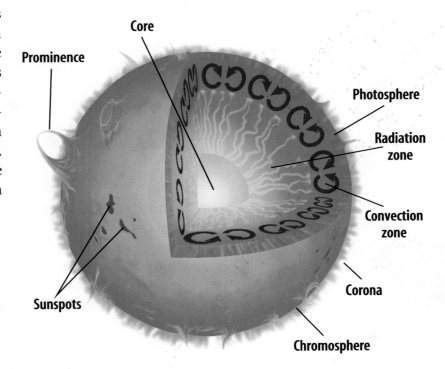

Figure 6 Energy produced in the Sun's core by fusion travels outward by radiation and convection. The Sun's atmosphere shines by the energy produced in the core.

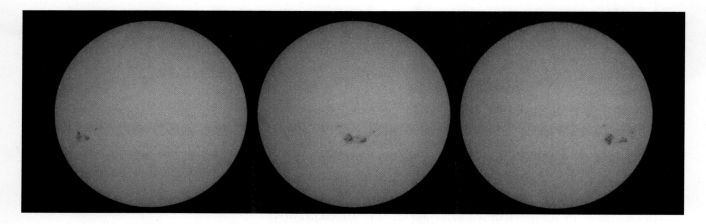

Figure 7 Sunspots are bright, but when viewed against the rest of the photosphere, they appear dark. Notice how these sunspots move as the Sun rotates.
Describe *the Sun's direction of rotation.*

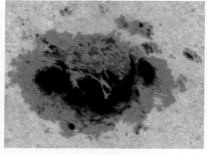

This is a close-up photo of a large sunspot.

Surface Features

From the viewpoint that you observe the Sun, its surface appears to be a smooth layer. But the Sun's surface has many features, including sunspots, prominences, flares, and CMEs.

Sunspots Areas of the Sun's surface that appear dark because they are cooler than surrounding areas are called **sunspots.** Ever since Galileo Galilei made drawings of sunspots, scientists have been studying them. Because scientists could observe the movement of individual sunspots, shown in **Figure 7,** they concluded that the Sun rotates. However, the Sun doesn't rotate as a solid body, as Earth does. It rotates faster at its equator than at its poles. Sunspots at the equator take about 25 days to complete one rotation. Near the poles, they take about 35 days.

Sunspots aren't permanent features on the Sun. They appear and disappear over a period of several days, weeks, or months. The number of sunspots increases and decreases in a fairly regular pattern called the sunspot, or solar activity, cycle. Times when many large sunspots occur are called sunspot maximums. Sunspot maximums occur about every 10 to 11 years. Periods of sunspot minimum occur in between.

Reading Check *What is a sunspot cycle?*

Prominences and Flares Sunspots are related to several features on the Sun's surface. The intense magnetic fields associated with sunspots might cause prominences, which are huge, arching columns of gas. Notice the huge prominence in **Figure 8.** Some prominences blast material from the Sun into space at speeds ranging from 600 km/s to more than 1,000 km/s.

Gases near a sunspot sometimes brighten suddenly, shooting outward at high speed. These violent eruptions are called solar flares. You can see a solar flare in **Figure 8.**

CMEs Coronal mass ejections (CMEs) occur when large amounts of electrically-charged gas are ejected suddenly from the Sun's corona. CMEs can occur as often as two or three times each day during a sunspot maximum.

CMEs present little danger to life on Earth, but they do have some effects. CMEs can damage satellites in orbit around Earth. They also can interfere with radio and power distribution equipment. CMEs often cause auroras. High-energy particles contained in CMEs and the solar wind are carried past Earth's magnetic field. This generates electric currents that flow toward Earth's poles. These electric currents ionize gases in Earth's atmosphere. When these ions recombine with electrons, they produce the light of an aurora, shown in **Figure 8.**

Topic: Space Weather
Visit blue.msscience.com for Web links to information about space weather and its effects.

Activity Record space weather conditions for several weeks. How does space weather affect Earth?

Figure 8 Features such as solar prominences and solar flares can reach hundreds of thousands of kilometers into space. CMEs are generated as magnetic fields above sunspot groups rearrange. CMEs can trigger events that produce auroras.

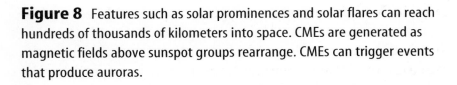

Solar prominence

Solar flare

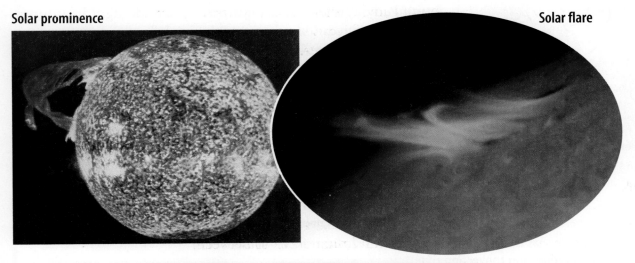

Aurora borealis, or northern lights

SECTION 2 The Sun

The Sun—An Average Star

The Sun is an average star. It is middle-aged, and its absolute magnitude is about average. It shines with a yellow light. Although the Sun is an average star, it is much closer to Earth than other stars. Light from the Sun reaches Earth in about eight minutes. Light from other stars takes from many years to many millions of years to reach Earth.

The Sun is unusual in one way. It is not close to any other stars. Most stars are part of a system in which two or more stars orbit each other. When two stars orbit each other, it is called a binary system. When three stars orbit each other, it is called a triple star system. The closest star system to the Sun—the Alpha Centauri system, including Proxima Centauri—is a triple star.

Stars also can move through space as a cluster. In a star cluster, many stars are relatively close, so the gravitational attraction among the stars is strong. Most star clusters are far from the solar system. They sometimes appear as a fuzzy patch in the night sky. The double cluster in the northern part of the constellation Perseus is shown in **Figure 9.** On a dark night in autumn, you can see the double cluster with binoculars, but you can't see its individual stars. The Pleiades star cluster can be seen in the constellation of Taurus in the winter sky. On a clear, dark night, you might be able to see seven of the stars in this cluster.

Figure 9 Most stars originally formed in large clusters containing hundreds, or even thousands, of stars.
Draw and label a sketch of the double cluster.

section 2 review

Summary

The Sun's Layers
- The Sun's interior has layers that include the core, radiation zone, and convection zone.

The Sun's Atmosphere
- The Sun's atmosphere includes the photosphere, chromosphere, and corona.

Surface Features
- The number of sunspots on the Sun varies in a 10- to 11-year cycle.
- Auroras occur when charged particles from the Sun interact with Earth's magnetic field.

The Sun—An Average Star
- The Sun is an average star, but it is much closer to Earth than any other star.

Self Check

1. **Explain** why the Sun is important for life on Earth.
2. **Describe** the sunspot cycle.
3. **Explain** why sunspots appear dark.
4. **Explain** why the Sun, which is an average star, appears so much brighter from Earth than other stars do.
5. **Think Critically** When a CME occurs on the Sun, it takes a couple of days for effects to be noticed on Earth. Explain.

Applying Skills

6. **Communicate** Make a sketch that shows the Sun's layers in your Science Journal. Write a short description of each layer.

Sunspots

Sunspots can be observed moving across the face of the Sun as it rotates. Measure the movement of sunspots, and use your data to determine the Sun's period of rotation.

Real-World Question

Can sunspot motion be used to determine the Sun's period of rotation?

Goals

- **Observe** sunspots and estimate their size.
- **Estimate** the rate at which sunspots move across the face of the Sun.

Materials

several books
piece of cardboard
drawing paper
refracting telescope
clipboard
small tripod
scissors

Safety Precautions

WARNING: *Handle scissors with care.*

Procedure

1. Find a location where the Sun can be viewed at the same time of day for a minimum of five days. **WARNING:** *Do not look directly at the Sun. Do not look through the telescope at the Sun. You could damage your eyes.*
2. If the telescope has a small finder scope attached, remove it or keep it covered.
3. Set up the telescope with the eyepiece facing away from the Sun, as shown. Align the telescope so that the shadow it casts on the ground is the smallest size possible. Cut and attach the cardboard as shown in the photo.
4. Use books to prop the clipboard upright. Point the eyepiece at the drawing paper.
5. Move the clipboard back and forth until you have the largest image of the Sun on the paper. Adjust the telescope to form a clear image. Trace the outline of the Sun on the paper.
6. Trace any sunspots that appear as dark areas on the Sun's image. Repeat this step at the same time each day for a week.
7. Using the Sun's diameter (approximately 1,390,000 km), estimate the size of the largest sunspots that you observed.
8. **Calculate** how many kilometers the sunspots move each day.
9. **Predict** how many days it will take for the same group of sunspots to return to the same position in which they appeared on day 1.

Conclude and Apply

1. What was the estimated size and rate of motion of the largest sunspots?
2. **Infer** how sunspots can be used to determine that the Sun's surface is not solid like Earth's surface.

Communicating Your Data

Compare your conclusions with those of other students in your class. **For more help, refer to the** Science Skill Handbook.

LAB **379**

section 3

Evolution of Stars

as you read

What You'll Learn
- **Describe** how stars are classified.
- **Compare** the Sun to other types of stars on the H-R diagram.
- **Describe** how stars evolve.

Why It's Important
Earth and your body contain elements that were made in stars.

Review Vocabulary
gravity: an attractive force between objects that have mass

New Vocabulary
- nebula
- giant
- white dwarf
- supergiant
- neutron star
- black hole

Classifying Stars

When you look at the night sky, all stars might appear to be similar, but they are quite different. Like people, they vary in age and size, but stars also vary in temperature and brightness.

In the early 1900s, Ejnar Hertzsprung and Henry Russell made some important observations. They noticed that, in general, stars with higher temperatures also have brighter absolute magnitudes.

Hertzsprung and Russell developed a graph, shown in **Figure 10,** to show this relationship. They placed temperatures across the bottom and absolute magnitudes up one side. A graph that shows the relationship of a star's temperature to its absolute magnitude is called a Hertzsprung-Russell (H-R) diagram.

The Main Sequence As you can see, stars seem to fit into specific areas of the graph. Most stars fit into a diagonal band that runs from the upper left to the lower right of the graph. This band, called the main sequence, contains hot, blue, bright stars in the upper left and cool, red, dim stars in the lower right. Yellow main sequence stars, like the Sun, fall in between.

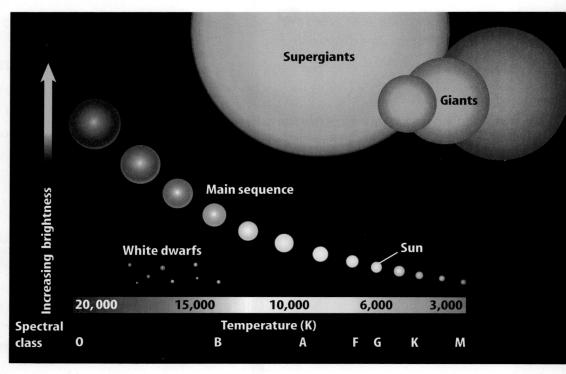

Figure 10 The relationships among a star's color, temperature, and brightness are shown in this H-R diagram. Stars in the upper left are hot, bright stars, and stars in the lower right are cool, dim stars.
Classify Which type of star shown in the diagram is the hottest, dimmest star?

Dwarfs and Giants About 90 percent of all stars are main sequence stars. Most of these are small, red stars found in the lower right of the H-R diagram. Among main sequence stars, the hottest stars generate the most light and the coolest ones generate the least. What about the ten percent of stars that are not part of the main sequence? Some of these stars are hot but not bright. These small stars are located on the lower left of the H-R diagram and are called white dwarfs. Other stars are extremely bright but not hot. These large stars on the upper right of the H-R diagram are called giants, or red giants, because they are usually red in color. The largest giants are called supergiants. **Figure 11** shows the supergiant, Antares—a star 300 times the Sun's diameter—in the constellation Scorpius. It is more than 11,000 times as bright as the Sun.

Reading Check *What kinds of stars are on the main sequence?*

How do stars shine?

For centuries, people were puzzled by the questions of what stars were made of and how they produced light. Many people had estimated that Earth was only a few thousand years old. The Sun could have been made of coal and shined for that long. However, when people realized that Earth was much older, they wondered what material possibly could burn for so many years. Early in the twentieth century, scientists began to understand the process that keeps stars shining for billions of years.

Generating Energy In the 1930s, scientists discovered reactions between the nuclei of atoms. They hypothesized that temperatures in the center of the Sun must be high enough to cause hydrogen to fuse to make helium. This reaction releases tremendous amounts of energy. Much of this energy is emitted as different wavelengths of light, including visible, infrared, and ultraviolet light. Only a tiny fraction of this light comes to Earth. During the fusion reaction, four hydrogen nuclei combine to create one helium nucleus. The mass of one helium nucleus is less than the mass of four hydrogen nuclei, so some mass is lost in the reaction.

Years earlier, in 1905, Albert Einstein had proposed a theory stating that mass can be converted into energy. This was stated as the famous equation $E = mc^2$. In this equation, E is the energy produced, m is the mass, and c is the speed of light. The small amount of mass "lost" when hydrogen atoms fuse to form a helium atom is converted to a large amount of energy.

Figure 11 Antares is a bright supergiant located 400 light-years from Earth. Although its temperature is only about 3,500 K, it is the 16th brightest star in the sky.

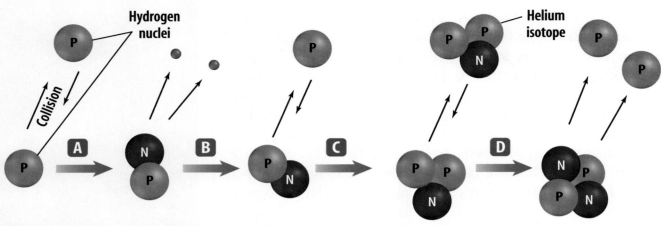

A Two protons (hydrogen nuclei) collide. One proton decays to a neutron, releasing subatomic particles and some energy.

B Another proton fuses with a proton and neutron to form an isotope of helium. Energy is given off again.

C Two helium isotopes collide with enough energy to fuse.

D A helium nucleus (two protons and two neutrons) forms as two protons break away. During the process, still more energy is released.

Figure 12 Fusion of hydrogen into helium occurs in a star's core. **Infer** what happens to the "lost" mass during this process.

Fusion Shown in **Figure 12,** fusion occurs in the cores of stars. Only in the core are temperatures high enough to cause atoms to fuse. Normally, they would repel each other, but in the core of a star where temperatures can exceed 15,000,000 K, atoms can move so fast that some of them fuse upon colliding.

Evolution of Stars

The H-R diagram explained a lot about stars. However, it also led to more questions. Many wondered why some stars didn't fit in the main sequence group and what happened when a star depleted its supply of hydrogen fuel. Today, scientists have theories about how stars evolve, what makes them different from one another, and what happens when they die. **Figure 13** illustrates the lives of different types of stars.

When hydrogen fuel is depleted, a star loses its main sequence status. This can take less than 1 million years for the brightest stars to many billions of years for the dimmest stars. The Sun has a main sequence life span of about 10 billion years. Half of its life span is still in the future.

Nebula Stars begin as a large cloud of gas and dust called a **nebula.** As the particles of gas and dust exert a gravitational force on each other, the nebula begins to contract. Gravitational forces cause instability within the nebula. The nebula can break apart into smaller and smaller pieces. Each piece eventually might collapse to form a star.

Science Online

Topic: Evolution of Stars
Visit blue.msscience.com for Web links to information about the evolution of stars.

Activity Make a three-circle Venn diagram to compare and contrast white dwarfs, neutron stars, and black holes.

A Star Is Born As the particles in the smaller pieces of nebula move closer together, the temperatures in each nebula piece increase. When the temperature inside the core of a nebula piece reaches 10 million K, fusion begins. The energy released radiates outward through the condensing ball of gas. As the energy radiates into space, stars are born.

Reading Check *How are stars born?*

Main Sequence to Giant Stars In the newly formed star, the heat from fusion causes pressure to increase. This pressure balances the attraction due to gravity. The star becomes a main sequence star. It continues to use its hydrogen fuel.

When hydrogen in the core of the star is depleted, a balance no longer exists between pressure and gravity. The core contracts, and temperatures inside the star increase. This causes the outer layers of the star to expand and cool. In this late stage of its life cycle, a star is called a **giant.**

After the core temperature reaches 100 million K, helium nuclei fuse to form carbon in the giant's core. By this time, the star has expanded to an enormous size, and its outer layers are much cooler than they were when it was a main sequence star. In about 5 billion years, the Sun will become a giant.

White Dwarfs After the star's core uses much of its helium, it contracts even more and its outer layers escape into space. This leaves behind the hot, dense core. At this stage in a star's evolution, it becomes a **white dwarf.** A white dwarf is about the size of Earth. Eventually, the white dwarf will cool and stop giving off light.

White Dwarf Matter The matter in white dwarf stars is more than 500,000 times as dense as the matter in Earth. In white dwarf matter, there are free electrons and atomic nuclei. The resistance of the electrons to pack together more provides pressure that keeps the star from collapsing. This state of matter is called electron degeneracy.

Figure 13 The life of a star depends on its mass. Massive stars eventually become neutron stars or black holes.
Explain *what happens to stars that are the size of the Sun.*

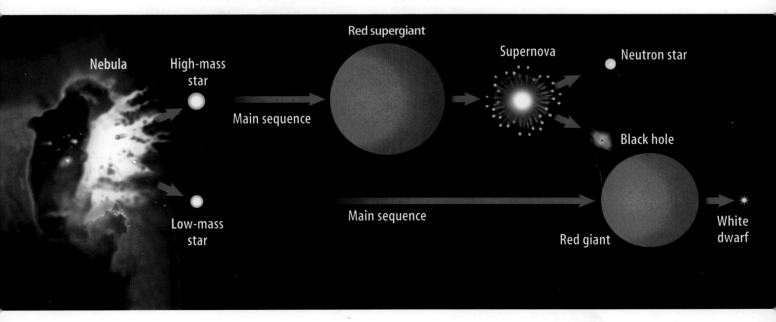

SECTION 3 Evolution of Stars **383**

INTEGRATE History

386 Supernova In 386 A.D., Chinese observers described a new star—a supernova—in the night sky. More recently, astronomers using the *Chandra X-ray Observatory* found evidence of a spinning neutron star, called a pulsar, in exactly the same location. Because of the Chinese account, astronomers better understand how neutron stars form and evolve.

Supergiants and Supernovas In stars that are more than about eight times more massive than the Sun, the stages of evolution occur more quickly and more violently. Look back at **Figure 13**. In massive stars, the core heats up to much higher temperatures. Heavier and heavier elements form by fusion, and the star expands into a **supergiant.** Eventually, iron forms in the core. Because of iron's atomic structure, it cannot release energy through fusion. The core collapses violently, and a shock wave travels outward through the star. The outer portion of the star explodes, producing a supernova. A supernova can be millions of times brighter than the original star was.

Neutron Stars If the collapsed core of a supernova is between about 1.4 and 3 times as massive as the Sun, it will shrink to approximately 20 km in diameter. Only neutrons can exist in the dense core, and it becomes a **neutron star.** Neutron stars are so dense that a teaspoonful would weigh more than 600 million metric tons in Earth's gravity. As dense as neutron stars are, they can contract only so far because the neutrons resist the inward pull of gravity.

Black Holes If the remaining dense core from a supernova is more than about three times more massive than the Sun, probably nothing can stop the core's collapse. Under these conditions, all of the core's mass collapses to a point. The gravity near this mass is so strong that nothing can escape from it, not even light. Because light cannot escape, the region is called a **black hole.** If you could shine a flashlight on a black hole, the light simply would disappear into it.

Reading Check *What is a black hole?*

Black holes, however, are not like giant vacuum cleaners, sucking in distant objects. A black hole has an event horizon, which is a region inside of which nothing can escape. If something—including light—crosses the event horizon, it will be pulled into the black hole. Beyond the event horizon, the black hole's gravity pulls on objects just as it would if the mass had not collapsed. Stars and planets can orbit around a black hole.

The photograph in **Figure 14** was taken by the *Hubble Space Telescope*. It shows a jet of gas streaming out of the center of galaxy M87. This jet of gas formed as matter flowed toward a black hole, and some of the gas was ejected along the polar axis.

Figure 14 The black hole at the center of galaxy M87 pulls matter into it at extremely high velocities. Some matter is ejected to produce a jet of gas that streams away from the center of the galaxy at nearly light speed.

Recycling Matter A star begins its life as a nebula, such as the one shown in **Figure 15.** Where does the matter in a nebula come from? Nebulas form partly from the matter that was once in other stars. A star ejects enormous amounts of matter during its lifetime. Some of this matter is incorporated into nebulas, which can evolve to form new stars. The matter in stars is recycled many times.

What about the matter created in the cores of stars and during supernova explosions? Are elements such as carbon and iron also recycled? These elements can become parts of new stars. In fact, spectrographs have shown that the Sun contains some carbon, iron, and other heavier elements. Because the Sun is an average, main sequence star, it is too young and its mass is too small to have formed these elements itself. The Sun condensed from material that was created in stars that died many billions of years ago.

Some elements condense to form planets and other bodies rather than stars. In fact, your body contains many atoms that were fused in the cores of ancient stars. Evidence suggests that the first stars formed from hydrogen and helium and that all the other elements have formed in the cores of stars or as stars explode.

Figure 15 Stars are forming in the Orion Nebula and other similar nebulae.
Describe *a star-forming nebula.*

section 3 review

Summary

Classifying Stars
- Most stars plot on the main sequence of an H-R diagram.
- As stars near the end of their lives, they move off of the main sequence.

How do stars shine?
- Stars shine because of a process called fusion.
- During fusion, nuclei of a lighter element merge to form a heavier element.

Evolution of Stars
- Stars form in regions of gas and dust called nebulae.
- Stars evolve differently depending on how massive they are.

Self Check

1. **Explain** how the Sun is different from other stars on the main sequence. How is it different from a giant star? How is it different from a white dwarf?
2. **Describe** how stars release energy.
3. **Outline** the past and probable future of the Sun.
4. **Define** a black hole.
5. **Think Critically** How can white dwarf stars be both hot and dim?

Applying Math

6. **Convert Units** A neutron star has a diameter of 20 km. One kilometer equals 0.62 miles. What is the neutron star's diameter in miles?

section 4

Galaxies and the Universe

as you read

What You'll Learn
- **Describe** the Sun's position in the Milky Way Galaxy.
- **Explain** that the same natural laws that apply to our solar system also apply in other galaxies.

Why It's Important
Studying the universe could determine whether life exists elsewhere.

Review Vocabulary
universe: the space that contains all known matter and energy

New Vocabulary
- galaxy
- big bang theory

Galaxies

If you enjoy science fiction, you might have read about explorers traveling through the galaxy. On their way, they visit planets around other stars and encounter strange alien beings. Although this type of space exploration is futuristic, it is possible to explore galaxies today. Using a variety of telescopes, much is being learned about the Milky Way and other galaxies.

A **galaxy** is a large group of stars, gas, and dust held together by gravity. Earth and the solar system are in a galaxy called the Milky Way. It might contain as many as one trillion stars. Countless other galaxies also exist. Each of these galaxies contains the same elements, forces, and types of energy that occur in Earth's solar system. Galaxies are separated by huge distances—often millions of light-years.

In the same way that stars are grouped together within galaxies, galaxies are grouped into clusters. The cluster that the Milky Way belongs to is called the Local Group. It contains about 45 galaxies of various sizes and types. The three major types of galaxies are spiral, elliptical, and irregular.

Spiral Galaxies Spiral galaxies are galaxies that have spiral arms that wind outward from the center. The arms consist of bright stars, dust, and gas. The Milky Way Galaxy, shown in **Figure 16,** is a spiral galaxy. The Sun and the rest of the solar system are located near the outer edge of the Milky Way Galaxy.

Spiral galaxies can be normal or barred. Arms in a normal spiral start close to the center of the galaxy. Barred spirals have spiral arms extending from a large bar of stars and gas that passes through the center of the galaxy.

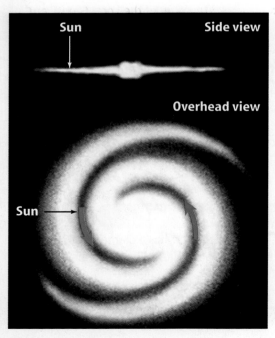

Figure 16 This illustration shows a side view and an overhead view of the Milky Way.
Describe where the Sun is in the Milky Way.

386 CHAPTER 13 Stars and Galaxies

Elliptical Galaxies A common type of galaxy is the elliptical galaxy. **Figure 17** shows an elliptical galaxy in the constellation Andromeda. These galaxies are shaped like large, three-dimensional ellipses. Many are football shaped, but others are round. Some elliptical galaxies are small, while others are so large that several galaxies the size of the Milky Way would fit inside one of them.

Irregular Galaxies The third type—an irregular galaxy—includes most of those galaxies that don't fit into the other categories. Irregular galaxies have many different shapes. They are smaller than the other types of galaxies. Two irregular galaxies called the Clouds of Magellan orbit the Milky Way. The Large Magellanic Cloud is shown in **Figure 18.**

Reading Check *How do the three different types of galaxies differ?*

Figure 17 This photo shows an example of an elliptical galaxy. **Identify** *the two other types of galaxies.*

The Milky Way Galaxy

The Milky Way might contain one trillion stars. The visible disk of stars shown in **Figure 16** is about 100,000 light-years across. Find the location of the Sun. Notice that it is located about 26,000 light-years from the galaxy's center in one of the spiral arms. In the galaxy, all stars orbit around a central region, or core. It takes about 225 million years for the Sun to orbit the center of the Milky Way.

The Milky Way often is classified as a normal spiral galaxy. However, recent evidence suggests that it might be a barred spiral. It is difficult to know for sure because astronomers have limited data about how the galaxy looks from the outside.

You can't see the shape of the Milky Way because you are located within one of its spiral arms. You can, however, see the Milky Way stretching across the sky as a misty band of faint light. You can see the brightest part of the Milky Way if you look low in the southern sky on a moonless summer night. All the stars you can see in the night sky belong to the Milky Way.

Like many other galaxies, the Milky Way has a supermassive black hole at its center. This black hole might be more than 2.5 million times as massive as the Sun. Evidence for the existence of the black hole comes from observing the orbit of a star near the galaxy's center. Additional evidence includes X-ray emissions detected by the *Chandra X-ray Observatory*. X rays are produced when matter spirals into a black hole.

Figure 18 The Large Magellanic Cloud is an irregular galaxy. It's a member of the Local Group, and it orbits the Milky Way.

Mini LAB

Measuring Distance in Space

Procedure

1. On a large sheet of **paper**, draw an overhead view of the Milky Way. If necessary, refer to **Figure 16**. Choose a scale to show distance in light-years.
2. Mark the approximate location of the solar system, which is about two-thirds of the way out on one of the spiral arms.
3. Now, draw a side view of the Milky Way Galaxy. Mark the position of the solar system.

Analysis

1. What scale did you use to represent distance on your model of the Milky Way?
2. The Andromeda Galaxy is about 2.9 million light-years from Earth. What scale distance would this represent?

Origin of the Universe

People long have wondered how the universe formed. Several models of its origin have been proposed. One model is the steady state theory. It suggests that the universe always has been the same as it is now. The universe always existed and always will. As the universe expands, new matter is created to keep the overall density of the universe the same or in a steady state. However, evidence indicates that the universe was much different in the past.

A second idea is called the oscillating model. In this model, the universe began with expansion. Over time, the expansion slowed and the universe contracted. Then the process began again, oscillating back and forth. Some scientists still hypothesize that the universe expands and contracts in a cycle.

A third model of how the universe formed is called the big bang theory. The universe started with a big bang and has been expanding ever since. This theory will be described later.

Expansion of the Universe

What does it sound like when a train is blowing its whistle while it travels past you? The whistle has a higher pitch as the train approaches you. Then the whistle seems to drop in pitch as the train moves away. This effect is called the Doppler shift. The Doppler shift occurs with light as well as with sound. **Figure 19** shows how the Doppler shift causes changes in the light coming from distant stars and galaxies. If a star is moving toward Earth, its wavelengths of light are compressed. If a star is moving away from Earth, its wavelengths of light are stretched.

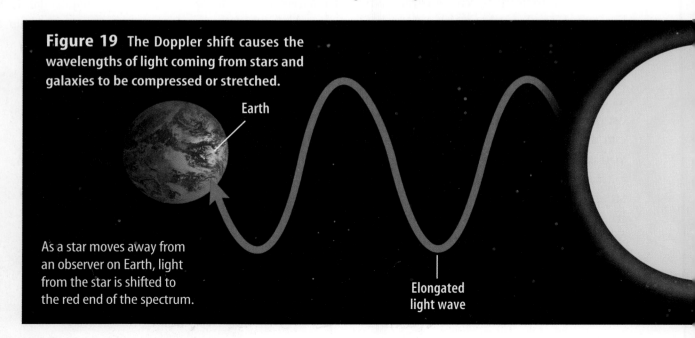

Figure 19 The Doppler shift causes the wavelengths of light coming from stars and galaxies to be compressed or stretched.

As a star moves away from an observer on Earth, light from the star is shifted to the red end of the spectrum.

The Doppler Shift Look at the spectrum of a star in **Figure 20A**. Note the position of the dark lines. How do they compare with the lines in **Figures 20B** and **20C**? They have shifted in position. What caused this shift? As you just read, when a star is moving toward Earth, its wavelengths of light are compressed, just as the sound waves from the train's whistle are. This causes the dark lines in the spectrum to shift toward the blue-violet end of the spectrum. A red shift in the spectrum occurs when a star is moving away from Earth. In a red shift, the dark lines shift toward the red end of the spectrum.

Red Shift In 1929, Edwin Hubble published an interesting fact about the light coming from most galaxies. When a spectrograph is used to study light from galaxies beyond the Local Group, a red shift occurs in the light. What does this red shift tell you about the universe?

Because all galaxies beyond the Local Group show a red shift in their spectra, they must be moving away from Earth. If all galaxies outside the Local Group are moving away from Earth, then the entire universe must be expanding. Remember the Launch Lab at the beginning of the chapter? The dots on the balloon moved apart as the model universe expanded. Regardless of which dot you picked, all the other dots moved away from it. In a similar way, galaxies beyond the Local Group are moving away from Earth.

Figure 20 **A** This spectrum shows dark absorption lines. **B** The dark lines shift toward the blue-violet end for a star moving toward Earth. **C** The lines shift toward the red end for a star moving away from Earth.

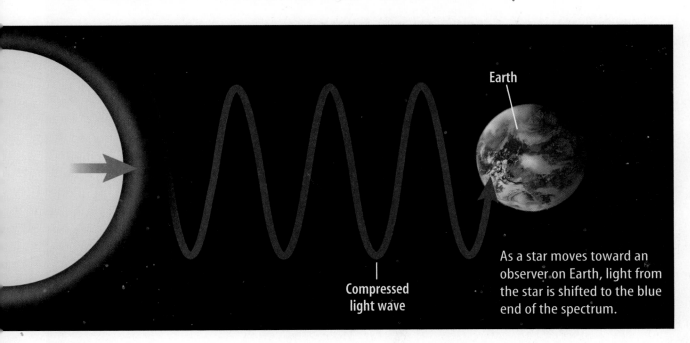

Compressed light wave

As a star moves toward an observer on Earth, light from the star is shifted to the blue end of the spectrum.

SECTION 4 Galaxies and the Universe

NATIONAL GEOGRAPHIC VISUALIZING THE BIG BANG THEORY

Figure 21

The big bang theory states that the universe probably began about 13.7 billion years ago with an enormous explosion. Even today, galaxies are rushing apart from this explosion.

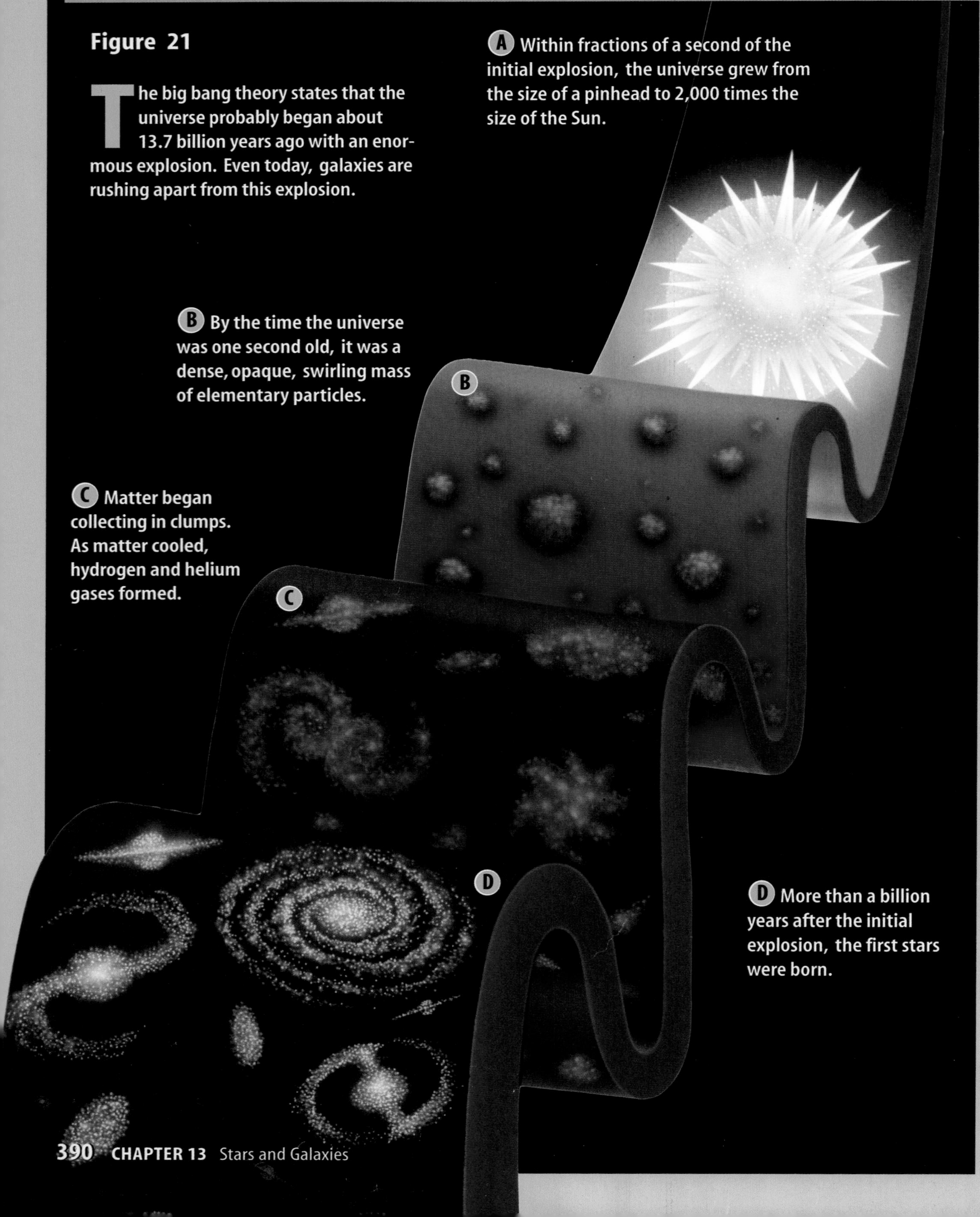

A Within fractions of a second of the initial explosion, the universe grew from the size of a pinhead to 2,000 times the size of the Sun.

B By the time the universe was one second old, it was a dense, opaque, swirling mass of elementary particles.

C Matter began collecting in clumps. As matter cooled, hydrogen and helium gases formed.

D More than a billion years after the initial explosion, the first stars were born.

The Big Bang Theory

When scientists determined that the universe was expanding, they developed a theory to explain their observations. The leading theory about the formation of the universe is called the **big bang theory**. **Figure 21** illustrates the big bang theory. According to this theory, approximately 13.7 billion years ago, the universe began with an enormous explosion. The entire universe began to expand everywhere at the same time.

Looking Back in Time The time-exposure photograph shown in **Figure 22** was taken by the *Hubble Space Telescope*. It shows more than 1,500 galaxies at distances of more than 10 billion light-years. These galaxies could date back to when the universe was no more than 1 billion years old. The galaxies are in various stages of development. One astronomer says that humans might be looking back to a time when the Milky Way was forming.

Whether the universe will expand forever or stop expanding is still unknown. If enough matter exists, gravity might halt the expansion, and the universe will contract until everything comes to a single point. However, studies of distant supernovae indicate that an energy, called dark energy, is causing the universe to expand faster. Scientists are trying to understand how dark energy might affect the fate of the universe.

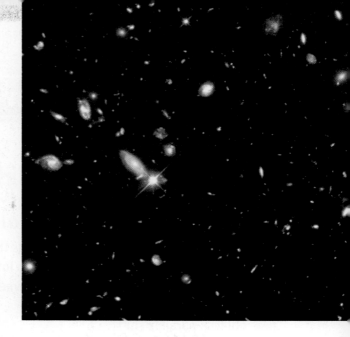

Figure 22 The light from the galaxies in this photo mosaic took billions of years to reach Earth.

section 4 review

Summary

Galaxies
- The three main types of galaxies are spiral, elliptical, and irregular.

The Milky Way Galaxy
- The Milky Way is a spiral galaxy and the Sun is about 26,000 light-years from its center.

Origin of the Universe
- Theories about how the universe formed include the steady state theory, the oscillating universe theory, and the big bang theory.

The Big Bang Theory
- This theory states that the universe began with an explosion about 13.7 billion years ago.

Self Check

1. **Describe** elliptical galaxies. How are they different from spiral galaxies?
2. **Identify** the galaxy that you live in.
3. **Explain** the Doppler shift.
4. **Explain** how all galaxies are similar.
5. **Think Critically** All galaxies outside the Local Group show a red shift. Within the Local Group, some show a red shift and some show a blue shift. What does this tell you about the galaxies in the Local Group?

Applying Skills

6. **Compare and contrast** the theories about the origin of the universe.

Design Your Own

Measuring Parallax

Goals
- **Design** a model to show how the distance from an observer to an object affects the object's parallax shift.
- **Describe** how parallax can be used to determine the distance to a star.

Possible Materials
meterstick
masking tape
metric ruler
pencil

Safety Precautions

WARNING: *Be sure to wear goggles to protect your eyes.*

● Real-World Question

Parallax is the apparent shift in the position of an object when viewed from two locations. How can you build a model to show the relationship between distance and parallax?

● Form a Hypothesis

State a hypothesis about how parallax varies with distance.

● Test Your Hypothesis

Make a Plan

1. As a group, agree upon and write your hypothesis statement.
2. **List** the steps you need to take to build your model. Be specific, describing exactly what you will do at each step.
3. **Devise** a method to test how distance from an observer to an object, such as a pencil, affects the parallax of the object.
4. **List** the steps you will take to test your hypothesis. Be specific, describing exactly what you will do at each step.
5. Read over your plan for the model to be used in this experiment.

392 **CHAPTER 13** Stars and Galaxies

Using Scientific Methods

6. How will you determine changes in observed parallax? Remember, these changes should occur when the distance from the observer to the object is changed.
7. You should measure shifts in parallax from several different positions. How will these positions differ?
8. How will you measure distances accurately and compare relative position shift?

Follow Your Plan

1. Make sure your teacher approves your plan before you start.
2. **Construct** the model your team has planned.
3. Carry out the experiment as planned.
4. While conducting the experiment, record any observations that you or other members of your group make in your Science Journal.

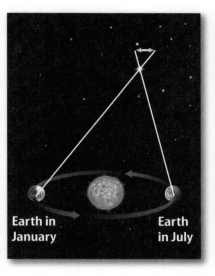

Earth in January

Earth in July

◯ Analyze Your Data

1. **Compare** what happened to the object when it was viewed with one eye closed, then the other.
2. At what distance from the observer did the object appear to shift the most?
3. At what distance did it appear to shift the least?

◯ Conclude and Apply

1. **Infer** what happened to the apparent shift of the object's location as the distance from the observer was increased or decreased.
2. **Describe** how astronomers might use parallax to study stars.

Communicating Your Data

Prepare a chart showing the results of your experiment. Share the chart with members of your class. **For more help, refer to the Science Skill Handbook.**

LAB **393**

SCIENCE Stats

Stars and Galaxies

Did you know...

...A star in Earth's galaxy explodes as a supernova about once a century. The most famous supernova of this galaxy occurred in 1054 and was recorded by the ancient Chinese and Koreans. The explosion was so powerful that it could be seen during the day, and its brightness lasted for weeks. Other major supernovas in the Milky Way that were observed from Earth occurred in 185, 386, 1006, 1181, 1572, and 1604.

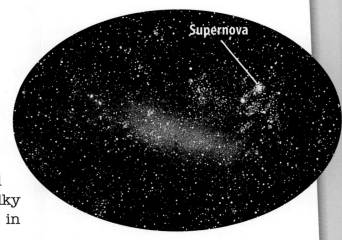
Supernova

...The large loops of material called solar prominences can extend more than 320,000 km above the Sun's surface. This is so high that two Jupiters and three Earths could fit under the arch.

...The red giant star Betelgeuse has a diameter larger than that of Earth's Sun. This gigantic star measures 450,520,000 km in diameter, while the Sun's diameter is a mere 1,390,176 km.

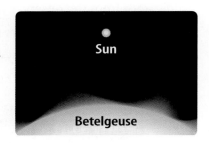
Sun
Betelgeuse

Applying Math Use words to express the number 450,520,000.

Write About It

Visit blue.msscience.com/science_stats to learn whether it might be possible for Earth astronauts to travel to the nearest stars. How long would such a trip take? What problems would have to be overcome? Write a brief report about what you find.

chapter 13 Study Guide

Reviewing Main Ideas

Section 1 Stars

1. Constellations are patterns of stars in the night sky. Some constellations can be seen all year. Other constellations are visible only during certain seasons.
2. Parallax is the apparent shift in the position of an object when viewed from two different positions. Parallax is used to find the distance to nearby stars.

Section 2 The Sun

1. The Sun is the closest star to Earth.
2. Sunspots are areas on the Sun's surface that are cooler and less bright than surrounding areas.

Section 3 Evolution of Stars

1. Stars are classified according to their position on the H-R diagram.
2. Low-mass stars end their lives as white dwarfs. High-mass stars become neutron stars or black holes.

Section 4 Galaxies and the Universe

1. A galaxy consists of stars, gas, and dust held together by gravity.
2. Earth's solar system is in the Milky Way, a spiral galaxy.
3. The universe is expanding. Scientists don't know whether the universe will expand forever or contract to a single point.

Visualizing Main Ideas

Copy and complete the following concept map that shows the evolution of a main sequence star with a mass similar to that of the Sun.

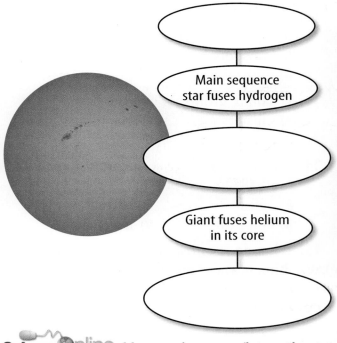

chapter 13 Review

Using Vocabulary

absolute magnitude p. 372
apparent magnitude p. 372
big bang theory p. 391
black hole p. 384
chromosphere p. 375
constellation p. 370
corona p. 375
galaxy p. 386
giant p. 383
light-year p. 373
nebula p. 382
neutron star p. 384
photosphere p. 375
sunspot p. 376
supergiant p. 384
white dwarf p. 383

Explain the difference between the terms in each of the following sets.

1. absolute magnitude—apparent magnitude
2. galaxy—constellation
3. giant—supergiant
4. chromosphere—photosphere
5. black hole—neutron star

Checking Concepts

Choose the word or phrase that best answers the question.

6. What is a measure of the amount of a star's light that is received on Earth?
   A) absolute magnitude
   B) apparent magnitude
   C) fusion
   D) parallax

7. What is higher for closer stars?
   A) absolute magnitude
   B) red shift
   C) parallax
   D) blue shift

8. What happens after a nebula contracts and its temperature increases to 10 million K?
   A) a black hole forms
   B) a supernova occurs
   C) fusion begins
   D) a white dwarf forms

9. Which of these has an event horizon?
   A) giant
   B) white dwarf
   C) black hole
   D) neutron star

10. What forms when the Sun fuses hydrogen?
    A) carbon
    B) oxygen
    C) iron
    D) helium

Use the illustration below to answer question 11.

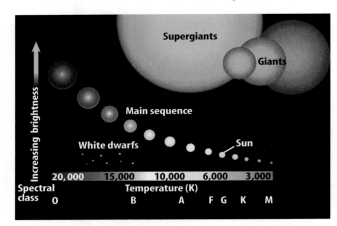

11. Which of the following best describes giant stars?
    A) hot, dim stars
    B) cool, dim stars
    C) hot, bright stars
    D) cool, bright stars

12. Which of the following are loops of matter flowing from the Sun?
    A) sunspots
    B) auroras
    C) coronas
    D) prominences

13. What are groups of galaxies called?
    A) clusters
    B) supergiants
    C) giants
    D) binary systems

14. Which galaxies are sometimes shaped like footballs?
    A) spiral
    B) elliptical
    C) barred
    D) irregular

15. What do scientists study to determine shifts in wavelengths of light?
    A) spectrum
    B) parallax
    C) corona
    D) nebula

chapter 13 Review

Thinking Critically

Use the table below to answer question 16.

Magnitude and Distance of Stars

Star	Apparent Magnitude	Absolute Magnitude	Distance in Light-Years
A	−26	4.8	0.00002
B	−1.5	1.4	8.7
C	0.1	4.4	4.3
D	0.1	−7.0	815
E	0.4	−5.9	520
F	1.0	−0.6	45

16. **Interpret Data** Use the table above to answer the following questions. *Hint: lower magnitude values are brighter than higher magnitude values.*
    a. Which star appears brightest from Earth?
    b. Which star would appear brightest from a distance of 10 light-years?
    c. Infer which star in the table above is the Sun.

17. **Infer** How do scientists know that black holes exist if these objects don't emit visible light?

18. **Recognize Cause and Effect** Why can parallax only be used to measure distances to stars that are relatively close to Earth?

19. **Compare and contrast** the Sun with other stars on the H-R diagram.

20. **Concept Map** Make a concept map showing the life history of a very large star.

21. **Make Models** Make a model of the Sun. Include all of the Sun's layers in your model.

Performance Activities

22. **Story** Write a short science-fiction story about an astronaut traveling through the universe. In your story, describe what the astronaut observes. Use as many vocabulary words as you can.

23. **Photomontage** Gather photographs of the aurora borealis from magazines and other sources. Use the photographs to create a photomontage. Write a caption for each photo.

Applying Math

24. **Travel to Vega** Vega is a star that is 26 light-years away. If a spaceship could travel at one-tenth the speed of light, how long would it take to reach this star?

Use the illustration below to answer question 25.

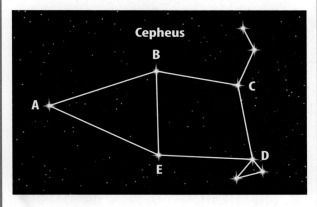

25. **Constellation Cepheus** The illustration above shows the constellation Cepheus. Answer the following questions about this contellation.
    a. Which of the line segments are nearly parallel?
    b. Which line segments are nearly perpendicular?
    c. Which angles are oblique?
    d. What geometric shape do the three stars at the left side of the drawing form?

Chapter 13 Standardized Test Practice

Part 1 Multiple Choice

Record your answers on the answer sheet provided by your teacher or on a sheet of paper.

Use the illustration below to answer question 1.

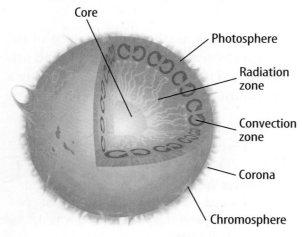

1. The illustration above shows the interior of which object?
   A. Earth   C. the Sun
   B. Saturn   D. the Moon

2. Which is a group of stars, gas, and dust held together by gravity?
   A. constellation   C. black hole
   B. supergiant   D. galaxy

3. The most massive stars end their lives as which type of object?
   A. black hole   C. neutron star
   B. white dwarf   D. black dwarf

4. In which galaxy does the Sun exist?
   A. Arp's galaxy   C. Milky Way galaxy
   B. Barnard's galaxy   D. Andromeda galaxy

Test-Taking Tip

Process of Elimination If you don't know the answer to a multiple-choice question, eliminate as many incorrect choices as possible. Mark your best guess from the remaining answers before moving on to the next question.

5. Which is the closest star to Earth?
   A. Sirius   C. Betelgeuse
   B. the Sun   D. the Moon

6. In which of the following choices are the objects ordered from smallest to largest?
   A. stars, galaxies, galaxy clusters, universe
   B. galaxy clusters, galaxies, stars, universe
   C. universe, galaxy clusters, galaxies, stars
   D. universe, stars, galaxies, galaxy clusters

Use the graph below to answer questions 7 and 8.

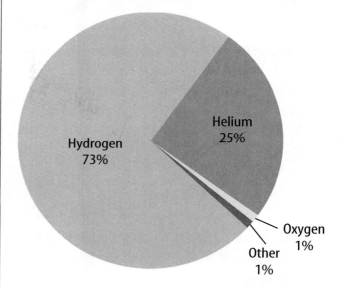

7. Which is the most abundant element in the Sun?
   A. helium
   B. hydrogen
   C. oxygen
   D. carbon

8. How will this circle graph change as the Sun ages?
   A. The hydrogen slice will get smaller.
   B. The hydrogen slice will get larger.
   C. The helium slice will get smaller.
   D. The circle graph will not change.

Standardized Test Practice

Part 2 Short Response/Grid In

Record your answers on the answer sheet provided by your teacher or on a sheet of paper.

9. How can events on the Sun affect Earth? Give one example.
10. How does a red shift differ from a blue shift?
11. How do astronomers know that the universe is expanding?
12. What is the main sequence?
13. What is a constellation?

Use the illustration below to answer questions 14–16.

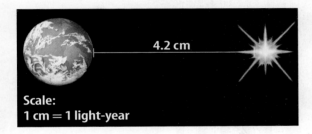

14. According to the illustration, how many light-years from Earth is Proxima Centauri?
15. How many years would it take for light from Proxima Centauri to get to Earth?
16. At this scale, how many centimeters would represent the distance to a star that is 100 light-years from Earth?
17. How can a star's color provide information about its temperature?
18. Approximately how long does it take light from the Sun to reach Earth? In general, how does this compare to the amount of time it takes light from all other stars to reach Earth?
19. How does the size, temperature, age, and brightness of the Sun compare to other stars in the Milky Way Galaxy?

Part 3 Open Ended

Record your answers on a sheet of paper.

Use the graph below to answer question 20.

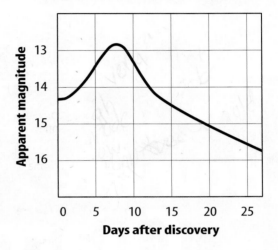

20. The graph above shows the brightness of a supernova that was observed from Earth in 1987. Describe how the brightness of this supernova changed through time. When was it brightest? What happened before May 20? What happened after May 20? How much did the brightness change?
21. Compare and contrast the different types of galaxies.
22. Write a detailed description of the Sun. What is it? What is it like?
23. Explain how parallax is used to measure the distance to nearby stars.
24. Why are some constellations visible all year? Why are other constellations only visible during certain seasons?
25. What are black holes? How do they form?
26. Explain the big bang theory.
27. What can be learned by studying the dark lines in a star's spectrum?

unit 5
Chemistry of Matter

How Are Charcoal & Celebrations Connected?

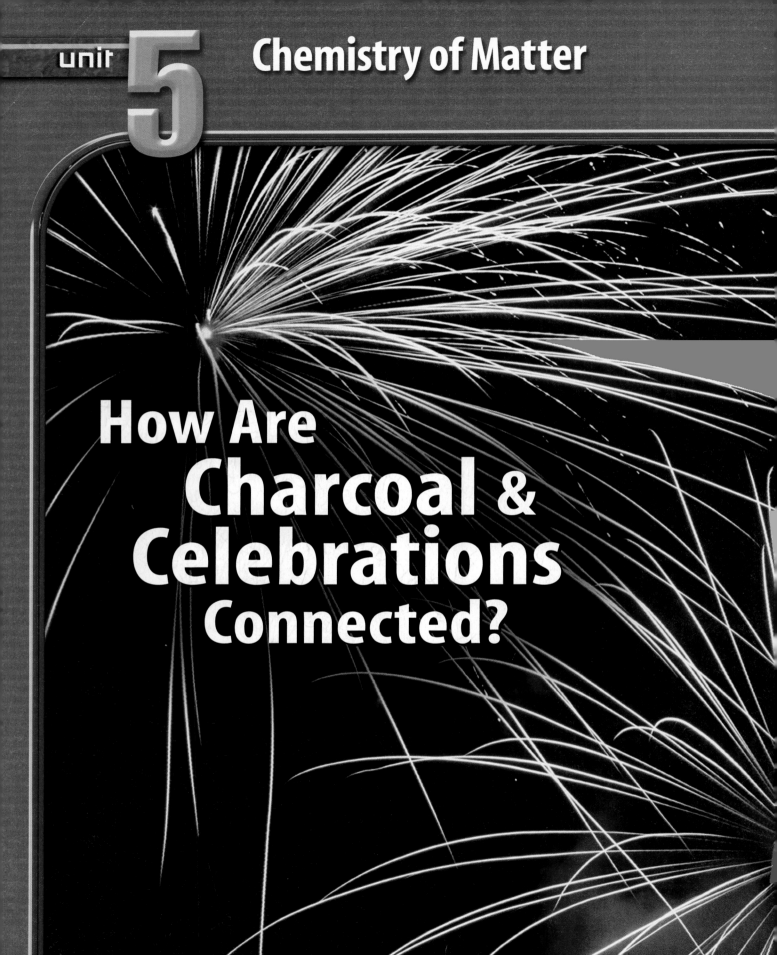

NATIONAL GEOGRAPHIC

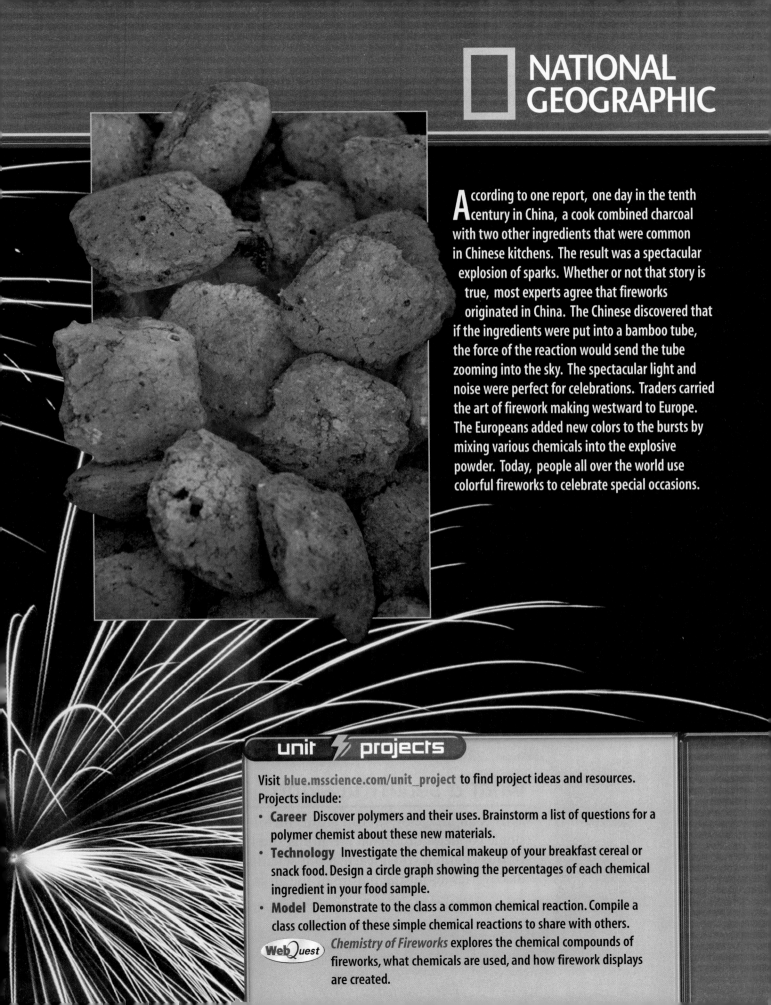

According to one report, one day in the tenth century in China, a cook combined charcoal with two other ingredients that were common in Chinese kitchens. The result was a spectacular explosion of sparks. Whether or not that story is true, most experts agree that fireworks originated in China. The Chinese discovered that if the ingredients were put into a bamboo tube, the force of the reaction would send the tube zooming into the sky. The spectacular light and noise were perfect for celebrations. Traders carried the art of firework making westward to Europe. The Europeans added new colors to the bursts by mixing various chemicals into the explosive powder. Today, people all over the world use colorful fireworks to celebrate special occasions.

unit projects

Visit blue.msscience.com/unit_project to find project ideas and resources.
Projects include:

- **Career** Discover polymers and their uses. Brainstorm a list of questions for a polymer chemist about these new materials.
- **Technology** Investigate the chemical makeup of your breakfast cereal or snack food. Design a circle graph showing the percentages of each chemical ingredient in your food sample.
- **Model** Demonstrate to the class a common chemical reaction. Compile a class collection of these simple chemical reactions to share with others.

WebQuest *Chemistry of Fireworks* explores the chemical compounds of fireworks, what chemicals are used, and how firework displays are created.

chapter 14

Inside the Atom

The BIG Idea

The model of the atom becomes more detailed as new information is learned.

SECTION 1
Models of the Atom
Main Idea Atoms contain protons and neutrons in a very small, dense nucleus and electrons in a larger area around the nucleus.

SECTION 2
The Nucleus
Main Idea For atoms of a given element, the number of protons is fixed, but the number of neutrons can vary.

What a beautiful sight!

This is an image of 48 iron atoms forming a "corral" around a single copper atom. What are atoms and how were they discovered? In this chapter, you'll learn about scientists and their amazing discoveries about the nature of the atom.

Science Journal Describe, based on your current knowledge, what an atom is.

Start-Up Activities

Model the Unseen

Have you ever had a wrapped birthday present that you couldn't wait to open? What did you do to try to figure out what was in it? The atom is like that wrapped present. You want to investigate it, but you cannot see it easily.

1. Your teacher will give you a piece of clay and some pieces of metal. Count the pieces of metal.
2. Bury these pieces in the modeling clay so they can't be seen.
3. Exchange clay balls with another group.
4. With a toothpick, probe the clay to find out how many pieces of metal are in the ball and what shape they are.
5. **Think Critically** In your Science Journal, sketch the shapes of the metal pieces as you identify them. How does the number of pieces you found compare with the number that were in the clay ball? How do their shapes compare?

Parts of the Atom Make the following Foldable to help you organize your thoughts and review the parts of an atom.

STEP 1 Collect two sheets of paper and layer them about 2 cm apart vertically. Keep the edges level.

STEP 2 Fold up the bottom edges of the paper to form four equal tabs.

STEP 3 Fold the papers and crease well to hold the tabs in place. Staple along the fold. Label the tabs *Atom, Electron, Proton,* and *Neutron* as shown.

Read and Write As you read the chapter, describe how each part of the atom was discovered and record other facts under the appropriate tabs.

Preview this chapter's content and activities at
blue.msscience.com

Get Ready to Read

Visualize

① Learn It! Visualize by forming mental images of the text as you read. Imagine how the text descriptions look, sound, feel, smell, or taste. Look for any pictures or diagrams on the page that may help you add to your understanding.

② Practice It! Read the following paragraph. As you read, use the underlined details to form a picture in your mind.

> The nuclear atom has a tiny nucleus tightly packed with positively charged electrons and neutral neutrons. Negatively charged electrons occupy the space surrounding the nucleus. The number of electrons in a neutral atom equals the number of protons in the atom.
>
> —*from page 411*

Based on the description above, try to visualize an atom. Now look at the diagram on page 411.
- How big is the nucleus?
- How many protons are in the atom?
- What is the charge of a proton and an electron?

③ Apply It! Read the chapter and list three subjects you were able to visualize. Make a rough sketch showing what you visualized.

Target Your Reading

Reading Tip: Forming your own mental images will help you remember what you read.

Use this to focus on the main ideas as you read the chapter.

① Before you read the chapter, respond to the statements below on your worksheet or on a numbered sheet of paper.
- Write an **A** if you **agree** with the statement.
- Write a **D** if you **disagree** with the statement.

② After you read the chapter, look back to this page to see if you've changed your mind about any of the statements.
- If any of your answers changed, explain why.
- Change any false statements into true statements.
- Use your revised statements as a study guide.

Science Online
Print out a worksheet of this page at blue.msscience.com

Before You Read A or D		Statement	After You Read A or D
	1	Early philosophers studied atoms by doing experiments.	
	2	Crooke determined that the beam he observed must have been light because it was bent by a magnet.	
	3	Rutherford expected the alpha particles to bounce back from the gold foil.	
	4	Atoms are made of mostly empty space.	
	5	Neutrons have no charge.	
	6	Electrons travel in very predictable paths around the nucleus.	
	7	All atoms of an element have the same number of protons and neutrons.	
	8	Atoms of one element can change into atoms of another element through radioactive decay.	
	9	Radioactive isotopes are too dangerous to be of any benefit to humans.	

section 1
Models of the Atom

as you read

What You'll Learn
- **Explain** how scientists discovered subatomic particles.
- **Explain** how today's model of the atom developed.
- **Describe** the structure of the nuclear atom.
- **Explain** that all matter is made up of atoms.

Why It's Important
Atoms make up everything in your world.

Review Vocabulary
matter: anything that has mass and takes up space

New Vocabulary
- element
- anode
- cathode
- electron
- alpha particle
- proton
- neutron
- electron cloud

First Thoughts

Do you like mysteries? Are you curious? Humans are curious. Someone always wants to know something that is not easy to detect or to see what can't be seen. For example, people began wondering about matter more than 2,500 years ago. Some of the early philosophers thought that matter was composed of tiny particles. They reasoned that you could take a piece of matter, cut it in half, cut the half piece in half again, and continue to cut again and again. Eventually, you wouldn't be able to cut any more. You would have only one particle left. They named these particles *atoms,* a term that means "cannot be divided." Another way to imagine this is to picture a string of beads like the one shown in **Figure 1.** If you keep dividing the string into pieces, you eventually come to one single bead.

Describing the Unseen Early philosophers didn't try to prove their theories by doing experiments as scientists now do. Their theories were the result of reasoning, debating, and discussion—not of evidence or proof. Today, scientists will not accept a theory that is not supported by experimental evidence. But even if these philosophers had experimented, they could not have proven the existence of atoms. People had not yet discovered much about what is now called chemistry, the study of matter. The kind of equipment needed to study matter was a long way from being invented. Even as recently as 500 years ago, atoms were still a mystery.

Figure 1 You can divide this string of beads in half, and in half again until you have one, indivisible bead. Like this string of beads, all matter can be divided until you reach one basic particle, the atom.

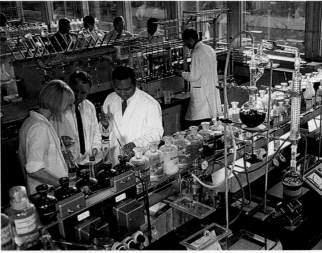

Figure 2 Even though the laboratories of the time were simple compared to those of today, incredible discoveries were made during the eighteenth century.

A Model of the Atom

A long period passed before the theories about the atom were developed further. Finally during the eighteenth century, scientists in laboratories, like the one on the left in **Figure 2,** began debating the existence of atoms once more. Chemists were learning about matter and how it changes. They were putting substances together to form new substances and taking substances apart to find out what they were made of. They found that certain substances couldn't be broken down into simpler substances. Scientists called these substances elements. An **element** is matter made of atoms of only one kind. For example, iron is an element made of iron atoms. Silver, another element, is made of silver atoms. Carbon, gold, and oxygen are other examples of elements.

Dalton's Concept John Dalton, an English schoolteacher in the early nineteenth century, combined the idea of elements with the earlier theory of the atom. He proposed the following ideas about matter: (1) Matter is made up of atoms, (2) atoms cannot be divided into smaller pieces, (3) all the atoms of an element are exactly alike, and (4) different elements are made of different kinds of atoms. Dalton pictured an atom as a hard sphere that was the same throughout, something like a tiny marble. A model like this is shown in **Figure 3.**

Scientific Evidence Dalton's theory of the atom was tested in the second half of the nineteenth century. In 1870, the English scientist William Crookes did experiments with a glass tube that had almost all the air removed from it. The glass tube had two pieces of metal called electrodes sealed inside. The electrodes were connected to a battery by wires.

Figure 3 Dalton pictured the atom as a hard sphere that was the same throughout.

SECTION 1 Models of the Atom **405**

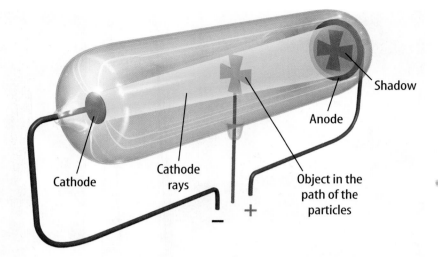

Figure 4 Crookes used a glass tube containing only a small amount of gas. When the glass tube was connected to a battery, something flowed from the negative electrode (cathode) to the positive electrode (anode).
Explain *if this unknown thing was light or a stream of particles.*

A Strange Shadow An electrode is a piece of metal that can conduct electricity. One electrode, called the **anode,** has a positive charge. The other, called the **cathode,** has a negative charge. In the tube that Crookes used, the metal cathode was a disk at one end of the tube. In the center of the tube was an object shaped like a cross, as you can see in **Figure 4.** When the battery was connected, the glass tube suddenly lit up with a greenish-colored glow. A shadow of the object appeared at the opposite end of the tube—the anode. The shadow showed Crookes that something was traveling in a straight line from the cathode to the anode, similar to the beam of a flashlight. The cross-shaped object was getting in the way of the beam and blocking it. This is similar to how a road crew uses a stencil to block paint from certain places on the road when they are marking lanes and arrows. You can see this in **Figure 5.**

Figure 5 Paint passing by a stencil is an example of what happened with Crookes's tube, the cathode ray, and the cross.

Cathode Rays Crookes hypothesized that the green glow in the tube was caused by rays, or streams of particles. These rays were called cathode rays because they were produced at the cathode. Crookes's tube is known as a cathode-ray tube, or CRT. **Figure 6** shows a CRT. They were used for TV and computer monitors for many years.

✓ Reading Check

What are cathode rays?

Discovering Charged Particles

The news of Crookes's experiments excited the scientific community of the time. But many scientists were not convinced that the cathode rays were streams of particles. Was the greenish glow light, or was it a stream of charged particles? In 1897, J.J. Thomson, an English physicist, tried to clear up the confusion. He placed a magnet beside the tube from Crookes's experiments. In **Figure 7,** you can see that the beam is bent in the direction of the magnet. Light cannot be bent by a magnet, so the beam couldn't be light. Therefore, Thomson concluded that the beam must be made up of charged particles of matter that came from the cathode.

The Electron Thomson then repeated the CRT experiment using different metals for the cathode and different gases in the tube. He found that the same charged particles were produced no matter what elements were used for the cathode or the gas in the tube. Thomson concluded that cathode rays are negatively charged particles of matter. How did Thomson know the particles were negatively charged? He knew that opposite charges attract each other. He observed that these particles were attracted to the positively charged anode, so he reasoned that the particles must be negatively charged.

These negatively charged particles are now called **electrons.** Thomson also inferred that electrons are a part of every kind of atom because they are produced by every kind of cathode material. Perhaps the biggest surprise that came from Thomson's experiments was the evidence that particles smaller than the atom do exist.

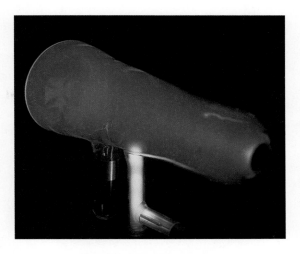

Figure 6 The cathode-ray tube got its name because the particles start at the cathode and travel to the anode. At one time, a CRT was in every TV and computer monitor.

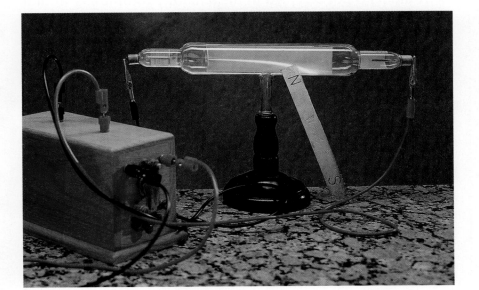

Figure 7 When a magnet was placed near a CRT, the cathode rays were bent. Since light is not bent by a magnet, Thomson determined that the cathode rays were made of charged particles.

Thomson's Atomic Model

Some of the questions posed by scientists were answered in light of Thomson's experiments. However, the answers inspired new questions. If atoms contain one or more negatively charged particles, then all matter, which is made of atoms, should be negatively charged as well. But all matter isn't negatively charged. Could it be that atoms also contain some positive charge? The negatively charged electrons and the unknown positive charge would then neutralize each other in the atom. Thomson came to this conclusion and included positive charge in his model of the atom.

Using his new findings, Thomson revised Dalton's model of the atom. Instead of a solid ball that was the same throughout, Thomson pictured a sphere of positive charge. The negatively charged electrons were spread evenly among the positive charge. This is modeled by the ball of clay shown in **Figure 8**. The positive charge of the clay is equal to the negative charge of the electrons. Therefore, the atom is neutral. It was later discovered that not all atoms are neutral. The number of electrons within an element can vary. If there is more positive charge than negative electrons, the atom has an overall positive charge. If there are more negative electrons than positive charge, the atom has an overall negative charge.

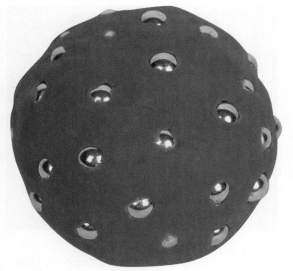

Figure 8 Modeling clay with ball bearings mixed through is another way to picture the J.J. Thomson atom. The clay contains all the positive charge of the atom. The ball bearings, which represent the negatively charged electrons, are mixed evenly in the clay.
Explain what made Thomson include positive particles in his atomic model?

 **Reading Check** *What particle did Thomson's model have scattered through it?*

Rutherford's Experiments

A model is not accepted in the scientific community until it has been tested and the tests support previous observations. In 1906, Ernest Rutherford and his coworkers began an experiment to find out if Thomson's model of the atom was correct. They wanted to see what would happen when they fired fast-moving, positively charged bits of matter, called alpha particles, at a thin film of a metal such as gold. Alpha particles, which come from unstable atoms, are positively charged, and so they are repelled by particles of matter which also have a positive charge.

Figure 9 shows how the experiment was set up. A source of alpha particles was aimed at a thin sheet of gold foil that was only 400 nm thick. The foil was surrounded by a fluorescent (floo REH sunt) screen that gave a flash of light each time it was hit by a charged particle.

Expected Results Rutherford was certain he knew what the results of this experiment would be. His prediction was that most of the speeding alpha particles would pass right through the foil and hit the screen on the other side, just like a bullet fired through a pane of glass. Rutherford reasoned that the thin, gold film did not contain enough matter to stop the speeding alpha particle or change its path. Also, there wasn't enough positive charge in any one place in Thomson's model to repel the alpha particle strongly. He thought that the positive charge in the gold atoms might cause a few minor changes in the path of the alpha particles. However, he assumed that this would only occur a few times.

That was a reasonable hypothesis because in Thomson's model, the positive charge is essentially neutralized by nearby electrons. Rutherford was so sure of what the results would be that he turned the work over to a graduate student.

The Model Fails Rutherford was shocked when his student rushed in to tell him that some alpha particles were veering off at large angles. You can see this in **Figure 9**. Rutherford expressed his amazement by saying, "It was about as believable as if you had fired a 15-inch shell at a piece of tissue paper, and it came back and hit you." How could such an event be explained? The positively charged alpha particles were moving with such high speed that it would take a large positive charge to cause them to bounce back. The mass and charges in Thomson's model of the atom were distributed evenly and could not have repelled an alpha particle.

Figure 9 In Rutherford's experiment, alpha particles bombarded the gold foil. Most particles passed right through the foil or veered slightly from a straight-line path, but some particles bounced right back. The path of a particle is shown by a flash of light when it hits the fluorescent screen.

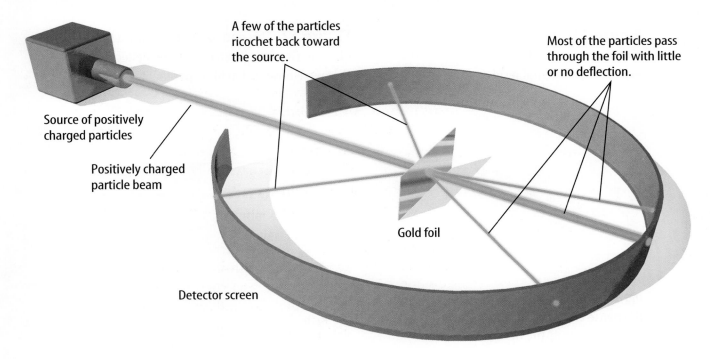

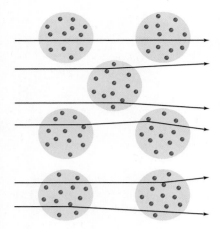

• Proton → Path of alpha particle

Figure 10 Rutherford thought that if the atom could be described by Thomson's model, as shown above, then only minor bends in the paths of the particles would have occurred.

A Model with a Nucleus

Now Rutherford and his team had to come up with an explanation for these unexpected results. They might have drawn diagrams like those in **Figure 10,** which uses Thomson's model and shows what Rutherford expected. Now and then, an alpha particle might be affected slightly by a positive charge in the atom and turn a bit off course. However, large changes in direction were not expected.

The Proton The actual results did not fit this model, so Rutherford proposed a new one, shown in **Figure 11.** He hypothesized that almost all the mass of the atom and all of its positive charge are crammed into an incredibly small region of space at the center of the atom called the nucleus. Eventually, his prediction was proved true. In 1920 scientists identified the positive charges in the nucleus as protons. A **proton** is a positively charged particle present in the nucleus of all atoms. The rest of each atom is empty space occupied by the atom's almost-massless electrons.

Reading Check *How did Rutherford describe his new model?*

Figure 12 shows how Rutherford's new model of the atom fits the experimental data. Most alpha particles could move through the foil with little or no interference because of the empty space that makes up most of the atom. However, if an alpha particle made a direct hit on the nucleus of a gold atom, which has 79 protons, the alpha particle would be strongly repelled and bounce back.

Figure 11 The nuclear model was new and helped explain experimental results.

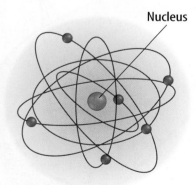

Rutherford's model included the dense center of positive charge known as the nucleus.

Figure 12 This nucleus that contained most of the mass of the atom caused the deflections that were observed in his experiment.

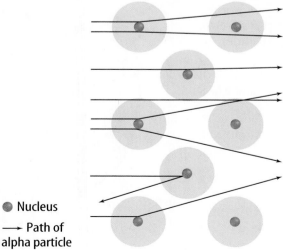

• Nucleus
→ Path of alpha particle

410 CHAPTER 14 Inside the Atom

The Neutron Rutherford's nuclear model was applauded as other scientists reviewed the results of the experiments. However, some data didn't fit. Once again, more questions arose and the scientific process continued. For instance, an atom's electrons have almost no mass. According to Rutherford's model, the only other particle in the atom was the proton. That meant that the mass of an atom should have been approximately equal to the mass of its protons. However, it wasn't. The mass of most atoms is at least twice as great as the mass of its protons. That left scientists with a dilemma and raised a new question. Where does the extra mass come from if only protons and electrons made up the atom?

It was proposed that another particle must be in the nucleus to account for the extra mass. The particle, which was later called the **neutron** (NEW trahn), would have the same mass as a proton and be electrically neutral. Proving the existence of neutrons was difficult though, because a neutron has no charge. Therefore, the neutron doesn't respond to magnets or cause fluorescent screens to light up. It took another 20 years before scientists were able to show by more modern experiments that atoms contain neutrons.

Reading Check *What particles are in the nucleus of the nuclear atom?*

The model of the atom was revised again to include the newly discovered neutrons in the nucleus. The nuclear atom, shown in **Figure 13,** has a tiny nucleus tightly packed with positively charged protons and neutral neutrons. Negatively charged electrons occupy the space surrounding the nucleus. The number of electrons in a neutral atom equals the number of protons in the atom.

Modeling the Nuclear Atom

Procedure
1. On a sheet of **paper,** draw a circle with a diameter equal to the width of the paper.
2. Small dots of **paper in two colors** will represent protons and neutrons. Using a dab of **glue** on each paper dot, make a model of the nucleus of the oxygen atom in the center of your circle. Oxygen has eight protons and eight neutrons.

Analysis
1. What particle is missing from your model of the oxygen atom?
2. How many of that missing particle should there be, and where should they be placed?

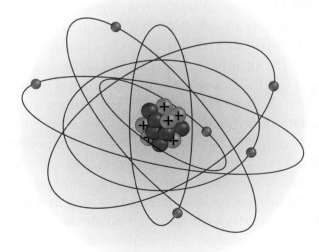

Figure 13 This atom of carbon, atomic number 6, has six protons and six neutrons in its nucleus. **Identify** *how many electrons are in the "empty" space surrounding the nucleus.*

SECTION 1 Models of the Atom **411**

Figure 14 If this Ferris wheel in London, with a diameter of 132 m, were the outer edge of the atom, the nucleus would be about the size of a single letter *o* on this page.

Protons Rutherford finally identified the particles of the nucleus as discrete positive charges of matter in 1919. Using alpha particles as bullets, he knocked hydrogen nuclei out of atoms from boron, fluorine, sodium, aluminum, phosphorus, and nitrogen. Rutherford named the hydrogen nuclei *protons*, which means "first" in Greek because protons were the first identified building blocks of the nuclei.

Size and Scale Drawings of the nuclear atom such as the one in **Figure 13** don't give an accurate representation of the extreme smallness of the nucleus compared to the rest of the atom. For example, if the nucleus were the size of a table-tennis ball, the atom would have a diameter of more than 2.4 km. Another way to compare the size of a nucleus with the size of the atom is shown in **Figure 14**. Perhaps now you can see better why in Rutherford's experiment, most of the alpha particles went directly through the gold foil without any interference from the gold atoms. Plenty of empty space allows the alpha particles an open pathway.

Further Developments

Even into the twentieth century, physicists were working on a theory to explain how electrons are arranged in an atom. It was natural to think that the negatively charged electrons are attracted to the positive nucleus in the same way the Moon is attracted to Earth. Then, electrons would travel in orbits around the nucleus. A physicist named Niels Bohr even calculated exactly what energy levels those orbits would represent for the hydrogen atom. His calculations explained experimental data found by other scientists. However, scientists soon learned that electrons are in constant, unpredictable motion and can't be described easily by an orbit. They determined that it was impossible to know the precise location of an electron at any particular moment. Their work inspired even more research and brainstorming among scientists around the world.

Electrons as Waves Physicists began to wrestle with explaining the unpredictable nature of electrons. Surely the experimental results they were seeing and the behavior of electrons could somehow be explained with new theories and models. The unconventional solution was to understand electrons not as particles, but as waves. This led to further mathematical models and equations that brought much of the experimental data together.

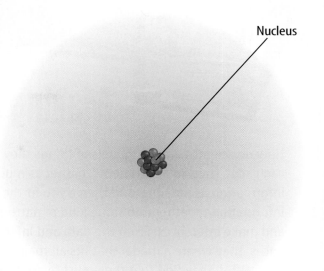

The Electron Cloud Model The new model of the atom allows for the somewhat unpredictable wave nature of electrons by defining a region where electrons are most likely to be found. Electrons travel in a region surrounding the nucleus, which is called the **electron cloud.** The current model for the electron cloud is shown in **Figure 15.** The electrons are more likely to be close to the nucleus rather than farther away because they are attracted to the positive charges of the protons. Notice the fuzzy outline of the electron cloud. Because the electrons could be anywhere, the cloud has no firm boundary. Interestingly, within the electron cloud, the electron in a hydrogen atom probably is found in the region Bohr calculated.

Figure 15 The electrons are more likely to be close to the nucleus rather than farther away, but they could be anywhere.

section 1 review

Summary

The Models of the Atom

- Some early philosophers believed all matter was made of small particles.
- John Dalton proposed that all matter is made of atoms that were hard spheres.
- J.J. Thomson showed that the particles in a CRT were negatively charged particles, later called electrons.
- Rutherford showed that a positive charge existed in a small region of the atom, which he called the nucleus.
- In order to explain the mass of an atom, the neutron was proposed as an uncharged particle with the same mass as a proton, located in the nucleus.
- Electrons are now believed to move about the nucleus in an electron cloud.

Self Check

1. **Explain** how the nuclear atom differs from the uniform sphere model of the atom.
2. **Determine** the number of electrons in a neutral atom with 49 protons.
3. **Think Critically** In Rutherford's experiment, why wouldn't the electrons in the atoms of the gold foil affect the paths of the alpha particles?
4. **Concept Map** Design and complete a concept map using all the words in the vocabulary list for this section. Add any other terms or words that will help create a complete diagram of the section and the concepts it contains.

Applying Math

5. **Solve One-Step Equations** The mass of an electron is 9.11×10^{-28} g. The mass of a proton is 1,836 times more than that of the electron. Calculate the mass of the proton in grams and convert that mass into kilograms.

Making a Model of the Invisible

How do scientists make models of things they can't see? They do experiments, gather as much information as possible, and then try to fit the information together into some kind of pattern and make inferences. From the data and inferences, they create a model that fits all their data. Often they find that they must revise their model when more data come to light.

● Real-World Question

How can you determine the inside structure of a box?

Goals
- **Observe** the motion of a marble inside a closed box.
- **Infer** the structure of the divisions inside the box.

Materials
sealed box
paper and pencil

● Procedure

1. **Record** the number of the box your teacher gives you. Don't take the lid off the box or look inside.
2. Lift the box. Tilt the box. Gently shake it. In your Science Journal, record all your observations. Make a sketch of the way you think the marble in the box is rolling.
3. Use your observations to infer what the inside of the box looks like.
4. **Compare** your inferences with those of students who have the same box as you do. Then you might want to make more observations or revise your inferences.
5. When you have gathered all the information, sketch your model in your Science Journal.
6. Open your box and compare your model with the actual inside structure of the box.

● Conclude and Apply

1. **Compare and Contrast** How did your model of the inside of the box compare with the actual inside?
2. **Draw Conclusions** Could you have used any other test to gather more information?
3. **Describe** how your notes in your Science Journal might be more helpful in this lab.
4. **Explain** how an observation is different from an inference.

Communicating Your Data

Create a data table and set of instructions that would help another student systematically test a new sealed box. **For more help, refer to the** Science Skill Handbook.

414 CHAPTER 14 Inside the Atom

section 2
The Nucleus

Identifying Numbers

The electron cloud model is a model of the average nuclear atom. But how does the nucleus in an atom of one element differ from the nucleus of an atom of another element? The atoms of different elements contain different numbers of protons. The **atomic number** of an element is the number of protons in the nucleus of an atom of that element. The smallest of the atoms, the hydrogen atom, has one proton in its nucleus, so hydrogen's atomic number is 1. Uranium, the heaviest naturally occurring element, has 92 protons. Its atomic number is 92. Atoms of an element are identified by the number of protons because this number never changes without changing the identity of the element.

Number of Neutrons The atomic number is the number of protons, but what about the number of neutrons in an atom's nucleus? A particular type of atom can have a varying number of neutrons in its nucleus. Most atoms of carbon have six neutrons. However, some carbon atoms have seven neutrons and some have eight, as you can see in **Figure 16.** They are all carbon atoms because they all have six protons. These three kinds of carbon atoms are called isotopes. **Isotopes** (I suh tohps) are atoms of the same element that have different numbers of neutrons. The isotopes of carbon are called carbon-12, carbon-13, and carbon-14. The numbers 12, 13, and 14 tell more about the nucleus of the isotopes. The combined masses of the protons and neutrons in an atom make up most of the mass of an atom.

as you read

What You'll Learn
- **Describe** the process of radioactive decay.
- **Explain** what is meant by half-life.
- **Describe** how radioactive isotopes are used.

Why It's Important
Radioactive elements are beneficial, but must be treated with caution.

Review Vocabulary
atom: the smallest particle of an element that retains all the properties of that element

New Vocabulary
- atomic number
- isotope
- mass number
- radioactive decay
- transmutation
- beta particle
- half-life

Figure 16 The three isotopes of carbon differ in the number of neutrons in each nucleus.

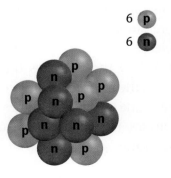
Carbon-12 nucleus

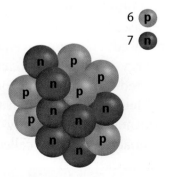
Carbon-13 nucleus

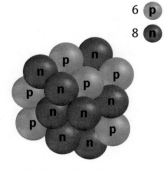
Carbon-14 nucleus

Table 1 Isotopes of Carbon

	Carbon-12	Carbon-13	Carbon-14
Mass number	12	13	14
Number of protons	6	6	6
Number of neutrons	6	7	8
Number of electrons	6	6	6
Atomic number	6	6	6

Mass Number The **mass number** of an isotope is the number of neutrons plus protons. **Table 1** shows the particles that make up each of the carbon isotopes. You can find the number of neutrons in an isotope by subtracting the atomic number from the mass number. For example, carbon-14 has a mass number of 14 and an atomic number of 6. The difference in these two numbers is 8, the number of neutrons in carbon-14.

Strong Nuclear Force When you need to hold something together, what do you use? Rubber bands, string, tape, or glue? What holds the protons and neutrons together in the nucleus of an atom? Because protons are positively charged, you might expect them to repel each other just as the north ends of two magnets tend to push each other apart. It is true that they normally would do just that. However, when they are packed together in the nucleus with the neutrons, an even stronger binding force takes over. That force is called the strong nuclear force. The strong nuclear force can hold the protons together only when they are as closely packed as they are in the nucleus of the atom.

Radioactive Decay

Many atomic nuclei are stable when they have about the same number of protons and neutrons. Carbon-12 is the most stable isotope of carbon. It has six protons and six neutrons. Some nuclei are unstable because they have too many or too few neutrons. This is especially true for heavier elements such as uranium and plutonium. In these nuclei, repulsion builds up. The nucleus releases particles and becomes more stable. When particles are released, energy is given off. The release of nuclear particles and energy is called **radioactive decay.** When the particles that are ejected from a nucleus include protons, the atomic number of the nucleus changes. When this happens, one element changes into another. The changing of one element into another through radioactive decay is called **transmutation.**

Topic: Radioactive Decay
Visit blue.msscience.com for Web links to information about radioactive decay.

Activity Explain how radioactive decay is used in home smoke detectors.

Reading Check *What occurs in radioactive decay?*

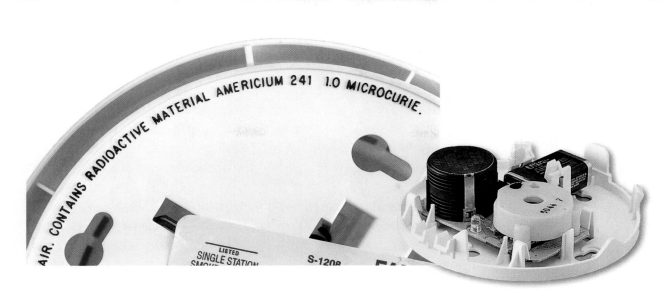

Loss of Alpha Particles Transmutation is occurring in most of your homes right now. **Figure 17** shows a smoke detector that makes use of radioactive decay. This device contains americium-241 (a muh RIH shee um), which undergoes transmutation by ejecting energy and an alpha particle. An **alpha particle** consists of two protons and two neutrons. Together, the energy and particles are called nuclear radiation. In the smoke detector, the fast-moving alpha particles enable the air to conduct an electric current. As long as the electric current is flowing, the smoke detector is silent. The alarm is triggered when the flow of electric current is interrupted by smoke entering the detector.

Changed Identity When americium expels an alpha particle, it's no longer americium. The atomic number of americium is 95, so americium has 95 protons. After the transmutation, it becomes the element that has 93 protons, neptunium. In **Figure 18,** notice that the mass and atomic numbers of neptunium and the alpha particle add up to the mass and atomic number of americium. All the nuclear particles of americium still exist after the transmutation.

Figure 17 This lifesaving smoke detector makes use of the radioactive isotope americium-241. The isotope is located inside the black, slotted chamber. When smoke particles enter the chamber, the alarm goes off.

Figure 18 Americium expels an alpha particle, which is made up of two protons and two neutrons. As a result, americium is changed into the element neptunium, which has two fewer protons than americium.

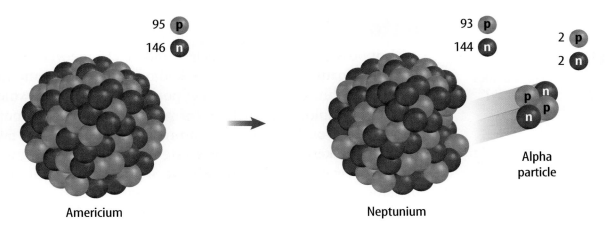

SECTION 2 The Nucleus **417**

Figure 19 Beta decay results in an element with an atomic number that is one greater than the original.

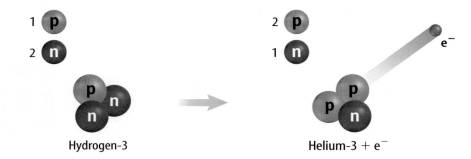

Hydrogen-3   Helium-3 + e⁻

Graphing Half-Life
Procedure
1. Make a table with three columns: *Number of Half-Lives, Days Passed,* and *Mass Remaining.*
2. Mark the table for six half-lives.
3. Thorium-234 has a half-life of 24 days. Fill in the second column with the total number of days after each half-life.
4. Begin with a 64-g sample of thorium and calculate the mass remaining after each half-life.
5. Make a graph with *Number of half-lives* on the *x*-axis and *Mass remaining* on the *y*-axis.

Analysis
1. During which half-life does the most thorium decay?
2. How much thorium was left by day 144?

Loss of Beta Particles Some elements undergo transmutations through a different process. Their nuclei emit an electron called a beta particle. A **beta particle** is a high-energy electron that comes from the nucleus, not from the electron cloud. However, the nucleus contains only protons and neutrons. How can it give off an electron? During this kind of transmutation, a neutron becomes unstable and splits into an electron and a proton. The electron, or beta particle, is released with a large amount of energy. The proton, however, remains in the nucleus.

Reading Check *What is a beta particle?*

Because a neutron has been changed into a proton, the nucleus of the element has an additional proton. Unlike the process of alpha decay, in beta decay the atomic number of the element that results is greater by one. **Figure 19** shows the beta decay of the hydrogen-3 nucleus. With two neutrons in its nucleus, hydrogen-3 is unstable. One neutron is converted to a proton and a beta particle by beta decay, and an isotope of helium is produced. The mass of the element stays almost the same because the mass of the electron that it loses is so small.

Rate of Decay

Is it possible to analyze a nucleus and determine when it will decay? Unfortunately, you cannot. Radioactive decay is random. It's like watching popcorn begin to pop. You can't predict which kernel will explode or when. But if you're an experienced popcorn maker, you might be able to predict how long it will take for half the kernels to pop. The rate of decay of a nucleus is measured by its half-life. The **half-life** of a radioactive isotope is the amount of time it takes for half of a sample of the element to decay.

Calculating Half-Life Decay Iodine-131 has a half-life of eight days. If you start with a sample of 4 g of iodine-131, after eight days you would have only 2 g of iodine-131 remaining. After 16 days, or two half-lives, half of the 2 g would have decayed and you would have only 1 g left. **Figure 20** illustrates this process.

The radioactive decay of unstable atoms goes on at a steady pace, unaffected by conditions such as weather, pressure, magnetic or electric fields, and even chemical reactions. Half-lives, which are different for each isotope, range in length from fractions of a second to billions of years.

February		1	2	3	4
	4 grams iodine-131				
5	6	7	8	9	10	11
			2 grams iodine-131			
12	13	14	15	16	17	18
				1 gram iodine-131		
19	20	21	22	23	24	25
					0.5 grams iodine-131	
26	27	28	1 March	2	3	4 ?

Figure 20 Half-life is the time it takes for one half of a sample to decay.
Calculate *how much of the sample you expect to find on March 4.*

Applying Math — Use Numbers

FIND HALF-LIVES Tritium has a half-life of 12.5 years. If you start with 20 g, how much tritium would be left after 50 years?

Solution

1 *This is what you know:*
- half-life = 12.5 years
- initial weight = 20 g

2 *This is what you need to find out:*
- the number of half-lives in 50 years
- final weight after 50 years

3 *This is the procedure you need to use:*
- Determine the number of half-lives.
  number of years/half life = number of half-lives
  50 years/12.5 years = 4 half-lives
- Determine the final weight.
  final weight = initial weight/$2^{\text{(number of half-lives)}}$
  final weight = 20 g/2^4 = 20 g/16 = 1.25 g

4 *Check your answer:* Substitute the number of half-lives and the final weight into the second equation and solve for initial weight. You should get the same initial weight.

Practice Problems

1. Carbon-14 has a half-life of 5,730 years. Starting with 100 g of carbon-14, how much would be left after 17,190 years?

2. Radon-222 has a half-life of 3.8 days. Starting with 50 g of radon-222, how much would be left after 19 days?

Science Online For more practice, visit blue.msscience.com/math_practice

Energy Conversion
Nuclear power plants convert the nuclear energy from the radioactive U-235 to electrical energy and heat energy. Research how the plants dispose of the heat energy, and infer what precautions they should take to prevent water pollution in the area.

Carbon Dating Scientists have found the study of radioactive decay useful in determining the age of artifacts and fossils. Carbon-14 is used to determine the age of dead animals, plants, and humans. The half-life of carbon-14 is 5,730 years. In a living organism, the amount of carbon-14 remains in constant balance with the levels of the isotope in the atmosphere or ocean. This balance occurs because living organisms take in and release carbon. For example, animals take in carbon from food such as plants and release carbon as carbon dioxide. While life processes go on, any carbon-14 nucleus that decays is replaced by another from the environment. When the plant or animal dies, the decaying nuclei no longer can be replaced.

When archaeologists find an ancient item, such as the one in **Figure 21,** they can find out how much carbon-14 it has and compare it with the amount of carbon-14 the animal would have had when it was alive. Knowing the half-life of carbon-14, they can then calculate when the animal lived.

When geologists want to determine the age of rocks, they cannot use carbon dating. Carbon dating is used only for things that have been alive. Instead, geologists examine the decay of uranium. Uranium-238 decays to lead-206 with a half-life of 4.5 billion years. By comparing the amount of uranium to lead, the scientist can determine the age of a rock. However, there is some disagreement in the scientific community about this method because some rocks might have had lead in them to start with. In addition, some of the isotopes could have migrated out of the rock over the years.

Figure 21 Using carbon-14 dating techniques, archaeologists can find out when an animal may have lived.

Disposal of Radioactive Waste Waste products from processes that involve radioactive decay are a problem because they can leave isotopes that still release radiation. This radioactive waste must be isolated from people and the environment because it continues to produce harmful radiation. Special disposal sites that can contain the radiation must be built to store this waste for long periods. One such site is in Carlsbad, New Mexico, where nuclear waste is buried 655 m below the surface of Earth.

Figure 22 Giant particle accelerators, such as this linear accelerator at Stanford, are needed to speed up particles until they are moving fast enough to cause an atomic transmutation.

Making Synthetic Elements

Scientists now create new elements by smashing atomic particles into a target element. Alpha and beta particles, for example, are accelerated in particle accelerators like the one in **Figure 22** to speeds fast enough that they can smash into a large nucleus and be absorbed on impact. The absorbed particle converts the target element into another element with a higher atomic number. The new element is called a synthetic element because it is made by humans. These artificial transmutations have created new elements that do not exist in nature. Elements with atomic numbers 93 to 112, and 114 have been made in this way.

Uses of Radioactive Isotopes The process of artificial transmutation has been adapted so that radioactive isotopes of normally stable elements can be used in hospitals and clinics using specially designed equipment. These isotopes, called tracer elements, are used to diagnose disease and to study environmental conditions. The radioactive isotope is introduced into a living system such as a person, animal, or plant. It then is followed by a device that detects radiation while it decays. These devices often present the results as a display on a screen or as a photograph. The isotopes chosen for medical purposes have short half-lives, which allows them to be used without the risk of exposing living organisms to prolonged radiation.

Topic: Isotopes in Medicine and Agriculture
Visit blue.msscience.com for Web links to information about the use of isotopes in medicine and agriculture.

Activity List the most commonly used radioactive elements and their isotopes used in medicine and agriculture.

SECTION 2 The Nucleus **421**

NATIONAL GEOGRAPHIC VISUALIZING TRACER ELEMENTS

Figure 23

Typically, we try to avoid radioactivity. However, very small amounts of radioactive substances, called radioisotopes or "tracer elements," can be used to diagnose disease. A healthy thyroid gland absorbs iodine to produce two metabolism-regulating hormones. To determine if a person's thyroid is functioning properly, a radioisotope thyroid scan can be performed. First, a radioactive isotope of iodine—iodine-131—is administered orally or by injection. The thyroid absorbs this isotope as it would regular iodine, and a device called a gamma camera is then used to detect the radiation that iodine-131 emits. A computer uses this information to create an image showing thyroid size and activity. Three thyroid images taken by a gamma camera are shown below.

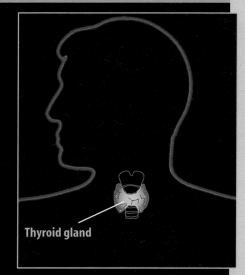
Thyroid gland

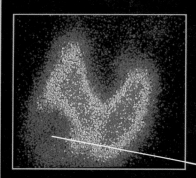

NORMAL
A healthy thyroid gland manufactures hormones that regulate a person's metabolism, including heart rate.

ENLARGED
Although rarely life threatening, an enlarged thyroid, or goiter—caused by too little iodine in the diet—can form a grapefruit-size lump in the neck.
— Goiter

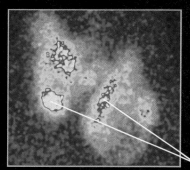

OVERACTIVE
In the condition known as hyperthyroidism, an overactive thyroid speeds up metabolism, causing weight loss and an increased heart rate.

— Areas of highest activity

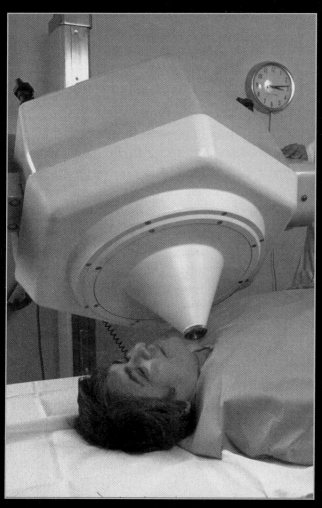

A gamma camera traces the location of iodine-131 during a thyroid scan procedure.

Medical Uses The isotope iodine-131 has been used to diagnose problems with the thyroid, a gland located at the base of the neck. This is discussed in **Figure 23.** Other radioactive isotopes are used to detect cancer, digestion problems, and circulation difficulties. Technetium-99 is a radioisotope with a half-life of 6 h that is used for tracing a variety of bodily processes. Tumors and fractures can be found because the isotope will show up as a stronger image wherever cells are growing rapidly.

Environmental Uses In the environment, tracers such as phosphorus-32 are injected into the root system of a plant. In the plant, the radioactive phosphorus behaves the same as the stable phosphorus would. A detector then is used to see how the plant uses phosphorus to grow and reproduce.

Radioisotopes also can be placed in pesticides and followed to see what impact the pesticide has as it moves through an ecosystem. Plants, streams, insects, and animals can be tested to see how far the pesticides travel and how long they last in the ecosystem. Fertilizers containing small amounts of radioactive isotopes are used to see how well plants absorb fertilizers. Water resources can be measured and traced using isotopes, as well. This technique has been used in many developing countries that are located in arid regions as they search for sources of water.

Cell Division in Tumors When a person has cancer, cells reproduce rapidly, causing a tumor. When radiation is focused directly on the tumor, it can slow or stop the cell division while leaving healthy, surrounding tissue largely unaffected. Find out more about radiation therapy and summarize your findings in your Science Journal.

section 2 review

Summary

Identifying Numbers
- The atomic number is the number of protons in the nucleus of an atom.
- The mass number is the total of protons and neutrons in the nucleus of an atom.
- Isotopes of an element have different numbers of neutrons.

Radioactivity
- Radioactive decay is the release of nuclear particles and energy.
- Transmutation is the change of one element into another through radioactive decay. One form of transmutation is the loss of an alpha particle and energy from the nucleus. Another is the loss of a beta particle from the nucleus.
- Half-life of a radioactive isotope is the amount of time it takes for half of a sample of the element to decay.

Self Check

1. **Define** the term *isotope*. What must you know to calculate the number of neutrons in an isotope of an element?
2. **Compare and contrast** two types of radioactive decay.
3. **Infer** Do all elements have half-lives? Why or why not?
4. **Explain** how radioactive isotopes are used to detect health problems.
5. **Think Critically** Suppose you had two samples of the same radioactive isotope. One sample had a mass of 25 g. The other had a mass of 50 g. Would the same number of particles be ejected from each sample in the first hour? Explain.

Applying Skills

6. **Make Models** You have learned how scientists used marbles, modeling clay, and a cloud to model the atom. Describe the materials you might use to create one of the atomic models described in the chapter.

LAB Design Your Own

Half-Life

Goals
- **Model** isotopes in a radioactive sample. For each half-life, determine the amount of change that occurs in the objects that represent the isotopes in the model.

Possible Materials
pennies
graph paper

Design an experiment to test the usefulness of half-life in predicting how much radioactive material still remains after a specific number of half-lives.

● Real-World Question

The decay rates of most radioactive isotopes range from milliseconds to billions of years. If you know the half-life of an isotope and the size of a sample of the isotope, can you predict how much will remain after a certain amount of time? Is it possible to predict when a specific atom will decay? How can you use pennies to create a model that will show the amount of a radioactive isotope remaining after specific numbers of half-lives?

● Form a Hypothesis

Using the definition of the term *half-life* and pennies to represent atoms, write a hypothesis that shows how half-life can be used to predict how much of a radioactive isotope will remain after a certain number of half-lives.

Using Scientific Methods

▶ Test Your Hypothesis

Make a Plan

1. With your group, write the hypothesis statement.
2. **Write** down the steps of the procedure you will use to test your hypothesis. Assume that each penny represents an atom in a radioactive sample. Each coin that lands heads up after flipping has decayed.
3. **List** the materials you will need.
4. In your Science Journal, make a data table with two columns. Label one Half-Life and the other Atoms Remaining.
5. **Decide** how you can use the pennies to represent the radioactive decay of an isotope.
6. **Determine** (a) what will represent one half-life in your model, and (b) how many half-lives you will investigate.
7. **Decide** (a) which variables your model will have, and (b) which variable will be represented on the *y*-axis of your graph and which will be represented on the *x*-axis.

Follow Your Plan

1. Make sure your teacher approves your plan and your data table before you start.
2. Carry out your plan and record your data carefully.

▶ Analyze Your Data

The relationship among the starting number of pennies, the number of pennies remaining *(y)*, and the number of half-lives *(x)* is shown in the following equation:

$$y = \frac{\text{(starting number of pennies)}}{2^x}$$

1. **Graph** this equation using a graphing calculator. Use your graph to find the number of pennies remaining after 2.5 half-lives.
2. **Compare** the results of your activity and your graph with those of other groups.

▶ Conclude and Apply

1. Is it possible to use your model to predict which individual atoms will decay during one half-life? Why or why not?
2. Can you predict the total number of atoms that will decay in one half-life? Explain.

Communicating Your Data

Display your data again using a bar graph. For more help, refer to the Science Skill Handbook.

LAB **425**

TIME SCIENCE AND HISTORY

SCIENCE CAN CHANGE THE COURSE OF HISTORY!

Pioneers in Radioactivity

A Surprise on a Cloudy Day

Most scientific discoveries are the result of meticulous planning. Others happen quite by accident. On a cloudy day in the spring of 1896, physicist Henri Becquerel was unable to complete the day's planned work requiring the sun as the primary energy source. Disappointed, he wrapped his experimental photographic plates and put them away in a darkened drawer along with some crystals containing uranium. Imagine Becquerel's surprise upon discovering that the covered plates had somehow been exposed in complete darkness! The unplanned discovery that uranium emits radiation ultimately led to a complete revision of theories about atomic structure and properties.

Marie Curie's Revolutionary Hypothesis

One year before this revolutionary event, physicist Wilhelm Roentgen discovered a type of ray that could penetrate flesh, yielding photographs of living people's bones. Were these "X" rays, as Roentgen named them, and the radiation emitted by uranium in any way related? Intrigued by these findings, scientist Marie Curie began studying uranium compounds. Her research led her to hypothesize that radiation is an atomic property of matter which causes atoms of some elements to emit radiation, changing into atoms of another element. Her revolutionary hypothesis challenged current beliefs that the atom was indivisible and unchangeable.

"The Miserable Old Shed"

Marie Curie's husband became interested in her research, shelving his own magnetism studies to partner with her. Together, in the laboratory she referred to as "the miserable old shed," they experimented with a uranium ore called pitchblende. Strangely, pitchblende proved to be more radioactive than pure uranium. The Curies hypothesized that one or more undiscovered radioactive elements must also be part of this ore. By eventually isolating the elements radium and polonium from pitchblende, they achieved the dream of every scientist of the day: adding elements to the periodic table. In 1903, Marie and Pierre Curie shared the Nobel prize in physics with Henri Becquerel for contributions made through radiation research. The first female recipient of a Nobel prize, Marie Curie was awarded a second Nobel in 1911 in chemistry for her work with radium and radium compounds.

Investigate Research the work of Ernest Rutherford, who won the Nobel prize in Chemistry in 1908. Use the link to the right to describe some of his discoveries dealing with transmutation, radiation, and atomic structure.

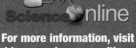

For more information, visit blue.msscience.com/time

chapter 14 Study Guide

Reviewing Main Ideas

Section 1 Models of the Atom

1. John Dalton proposed that an atom is a sphere of matter.
2. J.J. Thomson discovered that all atoms contain electrons.
3. Rutherford hypothesized that almost all the mass and all the positive charge of an atom is concentrated in an extremely tiny nucleus at the center of the atom.
4. Today's model of the atom has a concentrated nucleus containing the protons and neutrons surrounded by a cloud representing where the electrons are likely present.

Section 2 The Nucleus

1. The number of protons in the nucleus of an atom is its atomic number.
2. Isotopes are atoms of the same elements that have different numbers of neutrons. Each isotope has a different mass number.
3. An atom's nucleus is held together by the strong nuclear force.
4. Some nuclei decay by ejecting an alpha particle. Other nuclei decay by emitting a beta particle.
5. Half-life is a measure of the decay rate of a nucleus.

Visualizing Main Ideas

Copy and complete the following concept map about the parts of the atom.

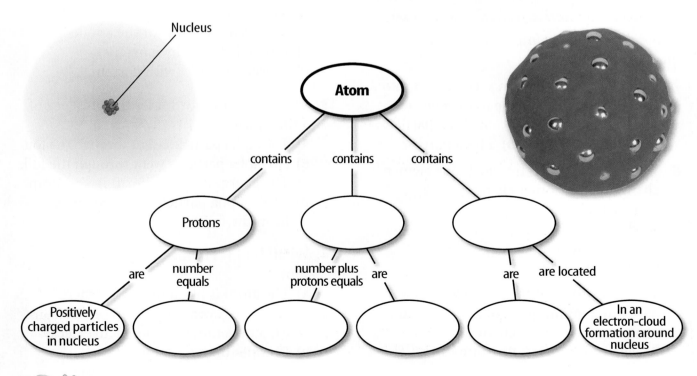

chapter 14 Review

Using Vocabulary

alpha particle p. 408	half-life p. 418
anode p. 406	isotope p. 415
atomic number p. 415	mass number p. 416
beta particle p. 418	neutron p. 411
cathode p. 406	proton p. 410
electron p. 407	radioactive decay p. 416
electron cloud p. 413	transmutation p. 416
element p. 405	

Each phrase below describes a science term from the list. Write the term that matches the phrase describing it.

1. a nuclear particle with no charge
2. a substance made up of only one type of atom
3. the number of protons and neutrons in the nucleus of an atom
4. a negatively charged particle
5. the release of nuclear particles and energy
6. the number of protons in an atom

Checking Concepts

Choose the word or phrase that best answers the question.

7. In beta decay, a neutron is converted into a proton and which of the following?
    A) an isotope   C) an alpha particle
    B) a nucleus   D) a beta particle

8. What is the process by which one element changes into another element?
    A) half-life
    B) chemical reaction
    C) chain reaction
    D) transmutation

9. What are atoms of the same element that have different numbers of neutrons called?
    A) protons   C) ions
    B) electrons D) isotopes

Use the illustration below to answer questions 10 and 11.

Boron nucleus

10. What is the atomic number equal to?
    A) energy levels   C) neutrons
    B) protons   D) nuclear particles

11. If the atomic number of boron is 5, boron-11 contains
    A) 11 electrons.
    B) five neutrons.
    C) five protons and six neutrons.
    D) six protons and five neutrons.

12. How did Thomson know that the glow in the CRT was from a stream of charged particles?
    A) It was green.
    B) It caused a shadow of the anode.
    C) It was deflected by a magnet.
    D) It occurred only with current.

13. Why did Rutherford infer the presence of a tiny nucleus?
    A) The alpha particles went through the foil.
    B) No alpha particles went through the foil.
    C) The charges were uniform in the atom.
    D) Some alpha particles bounced back from the foil.

14. What did J.J. Thomson's experiment show?
    A) The atom is like a uniform sphere.
    B) Cathode rays are made up of electrons.
    C) All atoms undergo radioactive decay.
    D) Isotopes undergo radioactive decay.

428 CHAPTER REVIEW   blue.msscience.com/vocabulary_puzzlemaker

chapter 14 Review

Thinking Critically

15. **Explain** how it is possible for two atoms of the same element to have different masses.

16. **Explain** Matter can't be created or destroyed, but could the amounts of some elements in Earth's crust decrease? Increase?

17. **Describe** why a neutral atom has the same number of protons and electrons.

18. **Compare and contrast** Dalton's model of the atom to today's model of the atom.

Use the figure below to answer question 19.

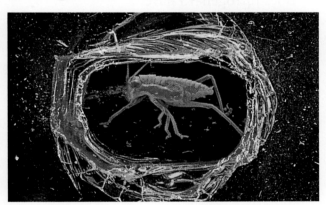

19. **Explain** how carbon-14 dating can provide the age of a dead animal or plant.

20. **Predict** If radium-226 releases an alpha particle, what is the mass number of the isotope formed?

21. **Concept Map** Make a concept map of the development of the theory of the atom.

22. **Predict** Given that the mass number of an isotope of mercury is 201, how many protons does it contain? Neutrons?

23. **Draw Conclusions** An experiment resulted in the release of a beta particle and an isotope of curium. What element was present at the beginning of this experiment?

Performance Activities

24. **Poster** Make a poster explaining one of the early models of the atom. Present this to your class.

25. **Game** Invent a game that illustrates radioactive decay.

Applying Math

26. **Half Life** A radioactive isotope has a half-life of two years. At the end of four years, how much of the original isotope remains?
    - **A)** one half
    - **B)** one fourth
    - **C)** one third
    - **D)** none

Use the graph below to answer question 27.

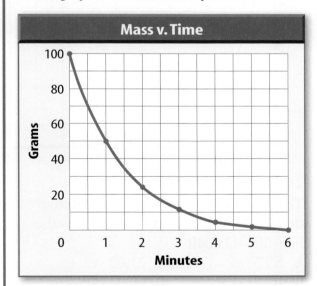

27. **Radioactive Decay** The radioactive decay of an isotope is plotted in the graph. What is the half-life of the isotope? How many grams of the isotope remain after three half-lives?

28. **Mass Number** An atom of rhodium-100 (^{100}Rh) has
    - **A)** 45 protons, 45 neutrons, 45 electrons.
    - **B)** 45 protons, 55 neutrons, 45 electrons.
    - **C)** 55 protons, 45 neutrons, 45 electrons.
    - **D)** 55 protons, 45 neutrons, 55 electrons.

Chapter 14 Standardized Test Practice

Part 1 | Multiple Choice

Record your answers on the answer sheet provided by your teacher or on a sheet of paper.

1. Which of the following is not an element?
   A. iron   C. steel
   B. carbon   D. oxygen

Use the graph below to answer questions 2 and 3.

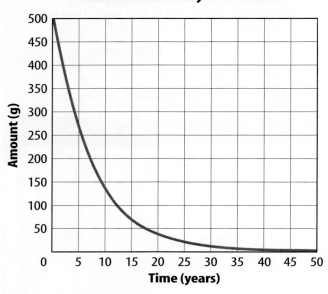

Radioactive Decay of Cobalt-60

2. The graph above shows the radioactive decay of 500 g of cobalt-60. What is the half-life of cobalt-60?
   A. 5.27 years   C. 21.08 years
   B. 10.54 years   D. 60.0 years

3. About how much of the original 500 g of cobalt-60 will be left after 20 years?
   A. 30 g   C. 90 g
   B. 60 g   D. 120 g

Test-Taking Tip

Use Recall Remember to recall any hands-on experience as you read the question. Base your answer on the information given on the test. What did you do in the pennies experiment?

Use the table below to answer questions 4 and 5.

Isotopes of Nitrogen		
Isotope	Mass Number	Number of Protons
Nitrogen-12	12	7
Nitrogen-13	13	7
Nitrogen-14	14	7
Nitrogen-15	15	7

4. The table above shows properties of some nitrogen isotopes. How many neutrons does nitrogen-15 have?
   A. 7   C. 14
   B. 8   D. 15

5. Which of the isotopes listed in the table would you expect to be the least stable?
   A. nitrogen-15   C. nitrogen-13
   B. nitrogen-14   D. nitrogen-12

6. Which of the following is the smallest?
   A. electron   C. proton
   B. nucleus   D. neutron

7. What is the heaviest naturally occurring element?
   A. Ac   C. Po
   B. Am   D. U

8. Ruthenium has an atomic number of 44 and a mass number of 101. How many protons does ruthenium have?
   A. 44   C. 88
   B. 57   D. 101

9. Which of the following could not be dated using carbon-14 dating?
   A. wooden bowl   C. bone fragments
   B. plant remains   D. rock tools

10. What is all matter made of?
    A. dust   C. atoms
    B. sun rays   D. metal alloys

Standardized Test Practice

Part 2 | Short Response/Grid In

Record your answers on the answer sheet provided by your teacher or on a sheet of paper.

11. What is an element?

12. What is the modern-day name for cathode rays?

Use the illustration below to answer questions 13 and 14.

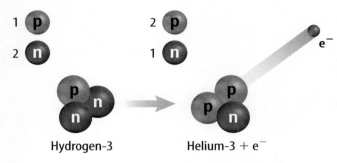

13. The figure above illustrates the beta decay of hydrogen-3 into helium-3 and an electron. What is a beta particle? From what part of the atom do beta particles originate?

14. Describe the transmutation that occurs during the beta decay shown in the illustration.

15. Describe Thomson's idea about the composition of an atom.

16. Are electrons more likely to be close to the nucleus or far away from the nucleus? Why?

17. Cesium-137 has a half-life of 30.3 years. If you start with 60 g, how much cesium would be left after 90.9 years?

18. Explain how the half-life of C-14 is used to date American Indian, Greek, and Roman artifacts, but cannot be used to date fossil remains from the Cretaceous Period.

Part 3 | Open Ended

Record your answers on a sheet of paper.

Use the illustration below to answer questions 19 and 20.

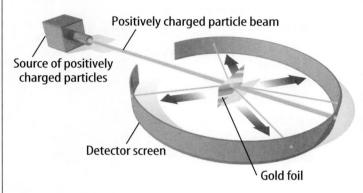

19. The illustration above shows Rutherford's gold foil experiment. Describe the setup shown. What result did Rutherford expect from his experiment?

20. What is the significance of the particles that reflected back from the gold foil? How did Rutherford explain his results?

21. Describe Dalton's ideas about the composition of matter, including the relationship between atoms and elements.

22. Describe the discovery of cathode rays.

23. Describe how was able to show Thomson that cathode rays were streams of particles, not light.

24. Some smoke detectors contain small radioactive sources. Explain how these detectors use radioactive decay to detect smoke.

25. Manganese-54 has a half-life of about 312 days. Draw a graph of the radioactive decay of a 600-g sample of manganese-54.

26. Describe the uses of radioactive elements in medicine, agriculture, and industry.

chapter 15

The Periodic Table

The BIG Idea

The periodic table provides information about all of the known elements.

SECTION 1
Introduction to the Periodic Table
Main Idea Elements are arranged in order of increasing atomic number on the periodic table.

SECTION 2
Representative Elements
Main Idea Representative elements within a group have similar properties.

SECTION 3
Transition Elements
Main Idea Transition elements are metals with a wide variety of uses.

Skyscrapers, Neon Lights, and the Periodic Table

Many cities have unique skylines. What is truly amazing is that everything in this photograph is made from 90 naturally occurring elements. In this chapter, you will learn more about the elements and the table that organizes them.

Science Journal Think of an element that you have heard about. Make a list of the properties you know and the properties you want to learn about.

Start-Up Activities

Make a Model of a Periodic Pattern

Every 29.5 days, the Moon begins to cycle through its phases from full moon to new moon and back again to full moon. Events that follow a predictable pattern are called periodic events. What other periodic events can you think of?

1. On a blank sheet of paper, make a grid with four squares across and four squares down.
2. Your teacher will give you 16 pieces of paper with different shapes and colors. Identify properties you can use to distinguish one piece of paper from another.
3. Place a piece of paper in each square on your grid. Arrange the pieces on the grid so that each column contains pieces that are similar.
4. Within each column, arrange the pieces to show a gradual change in their appearance.
5. **Think Critically** In your Science Journal, describe how the properties change in the rows across the grid and in the columns down the grid.

 Preview this chapter's content and activities at blue.msscience.com

FOLDABLES Study Organizer

Periodic Table Make the following Foldable to help you classify the elements in the periodic table as metals, nonmetals, and metalloids.

STEP 1 Fold a vertical sheet of paper from side to side. Make the front edge about 1.25 cm shorter than the back edge.

STEP 2 Turn lengthwise and fold into thirds.

STEP 3 Unfold and cut only the top layer along both folds to make three tabs. Label each tab as shown.

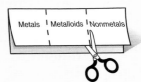

Find Main Ideas As you read the chapter, write information about the three types of elements under the appropriate tabs. Use the information in your Foldable to explain how metalloids have properties between those of metals and nonmetals.

Get Ready to Read

Make Connections

① Learn It! Make connections between what you read and what you already know. Connections can be based on personal experiences (text-to-self), what you have read before (text-to-text), or events in other places (text-to-world).

As you read, ask connecting questions. Are you reminded of a personal experience? Have you read about the topic before? Did you think of a person, a place, or an event in another part of the world?

② Practice It! Read the excerpt below and make connections to your own knowledge and experience.

Text-to-self: What metals do you use every day?

Text-to-text: What have you read about melting points before?

Text-to-world: Have you heard about mercury in the news or seen a mercury thermometer?

> If you look at the periodic table, you will notice it is color coded. The colors represent elements that are metals, nonmetals, or metalloids. With the exception of mercury, all the metals are solids, most with high melting points. A metal is an element that has luster, is a good conductor of heat and electricity, is malleable, and is ductile.
>
> —from page 438

③ Apply It! As you read this chapter, choose five words or phrases that make a connection to something you already know.

Target Your Reading

Use this to focus on the main ideas as you read the chapter.

① Before you read the chapter, respond to the statements below on your worksheet or on a numbered sheet of paper.
- Write an **A** if you **agree** with the statement.
- Write a **D** if you **disagree** with the statement.

② After you read the chapter, look back to this page to see if you've changed your mind about any of the statements.
- If any of your answers changed, explain why.
- Change any false statements into true statements.
- Use your revised statements as a study guide.

Reading Tip: Make connections with memorable events, places, or people in your life. The better the connection, the more likely you will be to remember it.

Science Online
Print out a worksheet of this page at blue.msscience.com

Before You Read A or D		Statement	After You Read A or D
	1	Scientists have discovered all the elements that could possibly exist.	
	2	The elements are arranged on the periodic table according to their atomic numbers and mass numbers.	
	3	Elements in a group have similar properties.	
	4	Metals are located on the right side of the periodic table.	
	5	When a new element is discovered, the IUPAC selects a name.	
	6	Only metals conduct electricity.	
	7	Noble gases rarely combine with other elements.	
	8	The transition elements contain metals, nonmetals, and metalloids.	
	9	Some elements are created in a lab.	

section 1

Introduction to the Periodic Table

as you read

What You'll Learn
- **Describe** the history of the periodic table.
- **Interpret** an element key.
- **Explain** how the periodic table is organized.

Why It's Important
The periodic table makes it easier for you to find information that you need about the elements.

Review Vocabulary
element: a substance that cannot be broken down into simpler substances

New Vocabulary
- period
- group
- representative element
- transition element
- metal
- nonmetal
- metalloid

Development of the Periodic Table

Early civilizations were familiar with a few of the substances now called elements. They made coins and jewelry from gold and silver. They also made tools and weapons from copper, tin, and iron. In the nineteenth century, chemists began to search for new elements. By 1830, they had isolated and named 55 different elements. The list continues to grow today.

Mendeleev's Table of Elements A Russian chemist, Dmitri Mendeleev (men duh LAY uhf), published the first version of his periodic table in the *Journal of the Russian Chemical Society* in 1869. His table is shown in **Figure 1.** When Mendeleev arranged the elements in order of increasing atomic mass, he began to see a pattern. Elements with similar properties fell into groups on the table. At that time, not all the elements were known. To make his table work, Mendeleev had to leave three gaps for missing elements. Based on the groupings in his table, he predicted the properties for the missing elements. Mendeleev's predictions spurred other chemists to look for the missing elements. Within 15 years, all three elements—gallium, scandium, and germanium—were discovered.

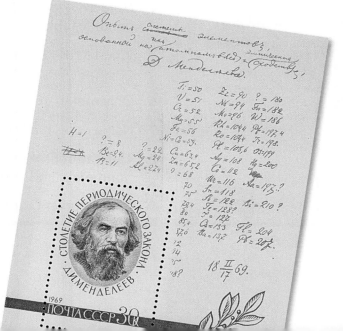

Figure 1 Mendeleev published his first periodic table in 1869. This postage stamp, with his table and photo, was issued in 1969 to commemorate the event. Notice the question marks that he used to mark his prediction of yet-undiscovered elements.

434 CHAPTER 15 The Periodic Table

Moseley's Contribution Although Mendeleev's table correctly organized most of the elements, a few elements seemed out of place. In the early twentieth century, the English physicist Henry Moseley, before age 27, realized that Mendeleev's table could be improved by arranging the elements according to atomic number rather than atomic mass. Moseley revised the periodic table by arranging the elements in order of increasing number of protons in the nucleus. With Moseley's table, it was clear how many elements still were undiscovered.

Today's Periodic Table

In the modern periodic table on the next page, the elements still are organized by increasing atomic number. The rows or periods are labeled 1–7. A **period** is a row of elements in the periodic table whose properties change gradually and predictably. The periodic table has 18 columns of elements. Each column contains a group, or family, of elements. A **group** contains elements that have similar physical or chemical properties.

Zones on the Periodic Table The periodic table can be divided into sections, as you can see in **Figure 2.** One section consists of the first two groups, Groups 1 and 2, and the elements in Groups 13–18. These eight groups are the **representative elements.** They include metals, metalloids, and nonmetals. The elements in Groups 3–12 are **transition elements.** They are all metals. Some transition elements, called the inner transition elements, are placed below the main table. These elements are called the lanthanide and actinide series because one series follows the element lanthanum, element 57, and the other series follows actinium, element 89.

Mini LAB

Designing a Periodic Table

Procedure
1. Collect **pens** and **pencils** from everyone in your class.
2. Decide which properties of the pens and pencils you will use to organize them into a periodic table. Consider properties such as color, mass, or length. Then create your table.

Analysis
1. Explain how your periodic table is similar to the periodic table of the elements.
2. If your classmates brought different pens or pencils to class tomorrow, how would you organize them on your periodic table?

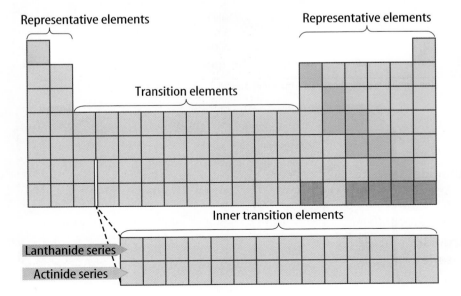

Figure 2 The periodic table is divided into sections. Traditionally, the lanthanides and actinides are placed below the table so that the table will not be as wide. These elements have similar properties. **Identify** *the section of the periodic table that contains only metal.*

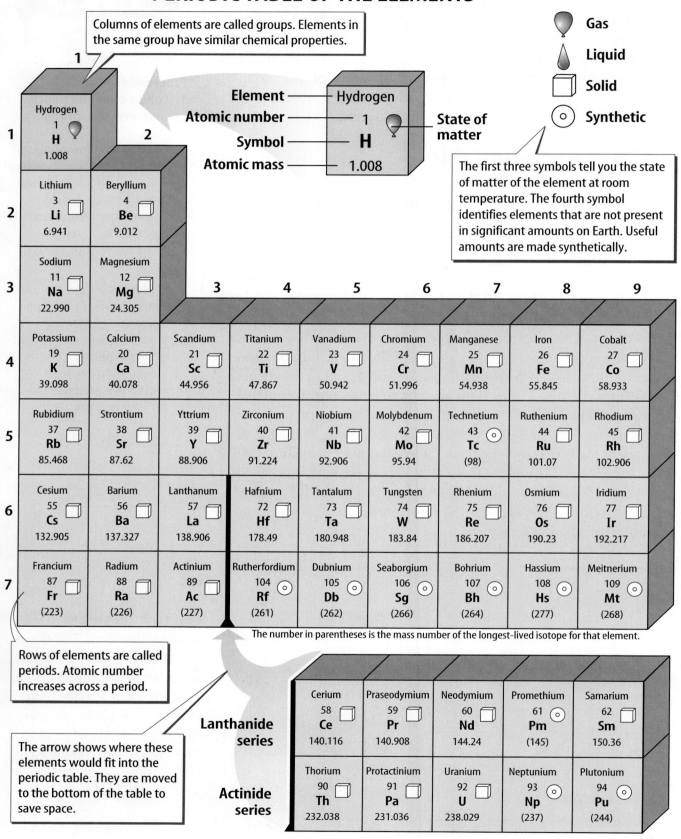

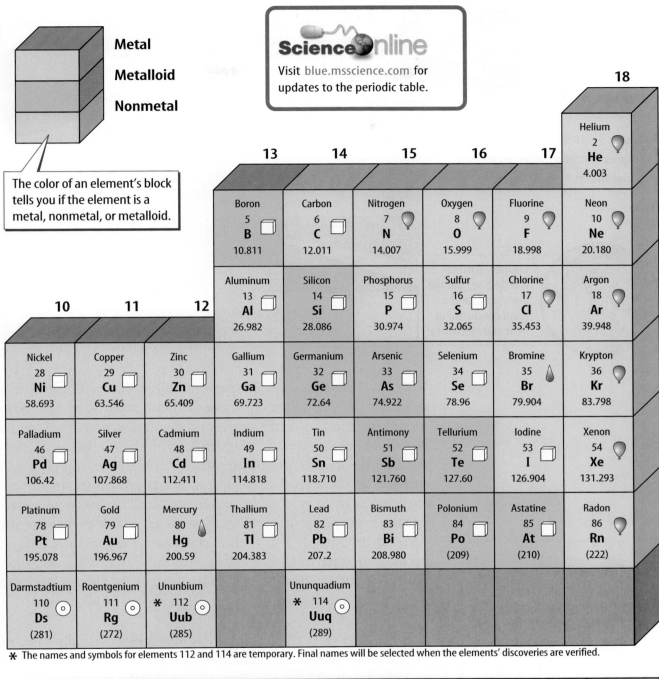

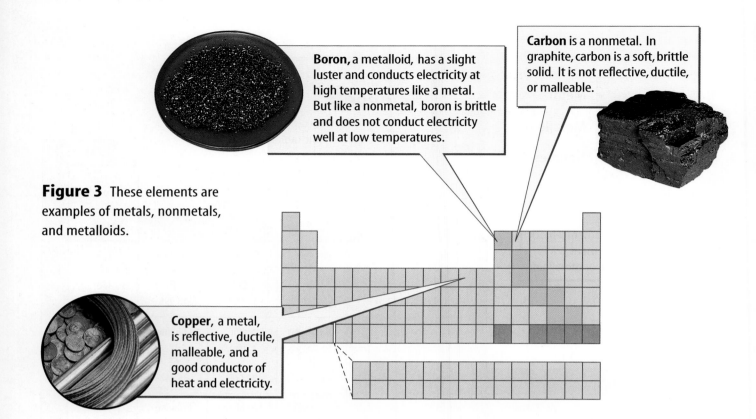

Figure 3 These elements are examples of metals, nonmetals, and metalloids.

Boron, a metalloid, has a slight luster and conducts electricity at high temperatures like a metal. But like a nonmetal, boron is brittle and does not conduct electricity well at low temperatures.

Carbon is a nonmetal. In graphite, carbon is a soft, brittle solid. It is not reflective, ductile, or malleable.

Copper, a metal, is reflective, ductile, malleable, and a good conductor of heat and electricity.

Metals If you look at the periodic table, you will notice it is color coded. The colors represent elements that are metals, nonmetals, or metalloids. Examples of a metal, a nonmetal, and a metalloid are illustrated in **Figure 3.** With the exception of mercury, all the metals are solids, most with high melting points. A **metal** is an element that has luster, is a good conductor of heat and electricity, is malleable, and is ductile. The ability to reflect light is a property of metals called luster. Many metals can be pressed or pounded into thin sheets or shaped into objects because they are malleable (MAL yuh bul). Metals are also ductile (DUK tul), which means that they can be drawn out into wires. Can you think of any items that are made of metals?

Nonmetals and Metalloids **Nonmetals** are usually gases or brittle solids at room temperature and poor conductors of heat and electricity. There are only 17 nonmetals, but they include many elements that are essential for life—carbon, sulfur, nitrogen, oxygen, phosphorus, and iodine.

The elements between metals and nonmetals on the periodic table are called metalloids (ME tuh loydz). As you might expect from the name, a **metalloid** is an element that shares some properties with metals and some with nonmetals. These elements also are called semimetals.

Science Online

Topic: Elements
Visit blue.msscience.com for Web links to information about how the periodic table was developed.

Activity Select an element and write about how, when, and by whom it was discovered.

Reading Check *How many elements are nonmetals?*

The Element Keys Each element is represented on the periodic table by a box called the element key. An enlarged key for hydrogen is shown in **Figure 4.** An element key shows you the name of the element, its atomic number, its symbol, and its average atomic mass. Element keys for elements that occur naturally on Earth include a logo that tells whether the element is a solid, a liquid, or a gas at room temperature. All the gases except hydrogen are on the right side of the table. They are marked with a balloon logo. Most of the other elements are solids at room temperature and are marked with a cube. Two elements on the periodic table are liquids at room temperature. Their logo is a drop. Elements that do not occur naturally on Earth are marked with a bull's-eye logo. These are synthetic elements.

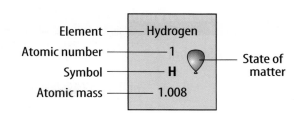

Figure 4 As you can see from the element key, a lot of information about an element is given on the periodic table.
Identify *the two elements that are liquids at room temperature.*

Applying Science

What does *periodic* mean in the periodic table?

Elements often combine with oxygen to form oxides and chlorine to form chlorides. For example, two hydrogen atoms combine with one oxygen atom to form oxide, H_2O or water. One sodium atom combines with one chlorine atom to form sodium chloride, NaCl or table salt. The location of an element on the periodic table is an indication of how it combines with other elements.

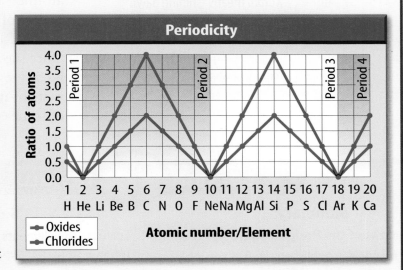

Identifying the Problem
The graph shows the number of oxygen atoms (red) and chlorine atoms (green) that will combine with the first 20 elements. What pattern do you see?

Solving the Problem
1. Find all of the elements in Group 1 on the graph. Do the same with the elements in Groups 14 and 18. What do you notice about their positions on the graph?
2. This relationship demonstrates one of the properties of a group of elements. Follow the elements in order on the periodic table and on the graph. Write a statement using the word *periodic* that describes what occurs with the elements and their properties.

SECTION 1 Introduction to the Periodic Table

Table 1 Chemical Symbols and Their Origins

Name	Symbol	Origin of Name
Mendelevium	Md	For Dimitri Mendeleev
Lead	Pb	The Latin name for lead is *plumbum*.
Thorium	Th	The Norse god of thunder is Thor.
Polonium	Po	For Poland, where Marie Curie, a famous scientist, was born
Hydrogen	H	From Greek words meaning "water former"
Mercury	Hg	*Hydrargyrum* means "liquid silver" in Greek.
Gold	Au	*Aurum* means "shining dawn" in Latin.
Unununium	Uuu	Named using the IUPAC naming system

Symbols for the Elements The symbols for the elements are either one- or two-letter abbreviations, often based on the element name. For example, V is the symbol for vanadium, and Sc is the symbol for scandium. Sometimes the symbols don't match the names. Examples are Ag for silver and Na for sodium. In those cases, the symbol might come from Greek or Latin names for the elements. Some elements are named for scientists such as Lise Meitner (meitnerium, Mt). Some are named for geographic locations such as France (francium, Fr).

Newly synthesized elements are given a temporary name and 3-letter symbol that is related to the element's atomic number. The International Union of Pure and Applied Chemistry (IUPAC) adopted this system in 1978. Once the discovery of the element is verified, the discoverers can choose a permanent name. **Table 1** shows the origin of some element names and symbols.

section 1 review

Summary

Development of the Periodic Table
- Dmitri Mendeleev published the first version of the periodic table in 1869.
- Mendeleev left three gaps on the periodic table for missing elements.
- Moseley arranged Mendeleev's table according to atomic number, not by atomic mass.

Today's Periodic Table
- The periodic table is divided into sections.
- A period is a row of elements whose properties change gradually and predictably.
- Groups 1 and 2 along with Groups 13–18 are called representative elements.
- Groups 3–12 are called transition elements.

Self Check

1. **Evaluate** the elements in period 4 to show how the physical state changes as the atomic number increases.
2. **Describe** where the metals, nonmetals, and metalloids are located in the periodic table.
3. **Classify** each of the following elements as metal, nonmetal, or metalloid: Fe, Li, B, Cl, Si, Na, and Ni.
4. **List** what an element key contains.
5. **Think Critically** How would the modern periodic table be different if elements were arranged by average atomic mass instead of by atomic number?

Applying Math

6. **Solve One-Step Equations** What is the difference in atomic mass of iodine and magnesium?

section 2
Representative Elements

Groups 1 and 2

Groups 1 and 2 are always found in nature combined with other elements. They're called active metals because of their readiness to form new substances with other elements. They are all metals except hydrogen, the first element in Group 1. Although hydrogen is placed in Group 1, it shares properties with the elements in Group 1 and Group 17.

Alkali Metals The Group 1 elements have a specific family name—**alkali metals.** All the alkali metals are silvery solids with low densities and low melting points. These elements increase in their reactivity, or tendency to combine with other substances, as you move from top to bottom on the periodic table. Some uses of the alkali metals are shown in **Figure 5**.

Alkali metals are found in many items. Lithium batteries are used in cameras. Sodium chloride is common table salt. Sodium and potassium, dietary requirements, are found in small quantities in potatoes and bananas.

as you read

What **You'll Learn**
- **Recognize** the properties of representative elements.
- **Identify** uses for the representative elements.
- **Classify** elements into groups based on similar properties.

Why **It's Important**
Many representative elements play key roles in your body, your environment, and in the things you use every day.

Review Vocabulary
atomic number: the number of protons in the nucleus of a given element

New Vocabulary
- alkali metal
- alkaline earth metal
- semiconductor
- halogen
- noble gas

Figure 5 These items contain alkali metals.

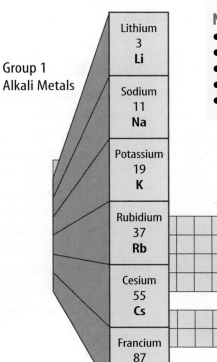

Group 1 Alkali Metals

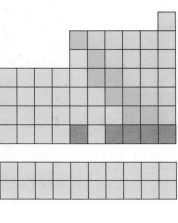

SECTION 2 Representative Elements **441**

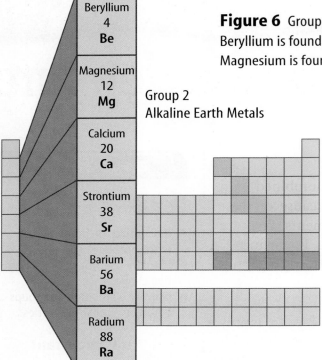

Figure 6 Group 2 elements are found in many things. Beryllium is found in the gems emerald and aquamarine. Magnesium is found in the chlorophyll of green plants.

Alkaline Earth Metals Next door to the alkali metals' family are their Group 2 neighbors, the **alkaline earth metals.** Each alkaline earth metal is denser and harder and has a higher melting point than the alkali metal in the same period. Alkaline earth metals are reactive, but not as reactive as the alkali metals. Some uses of the alkaline earth elements are shown in **Figure 6.**

 What are the names of the elements that are alkaline earth metals?

Groups 13 through 18

Notice on the periodic table that the elements in Groups 13–18 are not all solid metals like the elements of Groups 1 and 2. In fact, a single group can contain metals, nonmetals, and metalloids and have members that are solids, liquids, and gases.

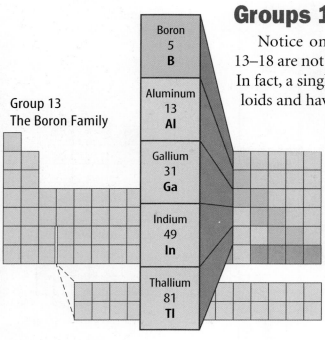

Group 13—The Boron Family The elements in Group 13 are all metals except boron, which is a brittle, black metalloid. This family of elements is used to make a variety of products. Cookware made with boron can be moved directly from the refrigerator into the oven without cracking. Aluminum is used to make soft-drink cans, cookware, siding for homes, and baseball bats. Gallium is a solid metal, but its melting point is so low that it will melt in your hand. It is used to make computer chips.

Group 14—The Carbon Group If you look at Group 14, you can see that carbon is a nonmetal, silicon and germanium are metalloids, and tin and lead are metals. The nonmetal carbon exists as an element in several forms. You're familiar with two of them—diamond and graphite. Carbon also is found in all living things. Carbon is followed by the metalloid silicon, an abundant element contained in sand. Sand contains ground-up particles of minerals such as quartz, which is composed of silicon and oxygen. Glass is an important product made from sand.

Silicon and its Group 14 neighbor, germanium, are metalloids. They are used in electronics as semiconductors. A **semiconductor** doesn't conduct electricity as well as a metal, but does conduct electricity better than a nonmetal. Silicon and small amounts of other elements are used for computer chips as shown in **Figure 7**.

Tin and lead are the two heaviest elements in Group 14. Lead is used in the apron, shown in **Figure 7,** to protect your torso during dental X rays. It also is used in car batteries, low-melting alloys, protective shielding around nuclear reactors, particle accelerators, X-ray equipment, and containers used for storing and transporting radioactive materials. Tin is used in pewter, toothpaste, and the coating on steel cans used for food.

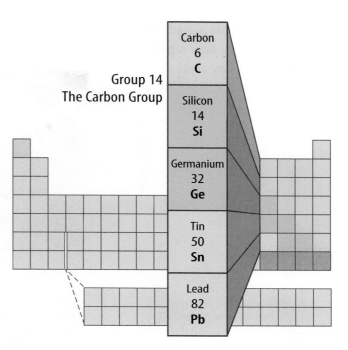

Figure 7 Members of Group 14 include one nonmetal, two metalloids, and two metals.

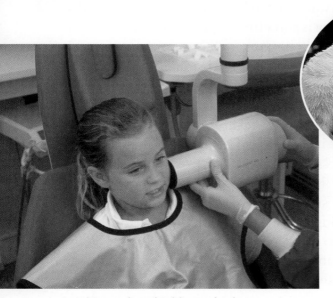

Lead is used to shield your body from unwanted X-ray exposure.

All living things contain carbon compounds.

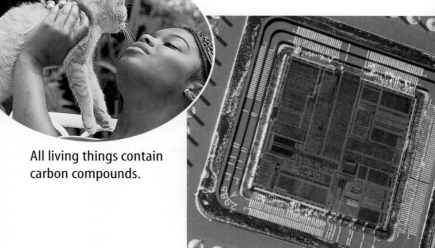

Silicon crystals are used to make computer chips.

SECTION 2 Representative Elements

Figure 8 Ammonia is used to make nylon, a tough, light fiber capable of replacing silk in many applications, including parachutes.

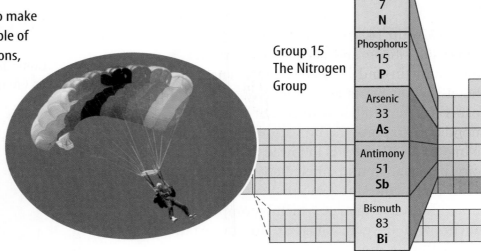

Group 15—The Nitrogen Group At the top of Group 15 are the two nonmetals—nitrogen and phosphorus. Nitrogen and phosphorus are required by living things and are used to manufacture various items. These elements also are parts of the biological materials that store genetic information and energy in living organisms. Although almost 80 percent of the air you breathe is nitrogen, you can't get the nitrogen your body needs by breathing nitrogen gas. Bacteria in the soil must first change nitrogen gas into substances that can be absorbed through the roots of plants. Then, by eating the plants, nitrogen becomes available to your body.

Farmers Each year farmers test their soil to determine the level of nutrients, the matter needed for plants to grow. The results of the test help the farmer decide how much nitrogen, phosphorus, and potassium to add to the fields. The additional nutrients increase the chance of having a successful crop.

Reading Check *Can your body obtain nitrogen by breathing air? Explain.*

Ammonia is a gas that contains nitrogen and hydrogen. When ammonia is dissolved in water, it can be used as a cleaner and disinfectant. Liquid ammonia is sometimes applied directly to soil as a fertilizer. Ammonia also can be converted into solid fertilizers. It also is used to freeze-dry food and as a refrigerant. Ammonia also is used to make nylon for parachutes, as shown in **Figure 8.**

The element phosphorus comes in two forms—white and red. White phosphorus is so active it can't be exposed to oxygen in the air or it will burst into flames. The heads of matches contain the less active red phosphorus, which ignites from the heat produced by friction when the match is struck. Phosphorous compounds are essential ingredients for healthy teeth and bones. Plants also need phosphorus, so it is one of the nutrients in most fertilizers. The fertilizer label in **Figure 9** shows the compounds of nitrogen and phosphorus that are used to give plants a synthetic supply of these elements.

Figure 9 Nitrogen and phosphorus are required for healthy green plants. This synthetic fertilizer label shows the nitrogen and phosphorous compounds that provide these.

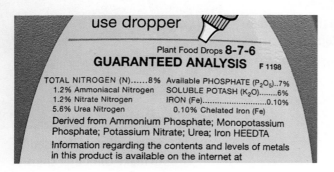

Group 16—The Oxygen Family The first two members of Group 16, oxygen and sulfur, are essential for life. The heavier members of the group, tellurium and polonium, are both metalloids.

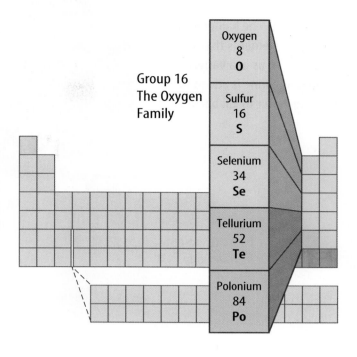

About 20 percent of Earth's atmosphere is the oxygen you breathe. Your body needs oxygen to release the energy from the foods you eat. Oxygen is abundant in Earth's rocks and minerals because it readily combines with other elements. Oxygen also is required for combustion to occur. Foam is used in fire fighting to keep oxygen away from the burning item, as shown in **Figure 10.** Ozone, a less common form of oxygen, is formed in the upper atmosphere through the action of electricity during thunderstorms. The presence of ozone is important because it shields living organisms from some harmful radiation from the Sun.

Sulfur is a solid, yellow nonmetal. Large amounts of sulfur are used to manufacture sulfuric acid, one of the most commonly used chemicals in the world. Sulfuric acid is a combination of sulfur, hydrogen, and oxygen. It is used in the manufacture of paints, fertilizers, detergents, synthetic fibers, and rubber.

Selenium conducts electricity when exposed to light, so it is used in solar cells, light meters, and photographic materials. Its most important use is as the light-sensitive component in photocopy machines. Traces of selenium are also necessary for good health.

Poison Buildup Arsenic disrupts the normal function of an organism by disrupting cellular metabolism. Because arsenic builds up in hair, forensic scientists can test hair samples to confirm or disprove a case of arsenic poisoning. Tests of Napoleon's hair suggest that he was poisoned with arsenic. Use reference books to find out who Napoleon I was and why someone might have wanted to poison him.

Figure 10 The foam used in aircraft fires forms a film of water over the burning fuel which suffocates the fire.

Figure 11 The halogens are a group of elements that are important to us in a variety of ways. Chlorine is added to drinking water to kill bacteria.

Iodine is needed by many systems in your body.

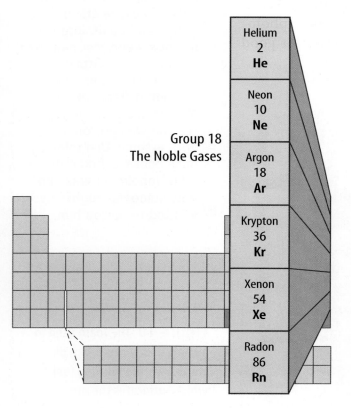

Group 17—The Halogen Group All the elements in Group 17 are nonmetals except for astatine, which is a radioactive metalloid. These elements are called **halogens**, which means "salt-former." Table salt, sodium chloride, is a substance made from sodium and chlorine. All of the halogens form similar salts with sodium and with the other alkali metals.

The halogen fluorine is the most reactive of the halogens in combining with other elements. Chlorine is less reactive than fluorine, and bromine is less reactive than chlorine. Iodine is the least reactive of the four nonmetals. **Figure 11** shows some uses of halogens.

Reading Check What do halogens form with the alkali metals?

Group 18—The Noble Gases The Group 18 elements are called the **noble gases.** This is because they rarely combine with other elements and are found only as uncombined elements in nature. Their reactivity is very low.

Helium is less dense than air, so it's great for all kinds of balloons, from party balloons to blimps that carry television cameras high above sporting events. Helium balloons, such as the one in **Figure 12,** lift instruments into the upper atmosphere to measure atmospheric conditions. Even though hydrogen is lighter than helium, helium is preferred for these purposes because helium will not burn.

446 CHAPTER 15 The Periodic Table

Uses for the Noble Gases The "neon" lights you see in advertising signs, like the one in **Figure 12,** can contain any of the noble gases, not just neon. Electricity is passed through the glass tubes that make up the sign. These tubes contain the noble gas, and the electricity causes the gas to glow. Each noble gas produces a unique color. Helium glows yellow, neon glows red-orange, and argon produces a bluish-violet color.

Argon, the most abundant of the noble gases on Earth, was first found in 1894. Krypton is used with nitrogen in ordinary lightbulbs because these gases keep the glowing filament from burning out. When a mixture of argon, krypton, and xenon is used, a bulb can last longer than bulbs that do not contain this mixture. Krypton lights are used to illuminate landing strips at airports, and xenon is used in strobe lights and was once used in photographic flash cubes.

At the bottom of the group is radon, a radioactive gas produced naturally as uranium decays in rocks and soil. If radon seeps into a home, the gas can be harmful because it continues to emit radiation. When people breathe the gas over a period of time, it can cause lung cancer.

Reading Check *Why are noble gases used in lights?*

Figure 12 Noble gases are used in many applications. Scientists use helium balloons to measure atmospheric conditions.

Each noble gas glows a different color when an electric current is passed through it.

section 2 review

Summary

Groups 1 and 2
- Groups 1 and 2 elements are always combined with other elements.
- The elements in Groups 1 and 2 are all metals except for hydrogen.
- Alkaline earth metals are not as active as the alkali metals.

Groups 13–18
- With Groups 13–18, a single group can contain metals, nonmetals, and metalloids.
- Nitrogen and phosphorus are required by living things.
- The halogen group will form salts with alkali metals.

Self Check

1. **Compare and contrast** the elements in Group 1 and the elements in Group 17.
2. **Describe** two uses for a member of each representative group.
3. **Identify** the group of elements that does not readily combine with other elements.
4. **Think Critically** Francium is a rare radioactive alkali metal at the bottom of Group 1. Its properties have not been studied carefully. Would you predict that francium would combine with water more or less readily than cesium?

Applying Skills

5. **Predict** how readily astatine would form a salt compared to the other elements in Group 17. Is there a trend for reactivity in this group?

SECTION 2 Representative Elements

section 3
Transition Elements

as you read

What You'll Learn
- **Identify** properties of some transition elements.
- **Distinguish** lanthanides from actinides.

Why It's Important
Transition elements provide the materials for many things including electricity in your home and steel for construction.

Review Vocabulary
mass number: the sum of neutrons and protons in the nucleus of an atom

New Vocabulary
- catalyst
- lanthanide
- actinide
- synthetic element

The Metals in the Middle

Groups 3–12 are called the transition elements and all of them are metals. Across any period from Group 3 through 12, the properties of the elements change less noticeably than they do across a period of representative elements.

Most transition elements are found combined with other elements in ores. A few transition elements such as gold and silver are found as pure elements.

The Iron Triad Three elements in period 4—iron, cobalt, and nickel—have such similar properties that they are known as the iron triad. These elements, among others, have magnetic properties. Industrial magnets are made from an alloy of nickel, cobalt, and aluminum. Nickel is used in batteries along with cadmium. Iron is a necessary part of hemoglobin, the substance that transports oxygen in the blood.

Iron also is mixed with other metals and with carbon to create a variety of steels with different properties. Structures such as bridges and skyscrapers, shown in **Figure 13,** depend upon steel for their strength.

Reading Check *Which metals make up the iron triad?*

Figure 13 These buildings and bridges have steel in their structure. **Explain** *why you think steel is used in their construction.*

The Iron Triad

| Iron 26 Fe | Cobalt 27 Co | Nickel 28 Ni |

Uses of Transition Elements Most transition metals have higher melting points than the representative elements. The filaments of lightbulbs, like the one in **Figure 14,** are made of tungsten, element 74. Tungsten has the highest melting point of any metal (3,410°C) and will not melt when a current passes through it.

Mercury, which has the lowest melting point of any metal (−39°C), is used in thermometers and in barometers. Mercury is the only metal that is a liquid at room temperatures. Like many of the heavy metals, mercury is poisonous to living beings. Therefore, mercury must be handled with care.

Chromium's name comes from the Greek word for color, *chroma,* and the element lives up to its name. Two substances containing chromium are shown in **Figure 15.** Many other transition elements combine to form substances with equally brilliant colors.

Ruthenium, rhodium, palladium, osmium, iridium, and platinum are sometimes called the platinum group because they have similar properties. They do not combine easily with other elements. As a result, they can be used as catalysts. A **catalyst** is a substance that can make something happen faster but is not changed itself. Other transition elements, such as nickel, zinc, and cobalt, can be used as catalysts. As catalysts, the transition elements are used to produce electronic and consumer goods, plastics, and medicines.

Figure 14 The transition metal tungsten is used in lightbulbs because of its high melting point.

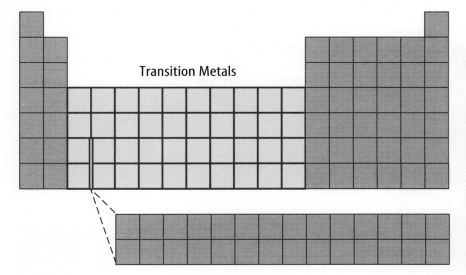

Figure 15 Transition metals are used in a variety of products.

Bright Lights Yttrium oxide (Y_2O_3) and europium oxide (Eu_2O_3) are used in color television screens to give a bright, natural color red. This blend of lanthanide elements will give off a red light when it's hit with a beam of electrons. Other compounds are used to make the additional colors required for a natural-looking picture.

Inner Transition Elements

There are two series of inner transition elements. The first series, from cerium to lutetium, is called the **lanthanides.** The lanthanides also are called the rare earths because at one time they were thought to be scarce. The lanthanides are usually found combined with oxygen in Earth's crust. The second series of elements, from thorium to lawrencium, is called the **actinides.**

Reading Check *What other name is used to refer to the lanthanides?*

The Lanthanides The lanthanides are soft metals that can be cut with a knife. The elements are so similar that they are hard to separate when they occur in the same ore, which they often do. Despite the name rare earth, the lanthanides are not as rare as originally thought. Earth's crust contains more cerium than lead. Cerium makes up 50 percent of an alloy called misch (MIHSH) metal. Flints in lighters, like the one in **Figure 16,** are made from misch metal. The other ingredients in flint are lanthanum, neodymium, and iron.

The Actinides All the actinides are radioactive. The nuclei of atoms of radioactive elements are unstable and decay to form other elements. Thorium, protactinium, and uranium are the only actinides that now are found naturally on Earth. Uranium is found in Earth's crust because its half-life is long—4.5 billion years. All other actinides are synthetic elements. **Synthetic elements** are made in laboratories and nuclear reactors. **Figure 17** shows how synthetic elements are made. The synthetic elements have many uses. Plutonium is used as a fuel in nuclear power plants. Americium is used in some home smoke detectors. Californium-252 is used to kill cancer cells.

Reading Check *What property do all actinides share?*

Figure 16 The flint in this lighter is called misch metal, which is about 50% cerium, 25% lanthanum, 15% neodymium, and 10% other rare earth metals and iron.

Lanthanide Series	58 Ce	59 Pr	60 Nd	61 Pm	62 Sm	63 Eu	64 Gd	65 Tb	66 Dy	67 Ho	68 Er	69 Tm	70 Yb	71 Lu
Actinide Series	90 Th	91 Pa	92 U	93 Np	94 Pu	95 Am	96 Cm	97 Bk	98 Cf	99 Es	100 Fm	101 Md	102 No	103 Lr

NATIONAL GEOGRAPHIC VISUALIZING SYNTHETIC ELEMENTS

Figure 17

No element heavier than uranium, with 92 protons and 146 neutrons, is typically found in nature. But by using a device called a particle accelerator, scientists can make synthetic elements with atomic numbers greater than that of uranium. Within the accelerator, atomic nuclei are made to collide at high speeds in the hope that some will fuse together to form new, heavier elements. These "heavy" synthetic elements are radioactive isotopes, some of which are so unstable that they survive only a fraction of a second before emitting radioactive particles and decaying into other, lighter elements.

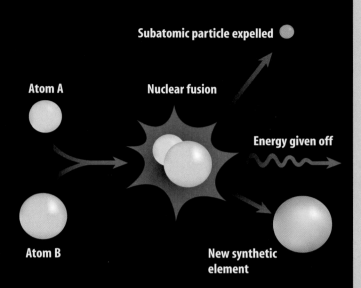

▲ When atoms collide in an accelerator, their nuclei may undergo a fusion reaction to form a new—and often short-lived—synthetic element. Energy and one or more subatomic particles typically are given off in the process.

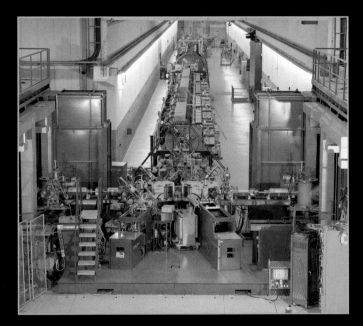

▲ Inside the airless vacuum chamber of a particle accelerator, such as this one in Hesse, Germany, streams of atoms move at incredibly high speeds.

▶ Recently, the IUPAC (International Union of Pure and Applied Chemistry) General Assembly confirmed the official name and symbol of element 110. Element 110 was previously known as Ununnilium and its symbol was Uun. The new name is darmstadtium and its symbol is Ds. Element 111 is expected to receive its official name and symbol in the near future.

SECTION 3 Transition Elements **451**

Topic: Health Risks
Visit blue.msscience.com for Web links to information about health risks due to mercury.

Activity Write a paragraph on how mercury can affect your health.

Dentistry and Dental Materials

Dentists have been using amalgam for over 150 years to fill cavities in decayed teeth. Amalgam, a mixture of silver, copper, tin, and mercury, is the familiar "silver filling." Because amalgam contains mercury, some people are concerned that the use of this type of filling may unnecessarily expose a person to mercury vapor. Today dentists have alternatives to amalgam. New composites, resins, and porcelains are being used to repair decayed, broken, or missing teeth. These new materials are strong, chemically resistant to body fluids, and can be altered to have the natural color of the tooth. Some of the new resins also contain fluoride that will protect the tooth from further decay. Many of these new materials would be useless without the development of new bonding agents. The new "glues" or bonding agents adhere the new materials to the natural tooth. These bonding agents must be strong and chemically resistant to body fluids.

Reading Check *Why are these new dental materials desirable for repairing teeth?*

Orthodontists are using new nickel and titanium alloys for the wires on braces. These wires have shape memory. The wires undergo a special heat treatment process to lock in their shapes. If the wires are forced out of their heat-treated shape, the wires will try to return to their original shape. Orthodontists have found these wires useful in straightening crooked teeth. How do you think these wires help to straighten crooked teeth?

section 3 review

Summary

Transition Elements
- Groups 3–12, which are transition elements, are all metals.
- Their properties change less than the representative elements.
- The elements in the iron triad are iron, cobalt, and nickel.

Inner Transition Elements
- The lanthanide series contains elements from cerium to lutetium.
- The lanthanides also are known as the rare earth elements.
- The actinide series contains elements from thorium to lawrencium.

Self Check

1. **State** how the elements in the iron triad differ from other transition metals.
2. **Explain** the major difference between the lanthanides and actinides.
3. **Explain** how mercury is used.
4. **Describe** how synthetic elements are made.
5. **Think Critically** Iridium and cadmium are both transition elements. Predict which element is toxic and which element is more likely to be a catalyst. Explain.

Applying Skills

6. **Form Hypotheses** How does the appearance of a burned-out lightbulb compare to a new lightbulb? What could explain the difference?

452 **CHAPTER 15** The Periodic Table

 blue.msscience.com/self_check_quiz

Metals and Nonmetals

Real-World Question

Metals on asteroids appear attractive for mining to space programs because the metals are essential for space travel. An asteroid could be processed to provide very pure iron and nickel. Valuable by-products would include cobalt, platinum, and gold. How can miners determine if an element is a metal or a nonmetal?

Goals
- **Describe** the appearance of metals and nonmetals.
- **Evaluate** the malleability or brittleness of metals and nonmetals
- **Observe** chemical reactions of metals and nonmetals with an acid and a base.

Materials (per group of 2–3 students)
10 test tubes with rack
10-mL graduated cylinder
forceps or tweezers
small hammer or mallet
dropper bottle of 0.5M HCl
dropper bottle of 0.1M CuCl$_2$
test-tube brush
marking pencil
25 g carbon
25 g silicon
25 g tin
25 g sulfur
25 g iron

Safety Precautions

Procedure

1. Copy data table into your Science Journal. Fill in data table as you complete the lab.
2. Describe in as much detail as possible the appearance of the sample, including color, luster, and state of matter.
3. Use the hammer or mallet to determine malleability or brittleness.
4. Label 5 test tubes #1–5. Place a 1-g sample of each element in separate test tubes. Add 5 mL of HCl to each tube. If bubbles form, this indicates a chemical reaction.
5. Repeat step 4, substituting HCl with CuCl$_2$. Do not discard the solutions immediately. Continue to observe for five minutes. Some of the changes may be slow. A chemical reaction is indicated by a change in appearance of the element.

Analyze Your Data

1. **Analyze Results** What characteristics distinguish metals from nonmetals?
2. **List** which elements you discovered to be metals.
3. **Describe** a metalloid. Are any of the elements tested a metalloid? If so, name them.

Conclude and Apply

1. **Explain** how the future might increase or decrease the need for selected elements.
2. **Infer** why discovering and mining metals on asteroids might be an important find.

Metal and Nonmetal Data

Element	Appearance	Malleable or Brittle	Reaction wth HCl	Reaction with CuCl$_2$
Carbon				
Silicon				
Tin		Do not write in this book.		
Sulfur				
Iron				

LAB 453

LAB Use the Internet

Health Risks from Heavy Metals

Goals
- **Organize** and synthesize information on a chemical or heavy metal thought to cause health problems in the area where you live.
- **Communicate** your findings to others in your class.

Data Source
Science Online

Visit blue.msscience.com/internet_lab for more information about health risks from heavy metals, hints on health risks, and data from other students.

Real-World Question

Many heavy metals are found naturally on the planet. People and animals are exposed to these metals every day. One way to reduce the exposure is to know as much as possible about the effects of chemicals on you and the environment. Do heavy metals and other chemicals pose a threat to the health of humans? Could health problems be caused by exposure to heavy metals such as lead, or a radioactive chemical element, such as radon? Is the incidence of these problems higher in one area than another?

Make a Plan

1. Read general information concerning heavy metals and other potentially hazardous chemicals.
2. Use the sites listed at the link to the left to research possible health problems in your area caused by exposure to chemicals or heavy metals. Do you see a pattern in the type of health risks that you found in your research?
3. Check the link to the left to see what others have learned.

Health Risk Data Table				
Location	Chemical or Heavy Metal	How People Come in Contact with Chemical	Potential Health Problem	Who Is Affected
		Do not write in this book.		

454 CHAPTER 15 The Periodic Table

Using Scientific Methods

▶ Follow Your Plan

1. Make sure your teacher approves your plan before you start.
2. **Research information** that can help you find out about health risks in your area.
3. **Organize** your information in a data table like the one shown.
4. **Write** a report in your Science Journal using the results of your research on heavy metals.
5. Post your data in the table provided at the link below.

▶ Analyze Your Data

1. **Evaluate** Did all your sources agree on the health risk of the chemical or heavy metal?
2. **Analyze** all your sources for possible bias. Are some sources more reliable than others?
3. **Explain** how the health risk differs for adults and children.
4. **Identify** the sources of the heavy metals in your area. Are the heavy metals still being deposited in your area?

▶ Conclude and Apply

1. **Analyze Results** Were the same substances found to be health risks in other parts of the country? From the data at the link below, try to predict what chemicals or heavy metals are health risks in different parts of the country.
2. **Determine** what information you think is the most important for the public to be aware of.
3. **Explain** what could be done to decrease the risk of the health problems you identified.

Communicating Your Data

Find this lab using the link below. **Post** your data in the table provided. **Compare** your data to those of other students. **Analyze** and look for patterns in the data.

Science online

blue.msscience.com/internet_lab

LAB **455**

Anansi Tries to Steal All the Wisdom in the World

A folktale, adapted by Matt Evans

The following African folktale about a spider named Anansi (or Anancy) is from the Ashanti people in Western Africa.

Anansi the spider knew that he was not wise... "I know... if I can get all of the wisdom in the village and put it in a hollow gourd... I would be the wisest of all!" So he set out to find a suitable gourd and then began his journey to collect the village's wisdom... He looked around and spotted a tall, tall tree. "Ah," he said to himself, "if I could hide my wisdom high in that tree, I would never have to worry about someone stealing it from me!"... He first took a cloth band and tied it around his waist. Then he tied the heavy gourd to the front of his belly where it would be safe. As he began to climb, however, the gourd full of wisdom kept getting in the way...

Soon Anansi's youngest son walked by... "But Father," said the son, "wouldn't it be much easier if you tied the gourd behind you instead of in front?"... Anansi moved the gourd so that it was behind him and proceeded up the tree with no problems at all. When he had reached the top, he cried out, "I walked all over and collected so much wisdom I am the wisest person ever, but still my baby son is wiser than me. Take back your wisdom!" He lifted the gourd high over his head and spilled its contents into the wind. The wisdom blew far and wide and settled across the land. And this is how wisdom came to the world.

Understanding Literature

Folktales The African folktale you have just read is called an animal-trickster tale. Trickster tales come from Africa, the Caribbean, and Latin American countries. Trickster tales portray a wily and cunning animal or human who at times bewilders the more powerful and at other times becomes a victim of his or her own schemes. Describe other kinds of folktales, such as fairy tales and tall tales

Respond to the Reading

1. Is Anansi a clever spider?
2. Why did Anansi scatter the wisdom he had collected?
3. **Linking Science and Writing** Write a folktale featuring an animal as a trickster.

INTEGRATE Chemistry Elements are classified in relation to one another in a periodic table. They also are classified in groups of elements that share similar characteristics. Thus, there is a group of elements known as the alkali metals, another called the halogens, and so on. This way of classifying elements is similar to the way in which folktales are classified. Trickster tales have similar characteristics such as a character that has certain traits, like cleverness, wit, cunning, and an ability to survive.

chapter 15 Study Guide

Reviewing Main Ideas

Section 1 — Introduction to the Periodic Table

1. When organized according to atomic number in a table, elements with similar properties occupy the same column and are called a group or family.
2. On the periodic table, the properties of the elements change gradually across a horizontal row called a period.
3. The periodic table can be divided into representative elements and transition elements.

Section 2 — Representative Elements

1. The groups on the periodic table are known also by other names. For instance, Group 17 is known as halogens.
2. Atoms of elements in Groups 1 and 2 readily combine with atoms of other elements.
3. Each element in Group 2 combines less readily than its neighbor in Group 1. Each alkaline earth metal is denser and has a higher melting point than the alkali metal in its period.
4. Sodium, potassium, magnesium, and calcium have important biological roles.

Section 3 — Transition Elements

1. The metals in the iron triad are found in a variety of places. Iron is found in blood and in the structure of skyscrapers.
2. Copper, silver, and gold are fairly unreactive, malleable elements.
3. The lanthanides are naturally occurring elements with similar properties.
4. The actinides are radioactive elements. All actinides except thorium, proactinium, and uranium are synthetic.

Visualizing Main Ideas

Copy and complete the following concept map on the periodic table.

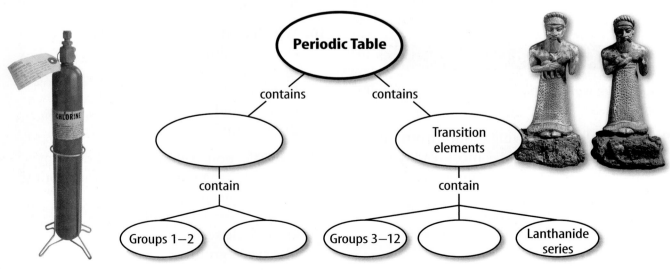

blue.msscience.com/interactive_tutor

Chapter 15 Review

Using Vocabulary

actinides p. 450
alkali metals p. 441
alkaline earth metals p. 442
catalyst p. 449
group p. 435
halogens p. 446
lanthanides p. 450
metal p. 438
metalloid p. 438
noble gases p. 446
nonmetal p. 438
period p. 435
representative element p. 435
semiconductor p. 443
synthetic elements p. 450
transition elements p. 435

Answer the following questions using complete sentences.

1. What is the difference between a group and a period?
2. What is the connection between a metalloid and a semiconductor?
3. What is a catalyst?
4. Arrange the terms *nonmetal, metal,* and *metalloid* according to increasing heat and electrical conductivity.
5. How is a metalloid like a metal? How is it different from a metal?
6. What are synthetic elements?
7. How are transition elements alike?
8. Why are some gases considered to be noble?

Checking Concepts

Choose the word or phrase that best answers the question.

9. Which of the following groups from the periodic table combines most readily with other elements to form compounds?
   A) transition metals
   B) alkaline earth metals
   C) alkali metals
   D) iron triad

10. Which element is NOT a part of the iron triad?
    A) nickel
    B) copper
    C) cobalt
    D) iron

11. Which element is located in Group 6, period 4?
    A) tungsten   C) titanium
    B) chromium   D) hafnium

12. Which element below is NOT a transition element?
    A) gold   C) silver
    B) calcium   D) copper

13. Several groups contain only metals. Which group contains only nonmetals?
    A) Group 1   C) Group 2
    B) Group 12   D) Group 18

14. Which of the following elements is likely to be contained in a substance with a brilliant yellow color?
    A) chromium   C) iron
    B) carbon   D) tin

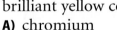

15. Which halogen is radioactive?
    A) astatine   C) bromine
    B) chlorine   D) iodine

16. Which of the following describes the element tellurium?
    A) alkali metal
    B) transition metal
    C) metalloid
    D) lanthanide

17. A brittle, non-conducting, solid might belong to which of the following groups?
    A) alkali metals
    B) alkaline earth metals
    C) actinide series
    D) oxygen group

chapter 15 Review

Thinking Critically

18. **Explain** why it is important that mercury be kept out of streams and waterways.

19. **Determine** If you were going to try to get the noble gas argon to combine with another element, would fluorine be a good choice for the other element? Explain.

Use the figure below to answer question 20.

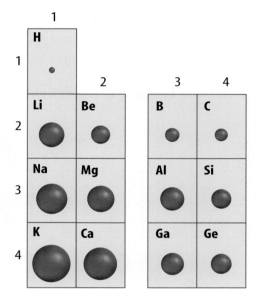

20. **Interpret Data** The periodic table shows trends across the rows and down the columns. In this portion of the periodic table, the relative size of the atom is represented by a ball. What trend can you see in this part of the table for relative size?

21. **Evaluate** It is theorized that some of the actinides beyond uranium were once present in Earth's crust. If this theory is true, how would their half-lives compare with the half-life of uranium, which is 4.5 billion years?

22. **Recognize Cause and Effect** Why do photographers work in low light when they work with materials containing selenium?

23. **Predict** How would life on Earth be different if the atmosphere were 80 percent oxygen and 20 percent nitrogen instead of the other way around?

24. **Compare and contrast** Na and Mg, which are in the same period, with F and Cl, which are in the same group.

Performance Activities

25. **Ask Questions** Research the contribution that Henry G. J. Moseley made to the development of the modern periodic table. Research the background and work of this scientist. Write your findings in the form of an interview.

Applying Math

26. **Elements at Room Temperature** Make a bar graph of the representative elements that shows how many of the elements are solids, liquids, and gases at room temperature.

27. **Calculate** Using the information that you collected in question 26, calculate the percentage of solids, liquids, and gases within the representative elements.

Use the figure below to answer question 28.

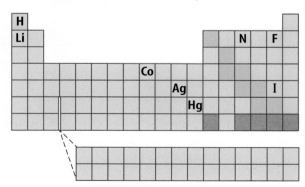

28. **Element Details** For each element shown, give the element's period and group number; whether the element is a metal or a nonmetal; and whether it is a solid, liquid, or gas at room temperature.

CHAPTER REVIEW **459**

Chapter 15 Standardized Test Practice

Part 1 Multiple Choice

Record your answers on the answer sheet provided by your teacher or on a sheet of paper.

1. Which statement about the periodic table is TRUE?
   A. Elements all occur naturally on Earth.
   B. Elements occur in the order in which they were discovered.
   C. Elements with similar properties occupy the same group.
   D. Elements are arranged in the order Mendeleev chose.

2. Which of these is NOT a property of metals?
   A. malleability
   B. luster
   C. ductility
   D. poor conductor of heat and electricity

Use the illustration below to answer questions 3 and 4.

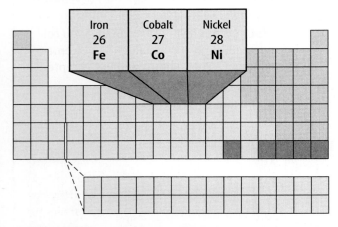

3. What name is given to these three elements which are used in processes that create steel and other metal mixtures?
   A. lanthanides   C. actinides
   B. the coin metals   D. the iron triad

Test-Taking Tip

The Best Answer Read all choices before answering the questions.

4. To which category do these elements belong?
   A. nonmetals
   B. transition elements
   C. noble gases
   D. representative metals

5. Which member of the boron family is used to make soft-drink cans, baseball bats, and siding for homes?
   A. aluminum   C. indium
   B. boron   D. gallium

Use the table below to answer questions 6 and 7.

6. Halogens are highly reactive nonmetals. Which group combines most readily with them?
   A. Group 1, alkali metals
   B. Group 2, alkaline earth metals
   C. Group 17, halogens
   D. Group 18, noble gases

7. Which alkali metal element is most reactive?
   A. Li   C. K
   B. Na   D. Cs

8. Many elements that are essential for life, including nitrogen, oxygen, and carbon, are part of what classification?
   A. nonmetals   C. metalloids
   B. metals   D. noble gases

Part 2 — Short Response/Grid In

Record your answers on the answer sheet provided by your teacher or on a sheet of paper.

9. Based on the information found in the periodic table, compare and contrast properties of the elements gold and silver.

10. Why don't the element symbols always match the name? Give two examples and describe the origin of each symbol.

Use the graph below to answer questions 11 and 12.

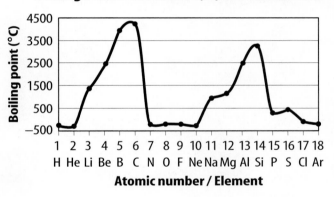

11. The data shows that boiling point is a periodic property. Explain what the term *periodic property* means.

12. Describe patterns evident in this data.

13. Describe the mixture used by dentists for the past 150 years to fill cavities in decayed teeth. Why do many dentists today use other materials to repair teeth?

14. Compare and contrast the periodic table that Mendeleev developed to the periodic table that Mosley organized.

15. Choose a representative element group and list the elements in that group. Then list three to four uses for those elements.

Part 3 — Open Ended

Record your answers on a sheet of paper.

16. What role does nitrogen play in the human body? Explain the importance of bacteria in the soil which change the form in which nitrogen naturally occurs.

17. Much of the wiring in houses is made from copper. What properties of copper make it ideal for this purpose?

18. Why do some homeowners check for the presence of the noble gas radon in their homes?

Use the graph below to answer questions 19 and 20.

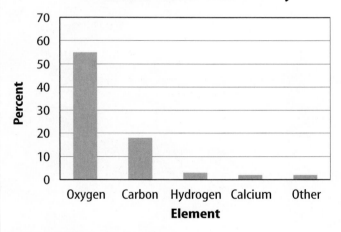

19. The graph above shows elements present in the greatest amounts in the human body. Use information from the periodic table to create a chart which shows properties of each element, including its symbol, atomic number, the group to which it belongs, and whether it is a metal, nonmetal, or metalloid.

20. One element shown here is an alkaline earth metal. Compare the properties of the elements in this family to those of the elements found in Group 1.

chapter 16

Atomic Structure and Chemical Bonds

The BIG Idea
An atom's structure affects how it bonds to other atoms.

SECTION 1
Why do atoms combine?
Main Idea When atoms combine, they become more stable.

SECTION 2
How Elements Bond
Main Idea Elements bond by transferring electrons or by sharing electrons.

The Noble Family

Blimps, city lights, and billboards, all have something in common—they use gases that are members of the same element family. In this chapter, you'll learn about the unique properties of element families. You'll also learn how electrons can be lost, gained, and shared by atoms to form chemical bonds.

Science Journal Write a sentence comparing household glue to chemical bonds.

Start-Up Activities

Model the Energy of Electrons

It's time to clean out your room—again. Where do all these things come from? Some are made of cloth and some of wood. The books are made of paper and an endless array of things are made of plastic. Fewer than 100 different kinds of naturally occurring elements are found on Earth. They combine to make all these different substances. What makes elements form chemical bonds with other elements?

1. Pick up a paper clip with a magnet. Touch that paper clip to another paper clip and pick it up.
2. Continue picking up paper clips this way until you have a strand of them and no more will attach.
3. Then, gently pull off the paper clips one by one.
4. **Think Critically** In your Science Journal, discuss which paper clip was easiest to remove and which was hardest. Was the clip that was easiest to remove closer to or farther from the magnet?

Chemical Bonds Make the following Foldable to help you classify information by diagramming ideas about chemical bonds.

STEP 1 Fold a vertical sheet of paper in half from top to bottom.

STEP 2 Fold in half from side to side with the fold at the top.

STEP 3 Unfold the paper once. Cut only the fold of the top flap to make two tabs.

STEP 4 Turn the paper vertically and label the tabs as shown.

Summarize As you read the chapter, identify the main ideas of bonding under the appropriate tabs. After you have read the chapter, explain the difference between polar covalent bonds and covalent bonds on the inside portion of your Foldable.

Preview this chapter's content and activities at
blue.msscience.com

463

Get Ready to Read

Questioning

① Learn It! Asking questions helps you to understand what you read. As you read, think about the questions you'd like answered. Often you can find the answer in the next paragraph or section. Learn to ask good questions by asking *who, what, when, where, why,* and *how*.

② Practice It! Read the following passage from Section 2.

> In medieval times, alchemists (AL kuh mists) were the first to explore the world of chemistry. Although many of them believed in magic and mystical transformations, alchemists did learn much about the properties of some elements. They even used symbols to represent them in chemical processes.
>
> —*from page 479*

Here are some questions you might ask about this paragraph:
- Who were alchemists?
- What was their contribution to chemistry?
- What symbols did alchemists use to represent these elements?
- How do modern element symbols compare to the symbols the alchemists used?

③ Apply It! As you read the chapter, look for answers to section headings that are in the form of questions.

Target Your Reading

Reading Tip
Test yourself. Create questions and then read to find answers to your own questions.

Use this to focus on the main ideas as you read the chapter.

① Before you read the chapter, respond to the statements below on your worksheet or on a numbered sheet of paper.
- Write an **A** if you **agree** with the statement.
- Write a **D** if you **disagree** with the statement.

② After you read the chapter, look back to this page to see if you've changed your mind about any of the statements.
- If any of your answers changed, explain why.
- Change any false statements into true statements.
- Use your revised statements as a study guide.

Science Online
Print out a worksheet of this page at blue.msscience.com

Before You Read A or D	Statement	After You Read A or D
	1 All matter, including solids like wood and steel, contains mostly empty space.	
	2 Scientists are able to pinpoint the exact location of an electron in an atom.	
	3 Electrons orbit a nucleus just like planets orbit the Sun.	
	4 The number of electrons in a neutral atom is the same as the atom's atomic number.	
	5 Noble gases react easily with other elements.	
	6 All elements transfer the same number of electrons when bonding with other elements.	
	7 Electrons in metals can move freely among all the ions in the metal.	
	8 Some atoms bond by sharing electrons between the two atoms.	
	9 Water molecules have two opposite ends, like poles in a magnet.	

section 1
Why do atoms combine?

as you read

What You'll Learn
- **Identify** how electrons are arranged in an atom.
- **Compare** the relative amounts of energy of electrons in an atom.
- **Compare** how the arrangement of electrons in an atom is related to its place in the periodic table.

Why It's Important
Chemical reactions take place all around you.

Review Vocabulary
atom: the smallest part of an element that keeps all the properties of that element

New Vocabulary
- electron cloud
- energy level
- electron dot diagram
- chemical bond

Atomic Structure

When you look at your desk, you probably see it as something solid. You might be surprised to learn that all matter, even solids like wood and metal contain mostly empty space. How can this be? The answer is that although there might be little or no space between atoms, a lot of empty space lies within each atom.

At the center of every atom is a nucleus containing protons and neutrons. This nucleus represents most of the atom's mass. The rest of the atom is empty except for the atom's electrons, which are extremely small compared with the nucleus. Although the exact location of any one electron cannot be determined, the atom's electrons travel in an area of space around the nucleus called the **electron cloud.**

To visualize an atom, picture the nucleus as the size of a penny. In this case, electrons would be smaller than grains of dust and the electron cloud would extend outward as far as 20 football fields.

Electrons You might think that electrons resemble planets circling the Sun, but they are very different, as you can see in **Figure 1.** First, planets have no charges, but the nucleus of an atom has a positive charge and electrons have negative charges.

Second, planets travel in predictable orbits—you can calculate exactly where one will be at any time. This is not true for electrons. Although electrons do travel in predictable areas, it is impossible to calculate the exact position of any one electron. Instead scientists use a mathematical model that predicts where an electron is most likely to be.

Figure 1 You can compare and contrast electrons with planets.

Planets travel in well-defined paths.

Electrons travel around the nucleus. However, their paths are not well-defined.

464 CHAPTER 16 Atomic Structure and Chemical Bonds

Element Structure Each element has a unique atomic structure consisting of a specific number of protons, neutrons, and electrons. The number of protons and electrons is always the same for a neutral atom of a given element. **Figure 2** shows a two-dimensional model of the electron structure of a lithium atom, which has three protons and four neutrons in its nucleus, and three electrons moving around its nucleus.

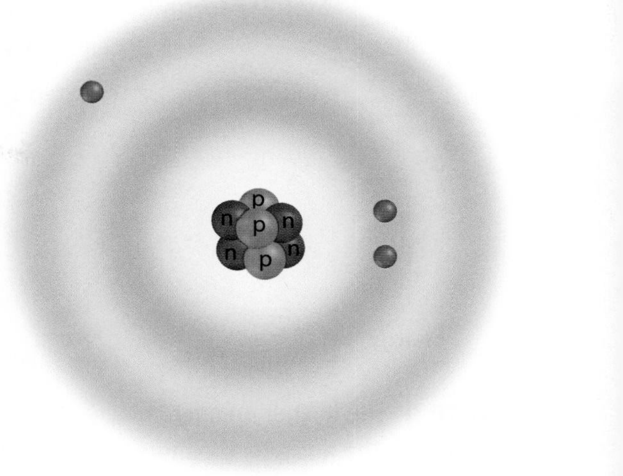

Electron Arrangement

The number and arrangement of electrons in the electron cloud of an atom are responsible for many of the physical and chemical properties of that element.

Electron Energy Although all the electrons in an atom are somewhere in the electron cloud, some electrons are closer to the nucleus than others. The different areas for an electron in an atom are called **energy levels**. **Figure 3** shows a model of what these energy levels might look like. Each level represents a different amount of energy.

Number of Electrons Each energy level can hold a maximum number of electrons. The farther an energy level is from the nucleus, the more electrons it can hold. The first energy level, energy level 1, can hold one or two electrons, the second, energy level 2, can hold up to eight, the third can hold up to 18, and the fourth energy level can hold a maximum of 32 electrons.

Figure 2 This neutral lithium atom has three positively charged protons, three negatively charged electrons, and four neutral neutrons.

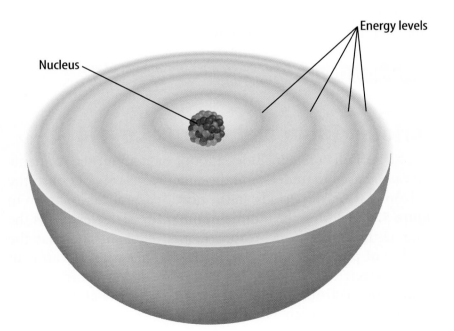

Figure 3 Electrons travel in three dimensions around the nucleus of an atom. The dark bands in this diagram show the energy levels where electrons are most likely to be found.
Identify the energy level that can hold the most electrons.

SECTION 1 Why do atoms combine?

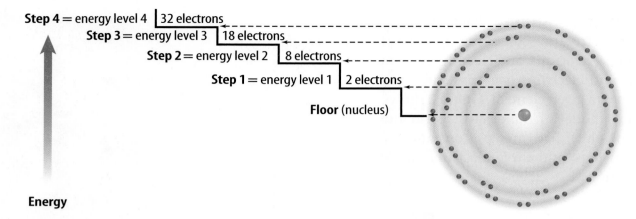

Figure 4 The farther an energy level is from the nucleus, the more electrons it can hold.
Identify *the energy level with the least energy and the energy level with the most energy.*

Energy Steps The stairway, shown in **Figure 4,** is a model that shows the maximum number of electrons each energy level can hold in the electron cloud. Think of the nucleus as being at floor level. Electrons within an atom have different amounts of energy, represented by energy levels. These energy levels are represented by the stairsteps in **Figure 4.** Electrons in the level closest to the nucleus have the lowest amount of energy and are said to be in energy level one. Electrons farthest from the nucleus have the highest amount of energy and are the easiest to remove. To determine the maximum number of electrons that can occupy an energy level, use the formula, $2n^2$, where n equals the number of the energy level.

Recall the Launch Lab at the beginning of the chapter. It took more energy to remove the paper clip that was closest to the magnet than it took to remove the one that was farthest away. That's because the closer a paper clip was to the magnet, the stronger the magnet's attractive force was on the clip. Similarly, the closer a negatively charged electron is to the positively charged nucleus, the more strongly it is attracted to the nucleus. Therefore, removing electrons that are close to the nucleus takes more energy than removing those that are farther away from the nucleus.

 *What determines the amount of energy an electron has?*

Topic: Electrons
Visit blue.msscience.com for Web links to information about electrons and their history.

Activity Research why scientists cannot locate the exact positions of an electron.

Periodic Table and Energy Levels

The periodic table includes a lot of data about the elements and can be used to understand the energy levels also. Look at the horizontal rows, or periods, in the portion of the table shown in **Figure 5.** Recall that the atomic number for each element is the same as the number of protons in that element and that the number of protons equals the number of electrons because an atom is electrically neutral. Therefore, you can determine the number of electrons in an atom by looking at the atomic number written above each element symbol.

Electron Configurations

If you look at the periodic table shown in **Figure 5**, you can see that the elements are arranged in a specific order. The number of electrons in a neutral atom of the element increases by one from left to right across a period. For example, the first period consists of hydrogen with one electron and helium with two electrons in energy level one. Recall from **Figure 4** that energy level one can hold up to two electrons. Therefore, helium's outer energy level is complete. Atoms with a complete outer energy level are stable. Therefore, helium is stable.

Reading Check What term is given to the rows of the periodic table?

The second period begins with lithium, which has three electrons—two in energy level one and one in energy level two. Lithium has one electron in its outer energy level. To the right of lithium is beryllium with two outer-level electrons, boron with three, and so on until you reach neon with eight.

Look again at **Figure 4.** You'll see that energy level two can hold up to eight electrons. Not only does neon have a complete outer energy level, but also this configuration of exactly eight electrons in an outer energy level is stable. Therefore, neon is stable. The third period elements fill their outer energy levels in the same manner, ending with argon. Although energy level three can hold up to 18 electrons, argon has eight electrons in its outer energy level—a stable configuration. Each period in the periodic table ends with a stable element.

Nobel Prize Winner
Ahmed H. Zewail is a professor of chemistry and physics and the director of the Laboratory for Molecular Sciences at the California Institute of Technology. He was awarded the 1999 Nobel Prize in Chemistry for his research. Zewail and his research team use lasers to record the making and breaking of chemical bonds.

Figure 5 This portion of the periodic table shows the electron configurations of some elements. Count the electrons in each element and notice how the number increases across a period.

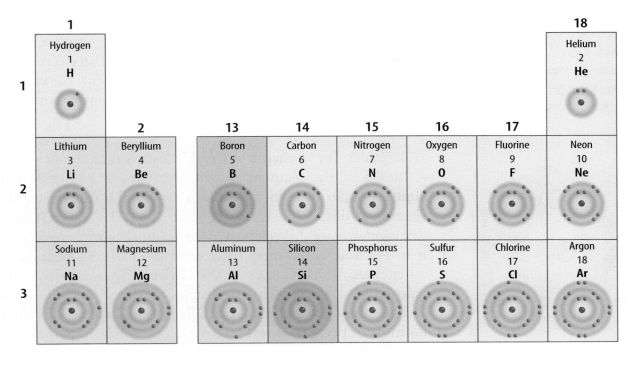

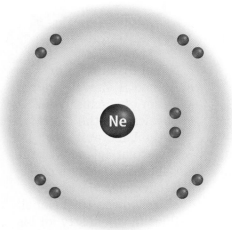

Figure 6 The noble gases are stable elements because their outer energy levels are complete or have a stable configuration of eight electrons like neon shown here.

Figure 7 The halogen element fluorine has seven electrons in its outer energy level. **Determine** *how many electrons the halogen family member bromine has in its outer energy level.*

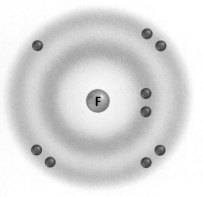

Element Families

Elements can be divided into groups, or families. Each column of the periodic table in **Figure 5** contains one element family. Hydrogen is usually considered separately, so the first element family begins with lithium and sodium in the first column. The second family starts with beryllium and magnesium in the second column, and so on. Just as human family members often have similar looks and traits, members of element families have similar chemical properties because they have the same number of electrons in their outer energy levels.

It was the repeating pattern of properties that gave Russian chemist Dmitri Mendeleev the idea for his first periodic table in 1869. While listening to his family play music, he noticed how the melody repeated with increasing complexity. He saw a similar repeating pattern in the elements and immediately wrote down a version of the periodic table that looks much as it does today.

Noble Gases Look at the structure of neon in **Figure 6**. Neon and the elements below it in Group 18 have eight electrons in their outer energy levels. Their energy levels are stable, so they do not combine easily with other elements. Helium, with two electrons in its lone energy level, is also stable. At one time these elements were thought to be completely unreactive, and therefore became known as the inert gases. When chemists learned that some of these gases can react, their name was changed to noble gases. They are still the most stable element group.

This stability makes possible one widespread use of the noble gases—to protect filaments in lightbulbs. Another use of noble gases is to produce colored light in signs. If an electric current is passed through them they emit light of various colors—orange-red from neon, lavender from argon, and yellowish-white from helium.

Halogens The elements in Group 17 are called the halogens. A model of the element fluorine in period 2 is shown in **Figure 7**. Like all members of this family, fluorine needs one electron to obtain a stable outer energy level. The easier it is for a halogen to gain this electron to form a bond, the more reactive it is. Fluorine is the most reactive of the halogens because its outer energy level is closest to the nucleus. The reactivity of the halogens decreases down the group as the outer energy levels of each element's atoms get farther from the nucleus. Therefore, bromine in period 4 is less reactive than fluorine in period 2.

Alkali Metals Look at the element family in Group 1 on the periodic table at the back of this book, called the alkali metals. The first members of this family, lithium and sodium, have one electron in their outer energy levels. You can see in **Figure 8** that potassium also has one electron in its outer level. Therefore, you can predict that the next family member, rubidium, does also. These electron arrangements are what determines how these metals react.

Reading Check *How many electrons do the alkali metals have in their outer energy levels?*

The alkali metals form compounds that are similar to each other. Alkali metals each have one outer energy level electron. It is this electron that is removed when alkali metals react. The easier it is to remove an electron, the more reactive the atom is. Unlike halogens, the reactivities of alkali metals increase down the group; that is, elements in the higher numbered periods are more reactive than elements in the lower numbered periods. This is because their outer energy levels are farther from the nucleus. Less energy is needed to remove an electron from an energy level that is farther from the nucleus than to remove one from an energy level that is closer to the nucleus. For this reason, cesium in period 6 loses an electron more readily and is more reactive than sodium in period 3.

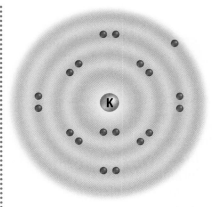

Figure 8 Potassium, like lithium and sodium, has only one electron in its outer level.

Applying Science

How does the periodic table help you identify properties of elements?

The periodic table displays information about the atomic structure of the elements. This information includes the properties, such as the energy level, of the elements. Can you identify an element if you are given information about its energy level? Use your ability to interpret the periodic table to find out.

Identifying the Problem

Recall that elements in a group in the periodic table contain the same number of electrons in their outer levels. The number of electrons increases by one from left to right across a period. Refer to **Figure 5.** Can you identify an unknown element or the group a known element belongs to?

Solving the Problem

1. An unknown element in Group 2 has a total number of 12 electrons and two electrons in its outer level. What is it?
2. Name the element that has eight electrons, six of which are in its outer level.
3. Silicon has a total of 14 electrons, four electrons in its outer level, and three energy levels. What group does silicon belong to?
4. Three elements have the same number of electrons in their outer energy levels. One is oxygen. Using the periodic table, what might the other two be?

SECTION 1 Why do atoms combine? **469**

Mini LAB

Drawing Electron Dot Diagrams

Procedure
1. Draw a periodic table that includes the first 18 elements—the elements from hydrogen through argon. Make each block a 3-cm square.
2. Fill in each block with the electron dot diagram of the element.

Analysis
1. What do you observe about the electron dot diagram of the elements in the same group?
2. Describe any changes you observe in the electron dot diagrams across a period.

Electron Dot Diagrams

You have read that the number of electrons in the outer energy level of an atom determines many of the chemical properties of the atom. Because these electrons are so important in determining the chemical properties of atoms, it can be helpful to make a model of an atom that shows only the outer electrons. A model like this can be used to show what happens to these electrons during reactions.

Drawing pictures of the energy levels and electrons in them takes time, especially when a large number of electrons are present. If you want to see how atoms of one element will react, it is handy to have an easier way to represent the atoms and the electrons in their outer energy levels. You can do this with electron dot diagrams. An **electron dot diagram** is the symbol for the element surrounded by as many dots as there are electrons in its outer energy level. Only the outer energy level electrons are shown because these are what determine how an element can react.

How to Write Them How do you know how many dots to make? For Groups 1 and 2, and 13–18, you can use the periodic table or the portion of it shown in **Figure 5**. Group 1 has one outer electron. Group 2 has two. Group 13 has three, Group 14, four, and so on to Group 18. All members of Group 18 have stable outer energy levels. From neon down, they have eight electrons. Helium has only two electrons, because that is all that its single energy level can hold.

The dots are written in pairs on four sides of the element symbol. Start by writing one dot on the top of the element symbol, then work your way around, adding dots to the right, bottom, and left. Add a fifth dot to the top to make a pair. Continue in this manner until you reach eight dots to complete the level.

The process can be demonstrated by writing the electron dot diagram for the element nitrogen. First, write N—the element symbol for nitrogen. Then, find nitrogen in the periodic table and see what group it is in. It's in Group 15, so it has five electrons in its outer energy level. The completed electron dot diagram for nitrogen can be seen in **Figure 9**.

The electron dot diagram for iodine can be drawn the same way. The completed diagram is shown on the right in **Figure 9**.

Figure 9 Electron dot diagrams show only the electrons in the outer energy level.
Explain why only the outer energy level electrons are shown.

Nitrogen contains five electrons in its outer energy level.

Iodine contains seven electrons in its outer energy level.

Figure 10 Some models are made by gluing pieces together. The glue that holds elements together in a chemical compound is the chemical bond.

Using Dot Diagrams Now that you know how to write electron dot diagrams for elements, you can use them to show how atoms bond with each other. A **chemical bond** is the force that holds two atoms together. Chemical bonds unite atoms in a compound much as glue unites the pieces of the model in **Figure 10.** Atoms bond with other atoms in such a way that each atom becomes more stable. That is, their outer energy levels will resemble those of the noble gases.

Reading Check *What is a chemical bond?*

section 1 review

Summary

Atom Structure
- At the center of the atom is the nucleus.
- Electrons exist in an area called the electron cloud.
- Electrons have a negative charge.

Electron Arrangement
- The different regions for an electron in an atom are called energy levels.
- Each energy level can hold a maximum number of electrons.

The Periodic Table
- The number of electrons is equal to the atomic number.
- The number of electrons in a neutral atom increases by one from left to right across a period.

Self Check

1. **Determine** how many electrons nitrogen has in its outer energy level. How many does bromine have?
2. **Solve** for the number of electrons that oxygen has in its first energy level. Second energy level?
3. **Identify** which electrons in oxygen have more energy, those in the first energy level or those in the second.
4. **Think Critically** Atoms in a group of elements increase in size as you move down the columns in the periodic table. Explain why this is so.

Applying Math

5. **Solve One-Step Equations** You can calculate the maximum number of electrons each energy level can hold using the formula $2n^2$. Calculate the number of electrons in the first five energy levels where n equals the number of energy levels.

Science Online blue.msscience.com/self_check_quiz

SECTION 1 Why do atoms combine? **471**

section 2

How Elements Bond

as you read

What You'll Learn
- **Compare and contrast** ionic and covalent bonds.
- **Distinguish** between compounds and molecules.
- **Identify** the difference between polar and nonpolar covalent bonds.
- **Interpret** chemical shorthand.

Why It's Important
Chemical bonds join the atoms in the materials you use every day.

Review Vocabulary
electron: a negatively charged particle that exists in an electron cloud around an atom's nucleus

New Vocabulary
- ion
- ionic bond
- compound
- metallic bond
- covalent bond
- molecule
- polar bond
- chemical formula

Ionic Bonds—Loss and Gain

When you put together the pieces of a jigsaw puzzle, they stay together only as long as you wish. When you pick up the completed puzzle, it falls apart. When elements are joined by chemical bonds, they do not readily fall apart. What would happen if suddenly the salt you were shaking on your fries separated into sodium and chlorine? Atoms form bonds with other atoms using the electrons in their outer energy levels. They have four ways to do this—by losing electrons, by gaining electrons, by pooling electrons, or by sharing electrons with another element.

Sodium is a soft, silvery metal as shown in **Figure 11**. It can react violently when added to water or to chlorine. What makes sodium so reactive? If you look at a diagram of its energy levels below, you will see that sodium has only one electron in its outer level. Removing this electron empties this level and leaves the completed level below. By removing one electron, sodium's electron configuration becomes the same as that of the stable noble gas neon.

Chlorine forms bonds in a way that is the opposite of sodium—it gains an electron. When chlorine accepts an electron, its electron configuration becomes the same as that of the noble gas argon.

Figure 11 Sodium and chlorine react, forming white crystalline sodium chloride.

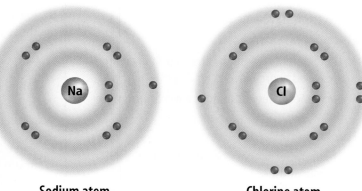

Sodium atom   Chlorine atom

Sodium is a silvery metal that can be cut with a knife. Chlorine is a greenish, poisonous gas.

Chlorine

If chlorine receives an electron from sodium, both atoms will become stable and a bond will form.

472 CHAPTER 16 Atomic Structure and Chemical Bonds

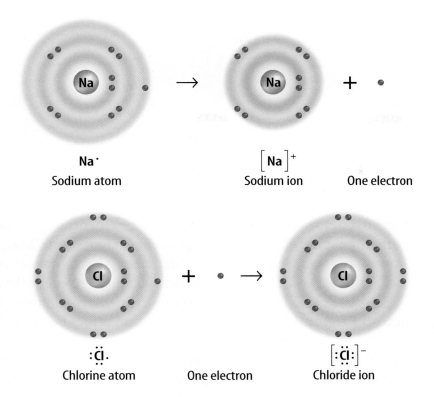

Figure 12 Ions form when elements lose or gain electrons. When sodium comes into contact with chlorine, an electron is transferred from the sodium atom to the chlorine atom. Na becomes a Na$^+$ ion. Cl becomes a Cl$^-$ ion.

Ions—A Question of Balance As you just learned, a sodium atom loses an electron and becomes more stable. But something else happens also. By losing an electron, the balance of electric charges changes. Sodium becomes a positively charged ion because there is now one fewer electron than there are protons in the nucleus. In contrast, chlorine becomes an ion by gaining an electron. It becomes negatively charged because there is one more electron than there are protons in the nucleus.

An atom that is no longer neutral because it has lost or gained an electron is called an **ion** (I ahn). A sodium ion is represented by the symbol Na$^+$ and a chloride ion is represented by the symbol Cl$^-$. **Figure 12** shows how each atom becomes an ion.

Bond Formation The positive sodium ion and the negative chloride ion are strongly attracted to each other. This attraction, which holds the ions close together, is a type of chemical bond called an **ionic bond**. In **Figure 13,** sodium and chloride ions form an ionic bond. The compound sodium chloride, or table salt, is formed. A **compound** is a pure substance containing two or more elements that are chemically bonded.

Ions When ions dissolve in water, they separate. Because of their positive and negative charges, the ions can conduct an electric current. If wires are placed in such a solution and the ends of the wires are connected to a battery, the positive ions move toward the negative terminal and the negative ions move toward the positive terminal. This flow of ions completes the circuit.

Figure 13 An ionic bond forms between atoms of opposite charges. **Describe** *how an atom becomes positive or negative.*

SECTION 2 How Elements Bond

Figure 14 Magnesium has two electrons in its outer energy level.

If one electron is lost to each of two chlorine atoms, magnesium chloride forms.

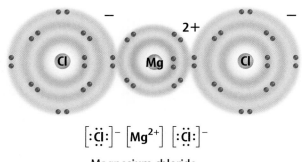

Magnesium chloride

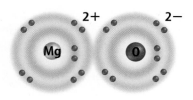

$[Mg^{2+}]\ [O^{2-}]$

Magnesium oxide

If both electrons are lost to one oxygen atom, magnesium oxide forms.

Determine *the electron arrangement for magnesium sulfide and calcium oxide.*

More Gains and Losses You have seen what happens when elements gain or lose one electron, but can elements lose or gain more than one electron? The element magnesium, Mg, in Group 2 has two electrons in its outer energy level. Magnesium can lose these two electrons and achieve a completed energy level. These two electrons can be gained by two chlorine atoms. As shown in **Figure 14,** a single magnesium ion represented by the symbol Mg^{2+} and two chloride ions are generated. The two negatively charged chloride ions are attracted to the positively charged magnesium ion forming ionic bonds. As a result of these bonds, the compound magnesium chloride ($MgCl_2$) is produced.

Some atoms, such as oxygen, need to gain two electrons to achieve stability. The two electrons released by one magnesium atom could be gained by a single atom of oxygen. When this happens, magnesium oxide (MgO) is formed, as shown in **Figure 14.** Oxygen can form similar compounds with any positive ion from Group 2.

Metallic Bonding—Pooling

You have just seen how metal atoms form ionic bonds with atoms of nonmetals. Metals can form bonds with other metal atoms, but in a different way. In a metal, the electrons in the outer energy levels of the atoms are not held tightly to individual atoms. Instead, they move freely among all the ions in the metal, forming a shared pool of electrons, as shown in **Figure 15. Metallic bonds** form when metal atoms share their pooled electrons. This bonding affects the properties of metals. For example, when a metal is hammered into sheets or drawn into a wire, it does not break. Instead, layers of atoms slide over one another. The pooled electrons tend to hold the atoms together. Metallic bonding also is the reason that metals conduct electricity well. The outer electrons in metal atoms readily move from one atom to the next to transmit current.

Figure 15 In metallic bonding, the outer electrons of the silver atoms are not attached to any one silver atom. This allows them to move and conduct electricity.

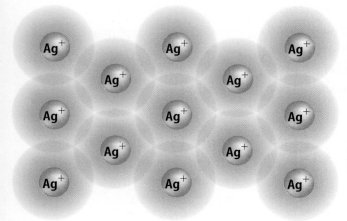

Covalent Bonds—Sharing

Some atoms are unlikely to lose or gain electrons because the number of electrons in their outer levels makes this difficult. For example, carbon has six protons and six electrons. Four of the six electrons are in its outer energy level. To obtain a more stable structure, carbon would either have to gain or lose four electrons. This is difficult because gaining and losing so many electrons takes so much energy. The alternative is sharing electrons.

The Covalent Bond Atoms of many elements become more stable by sharing electrons. The chemical bond that forms between nonmetal atoms when they share electrons is called a **covalent** (koh VAY luhnt) **bond**. Shared electrons are attracted to the nuclei of both atoms. They move back and forth between the outer energy levels of each atom in the covalent bond. So, each atom has a stable outer energy level some of the time. Covalently bonded compounds are called molecular compounds.

Reading Check *How do atoms form covalent bonds?*

The atoms in a covalent bond form a neutral particle, which contains the same numbers of positive and negative charges. The neutral particle formed when atoms share electrons is called a **molecule** (MAH lih kyewl). A molecule is the basic unit of a molecular compound. You can see how molecules form by sharing electrons in **Figure 16**. Notice that no ions are involved because no electrons are gained or lost. Crystalline solids, such as sodium chloride, are not referred to as molecules, because their basic units are ions, not molecules.

Mini LAB

Constructing a Model of Methane

Procedure
1. Using **circles of colored paper** to represent protons, neutrons, and electrons, build paper models of one carbon atom and four hydrogen atoms.
2. Use your models of atoms to construct a molecule of methane by forming covalent bonds. The methane molecule has four hydrogen atoms chemically bonded to one carbon atom.

Analysis
1. In the methane molecule, do the carbon and hydrogen atoms have the same arrangement of electrons as two noble gas elements? Explain your answer.
2. Does the methane molecule have a charge?

Try at Home

Figure 16 Covalent bonding is another way that atoms become more stable. Sharing electrons allows each atom to have a stable outer energy level. These atoms form a single covalent bond.

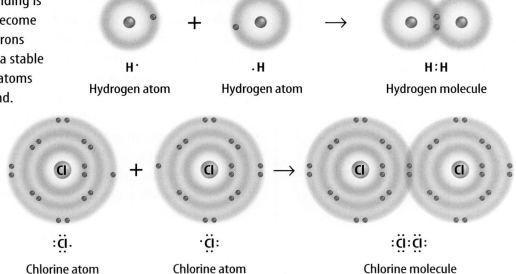

SECTION 2 How Elements Bond **475**

Figure 17 An atom can also form a covalent bond by sharing two or three electrons.

:C· + ·Ö: + ·Ö: → :Ö::C::Ö:

Carbon atom   Oxygen atoms   Carbon dioxide molecule

In carbon dioxide, carbon shares two electrons with each of two oxygen atoms forming two double bonds. Each oxygen atom shares two electrons with the carbon atom.

:N· + ·N: → :N⋮⋮N:

Nitrogen atoms   Nitrogen molecule

Each nitrogen atom shares three electrons in forming a triple bond.

Double and Triple Bonds Sometimes an atom shares more than one electron with another atom. In the molecule carbon dioxide, shown in **Figure 17,** each of the oxygen atoms shares two electrons with the carbon atom. The carbon atom shares two of its electrons with each oxygen atom. When two pairs of electrons are involved in a covalent bond, the bond is called a double bond. **Figure 17** also shows the sharing of three pairs of electrons between two nitrogen atoms in the nitrogen molecule. When three pairs of electrons are shared by two atoms, the bond is called a triple bond.

Reading Check *How many pairs of electrons are shared in a double bond?*

Polar and Nonpolar Molecules

You have seen how atoms can share electrons and that they become more stable by doing so, but do they always share electrons equally? The answer is no. Some atoms have a greater attraction for electrons than others do. Chlorine, for example, attracts electrons more strongly than hydrogen does. When a covalent bond forms between hydrogen and chlorine, the shared pair of electrons tends to spend more time near the chlorine atom than the hydrogen atom.

This unequal sharing makes one side of the bond more negative than the other, like poles on a battery. This is shown in **Figure 18.** Such bonds are called polar bonds. A **polar bond** is a bond in which electrons are shared unevenly. The bonds between the oxygen atom and hydrogen atoms in the water molecule are another example of polar bonds.

Figure 18 Hydrogen chloride is a polar covalent molecule.

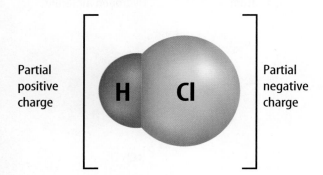

Partial positive charge   Partial negative charge

The Polar Water Molecule Water molecules form when hydrogen and oxygen share electrons. **Figure 19** shows how this sharing is unequal. The oxygen atom has a greater share of the electrons in each bond—the oxygen end of a water molecule has a slight negative charge and the hydrogen end has a slight positive charge. Because of this, water is said to be polar—having two opposite ends or poles like a magnet.

When they are exposed to a negative charge, the water molecules line up like magnets with their positive ends facing the negative charge. You can see how they are drawn to the negative charge on the balloon in **Figure 19.** Water molecules also are attracted to each other. This attraction between water molecules accounts for many of the physical properties of water.

Molecules that do not have these uneven charges are called nonpolar molecules. Because each element differs slightly in its ability to attract electrons, the only completely nonpolar bonds are bonds between atoms of the same element. One example of a nonpolar bond is the triple bond in the nitrogen molecule.

Like ionic compounds, some molecular compounds can form crystals, in which the basic unit is a molecule. Often you can see the pattern of the units in the shape of ionic and molecular crystals, as shown in **Figure 20.**

Topic: Polar Molecules
Visit blue.msscience.com for Web links to information about soaps and detergents.

Activity Oil and water are not soluble in one another. However, if you add a few grams of a liquid dish detergent, the oil will become soluble in the water. Instead of two layers, there will be only one. Explain why soap can help the oil become soluble in water.

Figure 19 Two hydrogen atoms share electrons with one oxygen atom, but the sharing is unequal. The electrons are more likely to be closer to the oxygen than the hydrogens. The space-saving model shows how the charges are separated or polarized. **Define** *the term* polar.

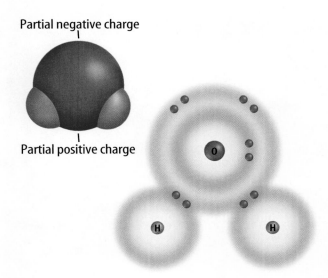

The positive ends of the water molecules are attracted to the negatively charged balloon, causing the stream of water to bend.

SECTION 2 How Elements Bond **477**

Chemical Shorthand

In medieval times, alchemists (AL kuh mists) were the first to explore the world of chemistry. Although many of them believed in magic and mystical transformations, alchemists did learn much about the properties of some elements. They even used symbols to represent them in chemical processes, some of which are shown in **Figure 21.**

Symbols for Atoms Modern chemists use symbols to represent elements, too. These symbols can be understood by chemists everywhere. Each element is represented by a one letter-, two letter-, or three-letter symbol. Many symbols are the first letters of the element's name, such as H for hydrogen and C for carbon. Others are the first letters of the element's name in another language, such as K for potassium, which stands for kalium, the Latin word for potassium.

Symbols for Compounds Compounds can be described using element symbols and numbers. For example, **Figure 22** shows how two hydrogen atoms join together in a covalent bond. The resulting hydrogen molecule is represented by the symbol H_2. The small 2 after the H in the formula is called a subscript. *Sub* means "below" and *script* means "write," so a subscript is a number that is written a little below a line of text. The subscript 2 means that two atoms of hydrogen are in the molecule.

Figure 21 Alchemists used elaborate symbols to describe elements and processes. Modern chemical symbols are letters that can be understood all over the world.

Figure 22 Chemical formulas show you the kind and number of atoms in a molecule. **Describe** *the term* subscript.

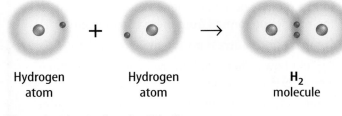

Hydrogen atom + Hydrogen atom → H_2 molecule

The subscript 2 after the H indicates that the hydrogen molecule contains two atoms of hydrogen.

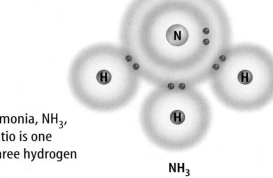

The formula for ammonia, NH_3, tells you that the ratio is one nitrogen atom to three hydrogen atoms.

NH_3

Chemical Formulas A **chemical formula** is a combination of chemical symbols and numbers that shows which elements are present in a compound and how many atoms of each element are present. When no subscript is shown, the number of atoms is understood to be one.

Reading Check *What is a chemical formula and what does it tell you about a compound?*

Now that you know a few of the rules for writing chemical formulas, you can look back at other chemical compounds shown earlier in this chapter and begin to predict their chemical formulas. A water molecule contains one oxygen atom and two hydrogen atoms, so its formula is H_2O. Ammonia, shown in **Figure 22,** is a covalent compound that contains one nitrogen atom and three hydrogen atoms. Its chemical formula is NH_3.

The black tarnish that forms on silver, shown in **Figure 23**, is a compound made up of the elements silver and sulfur in the proportion of two atoms of silver to one atom of sulfur. If alchemists knew the composition of silver tarnish, how might they have written a formula for the compound? The modern formula for silver tarnish is Ag_2S. The formula tells you that it is a compound that contains two silver atoms and one sulfur atom.

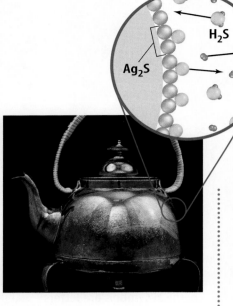

Figure 23 Silver tarnish is the compound silver sulfide, Ag_2S. The formula shows that two silver atoms are combined with one sulfur atom.

section 2 review

Summary

Four Types of Bonds
- Ionic bond is the attraction that holds ions close together.
- Metallic bonds form when metal atoms pool their electrons.
- Covalent bonds form when atoms share electrons.
- A polar covalent bond is a bond in which electrons are shared unevenly.

Chemical Shorthand
- Compounds can be described by using element symbols and numbers.
- A chemical formula is a combination of element symbols and numbers.

Self Check

1. **Determine** Use the periodic table to decide whether lithium forms a positive or negative ion. Does fluorine form a positive or negative ion? Write the formula for the compound formed from these two elements.
2. **Compare and contrast** polar and nonpolar bonds.
3. **Explain** how a chemical formula indicates the ratio of elements in a compound.
4. **Think Critically** Silicon has four electrons in its outer energy level. What type of bond is silicon most likely to form with other elements? Explain.

Applying Skills

5. **Predict** what type of bonds that will form between the following pairs of atoms: carbon and oxygen, potassium and bromine, fluorine and fluorine.

Ionic Compounds

Metals in Groups 1 and 2 often lose electrons and form positive ions. Nonmetals in Groups 16 and 17 often gain electrons and become negative ions. How can compounds form between these four groups of elements?

Real-World Question

How do different atoms combine with each other to form compounds?

Goals
- **Construct** models of electron gain and loss.
- **Determine** formulas for the ions and compounds that form when electrons are gained or lost.

Materials
paper (8 different colors)
tacks (2 different colors)
corrugated cardboard
scissors

Safety Precautions

Procedure

1. Cut colored-paper disks 7 cm in diameter to represent the elements Li, S, Mg, O, Ca, Cl, Na, and I. Label each disk with one symbol.
2. Lay circles representing the atoms Li and S side by side on cardboard.
3. Choose colored thumbtacks to represent the outer electrons of each atom. Place the tacks evenly around the disks to represent the outer electron levels of the elements.
4. Move electrons from the metal atom to the nonmetal atom so that both elements achieve noble gas arrangements of eight outer electrons. If needed, cut additional paper disks to add more atoms of one element.
5. Write the formula for each ion and the compound formed when you shift electrons.
6. Repeat steps 2 through 6 to combine Mg and O, Ca and Cl, and Na and I.

Conclude and Apply

1. **Draw** electron dot diagrams for all of the ions produced.
2. **Identify** the noble gas elements having the same electron arrangements as the ions you made in this lab.
3. **Analyze Results** Why did you have to use more than one atom in some cases? Why couldn't you take more electrons from one metal atom or add extra ones to a nonmetal atom?

Communicating Your Data

Compare your compounds and dot diagrams with those of other students in your class. **For more help, refer to the Science Skill Handbook.**

LAB **481**

Model and Invent

Atomic Structure

Goals
- **Design** a model of a chosen element.
- **Observe** the models made by others in the class and identify the elements they represent.

Possible Materials
magnetic board
rubber magnetic strips
candy-coated chocolates
scissors
paper
marker
coins

Safety Precautions

WARNING: *Never eat any food in the laboratory. Wash hands thoroughly.*

Real-World Question

As more information has become known about the structure of the atom, scientists have developed new models. Making your own model and studying the models of others will help you learn how protons, neutrons, and electrons are arranged in an atom. Can an element be identified based on a model that shows the arrangement of the protons, neutrons, and electrons of an atom? How will your group construct a model of an element that others will be able to identify?

Make A Model

1. Choose an element from periods 2 or 3 of the periodic table. How can you determine the number of protons, neutrons, and electrons in an atom given the atom's mass number?

2. How can you show the difference between protons and neutrons? What materials will you use to represent the electrons of the atom? How will you represent the nucleus?

3. How will you model the arrangement of electrons in the atom? Will the atom have a charge? Is it possible to identify an atom by the number of protons it has?

4. Make sure your teacher approves your plan before you proceed.

Test Your Model

1. **Construct** your model. Then record your observations in your Science Journal and include a sketch.
2. **Construct** another model of a different element.
3. **Observe** the models made by your classmates. Identify the elements they represent.

Analyze Your Data

1. **State** what elements you identified using your classmates' models.
2. **Identify** which particles always are present in equal numbers in a neutral atom.
3. **Predict** what would happen to the charge of an atom if one of the electrons were removed.
4. **Describe** what happens to the charge of an atom if two electrons are added. What happens to the charge of an atom if one proton and one electron are removed?
5. **Compare and contrast** your model with the electron cloud model of the atom. How is your model similar? How is it different?

Conclude and Apply

1. **Define** the minimum amount of information that you need to know in order to identify an atom of an element.
2. **Explain** If you made models of the isotopes boron-10 and boron-11, how would these models be different?

Communicating Your Data

Compare your models with those of other students. Discuss any differences you find among the models.

Science and Language Arts

"Baring the Atom's Mother Heart"
from Selu: Seeking the Corn-Mother's Wisdom
by Marilou Awiakta

Author Marilou Awiakta was raised near Oak Ridge National Laboratory, a nuclear research laboratory in Tennessee where her father worked. She is of Cherokee and Irish descent. This essay resulted from conversations the author had with writer Alice Walker. It details the author's concern with nuclear technology.

"What is the atom, Mother? Will it hurt us?"

I was nine years old. It was December 1945. Four months earlier, in the heat of an August morning—Hiroshima. Destruction. Death. Power beyond belief, released from something invisible[1]. Without knowing its name, I'd already felt the atoms' power in another form…

"What is the atom, Mother? Will it hurt us?"

"It can be used to hurt everybody, Marilou. It killed thousands[2] of people in Hiroshima and Nagasaki. But the atom itself. . . ? It's invisible, the smallest bit of matter. And it's in everything. Your hand, my dress, the milk you're drinking—. . .

. . . Mother already had taught me that beyond surface differences, everything is [connected]. It seemed natural for the atom to be part of this connection. At school, when I was introduced to Einstein's theory of relativity—that energy and matter are one—I accepted the concept easily.

1 can't see
2 10,500

Understanding Literature

Refrain Refrains are emotionally charged words or phrases that are repeated throughout a literary work and can serve a number of purposes. In this work, the refrain is when the author asks, "What is the atom, Mother? Will it hurt us?" Do you think the refrain helps the reader understand the importance of the atom?

Respond to the Reading

1. How did the author's mother explain the atom to her?
2. Is this a positive or negative explanation of the atom?
3. **Linking Science and Writing** Write a short poem about some element you learned about in this chapter.

 Nuclear fission, or splitting atoms, is the breakdown of an atom's nucleus. It occurs when a particle, such as a neutron, strikes the nucleus of a uranium atom, splitting the nucleus into two fragments, called fission fragments, and releasing two or three neutrons. These released neutrons ultimately cause a chain reaction by splitting more nuclei and releasing more neutrons. When it is uncontrolled, this chain reaction results in a devastating explosion.

chapter 16 Study Guide

Reviewing Main Ideas

Section 1 — Why do atoms combine?

1. The electrons in the electron cloud of an atom are arranged in energy levels.
2. Each energy level can hold a specific number of electrons.
3. The periodic table supplies a great deal of information about the elements.
4. The number of electrons in an atom increases across each period of the periodic table.
5. The noble gas elements are stable because their outer energy levels are stable.
6. Electron dot diagrams show the electrons in the outer energy level of an atom.

Section 2 — How Elements Bond

1. An atom can become stable by gaining, losing, or sharing electrons so that its outer energy level is full.
2. Ionic bonds form when a metal atom loses one or more electrons and a nonmetal atom gains one or more electrons.
3. Covalent bonds are created when two or more nonmetal atoms share electrons.
4. The unequal sharing of electrons results in a polar covalent bond.
5. A chemical formula indicates the kind and number of atoms in a compound.

Visualizing Main Ideas

Copy and complete the following concept map on types of bonds.

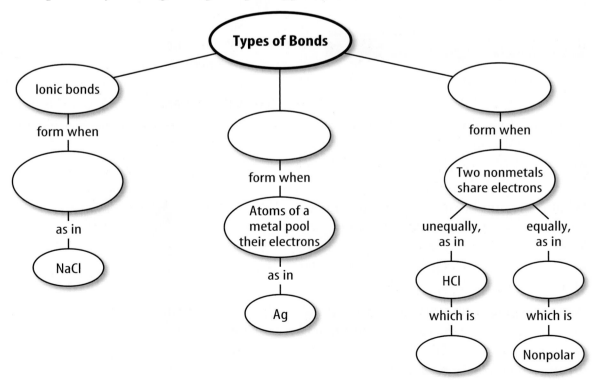

Chapter 16 Review

Using Vocabulary

chemical bond p. 471
chemical formula p. 480
compound p. 473
covalent bond p. 475
electron cloud p. 464
electron dot diagram p. 470
energy level p. 465
ion p. 473
ionic bond p. 473
metallic bond p. 474
molecule p. 475
polar bond p. 476

Distinguish between the terms in each of the following pairs.

1. ion—molecule
2. molecule—compound
3. electron dot diagram—ion
4. chemical formula—molecule
5. ionic bond—covalent bond
6. electron cloud—electron dot diagram
7. covalent bond—polar bond
8. compound—formula
9. metallic bond—ionic bond

Checking Concepts

Choose the word or phrase that best answers the question.

10. Which of the following is a covalently bonded molecule?
    A) Cl_2 C) Ne
    B) air   D) salt

11. What is the number of the group in which the elements have a stable outer energy level?
    A) 1   C) 16
    B) 13   D) 18

12. Which term describes the units that make up substances formed by ionic bonding?
    A) ions   C) acids
    B) molecules   D) atoms

13. Which of the following describes what is represented by the symbol Cl^-?
    A) an ionic compound
    B) a polar molecule
    C) a negative ion
    D) a positive ion

14. What happens to electrons in the formation of a polar covalent bond?
    A) They are lost.
    B) They are gained.
    C) They are shared equally.
    D) They are shared unequally.

15. Which of the following compounds is unlikely to contain ionic bonds?
    A) NaF   C) LiCl
    B) CO   D) $MgBr_2$

16. Which term describes the units that make up compounds with covalent bonds?
    A) ions   C) salts
    B) molecules   D) acids

17. In the chemical formula CO_2, the subscript 2 shows which of the following?
    A) There are two oxygen ions.
    B) There are two oxygen atoms.
    C) There are two CO_2 molecules.
    D) There are two CO_2 compounds.

Use the figure below to answer question 18.

18. Which is NOT true about the molecule H_2O?
    A) It contains two hydrogen atoms.
    B) It contains one oxygen atom.
    C) It is a polar covalent compound.
    D) It is an ionic compound.

chapter 16 Review

Thinking Critically

19. **Explain** why Groups 1 and 2 form many compounds with Groups 16 and 17.

Use the illustration below to answer questions 20 and 21.

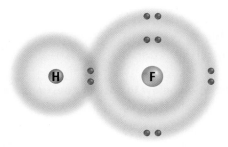

20. **Explain** what type of bond is shown here.

21. **Predict** In the HF molecule above, predict if the electrons are shared equally or unequally between the two atoms. Where do the electrons spend more of their time?

22. **Analyze** When salt dissolves in water, the sodium and chloride ions separate. Explain why this might occur.

23. **Interpret Data** Both cesium, in period 6, and lithium, in period 2, are in the alkali metals family. Cesium is more reactive. Explain this using the energy step diagram in **Figure 4**.

24. **Explain** Use the fact that water is a polar molecule to explain why water has a much higher boiling point than other molecules of its size.

25. **Predict** If equal masses of CuCl and $CuCl_2$ decompose into their components—copper and chlorine—predict which compound will yield more copper. Explain.

26. **Concept Map** Draw a concept map starting with the term *Chemical Bond* and use all the vocabulary words.

27. **Recognize Cause and Effect** A helium atom has only two electrons. Why does helium behave as a noble gas?

28. **Draw a Conclusion** A sample of an element can be drawn easily into wire and conducts electricity well. What kind of bonds can you conclude are present?

Performance Activities

29. **Display** Make a display featuring one of the element families described in this chapter. Include electronic structures, electron dot diagrams, and some compounds they form.

Applying Math

Use the table below to answer question 30.

Formulas of Compounds		
Compound	Number of Metal Atoms	Number of Nonmetal Atoms
Cu_2O		
Al_2S_3	Do not write in this book.	
NaF		
$PbCl_4$		

30. **Make and Use Tables** Fill in the second column of the table with the number of metal atoms in one unit of the compound. Fill in the third column with the number of atoms of the nonmetal in one unit.

31. **Molecules** What are the percentages of each atom for this molecule, K_2CO_3?

32. **Ionic Compounds** Lithium, as a positive ion, is written as Li^{1+}. Nitrogen, as a negative ion, is written as N^{3-}. In order for the molecule to be neutral, the plus and minus charges have to equal zero. How many lithium atoms are needed to make the charges equal to zero?

33. **Energy Levels** Calculate the maximum number of electrons in energy level 6.

Chapter 16 Standardized Test Practice

Part 1 | Multiple Choice

Record your answers on the answer sheet provided by your teacher or on a sheet of paper.

1. Sodium combines with fluorine to produce sodium fluoride (NaF), an active ingredient in toothpaste. In this form, sodium has the electron configuration of which other element?
   A. neon
   B. magnesium
   C. lithium
   D. chlorine

Use the illustration below to answer questions 2 and 3.

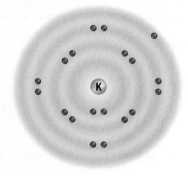

2. The illustration above shows the electron configuration for potassium. How many electrons does potassium need to gain or lose to become stable?
   A. gain 1
   B. gain 2
   C. lose 1
   D. lose 2

3. Potassium belongs to the Group 1 family of elements on the periodic table. What is the name of this group?
   A. halogens
   B. alkali metals
   C. noble gases
   D. alkaline metals

4. What type of bond connects the atoms in a molecule of nitrogen gas (N_2)?
   A. ionic
   B. single
   C. double
   D. triple

Use the illustration below to answer questions 5 and 6.

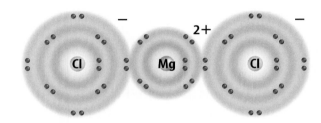

Magnesium chloride

5. The illustration above shows the electron distribution for magnesium chloride. Which of the following is the correct way to write the formula for magnesium chloride?
   A. Mg_2Cl
   B. $MgCl_2$
   C. $MgCl$
   D. Mg_2Cl_2

6. Which of the following terms best describes the type of bonding in magnesium chloride?
   A. ionic
   B. pooling
   C. metallic
   D. covalent

7. What is the maximum number of electrons in the third energy level?
   A. 8
   B. 16
   C. 18
   D. 24

Standardized Test Practice

Part 2 Short Response/Grid In

Record your answers on the answer sheet provided by your teacher or on a sheet of paper.

8. What is an electron cloud?

9. Explain what is wrong with the following statement: All covalent bonds between atoms are polar to some degree because each element differs slightly in its ability to attract electrons. Give an example to support your answer.

Use the illustration below to answer questions 10 and 11.

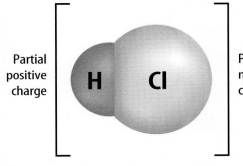

10. The illustration above shows how hydrogen and chlorine combine to form a polar molecule. Explain why the bond is polar.

11. What is the electron dot diagram for the molecule in the illustration?

12. What is the name of the family of elements in Group 17 of the periodic table?

13. Name two ways that electrons around a nucleus are different from planets circling the Sun.

14. Which family of elements used to be known as inert gases? Why was the name changed?

Test-Taking Tip

Take Your Time Stay focused during the test and don't rush, even if you notice that other students are finishing the test early.

Part 3 Open Ended

Record your answers on a sheet of paper.

15. Scientific experiments frequently require an oxygen-free environment. Such experiments often are performed in containers flooded with argon gas. Describe the arrangement of electrons in an argon atom. Why is argon often a good choice for these experiments?

16. Which group of elements is called the halogen elements? Describe their electron configurations and discuss their reactivity. Name two elements that belong to this group.

17. What is an ionic bond? Describe how sodium chloride forms an ionic bond.

18. Explain metallic bonding. What are some ways this affects the properties of metals?

19. Explain why polar molecules exist, but polar ionic compounds do not exist.

Use the illustration below to answer questions 20 and 21.

20. Explain what is happening in the photograph above. What would happen if the balloon briefly touched the water?

21. Draw a model showing the electron distribution for a water molecule. Explain how the position of the electrons causes the effect shown in your illustration.

chapter 17

Chemical Reactions

The BIG Idea

In chemical reactions, atoms in reactants are rearranged to form products with different chemical properties.

SECTION 1
Chemical Formulas and Equations
Main Idea Atoms are not created or destroyed in chemical reactions—they are just rearranged.

SECTION 2
Rates of Chemical Reactions
Main Idea Reaction rates are affected by several things, including temperature, concentration, surface area, inhibitors, and catalysts.

What chemical reactions happen at chemical plants?

Chemical plants like this one provide the starting materials for thousands of chemical reactions. The compact discs you listen to, personal items, such as shampoo and body lotion, and medicines all have their beginnings in a chemical plant.

Science Journal What additional types of products do you think are manufactured in a chemical plant?

Start-Up Activities

Identify a Chemical Reaction

You can see substances changing every day. Fuels burn, giving energy to cars and trucks. Green plants convert carbon dioxide and water into oxygen and sugar. Cooking an egg or baking bread causes changes too. These changes are called chemical reactions. In this lab you will observe a common chemical change.

WARNING: *Do not touch the test tube. It will be hot. Use extreme caution around an open flame. Point test tubes away from you and others.*

1. Place 3 g of sugar into a large test tube.
2. Carefully light a laboratory burner.
3. Using a test-tube holder, hold the bottom of the test tube just above the flame for 45 s or until something happens with the sugar.
4. Observe any change that occurs.
5. **Think Critically** Describe in your Science Journal the changes that took place in the test tube. What do you think happened to the sugar? Was the substance that remained in the test tube after heating the same as the substance you started with?

FOLDABLES Study Organizer

Chemical Reaction Make the following Foldable to help you understand chemical reactions.

STEP 1 Fold a vertical sheet of notebook paper in half lengthwise.

STEP 2 Cut along every third line of only the top layer to form tabs.

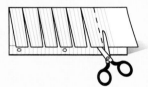

STEP 3 Label each tab.

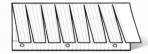

Research Information Before you read the chapter, write several questions you have about chemical reactions on the front of the tabs. As you read, add more questions. Under the tabs of your Foldable, write answers to the questions you recorded on the tabs.

Preview this chapter's content and activities at blue.msscience.com

Get Ready to Read

Make Predictions

① Learn It! A prediction is an educated guess based on what you already know. One way to predict while reading is to guess what you believe the author will tell you next. As you are reading, each new topic should make sense because it is related to the previous paragraph or passage.

② Practice It! Read the excerpt below from Section 1. Based on what you have read, make predictions about what you will read in the rest of the lesson. After you read Section 1, go back to your predictions to see if they were correct.

Predict what type of properties chemical changes affect.

Do you think melting is a chemical or physical change?

Predict what might happen to the atoms in water when it undergoes a chemical change.

> Matter can undergo two kinds of changes—physical and chemical. Physical changes in a substance affect only the physical properties, such as size and shape, or whether it is a solid, liquid, or gas. For example, when water freezes, its physical state changes from liquid to solid, but it is still water.
>
> —*from page 492*

③ Apply It! Before you read, skim the questions in the Chapter Review. Choose three questions and predict the answers.

Target Your Reading

Reading Tip

As you read, check the predictions you made to see if they were correct.

Use this to focus on the main ideas as you read the chapter.

1 Before you read the chapter, respond to the statements below on your worksheet or on a numbered sheet of paper.
- Write an **A** if you **agree** with the statement.
- Write a **D** if you **disagree** with the statement.

2 After you read the chapter, look back to this page to see if you've changed your mind about any of the statements.
- If any of your answers changed, explain why.
- Change any false statements into true statements.
- Use your revised statements as a study guide.

Science Online
Print out a worksheet of this page at blue.msscience.com

Before You Read A or D		Statement	After You Read A or D
	1	Burning is an example of a chemical change.	
	2	A chemical equation only tells the names of reactants and products.	
	3	When a substance burns, atoms disappear and new atoms are created.	
	4	When balancing a chemical equation, it's okay to change the subscripts of a chemical formula.	
	5	Some reactions release energy and some adsorb energy.	
	6	During chemical reactions, bonds in the reactants break and new bonds form.	
	7	Reactions that release energy do not need any energy to start the reaction.	
	8	Increasing temperature will speed up most chemical reactions.	

section 1

Chemical Formulas and Equations

as you read

What You'll Learn
- **Determine** whether or not a chemical reaction is occurring.
- **Determine** how to read and understand a balanced chemical equation.
- **Examine** some reactions that release energy and others that absorb energy.
- **Explain** the law of conservation of mass.

Why It's Important
Chemical reactions warm your home, cook your meals, digest your food, and power cars and trucks.

Review Vocabulary
atom: the smallest piece of matter that still retains the property of the element

New Vocabulary
- chemical reaction
- reactant
- product
- chemical equation
- endothermic reaction
- exothermic reaction

Physical or Chemical Change?

You can smell a rotten egg and see the smoke from a campfire. Signs like these tell you that a chemical reaction is taking place. Other evidence might be less obvious, but clues are always present to announce that a reaction is under way.

Matter can undergo two kinds of changes—physical and chemical. Physical changes in a substance affect only physical properties, such as its size and shape, or whether it is a solid, liquid, or gas. For example, when water freezes, its physical state changes from liquid to solid, but it's still water.

In contrast, chemical changes produce new substances that have properties different from those of the original substances. The rust on a bike's handlebars, for example, has properties different from those of the metal around it. Another example is the combination of two liquids that produce a precipitate, which is a solid, and a liquid. The reaction of silver nitrate and sodium chloride forms solid silver chloride and liquid sodium nitrate. A process that produces chemical change is a **chemical reaction.**

To compare physical and chemical changes, look at the newspaper shown in **Figure 1.** If you fold it, you change its size and shape, but it is still newspaper. Folding is a physical change. If you use it to start a fire, it will burn. Burning is a chemical change because new substances result. How can you recognize a chemical change? **Figure 2** shows what to look for.

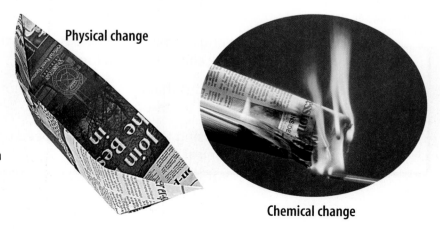

Physical change

Chemical change

Figure 1 Newspaper can undergo both physical and chemical changes.

492 CHAPTER 17 Chemical Reactions

NATIONAL GEOGRAPHIC VISUALIZING CHEMICAL REACTIONS

Figure 2

Chemical reactions take place when chemicals combine to form new substances. Your senses—sight, taste, hearing, smell, and touch—can help you detect chemical reactions in your environment.

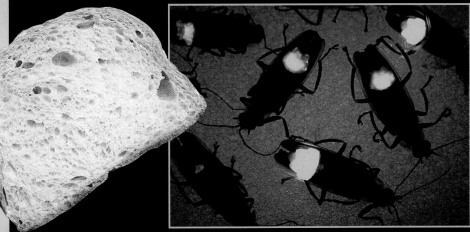

▲ **SIGHT** When you spot a firefly's bright glow, you are seeing a chemical reaction in progress—two chemicals are combining in the firefly's abdomen and releasing light in the process. The holes in a slice of bread are visible clues that sugar molecules were broken down by yeast cells in a chemical reaction that produces carbon dioxide gas. The gas caused the bread dough to rise.

▼ **TASTE** A boy grimaces after sipping milk that has gone sour due to a chemical reaction.

▲ **SMELL AND TOUCH** Billowing clouds of acrid smoke and waves of intense heat indicate that chemical reactions are taking place in this burning forest.

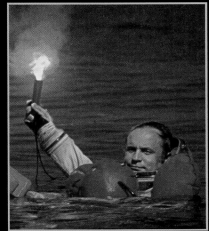

▲ **HEARING** A Russian cosmonaut hoists a flare into the air after landing in the ocean during a training exercise. The hissing sound of the burning flare is the result of a chemical reaction.

SECTION 1 Chemical Formulas and Equations **493**

Chemical Equations

To describe a chemical reaction, you must know which substances react and which substances are formed in the reaction. The substances that react are called the reactants (ree AK tunts). **Reactants** are the substances that exist before the reaction begins. The substances that form as a result of the reaction are called the **products.**

When you mix baking soda and vinegar, a vigorous chemical reaction occurs. The mixture bubbles and foams up inside the container, as you can see in **Figure 3.**

Baking soda and vinegar are the common names for the reactants in this reaction, but they also have chemical names. Baking soda is the compound sodium hydrogen carbonate (often called sodium bicarbonate), and vinegar is a solution of acetic (uh SEE tihk) acid in water. What are the products? You saw bubbles form when the reaction occurred, but is that enough of a description?

Describing What Happens Bubbles tell you that a gas has been produced, but they don't tell you what kind of gas. Are bubbles of gas the only product, or do some atoms from the vinegar and baking soda form something else? What goes on in the chemical reaction can be more than what you see with your eyes. Chemists try to find out which reactants are used and which products are formed in a chemical reaction. Then, they can write it in a shorthand form called a chemical equation. A **chemical equation** tells chemists at a glance the reactants, products, and proportions of each substance present. Some equations also tell the physical state of each substance.

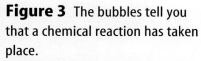 *What does a chemical equation tell chemists?*

Figure 3 The bubbles tell you that a chemical reaction has taken place.
Predict *how you might find out whether a new substance has formed.*

Table 1 Reactions Around the Home

Reactants		Products
Baking soda + Vinegar	→	Gas + White solid
Charcoal + Oxygen	→	Ash + Gas + Heat
Iron + Oxygen + Water	→	Rust
Silver + Hydrogen sulfide	→	Black tarnish + Gas
Gas (kitchen range) + Oxygen	→	Gas + Heat
Sliced apple + Oxygen	→	Apple turns brown

Using Words One way you can describe a chemical reaction is with an equation that uses words to name the reactants and products. The reactants are listed on the left side of an arrow, separated from each other by plus signs. The products are placed on the right side of the arrow, also separated by plus signs. The arrow between the reactants and products represents the changes that occur during the chemical reaction. When reading the equation, the arrow is read as *produces*.

You can begin to think of processes as chemical reactions even if you do not know the names of all the substances involved. **Table 1** can help you begin to think like a chemist. It shows the word equations for chemical reactions you might see around your home. See how many other reactions you can find. Look for the signs you have learned that indicate a reaction might be taking place. Then, try to write them in the form shown in the table.

Using Chemical Names Many chemicals used around the home have common names. For example, acetic acid dissolved in water is called vinegar. Some chemicals, such as baking soda, have two common names—it also is known as sodium bicarbonate. However, chemical names are usually used in word equations instead of common names. In the baking soda and vinegar reaction, you already know the chemical names of the reactants—sodium hydrogen carbonate and acetic acid. The names of the products are sodium acetate, water, and carbon dioxide. The word equation for the reaction is as follows.

Acetic acid + Sodium hydrogen carbonate → Sodium acetate + Water + Carbon dioxide

Autumn Leaves A color change can indicate a chemical reaction. When leaves change colors in autumn, the reaction may not be what you expect. The bright yellow and orange are always in the leaves, but masked by green chlorophyll. When the growth season ends, more chlorophyll is broken down than produced. The orange and yellow colors become visible.

Mini LAB

Observing the Law of Conservation of Mass

Procedure

1. Place a piece of **steel wool** into a **medium test tube**. Seal the end of the test tube with a **balloon**.
2. Find the mass.
3. Using a test-tube holder, heat the bottom of the tube for two minutes in a **hot water bath** provided by your teacher. Allow the tube to cool completely.
4. Find the mass again.

Analysis

1. What did you observe that showed a chemical reaction took place?
2. Compare the mass before and after the reaction.
3. Why was it important for the test tube to be sealed?

Using Formulas The word equation for the reaction of baking soda and vinegar is long. That's why chemists use chemical formulas to represent the chemical names of substances in the equation. You can convert a word equation into a chemical equation by substituting chemical formulas for the chemical names. For example, the chemical equation for the reaction between baking soda and vinegar can be written as follows:

$$CH_3COOH + NaHCO_3 \rightarrow CH_3COONa + H_2O + CO_2$$

Acetic acid (vinegar)   Sodium hydrogen carbonate (baking soda)   Sodium acetate   Water   Carbon dioxide

Subscripts When you look at chemical formulas, notice the small numbers written to the right of the atoms. These numbers, called subscripts, tell you the number of atoms of each element in that compound. For example, the subscript 2 in CO_2 means that each molecule of carbon dioxide has two oxygen atoms. If an atom has no subscript, it means that only one atom of that element is in the compound, so carbon dioxide has only one carbon atom.

Conservation of Mass

What happens to the atoms in the reactants when they are converted into products? According to the law of conservation of mass, the mass of the products must be the same as the mass of the reactants in that chemical reaction. This principle was first stated by the French chemist Antoine Lavoisier (1743–1794), who is considered the first modern chemist. Lavoisier used logic and scientific methods to study chemical reactions. He proved by his experiments that nothing is lost or created in chemical reactions.

He showed that chemical reactions are much like mathematical equations. In math equations, the right and left sides of the equation are numerically equal. Chemical equations are similar, but it is the number and kind of atoms that are equal on the two sides. Every atom that appears on the reactant side of the equation also appears on the product side, as shown in **Figure 4**. Atoms are never lost or created in a chemical reaction; however, they do change partners.

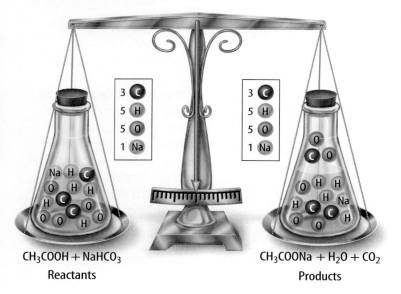

Figure 4 The law of conservation of mass states that the number and kind of atoms must be equal for products and reactants.

$CH_3COOH + NaHCO_3$
Reactants

$CH_3COONa + H_2O + CO_2$
Products

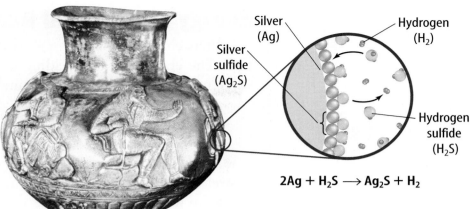

Figure 5 Keeping silver bright takes frequent polishing, especially in homes heated by gas. Sulfur compounds found in small concentrations in natural gas react with silver, forming black silver sulfide, Ag_2S.

$$2Ag + H_2S \longrightarrow Ag_2S + H_2$$

Balancing Chemical Equations

When you write the chemical equation for a reaction, you must observe the law of conservation of mass. Look back at **Figure 4.** It shows that when you count the number of carbon, hydrogen, oxygen, and sodium atoms on each side of the arrow in the equation, you find equal numbers of each kind of atom. This means the equation is balanced and the law of conservation of mass is observed.

Not all chemical equations are balanced so easily. For example, silver tarnishes, as in **Figure 5,** when it reacts with sulfur compounds in the air, such as hydrogen sulfide. The following unbalanced equation shows what happens when silver tarnishes.

$$\underset{\text{Silver}}{Ag} + \underset{\text{Hydrogen sulfide}}{H_2S} \rightarrow \underset{\text{Silver sulfide}}{Ag_2S} + \underset{\text{Hydrogen}}{H_2}$$

Count the Atoms Count the number of atoms of each type in the reactants and in the products. The same numbers of hydrogen and sulfur atoms are on each side, but one silver atom is on the reactant side and two silver atoms are on the product side. This cannot be true. A chemical reaction cannot create a silver atom, so this equation does not represent the reaction correctly. Place a 2 in front of the reactant Ag and check to see if the equation is balanced. Recount the number of atoms of each type.

$$2Ag + H_2S \rightarrow Ag_2S + H_2$$

The equation is now balanced. There are an equal number of silver atoms in the reactants and the products. When balancing chemical equations, numbers are placed before the formulas as you did for Ag. These are called coefficients. However, never change the subscripts written to the right of the atoms in a formula. Changing these numbers changes the identity of the compound.

Topic: Chemical Equations
Visit blue.msscience.com for Web links to information about chemical equations and balancing them.

Activity Find a chemical reaction that takes place around your home or school. Write a chemical equation describing it.

Energy in Chemical Reactions

Often, energy is released or absorbed during a chemical reaction. The energy for the welding torch in **Figure 6** is released when hydrogen and oxygen combine to form water.

$$2H_2 + O_2 \rightarrow 2H_2O + \text{energy}$$

Energy Released Where does this energy come from? To answer this question, think about the chemical bonds that break and form when atoms gain, lose, or share electrons. When such a reaction takes place, bonds break in the reactants and new bonds form in the products. In reactions that release energy, the products are more stable, and their bonds have less energy than those of the reactants. The extra energy is released in various forms—light, sound, and thermal energy.

Applying Math — Balancing Equations

CONSERVING MASS Methane and oxygen react to form carbon dioxide and water. You can see how mass is conserved by balancing the equation: $CH_4 + O_2 \rightarrow CO_2 + H_2O$.

Solution

1 *This is what you know:*

The number of atoms of C, H, and O in reactants and products.

2 *This is what you need to do:*

Make sure that the reactants and products have equal numbers of atoms of each element. Start with the reactant having the greatest number of different elements.

Reactants	Products	Action
$CH_4 + O_2$ have 4 H atoms	$CO_2 + H_2O$ have 2 H atoms	Need 2 more H atoms in Products. Multiply H_2O by 2 to give 4 H atoms
$CH_4 + O_2$ have 2 O atoms	$CO_2 + 2H_2O$ have 4 O atoms	Need 2 more O atoms in Reactants. Multiply O_2 by 2 to give 4 O atoms

The balanced equation is $CH_4 + 2O_2 \rightarrow CO_2 + 2H_2O$.

3 *Check your answer:*

Count the carbons, hydrogens, and oxygens on each side.

Practice Problems

1. Balance the equation $Fe_2O_3 + CO \rightarrow Fe_3O_4 + CO_2$.
2. Balance the equation $Al + I_2 \rightarrow AlI_3$.

For more practice, visit
blue.msscience.com/
math_practice

Figure 6 This welding torch burns hydrogen and oxygen to produce temperatures above 3,000°C. It can even be used underwater.
Identify the products of this chemical reaction.

Energy Absorbed What happens when the reverse situation occurs? In reactions that absorb energy, the reactants are more stable, and their bonds have less energy than those of the products.

$$2H_2O + \text{energy} \rightarrow 2H_2 + O_2$$
$$\text{Water} \qquad\qquad\qquad \text{Hydrogen} \quad \text{Oxygen}$$

In this reaction the extra energy needed to form the products can be supplied in the form of electricity, as shown in **Figure 7**.

Reactions can release or absorb several kinds of energy, including electricity, light, sound, and thermal energy. When thermal energy is gained or lost in reactions, special terms are used. **Endothermic** (en doh THUR mihk) **reactions** absorb thermal energy. **Exothermic** (ek soh THUR mihk) **reactions** release thermal energy. The root word *therm* refers to heat, as it does in thermos bottles and thermometers.

Energy Released Several types of reactions that release thermal energy. Burning is an exothermic chemical reaction in which a substance combines with oxygen to produce thermal energy along with light, carbon dioxide, and water.

Reading Check What type of chemical reaction is burning?

Rapid Release Sometimes energy is released rapidly. For example, charcoal lighter fluid combines with oxygen in the air and produces enough thermal energy to ignite a charcoal fire within a few minutes.

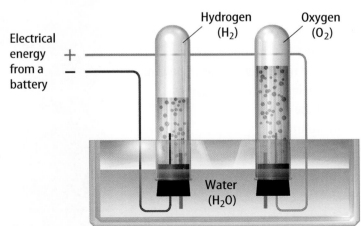

Figure 7 Electrical energy is needed to break water into its components. This is the reverse of the reaction that takes place in the welding torch shown in **Figure 6**.

SECTION 1 Chemical Formulas and Equations **499**

Figure 8 Two exothermic reactions are shown. The charcoal fire to cook the food was started when lighter fluid combined rapidly with oxygen in air. The iron in the wheelbarrow combined slowly with oxygen in the air to form rust.

Fast reaction

Slow reaction

Slow Release Other materials also combine with oxygen but release thermal energy so slowly that you cannot see or feel it happen. This is the case when iron combines with oxygen in the air to form rust. The slow release of thermal energy from a reaction also is used in heat packs that can keep your hands warm for several hours. Fast and slow energy release are compared in **Figure 8.**

Energy Absorbed Some chemical reactions and physical processes need to have thermal energy added before they can proceed. An example of an endothermic physical process that absorbs thermal energy is the cold pack shown in **Figure 9.**

The heavy plastic cold pack holds ammonium nitrate and water. The two substances are separated by a plastic divider. When you squeeze the bag, you break the divider so that the ammonium nitrate dissolves in the water. The dissolving process absorbs thermal energy, which must come from the surrounding environment—the surrounding air or your skin after you place the pack on the injury.

Figure 9 The thermal energy needed to dissolve the ammonium nitrate in this cold pack comes from the surrounding environment.

500 **CHAPTER 17** Chemical Reactions

Energy in the Equation The word *energy* often is written in equations as either a reactant or a product. Energy written as a reactant helps you think of energy as a necessary ingredient for the reaction to take place. For example, electrical energy is needed to break up water into hydrogen and oxygen. It is important to know that energy must be added to make this reaction occur.

Similarly, in the equation for an exothermic reaction, the word *energy* often is written along with the products. This tells you that energy is released. You include energy when writing the reaction that takes place between oxygen and methane in natural gas when you cook on a gas range, as shown in **Figure 10.** This thermal energy cooks your food.

Figure 10 Energy from a chemical reaction is used to cook.
Determine *if energy is used as a reactant or a product in this reaction.*

$$CH_4 + 2O_2 \rightarrow CO_2 + 2H_2O + energy$$
Methane Oxygen   Carbon   Water
                    dioxide

Although it is not necessary, writing the word *energy* can draw attention to an important aspect of the equation.

section 1 review

Summary

Physical or Chemical Change?
- Matter can undergo physical and chemical changes.
- A chemical reaction produces chemical changes.

Chemical Equations
- A chemical equation describes a chemical reaction.
- Chemical formulas represent chemical names for substances.
- A balanced chemical equation has the same number of atoms of each kind on both sides of the equation.

Energy in Chemical Reactions
- Endothermic reactions absorb thermal energy.
- Exothermic reactions release thermal energy.

Self Check

1. **Determine** if each of these equations is balanced. Why or why not?
   a. $Ca + Cl_2 \rightarrow CaCl_2$
   b. $Zn + Ag_2S \rightarrow ZnS + Ag$
2. **Describe** what evidence might tell you that a chemical reaction has occurred.
3. **Think Critically** After a fire, the ashes have less mass and take up less space than the trees and vegetation before the fire. How can this be explained in terms of the law of conservation of mass?

Applying Math

4. **Calculate** The equation for the decomposition of silver oxide is $2Ag_2O \rightarrow 4Ag + O_2$. Set up a ratio to calculate the number of oxygen molecules released when 1 g of silver oxide is broken down. There are 2.6×10^{21} molecules in 1 g of silver oxide.

SECTION 1 Chemical Formulas and Equations

section 2
Rates of Chemical Reactions

as you read

What You'll Learn
- **Determine** how to describe and measure the speed of a chemical reaction.
- **Identify** how chemical reactions can be sped up or slowed down.

Why It's Important
Speeding up useful reactions and slowing down destructive ones can be helpful.

Review Vocabulary
state of matter: physical property that is dependent on temperature and pressure and occurs in four forms—solid, liquid, gas, or plasma

New Vocabulary
- activation energy
- rate of reaction
- concentration
- inhibitor
- catalyst
- enzyme

How Fast?

Fireworks explode in rapid succession on a summer night. Old copper pennies darken slowly while they lie forgotten in a drawer. Cooking an egg for two minutes instead of five minutes makes a difference in the firmness of the yolk. The amount of time you leave coloring solution on your hair must be timed accurately to give the color you want. Chemical reactions are common in your life. However, notice from these examples that time has something to do with many of them. As you can see in **Figure 11,** not all chemical reactions take place at the same rate.

Some reactions, such as fireworks or lighting a campfire, need help to get going. You may also notice that others seem to start on their own. In this section, you will also learn about factors that make reactions speed up or slow down once they get going.

Figure 11 Reaction speeds vary greatly. Fireworks are over in a few seconds. However, the copper coating on pennies darkens slowly as it reacts with substances it touches.

502 CHAPTER 17 Chemical Reactions

Activation Energy—Starting a Reaction

Before a reaction can start, molecules of the reactants have to bump into each other, or collide. This makes sense because to form new chemical bonds, atoms have to be close together. But, not just any collision will do. The collision must be strong enough. This means the reactants must smash into each other with a certain amount of energy. Anything less, and the reaction will not occur. Why is this true?

To form new bonds in the product, old bonds must break in the reactants, and breaking bonds takes energy. To start any chemical reaction, a minimum amount of energy is needed. This energy is called the **activation energy** of the reaction.

Science Online

Topic: Olympic Torch
Visit blue.msscience.com for Web links to information about the Olympic Torch.

Activity With each new Olympics, the host city devises a new Olympic Torch. Research the process that goes into developing the torch and the fuel it uses.

Reading Check *What term describes the minimum amount of energy needed to start a reaction?*

What about reactions that release energy? Is there an activation energy for these reactions too? Yes, even though they release energy later, these reactions also need enough energy to start.

One example of a reaction that needs energy to start is the burning of gasoline. You have probably seen movies in which a car plunges over a cliff, lands on the rocks below, and suddenly bursts into flames. But if some gasoline is spilled accidentally while filling a gas tank, it probably will evaporate harmlessly in a short time.

Why doesn't this spilled gasoline explode as it does in the movies? The reason is that gasoline needs energy to start burning. That is why there are signs at filling stations warning you not to smoke. Other signs advise you to turn off the ignition, not to use mobile phones, and not to reenter the car until fueling is complete.

This is similar to the lighting of the Olympic Cauldron, as shown in **Figure 12.** Cauldrons designed for each Olympics contain highly flammable materials that cannot be extinguished by high winds or rain. However, they do not ignite until the opening ceremonies when a runner lights the cauldron using a flame that was kindled in Olympia, Greece, the site of the original Olympic Games.

Figure 12 Most fuels need energy to ignite. The Olympic Torch, held by Cathy Freeman in the 2000 Olympics, provided the activation energy required to light the fuel in the cauldron.

Reaction Rate

Many physical processes are measured in terms of a rate. A rate tells you how much something changes over a given period of time. For example, the rate or speed at which you run or ride your bike is the distance you move divided by the time it took you to move that distance. You may jog at a rate of 8 km/h.

Chemical reactions have rates, too. The **rate of reaction** tells how fast a reaction occurs after it has started. To find the rate of a reaction, you can measure either how quickly one of the reactants is consumed or how quickly one of the products is created, as in **Figure 13.** Both measurements tell how the amount of a substance changes per unit of time.

Figure 13 The diminishing amount of wax in this candle as it burns indicates the rate of the reaction.

Reading Check *What can you measure to determine the rate of a reaction?*

Reaction rate is important in industry because the faster the product can be made, the less it usually costs. However, sometimes fast rates of reaction are undesirable such as the rates of reactions that cause fruit to ripen. The slower the reaction rate, the longer the food will stay edible. What conditions control the reaction rate, and how can the rate be changed?

Temperature Changes Rate You can slow the ripening of some fruits by putting them in a refrigerator, as in **Figure 14.** Ripening is caused by a series of chemical reactions. Lowering the temperature of the fruit slows the rates of these reactions.

Figure 14 Tomatoes are often picked green and then held in refrigerated storage until they can be delivered to grocery stores.

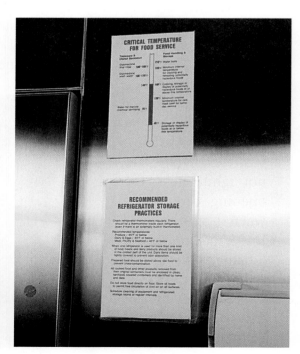

INTEGRATE Health Meat and fish decompose faster at higher temperatures, producing toxins that can make you sick. Keeping these foods chilled slows the decomposition process. Bacteria grow faster at higher temperatures, too, so they reach dangerous levels sooner. Eggs may contain such bacteria, but the heat required to cook eggs also kills bacteria, so hard-cooked eggs are safer to eat than soft-cooked or raw eggs.

Temperature Affects Rate Most chemical reactions speed up when temperature increases. This is because atoms and molecules are always in motion, and they move faster at higher temperatures, as shown in **Figure 15**. Faster molecules collide with each other more often and with greater energy than slower molecules do, so collisions are more likely to provide enough energy to break the old bonds. This is the activation energy.

The high temperature inside an oven speeds up the chemical reactions that turn a liquid cake batter into a more solid, spongy cake. This works the other way, too. Lowering the temperature slows down most reactions. If you set the oven temperature too low, your cake will not bake properly.

Concentration Affects Rate The closer reactant atoms and molecules are to each other, the greater the chance of collisions between them and the faster the reaction rate. It's like the situation shown in **Figure 16**. When you try to walk through a crowded train station, you're more likely to bump into other people than if the station were not so crowded. The amount of substance present in a certain volume is called the **concentration** of that substance. If you increase the concentration, you increase the number of particles of a substance per unit of volume.

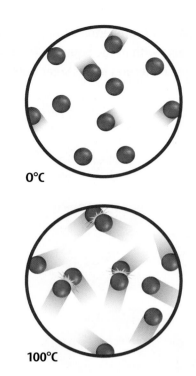

Figure 15 Molecules collide more frequently at higher temperatures than at lower temperatures. This means they are more likely to react.

Collisions are more frequent in a concentrated solution.

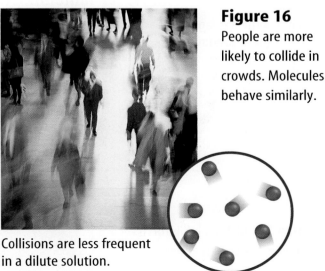

Collisions are less frequent in a dilute solution.

Figure 16 People are more likely to collide in crowds. Molecules behave similarly.

SECTION 2 Rates of Chemical Reactions **505**

Figure 17 Iron atoms trapped inside this steel beam cannot react with oxygen quickly. More iron atoms are exposed to oxygen molecules in this steel wool, so the reaction speeds up.

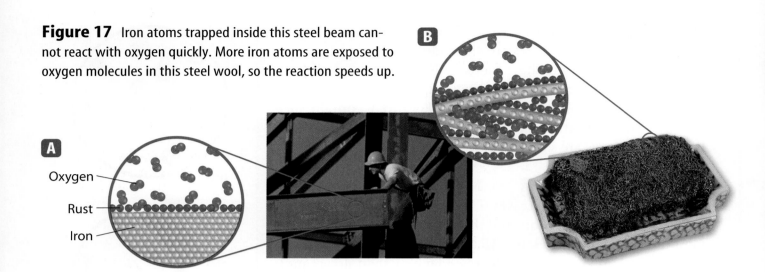

Mini LAB

Identifying Inhibitors

Procedure
1. Look at the ingredients listed on **packages of cereals** and **crackers** in your kitchen.
2. Note the preservatives listed. These are chemical inhibitors.
3. Compare the date on the box with the approximate date the box was purchased to estimate shelf life.

Analysis
1. What is the average shelf life of these products?
2. Why is increased shelf life of such products important?

Surface Area Affects Rate The exposed surface area of reactant particles also affects how fast the reaction can occur. You can quickly start a campfire with small twigs, but starting a fire with only large logs would probably not work.

Only the atoms or molecules in the outer layer of the reactant material can touch the other reactants and react. **Figure 17A** shows that when particles are large, most of the iron atoms are stuck inside and can't react. In **Figure 17B,** more of the reactant atoms are exposed to the oxygen and can react.

Slowing Down Reactions

Sometimes reactions occur too quickly. For example, food and medications can undergo chemical reactions that cause them to spoil or lose their effectiveness too rapidly. Luckily, these reactions can be slowed down.

A substance that slows down a chemical reaction is called an **inhibitor.** An inhibitor makes the formation of a certain amount of product take longer. Some inhibitors completely stop reactions. Many cereals and cereal boxes contain the compound butylated hydroxytoluene, or BHT. The BHT slows the spoiling of the cereal and increases its shelf life.

Figure 18 BHT, an inhibitor, is found in many cereals and cereal boxes.

Speeding Up Reactions

Is it possible to speed up a chemical reaction? Yes, you can add a catalyst (KAT uh lihst). A **catalyst** is a substance that speeds up a chemical reaction. Catalysts do not appear in chemical equations, because they are not changed permanently or used up. A reaction using a catalyst will not produce more product than a reaction without a catalyst, but it will produce the same amount of product faster.

Reading Check *What does a catalyst do in a chemical reaction?*

How does a catalyst work? Many catalysts speed up reaction rates by providing a surface for the reaction to take place. Sometimes the reacting molecules are held in a particular position that favors reaction. Other catalysts reduce the activation energy needed to start the reaction.

Catalytic Converters Catalysts are used in the exhaust systems of cars and trucks to aid fuel combustion. The exhaust passes through the catalyst, often in the form of beads coated with metals such as platinum or rhodium. Catalysts speed the reactions that change incompletely burned substances that are harmful, such as carbon monoxide, into less harmful substances, such as carbon dioxide. Similarly, hydrocarbons are changed into carbon dioxide and water. The result of these reactions is cleaner air. These reactions are shown in **Figure 19.**

Breathe Easy The Clean Air Act of 1970 required the reduction of 90 percent of automobile tailpipe emissions. The reduction of emissions included the amount of hydrocarbons and carbon monoxide released. Automakers needed to develop technology to meet this new standard. After much hard work, the result of this legislation was the introduction of the catalytic converter in 1975.

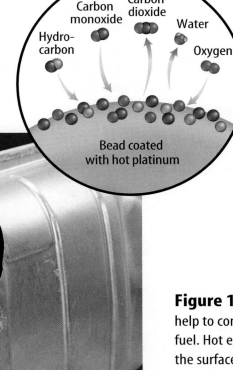

Figure 19 Catalytic converters help to complete combustion of fuel. Hot exhaust gases pass over the surfaces of metal-coated beads. On the surface of the beads, carbon monoxide and hydrocarbons are converted to CO_2 and H_2O.

SECTION 2 Rates of Chemical Reactions

Figure 20 The enzymes in meat tenderizer break down protein in meat, making it more tender.

Enzymes Are Specialists Some of the most effective catalysts are at work in thousands of reactions that take place in your body. These catalysts, called **enzymes,** are large protein molecules that speed up reactions needed for your cells to work properly. They help your body convert food to fuel, build bone and muscle tissue, convert extra energy to fat, and even produce other enzymes.

These are complex reactions. Without enzymes, they would occur at rates that are too slow to be useful or they would not occur at all. Enzymes make it possible for your body to function. Like other catalysts, enzymes function by positioning the reacting molecules so that their structures fit together properly. Enzymes are a kind of chemical specialist—enzymes exist to carry out each type of reaction in your body.

Other Uses Enzymes work outside your body, too. One class of enzymes, called proteases (PROH tee ay ses), specializes in protein reactions. They work within cells to break down large, complex molecules called proteins. The meat tenderizer shown in **Figure 20** contains proteases that break down protein in meat, making it more tender. Contact lens cleaning solutions also contain proteases that break down proteins from your eyes that can collect on your lenses and cloud your view.

section 2 review

Summary

Chemical Reactions
- To form new bonds in the product, old bonds must break in the reactants. This takes energy.
- Activation energy is the minimum quantity of energy needed to start a reaction.

Reaction Rate
- The rate of reaction tells you how fast a reaction occurs.
- Temperature, concentration, and surface area affect the rate of reaction.

Inhibitors and Catalysts
- Inhibitors slow down reactions. Catalysts speed up reactions.
- Enzymes are catalysts that speed up or slow down reactions for your cells.

Self Check

1. **Describe** how you can measure reaction rates.
2. **Explain** in the general reaction A + B + energy → C, how the following will affect the reaction rate.
   a. increasing the temperature
   b. decreasing the reactant concentration
3. **Describe** how catalysts work to speed up chemical reactions.
4. **Think Critically** Explain why a jar of spaghetti sauce can be stored for weeks on the shelf in the market but must be placed in the refrigerator after it is opened.

Applying Math

5. **Solve One-Step Equations** A chemical reaction is proceeding at a rate of 2 g of product every 45 s. How long will it take to obtain 50 g of product?

Physical or Chemical Change?

Real-World Question

Matter can undergo two kinds of changes—physical and chemical. A physical change affects only physical properties. When a chemical change takes place, a new product is produced. How can a scientist tell if a chemical change took place?

Goals
- **Determine** if a physical or chemical change took place.

Materials
500-mL Erlenmeyer flask
100-mL graduated cylinder
one-hole stopper with 15-cm length of glass tube inserted
1,000-mL beaker
45-cm length of rubber (or plastic) tubing
stopwatch or clock with second hand
weighing dish
balance
baking soda
vinegar

Safety Precautions

WARNING: *Vinegar (acetic acid) may cause skin and eye irritation.*

Procedure

1. Add about 300 mL of water to the 500-mL Erlenmeyer flask.
2. Weigh 5 g of baking soda. Carefully pour the baking soda into the flask. Swirl the flask until the solution is clear.
3. Insert the rubber stopper with the glass tubing into the flask.
4. Add about 600 mL of water to the 1,000-mL beaker.

5. Attach one end of the rubber tubing to the top of the glass tubing. Place the other end of the rubber tubing in the beaker. Be sure the rubber tubing remains under the water.
6. Remove the stopper from the flask. Carefully add 80 mL of vinegar to the flask. Replace the stopper.
7. Count the number of bubbles coming into the beaker for 20 s. Repeat this two more times.
8. Record your data in your Science Journal.

Conclude and Apply

1. **Describe** what you observed in the flask after the acid was added to the baking soda solution.
2. **Classify** Was this a physical or chemical change? How do you know?
3. **Analyze Results** Was this process endothermic or exothermic?
4. **Calculate** the average reaction rate based on the number of bubbles per second.

Communicating Your Data

Compare your results with those of other students in your class.

Design Your Own

Exothermic or Endothermic?

Goals
- **Design** an experiment to test whether a reaction is exothermic or endothermic.
- **Measure** the temperature change caused by a chemical reaction.

Possible Materials
test tubes (8)
test-tube rack
3% hydrogen peroxide solution
raw liver
raw potato
thermometer
stopwatch
clock with second hand
25-mL graduated cylinder

Safety Precautions

WARNING: *Hydrogen peroxide can irritate skin and eyes and damage clothing.* Be careful when handling glass thermometers. Test tubes containing hydrogen peroxide should be placed and kept in racks. Dispose of materials as directed by your teacher. Wash your hands when you complete this lab.

▶ Real-World Question

Energy is always a part of a chemical reaction. Some reactions need a constant supply of energy to proceed. Other reactions release energy into the environment. What evidence can you find to show that a reaction between hydrogen peroxide and liver or potato is exothermic or endothermic?

▶ Form a Hypothesis

Make a hypothesis that describes how you can use the reactions between hydrogen peroxide and liver or potato to determine whether a reaction is exothermic or endothermic.

▶ Test Your Hypothesis

Make a Plan

1. As a group, look at the list of materials. Decide which procedure you will use to test your hypothesis, and which measurements you will make.
2. **Decide** how you will detect the heat released to the environment during the reaction. Determine how many measurements you will need to make during a reaction.
3. You will get more accurate data if you repeat each experiment several times. Each repeated experiment is called a trial. Use the average of all the trials as your data for supporting your hypothesis.
4. **Decide** what the variables are and what your control will be.
5. **Copy** the data table in your Science Journal before you begin to carry out your experiment.

Follow Your Plan

1. Make sure your teacher approves your plan before you start.
2. Carry out your plan.
3. **Record** your measurements immediately in your data table.
4. **Calculate** the averages of your trial results and record them in your Science Journal.

510 CHAPTER 17 Chemical Reactions

Using Scientific Methods

▶ Analyze Your Data

1. Can you infer that a chemical reaction took place? What evidence did you observe to support this?
2. **Identify** what the variables were in this experiment.
3. **Identify** the control.

Temperature After Adding Liver/Potato				
Trial	Temperature After Adding Liver (°C)		Temperature After Adding Potato (°C)	
	Starting	After ____ min	Starting	After ____ min
1				
2	Do not write in this book.			
3				
4				

▶ Conclude and Apply

1. Do your observations allow you to distinguish between an exothermic reaction and an endothermic reaction? Use your data to explain your answer.
2. Where do you think that the energy involved in this experiment came from? Explain your answer.

Communicating Your Data

Compare the results obtained by your group with those obtained by other groups. Are there differences? **Explain** how these might have occurred.

LAB **511**

TIME SCIENCE AND HISTORY

SCIENCE CAN CHANGE THE COURSE OF HISTORY!

Synthetic Diamonds

Natural Diamond

Almost the Real Thing

 Synthetic Diamond

Diamonds are the most dazzling, most dramatic, most valuable natural objects on Earth. Strangely, these beautiful objects are made of carbon, the same material graphite—the stuff found in pencils—is made of. So why is a diamond hard and clear and graphite soft and black? A diamond's hardness is a result of how strongly its atoms are linked. What makes a diamond transparent is the way its crystals are arranged. The carbon in a diamond is almost completely pure, with trace amounts of boron and nitrogen in it. These elements account for the many shades of color found in diamonds.

A diamond is the hardest naturally occurring substance on Earth. It's so hard, only a diamond can scratch another diamond. Diamonds are impervious to heat and household chemicals. Their crystal structure allows them to be split (or crushed) along particular lines.

Diamonds are made when carbon is squeezed at high pressures and temperatures in Earth's upper mantle, about 150 km beneath the surface. At that depth, the temperature is about 1,400°C, and the pressure is about 55,000 atmospheres greater than the pressure at sea level.

As early as the 1850s, scientists tried to convert graphite into diamonds. It wasn't until 1954 that researchers produced the first synthetic diamonds by compressing carbon under extremely high pressure and heat. Scientists converted graphite powder into tiny diamond crystals using pressure of more than 68,000 atm, and a temperature of about 1,700°C for about 16 hours.

Synthetic diamonds are human-made, but they're not fake. They have all the properties of natural diamonds, from hardness to excellent heat conductivity. Experts claim to be able to detect synthetics because they contain tiny amounts of metal (used in their manufacturing process) and have a different luminescence than natural diamonds. In fact, most synthetics are made for industrial use. One major reason is that making small synthetic diamonds is cheaper than finding small natural ones. The other reason is that synthetics can be made to a required size and shape. Still, if new techniques bring down the cost of producing large, gem-quality synthetic diamonds, they may one day compete with natural diamonds as jewelry.

Research Investigate the history of diamonds—natural and synthetic. Explain the differences between them and their uses. Share your findings with the class.

For more information, visit blue.msscience.com/time

Chapter 17 Study Guide

Reviewing Main Ideas

Section 1: Formulas and Chemical Equations

1. Chemical reactions often cause observable changes, such as a change in color or odor, a release or absorption of heat or light, or a release of gas.

2. A chemical equation is a shorthand method of writing what happens in a chemical reaction. Chemical equations use symbols to represent the reactants and products of a reaction, and sometimes show whether energy is produced or absorbed.

3. The law of conservation of mass requires a balanced chemical reaction that contains the same number of atoms of each element in the products as in the reactants. This is true in every balanced chemical equation.

Section 2: Rates of Chemical Reactions

1. The rate of reaction is a measure of how quickly a reaction occurs.

2. All reactions have an activation energy—a certain minimum amount of energy required to start the reaction.

3. The rate of a chemical reaction can be influenced by the temperature, the concentration of the reactants, and the exposed surface area of the reactant particles.

4. Catalysts can speed up a reaction without being used up. Inhibitors slow down the rate of reaction.

5. Enzymes are protein molecules that act as catalysts in your body's cells.

Visualizing Main Ideas

Copy and complete the following concept map on chemical reactions.

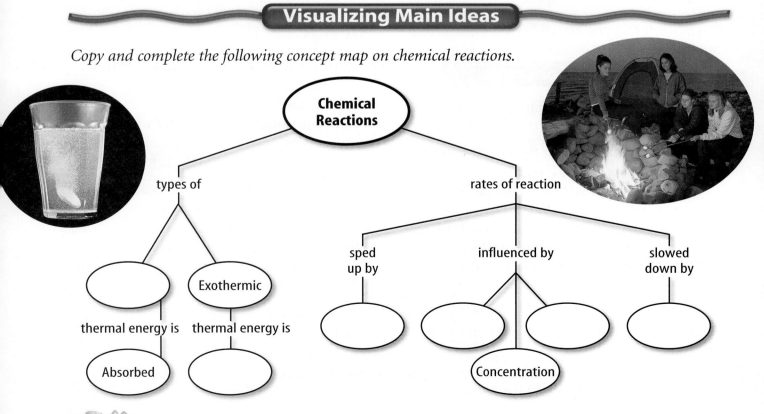

Chapter 17 Review

Using Vocabulary

activation energy p. 503
catalyst p. 507
chemical equation p. 494
chemical reaction p. 492
concentration p. 505
endothermic reaction p. 499
enzyme p. 508
exothermic reaction p. 499
inhibitor p. 506
product p. 494
rate of reaction p. 504
reactant p. 494

Explain the differences between the vocabulary terms in each of the following sets.

1. exothermic reaction—endothermic reaction
2. activation energy—rate of reaction
3. reactant—product
4. catalyst—inhibitor
5. concentration—rate of reaction
6. chemical equation—reactant
7. inhibitor—product
8. catalyst—chemical equation
9. rate of reaction—enzyme

Checking Concepts

Choose the word or phrase that best answers the question.

10. Which statement about the law of conservation of mass is NOT true?
    A) The mass of reactants must equal the mass of products.
    B) All the atoms on the reactant side of an equation are also on the product side.
    C) The reaction creates new types of atoms.
    D) Atoms are not lost, but are rearranged.

11. To slow down a chemical reaction, what should you add?
    A) catalyst   C) inhibitor
    B) reactant   D) enzyme

12. Which of these is a chemical change?
    A) Paper is shredded.
    B) Liquid wax turns solid.
    C) A raw egg is broken.
    D) Soap scum forms.

13. Which of these reactions releases thermal energy?
    A) unbalanced   C) exothermic
    B) balanced   D) endothermic

14. A balanced chemical equation must have the same number of which of these on both sides of the equation?
    A) atoms   C) molecules
    B) reactants   D) compounds

15. What does NOT affect reaction rate?
    A) balancing   C) surface area
    B) temperature   D) concentration

16. Which is NOT a balanced equation?
    A) $CuCl_2 + H_2S \rightarrow CuS + 2HCl$
    B) $AgNO_3 + NaI \rightarrow AgI + NaNO_3$
    C) $2C_2H_6 + 7O_2 \rightarrow 4CO_2 + 6H_2O$
    D) $MgO + Fe \rightarrow Fe_2O_3 + Mg$

17. Which is NOT evidence that a chemical reaction has occurred?
    A) Milk tastes sour.
    B) Steam condenses on a cold window.
    C) A strong odor comes from a broken egg.
    D) A slice of raw potato darkens.

18. Which of the following would decrease the rate of a chemical reaction?
    A) increase the temperature
    B) reduce the concentration of a reactant
    C) increase the concentration of a reactant
    D) add a catalyst

19. Which of these describes a catalyst?
    A) It is a reactant.
    B) It speeds up a reaction.
    C) It appears in the chemical equation.
    D) It can be used in place of an inhibitor.

blue.msscience.com/vocabulary_puzzlemaker

chapter 17 Review

Thinking Critically

20. **Cause and Effect** Pickled cucumbers remain edible much longer than fresh cucumbers do. Explain.

21. **Analyze** A beaker of water in sunlight becomes warm. Has a chemical reaction occurred? Explain.

22. **Distinguish** if $2Ag + S$ is the same as Ag_2S. Explain.

23. **Infer** Apple slices can be kept from browning by brushing them with lemon juice. Infer what role lemon juice plays in this case.

24. **Draw a Conclusion** Chili can be made using ground meat or chunks of meat. Which would you choose, if you were in a hurry? Explain.

Use the graph below to answer question 25.

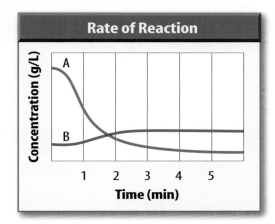

Rate of Reaction

25. **Interpret Scientific Illustrations** The two curves on the graph represent the concentrations of compounds A (blue) and B (red) during a chemical reaction.
    a. Which compound is a reactant?
    b. Which compound is a product?
    c. During which time period is the concentration of the reactant changing most rapidly?

26. **Form a Hypothesis** You are cleaning out a cabinet beneath the kitchen sink and find an unused steel wool scrub pad that has rusted completely. Will the remains of this pad weigh more or less than when it was new? Explain.

Performance Activities

27. **Poster** Make a list of the preservatives in the food you eat in one day. Present your findings to your class in a poster.

Applying Math

Use the graph below to answer question 28.

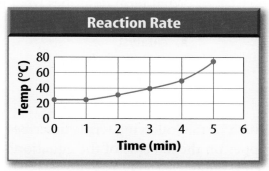

Reaction Rate

28. **Reaction Rates** In the reaction graph above, how long does it take the reaction to reach 50°C?

29. **Chemical Equation** In the following chemical equation, $3Na + AlCl_3 \rightarrow 3NaCl + Al$, how many aluminum molecules will be produced if you have 30 molecules of sodium?

30. **Catalysis** A zinc catalyst is used to reduce the reaction time by 30%. If the normal time for the reaction to finish is 3 h, how long will it take with the catalyst?

31. **Molecules** Silver has 6.023×10^{23} molecules per 107.9 g. How many molecules are there if you have
    a. 53.95 g?
    b. 323.7 g?
    c. 10.79 g?

blue.msscience.com/chapter_review

chapter 17 Standardized Test Practice

Part 1 Multiple Choice

Record your answers on the answer sheet provided by your teacher or on a sheet of paper.

Use the photo below to answer questions 1 and 2.

1. The photograph shows the reaction of copper (Cu) with silver nitrate (AgNO$_3$) to produce copper nitrate (Cu(NO$_3$)$_2$) and silver (Ag). The chemical equation that describes this reaction is the following:

   $$2AgNO_3 + Cu \rightarrow Cu(NO_3)_2 + 2Ag$$

   What term describes what is happening in the reaction?
   A. catalyst
   B. chemical change
   C. inhibitor
   D. physical change

2. Which of the following terms describes the copper on the left side of the equation?
   A. reactant   C. enzyme
   B. catalyst   D. product

3. Which term best describes a chemical reaction that absorbs thermal energy?
   A. catalytic   C. endothermic
   B. exothermic   D. acidic

4. What should be balanced in a chemical equation?
   A. compounds   C. molecules
   B. atoms   D. molecules and atoms

Test-Taking Tip

Read All Questions Never skip a question. If you are unsure of an answer, mark your best guess on another sheet of paper and mark the question in your test booklet to remind you to come back to it at the end of the test.

Use the photo below to answer questions 5 and 6.

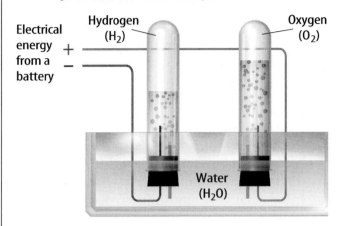

5. The photograph above shows a demonstration of electrolysis, in which water is broken down into hydrogen and oxygen. Which of the following is the best way to write the chemical equation for this process?
   A. $H_2O + \text{energy} \rightarrow H_2 + O_2$
   B. $H_2O + \text{energy} \rightarrow 2H_2 + O_2$
   C. $2H_2O + \text{energy} \rightarrow 2H_2 + O_2$
   D. $2H_2O + \text{energy} \rightarrow 2H_2 + 2O_2$

6. For each atom of hydrogen that is present before the reaction begins, how many atoms of hydrogen are present after the reaction?
   A. 1   C. 4
   B. 2   D. 8

7. What is the purpose of an inhibitor in a chemical reaction?
   A. decrease the shelf life of food
   B. increase the surface area
   C. decrease the speed of a chemical reaction
   D. increase the speed of a chemical reaction

516 STANDARDIZED TEST PRACTICE

Standardized Test Practice

Part 2 Short Response/Grid In

Record your answers on the answer sheet provided by your teacher or on a sheet of paper.

8. If the volume of a substance changes but no other properties change, is this a physical or a chemical change? Explain.

Use the equation below to answer question 9.

$$CaCl_2 + 2AgNO_3 \rightarrow 2\,\boxed{} + Ca(NO_3)_2$$

9. When solutions of calcium chloride ($CaCl_2$) and silver nitrate ($AgNO_3$) are mixed, calcium nitrate ($Ca(NO_3)_2$) and a white precipitate, or residue, form. Determine the chemical formula of the precipitate.

Use the illustration below to answer questions 10 and 11.

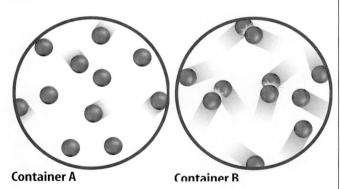

Container A   Container B

10. The figure above demonstrates the movement of atoms at temperatures of 0°C and 100°C. What would happen to the movement of the atoms if the temperature dropped far below 0°C?

11. Describe how the difference in the movement of the atoms at two different temperatures affects the rate of most chemical reactions.

12. Is activation energy needed for reactions that release energy? Explain.

Part 3 Open Ended

Record your answers on a sheet of paper.

Use the illustration below to answer questions 13 and 14.

13. The photograph above shows a forest fire that began when lightning struck a tree. Describe the chemical reaction that occurs when trees burn. Is the reaction endothermic or exothermic? What does this mean? Why does this cause a forest fire to spread?

14. The burning of logs in a forest fire is a chemical reaction. What prevents this chemical reaction from occurring when there is no lightning to start a fire?

15. Explain how the surface area of a material can affect the rate at which the material reacts with other substances. Give an example to support your answer.

16. One of the chemical reactions that occurs in the formation of glass is the combining of calcium carbonate ($CaCO_3$) and silica (SiO_2) to form calcium silicate ($CaSiO_3$) and carbon dioxide (CO_2):

$$CaCO_3 + SiO_2 \rightarrow CaSiO_3 + CO_2$$

Describe this reaction using the names of the chemicals. Discuss which bonds are broken and how atoms are rearranged to form new bonds.

unit 6
Motion, Forces, and Energy

How Are
City Streets &
Zebra Mussels
Connected?

As long as people and cargo have traveled the open seas, ships have taken on extra deadweight, or ballast. Ballast helps adjust the ships' depth in the water and counteracts uneven cargo loads. For centuries, ballast was made up of solid materials—usually rocks, bricks, or sand. These materials had to be unloaded by hand when heavier cargo was taken on board. Many port cities used the discarded ballast stones to pave their dirt roads. The new streets were called "cobblestone." By the mid-1800s, shipbuilding and pump technology had improved, so that taking on and flushing out large quantities of water was relatively easy. Water began to replace rocks and sand as ballast. This new form of deadweight often contained living creatures. When the tanks were flushed at the ships' destinations, those sea creatures would be expelled as well. One such creature was the zebra mussel, which is believed to have been introduced to North America in the 1980s as ballast on a cargo ship. Since then, the population of zebra mussels in the Great Lakes has increased rapidly. Some native species in these lakes are now being threatened by the invading zebra mussel population.

unit projects

Visit blue.msscience.com/unit_project to find project ideas and resources.
Projects include:

- **History** Write a 60-second Moment in History on the life and scientific contributions of Sir Isaac Newton.
- **Technology** Dissect gears in clocks and explore how clocks work. Design a flow chart of the system where every minute counts.
- **Model** Design a tower system that will keep a ball moving down the track using limited supplies. This time, slower is better!

WebQuest *Roller Coaster Physics* is an investigation of acceleration, laws of motion, gravity, and coaster design. Create your own roller coaster.

chapter 18

Motion and Momentum

The BIG Idea

The motion of an object can be described by its velocity.

**SECTION 1
What is motion?**
Main Idea Motion is a change in position.

**SECTION 2
Acceleration**
Main Idea Acceleration occurs when an object speeds up, slows down, or changes direction.

**SECTION 3
Momentum**
Main Idea In a collision, momentum can be transferred from one object to another.

Springing into Action

The hunt might just be over for this pouncing leopard. A leopard can run as fast as 60 km/h over short distances and can jump as high as 3 m. However, a leopard must be more than fast and strong. To catch its prey, it must also be able to change its speed and direction quickly.

Science Journal Describe how your motion changed as you moved from your school's entrance to your classroom.

Start-Up Activities

Motion After a Collision

How is it possible for a 70-kg football player to knock down a 110-kg football player? The smaller player usually must be running faster. Mass makes a difference when two objects collide, but the speed of the objects also matters. Explore the behavior of colliding objects during this lab.

1. Space yourself about 2 m away from a partner. Slowly roll a baseball on the floor toward your partner, and have your partner roll a baseball quickly into your ball.
2. Have your partner slowly roll a baseball as you quickly roll a tennis ball into the baseball.
3. Roll two tennis balls toward each other at the same speed.
4. **Think Critically** In your Science Journal, describe how the motion of the balls changed after the collisions, including the effects of speed and type of ball.

Motion and Momentum Make the following Foldable to help you understand the vocabulary terms in this chapter.

STEP 1 Fold a vertical sheet of notebook paper from side to side.

STEP 2 Cut along every third line of only the top layer to form tabs.

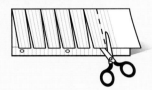

STEP 3 Label each tab.

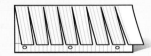

Build Vocabulary As you read the chapter, list the vocabulary words about motion and momentum on the tabs. As you learn the definitions, write them under the tab for each vocabulary word.

 Preview this chapter's content and activities at blue.msscience.com

Get Ready to Read

Summarize

① Learn It! Summarizing helps you organize information, focus on main ideas, and reduce the amount of information to remember. To summarize, restate the important facts in a short sentence or paragraph. Be brief and do not include too many details.

② Practice It! Read the text on page 535 labeled *Conservation of Momentum*. Then read the summary below and look at the important facts from that passage.

Important Facts

- A collision doesn't change the total momentum of a group of objects.
- Friction is an outside force that causes the total momentum to decrease.
- Objects can bounce off each other when they collide.
- Objects can stick together when they collide.

Summary

The total momentum of a group of objects remains constant unless an outside force acts on the objects.

③ Apply It! Practice summarizing as you read this chapter. Stop after each section and write a brief summary.

Target Your Reading

Reading Tip
Reread your summary to make sure you didn't change the author's original meaning or ideas.

Use this to focus on the main ideas as you read the chapter.

1 Before you read the chapter, respond to the statements below on your worksheet or on a numbered sheet of paper.
- Write an **A** if you **agree** with the statement.
- Write a **D** if you **disagree** with the statement.

2 After you read the chapter, look back to this page to see if you've changed your mind about any of the statements.
- If any of your answers changed, explain why.
- Change any false statements into true statements.
- Use your revised statements as a study guide.

Science Online
Print out a worksheet of this page at blue.msscience.com

Before You Read A or D		Statement	After You Read A or D
	1	Distance traveled and displacement are always equal.	
	2	When an object changes direction, it is accelerating.	
	3	A horizontal line on a distance-time graph means the speed is zero.	
	4	If two objects are moving at the same speed, the heavier object is harder to stop.	
	5	The instantaneous speed of an object is always equal to its average speed.	
	6	Momentum equals mass divided by velocity.	
	7	Speed always is measured in kilometers per hour.	
	8	An object's momentum increases if its speed increases.	
	9	If a car is accelerating, its speed must be increasing.	
	10	Speed and velocity are the same thing.	

522 B

section 1

What is motion?

as you read

What You'll Learn
- **Define** distance, speed, and velocity.
- **Graph** motion.

Why It's Important
The different motions of objects you see every day can be described in the same way.

Review Vocabulary
meter: SI unit of distance, abbreviated m; equal to 39.37 in

New Vocabulary
- speed
- average speed
- instantaneous speed
- velocity

Matter and Motion

All matter in the universe is constantly in motion, from the revolution of Earth around the Sun to electrons moving around the nucleus of an atom. Leaves rustle in the wind. Lava flows from a volcano. Bees move from flower to flower as they gather pollen. Blood circulates through your body. These are all examples of matter in motion. How can the motion of these different objects be described?

Changing Position

To describe an object in motion, you must first recognize that the object is in motion. Something is in motion if it is changing position. It could be a fast-moving airplane, a leaf swirling in the wind, or water trickling from a hose. Even your school, attached to Earth, is moving through space. When an object moves from one location to another, it is changing position. The runners shown in **Figure 1** sprint from the start line to the finish line. Their positions change, so they are in motion.

Figure 1 These sprinters are in motion because their positions change.

522 CHAPTER 18 Motion and Momentum

Relative Motion Determining whether something changes position requires a point of reference. An object changes position if it moves relative to a reference point. To visualize this, picture yourself competing in a 100-m dash. You begin just behind the start line. When you pass the finish line, you are 100 m from the start line. If the start line is your reference point, then your position has changed by 100 m relative to the start line, and motion has occurred. Look at **Figure 2.** How can you determine that the dog has been in motion?

 *How do you know if an object has changed position?*

Distance and Displacement Suppose you are to meet your friends at the park in five minutes. Can you get there on time by walking, or should you ride your bike? To help you decide, you need to know the distance you will travel to get to the park. This distance is the length of the route you will travel from your house to the park.

Suppose the distance you traveled from your house to the park was 200 m. When you get to the park, how would you describe your location? You could say that your location was 200 m from your house. However, your final position depends on both the distance you travel and the direction. Did you go 200 m east or west? To describe your final position exactly, you also would have to tell the direction from your starting point. To do this, you would specify your displacement. Displacement includes the distance between the starting and ending points and the direction in which you travel. **Figure 3** shows the difference between distance and displacement.

Figure 2 Motion occurs when something changes position relative to a reference point.
Explain *how the position of the dog changed.*

Figure 3 Distance is how far you have walked. Displacement is the direction and difference in position between your starting and ending points.

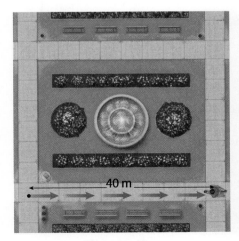

Distance: 40 m
Displacement: 40 m east

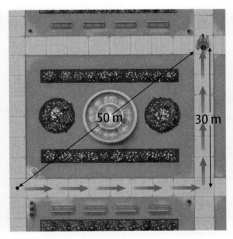

Distance: 70 m
Displacement: 50 m northeast

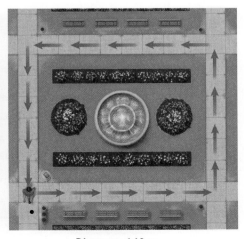

Distance: 140 m
Displacement: 0 m

SECTION 1 What is motion? **523**

Animal Speeds Different animals can move at different top speeds. What are some of the fastest animals? Research the characteristics that help animals run, swim, or fly at high speed.

Speed

To describe motion, you usually want to describe how fast something is moving. The faster something is moving, the greater the distance it can travel in a unit of time, such as one second or one hour. **Speed** is the distance an object travels in a unit of time. For example, an object with a speed of 5 m/s can travel 5 m in 1 s. Speed can be calculated from this equation:

Speed Equation

$$\text{speed (in meters/second)} = \frac{\text{distance (in meters)}}{\text{time (in seconds)}}$$

$$s = \frac{d}{t}$$

The unit for speed is the unit of distance divided by the unit of time. In SI units, speed is measured in units of m/s—meters per second. However, speed can be calculated using other units such as kilometers for distance and hours for time.

Applying Math — Solve a Simple Equation

SPEED OF A SWIMMER Calculate the speed of a swimmer who swims 100 m in 56 s.

Solution

1 *This is what you know:*
- distance: $d = 100$ m
- time: $t = 56$ s

2 *This is what you need to know:*
speed: $s = ?$ m/s

3 *This is the procedure you need to use:*
Substitute the known values for distance and time into the speed equation and calculate the speed:
$$s = \frac{d}{t} = \frac{100 \text{ m}}{56 \text{ s}} = \frac{100}{56} \frac{\text{m}}{\text{s}} = 1.8 \text{ m/s}$$

4 *Check your answer:*
Multiply your answer by the time. You should get the distance that was given.

Practice Problems

1. A runner completes a 400-m race in 43.9 s. In a 100-m race, he finishes in 10.4 s. In which race was his speed faster?
2. A passenger train travels from Boston to New York, a distance of 350 km, in 3.5 h. What is the train's speed?

For more practice, visit blue.msscience.com/math_practice

Average Speed A car traveling in city traffic might have to speed up and slow down many times. How could you describe the speed of an object whose speed is changing? One way is to determine the object's average speed between where it starts and stops. The speed equation on the previous page can be used to calculate the average speed. **Average speed** is found by dividing the total distance traveled by the total time taken.

Reading Check *How is average speed calculated?*

Instantaneous Speed An object in motion can change speeds many times as it speeds up or slows down. The speed of an object at one instant of time is the object's **instantaneous speed.** To understand the difference between average and instantaneous speeds, think about walking to the library. If it takes you 0.5 h to walk 2 km to the library, your average speed would be as follows:

$$s = \frac{d}{t}$$
$$= \frac{2 \text{ km}}{0.5 \text{ h}} = 4 \text{ km/h}$$

However, you might not have been moving at the same speed throughout the trip. At a crosswalk, your instantaneous speed might have been 0 km/h. If you raced across the street, your speed might have been 7 km/h. If you were able to walk at a steady rate of 4 km/h during the entire trip, you would have moved at a constant speed. Average speed, instantaneous speed, and constant speed are illustrated in **Figure 4.**

Mini LAB

Measuring Average Speed

Procedure
1. Choose two points, such as two doorways, and mark each with a small piece of **masking tape.**
2. Measure the distance between the two points.
3. Use a **watch, clock,** or **timer** that indicates seconds to time yourself walking from one mark to the other.
4. Time yourself walking slowly, walking safely and quickly, and walking with a varying speed; for example, slow/fast/slow.

Analysis
1. Calculate your average speed in each case.
2. Predict how long it would take you to walk 100 m slowly, at your normal speed, and quickly.

Try at Home

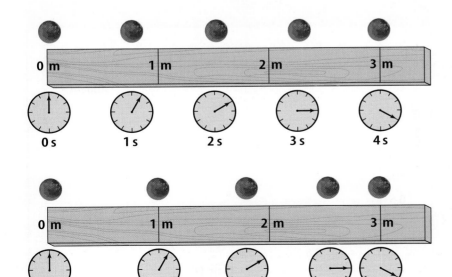

Figure 4 The average speed of each ball is the same from 0 s to 4 s.

The top ball is moving at a constant speed. In each second, the ball moves the same distance.

The bottom ball has a varying speed. Its instantaneous speed is fastest between 0 s and 1 s, slower between 2 s and 3 s, and slowest between 3 s and 4 s.

SECTION 1 What is motion? **525**

Topic: Land Speed Record
Visit blue.msscience.com for Web links to information about how the land speed record has changed over the past century.

Activity Make a graph showing the increase in the land speed over time.

Graphing Motion

You can represent the motion of an object with a distance-time graph. For this type of graph, time is plotted on the horizontal axis and distance is plotted on the vertical axis. **Figure 5** shows the motion of two students who walked across a classroom plotted on a distance-time graph.

Distance-Time Graphs and Speed A distance-time graph can be used to compare the speeds of objects. Look at the graph shown in **Figure 5**. According to the graph, after 1 s student A traveled 1 m. Her average speed during the first second is as follows:

$$\text{speed} = \frac{\text{distance}}{\text{time}} = \frac{1 \text{ m}}{1 \text{ s}} = 1 \text{ m/s}$$

Student B, however, traveled only 0.5 m in the first second. His average speed is

$$\text{speed} = \frac{\text{distance}}{\text{time}} = \frac{0.5 \text{ m}}{1 \text{ s}} = 0.5 \text{ m/s}$$

So student A traveled faster than student B. Now compare the steepness of the lines on the graph in **Figure 5**. The line representing the motion of student A is steeper than the line for student B. A steeper line on the distance-time graph represents a greater speed. A horizontal line on the distance-time graph means that no change in position occurs. In that case, the speed of the object is zero.

Figure 5 The motion of two students walking across a classroom is plotted on this distance-time graph.
Use the graph to determine which student had the faster average speed.

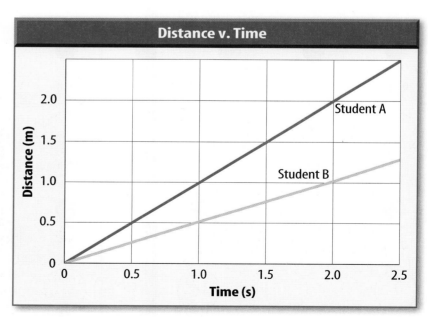

Velocity

The motion of an object also depends on the direction in which the object is moving. The direction of an object's motion can be described with its velocity. The **velocity** of an object is the speed of the object and the direction of its motion. For example, if a car is moving west with a speed of 80 km/h, the car's velocity is 80 km/h west. The velocity of an object is sometimes represented by an arrow. The arrow points in the direction in which the object is moving. **Figure 6,** uses arrows to show the velocities of two hikers.

The velocity of an object can change if the object's speed changes, its direction of motion changes, or they both change. For example, suppose a car is traveling at a speed of 40 km/h north and then turns left at an intersection and continues on with a speed of 40 km/h. The speed of the car is constant at 40 km/h, but the velocity changes from 40 km/h north to 40 km/h west. Why can you say the velocity of a car changes as it comes to a stop at an intersection?

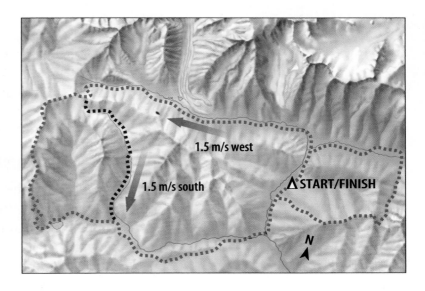

Figure 6 The arrows show the velocities of two hikers. Although the hikers have the same speed, they have different velocities because they are moving in different directions.

section 1 review

Summary

Changing Position
- An object is in motion if it changes position relative to a reference point.
- Motion can be described by distance, speed, displacement, and velocity, where displacement and velocity also include direction.

Speed and Velocity
- The speed of an object can be calculated by dividing the distance traveled by the time needed to travel the distance.
- For an object traveling at constant speed, its average speed is the same as its instantaneous speed.
- The velocity of an object is the speed of the object and its direction of motion.

Graphing Motion
- A line on a distance-time graph becomes steeper as an object's speed increases.

Self Check

1. **Identify** the two pieces of information you need to know the velocity of an object.
2. **Make and Use Graphs** You walk forward at 1.5 m/s for 8 s. Your friend decides to walk faster and starts out at 2.0 m/s for the first 4 s. Then she slows down and walks forward at 1.0 m/s for the next 4 s. Make a distance-time graph of your motion and your friend's motion. Who walked farther?
3. **Think Critically** A bee flies 25 m north of the hive, then 10 m east, 5 m west, and 10 m south. How far north and east of the hive is it now? Explain how you calculated your answer.

Applying Math

4. **Calculate** the average velocity of a dancer who moves 5 m toward the left of the stage over the course of 15 s.
5. **Calculate Travel Time** An airplane flew a distance of 650 km at an average speed of 300 km/h. How much time did the flight take?

section 2
Acceleration

as you read

What You'll Learn
- **Define** acceleration.
- **Predict** what effect acceleration will have on motion.

Why It's Important
Whenever the motion of an object changes, it is accelerating.

Review Vocabulary
kilogram: SI unit of mass, abbreviated kg; equal to approximately 2.2 lbs

New Vocabulary
- acceleration

Figure 7 The toy car is accelerating because its speed is increasing.

Acceleration and Motion

When you watch the first few seconds of a liftoff, a rocket barely seems to move. With each passing second, however, you can see it move faster until it reaches an enormous speed. How could you describe the change in the rocket's motion? When an object changes its motion, it is accelerating. **Acceleration** is the change in velocity divided by the time it takes for the change to occur.

Like velocity, acceleration has a direction. If an object speeds up, the acceleration is in the direction that the object is moving. If an object slows down, the acceleration is opposite to the direction that the object is moving. What if the direction of the acceleration is at an angle to the direction of motion? Then the direction of motion will turn toward the direction of the acceleration.

Speeding Up You get on a bicycle and begin to pedal. The bike moves slowly at first, but speeds up as you keep pedaling. Recall that the velocity of an object is the speed of an object and its direction of motion. Acceleration occurs whenever the velocity of an object changes. Because the bike's speed is increasing, the velocity of the bike is changing. As a result, the bike is accelerating.

For example, the toy car in **Figure 7** is accelerating because it is speeding up. The speed of the car is 10 cm/s after 1s, 20 cm/s after 2s, and 30 cm/s after 3s. Here the direction of the car's acceleration is in the same direction as the car's velocity—to the right.

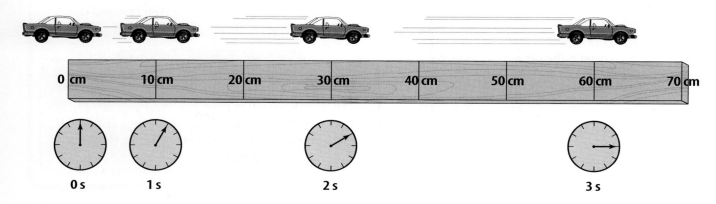

528 CHAPTER 18 Motion and Momentum

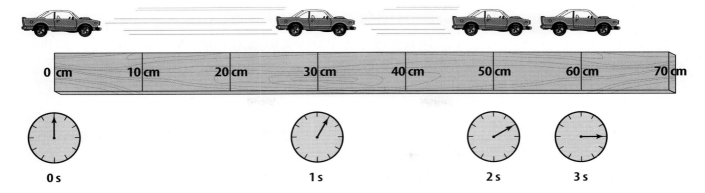

Slowing Down Now suppose you are biking at a speed of 4 m/s and you apply the brakes. This causes you to slow down. When you slow down, your velocity changes because your speed decreases. This means that acceleration occurs when an object slows down, as well as when it speeds up. The car in **Figure 8** is slowing down. During each time interval, the car travels a smaller distance, so its speed is decreasing.

In both of these examples, speed is changing, so acceleration is occurring. Because speed is decreasing in the second example, the direction of the acceleration is opposite to the direction of motion. Any time an object slows down, its acceleration is in the direction opposite to the direction of its motion.

Changing Direction The velocity of an object also changes if the direction of motion changes. Then the object doesn't move in a straight line, but instead moves in a curved path. The object is accelerating because its velocity is changing. In this case the direction of acceleration is at an angle to the direction of motion.

Again imagine yourself riding a bicycle. When you turn the handlebars, the bike turns. Because the direction of the bike's motion has changed, the bike has accelerated. The acceleration is in the direction that the bicycle turned.

Figure 9 shows another example of an object that is accelerating. The ball starts moving upward, but its direction of motion changes as its path turns downward. Here the acceleration is downward. The longer the ball accelerates, the more its path turns toward the direction of acceleration.

Reading Check What are three ways to accelerate?

Figure 8 The car is moving to the right but accelerating to the left. In each time interval, it covers less distance and moves more slowly.
Determine how the car's velocity is changing.

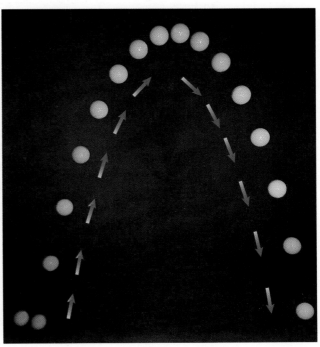

Figure 9 The ball starts out by moving forward and upward, but the acceleration is downward, so the ball's path turns in that direction.

Calculating Acceleration

If an object is moving in only one direction, its acceleration can be calculated using this equation.

Acceleration Equation

acceleration (in m/s^2) = $\dfrac{(\text{final speed (in m/s)} - \text{initial speed (in m/s)})}{\text{time (in s)}}$

$$a = \dfrac{(s_f - s_i)}{t}$$

In this equation, time is the length of time over which the motion changes. In SI units, acceleration has units of meters per second squared (m/s^2).

Applying Math — Solve a Simple Equation

ACCELERATION OF A BUS Calculate the acceleration of a bus whose speed changes from 6 m/s to 12 m/s over a period of 3 s.

Solution

1 *This is what you know:*
- initial speed: $s_i = 6$ m/s
- final speed: $s_f = 12$ m/s
- time: $t = 3$ s

2 *This is what you need to know:* acceleration: $a = ?$ m/s^2

3 *This is the procedure you need to use:* Substitute the known values of initial speed, final speed and time in the acceleration equation and calculate the acceleration:

$$a = \dfrac{(s_f - s_i)}{t} = \dfrac{(12 \text{ m/s} - 6 \text{ m/s})}{3 \text{ s}} = 6\dfrac{\text{m}}{\text{s}} \times \dfrac{1}{3 \text{ s}} = 2 \text{ m/s}^2$$

4 *Check your answer:* Multiply the calculated acceleration by the known time. Then add the known initial speed. You should get the final speed that was given.

Practice Problems

1. Find the acceleration of a train whose speed increases from 7 m/s to 17 m/s in 120 s.
2. A bicycle accelerates from rest to 6 m/s in 2 s. What is the bicycle's acceleration?

For more practice, visit blue.msscience.com/math_practice

Figure 10 When skidding to a stop, you are slowing down. This means you have a negative acceleration.

Positive and Negative Acceleration An object is accelerating when it speeds up, and the acceleration is in the same direction as the motion. An object also is accelerating when it slows down, but the acceleration is in the direction opposite to the motion, such as the bicycle in **Figure 10**. How else is acceleration different when an object is speeding up and slowing down?

Suppose you were riding your bicycle in a straight line and increased your speed from 4 m/s to 6 m/s in 5 s. You could calculate your acceleration from the equation on the previous page.

$$a = \frac{(s_f - s_i)}{t}$$
$$= \frac{(6 \text{ m/s} - 4 \text{ m/s})}{5 \text{ s}} = \frac{+2 \text{ m/s}}{5 \text{ s}}$$
$$= +0.4 \text{ m/s}^2$$

When you speed up, your final speed always will be greater than your initial speed. So subtracting your initial speed from your final speed gives a positive number. As a result, your acceleration is positive when you are speeding up.

Suppose you slow down from a speed of 4 m/s to 2 m/s in 5 s. Now the final speed is less than the initial speed. You could calculate your acceleration as follows:

$$a = \frac{(s_f - s_i)}{t}$$
$$= \frac{(2 \text{ m/s} - 4 \text{ m/s})}{5 \text{ s}} = \frac{-2 \text{ m/s}}{5 \text{ s}}$$
$$= -0.4 \text{ m/s}^2$$

Because your final speed is less than your initial speed, your acceleration is negative when you slow down.

Modeling Acceleration

Procedure
1. Use **masking tape** to lay a course on the floor. Mark a starting point and place marks along a straight path at 10 cm, 40 cm, 90 cm, 160 cm, and 250 cm from the start.
2. Clap a steady beat. On the first beat, the person walking the course should be at the starting point. On the second beat, the walker should be on the first mark, and so on.

Analysis
1. Describe what happens to your speed as you move along the course. Infer what would happen if the course were extended farther.
2. Repeat step 2, starting at the other end. Are you still accelerating? Explain.

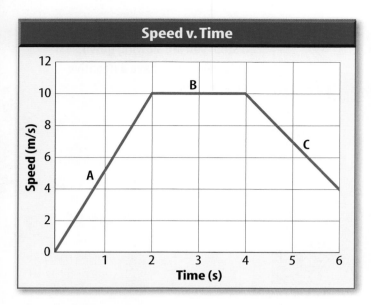

Figure 11 A speed-time graph can be used to find acceleration. When the line rises, the object is speeding up. When the line falls, the object is slowing down.
Infer *what acceleration a horizontal line represents.*

Graphing Accelerated Motion The motion of an object that is moving in a single direction can be shown with a graph. For this type of graph, speed is plotted on the vertical axis and time on the horizontal axis. Take a look at **Figure 11**. On section A of the graph, the speed increases from 0 m/s to 10 m/s during the first 2 s, so the acceleration is +5 m/s². The line in section A slopes upward to the right. An object that is speeding up will have a line on a speed-time graph that slopes upward.

Now look at section C. Between 4 s and 6 s, the object slows down from 10 m/s to 4 m/s. The acceleration is −3 m/s². On the speed-time graph, the line in section C is sloping downward to the right. An object that is slowing down will have a line on a speed-time graph that slopes downward.

On section B, where the line is horizontal, the change in speed is zero. So a horizontal line on the speed-time graph represents an acceleration of zero or constant speed.

section 2 review

Summary

Acceleration and Motion
- Acceleration is the change in velocity divided by the time it takes to make the change. Acceleration has direction.
- Acceleration occurs whenever an object speeds up, slows down, or changes direction.

Calculating Acceleration
- For motion in a single direction, acceleration can be calculated from this equation:

$$a = \frac{s_f - s_i}{t}$$

- If an object is speeding up, its acceleration is positive; if an object is slowing down, its acceleration is negative.
- On a speed-time graph, a line sloping upward represents increasing speed, a line sloping downward represents decreasing speed, and a horizontal line represents zero acceleration or constant speed.

Self Check

1. **Compare and contrast** speed, velocity, and acceleration.
2. **Infer** the motion of a car whose speed-time graph shows a horizontal line, followed by a straight line that slopes downward to the bottom of the graph.
3. **Think Critically** You start to roll backward down a hill on your bike, so you use the brakes to stop your motion. In what direction did you accelerate?

Applying Math

4. **Calculate** the acceleration of a runner who accelerates from 0 m/s to 3 m/s in 12 s.
5. **Calculate Speed** An object falls with an acceleration of 9.8 m/s². What is its speed after 2 s?
6. **Make and Use a Graph** A sprinter had the following speeds at different times during a race: 0 m/s at 0 s, 4 m/s at 2 s, 7 m/s at 4 s, 10 m/s at 6 s, 12 m/s at 8 s, and 10 m/s at 10 s. Plot these data on a speed-time graph. During what time intervals is the acceleration positive? Negative? Is the acceleration ever zero?

section 3
Momentum

Collisions

A collision occurs when a moving object collides with other objects. What happens when a cue ball collides with another ball in a game of pool? The answer is that the velocities of the two balls change. The collision can change the speed of each ball, the direction of motion of each ball, or both. When a collision occurs, changes in motion of the colliding objects depend on their masses and their velocities before the collision.

Mass and Inertia

The mass of an object affects how easy it is to change its motion. **Mass** is the amount of matter in an obejct. Imagine a person rushing toward you. To stop the person, you would have to push on him or her. However, you would have to push harder to stop an adult than to stop a child. The child would be easier to stop because it has less mass than the adult. The more mass an object has, the harder it is to change its motion. In **Figure 12,** the tennis ball has more mass than the table-tennis ball. A big racquet rather than a small paddle is used to change its motion. The tendency of an object to resist a change in its motion is called **inertia**. The amount of resistance to a change in motion increases as an object's mass increases.

Reading Check *What is inertia?*

as you read

What **You'll Learn**
- **Define** momentum.
- **Explain** why momentum might not be conserved after a collision.
- **Predict** motion using the law of conservation of momentum.

Why **It's Important**
Objects in motion have momentum. The motion of objects after they collide depends on their momentum.

Review Vocabulary
triple-beam balance: scientific instrument used to measure mass precisely by comparing the mass of a sample to known masses

New Vocabulary
- mass
- inertia
- momentum
- law of conservation of momentum

Figure 12 A tennis ball has more mass than a table-tennis ball. The tennis ball must be hit harder than the table-tennis ball to change its velocity by the same amount.

Forensics and Momentum Forensic investigations of accidents and crimes often involve determining the momentum of an object. For example, the law of conservation of momentum sometimes is used to reconstruct the motion of vehicles involved in a collision. Research other ways momentum is used in forensic investigations.

Momentum

You know that the faster a bicycle moves, the harder it is to stop. Just as increasing the mass of an object makes it harder to stop, so does increasing the speed or velocity of the object. The **momentum** of an object is a measure of how hard it is to stop the object, and it depends on the object's mass and velocity. Momentum is usually symbolized by p.

Momentum Equation

momentum (in kg·m/s) = **mass** (in kg) × **velocity** (in m/s)

$$p = mv$$

Mass is measured in kilograms and velocity has units of meters per second, so momentum has units of kilograms multiplied by meters per second (kg·m/s). Also, because velocity includes a direction, momentum has a direction that is the same as the direction of the velocity.

Reading Check *Explain how an object's momentum changes as its velocity changes.*

Applying Math — Solve a Simple Equation

MOMENTUM OF A BICYCLE Calculate the momentum of a 14-kg bicycle traveling north at 2 m/s.

Solution

1. *This is what you know:*
   - mass: $m = 14$ kg
   - velocity: $v = 2$ m/s north

2. *This is what you need to find:*
   momentum: $p = ?$ kg·m/s

3. *This is the procedure you need to use:*
   Substitute the known values of mass and velocity into the momentum equation and calculate the momentum:
   $p = mv = (14$ kg$)(2$ m/s north$) = 28$ kg·m/s north

4. *Check your answer:*
   Divide the calculated momentum by the mass of the bicycle. You should get the velocity that was given.

Practice Problems

1. A 10,000-kg train is traveling east at 15 m/s. Calculate the momentum of the train.
2. What is the momentum of a car with a mass of 900 kg traveling north at 27 m/s?

For more practice, visit blue.msscience.com/math_practice

Conservation of Momentum

If you've ever played billiards, you know that when the cue ball hits another ball, the motions of both balls change. The cue ball slows down and may change direction, so its momentum decreases. Meanwhile, the other ball starts moving, so its momentum increases.

In any collision, momentum is transferred from one object to another. Think about the collision between two billiard balls. If the momentum lost by one ball equals the momentum gained by the other ball, then the total amount of momentum doesn't change. When the total momentum of a group of objects doesn't change, momentum is conserved.

The Law of Conservation of Momentum According to the **law of conservation of momentum,** the total momentum of a group of objects remains constant unless outside forces act on the group. The moving cue ball and the other billiard balls in **Figure 13** are a group of objects. The law of conservation of momentum means that collisions between these objects don't change the total momentum of all the objects in the group.

Only an outside force, such as friction between the billiard balls and the table, can change the total momentum of the group of objects. Friction will cause the billiard balls to slow down as they roll on the table and the total momentum will decrease.

Types of Collisions Objects can collide with each other in different ways. **Figure 14** shows two examples. Sometimes objects will bounce off each other like the bowling ball and bowling pins. In other collisions, objects will collide and stick to each other, as when one football player tackles another.

Figure 13 When the cue ball hits the other billiard balls, it slows down because it transfers some of its momentum to the other billiard balls.
Predict what would happen to the speed of the cue ball if all of its momentum were transferred to the other billiard balls.

Figure 14 When objects collide, they can bounce off each other, or they can stick to each other.

When the bowling ball hits the pins, the ball and the pins bounce off each other. The momentum of the ball and the pins changes during the collision.

When one player tackles the other, they stick together. The momentum of each player changes during the collision.

Before the student on skates and the backpack collide, she is not moving.

After the collision, the student and the backpack move together at a slower speed than the backpack had before the collision.

Figure 15 Momentum is transferred from the backpack to the student.

Using Momentum Conservation The law of momentum conservation can be used to predict the velocity of objects after they collide. To use the law of conservation of momentum, assume that the total momentum of the colliding objects doesn't change.

For example, imagine being on skates when someone throws a backpack to you, as in **Figure 15.** The law of conservation of momentum can be used to find your velocity after you catch the backpack. Suppose a 2-kg backpack initially has a velocity of 5 m/s east. Your mass is 48 kg, and initially you're at rest. Then the total initial momentum is:

$$\begin{aligned}\text{total momentum} &= \text{momentum of backpack} + \text{your momentum} \\ &= 2 \text{ kg} \times 5 \text{ m/s east} + 48 \text{ kg} \times 0 \text{ m/s} \\ &= 10 \text{ kg} \cdot \text{m/s east}\end{aligned}$$

After the collision, the total momentum remains the same, and only one object is moving. Its mass is the sum of your mass and the mass of the backpack. You can use the equation for momentum to find the final velocity

$$\begin{aligned}\text{total momentum} &= (\text{mass of backpack} + \text{your mass}) \times \text{velocity} \\ 10 \text{ kg} \cdot \text{m/s east} &= (2 \text{ kg} + 48 \text{ kg}) \times \text{velocity} \\ 10 \text{ kg} \cdot \text{m/s east} &= (50 \text{ kg}) \times \text{velocity} \\ 0.2 \text{ m/s east} &= \text{velocity}\end{aligned}$$

This is your velocity right after you catch the backpack. The velocity of you and the backpack together is much smaller than the initial velocity of the backpack. **Figure 16** shows the results of collisions between two objects that don't stick to each other.

Topic: Collisions
Visit blue.msscience.com for Web links to information about collisions between objects with different masses.

Activity Draw diagrams showing the results of collisions between a bowling ball and a tennis ball if they are moving in the same direction and if they are in opposite directions.

NATIONAL GEOGRAPHIC VISUALIZING CONSERVATION OF MOMENTUM

Figure 16

The law of conservation of momentum can be used to predict the results of collisions between different objects, whether they are subatomic particles smashing into each other at enormous speeds, or the collisions of marbles, as shown on this page. What happens when one marble hits another marble initially at rest? The results of the collisions depend on the masses of the marbles.

A Here, a less massive marble strikes a more massive marble that is at rest. After the collision, the smaller marble bounces off in the opposite direction. The larger marble moves in the same direction that the small marble was initially moving.

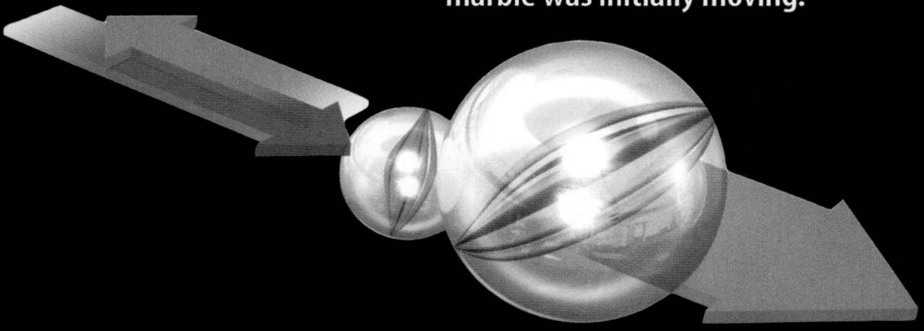

B Here, the large marble strikes the small marble that is at rest. After the collision, both marbles move in the same direction. The less massive marble always moves faster than the more massive one.

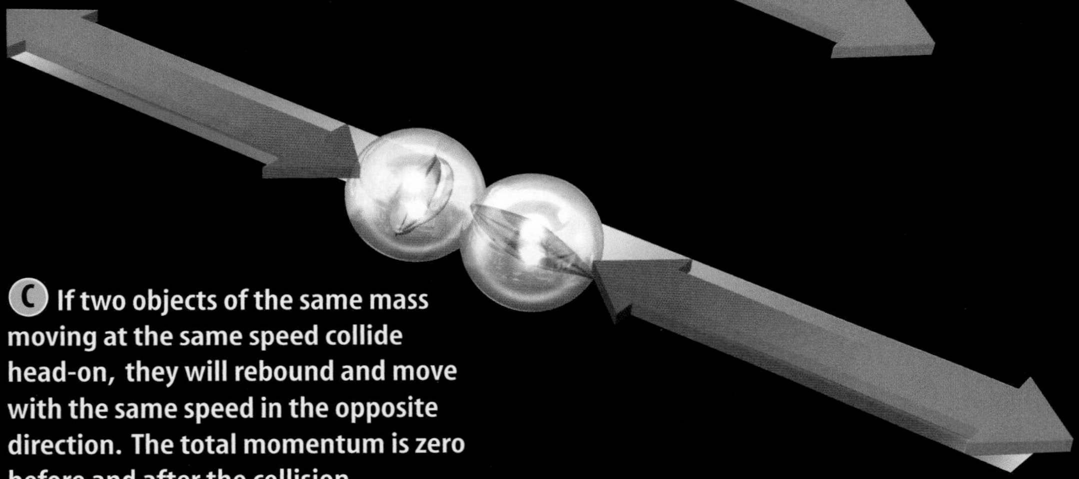

C If two objects of the same mass moving at the same speed collide head-on, they will rebound and move with the same speed in the opposite direction. The total momentum is zero before and after the collision.

SECTION 3 Momentum

Figure 17 When bumper cars collide, they bounce off each other, and momentum is transferred.

Colliding and Bouncing Off In some collisions, the objects involved, like the bumper cars in **Figure 17**, bounce off each other. The law of conservation of momentum can be used to determine how these objects move after they collide.

For example, suppose two identical objects moving with the same speed collide head on and bounce off. Before the collision, the momentum of each object is the same, but in opposite directions. So the total momentum before the collision is zero. If momentum is conserved, the total momentum after the collision must be zero also. This means that the two objects must move in opposite directions with the same speed after the collision. Then the total momentum once again is zero.

section 3 review

Summary

Inertia and Momentum
- Inertia is the tendency of an object to resist a change in motion.
- The momentum of an object in motion is related to how hard it is to stop the object, and can be calculated from the following equation:
  $$p = mv$$
- The momentum of an object is in the same direction as its velocity.

The Law of Conservation of Momentum
- According to the law of conservation of momentum, the total momentum of a group of objects remains constant unless outside forces act on the group.
- When objects collide, they can bounce off each other or stick together.

Self Check

1. **Explain** how momentum is transferred when a golfer hits a ball with a golf club.
2. **Determine** if the momentum of an object moving in a circular path at constant speed is constant.
3. **Explain** why the momentum of a billiard ball rolling on a billiard table changes.
4. **Think Critically** Two identical balls move directly toward each other with equal speeds. How will the balls move if they collide and stick together?

Applying Math

5. **Momentum** What is the momentum of a 0.1-kg mass moving with a velocity of 5 m/s west?
6. **Momentum Conservation** A 1-kg ball moving at 3 m/s east strikes a 2-kg ball and stops. If the 2-kg ball was initially at rest, find its velocity after the collision.

blue.msscience.com/self_check_quiz

Collisions

A collision occurs when a baseball bat hits a baseball or a tennis racket hits a tennis ball. What would happen if you hit a baseball with a table-tennis paddle or a table-tennis ball with a baseball bat? How do the masses of colliding objects change the results of collisions?

Real-World Question

How does changing the size and number of objects in a collision affect the collision?

Goals

- **Compare and contrast** different collisions.
- **Determine** how the velocities after a collision depend on the masses of the colliding objects.

Materials

small marbles (5)   metersticks (2)
large marbles (2)   tape

Safety Precautions

Procedure

1. Tape the metersticks next to each other, slightly farther apart than the width of the large marbles. This limits the motion of the marbles to nearly a straight line.
2. Place a small target marble in the center of the track formed by the metersticks. Place another small marble at one end of the track. Flick the small marble toward the target marble. Describe the collision.
3. Repeat step 2, replacing the two small marbles with the two large marbles.
4. Repeat step 2, replacing the small shooter marble with a large marble.
5. Repeat step 2, replacing the small target marble with a large marble.
6. Repeat step 2, replacing the small target marble with four small marbles that are touching.
7. Place two small marbles at opposite ends of the metersticks. Shoot the marbles toward each other and describe the collision.
8. Place two large marbles at opposite ends of the metersticks. Shoot the marbles toward each other and describe the collision.
9. Place a small marble and a large marble at opposite ends of the metersticks. Shoot the marbles toward each other and describe the collision.

Conclude and Apply

1. **Describe** In which collisions did the shooter marble change direction? How did the mass of the target marble compare with the mass of the shooter marble in these collisions?
2. **Describe** how the velocity of the shooter marble changed when the target marble had the same mass and was at rest.

Communicating Your Data

Make a chart showing your results. You might want to make before-and-after sketches, with short arrows to show slow movement and long arrows to show fast movement.

LAB **539**

Design Your Own

Car Safety Testing

Goals
- **Construct** a fast car.
- **Design** a safe car that will protect a plastic egg from the effects of inertia when the car crashes.

Possible Materials
insulated foam meat trays or fast food trays
insulated foam cups
straws, narrow and wide
straight pins
tape
plastic eggs

Safety Precautions

WARNING: *Protect your eyes from possible flying objects.*

● Real-World Question

Imagine that you are a car designer. How can you create an attractive, fast car that is safe? When a car crashes, the passengers have inertia that can keep them moving. How can you protect the passengers from stops caused by sudden, head-on impacts?

● Form a Hypothesis

Develop a hypothesis about how to design a car to deliver a plastic egg quickly and safely through a race course and a crash at the end.

● Test Your Hypothesis

Make a Plan

1. Be sure your group has agreed on the hypothesis statement.
2. **Sketch** the design for your car. List the materials you will need. Remember that to make the car move smoothly, narrow straws will have to fit into the wider straws.

540 CHAPTER 18 Motion and Momentum

Using Scientific Methods

3. As a group, make a detailed list of the steps you will take to test your hypothesis.
4. Gather the materials you will need to carry out your experiment.

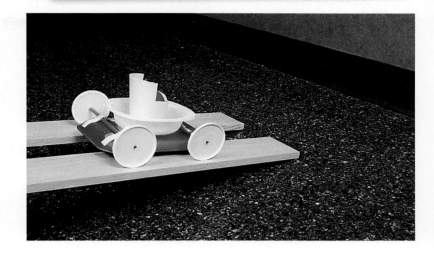

Follow Your Plan

1. Make sure your teacher approves your plan before you start. Include any changes suggested by your teacher in your plans.
2. Carry out the experiment as planned.
3. **Record** any observations that you made while doing your experiment. Include suggestions for improving your design.

Analyze Your Data

1. **Compare** your car design to the designs of the other groups. What made the fastest car fast? What slowed the slowest car?
2. **Compare** your car's safety features to those of the other cars. What protected the eggs the best? How could you improve the unsuccessful designs?
3. **Predict** What effect would decreasing the speed of your car have on the safety of the egg?

Conclude and Apply

1. **Summarize** How did the best designs protect the egg?
2. **Apply** If you were designing cars, what could you do to better protect passengers from sudden stops?

Communicating Your Data

Write a descriptive paragraph about ways a car could be designed to protect its passengers effectively. Include a sketch of your ideas.

Oops! Accidents in Science

SOMETIMES GREAT DISCOVERIES HAPPEN BY ACCIDENT!

What Goes Around Comes Around
The Story of Boomerangs

Imagine a group gathered on a flat, yellow plain on the Australian Outback. One youth steps forward and, with the flick of an arm, sends a long, flat, angled stick soaring and spinning into the sky. The stick's path curves until it returns right back into the thrower's hand. Thrower after thrower steps forward, and the contest goes on all afternoon.

This contest involved throwing boomerangs—elegantly curved sticks. Because of how boomerangs are shaped, they always return to the thrower's hand

This amazing design is over 15,000 years old. Scientists believe that boomerangs developed from simple clubs thrown to stun and kill animals for food. Differently shaped clubs flew in different ways. As the shape of the club was refined, people probably started throwing them for fun too. In fact, today, using boomerangs for fun is still a popular sport, as world-class throwers compete in contests of strength and skill.

Boomerangs come in several forms, but all of them have several things in common. First a boomerang is shaped like an airplane's wing: flat on one side and curved on the other. Second, boomerangs are angled, which makes them spin as they fly. These two features determine the aerodynamics that give the boomerang its unique flight path.

From its beginning as a hunting tool to its use in today's World Boomerang Championships, the boomerang has remained a source of fascination for thousands of years.

Design Boomerangs are made from various materials. Research to find instructions for making boomerangs. After you and your friends build some boomerangs, have a competition of your own.

For more information, visit blue.msscience.com/oops

chapter 18 Study Guide

Reviewing Main Ideas

Section 1 What is motion?

1. The position of an object depends on the reference point that is chosen.
2. An object is in motion if the position of the object is changing.
3. The speed of an object equals the distance traveled divided by the time:
$$s = \frac{d}{t}$$
4. The velocity of an object includes the speed and the direction of motion.
5. The motion of an object can be represented on a speed-time graph.

Section 2 Acceleration

1. Acceleration is a measure of how quickly velocity changes. It includes a direction.
2. An object is accelerating when it speeds up, slows down, or turns.
3. When an object moves in a straight line, its acceleration can be calculated by
$$a = \frac{(s_f - s_i)}{t}$$

Section 3 Momentum

1. Momentum equals the mass of an object times its velocity:
$$p = mv$$
2. Momentum is transferred from one object to another in a collision.
3. According to the law of conservation of momentum, the total amount of momentum of a group of objects doesn't change unless outside forces act on the objects.

Visualizing Main Ideas

Copy and complete the following table on motion.

Describing Motion

Quantity	Definition	Direction
Distance	length of path traveled	no
Displacement	direction and change in position	
Speed		no
Velocity	rate of change in position and direction	
Acceleration		
Momentum		yes

chapter 18 Review

Using Vocabulary

acceleration p. 528
average speed p. 525
inertia p. 533
instantaneous speed p. 525
law of conservation
  of momentum p. 535
mass p. 533
momentum p. 534
speed p. 524
velocity p. 527

Explain the relationship between each pair of terms.

1. speed—velocity
2. velocity—acceleration
3. velocity—momentum
4. momentum—law of conservation of momentum
5. mass—momentum
6. mass—inertia
7. momentum—inertia
8. average speed—instantaneous speed

Checking Concepts

Choose the word or phrase that best answers the question.

9. What measures the quantity of matter?
   A) speed
   B) weight
   C) acceleration
   D) mass

10. Which of the following objects is NOT accelerating?
    A) a jogger moving at a constant speed
    B) a car that is slowing down
    C) Earth orbiting the Sun
    D) a car that is speeding up

11. Which of the following equals speed?
    A) acceleration/time
    B) (change in velocity)/time
    C) distance/time
    D) displacement/time

12. Which of these is an acceleration?
    A) 5 m east
    B) 15 m/s east
    C) 52 m/s^2 east
    D) 32 s^2 east

13. Resistance to a change in motion increases when which of these increases?
    A) velocity
    B) speed
    C) instantaneous speed
    D) mass

14. What is 18 cm/h north an example of?
    A) speed
    B) velocity
    C) acceleration
    D) momentum

15. Which is true when the velocity and the acceleration of an object are in the same direction?
    A) The object's speed is constant.
    B) The object changes direction.
    C) The object speeds up.
    D) The object slows down.

16. Which of the following equals the change in velocity divided by the time?
    A) speed
    B) displacement
    C) momentum
    D) acceleration

17. You travel to a city 200 km away in 2.5 hours. What is your average speed in km/h?
    A) 180 km/h
    B) 12.5 km/h
    C) 80 km/h
    D) 500 km/h

18. A cue ball hits another billiard ball and slows down. Why does the speed of the cue ball decrease?
    A) The cue ball's momentum is positive.
    B) The cue ball's momentum is negative.
    C) Momentum is transferred to the cue ball.
    D) Momentum is transferred from the cue ball.

chapter 18 Review

Thinking Critically

19. **Explain** You run 100 m in 25 s. If you later run the same distance in less time, explain if your average speed increase or decrease.

Use the graph below to answer questions 20 and 21.

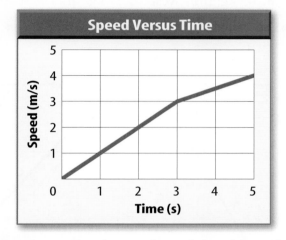

20. **Compare** For the motion of the object plotted on the speed-time graph above, how does the acceleration between 0 s and 3 s compare to the acceleration between 3 s and 5 s?

21. **Calculate** the acceleration of the object over the time interval from 0 s to 3 s.

22. **Infer** The molecules in a gas are often modeled as small balls. If the molecules all have the same mass, infer what happens if two molecules traveling at the same speed collide head on.

23. **Calculate** What is your displacement if you walk 100 m north, 20 m east, 30 m south, 50 m west, and then 70 m south?

24. **Infer** You are standing on ice skates and throw a basketball forward. Infer how your velocity after you throw the basketball compares with the velocity of the basketball.

25. **Determine** You throw a ball upward and then it falls back down. How does the velocity of the ball change as it rises and falls?

Use the graph below to answer question 26.

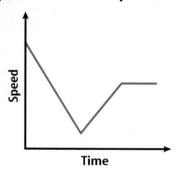

26. **Make and Use Graphs** The motion of a car is plotted on the speed-time graph above. Over which section of the graph is the acceleration of the car zero?

Performance Activities

27. **Demonstrate** Design a racetrack and make rules that specify the types of motion allowed. Demonstrate how to measure distance, measure time, and calculate speed accurately.

Applying Math

28. **Velocity of a Ball** Calculate the velocity of a 2-kg ball that has a momentum of 10 kg · m/s west.

29. **Distance Traveled** A car travels for a half hour at a speed of 40 km/h. How far does the car travel?

Use the graph below to answer question 30.

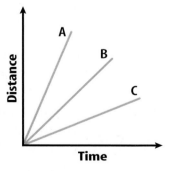

30. **Speed** From the graph determine which object is moving the fastest and which is moving the slowest.

Chapter 18 Standardized Test Practice

Part 1 Multiple Choice

Record your answers on the answer sheet provided by your teacher or on a sheet of paper.

1. What is the distance traveled divided by the time taken to travel that distance?
   A. acceleration   C. speed
   B. velocity   D. inertia

2. Sound travels at a speed of 330 m/s. How long does it take for the sound of thunder to travel 1,485 m?
   A. 45 s   C. 4,900 s
   B. 4.5 s   D. 0.22 s

Use the figure below to answer questions 3 and 4.

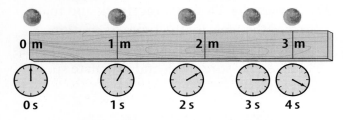

3. During which time period is the ball's average speed the fastest?
   A. between 0 and 1 s
   B. between 1 and 2 s
   C. between 2 and 3 s
   D. between 3 and 4 s

4. What is the average speed of the ball?
   A. 0.75 m/s   C. 10 m/s
   B. 1 m/s   D. 1.3 m/s

5. A car accelerates from 15 m/s to 30 m/s in 3.0 s. What is the car's acceleration?
   A. 10 m/s^2 C. 15 m/s^2
   B. 25 m/s^2 D. 5.0 m/s^2

6. Which of the following can occur when an object is accelerating?
   A. It speeds up.   C. It changes direction.
   B. It slows down.   D. all of the above

7. What is the momentum of a 21-kg bicycle traveling west at 3.0 m/s?
   A. 7 kg · m/s west   C. 18 kg · m/s west
   B. 63 kg · m/s west   D. 24 kg · m/s west

Use the figure below to answer questions 8–10.

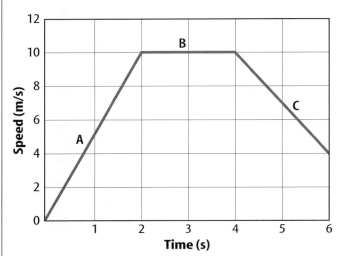

8. What is the acceleration between 0 and 2 s?
   A. 10 m/s^2 C. 0 m/s^2
   B. 5 m/s^2 D. −5 m/s^2

9. During what time period does the object have a constant speed?
   A. between 1 and 2 s
   B. between 2 and 3 s
   C. between 4 and 5 s
   D. between 5 and 6 s

10. What is the acceleration between 4 and 6 s?
    A. 10 m/s^2 C. 6 m/s^2
    B. 4 m/s^2 D. −3 m/s^2

11. An acorn falls from the top of an oak and accelerates at 9.8 m/s^2. It hits the ground in 1.5 s. What is the speed of the acorn when it hits the ground?
    A. 9.8 m/s   C. 15 m/s
    B. 20 m/s   D. 30 m/s

Standardized Test Practice

Part 2 Short Response/Grid In

Record your answers on the answer sheet provided by your teacher or on a sheet of paper.

12. Do two objects that have the same mass always have the same momentum? Why or why not?

13. What is the momentum of a 57 kg cheetah running north at 27 m/s?

14. A sports car and a moving van are traveling at a speed of 30 km/h. Which vehicle will be easier to stop? Why?

Use the figure below to answer questions 15 and 16.

15. What happens to the momentum of the bowling ball when it hits the pins?

16. What happens to the speed of the ball and the speed of the pins?

17. What is the speed of a race horse that runs 1,500 m in 125 s?

18. A car travels for 5.5 h at an average speed of 75 km/h. How far did it travel?

19. If the speedometer on a car indicates a constant speed, can you be certain the car is not accelerating? Explain.

20. A girl walks 2 km north, then 2 km east, then 2 km south, then 2 km west. What distance does she travel? What is her displacement?

Part 3 Open Ended

Record your answers on a sheet of paper.

Use the figure below to answer questions 21 and 22.

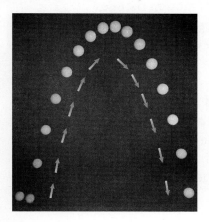

21. Describe the motion of the ball in terms of its speed, velocity, and acceleration.

22. During which part of its path does the ball have positive acceleration? During which part of its path does it have negative acceleration? Explain.

23. Describe what will happen when a baseball moving to the left strikes a bowling ball that is at rest.

24. A girl leaves school at 3:00 and starts walking home. Her house is 2 km from school. She gets home at 3:30. What was her average speed? Do you know her instantaneous speed at 3:15? Why or why not?

25. Why is it dangerous to try to cross a railroad track when a very slow-moving train is approaching?

Test-Taking Tip

Look for Missing Information Questions sometimes will ask about missing information. Notice what is missing as well as what is given.

chapter 19

Force and Newton's Laws

The BIG Idea

An object's motion changes if the forces acting on the object are unbalanced.

SECTION 1
Newton's First Law
Main Idea If the net force on an object is zero, the motion of the object does not change.

SECTION 2
Newton's Second Law
Main Idea An object's acceleration equals the net force divided by the mass.

SECTION 3
Newton's Third Law
Main Idea Forces act in equal but opposite pairs.

Moving at a Crawl

This enormous vehicle is a crawler that moves a space shuttle to the launch pad. The crawler and space shuttle together have a mass of about 7,700,000 kg. To move the crawler at a speed of about 1.5 km/h requires a force of about 10,000,000 N. This force is exerted by 16 electric motors in the crawler.

Science Journal Describe three examples of pushing or pulling an object. How did the object move?

Start-Up Activities

Forces and Motion

Imagine being on a bobsled team speeding down an icy run. Forces are exerted on the sled by the ice, the sled's brakes and steering mechanism, and gravity. Newton's laws predict how these forces cause the bobsled to turn, speed up, or slow down. Newton's Laws tell how forces cause the motion of any object to change.

1. Lean two metersticks parallel, less than a marble width apart on three books as shown on the left. This is your ramp.
2. Tap a marble so it rolls up the ramp. Measure how far up the ramp it travels before rolling back.
3. Repeat step 2 using two books, one book, and zero books. The same person should tap with the same force each time.
4. **Think Critically** Make a table to record the motion of the marble for each ramp height. What would happen if the ramp were perfectly smooth and level?

Newton's Laws Make the following Foldable to help you organize your thoughts about Newton's laws.

STEP 1 Fold a sheet of paper in half lengthwise. Make the back edge about 5 cm longer than the front edge.

STEP 2 Turn the paper so the fold is on the bottom. Then fold it into thirds.

STEP 3 Unfold and cut only the top layer along both folds to make three tabs.

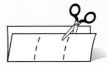

STEP 4 Label the foldable as shown.

Make a Concept Map As you read the chapter, record what you learn about each of Newton's laws in your concept map.

Preview this chapter's content and activities at
blue.msscience.com

Get Ready to Read

Compare and Contrast

1) Learn It! Good readers compare and contrast information as they read. This means they look for similarities and differences to help them to remember important ideas. Look for signal words in the text to let you know when the author is comparing or contrasting.

Compare and Contrast Signal Words	
Compare	**Contrast**
as	but
like	or
likewise	unlike
similarly	however
at the same time	although
in a similar way	on the other hand

2) Practice It! Read the excerpt below and notice how the author uses contrast signal words to describe the differences between weight and mass.

> When you stand on a bathroom scale, you are measuring the pull of the Earth's gravity — a force. **However,** mass is the amount of matter in an object, and doesn't depend on location. Weight will vary with location, **but** mass will remain constant.
>
> —*from page 558*

3) Apply It! Compare and contrast sliding friction on page 554 and air resistance on page 561.

Target Your Reading

Reading Tip

As you read, use other skills, such as summarizing and connecting, to help you understand comparisons and contrasts.

Use this to focus on the main ideas as you read the chapter.

1 Before you read the chapter, respond to the statements below on your worksheet or on a numbered sheet of paper.
- Write an **A** if you **agree** with the statement.
- Write a **D** if you **disagree** with the statement.

2 After you read the chapter, look back to this page to see if you've changed your mind about any of the statements.
- If any of your answers changed, explain why.
- Change any false statements into true statements.
- Use your revised statements as a study guide.

Science Online
Print out a worksheet of this page at blue.msscience.com

Before You Read A or D		Statement	After You Read A or D
	1	If an object is moving, unbalanced forces are acting on the object.	
	2	When you jump up into the air, the ground exerts a force on you.	
	3	A force is a push or a pull.	
	4	Gravity does not pull on astronauts while in orbit around Earth.	
	5	Objects must be touching each other to apply forces on one another.	
	6	An object traveling in a circle at a constant speed is not accelerating.	
	7	Action and reaction force pairs cancel each other because they are equal in size but opposite in direction.	
	8	Gravity pulls on all objects that have mass.	
	9	An object at rest can have forces acting on it.	

550 B

section 1
Newton's First Law

as you read

What You'll Learn
- **Distinguish** between balanced and net forces.
- **Describe** Newton's first law of motion.
- **Explain** how friction affects motion.

Why It's Important
Forces can cause the motion of objects to change.

Review Vocabulary
velocity: the speed and direction of a moving object

New Vocabulary
- force
- net force
- balanced forces
- unbalanced forces
- Newton's first law of motion
- friction

Force

A soccer ball sits on the ground, motionless, until you kick it. Your science book sits on the table until you pick it up. If you hold your book above the ground, then let it go, gravity pulls it to the floor. In every one of these cases, the motion of the ball or book was changed by something pushing or pulling on it. An object will speed up, slow down, or turn only if something is pushing or pulling on it.

A **force** is a push or a pull. Examples of forces are shown in **Figure 1.** Think about throwing a ball. Your hand exerts a force on the ball, and the ball accelerates forward until it leaves your hand. After the ball leaves your hand, the force of gravity causes its path to curve downward. When the ball hits the ground, the ground exerts a force, stopping the ball.

A force can be exerted in different ways. For instance, a paper clip can be moved by the force a magnet exerts, the pull of Earth's gravity, or the force you exert when you pick it up. These are all examples of forces acting on the paper clip.

Figure 1 A force is a push or a pull.

The magnet on the crane pulls the pieces of scrap metal upward.

This golf club exerts a force by pushing on the golf ball.

550 CHAPTER 19 Force and Newton's Laws

This door is not moving because the forces exerted on it are equal and in opposite directions.

The door is closing because the force pushing the door closed is greater than the force pushing it open.

Figure 2 When the forces on an object are balanced, no change in motion occurs. A change in motion occurs only when the forces acting on an object are unbalanced.

Combining Forces More than one force can act on an object at the same time. If you hold a paper clip near a magnet, you, the magnet, and gravity all exert forces on the paper clip. The combination of all the forces acting on an object is the **net force.** When more than one force acts on an object, the net force determines how the motion of an object changes. If the motion of an object changes, its velocity changes. A change in velocity means the object is accelerating.

How do forces combine to form the net force? If the forces are in the same direction, they add together to form the net force. If two forces are in opposite directions, then the net force is the difference between the two forces, and it is in the direction of the larger force.

Balanced and Unbalanced Forces A force can act on an object without causing it to accelerate if other forces cancel the push or pull of the force. Look at **Figure 2.** If you and your friend push on a door with the same force in opposite directions, the door does not move. Because you both exert forces of the same size in opposite directions on the door, the two forces cancel each other. Two or more forces exerted on an object are **balanced forces** if their effects cancel each other and they do not change the object's velocity. If the forces on an object are balanced, the net force is zero. If the net force is not zero, the forces are **unbalanced forces.** Then the effects of the forces don't cancel, and the object's velocity changes.

Biomechanics Whether you run, jump, or sit, forces are being exerted on different parts of your body. Biomechanics is the study of how the body exerts forces and how it is affected by forces acting on it. Research how biomechanics has been used to reduce job-related injuries. Write a paragraph on what you've learned in your Science Journal.

SECTION 1 Newton's First Law **551**

Newton's First Law of Motion

If you stand on a skateboard and someone gives you a push, then you and your skateboard will start moving. You will begin to move when the force was applied. An object at rest—like you on your skateboard—remains at rest unless an unbalanced force acts on it and causes it to move.

Because a force had to be applied to make you move when you and your skateboard were at rest, you might think that a force has to be applied continually to keep an object moving. Surprisingly, this is not the case. An object can be moving even if the net force acting on it is zero.

The Italian scientist Galileo Galilei, who lived from 1564 to 1642, was one of the first to understand that a force doesn't need to be constantly applied to an object to keep it moving.

Galileo's ideas helped Isaac Newton to better understand the nature of motion. Newton, who lived from 1642 to 1727, explained the motion of objects in three rules called Newton's laws of motion.

Newton's first law of motion describes how an object moves when the net force acting on it is zero. According to **Newton's first law of motion,** if the net force acting on an object is zero, the object remains at rest, or if the object is already moving, continues to move in a straight line with constant speed.

Friction

Galileo realized the motion of an object doesn't change until an unbalanced force acts on it. Every day you see moving objects come to a stop. The force that brings nearly everything to a stop is **friction,** which is the force that acts to resist sliding between two touching surfaces, as shown in **Figure 3.** Friction is why you never see objects moving with constant velocity unless a net force is applied. Friction is the force that eventually brings your skateboard to a stop unless you keep pushing on it. Friction also acts on objects that are sliding or moving through substances such as air or water.

Figure 3 When two objects in contact try to slide past each other, friction keeps them from moving or slows them down.

Friction slows down this sliding baseball player.

Friction Opposes Sliding Although several different forms of friction exist, they all have one thing in common. If two objects are in contact, frictional forces always try to prevent one object from sliding on the other object. If you rub your hand against a tabletop, you can feel the friction push against the motion of your hand. If you rub the other way, you can feel the direction of friction change so it is again acting against your hand's motion.

Reading Check *What do the different forms of friction have in common?*

Older Ideas About Motion It took a long time to understand motion. One reason was that people did not understand the behavior of friction and that friction was a force. Because moving objects eventually come to a stop, people thought the natural state of an object was to be at rest. For an object to be in motion, something always had to be pushing or pulling it to keep the object moving. As soon as the force stopped, the object would stop moving.

Galileo understood that an object in constant motion is as natural as an object at rest. It was usually friction that made moving objects slow down and eventually come to a stop. To keep an object moving, a force had to be applied to overcome the effects of friction. If friction could be removed, an object in motion would continue to move in a straight line with constant speed. **Figure 4** shows motion where there is almost no friction.

Science Online

Topic: Galileo and Newton
Visit blue.msscience.com for Web links to information about the lives of Galileo and Newton.

Activity Make a time line showing important events in the lives of either Galileo or Newton.

Figure 4 In an air hockey game, the puck floats on a layer of air, so that friction is almost eliminated. As a result, the puck moves in a straight line with nearly constant speed after it's been hit.
Infer *how the puck would move if there was no layer of air.*

SECTION 1 Newton's First Law **553**

Mini LAB

Observing Friction

Procedure
1. Lay a **bar of soap**, a **flat eraser**, and a **key** side by side on one end of a **hard-sided notebook.**
2. At a constant rate, slowly lift the end of notebook with objects on it. Note the order in which the objects start sliding.

Analysis
1. For which object was static friction the greatest? For which object was it the smallest? Explain, based on your observations.
2. Which object slid the fastest? Which slid the slowest? Explain why there is a difference in speed.
3. How could you increase and decrease the amount of friction between two materials?

Try at Home

Static Friction If you've ever tried pushing something heavy, like a refrigerator, you might have discovered that nothing happened at first. Then as you push harder and harder, the object suddenly will start to move. When you first start to push, friction between the heavy refrigerator and the floor opposes the force you are exerting and the net force is zero. The type of friction that prevents an object from moving when a force is applied is called static friction.

Static friction is caused by the attraction between the atoms on the two surfaces that are in contact. This causes the surfaces to stick or weld together where they are in contact. Usually, as the surface gets rougher and the object gets heavier, the force of static friction will be larger. To move the object, you have to exert a force large enough to break the bonds holding two surfaces together.

Sliding Friction While static friction acts on an object at rest, sliding friction slows down an object that slides. When you push a box over the floor, sliding friction acts in the direction opposite to the motion of the box. If you stop pushing, sliding friction causes the box to stop. Sliding friction is due to the microscopic roughness of two openers, as shown in **Figure 5.** The surfaces tend to stick together where they touch. The bonds between the surfaces are broken and form again as the surfaces slide past each other. This causes sliding friction. **Figure 6** shows that sliding friction is produced when the brake pad in a bicycle's brakes rub against the wheel.

Reading Check *What is the difference between static friction and sliding friction?*

Figure 5 Microscopic roughness, even on surfaces that seem smooth, such as the tray and metal shelf, causes sliding friction.

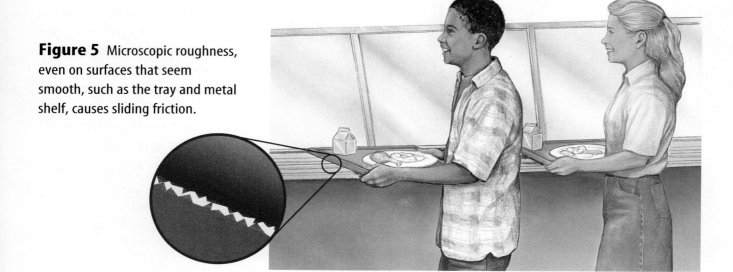

Figure 6 Sliding friction and rolling friction can act on a bicycle.

Sliding friction between the brake pads and the wheel causes the wheel to slow down.

Rolling friction acts between the ground and the bicycle tires when the wheels are rolling.

Rolling Friction If you're coasting on a bicycle or on a skateboard, you slow down and eventually stop because of another type of frictional force. Rolling friction occurs when an object rolls across a surface. It is rolling friction between the bicycle tires and the ground, as in **Figure 6,** that slows a moving bicycle.

The size of the rolling friction force due usually is much less than the force of sliding friction between the same surfaces. This is why it takes less force to pull a box on a wagon or cart with wheels, than to drag the box along the ground. Rolling friction between the wheels and the ground is less than the sliding friction between the box and the ground.

section 1 review

Summary

Force
- A force is a push or a pull.
- The net force on an object is the combination of all the forces acting on the object.
- The forces acting on an object can be balanced or unbalanced. If the forces are balanced, the net force is zero.

Newton's First Law of Motion
- If the net force on an object at rest is zero, the object remains at rest, or if the object is moving, it continues moving in a straight line with constant speed.

Friction
- Friction is the force that acts to resist sliding between two surfaces that are touching.
- Three types of friction are static friction, sliding friction, and rolling friction.

Self Check

1. **Explain** whether a force is acting on a car that is moving at 20 km/h and turns to the left.
2. **Describe** the factors that cause static friction between two surfaces to increase.
3. **Discuss** why friction made it difficult to discover Newton's first law of motion.
4. **Discuss** whether an object can be moving if the net force acting on the object is zero.
5. **Think Critically** For the following actions, explain whether the forces involved are balanced or unbalanced.
   a. You push a box until it moves.
   b. You push a box but it doesn't move.
   c. You stop pushing a box and it slows down.

Applying Skills

6. **Compare and contrast** static, sliding, and rolling friction.

section 2
Newton's Second Law

as you read

What You'll Learn
- **Explain** Newton's second law of motion.
- **Explain** why the direction of force is important.

Why It's Important
Newton's second law of motion explains how changes in motion can be calculated.

Review Vocabulary
acceleration: the change in velocity divided by the time over which the change occurred

New Vocabulary
- Newton's second law of motion
- weight
- center of mass

Force and Acceleration

When you go shopping in a grocery store and push a cart, you exert a force to make the cart move. If you want to slow down or change the direction of the cart, a force is required to do this, as well. Would it be easier for you to stop a full or empty grocery cart suddenly, as in **Figure 7?** When the motion of an object changes, the object is accelerating. Acceleration occurs any time an object speeds up, slows down, or changes its direction of motion.

Newton's second law of motion connects the net force on an object, its mass, and its acceleration. According to **Newton's second law of motion,** the acceleration of an object equals the net force divided by the mass and is in the direction of the net force. The acceleration can be calculated using this equation:

Newton's Second Law Equation

$$\text{acceleration (in meters/second}^2\text{)} = \frac{\text{net force (in newtons)}}{\text{mass (in kilograms)}}$$

$$a = \frac{F_{net}}{m}$$

In this equation, a is the acceleration, m is the mass, and F_{net} is the net force. If both sides of the above equation are multiplied by the mass, the equation can be written this way:

$$F_{net} = ma$$

Reading Check *What is Newton's second law?*

Figure 7 The force needed to change the motion of an object depends on its mass.
Predict *which grocery cart would be easier to stop.*

Units of Force Force is measured in newtons, abbreviated N. Because the SI unit for mass is the kilogram (kg) and acceleration has units of meters per second squared (m/s²), 1 N also is equal to 1 kg·m/s². In other words, to calculate a force in newtons from the equation shown on the prior page, the mass must be given in kg and the acceleration in m/s².

Gravity

One force that you are familiar with is gravity. Whether you're coasting down a hill on a bike or a skateboard or jumping into a pool, gravity is pulling you downward. Gravity also is the force that causes Earth to orbit the Sun and the Moon to orbit Earth.

What is gravity? The force of gravity exists between any two objects that have mass. Gravity always is attractive and pulls objects toward each other. A gravitational attraction exists between you and every object in the universe that has mass. However, the force of gravity depends on the mass of the objects and the distance between them. The gravitational force becomes weaker the farther apart the objects are and also decreases as the masses of the objects involved decrease.

For example, there is a gravitational force between you and the Sun and between you and Earth. The Sun is much more massive than Earth, but is so far away that the gravitational force between you and the Sun is too weak to notice. Only Earth is close enough and massive enough to exert a noticeable gravitational force on you. The force of gravity between you and Earth is about 1,650 times greater than between you and the Sun.

Newton and Gravity
Isaac Newton was the first to realize that gravity—the force that made objects fall to Earth—was also the force that caused the Moon to orbit Earth and the planets to orbit the Sun. In 1687, Newton published a book that included the law of universal gravitation. This law showed how to calculate the gravitational force between any two objects. Using the law of universal gravitation, astronomers were able to explain the motions of the planets in the solar system, as well as the motions of distant stars and galaxies.

Weight When you stand on a bathroom scale, what are you measuring? The **weight** of an object is the size of the gravitational force exerted on an object. Your weight on Earth is the gravitational force between you and Earth. On Earth, weight is calculated from this equation:

$$W = m\,(9.8 \text{ m/s}^2)$$

In this equation, W is the weight in N, and m is the mass in kg. Your weight would change if you were standing on a planet other than Earth, as shown in **Table 1.** Your weight on a different planet would be the gravitational force between you and the planet.

Table 1 Weight of 60-kg Person on Different Planets		
Place	Weight in Newtons if Your Mass were 60 kg	Percent of Your Weight on Earth
Mars	221	37.7
Earth	588	100.0
Jupiter	1,390	236.4
Pluto	35	5.9

Figure 8 The sled speeds up when the net force on the sled is in the same direction as the sled's velocity.

Figure 9 The sled slows down when the net force on the sled is in the direction opposite to the sled's velocity.

Weight and Mass Weight and mass are different. Weight is a force, just like the push of your hand is a force, and is measured in newtons. When you stand on a bathroom scale, you are measuring the pull of Earth's gravity—a force. However, mass is the amount of matter in an object, and doesn't depend on location. Weight will vary with location, but mass will remain constant. A book with a mass of 1 kg has a mass of 1 kg on Earth or on Mars. However, the weight of the book would be different on Earth and Mars. The two planets would exert a different gravitational force on the book.

Using Newton's Second Law

The second law of motion tells how to calculate the acceleration of an object if the object's mass and the forces acting on it are known. Recall that the acceleration equals the change in velocity divided by the change in time. If an object's acceleration is known, the change in its velocity can be determined

Speeding Up When does an unbalanced force cause an object to speed up? If an object is moving, a net force applied in the same direction as the object is moving causes the object to speed up. For exmple, in **Figure 8** the applied force is in the same direction as the sled's velocity. This makes the sled speed up and its velocity increase.

The net force on a ball falling to the ground is downward. This force is in the same direction that the ball is moving. Because the net force on the ball is in the same direction as the ball's velocity, the ball speeds up as it falls.

Slowing Down If the net force on an object is in the direction opposite to the object's velocity, the object slows down. In **Figure 9,** the force of sliding friction becomes larger when the boy puts his feet in the snow. The net force on the sled is the combination of gravity and sliding friction. When the sliding friction force becomes large enough, the net force is opposite to the sled's velocity. This causes the sled to slow down.

Calculating Acceleration Newton's second law of motion can be used to calculate acceleration. For example, suppose you pull a 10-kg box so that the net force on the box is 5 N. The acceleration can be found as follows:

$$a = \frac{F_{net}}{m} = \frac{5 \text{ N}}{10 \text{ kg}} = 0.5 \text{ m/s}^2$$

The box keeps accelerating as long as you keep pulling on it. The acceleration does not depend on how fast the box is moving. It depends only on the net force and the mass of the box.

Applying Math — Solve a Simple Equation

ACCELERATION OF A CAR A net force of 4,500 N acts on a car with a mass of 1,500 kg. What is the acceleration of the car?

Solution

1 *This is what you know:*
- net force: $F_{net} = 4{,}500$ N
- mass: $m = 1{,}500$ kg

2 *This is what you need to find:*
acceleration: $a = ?$ m/s^2

3 *This is the procedure you need to use:* Substitute the known values for net force and mass into the equation for Newton's second law of motion to calculate the acceleration:

$$a = \frac{F_{net}}{m} = \frac{4{,}500 \text{ N}}{1{,}500 \text{ kg}} = 3.0 \, \frac{\text{N}}{\text{kg}} = 3.0 \text{ m/s}^2$$

4 *Check your answer:* Multiply your answer by the mass, 1,500 kg. The result should be the given net force, 4,500 N.

Practice Problems

1. A book with a mass of 2.0 kg is pushed along a table. If the net force on the book is 1.0 N, what is the book's acceleration?

2. A baseball has a mass of 0.15 kg. What is the net force on the ball if its acceleration is 40.0 m/s^2?

Science Online — For more practice visit blue.msscience.com/math_practice

Figure 10 The force due to gravity on the ball is at an angle to the ball's velocity. This causes the ball to move in a curved path. **Infer** *how the ball would move if it were thrown horizontally.*

Turning Sometimes the net force on an object and the object's velocity, are neither in the same direction nor in the opposite direction. Then the object will move in a curved path, instead of a straight line.

When you shoot a basketball, it doesn't continue to move in a straight line after it leaves your hand. Instead the ball starts to curve downward, as shown in **Figure 10.** Gravity pulls the ball downward, so the net force on the ball is at an angle to the ball's velocity. This causes the ball to move in a curved path.

Circular Motion

A rider on a merry-go-round ride moves in a circle. This type of motion is called circular motion. If you are in circular motion, your direction of motion is constantly changing. This means you are constantly accelerating. According to Newton's second law of motion, if you are constantly accelerating, there must be a non-zero net force acting on you the entire time.

To cause an object to move in circular motion with constant speed, the net force on the object must be at right angles to the velocity. When an object moves in circular motion, the net force on the object is called centripetal force. The direction of the centripetal force is toward the center of the object's circular path.

Satellite Motion Objects that orbit Earth are satellites of Earth. Some satellites go around Earth in nearly circular orbits. The centripetal force is due to gravity between the satellite and Earth. Gravity causes the net force on the satellite to always point toward Earth, which is center of the satellite's orbit. Why doesn't a satellite fall to Earth like a baseball does? Actually, a satellite is falling to Earth just like a baseball.

Suppose Earth were perfectly smooth and you throw a baseball horizontally. Gravity pulls the baseball downward so it travels in a curved path. If the baseball is thrown faster, its path is less curved, and it travels farther before it hits the ground. If the baseball were traveling fast enough, as it fell, its curved path would follow the curve of Earth's surface as shown in **Figure 11.** Then the baseball would never hit the ground. Instead, it would continue to fall around Earth.

Satellites in orbit are being pulled toward Earth just as baseballs are. The difference is that satellites are moving so fast horizontally that Earth's surface curves downward at the same rate that the satellites are falling downward. The speed at which a object must move to go into orbit near Earth's surface is about 8 km/s, or about 29,000 km/h.

Figure 11 The faster a ball is thrown, the farther it travels before gravity pulls it to Earth. If the ball is traveling fast enough, Earth's surface curves away from it as fast as it falls downward. Then the ball never hits the ground.

Air Resistance

Whether you are walking, running, or biking, air is pushing against you. This push is air resistance. Air resistance is a form of friction that acts to slow down any object moving in the air. Air resistance is a force that gets larger as an object moves faster. Air resistance also depends on the shape of an object. A piece of paper crumpled into a ball falls faster than a flat piece of paper falls.

When an object falls it speeds up as gravity pulls it downward. At the same time, the force of air resistance pushing up on the object is increasing as the object moves faster. Finally, the upward air resistance force becomes large enough to equal the downward force of gravity.

When the air resistance force equals the weight, the net force on the object is zero. By Newton's second law, the object's acceleration then is zero, and its speed no longer increases. When air resistance balances the force of gravity, the object falls at a constant speed called the terminal velocity.

Figure 12 The wrench is spinning as it slides across the table. The center of mass of the wrench, shown by the dots, moves as if the net force is acting at that point.

Center of Mass

When you throw a stick, the motion of the stick might seem to be complicated. However, there is one point on the stick, called the center of mass, that moves in a smooth path. The **center of mass** is the point in an object that moves as if all the object's mass were concentrated at that point. For a symmetrical object, such as a ball, the center of mass is at the object's center. However, for any object the center of mass moves as if the net force is being applied there.

Figure 12 shows how the center of mass of a wrench moves as it slides across a table. The net force on the wrench is the force of friction between the wrench and the table. This causes the center of mass to move in a straight line with decreasing speed.

section 2 review

Summary

Force and Acceleration
- According to Newton's second law, the net force on an object, its mass, and its acceleration are related by
$$F_{net} = ma$$

Gravity
- The force of gravity between any two objects is always attractive and depends on the masses of the objects and the distance between them.

Using Newton's Second Law
- A moving object speeds up if the net force is in the direction of the motion.
- A moving object slows down if the net force is in the direction opposite to the motion.
- A moving object turns if the net force is at an angle to the direction of motion.

Circular Motion
- In circular motion with constant speed, the net force is called the centripetal force and points toward the center of the circular path.

Self Check

1. **Make a diagram** showing the forces acting on a coasting bike rider traveling at 25 km/h on a flat roadway.
2. **Analyze** how your weight would change with time if you were on a space ship traveling away from Earth toward the Moon.
3. **Explain** how the force of air resistance depends on an object's speed.
4. **Infer** the direction of the net force acting on a car as it slows down and turns right.
5. **Think Critically** Three students are pushing on a box. Under what conditions will the motion of the box change?

Applying Math

6. **Calculate Net Force** A car has a mass of 1,500 kg. If the car has an acceleration of 2.0 m/s^2, what is the net force acting on the car?
7. **Calculate Mass** During a softball game, a softball is struck by a bat and has an acceleration of 1,500 m/s^2. If the net force exerted on the softball by the bat is 300 N, what is the softball's mass?

section 3
Newton's Third Law

Action and Reaction

Newton's first two laws of motion explain how the motion of a single object changes. If the forces acting on the object are balanced, the object will remain at rest or stay in motion with constant velocity. If the forces are unbalanced, the object will accelerate in the direction of the net force. Newton's second law tells how to calculate the acceleration, or change in motion, of an object if the net force acting on it is known.

Newton's third law describes something else that happens when one object exerts a force on another object. Suppose you push on a wall. It may surprise you to learn that if you push on a wall, the wall also pushes on you. According to **Newton's third law of motion**, forces always act in equal but opposite pairs. Another way of saying this is for every action, there is an equal but opposite reaction. This means that when you push on a wall, the wall pushes back on you with a force equal in strength to the force you exerted. When one object exerts a force on another object, the second object exerts the same size force on the first object, as shown in **Figure 13**.

as you read

What You'll Learn
- **Identify** the relationship between the forces that objects exert on each other.

Why It's Important
Newton's third law can explain how birds fly and rockets move.

Review Vocabulary
force: a push or a pull

New Vocabulary
- Newton's third law of motion

Figure 13 The car jack is pushing up on the car with the same amount of force with which the car is pushing down on the jack.
Identify *the other force acting on the car.*

Figure 14 In this collision, the first car exerts a force on the second. The second exerts the same force in the opposite direction on the first car.
Explain whether both cars will have the same acceleration.

Topic: How Birds Fly
Visit blue.msscience.com for Web links to information about how birds and other animals fly.

Activity Make a diagram showing the forces acting on a bird as it flies.

Figure 15 When the child pushes against the wall, the wall pushes against the child.

Action and Reaction Forces Don't Cancel The forces exerted by two objects on each other are often called an action-reaction force pair. Either force can be considered the action force or the reaction force. You might think that because action-reaction forces are equal and opposite that they cancel. However, action and reaction force pairs don't cancel because they act on different objects. Forces can cancel only if they act on the same object.

For example, imagine you're driving a bumper car and are about to bump a friend in another car, as shown in **Figure 14.** When the two cars collide, your car pushes on the other car. By Newton's third law, that car pushes on your car with the same force, but in the opposite direction. This force causes you to slow down. One force of the action-reaction force pair is exerted on your friend's car, and the other force of the force pair is exerted on your car. Another example of an action-reaction pair is shown in **Figure 15.**

You constantly use action-reaction force pairs as you move about. When you jump, you push down on the ground. The ground then pushes up on you. It is this upward force that pushes you into the air. **Figure 16** shows some examples of how Newton's laws of motion are demonstrated in sporting events.

 Birds and other flying creatures also use Newton's third law. When a bird flies, its wings push in a downward and a backward direction. This pushes air downward and backward. By Newton's third law, the air pushes back on the bird in the opposite directions—upward and forward. This force keeps a bird in the air and propels it forward.

564 CHAPTER 19 Force and Newton's Laws

NATIONAL GEOGRAPHIC VISUALIZING NEWTON'S LAWS IN SPORTS

Figure 16

Although it is not obvious, Newton's laws of motion are demonstrated in sports activities all the time. According to the first law, if an object is in motion, it moves in a straight line with constant speed unless a net force acts on it. If an object is at rest, it stays at rest unless a net force acts on it. The second law states that a net force acting on an object causes the object to accelerate in the direction of the force. The third law can be understood this way—for every action force, there is an equal and opposite reaction force.

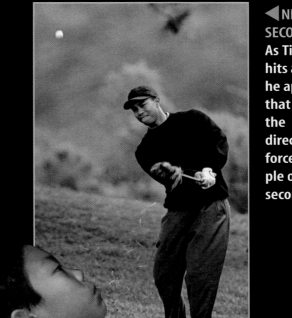

◀ **NEWTON'S SECOND LAW** As Tiger Woods hits a golf ball, he applies a force that will drive the ball in the direction of that force—an example of Newton's second law.

▲ **NEWTON'S FIRST LAW** According to Newton's first law, the diver does not move in a straight line with constant speed because of the force of gravity.

▶ **NEWTON'S THIRD LAW** Newton's third law applies even when objects do not move. Here a gymnast pushes downward on the bars. The bars push back on the gymnast with an equal force.

SECTION 3 Newton's Third Law

Figure 17 The force of the ground on your foot is equal and opposite to the force of your foot on the ground. If you push back harder, the ground pushes forward harder.
Determine *In what direction does the ground push on you if you are standing still?*

Change in Motion Depends on Mass Often you might not notice the effects of the forces in an action-reaction pair. This can hapen if one of the objects involved is much more massive than the other. Then the massice object might seem to remain motionless when one of the action-reaction forces acts on it. For example, when you walk forward, as in **Figure 17,** you push backward on the ground. Earth then pushes you forward with the same size force. Because Earth's mass is so large, the force you exert causes Earth to have an extremely small acceleration. This acceleration is so small that Earth's change in motion is undetectable as you walk.

A Rocket Launch The launching of a space shuttle is a spectacular example of Newton's third law. Three rocket engines supply the force, called thrust, that lifts the rocket. When the rocket fuel is ignited, a hot gas is produced. The gas molecules collide with the inside of the engine, as shown in **Figure 18.** As these collisions occur, the engine pushes the hot gases downward. According to Newton's third law of motion, the hot gases push upward on the engine. The upward force exerted by the gases on the rocket propels the rocket upward.

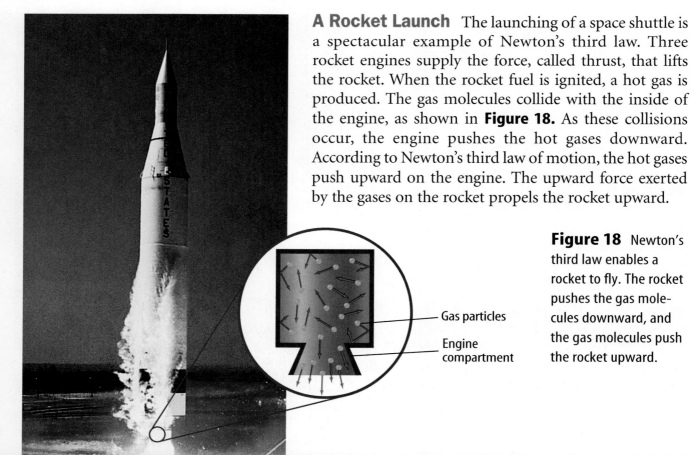

Figure 18 Newton's third law enables a rocket to fly. The rocket pushes the gas molecules downward, and the gas molecules push the rocket upward.

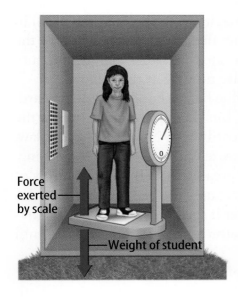
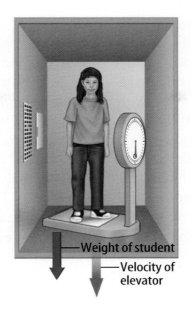

Figure 19 Whether you are standing on Earth or falling, the force of Earth's gravity on you doesn't change. However, your weight measured by a scale would change.

Weightlessness

You might have seen pictures of astronauts floating inside a space shuttle as it orbits Earth. The astronauts are said to be weightless, as if Earth's gravity were no longer pulling on them. Yet the force of gravity on the shuttle is almost 90 percent as large as at Earth's surface. Newton's laws of motion can explain why the astronauts float as if there were no forces acting on them.

Measuring Weight Think about how you measure your weight. When you stand on a scale, your weight pushes down on the scale. This causes the scale pointer to point to your weight. At the same time, by Newton's third law the scale pushes up on you with a force equal to your weight, as shown in **Figure 19**. This force balances the downward pull of gravity on you.

Free Fall and Weightlessness Now suppose you were standing on a scale in an elevator that is falling, as shown in **Figure 19**. A falling object is in free fall when the only force acting on the object is gravity. Inside the free-falling elevator, you and the scale are both in free fall. Because the only force acting on you is gravity, the scale no longer is pushing up on you. According to Newton's third law, you no longer push down on the scale. So the scale pointer stays at zero and you seem to be weightless. Weightlessness is the condition that occurs in free fall when the weight of an object seems to be zero.

However, you are not really weightless in free fall because Earth is still pulling down on you. With nothing to push up on you, such as your chair, you would have no sensation of weight.

Measuring Force Pairs

Procedure
1. Work in pairs. Each person needs a **spring scale**.
2. Hook the two scales together. Each person should pull back on a scale. Record the two readings. Pull harder and record the two readings.
3. Continue to pull on both scales, but let the scales move toward one person. Do the readings change?
4. Try to pull in such a way that the two scales have different readings.

Analysis
1. What can you conclude about the pair of forces in each situation?
2. Explain how this experiment demonstrates Newton's third law.

SECTION 3 Newton's Third Law **567**

Figure 20 These oranges seem to be floating because they are falling around Earth at the same speed as the space shuttle and the astronauts. As a result, they aren't moving relative to the astronauts in the cabin.

Weightlessness in Orbit To understand how objects move in the orbiting space shuttle, imagine you were holding a ball in the free-falling elevator. If you let the ball go, the position of the ball relative to you and the elevator wouldn't change, because you, the ball, and the elevator are moving at the same speed.

However, suppose you give the ball a gentle push downward. While you are pushing the ball, this downward force adds to the downward force of gravity. According to Newton's second law, the acceleration of the ball increases. So while you are pushing, the acceleration of the ball is greater than the acceleration of both you and the elevator. This causes the ball to speed up relative to you and the elevator. After it speeds up, it continues moving faster than you and the elevator, and it drifts downward until it hits the elevator floor.

When the space shuttle orbits Earth, the shuttle and all the objects in it are in free fall. They are falling in a curved path around Earth, instead of falling straight downward. As a result, objects in the shuttle appear to be weightless, as shown in **Figure 20.** A small push causes an object to drift away, just as a small downward push on the ball in the free-falling elevator caused it to drift to the floor.

section 3 review

Summary

Action and Reaction

- According to Newton's third law, when one object exerts a force on another object, the second object exerts the same size force on the first object.
- Either force in an action-reaction force pair can be the action force or the reaction force.
- Action and reaction force pairs don't cancel because they are exerted on different objects.
- When action and reaction forces are exerted by two objects, the accelerations of the objects depend on the masses of the objects.

Weightlessness

- A falling object is in free fall if the only force acting on it is gravity.
- Weightlessness occurs in free fall when the weight of an object seems to be zero.
- Objects orbiting Earth appear to be weightless because they are in free fall in a curved path around Earth.

Self Check

1. **Evaluate** the force a skateboard exerts on you if your mass is 60 kg and you push on the skateboard with a force of 60 N.
2. **Explain** why you move forward and a boat moves backward when you jump from a boat to a pier.
3. **Describe** the action and reaction forces when a hammer hits a nail.
4. **Infer** You and a child are on skates and you give each other a push. If the mass of the child is half your mass, who has the greater acceleration? By what factor?
5. **Think Critically** Suppose you are walking in an airliner in flight. Use Newton's third law to describe the effect of your walk on the motion on the airliner.

Applying Math

6. **Calculate Acceleration** A person standing in a canoe exerts a force of 700 N to throw an anchor over the side. Find the acceleration of the canoe if the total mass of the canoe and the person is 100 kg.

BALLOON *RACES*

▶ *Real-World Question*

The motion of a rocket lifting off a launch pad is determined by Newton's laws of motion. Here you will make a balloon rocket that is powered by escaping air. How do Newton's laws of motion explain the motion of balloon rockets?

Goals
- **Measure** the speed of a balloon rocket.
- **Describe** how Newton's laws explain a rocket's motion.

Materials
balloons
drinking straws
string
tape
meterstick
stopwatch
*clock
*Alternate materials

Safety Precautions

▶ *Procedure*

1. Make a rocket path by threading a string through a drinking straw. Run the string across the classroom and fasten at both ends.
2. Blow up a balloon and hold it tightly at the end to prevent air from escaping. Tape the balloon to the straw on the string.
3. Release the balloon so it moves along the string. Measure the distance the balloon travels and the time it takes.
4. Repeat steps 2 and 3 with different balloons.

▶ *Conclude and Apply*

1. **Compare and contrast** the distances traveled. Which rocket went the greatest distance?
2. **Calculate** the average speed for each rocket. Compare and contrast them. Which rocket has the greatest average speed?
3. **Infer** which aspects of these rockets made them travel far or fast.
4. **Draw** a diagram showing all the forces acting on a balloon rocket.
5. Use Newton's laws of motion to explain the motion of a balloon rocket from launch until it comes to a stop.

*C*ommunicating **Your Data**

Discuss with classmates which balloon rocket traveled the farthest. Why? **For more help, refer to the** Science Skill Handbook.

LAB Design Your Own

MODELING MOTION IN TWO DIRECTIONS

▶ Real-World Question

When you move a computer mouse across a mouse pad, how does the rolling ball tell the computer cursor to move in the direction that you push the mouse? Inside the housing for the mouse's ball are two or more rollers that the ball rubs against as you move the mouse. They measure up-and-down and back-and-forth motions. The motion of the cursor on the screen is based on the movement of the up-and-down rollers and the back-and-forth rollers. Can any object be moved along a path by a series of motions in only two directions?

Goals
- **Move** the skid across the ground using two forces.
- **Measure** how fast the skid can be moved.
- **Determine** how smoothly the direction can be changed.

Possible Materials
masking tape
stopwatch
*watch or clock with a second hand
meterstick
*metric tape measure
spring scales marked in newtons (2)
plastic lid
golf ball
*tennis ball
*Alternate materials

Safety Precautions

▶ Form a Hypothesis

How can you combine forces to move in a straight line, along a diagonal, or around corners? Place a golf ball on something that will slide, such as a plastic lid. The plastic lid is called a skid. Lay out a course to follow on the floor. Write a plan for moving your golf ball along the path without having the golf ball roll away.

▶ Test Your Hypothesis

Make a Plan

1. Lay out a course that involves two directions, such as always moving forward or left.
2. Attach two spring scales to the skid. One always will pull straight forward. One always will pull to one side. You cannot turn the skid. If one scale is pulling toward the door of your classroom, it always must pull in that direction. (It can pull with zero force if needed, but it can't push.)
3. How will you handle movements along diagonals and turns?
4. How will you measure speed?

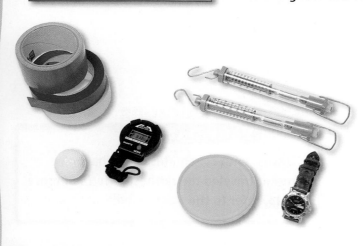

570 CHAPTER 19 Force and Newton's Laws

Using Scientific Methods

5. **Experiment** with your skid. How hard do you have to pull to counteract sliding friction at a given speed? How fast can you accelerate? Can you stop suddenly without spilling the golf ball, or do you need to slow down?

6. **Write** a plan for moving your golf ball along the course by pulling only forward or to one side. Be sure you understand your plan and have considered all the details.

Follow Your Plan

1. Make sure your teacher approves your plan before you start.
2. Move your golf ball along the path.
3. Modify your plan, if needed.
4. **Organize** your data so they can be used to run your course and write them in your Science Journal.
5. **Test** your results with a new route.

● Analyze Your Data

1. What was the difference between the two routes? How did this affect the forces you needed to use on the golf ball?
2. How did you separate and control variables in this experiment?
3. Was your hypothesis supported? Explain.

● Conclude and Apply

1. What happens when you combine two forces at right angles?
2. If you could pull on all four sides (front, back, left, right) of your skid, could you move anywhere along the floor? Make a hypothesis to explain your answer.

Communicating Your Data

Compare your conclusions with those of other students in your class. **For more help, refer to the** Science Skill Handbook.

LAB **571**

TIME SCIENCE AND Society
SCIENCE ISSUES THAT AFFECT YOU!

Air Bag Safety

After complaints and injuries, air bags in cars are helping all passengers

The car in front of yours stops suddenly. You hear the crunch of car against car and feel your seat belt grab you. Your mom is covered with, not blood, thank goodness, but with a big white cloth. Your seat belts and air bags worked perfectly.

Popcorn in the Dash

Air bags have saved more than a thousand lives since 1992. They are like having a giant popcorn kernel in the dashboard that pops and becomes many times its original size. But unlike popcorn, an air bag is triggered by impact, not temperature. In a crash, a chemical reaction produces a gas that expands in a split second, inflating a balloonlike bag to cushion the driver and possibly the front-seat passenger. The bag deflates quickly so it doesn't trap people in the car.

Newton and the Air Bag

When you're traveling in a car, you move with it at whatever speed it is going. According to Newton's first law, you are the object in motion, and you will continue in motion unless acted upon by a force, such as a car crash.

Unfortunately, a crash stops the car, but it doesn't stop you, at least, not right away. You continue moving forward if your car doesn't have air bags or if you haven't buckled your seat belt. You stop when you strike the inside of the car. You hit the dashboard or steering wheel while traveling at the speed of the car. When an air bag inflates, you come to a stop more slowly, which reduces the force that is exerted on you.

A test measures the speed at which an air bag deploys.

Measure Hold a paper plate 26 cm in front of you. Use a ruler to measure the distance. That's the distance drivers should have between the chest and the steering wheel to make air bags safe. Inform adult drivers in your family about this safety distance.

For more information, visit blue.msscience.com/time

chapter 19 Study Guide

Reviewing Main Ideas

Section 1 Newton's First Law

1. A force is a push or a pull.
2. Newton's first law states that objects in motion tend to stay in motion and objects at rest tend to stay at rest unless acted upon by a nonzero net force.
3. Friction is a force that resists motion between surfaces that are touching each other.

Section 2 Newton's Second Law

1. Newton's second law states that an object acted upon by a net force will accelerate in the direction of this force.
2. The acceleration due to a net force is given by the equation $a = F_{net}/m$.

3. The force of gravity between two objects depends on their masses and the distance between them.
4. In circular motion, a force pointing toward the center of the circle acts on an object.

Section 3 Newton's Third Law

1. According to Newton's third law, the forces two objects exert on each other are always equal but in opposite directions.
2. Action and reaction forces don't cancel because they act on different objects.
3. Objects in orbit appear to be weightless because they are in free fall around Earth.

Visualizing Main Ideas

Copy and complete the following concept map on Newton's laws of motion.

chapter 19 Review

Using Vocabulary

balanced forces p. 551
center of mass p. 562
force p. 550
friction p. 552
net force p. 551
Newton's first law of motion p. 552
Newton's second law of motion p. 556
Newton's third law of motion p. 563
unbalanced forces p. 551
weight p. 557

Explain the differences between the terms in the following sets.

1. force—inertia—weight
2. Newton's first law of motion—Newton's third law of motion
3. friction—force
4. net force—balanced forces
5. weight—weightlessness
6. balanced forces—unbalanced forces
7. friction—weight
8. Newton's first law of motion—Newton's second law of motion
9. friction—unbalanced force
10. net force—Newton's third law of motion

Checking Concepts

Choose the word or phrase that best answers the question.

11. Which of the following changes when an unbalanced force acts on an object?
    A) mass   C) inertia
    B) motion   D) weight

12. Which of the following is the force that slows a book sliding on a table?
    A) gravity
    B) static friction
    C) sliding friction
    D) inertia

Use the illustration below to answer question 13.

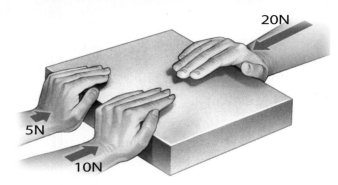

13. Two students are pushing on the left side of a box and one student is pushing on the right. The diagram above shows the forces they exert. Which way will the box move?
    A) up   C) down
    B) left   D) right

14. What combination of units is equivalent to the newton?
    A) m/s^2 C) kg·m/s^2
    B) kg·m/s   D) kg/m

15. Which of the following is a push or a pull?
    A) force   C) acceleration
    B) momentum   D) inertia

16. An object is accelerated by a net force in which direction?
    A) at an angle to the force
    B) in the direction of the force
    C) in the direction opposite to the force
    D) Any of these is possible.

17. You are riding on a bike. In which of the following situations are the forces acting on the bike balanced?
    A) You pedal to speed up.
    B) You turn at constant speed.
    C) You coast to slow down.
    D) You pedal at constant speed.

18. Which of the following has no direction?
    A) force   C) weight
    B) acceleration D) mass

chapter 19 Review

Thinking Critically

19. **Explain** why the speed of a sled increases as it moves down a snow-covered hill, even though no one is pushing on the sled.

20. **Explain** A baseball is pitched east at a speed of 40 km/h. The batter hits it west at a speed of 40 km/h. Did the ball accelerate?

21. **Form a Hypothesis** Frequently, the pair of forces acting between two objects are not noticed because one of the objects is Earth. Explain why the force acting on Earth isn't noticed.

22. **Identify** A car is parked on a hill. The driver starts the car, accelerates until the car is driving at constant speed, drives at constant speed, and then brakes to put the brake pads in contact with the spinning wheels. Explain how static friction, sliding friction, rolling friction, and air resistance are acting on the car.

23. **Draw Conclusions** You hit a hockey puck and it slides across the ice at nearly a constant speed. Is a force keeping it in motion? Explain.

24. **Infer** Newton's third law describes the forces between two colliding objects. Use this connection to explain the forces acting when you kick a soccer ball.

25. **Recognize Cause and Effect** Use Newton's third law to explain how a rocket accelerates upon takeoff.

26. **Predict** Two balls of the same size and shape are dropped from a helicopter. One ball has twice the mass of the other ball. On which ball will the force of air resistance be greater when terminal velocity is reached?

Use the figure below to answer question 27.

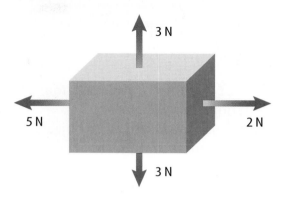

27. **Interpreting Scientific Illustrations** Is the force on the box balanced? Explain.

Performance Activities

28. **Oral Presentation** Research one of Newton's laws of motion and compose an oral presentation. Provide examples of the law. You might want to use a visual aid.

29. **Writing in Science** Create an experiment that deals with Newton's laws of motion. Document it using the following subject heads: *Title of Experiment, Partners' Names, Hypothesis, Materials, Procedures, Data, Results,* and *Conclusion.*

Applying Math

30. **Acceleration** If you exert a net force of 8 N on a 2-kg object, what will its acceleration be?

31. **Force** You push against a wall with a force of 5 N. What is the force the wall exerts on your hands?

32. **Net Force** A 0.4-kg object accelerates at 2 m/s^2. Find the net force.

33. **Friction** A 2-kg book is pushed along a table with a force of 4 N. Find the frictional force on the book if the book's acceleration is 1.5 m/s^2.

Chapter 19 Standardized Test Practice

Part 1 Multiple Choice

Record your answers on the answer sheet provided by your teacher or on a sheet of paper.

1. Which of the following descriptions of gravitational force is *not* true?
   A. It depends on the mass of objects.
   B. It is a repulsive force.
   C. It depends on the distance between objects.
   D. It exists between all objects.

Use the table below to answer questions 2 and 3.

Mass of Common Objects	
Object	Mass (g)
Cup	380
Book	1,100
Can	240
Ruler	25
Stapler	620

2. Which object would have an acceleration of 0.89 m/s² if you pushed on it with a force of 0.55 N?
   A. book
   B. can
   C. ruler
   D. stapler

3. Which object would have the greatest acceleration if you pushed on it with a force of 8.2 N?
   A. can
   B. stapler
   C. ruler
   D. book

Test-Taking Tip

Check Symbols Be sure you understand all symbols on a table or graph before answering any questions about the table or graph.

Question 3 The mass of the objects are given in grams, but the force is given in newtons which is a kg·m/s². The mass must be converted from grams to kilograms.

4. What is the weight of a book that has a mass of 0.35 kg?
   A. 0.036 N
   B. 3.4 N
   C. 28 N
   D. 34 N

5. If you swing an object on the end of a string around in a circle, the string pulls on the object to keep it moving in a circle. What is the name of this force?
   A. inertial
   B. centripetal
   C. resistance
   D. gravitational

6. What is the acceleration of a 1.4-kg object if the gravitational force pulls downward on it, but air resistance pushes upward on it with a force of 2.5 N?
   A. 11.6 m/s², downward
   B. 11.6 m/s², upward
   C. 8.0 m/s², downward
   D. 8.0 m/s², upward

Use the figure below to answer questions 7 and 8.

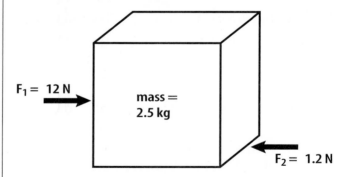

7. The figure above shows the horizontal forces that act on a box that is pushed from the left with a force of 12 N. What force is resisting the horizontal motion in this illustration?
   A. friction
   B. gravity
   C. inertia
   D. momentum

8. What is the acceleration of the box?
   A. 27 m/s²
   B. 4.8 m/s²
   C. 4.3 m/s²
   D. 0.48 m/s²

Part 2 Short Response/Grid In

Record your answers on the answer sheet provided by your teacher or on a sheet of paper.

9. A skater is coasting along the ice without exerting any apparent force. Which law of motion explains the skater's ability to continue moving?

10. After a soccer ball is kicked into the air, what force or forces are acting on it?

11. What is the force on an 8.55-kg object that accelerates at 5.34 m/s².

Use the figure below to answer questions 12 and 13.

12. Two bumper cars collide and then move away from each other. How do the forces the bumper cars exert on each other compare?

13. After the collision, determine whether both bumper cars will have the same acceleration.

14. Does acceleration depend on the speed of an object? Explain.

15. An object acted on by a force of 2.8 N has an acceleration of 3.6 m/s². What is the mass of the object?

16. What is the acceleration a 1.4-kg object falling through the air if the force of air resistance on the object is 2.5 N?

17. Name three ways you could accelerate if you were riding a bicycle.

Part 3 Open Ended

Record your answers on a sheet of paper.

18. When astronauts orbit Earth, they float inside the spaceship because of weightlessness. Explain this effect.

19. Describe how satellites are able to remain in orbit around Earth.

Use the figure below to answer questions 20 and 21.

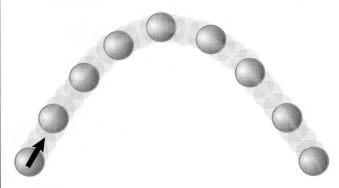

20. The figure above shows the path a ball thrown into the air follows. What causes the ball to move along a curved path?

21. What effect would throwing the ball harder have on the ball's path? Explain.

22. How does Newton's second law determine the motion of a book as you push it across a desktop?

23. A heavy box sits on a sidewalk. If you push against the box, the box moves in the direction of the force. If the box is replaced with a ball of the same mass, and you push with the same force against the ball, will it have the same acceleration as the box? Explain.

24. According to Newton's third law of motion, a rock sitting on the ground pushes against the ground, and the ground pushes back against the rock with an equal force. Explain why this force doesn't cause the rock to accelerate upward from the ground according to Newton's second law.

Work and Simple Machines

The BIG Idea
A machine makes doing a job easier.

SECTION 1
Work and Power
Main Idea Work is done when a force causes an object to move in the same direction as the force.

SECTION 2
Using Machines
Main Idea A machine can change the force needed to do a job.

SECTION 3
Simple Machines
Main Idea There are six types of simple machines.

Heavy Lifting

It took the ancient Egyptians more than 100 years to build the pyramids without machines like these. But now, even tall skyscrapers can be built in a few years. Complex or simple, machines have the same purpose. They make doing a job easier.

Science Journal Describe three machines you used today, and how they made doing a task easier.

Start-Up Activities

Compare Forces

Two of the world's greatest structures were built using different tools. The Great Pyramid at Giza in Egypt was built nearly 5,000 years ago using blocks of limestone moved into place by hand with ramps and levers. In comparison, the Sears Tower in Chicago was built in 1973 using tons of steel that were hoisted into place by gasoline-powered cranes. How do machines such as ramps, levers, and cranes change the forces needed to do a job?

1. Place a ruler on an eraser. Place a book on one end of the ruler.
2. Using one finger, push down on the free end of the ruler to lift the book.
3. Repeat the experiment, placing the eraser in various positions beneath the ruler. Observe how much force is needed in each instance to lift the book.
4. **Think Critically** In your Science Journal, describe your observations. How did changing the distance between the book and the eraser affect the force needed to lift the book?

Simple Machines Many of the devices that you use every day are simple machines. Make the following Foldable to help you understand the characteristics of simple machines.

STEP 1 Draw a mark at the midpoint of a sheet of paper along the side edge. Then **fold** the top and bottom edges in to touch the midpoint.

STEP 2 Fold in half from side to side.

STEP 3 Turn the paper vertically. Open and cut along the inside fold lines to form four tabs.

STEP 4 Label the tabs *Inclined Plane, Lever, Wheel and Axle,* and *Pulley.*

Read for Main Ideas As you read the chapter, list the characteristics of inclined planes, levers, wheels and axles, and pulleys under the appropriate tab.

Preview this chapter's content and activities at blue.msscience.com

Get Ready to Read

Questions and Answers

① Learn It! Knowing how to find answers to questions will help you on reviews and tests. Some answers can be found in the textbook, while other answers require you to go beyond the textbook. These answers might be based on knowledge you already have or things you have experienced.

② Practice It! Read the excerpt below. Answer the following questions and then discuss them with a partner.

> Did you use a machine today? When you think of a machine, you might think of a device, such as a car, with many moving parts powered by an engine or an electric motor. But if you used a pair of scissors or a broom, or cut your food with a knife, you used a machine. A machine is simply a device that makes doing work easier. Even a sloping surface can be a machine.
>
> —*from page 586*

- Describe how using a broom makes cleaning a floor easier.
- How is pushing a box up a smooth ramp easier than lifting the box upward?
- Why does a screwdriver make it easier to tighten a screw?

③ Apply It! Look at some questions in the text. Which questions can be answered directly from the text? Which require you to go beyond the text?

Target Your Reading

Reading Tip: As you read, keep track of questions you answer in the chapter. This will help you remember what you read.

Use this to focus on the main ideas as you read the chapter.

1 Before you read the chapter, respond to the statements below on your worksheet or on a numbered sheet of paper.
- Write an **A** if you **agree** with the statement.
- Write a **D** if you **disagree** with the statement.

2 After you read the chapter, look back to this page to see if you've changed your mind about any of the statements.
- If any of your answers changed, explain why.
- Change any false statements into true statements.
- Use your revised statements as a study guide.

Science Online
Print out a worksheet of this page at blue.msscience.com

Before You Read A or D		Statement	After You Read A or D
	1	Friction is caused by atoms or molecules of one object bonding to atoms or molecules in another object.	
	2	Power measures how fast work is done.	
	3	The fulcrum of a lever is always between the input force and the output force.	
	4	Efficiency is the ratio of output work to input work.	
	5	When you do work on an object, you transfer energy to the object.	
	6	A car is a combination of simple machines.	
	7	A wedge and a screw are both types of inclined planes.	
	8	Work is done anytime a force is applied.	
	9	Mechanical advantage can never be less than 1.	

section 1

Work and Power

as you read

What You'll Learn
- **Recognize** when work is done.
- **Calculate** how much work is done.
- **Explain** the relation between work and power.

Why It's Important
If you understand work, you can make your work easier.

Review Vocabulary
force: a push or a pull

New Vocabulary
- work
- power

What is work?

What does the term *work* mean to you? You might think of household chores; a job at an office, a factory, a farm; or the homework you do after school. In science, the definition of work is more specific. **Work** is done when a force causes an object to move in the same direction that the force is applied.

Can you think of a way in which you did work today? Maybe it would help to know that you do work when you lift your books, turn a doorknob, raise window blinds, or write with a pen or pencil. You also do work when you walk up a flight of stairs or open and close your school locker. In what other ways do you do work every day?

Work and Motion Your teacher has asked you to move a box of books to the back of the classroom. Try as you might, though, you just can't budge the box because it is too heavy. Although you exerted a force on the box and you feel tired from it, you have not done any work. In order for you to do work, two things must occur. First, you must apply a force to an object. Second, the object must move in the same direction as your applied force. You do work on an object only when the object moves as a result of the force you exert. The girl in **Figure 1** might think she is working by holding the bags of groceries. However, if she is not moving, she is not doing any work because she is not causing something to move.

 *To do work, how must a force make an object move?*

Figure 1 This girl is holding bags of groceries, yet she isn't doing any work. **Explain** *what must happen for work to be done.*

Figure 2 To do work, an object must move in the direction a force is applied.

The boy's arms do work when they exert an upward force on the basket and the basket moves upward.

The boy's arms still exert an upward force on the basket. But when the boy walks forward, no work is done by his arms.

Applying Force and Doing Work Picture yourself lifting the basket of clothes in **Figure 2.** You can feel your arms exerting a force upward as you lift the basket, and the basket moves upward in the direction of the force your arms applied. Therefore, your arms have done work. Now, suppose you carry the basket forward. You can still feel your arms applying an upward force on the basket to keep it from falling, but now the basket is moving forward instead of upward. Because the direction of motion is not in the same direction of the force applied by your arms, no work is done by your arms.

Force in Two Directions Sometimes only part of the force you exert moves an object. Think about what happens when you push a lawn mower. You push at an angle to the ground as shown in **Figure 3.** Part of the force is to the right and part of the force is downward. Only the part of the force that is in the same direction as the motion of the mower—to the right—does work.

Figure 3 When you exert a force at an angle, only part of your force does work—the part that is in the same direction as the motion of the object.

SECTION 1 Work and Power

James Prescott Joule This English physicist experimentally verified the law of conservation of energy. He showed that various forms of energy—mechanical, electrical, and thermal—are essentially the same and can be converted one into another. The SI unit of energy and work, the joule, is named after him. Research the work of Joule and write what you learn in your Science Journal.

Calculating Work

Work is done when a force makes an object move. More work is done when the force is increased or the object is moved a greater distance. Work can be calculated using the work equation below. In SI units, the unit for work is the joule, named for the nineteenth-century scientist James Prescott Joule.

Work Equation
work (in joules) = force (in newtons) × distance (in meters)
$$W = Fd$$

Work and Distance Suppose you give a book a push and it slides across a table. To calculate the work you did, the distance in the above equation is not the distance the book moved. The distance in the work equation is the distance an object moves while the force is being applied. So the distance in the work equation is the distance the book moved while you were pushing.

Applying Math — Solve a One-Step Equation

CALCULATING WORK A painter lifts a can of paint that weighs 40 N a distance of 2 m. How much work does she do? *Hint: to lift a can weighing 40 N, the painter must exert a force of 40 N.*

Solution

1. *This is what you know:*
   - force: $F = 40$ N
   - distance: $d = 2$ m

2. *This is what you need to find out:*
   work: $W = ?$ J

3. *This is the procedure you need to use:*
   Substitute the known values $F = 40$ N and $d = 2$ m into the work equation:
   $$W = Fd = (40 \text{ N})(2 \text{ m}) = 80 \text{ N} \cdot \text{m} = 80 \text{ J}$$

4. *Check your answer:*
   Check your answer by dividing the work you calculated by the distance given in the problem. The result should be the force given in the problem.

Practice Problems

1. As you push a lawn mower, the horizontal force is 300 N. If you push the mower a distance of 500 m, how much work do you do?

2. A librarian lifts a box of books that weighs 93 N a distance of 1.5 m. How much work does he do?

For more practice, visit blue.msscience.com

What is power?

What does it mean to be powerful? Imagine two weightlifters lifting the same amount of weight the same vertical distance. They both do the same amount of work. However, the amount of power they use depends on how long it took to do the work. **Power** is how quickly work is done. The weightlifter who lifted the weight in less time is more powerful.

Calculating Power Power can be calculated by dividing the amount of work done by the time needed to do the work.

Power Equation

$$\text{power (in watts)} = \frac{\text{work (in joules)}}{\text{time (in seconds)}}$$

$$P = \frac{W}{t}$$

In SI units, the unit of power is the watt, in honor of James Watt, a nineteenth-century British scientist who invented a practical version of the steam engine.

Mini LAB

Work and Power

Procedure
1. Weigh yourself on a **scale**.
2. Multiply your weight in pounds by 4.45 to convert your weight to newtons.
3. Measure the vertical height of a **stairway**. **WARNING:** *Make sure the stairway is clear of all objects.*
4. Time yourself walking slowly and quickly up the stairway.

Analysis
Calculate and compare the work and power in each case.

Try at Home

Applying Math — Solve a One-Step Equation

CALCULATING POWER You do 200 J of work in 12 s. How much power did you use?

Solution

❶ *This is what you know:*
- work: $W = 200$ J
- time: $t = 12$ s

❷ *This is what you need to find out:*
- power: $P = ?$ watts

❸ *This is the procedure you need to use:*

Substitute the known values $W = 200$ J and $t = 12$ s into the power equation:

$$P = \frac{W}{t} = \frac{200 \text{ J}}{12 \text{ s}} = 17 \text{ watts}$$

❹ *Check your answer:*

Check your answer by multiplying the power you calculated by the time given in the problem. The result should be the work given in the problem.

Practice Problems

1. In the course of a short race, a car does 50,000 J of work in 7 s. What is the power of the car during the race?
2. A teacher does 140 J of work in 20 s. How much power did he use?

 For more practice, visit blue.msscience.com

Topic: James Watt
Visit blue.msscience.com for Web links to information about James Watt and his steam engine.

Activity Draw a diagram showing how his steam engine worked.

Work and Energy If you push a chair and make it move, you do work on the chair and change its energy. Recall that when something is moving it has energy of motion, or kinetic energy. By making the chair move, you increase its kinetic energy.

You also change the energy of an object when you do work and lift it higher. An object has potential energy that increases when it is higher above Earth's surface. By lifting an object, you do work and increase its potential energy.

Power and Energy When you do work on an object you increase the energy of the object. Because energy can never be created or destroyed, if the object gains energy then you must lose energy. When you do work on an object you transfer energy to the object, and your energy decreases. The amount of work done is the amount of energy transferred. So power is also equal to the amount of energy transferred in a certain amount of time.

Sometimes energy can be transferred even when no work is done, such as when heat flows from a warm to a cold object. In fact, there are many ways energy can be transferred even if no work is done. Power is always the rate at which energy is transferred, or the amount of energy transferred divided by the time needed.

section 1 review

Summary

What is work?

- Work is done when a force causes an object to move in the same direction that the force is applied.
- If the movement caused by a force is at an angle to the direction the force is applied, only the part of the force in the direction of motion does work.
- Work can be calculated by multiplying the force applied by the distance:
  $$W = Fd$$
- The distance in the work equation is the distance an object moves while the force is being applied.

What is power?

- Power is how quickly work is done. Something is more powerful if it can do a given amount of work in less time.
- Power can be calculated by dividing the work done by the time needed to do the work:
  $$P = \frac{W}{t}$$

Self Check

1. **Describe** a situation in which work is done on an object.
2. **Evaluate** which of the following situations involves more power: 200 J of work done in 20 s or 50 J of work done in 4 s? Explain your answer.
3. **Determine** two ways power can be increased.
4. **Calculate** how much power, in watts, is needed to cut a lawn in 50 min if the work involved is 100,000 J.
5. **Think Critically** Suppose you are pulling a wagon with the handle at an angle. How can you make your task easier?

Applying Math

6. **Calculate Work** How much work was done to lift a 1,000-kg block to the top of the Great Pyramid, 146 m above ground?
7. **Calculate Work Done by an Engine** An engine is used to lift a beam weighing 9,800 N up to 145 m. How much work must the engine do to lift this beam? How much work must be done to lift it 290 m?

584 CHAPTER 20 Work and Simple Machines

Building the Pyramids

Imagine moving 2.3 million blocks of limestone, each weighing more than 1,000 kg. That is exactly what the builders of the Great Pyramid at Giza did. Although no one knows for sure exactly how they did it, they probably pulled the blocks most of the way.

Work Done Using Different Ramps		
Distance (cm)	Force (N)	Work (J)
Do not write in this book.		

Real-World Question

How is the force needed to lift a block related to the distance it travels?

Goals
- **Compare** the force needed to lift a block with the force needed to pull it up a ramp.

Materials
wood block
tape
spring scale
ruler
thin notebooks
meterstick
several books

Safety Precautions

Procedure

1. Stack several books together on a tabletop to model a half-completed pyramid. Measure the height of the books in centimeters. Record the height on the first row of the data table under *Distance*.
2. Use the wood block as a model for a block of stone. Use tape to attach the block to the spring scale.
3. Place the block on the table and lift it straight up the side of the stack of books until the top of the block is even with the top of the books. Record the force shown on the scale in the data table under *Force*.
4. **Arrange** a notebook so that one end is on the stack of books and the other end is on the table. Measure the length of the notebook and record this length as distance in the second row of the data table under *Distance*.
5. **Measure** the force needed to pull the block up the ramp. Record the force in the data table.
6. Repeat steps 4 and 5 using a longer notebook to make the ramp longer.
7. **Calculate** the work done in each row of the data table.

Conclude and Apply

1. **Evaluate** how much work you did in each instance.
2. **Determine** what happened to the force needed as the length of the ramp increased.
3. **Infer** How could the builders of the pyramids have designed their task to use less force than they would lifting the blocks straight up? Draw a diagram to support your answer.

Communicating Your Data

Add your data to that found by other groups. **For more help, refer to the** Science Skill Handbook.

LAB **585**

SECTION 2

Using Machines

as you read

What You'll Learn
- **Explain** how a machine makes work easier.
- **Calculate** the mechanical advantages and efficiency of a machine.
- **Explain** how friction reduces efficiency.

Why It's Important
Machines can't change the amount of work you need to do, but they can make doing work easier.

Review Vocabulary
friction: force that opposes motion between two touching surfaces

New Vocabulary
- input force
- output force
- mechanical advantage
- efficiency

What is a machine?

Did you use a machine today? When you think of a machine you might think of a device, such as a car, with many moving parts powered by an engine or an electric motor. But if you used a pair of scissors or a broom, or cut your food with a knife, you used a machine. A machine is simply a device that makes doing work easier. Even a sloping surface can be a machine.

Mechanical Advantage

Even though machines make work easier, they don't decrease the amount of work you need to do. Instead, a machine changes the way in which you do work. When you use a machine, you exert a force over some distance. For example, you exert a force to move a rake or lift the handles of a wheelbarrow. The force that you apply on a machine is the **input force.** The work you do on the machine is equal to the input force times the distance over which your force is applied. The work that you do on the machine is the input work.

The machine also does work by exerting a force to move an object over some distance. A rake, for example, exerts a force to move leaves. Sometimes this force is called the resistance force because the machine is trying to overcome some resistance. The force that the machine applies is the **output force.** The work that the machine does is the output work. **Figure 4** shows how a machine transforms input work to output work.

When you use a machine, the output work can never be greater than the input work. So what is the advantage of using a machine? A machine makes work easier by changing the amount of force you need to exert, the distance over which the force is exerted, or the direction in which you exert your force.

Figure 4 No matter what type of machine is used, the output work is never greater than the input work.

586 CHAPTER 20 Work and Simple Machines

Changing Force Some machines make doing a job easier by reducing the force you have to apply to do the job. For this type of machine the output force is greater than the input force. How much larger the output force is compared to the input force is the **mechanical advantage** of the machine. The mechanical advantage of a machine is the ratio of the output force to the input force and can be calculated from this equation:

Mechanical Advantage Equation

$$\text{mechanical advantage} = \frac{\text{output force (in newtons)}}{\text{input force (in newtons)}}$$

$$MA = \frac{F_{out}}{F_{in}}$$

Mechanical advantage does not have any units, because it is the ratio of two numbers with the same units.

Topic: Historical Tools
Visit blue.msscience.com for Web links to information about early types of tools and how they took advantage of simple machines.

Activity Write a paragraph describing how simple machines were used to design early tools.

Applying Math — Solve a One-Step Equation

CALCULATING MECHANICAL ADVANTAGE To pry the lid off a paint can, you apply a force of 50 N to the handle of the screwdriver. What is the mechanical advantage of the screwdriver if it applies a force of 500 N to the lid?

Solution

① *This is what you know:*
- input force: F_{in} = 50 N
- output force: F_{out} = 500 N

② *This is what you need to find out:* mechanical advantage: MA = ?

③ *This is the procedure you need to use:* Substitute the known values F_{in} = 50 N and F_{out} = 500 N into the mechanical advantage equation:

$$MA = \frac{F_{out}}{F_{in}} = \frac{500 \text{ N}}{50 \text{ N}} = 10$$

④ *Check your answer:* Check your answer by multiplying the mechanical advantage you calculated by the input force given in the problem. The result should be the output force given in the problem.

Practice Problems

1. To open a bottle, you apply a force of 50 N to the bottle opener. The bottle opener applies a force of 775 N to the bottle cap. What is the mechanical advantage of the bottle opener?

2. To crack a pecan, you apply a force of 50 N to the nutcracker. The nutcracker applies a force of 750 N to the pecan. What is the mechanical advantage of the nutcracker?

 For more practice, visit blue.msscience.com

SECTION 2 Using Machines

Figure 5 Changing the direction or the distance that a force is applied can make a task easier.

Sometimes it is easier to exert your force in a certain direction. This boy would rather pull down on the rope to lift the flag than to climb to the top of the pole and pull up.

When you rake leaves, you move your hands a short distance, but the end of the rake moves over a longer distance.

Changing Distance Some machines allow you to exert your force over a shorter distance. In these machines, the output force is less than the input force. The rake in **Figure 5** is this type of machine. You move your hands a small distance at the top of the handle, but the bottom of the rake moves a greater distance as it moves the leaves. The mechanical advantage of this type of machine is less than one because the output force is less than the input force.

Changing Direction Sometimes it is easier to apply a force in a certain direction. For example, it is easier to pull down on the rope in **Figure 5** than to pull up on it. Some machines enable you to change the direction of the input force. In these machines neither the force nor the distance is changed. The mechanical advantage of this type of machine is equal to one because the output force is equal to the input force. The three ways machines make doing work easier are summarized in **Figure 6.**

Figure 6 Machines are useful because they can increase force, increase distance, or change the direction in which a force is applied.

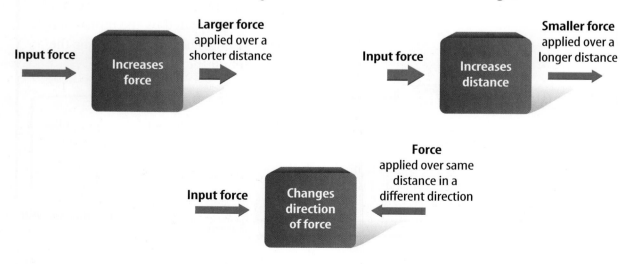

588 CHAPTER 20 Work and Simple Machines

Efficiency

A machine can't make the output work greater than the input work. In fact, for a real machine, the output work is always less than the input work. In a real machine, there is friction as parts of the machine move. Friction converts some of the input work into heat, so that the output work is reduced. The **efficiency** of a machine is the ratio of the output work to the input work, and can be calculated from this equation:

Efficiency Equation

$$\text{efficiency (in percent)} = \frac{\text{output work (in joules)}}{\text{input work (in joules)}} \times 100\%$$

$$\mathit{eff} = \frac{W_{out}}{W_{in}} \times 100\%$$

If the amount of friction in the machine is reduced, the efficiency of the machine increases.

Body Temperature Chemical reactions that enable your muscles to move also produce heat that helps maintain your body temperature. When you shiver, rapid contraction and relaxation of muscle fibers produces a large amount of heat that helps raise your body temperature. This causes the efficiency of your muscles to decrease as more energy is converted into heat.

Applying Math — Solve a One-Step Equation

CALCULATING EFFICIENCY Using a pulley system, a crew does 7,500 J of work to load a box that requires 4,500 J of work. What is the efficiency of the pulley system?

Solution

1 *This is what you know:*
- input work: $W_{in} = 7{,}500$ J
- output work: $W_{out} = 4{,}500$ J

2 *This is what you need to find out:* efficiency: $\mathit{eff} = ?\,\%$

3 *This is the procedure you need to use:* Substitute the known values $W_{in} = 7{,}500$ J and $W_{out} = 4{,}500$ J into the efficiency equation:

$$\mathit{eff} = \frac{W_{out}}{W_{in}} = \frac{4{,}500 \text{ J}}{7{,}500 \text{ J}} \times 100\% = 60\%$$

4 *Check your answer:* Check your answer by dividing the efficiency by 100% and then multiplying your answer times the work input. The product should be the work output given in the problem.

Practice Problems

1. You do 100 J of work in pulling out a nail with a claw hammer. If the hammer does 70 J of work, what is the hammer's efficiency?
2. You do 150 J of work pushing a box up a ramp. If the ramp does 105 J of work, what is the efficiency of the ramp?

 For more practice, visit blue.msscience.com

SECTION 2 Using Machines

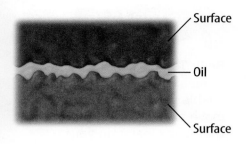

Figure 7 Lubrication can reduce the friction between two surfaces. Two surfaces in contact can stick together where the high spots on each surface come in contact. Adding oil or another lubricant separates the surface so that fewer high spots make contact.

Friction To help understand friction, imagine pushing a heavy box up a ramp. As the box begins to move, the bottom surface of the box slides across the top surface of the ramp. Neither surface is perfectly smooth—each has high spots and low spots, as shown in **Figure 7**.

As the two surfaces slide past each other, high spots on the two surfaces come in contact. At these contact points, shown in **Figure 7,** atoms and molecules can bond together. This makes the contact points stick together. The attractive forces between all the bonds in the contact points added together is the frictional force that tries to keep the two surfaces from sliding past each other.

To keep the box moving, a force must be applied to break the bonds between the contact points. Even after these bonds are broken and the box moves, new bonds form as different parts of the two surfaces come into contact.

Friction and Efficiency One way to reduce friction between two surfaces is to add oil. **Figure 7** shows how oil fills the gaps between the surfaces, and keeps many of the high spots from making contact. Because there are fewer contact points between the surfaces, the force of friction is reduced. More of the input work then is converted to output work by the machine.

section 2 review

Summary

What is a machine?
- A machine is a device that makes doing work easier.
- A machine can make doing work easier by reducing the force exerted, changing the distance over which the force is exerted, or changing the direction of the force.
- The output work done by a machine can never be greater than the input work done on the machine.

Mechanical Advantage and Efficiency
- The mechanical advantage of a machine is the number of times the machine increases the input force:
$$MA = \frac{F_{out}}{F_{in}}$$
- The efficiency of a machine is the ratio of the output work to the input work:
$$eff = \frac{W_{out}}{W_{in}} \times 100\%$$

Self Check

1. **Identify** three specific situations in which machines make work easier.
2. **Infer** why the output force exerted by a rake must be less than the input force.
3. **Explain** how the efficiency of an ideal machine compares with the efficiency of a real machine.
4. **Explain** how friction reduces the efficiency of machines.
5. **Think Critically** Can a machine be useful even if its mechanical advantage is less than one? Explain and give an example.

Applying Math

6. **Calculate Efficiency** Find the efficiency of a machine if the input work is 150 J and the output work is 90 J.
7. **Calculate Mechanical Advantage** To lift a crate, a pulley system exerts a force of 2,750 N. Find the mechanical advantage of the pulley system if the input force is 250 N.

section 3

Simple Machines

What is a simple machine?

What do you think of when you hear the word *machine?* Many people think of machines as complicated devices such as cars, elevators, or computers. However, some machines are as simple as a hammer, shovel, or ramp. A **simple machine** is a machine that does work with only one movement. The six simple machines are the inclined plane, lever, wheel and axle, screw, wedge, and pulley. A machine made up of a combination of simple machines is called a **compound machine.** A can opener is a compound machine. The bicycle in **Figure 8** is a familiar example of another compound machine.

Inclined Plane

Ramps might have enabled the ancient Egyptians to build their pyramids. To move limestone blocks weighing more than 1,000 kg each, archaeologists hypothesize that the Egyptians built enormous ramps. A ramp is a simple machine known as an inclined plane. An **inclined plane** is a flat, sloped surface. Less force is needed to move an object from one height to another using an inclined plane than is needed to lift the object. As the inclined plane becomes longer, the force needed to move the object becomes smaller.

as you read

What You'll Learn
- **Distinguish** among the different simple machines.
- **Describe** how to find the mechanical advantage of each simple machine.

Why It's Important
All machines, no matter how complicated, are made of simple machines.

Review Vocabulary
compound: made of separate pieces or parts

New Vocabulary
- simple machine
- compound machine
- inclined plane
- wedge
- screw
- lever
- wheel and axle
- pulley

Figure 8 Devices that use combinations of simple machines, such as this bicycle, are called compound machines.

SECTION 3 Simple Machines

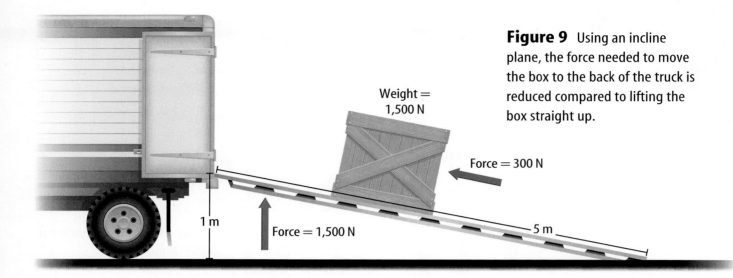

Figure 9 Using an incline plane, the force needed to move the box to the back of the truck is reduced compared to lifting the box straight up.

Using Inclined Planes Imagine having to lift a box weighing 1,500 N to the back of a truck that is 1 m off the ground. You would have to exert a force of 1,500 N, the weight of the box, over a distance of 1 m, which equals 1,500 J of work. Now suppose that instead you use a 5-m-long ramp, as shown in **Figure 9.** The amount of work you need to do does not change. You still need to do 1,500 J of work. However, the distance over which you exert your force becomes 5 m. You can calculate the force you need to exert by dividing both sides of the equation for work by distance.

$$\text{Force} = \frac{\text{work}}{\text{distance}}$$

If you do 1,500 J of work by exerting a force over 5 m, the force is only 300 N. Because you exert the input force over a distance that is five times as long, you can exert a force that is five times less.

The mechanical advantage of an inclined plane is the length of the inclined plane divided by its height. In this example, the ramp has a mechanical advantage of 5.

Wedge An inclined plane that moves is called a **wedge.** A wedge can have one or two sloping sides. The knife shown in **Figure 10** is an example of a wedge. An axe and certain types of doorstops are also wedges. Just as for an inclined plane, the mechanical advantage of a wedge increases as it becomes longer and thinner.

Figure 10 This chef's knife is a wedge that slices through food.

Figure 11 Wedge-shaped teeth help tear food.

Your front teeth help tear an apple apart.

The wedge-shaped teeth of this *Tyrannosaurus rex* show that it was a carnivore.

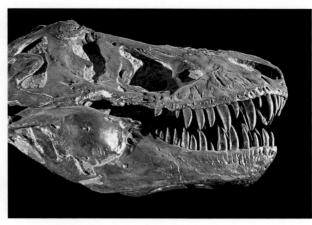

Wedges in Your Body You have wedges in your body. The bite marks on the apple in **Figure 11** show how your front teeth are wedge shaped. A wedge changes the direction of the applied effort force. As you push your front teeth into the apple, the downward effort force is changed by your teeth into a sideways force that pushes the skin of the apple apart.

The teeth of meat eaters, or carnivores, are more wedge shaped than the teeth of plant eaters, or herbivores. The teeth of carnivores are used to cut and rip meat, while herbivores' teeth are used for grinding plant material. By examining the teeth of ancient animals, such as the dinosaur in **Figure 11**, scientists can determine what the animal ate when it was living.

The Screw Another form of the inclined plane is a screw. A **screw** is an inclined plane wrapped around a cylinder or post. The inclined plane on a screw forms the screw threads. Just like a wedge changes the direction of the effort force applied to it, a screw also changes the direction of the applied force. When you turn a screw, the force applied is changed by the threads to a force that pulls the screw into the material. Friction between the threads and the material holds the screw tightly in place. The mechanical advantage of the screw is the length of the inclined plane wrapped around the screw divided by the length of the screw. The more tightly wrapped the threads are, the easier it is to turn the screw. Examples of screws are shown in **Figure 12.**

Reading Check *How are screws related to the inclined plane?*

Figure 12 The thread around a screw is an inclined plane. Many familiar devices use screws to make work easier.

SECTION 3 Simple Machines **593**

Figure 13 The mechanical advantage of a lever changes as the position of the fulcrum changes. The mechanical advantage increases as the fulcrum is moved closer to the output force.

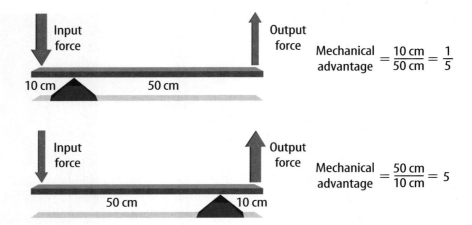

Figure 14 A faucet handle is a wheel and axle. A wheel and axle is similar to a circular lever. The center is the fulcrum, and the wheel and axle turn around it. **Explain** *how you can increase the mechanical advantage of a wheel and axle.*

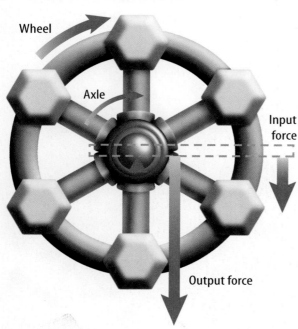

Lever

You step up to the plate. The pitcher throws the ball and you swing your lever to hit the ball? That's right! A baseball bat is a type of simple machine called a lever. A **lever** is any rigid rod or plank that pivots, or rotates, about a point. The point about which the lever pivots is called a fulcrum.

The mechanical advantage of a lever is found by dividing the distance from the fulcrum to the input force by the distance from the fulcrum to the output force, as shown in **Figure 13**. When the fulcrum is closer to the output force than the input force, the mechanical advantage is greater than one.

Levers are divided into three classes according to the position of the fulcrum with respect to the input force and output force. **Figure 15** shows examples of three classes of levers.

Wheel and Axle

Do you think you could turn a doorknob easily if it were a narrow rod the size of a pencil? It might be possible, but it would be difficult. A doorknob makes it easier for you to open a door because it is a simple machine called a wheel and axle. A **wheel and axle** consists of two circular objects of different sizes that are attached in such a way that they rotate together. As you can see in **Figure 14,** the larger object is the wheel and the smaller object is the axle.

The mechanical advantage of a wheel and axle is usually greater than one. It is found by dividing the radius of the wheel by the radius of the axle. For example, if the radius of the wheel is 12 cm and the radius of the axle is 4 cm, the mechanical advantage is 3.

NATIONAL GEOGRAPHIC VISUALIZING LEVERS

Figure 15

Levers are among the simplest of machines, and you probably use them often in everyday life without even realizing it. A lever is a bar that pivots around a fixed point called a fulcrum. As shown here, there are three types of levers—first class, second class, and third class. They differ in where two forces—an input force and an output force—are located in relation to the fulcrum.

▲ Fulcrum
▼ Input force
▲ Output force

In a first-class lever, the fulcrum is between the input force and the output force. First-class levers, such as scissors and pliers, multiply force or distance depending on where the fulcrum is placed. They always change the direction of the input force, too.

First-class lever

In a second-class lever, such as a wheelbarrow, the output force is between the input force and the fulcrum. Second-class levers always multiply the input force but don't change its direction.

Second-class lever

Third-class lever

In a third-class lever, such as a baseball bat, the input force is between the output force and the fulcrum. For a third-class lever, the output force is less than the input force, but is in the same direction.

SECTION 3 Simple Machines **595**

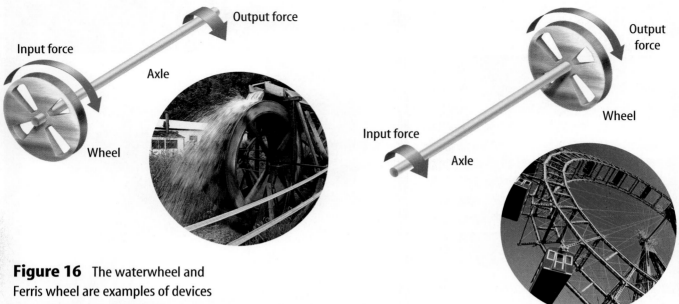

Figure 16 The waterwheel and Ferris wheel are examples of devices that rely on a wheel and axle.
Compare and contrast waterwheels and Ferris wheels in terms of wheels and axles.

Using Wheels and Axles In some devices, the input force is used to turn the wheel and the output force is exerted by the axle. Because the wheel is larger than the axle, the mechanical advantage is greater than one. So the output force is greater than the input force. A doorknob, a steering wheel, and a screwdriver are examples of this type of wheel and axle.

In other devices, the input force is applied to turn the axle and the output force is exerted by the wheel. Then the mechanical advantage is less than one and the output force is less than the input force. A fan and a ferris wheel are examples of this type of wheel and axle. **Figure 16** shows an example of each type of wheel and axle.

Pulley

To raise a sail, a sailor pulls down on a rope. The rope uses a simple machine called a pulley to change the direction of the force needed. A **pulley** consists of a grooved wheel with a rope or cable wrapped over it.

Fixed Pulleys Some pulleys, such as the one on a sail, a window blind, or a flagpole, are attached to a structure above your head. When you pull down on the rope, you pull something up. This type of pulley, called a fixed pulley, does not change the force you exert or the distance over which you exert it. Instead, it changes the direction in which you exert your force, as shown in **Figure 17**. The mechanical advantage of a fixed pulley is 1.

 *How does a fixed pulley affect the input force?*

Observing Pulleys
Procedure
1. Obtain two **broomsticks**. Tie a 3-m-long **rope** to the middle of one stick. Wrap the rope around both sticks four times.
2. Have two students pull the broomsticks apart while a third pulls on the rope.
3. Repeat with two wraps of rope.

Analysis
1. Compare the results.
2. Predict whether it will be easier to pull the broomsticks together with ten wraps of rope.

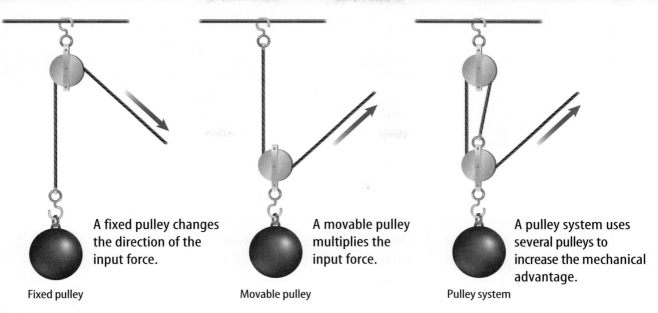

Figure 17 Pulleys can change force and direction.

Movable Pulleys Another way to use a pulley is to attach it to the object you are lifting, as shown in **Figure 17**. This type of pulley, called a movable pulley, allows you to exert a smaller force to lift the object. The mechanical advantage of a movable pulley is always 2.

More often you will see combinations of fixed and movable pulleys. Such a combination is called a pulley system. The mechanical advantage of a pulley system is equal to the number of sections of rope pulling up on the object. For the pulley system shown in **Figure 17** the mechanical advantage is 3.

section 3 review

Summary

Simple and Compound Machines
- A simple machine is a machine that does work with only one movement.
- A compound machine is made from a combination of simple machines.

Types of Simple Machines
- An inclined plane is a flat, sloped surface.
- A wedge is an inclined plane that moves.
- A screw is an inclined plane that is wrapped around a cylinder or post.
- A lever is a rigid rod that pivots around a fixed point called the fulcrum.
- A wheel and axle consists of two circular objects of different sizes that rotate together.
- A pulley is a grooved wheel with a rope or cable wrapped over it.

Self Check

1. **Determine** how the mechanical advantage of a ramp changes as the ramp becomes longer.
2. **Explain** how a wedge changes an input force.
3. **Identify** the class of lever for which the fulcrum is between the input force and the output force.
4. **Explain** how the mechanical advantage of a wheel and axle change as the size of the wheel increases.
5. **Think Critically** How are a lever and a wheel and axle similar?

Applying Math

6. **Calculate Length** The Great Pyramid is 146 m high. How long is a ramp from the top of the pyramid to the ground that has a mechanical advantage of 4?
7. **Calculate Force** Find the output force exerted by a moveable pulley if the input force is 50 N.

Design Your Own

Pulley Power

Real-World Question

Imagine how long it might have taken to build the Sears Tower in Chicago without the aid of a pulley system attached to a crane. Hoisting the 1-ton I beams to a maximum height of 110 stories required large lifting forces and precise control of the beam's movement.

Construction workers also use smaller pulleys that are not attached to cranes to lift supplies to where they are needed. Pulleys are not limited to construction sites. They also are used to lift automobile engines out of cars, to help load and unload heavy objects on ships, and to lift heavy appliances and furniture. How can you use a pulley system to reduce the force needed to lift a load?

Goals
- **Design** a pulley system.
- **Measure** the mechanical advantage and efficiency of the pulley system.

Possible Materials
single- and multiple-pulley systems
nylon rope
steel bar to support the pulley system
meterstick
*metric tape measure
variety of weights to test pulleys
force spring scale
brick
*heavy book
balance
*scale
*Alternate materials

Safety Precautions

WARNING: *The brick could be dangerous if it falls. Keep your hands and feet clear of it.*

Form a Hypothesis

Write a hypothesis about how pulleys can be combined to make a system of pulleys to lift a heavy load, such as a brick. Consider the efficiency of your system.

Test Your Hypothesis

Make a Plan

1. Decide how you are going to support your pulley system. What materials will you use?
2. How will you measure the effort force and the resistance force? How will you determine the mechanical advantage? How will you measure efficiency?
3. **Experiment** by lifting small weights with a single pulley, double pulley, and so on. How efficient are the pulleys? In what ways can you increase the efficiency of your setup?

CHAPTER 20 Work and Simple Machines

Using Scientific Methods

4. Use the results of step 3 to design a pulley system to lift the brick. Draw a diagram of your design. Label the different parts of the pulley system and use arrows to indicate the direction of movement for each section of rope.

Follow Your Plan

1. Make sure your teacher approves your plan before you start.
2. Assemble the pulley system you designed. You might want to test it with a smaller weight before attaching the brick.
3. **Measure** the force needed to lift the brick. How much rope must you pull to raise the brick 10 cm?

Analyze Your Data

1. **Calculate** the ideal mechanical advantage of your design.
2. **Calculate** the actual mechanical advantage of the pulley system you built.
3. **Calculate** the efficiency of your pulley system.
4. How did the mechanical advantage of your pulley system compare with those of your classmates?

Conclude and Apply

1. **Explain** how increasing the number of pulleys increases the mechanical advantage.
2. **Infer** How could you modify the pulley system to lift a weight twice as heavy with the same effort force used here?
3. **Compare** this real machine with an ideal machine.

Show your design diagram to the class. Review the design and point out good and bad characteristics of your pulley system. **For more help, refer to the** Science Skill Handbook.

TIME SCIENCE AND Society

SCIENCE ISSUES THAT AFFECT YOU!

Bionic People

Artificial limbs can help people lead normal lives

People in need of transplants usually receive human organs. But many people's medical problems can only be solved by receiving artificial body parts. These synthetic devices, called prostheses, are used to replace anything from a heart valve to a knee joint. Bionics is the science of creating artificial body parts. A major focus of bionics is the replacement of lost limbs. Through accident, birth defect, or disease, people sometimes lack hands or feet, or even whole arms or legs. For centuries, people have used prostheses to replace limbs. In the past, physically challenged people used devices like peg legs or artificial arms that ended in a pair of hooks. These prostheses didn't do much to replace lost functions of arms and legs.

The knowledge that muscles respond to electricity has helped create more effective prostheses. One such prostheses is the myoelectric arm. This battery-powered device connects muscle nerves in an amputated arm to a sensor.

The sensor detects when the arm tenses, then transmits the signal to an artificial hand, which opens or closes. New prosthetic hands even give a sense of touch, as well as cold and heat.

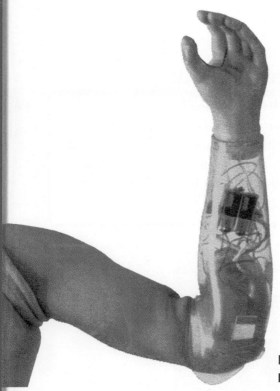

Myoelectric arms make life easier for people who have them.

Research Use your school's media center to find other aspects of robotics such as walking machines or robots that perform planetary exploration. What are they used for? How do they work? You could take it one step further and learn about cyborgs. Report to the class.

For more information, visit blue.msscience.com

Chapter 20 Study Guide

Reviewing Main Ideas

Section 1 Work and Power

1. Work is done when a force exerted on an object causes the object to move.
2. A force can do work only when it is exerted in the same direction as the object moves.
3. Work is equal to force times distance, and the unit of work is the joule.
4. Power is the rate at which work is done, and the unit of power is the watt.

Section 2 Using Machines

1. A machine can change the size or direction of an input force or the distance over which it is exerted.
2. The mechanical advantage of a machine is its output force divided by its input force.

Section 3 Simple Machines

1. A machine that does work with only one movement is a simple machine. A compound machine is a combination of simple machines.
2. Simple machines include the inclined plane, lever, wheel and axle, screw, wedge, and pulley.
3. Wedges and screws are inclined planes.
4. Pulleys can be used to multiply force and change direction.

Visualizing Main Ideas

Copy and complete the following concept map on simple machines.

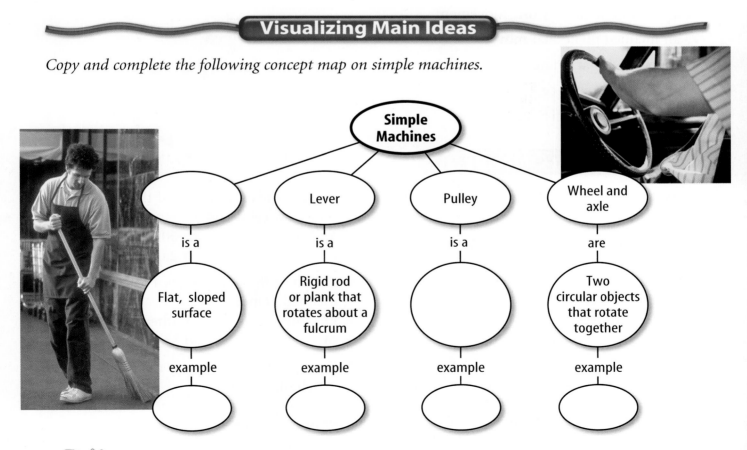

Chapter 20 Review

Using Vocabulary

compound machine p. 591
efficiency p. 589
inclined plane p. 591
input force p. 586
lever p. 594
mechanical advantage p. 587
output force p. 586
power p. 583
pulley p. 596
screw p. 593
simple machine p. 591
wedge p. 592
wheel and axle p. 594
work p. 580

Each phrase below describes a vocabulary word. Write the vocabulary word that matches the phrase describing it.

1. percentage of work in to work out
2. force put into a machine
3. force exerted by a machine
4. two rigidly attached wheels
5. output force divided by input force
6. a machine with only one movement
7. an inclined plane that moves
8. a rigid rod that rotates about a fulcrum
9. a flat, sloped surface
10. amount of work divided by time

Checking Concepts

Choose the word or phrase that best answers the question.

11. Which of the following is a requirement for work to be done?
    A) Force is exerted.
    B) Object is carried.
    C) Force moves an object.
    D) Machine is used.

12. How much work is done when a force of 30 N moves an object a distance of 3 m?
    A) 3 J
    B) 10 J
    C) 30 J
    D) 90 J

13. How much power is used when 600 J of work are done in 10 s?
    A) 6 W
    B) 60 W
    C) 600 W
    D) 610 W

14. Which is a simple machine?
    A) baseball bat
    B) bicycle
    C) can opener
    D) car

15. Mechanical advantage can be calculated by which of the following expressions?
    A) input force/output force
    B) output force/input force
    C) input work/output work
    D) output work/input work

16. What is the ideal mechanical advantage of a machine that changes only the direction of the input force?
    A) less than 1
    B) zero
    C) 1
    D) greater than 1

Use the illustration below to answer question 17.

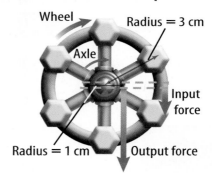

17. What is the output force if the input force on the wheel is 100 N?
    A) 5 N
    B) 200 N
    C) 500 N
    D) 2,000 N

18. Which of the following is a form of the inclined plane?
    A) pulley
    B) screw
    C) wheel and axle
    D) lever

19. For a given input force, a ramp increases which of the following?
    A) height
    B) output force
    C) output work
    D) efficiency

chapter 20 Review

Thinking Critically

Use the illustration below to answer question 20.

20. **Evaluate** Would a 9-N force applied 2 m from the fulcrum lift the weight? Explain.

21. **Explain** why the output work for any machine can't be greater than the input work.

22. **Explain** A doorknob is an example of a wheel and axle. Explain why turning the knob is easier than turning the axle.

23. **Infer** On the Moon, the force of gravity is less than on Earth. Infer how the mechanical advantage of an inclined plane would change if it were on the Moon, instead of on Earth.

24. **Make and Use Graphs** A pulley system has a mechanical advantage of 5. Make a graph with the input force on the *x*-axis and the output force on the *y*-axis. Choose five different values of the input force, and plot the resulting output force on your graph.

Use the diagram below to answer question 25.

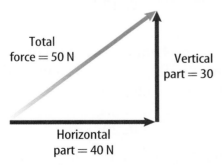

25. **Work** The diagram above shows a force exerted at an angle to pull a sled. How much work is done if the sled moves 10 m horizontally?

Performance Activities

26. **Identify** You have levers in your body. Your muscles and tendons provide the input force. Your joints act as fulcrums. The output force is used to move everything from your head to your hands. Describe and draw any human levers you can identify.

27. **Display** Make a display of everyday devices that are simple and compound machines. For devices that are simple machines, identify which simple machine it is. For compound machines, identify the simple machines that compose it.

Applying Math

28. **Mechanical Advantage** What is the mechanical advantage of a 6-m long ramp that extends from a ground-level sidewalk to a 2-m high porch?

29. **Input Force** How much input force is required to lift an 11,000-N beam using a pulley system with a mechanical advantage of 20?

30. **Efficiency** The input work done on a pulley system is 450 J. What is the efficiency of the pulley system if the output work is 375 J?

Use the table below to answer question 31.

Output Force Exerted by Machines		
Machine	Input Force (N)	Output Force (N)
A	500	750
B	300	200
C	225	225
D	800	1,100
E	75	110

31. **Mechanical Advantage** According to the table above, which of the machines listed has the largest mechanical advantage?

Chapter 20 Standardized Test Practice

Part 1 Multiple Choice

Record your answers on the answer sheet provided by your teacher or on a sheet of paper.

1. The work done by a boy pulling a snow sled up a hill is 425 J. What is the power expended by the boy if he pulls on the sled for 10.5 s?
   A. 24.7 W
   B. 40.5 W
   C. 247 W
   D. 4460 W

Use the illustration below to answer questions 2 and 3.

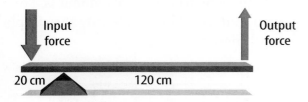

2. What is the mechanical advantage of the lever shown above?
   A. $\frac{1}{6}$
   B. $\frac{1}{2}$
   C. 2
   D. 6

3. What would the mechanical advantage of the lever be if the triangular block were moved to a position 35 cm from the edge of the output force side of the plank?
   A. $\frac{1}{4}$
   B. $\frac{1}{3}$
   C. 3
   D. 4

4. Which of the following causes the efficiency of a machine to be less than 100%?
   A. work
   B. power
   C. mechanical advantage
   D. friction

> **Test-Taking Tip**
>
> **Simplify Diagrams** Write directly on complex charts, such as a Punnett square.

Use the illustration below to answer questions 5 and 6.

5. The pulley system in the illustration above uses several pulleys to increase the mechanical advantage. What is the mechanical advantage of this system?
   A. 1
   B. 2
   C. 3
   D. 4

6. Suppose the lower pulley was removed so that the object was supported only by the upper pulley. What would the mechanical advantage be?
   A. 0
   B. 1
   C. 2
   D. 3

7. You push a shopping cart with a force of 12 N for a distance of 1.5 m. You stop pushing the cart, but it continues to roll for 1.1 m. How much work did you do?
   A. 8.0 J
   B. 13 J
   C. 18 J
   D. 31 J

8. What is the mechanical advantage of a wheel with a radius of 8.0 cm connected to an axle with a radius of 2.5 cm?
   A. 0.31
   B. 2.5
   C. 3.2
   D. 20

9. You push a 5-kg box across the floor with a force of 25 N. How far do you have to push the box to do 63 J of work?
   A. 0.40 m
   B. 1.6 m
   C. 2.5 m
   D. 13 m

Part 2 Short Response/Grid In

Record your answers on the answer sheet provided by your teacher or on a sheet of paper.

10. What is the name of the point about which a lever rotates?

11. Describe how you can determine the mechanical advantage of a pulley or a pulley system.

Use the figure below to answer questions 12 and 13.

12. What type of simple machine is the tip of the dart in the photo above?

13. Would the mechanical advantage of the dart tip change if the tip were longer and thinner? Explain.

14. How much energy is used by a 75-W lightbulb in 15 s?

15. The input and output forces are applied at the ends of the lever. If the lever is 3 m long and the output force is applied 1 m from the fulcrum, what is the mechanical advantage?

16. Your body contains simple machines. Name one part that is a wedge and one part that is a lever.

17. Explain why applying a lubricant, such as oil, to the surfaces of a machine causes the efficiency of the machine to increase.

18. Apply the law of conservation of energy to explain why the output work done by a real machine is always less than the input work done on the machine.

Part 3 Open Ended

Record your answers on a sheet of paper.

19. The output work of a machine can never be greater than the input work. However, the advantage of using a machine is that it makes work easier. Describe and give an example of the three ways a machine can make work easier.

20. A wheel and axle may have a mechanical advantage that is either greater than 1 or less than 1. Describe both types and give some examples of each.

21. Draw a sketch showing the cause of friction as two surfaces slide past each other. Explain your sketch, and describe how lubrication can reduce the friction between the two surfaces.

22. Draw the two types of simple pulleys and an example of a combination pulley. Draw arrows to show the direction of force on your sketches.

Use the figure below to answer question 23.

23. Identify two simple machines in the photo above and describe how they make riding a bicycle easier.

24. Explain why the mechanical advantage of an inclined plane can never be less than 1.

chapter 21

Thermal Energy

The BIG Idea
Thermal energy flows from areas of higher temperature to areas of lower temperature.

SECTION 1
Temperature and Thermal Energy
Main Idea Atoms and molecules in an object are moving in all directions with different speeds.

SECTION 2
Heat
Main Idea Thermal energy moves by conduction, convection, and radiation.

SECTION 3
Engines and Refrigerators
Main Idea Engines convert thermal energy to mechanical energy; refrigerators transfer thermal energy from one place to another.

Fastest to the Finish Line
In order to reach an extraordinary speed in a short distance, this dragster depends on more than an aerodynamic design. Its engine must transform the thermal energy produced by burning fuel to mechanical energy, which propels the dragster down the track.

Science Journal Describe five things that you do to make yourself feel warmer or cooler.

Start-Up Activities

Measuring Temperature

When you leave a glass of ice water on a kitchen table, the ice gradually melts and the temperature of the water increases. What is temperature, and why does the temperature of the ice water increase? In this lab you will explore one way of determining temperature.

1. Obtain three pans. Fill one pan with lukewarm water. Fill a second pan with cold water and crushed ice. Fill a third pan with very warm tap water. Label each pan.
2. Soak one of your hands in the warm water for one minute. Remove your hand from the warm water and put it in the lukewarm water. Does the lukewarm water feel cool or warm?
3. Now soak your hand in the cold water for one minute. Remove your hand from the cold water and place it in the lukewarm water. Does the lukewarm water feel cool or warm?
4. **Think Critically** Write a paragraph in your Science Journal discussing whether your sense of touch would make a useful thermometer.

Thermal Energy Make the following Foldable to help you identify how thermal energy, heat, and temperature are related.

STEP 1 Fold a vertical piece of paper into thirds.

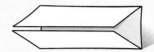

STEP 2 Turn the paper horizontally. Unfold and label the three columns as shown.

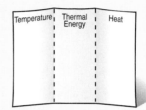

Read for Main Ideas Before you read the chapter, write down what you know about temperature, thermal energy, and heat on the appropriate tab. As you read, add to and correct what you wrote. Write what you have learned about the relationship between heat and thermal energy on the back of your Foldable.

Preview this chapter's content and activities at blue.msscience.com

Get Ready to Read

Identify the Main Idea

① Learn It! Main ideas are the most important ideas in a paragraph, a section, or a chapter. Supporting details are facts or examples that explain the main idea. Understanding the main idea allows you to grasp the whole picture.

② Picture It! Read the following paragraph. Draw a graphic organizer like the one below to show the main idea and supporting details.

> When you heat a pot of water on a stove, thermal energy can be transferred through the water by a process other than conduction and radiation. In a gas or liquid, atoms or molecules can move much more easily than they can in a solid. As a result, these particles can travel from one place to another, carrying their energy with them. This transfer of thermal energy by the movement of atoms or molecules from one part of a material to another is called convection.
>
> —from page 614

```
                    Main Idea
           ┌───────────┼───────────┐
           ▼           ▼           ▼
      Supporting   Supporting   Supporting
       Details      Details      Details
```

③ Apply It! Pick a paragraph from another section of this chapter and diagram the main ideas as you did above.

Target Your Reading

Reading Tip
The main idea is often the first sentence in a paragraph, but not always.

Use this to focus on the main ideas as you read the chapter.

1 Before you read the chapter, respond to the statements below on your worksheet or on a numbered sheet of paper.
- Write an **A** if you **agree** with the statement.
- Write a **D** if you **disagree** with the statement.

2 After you read the chapter, look back to this page to see if you've changed your mind about any of the statements.
- If any of your answers changed, explain why.
- Change any false statements into true statements.
- Use your revised statements as a study guide.

Science Online
Print out a worksheet of this page at blue.msscience.com

Before You Read A or D		Statement	After You Read A or D
	1	Temperature depends on the kinetic energy of the molecules in a material.	
	2	Heat engines can convert energy from one form to another.	
	3	Objects cannot have a temperature below zero on the Celsius scale.	
	4	In a refrigerator, the coolant gas gets cooler as it is compressed.	
	5	A conductor is any material that easily transfers thermal energy.	
	6	Energy is created by an engine.	
	7	Thermal energy from the Sun reaches Earth by conduction through space.	
	8	A car's engine converts thermal energy to mechanical energy.	
	9	Thermal energy always moves from colder objects to warmer objects.	

608 B

section 1

Temperature and Thermal Energy

as you read

What You'll Learn
- **Explain** how temperature is related to kinetic energy.
- **Describe** three scales used for measuring temperature.
- **Define** thermal energy.

Why It's Important
The movement of thermal energy toward or away from your body determines whether you feel too cold, too hot, or just right.

Review Vocabulary
kinetic energy: energy a moving object has that increases as the speed of the object increases

New Vocabulary
- temperature
- thermal energy

What is temperature?

Imagine it's a hot day and you jump into a swimming pool to cool off. When you first hit the water, you might think it feels cold. Perhaps someone else, who has been swimming for a few minutes, thinks the water feels warm. When you swim in water, touch a hot pan, or swallow a cold drink, your sense of touch tells you whether something is hot or cold. However, the words *cold*, *warm*, and *hot* can mean different things to different people.

Temperature How hot or cold something feels is related to its temperature. To understand temperature, think of a glass of water sitting on a table. The water might seem perfectly still, but water is made of molecules that are in constant, random motion. Because these molecules are always moving, they have energy of motion, or kinetic energy.

However, water molecules in random motion don't all move at the same speed. Some are moving faster and some are moving slower. **Temperature** is a measure of the average value of the kinetic energy of the molecules in random motion. The more kinetic energy the molecules have, the higher the temperature. Molecules have more kinetic energy when they are moving faster. So the higher the temperature, the faster the molecules are moving, as shown in **Figure 1**.

Figure 1 The temperature of a substance depends on how fast its molecules are moving. Water molecules are moving faster in the hot water on the left than in the cold water on the right.

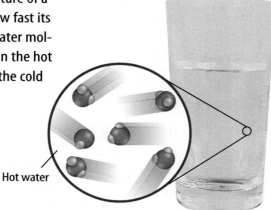

Thermal Expansion It wasn't an earthquake that caused the sidewalk to buckle in **Figure 2.** Hot weather caused the concrete to expand so much that it cracked, and the pieces squeezed each other upward. When the temperature of an object is increased, its molecules speed up and tend to move farther apart. This causes the object to expand. When the object is cooled, its molecules slow down and move closer together. This causes the object to shrink, or contract.

Almost all substances expand when they are heated and contract when they are cooled. The amount of expansion or contraction depends on the type of material and the change in temperature. For example, liquids usually expand more than solids. Also, the greater the change in temperature, the more an object expands or contracts.

Figure 2 Most objects expand as their temperatures increase. Pieces of this concrete sidewalk forced each other upward when the concrete expanded on a hot day.

 Why do materials expand when their temperatures increase?

Measuring Temperature

The temperature of an object depends on the average kinetic energy of all the molecules in an object. However, molecules are so small and objects contain so many of them, that it is impossible to measure the kinetic energy of all the individual molecules.

A more practical way to measure temperature is to use a thermometer. Thermometers usually use the expansion and contraction of materials to measure temperature. One common type of thermometer uses a glass tube containing a liquid. When the temperature of the liquid increases, it expands so that the height of the liquid in the tube depends on the temperature.

Temperature Scales To be able to give a number for the temperature, a thermometer has to have a temperature scale. Two common temperature scales are the Fahrenheit and Celsius scales, shown in **Figure 3.**

On the Fahrenheit scale, the freezing point of water is given the temperature 32°F and the boiling point 212°F. The space between the boiling point and the freezing point is divided into 180 equal degrees. The Fahrenheit scale is used mainly in the United States.

On the Celsius temperature scale, the freezing point of water is given the temperature 0°C and the boiling point is given the temperature 100°C. Because there are only 100 Celsius degrees between the boiling and freezing point of water, Celsius degrees are bigger than Fahrenheit degrees.

Figure 3 The Fahrenheit and Celsius scales are commonly used temperature scales.

SECTION 1 Temperature and Thermal Energy **609**

Converting Fahrenheit and Celsius You can convert temperatures back and forth between the two temperature scales by using the following equations.

Temperature Conversion Equations

To convert temperature in °F to °C: $°C = (\frac{5}{9})(°F - 32)$

To convert temperature in °C to °F: $°F = (\frac{9}{5})(°C) + 32$

For example, to convert 68°F to degrees Celsius, first subtract 32, multiply by 5, then divide by 9. The result is 20°C.

The Kelvin Scale Another temperature scale that is sometimes used is the Kelvin scale. On this scale, 0 K is the lowest temperature an object can have. This temperature is known as absolute zero. The size of a degree on the Kelvin scale is the same as on the Celsius scale. You can change from Celsius degrees to Kelvin degrees by adding 273 to the Celsius temperature.

$$K = °C + 273$$

Applying Math — Solving a Simple Equation

CONVERTING TO CELSIUS On a hot summer day, a Fahrenheit thermometer shows the temperature to be 86°F. What is this temperature on the Celsius scale?

Solution

1 *This is what you know:* Fahrenheit temperature: °F = 86

2 *This is what you need to find:* Celsius temperature: °C

3 *This is the procedure you need to use:* Substitute the Fahrenheit temperature into the equation that converts temperature in °F to °C.
$°C = (\frac{5}{9})(°F - 32) = \frac{5}{9}(86 - 32) = \frac{5}{9}(54) = 30°C$

4 *Check the answer:* Add 32 to your answer and multiply by 9/5. The result should be the given Fahrenheit temperature.

Practice Problems

1. A student's body temperature is 98.6°F. What is this temperature on the Celsius scale?
2. A temperature of 57°C was recorded in 1913 at Death Valley, California. What is this temperature on the Fahrenheit scale?

For more practice visit blue.msscience.com

Thermal Energy

The temperature of an object is related to the average kinetic energy of molecules in random motion. But molecules also have potential energy. Potential energy is energy that the molecules have that can be converted into kinetic energy. The sum of the kinetic and potential energy of all the molecules in an object is the **thermal energy** of the object.

The Potential Energy of Molecules When you hold a ball above the ground, it has potential energy. When you drop the ball, its potential energy is converted into kinetic energy as the ball falls toward Earth. It is the attractive force of gravity between Earth and the ball that gives the ball potential energy.

The molecules in a material also exert attractive forces on each other. As a result, the molecules in a material have potential energy. As the molecules get closer together or farther apart, their potential energy changes.

Increasing Thermal Energy Temperature and thermal energy are different. Suppose you have two glasses filled with the same amount of milk, and at the same temperature. If you pour both glasses of milk into a pitcher, as shown in **Figure 4,** the temperature of the milk won't change. However, because there are more molecules of milk in the pitcher than in either glass, the thermal energy of the milk in the pitcher is greater than the thermal energy of the milk in either glass.

Figure 4 At the same temperature, the larger volume of milk in the pitcher has more thermal energy than the smaller volumes of milk in either glass.

section 1 review

Summary

Temperature
- Temperature is related to the average kinetic energy of the molecules an object contains.
- Most materials expand when their temperatures increase.

Measuring Temperature
- On the Celsius scale the freezing point of water is 0°C and the boiling point is 100°C.
- On the Fahrenheit scale the freezing point of water is 32°F and the boiling point is 212°F.

Thermal Energy
- The thermal energy of an object is the sum of the kinetic and potential energy of all the molecules in an object.

Self Check

1. **Explain** the difference between temperature and thermal energy. How are they related?
2. **Determine** which temperature is always larger—an object's Celsius temperature or its Kelvin temperature.
3. **Explain** how kinetic energy and thermal energy are related.
4. **Describe** how a thermometer uses the thermal expansion of a material to measure temperature.

Applying Math

5. **Convert Temperatures** A turkey cooking in an oven will be ready when the internal temperature reaches 180°F. Convert this temperature to °C and K.

section 2
Heat

Heat and Thermal Energy

It's the heat of the day. Heat the oven to 375°F. A heat wave has hit the Midwest. You've often heard the word *heat*, but what is it? Is it something you can see? **Heat** is the transfer of thermal energy from one object to another when the objects are at different temperatures. The amount of thermal energy that is transferred when two objects are brought into contact depends on the difference in temperature between the objects.

For example, no thermal energy is transferred when two pots of boiling water are touching, because the water in both pots is at the same temperature. However, thermal energy is transferred from the pot of hot water in **Figure 5** that is touching a pot of cold water. The hot water cools down and the cold water gets hotter. Thermal energy continues to be transferred until both objects are the same temperature.

Transfer of Thermal Energy When thermal energy is transferred, it always moves from warmer to cooler objects. Thermal energy never flows from a cooler object to a warmer object. The warmer object loses thermal energy and becomes cooler as the cooler object gains thermal energy and becomes warmer. This process of thermal energy transfer can occur in three ways—by conduction, radiation, or convection.

as you read

What You'll Learn
- **Explain** the difference between thermal energy and heat.
- **Describe** three ways thermal energy is transferred.
- **Identify** materials that are insulators or conductors.

Why It's Important
To keep you comfortable, the flow of thermal energy into and out of your house must be controlled.

Review Vocabulary
electromagnetic wave: a wave produced by vibrating electric charges that can travel in matter and empty space

New Vocabulary
- heat
- conduction
- radiation
- convection
- conductor
- specific heat
- thermal pollution

Figure 5 Thermal energy is transferred only when two objects are at different temperatures. Thermal energy always moves from the warmer object to the cooler object.

612 CHAPTER 21 Thermal Energy

Conduction

When you eat hot pizza, you experience conduction. As the hot pizza touches your mouth, thermal energy moves from the pizza to your mouth. This transfer of thermal energy by direct contact is called conduction. **Conduction** occurs when the particles in a material collide with neighboring particles.

Imagine holding an ice cube in your hand, as in **Figure 6.** The faster-moving molecules in your warm hand bump against the slower-moving molecules in the cold ice. In these collisions, energy is passed from molecule to molecule. Thermal energy flows from your warmer hand to the colder ice, and the slow-moving molecules in the ice move faster. As a result, the ice becomes warmer and its temperature increases. Molecules in your hand move more slowly as they lose thermal energy, and your hand becomes cooler.

Conduction usually occurs most easily in solids and liquids, where atoms and molecules are close together. Then atoms and molecules need to move only a short distance before they bump into one another. As a result, thermal energy is transferred more rapidly by conduction in solids and liquids than in gases.

 *Why does conduction occur more easily in solids and liquids than in gases?*

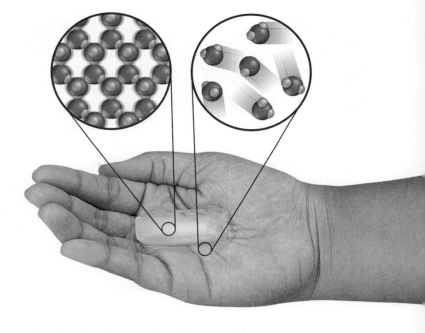

Figure 6 An ice cube in your hand melts because of conduction.

Radiation

On a clear day, you walk outside and feel the warmth of the Sun. How does this transfer of thermal energy occur? Thermal energy transfer does not occur by conduction because almost no matter exists between the Sun and Earth. Instead, thermal energy is transferred from the Sun to Earth by radiation. Thermal energy transfer by **radiation** occurs when energy is transferred by electromagnetic waves. These waves carry energy through empty space, as well as through matter. The transfer of thermal energy by radiation can occur in empty space, as well as in solids, liquids, and gases.

The Sun is not the only source of radiation. All objects emit electromagnetic radiation, although warm objects emit more radiation than cool objects. The warmth you feel when you sit next to a fireplace is due to the thermal energy transferred by radiation from the fire to your skin.

Mini LAB

Comparing Rates of Melting

Procedure
1. Prepare ice water by filling a **glass** with ice and then adding water. Let the glass sit until all the ice melts.
2. Place an ice cube in a **coffee cup.**
3. Place a similar-sized ice cube in another **coffee cup** and add ice water to a depth of about 1 cm.
4. Time how long it takes both ice cubes to melt.

Analysis
1. Which ice cube melted fastest? Why?
2. Is air or water a better insulator? Explain.

Try at Home

Convection

When you heat a pot of water on a stove, thermal energy can be transferred through the water by a process other than conduction and radiation. In a gas or liquid, atoms or molecules can move much more easily than they can in a solid. As a result, these particles can travel from one place to another, carrying their energy along with them. This transfer of thermal energy by the movement of atoms or molecules from one part of a material to another is called **convection.**

Transferring Thermal Energy by Convection As a pot of water is heated, thermal energy is transferred by convection. First, thermal energy is transferred to the water molecules at the bottom of the pot from the stove. These water molecules move faster as their thermal energy increases. The faster-moving molecules tend to be farther apart than the slower-moving molecules in the cooler water above. This causes the warm water to be less dense than the cooler water. As a result, the warm water rises and is replaced at the bottom of the pot by cooler water. The cooler water is heated, rises, and the cycle is repeated until all the water in the pan is at the same temperature.

Natural Convection Natural convection occurs when a warmer, less dense fluid is pushed away by a cooler, denser fluid. For example, imagine the shore of a lake. During the day, the water is cooler than the land. As shown in **Figure 7,** air above the warm land is heated by conduction. When the air gets hotter, its particles move faster and get farther from each other, making the air less dense. The cooler, denser air from over the lake flows in over the land, pushing the less dense air upward. You feel this movement of incoming cool air as wind. The cooler air then is heated by the land and also begins to rise.

Figure 7 Wind movement near a lake or ocean can result from natural convection.

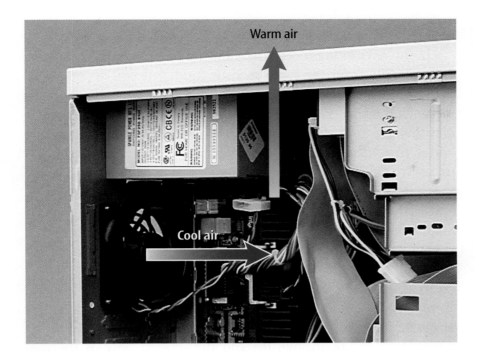

Figure 8 This computer uses forced convection to keep the electronic components surrounded by cooler air.
Identify *another example of forced convection.*

Forced Convection Sometimes convection can be forced. Forced convection occurs when an outside force pushes a fluid, such as air or water, to make it move and transfer thermal energy. A fan is one type of device that is used to move air. For example, computers use fans to keep their electronic components from getting too hot, which can damage them. The fan blows cool air onto the hot electronic components, as shown in **Figure 8**. Thermal energy from the electronic components is transferred to the air around them by conduction. The warm air is pushed away as cool air rushes in. The hot components then continue to lose thermal energy as the fan blows cool air over them.

Thermal Conductors

Why are cooking pans usually made of metal? Why does the handle of a metal spoon in a bowl of hot soup become warm? The answer to both questions is that metal is a good conductor. A **conductor** is any material that easily transfers thermal energy. Some materials are good conductors because of the types of atoms or chemical compounds they contain.

Reading Check *What is a conductor?*

Remember that an atom has a nucleus surrounded by one or more electrons. Certain materials, such as metals, have some electrons that are not held tightly by the nucleus and are freer to move around. These loosely held electrons can bump into other atoms and help transfer thermal energy. The best conductors of thermal energy are metals such as gold and copper.

Observing Convection

Procedure

1. Fill a **250-mL beaker** with room-temperature **water** and let it stand undisturbed for at least 1 min.
2. Using a **hot plate**, heat a small amount of water in a **50-mL beaker** until it is almost boiling.
   WARNING: *Do not touch the heated hot plate.*
3. Carefully place a **penny** into the hot water and let it stand for about 1 min.
4. Take the penny out of the hot water with **metal tongs** and place it on a table. Immediately place the 250-mL beaker on the penny.
5. Using a **dropper**, gently place one drop of **food coloring** on the bottom of the 250-mL beaker of water.
6. Observe what happens in the beaker for several minutes.

Analysis
What happened when you placed the food coloring in the 250-mL beaker? Why?

SECTION 2 Heat **615**

Animal Insulation
To survive in its arctic environment, a polar bear needs good insulation against the cold. Underneath its fur, a polar bear has 10 cm of insulating blubber. Research how animals in polar regions are able to keep themselves warm. Summarize the different ways in your Science Journal.

Thermal Insulators

If you're cooking food, you want the pan to conduct thermal energy easily from the stove to your food, but you do not want the handle of the pan to become hot. An insulator is a material in which thermal energy doesn't flow easily. Most pans have handles that are made from insulators. Liquids and gases are usually better insulators than solids are. Air is a good insulator, and many insulating materials contain air spaces that reduce the transfer of thermal energy by conduction within the material. Materials that are good conductors, such as metals, are poor insulators, and poor conductors are good insulators.

Houses and buildings are made with insulating materials to reduce the flow of thermal energy between the inside and outside. Fluffy insulation like that shown in **Figure 9** is put in the walls. Some windows have double layers of glass that sandwich a layer of air or other insulating gas. This reduces the outward flow of thermal energy in the winter and the inward flow of thermal energy in the summer.

Heat Absorption

On a hot day, you can walk barefoot across the lawn, but the asphalt pavement of a street is too hot to walk on. Why is the pavement hotter than the grass? The change in temperature of an object as it is heated depends on the material it is made of.

Figure 9 The insulation in houses and buildings helps reduce the transfer of thermal energy between the air inside and air outside.

Specific Heat The temperature change of a material as it is heated depends on the specific heat of the material. The **specific heat** of a material is the amount of thermal energy needed to raise the temperature of 1 kg of the material by 1°C.

More thermal energy is needed to change the temperature of a material with a high specific heat than one with a low specific heat. For example, sand on a beach has a lower specific heat than water. During the day, radiation from the Sun warms the sand and the water. Because of its lower specific heat, the sand heats up faster than the water. At night, however, the sand feels cool and the water feels warmer. The temperature of the water changes more slowly than the temperature of the sand as they both lose thermal energy to the cooler night air.

616 CHAPTER 21 Thermal Energy

Thermal Pollution

INTEGRATE Life Science Some electric power plants and factories that use water for cooling produce hot water as a by-product. If this hot water is released into an ocean, lake, or river, it will raise the temperature of the water nearby. This increase in the temperature of a body of water caused by adding warmer water is called **thermal pollution.** Rainwater that is heated after it falls on warm roads or parking lots also can cause thermal pollution if it runs off into a river or lake.

Effects of Thermal Pollution Increasing the water temperature causes fish and other aquatic organisms to use more oxygen. Because warmer water contains less dissolved oxygen than cooler water, some organisms can die due to a lack of oxygen. Also, in warmer water, many organisms become more sensitive to chemical pollutants, parasites, and diseases.

Reducing Thermal Pollution Thermal pollution can be reduced by cooling the warm water produced by factories, power plants, and runoff before it is released into a body of water. Cooling towers like the ones shown in **Figure 10** are used to cool the water used by some power plants and factories.

Figure 10 This power plant uses cooling towers to cool the warm water produced by the power plant.

section 2 review

Summary

Heat and Thermal Energy
- Heat is the transfer of thermal energy due to a temperature difference.
- Thermal energy always moves from a higher temperature to a lower temperature.

Conduction, Radiation, and Convection
- Conduction is the transfer of thermal energy when substances are in direct contact.
- Radiation is the transfer of thermal energy by electromagnetic waves.
- Convection is the transfer of thermal energy by the movement of matter.

Thermal Conductors and Specific Heat
- A thermal conductor is a material in which thermal energy moves easily.
- The specific heat of a substance is the amount of thermal energy needed to raise the temperature of 1 kg of the substance by 1°C.

Self Check

1. **Explain** why materials such as plastic foam, feathers, and fur are poor conductors of thermal energy.
2. **Explain** why the sand on a beach cools down at night more quickly than the ocean water.
3. **Infer** If a substance can contain thermal energy, can a substance also contain heat?
4. **Describe** how thermal energy is transferred from one place to another by convection.
5. **Explain** why a blanket keeps you warm.
6. **Think Critically** In order to heat a room evenly, should heating vents be placed near the floor or near the ceiling of the room? Explain.

Applying Skills

7. **Design an Experiment** to determine whether wood or iron is a better thermal conductor. Identify the dependent and independent variables in your experiment.

Science Online blue.msscience.com/self_check_quiz

Heating Up and Cooling Down

Do you remember how long it took for a cup of hot chocolate to cool before you could take a sip? The hotter the chocolate, the longer it seemed to take to cool.

Real-World Question

How does the temperature of a liquid affect how quickly it warms or cools?

Goals

- Measure the temperature change of water at different temperatures.
- Infer how the rate of heating or cooling depends on the initial water temperature.

Materials

thermometers (5)
400-mL beakers (5)
stopwatch
*watch with second hand
hot plate
*Alternate materials

Safety Precautions

WARNING: *Do not use mercury thermometers. Use caution when heating with a hot plate. Hot and cold glass appears the same.*

Procedure

1. Make a data table to record the temperature of water in five beakers every minute from 0 to 10 min.
2. Fill one beaker with 100 mL of water. Place the beaker on a hot plate and bring the water to a boil. Carefully remove the hot beaker from the hot plate.

3. Record the water temperature at minute 0, and then every minute for 10 min.
4. Repeat step 3 starting with hot tap water, cold tap water, refrigerated water, and ice water with the ice removed.

Conclude and Apply

1. **Graph** your data. **Plot and label** lines for all five beakers on one graph.
2. **Calculate** the rate of heating or cooling for the water in each beaker by subtracting the initial temperature of the water from the final temperature and then dividing this answer by 10 min.
3. **Infer** from your results how the difference between room temperature and the initial temperature of the water affected the rate at which it heated up or cooled down.

Communicating Your Data

Share your data and graphs with other classmates and explain any differences among your data.

section 3
Engines and Refrigerators

Heat Engines

The engines used in cars, motorcycles, trucks, and other vehicles, like the one shown in **Figure 11,** are heat engines. A **heat engine** is a device that converts thermal energy into mechanical energy. Mechanical energy is the sum of the kinetic and potential energy of an object. The heat engine in a car converts thermal energy into mechanical energy when it makes the car move faster, causing the car's kinetic energy to increase.

Forms of Energy There are other forms of energy besides thermal energy and mechanical energy. For example, chemical energy is energy stored in the chemical bonds between atoms. Radiant energy is the energy carried by electromagnetic waves. Nuclear energy is energy stored in the nuclei of atoms. Electrical energy is the energy carried by electric charges as they move in a circuit. Devices such as heat engines convert one form of energy into other useful forms.

The Law of Conservation of Energy When energy is transformed from one form to another, the total amount of energy doesn't change. According to the law of conservation of energy, energy cannot be created or destroyed. Energy only can be transformed from one form to another. No device, including a heat engine, can produce energy or destroy energy.

as you read

What **You'll Learn**
- **Describe** what a heat engine does.
- **Explain** that energy can exist in different forms, but is never created or destroyed.
- **Describe** how an internal combustion engine works.
- **Explain** how refrigerators move thermal energy.

Why **It's Important**

Heat engines enable you to travel long distances.

Review Vocabulary
work: a way of transferring energy by exerting a force over a distance

New Vocabulary
- heat engine
- internal combustion engine

Figure 11 The engine in this earth mover transforms thermal energy into mechanical energy that can perform useful work.

Figure 12 Internal combustion engines are found in many tools and machines.

Topic: Automobile Engines
Visit blue.msscience.com for Web links to information on how internal combustion engines were developed for use in cars.

Activity Make a time line showing the five important events in the development of the automobile engine.

Internal Combustion Engines The heat engine you are probably most familiar with is the internal combustion engine. In **internal combustion engines,** the fuel burns in a combustion chamber inside the engine. Many machines, including cars, airplanes, buses, boats, trucks, and lawn mowers, use internal combustion engines, as shown in **Figure 12.**

Most cars have an engine with four or more combustion chambers, or cylinders. Usually the more cylinders an engine has, the more power it can produce. Each cylinder contains a piston that can move up and down. A mixture of fuel and air is injected into a combustion chamber and ignited by a spark. When the fuel mixture is ignited, it burns explosively and pushes the piston down. The up-and-down motion of the pistons turns a rod called a crankshaft, which turns the wheels of the car. **Figure 13** shows how an internal combustion engine converts thermal energy to mechanical energy in a process called the four-stroke cycle.

Several kinds of internal combustion engines have been designed. In diesel engines, the air in the cylinder is compressed to such a high pressure that the highly flammable fuel ignites without the need for a spark plug. Many lawn mowers use a two-stroke gasoline engine. The first stroke is a combination of intake and compression. The second stroke is a combination of power and exhaust.

Reading Check *How does the burning of fuel mixture cause a piston to move?*

NATIONAL GEOGRAPHIC VISUALIZING THE FOUR-STROKE CYCLE

Figure 13

Most modern cars are powered by fuel-injected internal combustion engines that have a four-stroke combustion cycle. Inside the engine, thermal energy is converted into mechanical energy as gasoline is burned under pressure inside chambers known as cylinders. The steps in the four-stroke cycle are shown here.

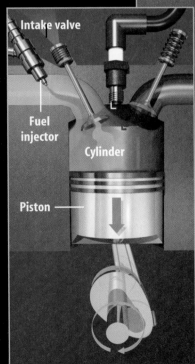

COMPRESSION STROKE
Fuel-air mixture

POWER STROKE
Spark plug
Crankshaft

EXHAUST STROKE
Exhaust valve
Exhaust gases

INTAKE STROKE
Intake valve
Fuel injector
Cylinder
Piston

A During the intake stroke, the piston inside the cylinder moves downward. As it does, air fills the cylinder through the intake valve, and a mist of fuel is injected into the cylinder.

B The piston moves up, compressing the fuel-air mixture.

C At the top of the compression stroke, a spark ignites the fuel-air mixture. The hot gases that are produced expand, pushing the piston down and turning the crankshaft.

D The exhaust valve opens as the piston moves up, pushing the exhaust gases out of the cylinder.

INTEGRATE Career

Mechanical Engineering People who design engines and machines are mechanical engineers. Some mechanical engineers study ways to maximize the transformation of useful energy during combustion—the transformation of energy from chemical form to mechanical form.

Refrigerators

If thermal energy will only flow from something that is warm to something that is cool, how can a refrigerator be cooler inside than the air in the kitchen? A refrigerator is a heat mover. It absorbs thermal energy from the food inside the refrigerator. Then it carries the thermal energy to outside the refrigerator, where it is transferred to the surrounding air.

A refrigerator contains a material called a coolant that is pumped through pipes inside and outside the refrigerator. The coolant is the substance that carries thermal energy from the inside to the outside of the refrigerator.

Absorbing Thermal Energy Figure 14 shows how a refrigerator operates. Liquid coolant is forced up a pipe toward the freezer unit. The liquid passes through an expansion valve where it changes into a gas. When it changes into a gas, it becomes cold. The cold gas passes through pipes around the inside of the refrigerator. Because the coolant gas is so cold, it absorbs thermal energy from inside the refrigerator, and becomes warmer.

Releasing Thermal Energy However, the gas is still colder than the outside air. So, the thermal energy absorbed by the coolant cannot be transferred to the air. The coolant gas then passes through a compressor that compresses the gas. When the gas is compressed, it becomes warmer than room temperature. The gas then flows through the condenser coils, where thermal energy is transferred to the cooler air in the room. As the coolant gas cools, it changes into a liquid. The liquid is pumped through the expansion valve, changes into a gas, and the cycle is repeated.

Figure 14 A refrigerator uses a coolant to move thermal energy from inside to outside the refrigerator. The compressor supplies the energy that enables the coolant to transfer thermal energy to the room.
Diagram how the temperature of the coolant changes as it moves in a refrigerator.

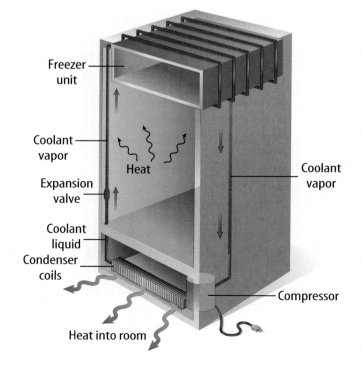

Air Conditioners Most air conditioners cool in the same way that a refrigerator does. You've probably seen air-conditioning units outside of many houses. As in a refrigerator, thermal energy from inside the house is absorbed by the coolant within pipes inside the air conditioner. The coolant then is compressed by a compressor, and becomes warmer. The warmed coolant travels through pipes that are exposed to the outside air. Here the thermal energy is transferred to the outside air.

Heat Pumps Some buildings use a heat pump for heating and cooling. Like an air conditioner or refrigerator, a heat pump moves thermal energy from one place to another. In heating mode, shown in **Figure 15,** the coolant absorbs thermal energy through the outside coils. The coolant is warmed when it is compressed and transfers thermal energy to the house through the inside coils. When a heat pump is used for cooling, it removes thermal energy from the indoor air and transfers it outdoors.

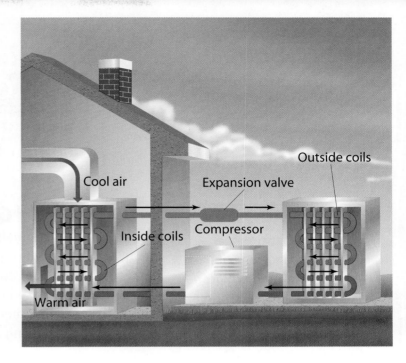

Figure 15 A heat pump heats a building by absorbing thermal energy from the outside air and transferring thermal energy to the cooler air inside.

section 3 review

Summary

Heat Engines and Energy
- A heat engine is a device that converts thermal energy into mechanical energy.
- Energy cannot be created or destroyed. It only can be transformed from one form to another.
- An internal combustion engine is a heat engine that burns fuel in a combustion chamber inside the engine.

Refrigerators and Heat Pumps
- A refrigerator uses a coolant to transfer thermal energy to outside the refrigerator.
- The coolant gas absorbs thermal energy from inside the refrigerator.
- Compressing the coolant makes it warmer than the air outside the refrigerator.
- A heat pump heats by absorbing thermal energy from the air outside, and transferring it inside a building.

Self Check

1. **Diagram** the movement of coolant and the flow of heat when a heat pump is used to cool a building.
2. **Explain** why diesel engines don't use spark plugs.
3. **Identify** the source of thermal energy in an internal combustion engine.
4. **Determine** whether you could cool a kitchen by keeping the refrigerator door open.
5. **Describe** how a refrigerator uses a coolant to keep the food compartment cool.
6. **Think Critically** Explain how an air conditioner could also be used to heat a room.

Applying Skills

7. **Make a Concept Map** Make an events-chain concept map showing the sequence of steps in a four-stroke cycle.

Design Your Own

Comparing Thermal Insulators

Goals
- **Predict** the temperature change of a hot drink in various types of containers over time.
- **Design** an experiment to test the hypothesis and collect data that can be graphed.
- **Interpret** the data.

Possible Materials
hot plate
large beaker
water
100-mL graduated cylinder
alcohol thermometers
various beverage containers
material to cover the containers
stopwatch
tongs
thermal gloves or mitts

Safety Precautions

WARNING: *Use caution when heating liquids. Use tongs or thermal gloves when handling hot materials. Hot and cold glass appear the same. Treat thermometers with care and keep them away from the edges of tables.*

● Real-World Question
Insulated beverage containers are used to reduce the flow of thermal energy. What kinds of containers do you most often drink from? Aluminum soda cans? Paper, plastic, or foam cups? Glass containers? In this investigation, compare how well several different containers reduce the transfer of thermal energy. Which types of containers are most effective at keeping a hot drink hot?

● Form a Hypothesis
Predict the temperature change of a hot liquid in several containers made of different materials over a time interval.

● Test Your Hypothesis

Make a Plan

1. **Decide** what types of containers you will test. Design an experiment to test your hypothesis. This is a group activity, so make certain that everyone gets to contribute to the discussion.

624 CHAPTER 21 Thermal Energy

Using Scientific Methods

2. **List** the materials you will use in your experiment. Describe exactly how you will use these materials. Which liquid will you test? What will be its starting temperature? How will you cover the hot liquids in the container? What material will you use as a cover?
3. **Identify** the variables and controls in your experiment.
4. **Design** a data table in your Science Journal to record the observations you make.

Follow Your Plan

1. Have your teacher approve the steps of your experiment and your data table before you start.
2. To see the pattern of how well various containers retain thermal energy, you will need to graph your data. What kind of graph will you use? Make certain you take enough measurements during the experiment to make your graph.
3. The time intervals between measurements should be the same. Be sure to keep track of time as the experiment goes along. For how long will you measure the temperature?
4. Carry out your investigation and record your observations.

ⓘ Analyze Your Data

1. **Graph** your data. Use one graph to show the data collected from all your containers. Label each line on your graph.
2. **Interpret Data** How can you tell by looking at your graphs which containers best reduce the flow of thermal energy?
3. **Evaluate** Did the water temperature change as you had predicted? Use your data and graph to explain your answers.

ⓘ Conclude and Apply

1. **Explain** why the rate of temperature change varies among the containers. Did the size of the containers affect the rate of cooling?
2. **Conclude** which containers were the best insulators.

Compare your data and graphs with other classmates and explain any differences in your results or conclusions.

LAB 625

TIME SCIENCE AND Society
SCIENCE ISSUES THAT AFFECT YOU!

The Heat Is On

You may live far from water, but still live on an island—a heat island

Think about all the things that are made of asphalt and concrete in a city. As far as the eye can see, there are buildings and parking lots, sidewalks and streets. The combined effect of these paved surfaces and towering structures can make a city sizzle in the summer. There's even a name for this effect. It's called the heat island effect.

Hot Times

You can think of a city as an island surrounded by an ocean of green trees and other vegetation. In the midst of those green trees, the air can be up to 8°C cooler than it is downtown. During the day in rural areas, the Sun's energy is absorbed by plants and soil. Some of this energy causes water to evaporate, so less energy is available to heat the surroundings. This keeps the temperature lower.

Higher temperatures aren't the only problems caused by heat islands. People crank up their air conditioners for relief, so the use of energy skyrockets. Also, the added heat speeds up the rates of chemical reactions in the atmosphere. Smog is due to chemical reactions caused by the interaction of sunlight and vehicle emissions. So hotter air means more smog. And more smog means more health problems.

Cool Cures

Several U.S. cities are working with NASA scientists to come up with a cure for the summertime blues. For instance, dark materials absorb thermal energy more efficiently than light materials. So painting buildings, especially roofs, white can reduce temperature and save on cooling bills.

Dark materials, such as asphalt, absorb more thermal energy than light materials. In extreme heat, it's even possible to fry an egg on dark pavement!

Design and Research Visit the Web Site to the right to research NASA's Urban Heat Island Project. What actions are cities taking to reduce the heat-island effect? Design a city area that would help reduce this effect.

For more information, visit blue.msscience.com

Chapter 21 Study Guide

Reviewing Main Ideas

Section 1 — Temperature and Thermal Energy

1. Molecules of matter are moving constantly. Temperature is related to the average value of the kinetic energy of the molecules.
2. Thermometers measure temperature. Three common temperature scales are the Celsius, Fahrenheit, and Kelvin scales.
3. Thermal energy is the total kinetic and potential energy of the particles in matter.

Section 2 — Heat

1. Heat is thermal energy that is transferred from a warmer object to a colder object.
2. Thermal energy can be transferred by conduction, convection, and radiation.
3. A material that easily transfers thermal energy is a conductor. Thermal energy doesn't flow easily in an insulator.
4. The specific heat of a material is the thermal energy needed to change the temperature of 1 kg of the material 1°C.
5. Thermal pollution occurs when warm water is added to a body of water.

Section 3 — Engines and Refrigerators

1. A device that converts thermal energy into mechanical energy is an engine.
2. In an internal combustion engine, fuel is burned in combustion chambers inside the engine using a four-stroke cycle.
3. Refrigerators and air conditioners use a coolant to move thermal energy.

Visualizing Main Ideas

Copy and complete the following cycle map about the four-stroke cycle.

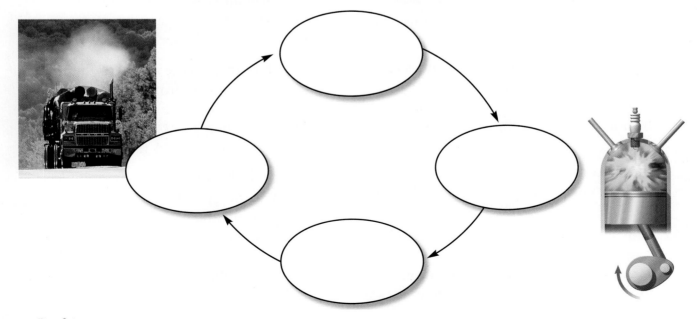

CHAPTER STUDY GUIDE 627

Chapter 21 Review

Using Vocabulary

conduction p. 613
conductor p. 615
convection p. 614
heat p. 612
heat engine p. 619
internal combustion engine p. 620
radiation p. 613
specific heat p. 616
temperature p. 608
thermal energy p. 611
thermal pollution p. 617

Explain the differences in the vocabulary words given below. Then explain how the words are related. Use complete sentences in your answers.

1. internal combustion engine—heat engine
2. temperature—thermal energy
3. thermal energy—thermal pollution
4. conduction—convection
5. conduction—thermal energy
6. thermal energy—specific heat
7. conduction—radiation
8. convection—radiation
9. conductor—thermal energy

Checking Concepts

Choose the word or phrase that best answers the question.

10. What source of thermal energy does an internal combustion engine use?
    A) steam
    B) hot water
    C) burning fuel
    D) refrigerant

11. What happens to most materials when they become warmer?
    A) They contract.
    B) They float.
    C) They vaporize.
    D) They expand.

12. Which occurs if two objects at different temperatures are in contact?
    A) convection
    B) radiation
    C) condensation
    D) conduction

13. Which of the following describes the thermal energy of particles in a substance?
    A) average value of all kinetic energy
    B) total value of all kinetic energy
    C) total value of all kinetic and potential energy
    D) average value of all kinetic and potential energy

14. Thermal energy moving from the Sun to Earth is an example of which process?
    A) convection
    B) expansion
    C) radiation
    D) conduction

15. Many insulating materials contain spaces filled with air because air is what type of material?
    A) conductor
    B) coolant
    C) radiator
    D) insulator

16. A recipe calls for a cake to be baked at a temperature of 350°F. What is this temperature on the Celsius scale?
    A) 162°C
    B) 177°C
    C) 194°C
    D) 212°C

17. Which of the following is true?
    A) Warm air is less dense than cool air.
    B) Warm air is as dense as cool air.
    C) Warm air has no density.
    D) Warm air is denser than cool air.

18. Which of these is the name for thermal energy that moves from a warmer object to a cooler one?
    A) kinetic energy
    B) specific heat
    C) heat
    D) temperature

19. Which of the following is an example of thermal energy transfer by conduction?
    A) water moving in a pot of boiling water
    B) warm air rising from hot pavement
    C) the warmth you feel sitting near a fire
    D) the warmth you feel holding a cup of hot cocoa

chapter 21 Review

Thinking Critically

20. **Infer** When you heat water in a pan, the surface gets hot quickly, even though you are applying heat to the bottom of the water. Explain.

21. **Explain** why several layers of clothing often keep you warmer than a single layer.

22. **Identify** The phrase "heat rises" is sometimes used to describe the movement of thermal energy. For what type of materials is this phrase correct? Explain.

23. **Describe** When a lightbulb is turned on, the electric current in the filament causes the filament to become hot and glow. If the filament is surrounded by a gas, describe how thermal energy is transferred from the filament to the air outside the bulb.

24. **Design an Experiment** Some colors of clothing absorb radiation better than other colors. Design an experiment that will test various colors by placing them in the hot Sun for a period of time.

25. **Explain** Concrete sidewalks usually are made of slabs of concrete. Why do the concrete slabs have a space between them?

26. **Concept Map** Copy and complete the following concept map on convection in a liquid.

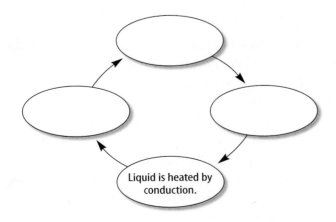

27. **Explain** A winter jacket is lined with insulating material that contains air spaces. How would the insulating properties of the jacket change if the insulating material in the jacket becomes wet? Explain.

28. **Compare** Two glasses of water are poured into a pitcher. If the temperature of the water in both glasses was the same before and after they were mixed, describe how the thermal energy of the water in the pitcher compares to the water in the glasses.

Performance Activities

29. **Poll** In the United States, the Fahrenheit temperature scale is used most often. Some people feel that Americans should switch to the Celsius scale. Take a poll of at least 20 people. Find out if they feel the switch to the Celsius scale should be made. Make a list of reasons people give for or against changing.

Applying Math

30. **Temperature Order** List the following temperatures from coldest to warmest: 80° C, 200 K, 50° F.

31. **Temperature Change** The high temperature on a summer day is 88°F and the low temperature is 61°F. What is the difference between these two temperatures in degrees Celsius?

32. **Global Temperature** The average global temperature is 286 K. Convert this temperature to degrees Celsius.

33. **Body Temperature** A doctor measures a patient's temperature at 38.4°C. Convert this temperature to degrees Fahrenheit.

Chapter 21 Standardized Test Practice

Part 1 Multiple Choice

Record your answers on the answer sheet provided by your teacher or on a sheet of paper.

Use the photo below to answer questions 1 and 2.

1. The temperatures of the two glasses of water shown in the photograph above are 30°C and 0°C. Which of the following is a correct statement about the two glasses of water?
   A. The cold water has a higher average kinetic energy.
   B. The warmer water has lower thermal energy.
   C. The molecules of the cold water move faster.
   D. The molecules of the warmer water have more kinetic energy.

2. The difference in temperature of the two glasses of water is 30°C. What is their difference in temperature on the Kelvin scale?
   A. 30 K   C. 243 K
   B. 86 K   D. 303 K

3. Which of the following describes a refrigerator?
   A. heat engine   C. heat mover
   B. heat pump   D. conductor

Test-Taking Tip

Avoid rushing on test day. Prepare your clothes and test supplies the night before. Wake up early and arrive at school on time on test day.

4. Which of the following is not a step in the four-stroke cycle of internal combustion engines?
   A. compression   C. idling
   B. exhaust   D. power

Use the table below to answer question 5.

Material	Specific Heat (J/kg °C)
aluminum	897
copper	385
lead	129
nickel	444
zinc	388

5. A sample of each of the metals in the table above is formed into a 50-g cube. If 100 J of thermal energy are applied to each of the samples, which metal would change temperature by the greatest amount?
   A. aluminum   C. lead
   B. copper   D. nickel

6. An internal combustion engine converts thermal energy to which of the following forms of energy?
   A. chemical   C. radiant
   B. mechanical   D. electrical

7. Which of the following is a statement of the law of conservation of energy?
   A. Energy never can be created or destroyed.
   B. Energy can be created, but never destroyed.
   C. Energy can be destroyed, but never created.
   D. Energy can be created and destroyed when it changes form.

Standardized Test Practice

Part 2 Short Response/Grid In

Record your answers on the answer sheet provided by your teacher or on a sheet of paper.

8. If you add ice to a glass of room-temperature ice, does the water warm the ice or does the ice cool the water? Explain.

9. Strong winds that occur during a thunderstorm are the result of temperature differences between neighboring air masses. Would you expect the warmer or the cooler air mass to rise above the other?

10. A diesel engine uses a different type of fuel than the fuel used in a gasoline engine. Explain why.

11. What are the two main events that occur while the cylinder moves downward during the intake stroke of an internal combustion engine's four-stroke cycle?

Use the photo below to answer questions 12 and 13.

12. Why are cooking pots like the one in the photograph above often made of metal? Why isn't the handle made of metal?

13. When heating water in the pot, electrical energy from the cooking unit is changed to what other type of energy?

Part 3 Open Ended

Record your answers on a sheet of paper.

Use the illustration below to answer questions 14 and 15.

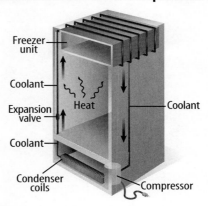

14. The illustration above shows the parts of a refrigerator and how coolant flows through the refrigerator. Explain how thermal energy is transferred to the coolant inside the refrigerator and then transferred from the coolant to the outside air.

15. What are the functions of the expansion valve, the condenser coils, and the compressor in the illustration?

16. Define convection. Explain the difference between natural and forced convection, and give an example of each.

17. Draw a sketch with arrows showing how conduction, convection, and radiation affect the movement and temperature of air near an ocean.

18. Define temperature and explain how it is related to the movement of molecules in a substance.

19. Explain what makes some materials good thermal conductors.

20. You place a cookie sheet in a hot oven. A few minutes later you hear a sound as the cookie sheet bends slightly. Explain what causes this.

unit 7
Physical Interactions

How Are Radar & Popcorn Connected?

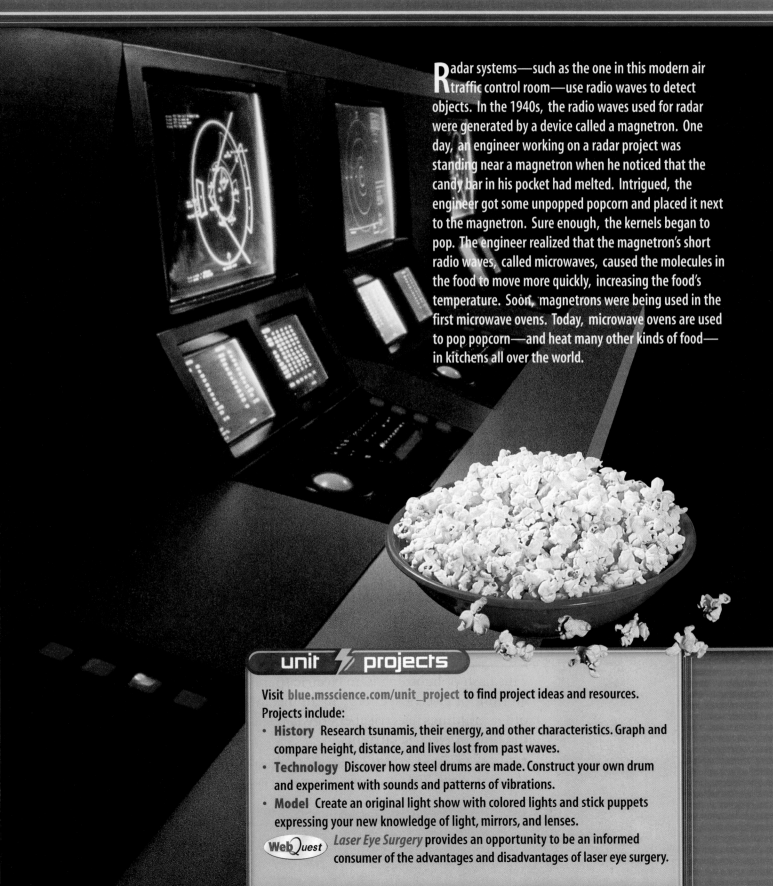

NATIONAL GEOGRAPHIC

Radar systems—such as the one in this modern air traffic control room—use radio waves to detect objects. In the 1940s, the radio waves used for radar were generated by a device called a magnetron. One day, an engineer working on a radar project was standing near a magnetron when he noticed that the candy bar in his pocket had melted. Intrigued, the engineer got some unpopped popcorn and placed it next to the magnetron. Sure enough, the kernels began to pop. The engineer realized that the magnetron's short radio waves, called microwaves, caused the molecules in the food to move more quickly, increasing the food's temperature. Soon, magnetrons were being used in the first microwave ovens. Today, microwave ovens are used to pop popcorn—and heat many other kinds of food— in kitchens all over the world.

unit projects

Visit blue.msscience.com/unit_project to find project ideas and resources.
Projects include:
- **History** Research tsunamis, their energy, and other characteristics. Graph and compare height, distance, and lives lost from past waves.
- **Technology** Discover how steel drums are made. Construct your own drum and experiment with sounds and patterns of vibrations.
- **Model** Create an original light show with colored lights and stick puppets expressing your new knowledge of light, mirrors, and lenses.

WebQuest *Laser Eye Surgery* provides an opportunity to be an informed consumer of the advantages and disadvantages of laser eye surgery.

chapter 22

Electricity

The BIG Idea

Electrical energy can be converted into other forms of energy when electric charges flow in a circuit.

SECTION 1
Electric Charge
Main Idea Electric charges are positive or negative and exert forces on each other.

SECTION 2
Electric Current
Main Idea A battery produces an electric field in a closed circuit that causes electric charges to flow.

SECTION 3
Electric Circuits
Main Idea Electrical energy can be transferred to devices connected in an electric circuit.

A Blast of Energy

This flash of lightning is an electric spark that releases an enormous amount of electrical energy in an instant. However, in homes and other buildings, electrical energy is released in a controlled way by the flow of electric currents.

Science Journal Write a paragraph describing a lightning flash you have seen. Include information about the weather conditions at the time.

Start-Up Activities

Observing Electric Forces

No computers? No CD players? No video games? Can you imagine life without electricity? Electricity also provides energy that heats and cools homes and produces light. The electrical energy that you use every day is produced by the forces that electric charges exert on each other.

1. Inflate a rubber balloon.
2. Place small bits of paper on your desktop and bring the balloon close to the bits of paper. Record your observations.
3. Charge the balloon by holding it by the knot and rubbing it on your hair or a piece of wool.
4. Bring the balloon close to the bits of paper. Record your observations.
5. Charge two balloons using the procedure in step 3. Hold each balloon by its knot and bring the balloons close to each other. Record your observations.
6. **Think Critically** Compare and contrast the force exerted on the bits of paper by the charged balloon and the force exerted by the two charged balloons on each other.

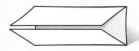

Electricity Make the following Foldable to help you understand the terms *electric charge*, *electric current*, and *electric circuit*.

STEP 1 Fold the top of a vertical piece of paper down and the bottom up to divide the paper into thirds.

STEP 2 Turn the paper horizontally; **unfold and label** the three columns as shown.

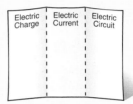

Read and Write Before you read the chapter, write a definition of electric charge, electric current, and electric circuit in the appropriate column. As you read the chapter, correct your definition and add additional information about each term.

Preview this chapter's content and activities at
blue.msscience.com

635

Get Ready to Read

Make Predictions

① Learn It! A prediction is an educated guess based on what you already know. One way to predict while reading is to guess what you believe the author will tell you next. As you are reading, each new topic should make sense because it is related to the previous paragraph or passage.

② Practice It! Read the excerpt below from Section 1. Based on what you have read, make predictions about what you will read in the rest of the lesson. After you read Section 1, go back to your predictions to see if they were correct.

> Predict how the electric force depends on the amount of charge.

> Predict how the electric force would change as charged objects get farther apart.

> Can you predict how the electric force between two electrons changes as the electrons get closer together?

The electric force between two charged objects depends on the distance between them and the **amount of charge** on each object. The electric force between two charged objects gets stronger **as the charges get closer together**. **A positive and a negative charge** are attracted to each other more strongly if they are closer together.

—from page 639

③ Apply It! Before you read, skim the questions in the Chapter Review. Choose three questions and predict the answers.

Target Your Reading

Reading Tip: As you read, check the predictions you made to see if they were correct.

Use this to focus on the main ideas as you read the chapter.

① Before you read the chapter, respond to the statements below on your worksheet or on a numbered sheet of paper.
- Write an **A** if you **agree** with the statement.
- Write a **D** if you **disagree** with the statement.

② After you read the chapter, look back to this page to see if you've changed your mind about any of the statements.
- If any of your answers changed, explain why.
- Change any false statements into true statements.
- Use your revised statements as a study guide.

Science Online
Print out a worksheet of this page at blue.msscience.com

Before You Read A or D		Statement	After You Read A or D
	1	Atoms become ions by gaining or losing electrons.	
	2	It is safe to take shelter under a tree during a lightning storm.	
	3	Electric current can follow only one path in a parallel circuit.	
	4	Electrons flow in straight lines through conducting wires.	
	5	Batteries produce electrical energy through nuclear reactions.	
	6	The force between electric charges always is attractive.	
	7	Electrical energy can be transformed into other forms of energy.	
	8	If the voltage in a circuit doesn't change, the current increases if the resistance decreases.	
	9	Electric charges must be touching to exert forces on each other.	

section 1
Electric Charge

as you read

What You'll Learn
- **Describe** how objects can become electrically charged.
- **Explain** how an electric charge affects other electric charges.
- **Distinguish** between electric conductors and insulators.
- **Describe** how electric discharges such as lightning occur.

Why It's Important
All electric phenomena result from the forces electric charges exert on each other.

Review Vocabulary
force: the push or pull one object exerts on another

New Vocabulary
- ion
- static charge
- electric force
- electric field
- insulator
- conductor
- electric discharge

Electricity

You can't see, smell, or taste electricity, so it might seem mysterious. However, electricity is not so hard to understand when you start by thinking small—very small. All solids, liquids, and gases are made of tiny particles called atoms. Atoms, as shown in **Figure 1,** are made of even smaller particles called protons, neutrons, and electrons. Protons and neutrons are held together tightly in the nucleus at the center of an atom, but electrons swarm around the nucleus in all directions. Protons and electrons have electric charge, but neutrons have no electric charge.

Positive and Negative Charge There are two types of electric charge—positive and negative. Protons have a positive charge, and electrons have a negative charge. The amount of negative charge on an electron is exactly equal to the amount of positive charge on a proton. Because atoms have equal numbers of protons and electrons, the amount of positive charge on all the protons in the nucleus of an atom is balanced by the negative charge on all the electrons moving around the nucleus. Therefore, atoms are electrically neutral, which means they have no overall electric charge.

An atom becomes negatively charged when it gains extra electrons. If an atom loses electrons it becomes positively charged. A positively or negatively charged atom is called an **ion** (I ahn).

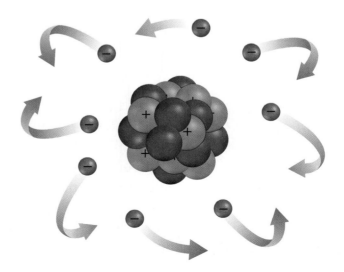

Figure 1 An atom is made of positively charged protons (orange), negatively charged electrons (red), and neutrons (blue) with no electric charge.
Identify *where the protons and neutrons are located in an atom.*

636 CHAPTER 22 Electricity

Figure 2 Rubbing can move electrons from one object to another. Hair holds electrons more loosely than the balloon holds them. As a result, electrons are moved from the hair to the balloon when the two make contact. **Infer** *which object has become positively charged and which has become negatively charged.*

Electrons Move in Solids Electrons can move from atom to atom and from object to object. Rubbing is one way that electrons can be transferred. If you have ever taken clinging clothes from a clothes dryer, you have seen what happens when electrons are transferred from one object to another.

Suppose you rub a balloon on your hair. The atoms in your hair hold their electrons more loosely than the atoms on the balloon hold theirs. As a result, electrons are transferred from the atoms in your hair to the atoms on the surface of the balloon, as shown in **Figure 2.** Because your hair loses electrons, it becomes positively charged. The balloon gains electrons and becomes negatively charged. Your hair and the balloon become attracted to one another and make your hair stand on end. This imbalance of electric charge on an object is called a **static charge.** In solids, static charge is due to the transfer of electrons between objects. Protons cannot be removed easily from the nucleus of an atom and usually do not move from one object to another.

Reading Check *How does an object become electrically charged?*

Ions Move in Solutions Sometimes, the movement of charge can be caused by the movement of ions instead of the movement of electrons. Table salt—sodium chloride—is made of sodium ions and chloride ions that are fixed in place and cannot move through the solid. However, when salt is dissolved in water, the sodium and chloride ions break apart and spread out evenly in the water, forming a solution, as shown in **Figure 3.** Now the positive and negative ions are free to move. Solutions containing ions play an important role in enabling different parts of your body to communicate with each other. **Figure 4** shows how a nerve cell uses ions to transmit signals. These signals moving throughout your body enable you to sense, move, and even think.

Figure 3 When table salt (NaCl) dissolves in water, the sodium ions and chloride ions break apart. These ions now are able to carry electric energy.

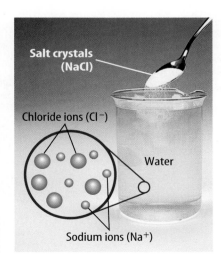

SECTION 1 Electric Charge **637**

NATIONAL GEOGRAPHIC VISUALIZING NERVE IMPULSES

Figure 4

The control and coordination of all your bodily functions involves signals traveling from one part of your body to another through nerve cells. Nerve cells use ions to transmit signals from one nerve cell to another.

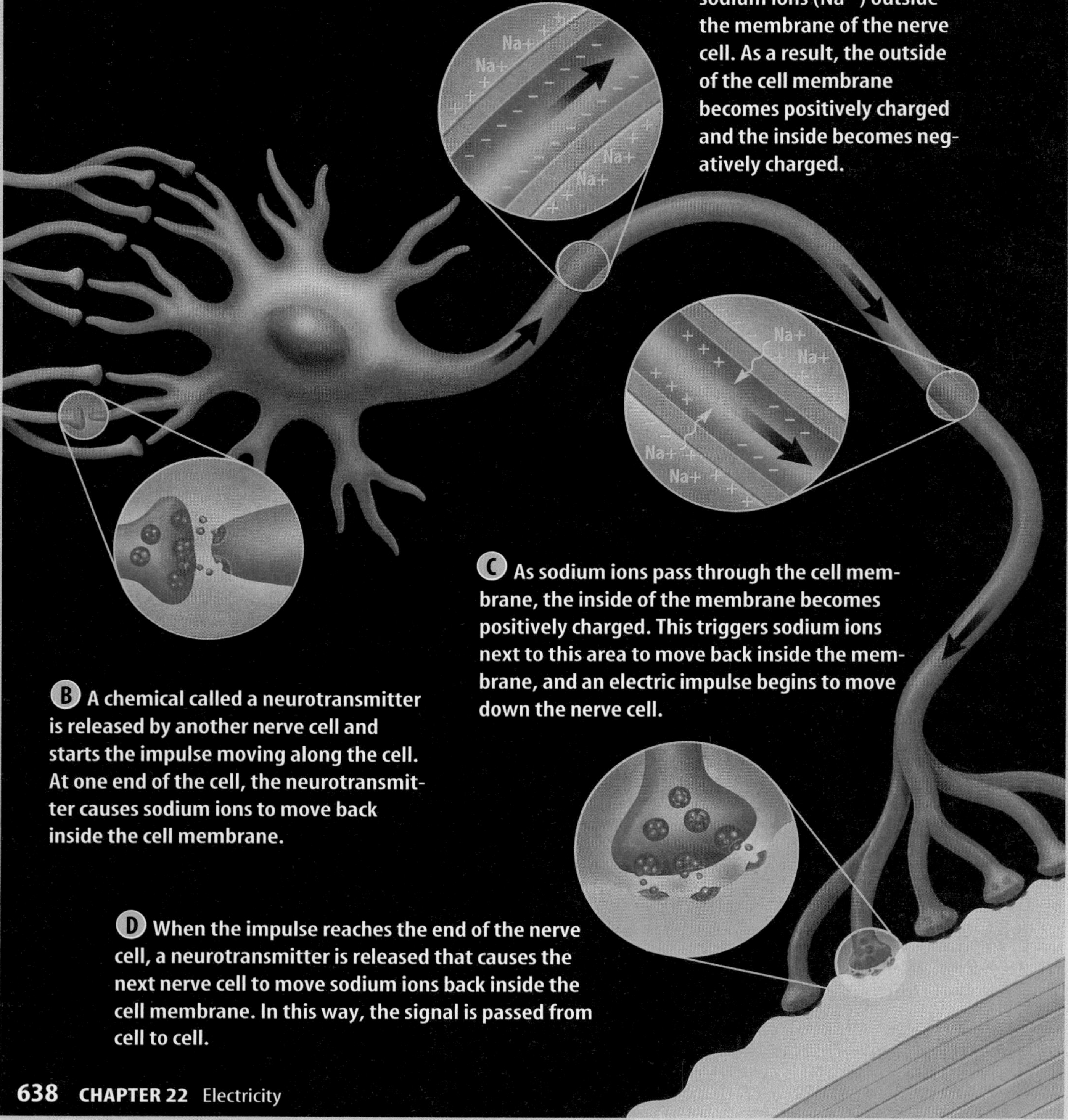

A When a nerve cell is not transmitting a signal, it moves positively charged sodium ions (Na^+) outside the membrane of the nerve cell. As a result, the outside of the cell membrane becomes positively charged and the inside becomes negatively charged.

C As sodium ions pass through the cell membrane, the inside of the membrane becomes positively charged. This triggers sodium ions next to this area to move back inside the membrane, and an electric impulse begins to move down the nerve cell.

B A chemical called a neurotransmitter is released by another nerve cell and starts the impulse moving along the cell. At one end of the cell, the neurotransmitter causes sodium ions to move back inside the cell membrane.

D When the impulse reaches the end of the nerve cell, a neurotransmitter is released that causes the next nerve cell to move sodium ions back inside the cell membrane. In this way, the signal is passed from cell to cell.

Unlike charges attract.

Like charges repel. Like charges repel.

Figure 5 A positive charge and a negative charge attract each other. Two positive charges repel each other, as do two negative charges.

Electric Forces

The electrons in an atom swarm around the nucleus. What keeps these electrons close to the nucleus? The positively charged protons in the nucleus exert an attractive electric force on the negatively charged electrons. All charged objects exert an **electric force** on each other. The electric force between two charges can be attractive or repulsive, as shown in **Figure 5.** Objects with the same type of charge repel one another and objects with opposite charges attract one another. This rule is often stated as "like charges repel, and unlike charges attract."

The electric force between two charged objects depends on the distance between them and the amount of charge on each object. The electric force between two electric charges gets stronger as the charges get closer together. A positive and a negative charge are attracted to each other more strongly if they are closer together. Two like charges are pushed away more strongly from each other the closer they are. The electric force between two objects that are charged, such as two balloons that have been rubbed on wool, increases if the amount of charge on at least one of the objects increases.

 *How does the electric force between two charged objects depend on the distance between them?*

Electric Fields You might have noticed examples of how charged objects don't have to be touching to exert an electric force on each other. For instance, two charged balloons push each other apart even though they are not touching. How are charged objects able to exert forces on each other without touching?

Electric charges exert a force on each other at a distance through an **electric field** that exists around every electric charge. **Figure 6** shows the electric field around a positive and a negative charge. An electric field gets stronger as you get closer to a charge, just as the electric force between two charges becomes greater as the charges get closer together.

Figure 6 The lines with arrowheads represent the electric field around charges. The direction of each arrow is the direction a positive charge would move if it were placed in the field.

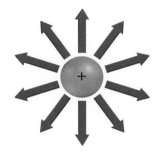

The electric field arrows point away from a positive charge.

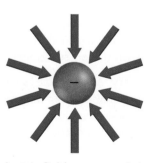

The electric field arrows point toward a negative charge.
Explain *why the electric field arrows around a negative charge are in the opposite direction of the arrows around a positive charge.*

SECTION 1 Electric Charge **639**

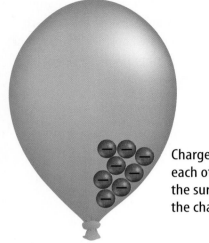

Charges placed on an insulator repel each other but cannot move easily on the surface of the insulator. As a result, the charges remain in one place.

Figure 7 Electric charges move more easily through conductors than through insulators.

The three wires in this electric cable are made of copper, which is a conductor. The wires are covered with plastic insulation that keeps the copper wires from touching each other.

Insulators and Conductors

Rubbing a balloon on your hair transfers electrons from your hair to the balloon. However, only the part of the balloon that was rubbed on your hair becomes charged, because electrons cannot move easily through rubber. As a result, the electrons that were rubbed onto the balloon tend to stay in one place, as shown in **Figure 7.** A material in which electrons cannot move easily from place to place is called an **insulator.** Examples of insulators are plastic, wood, glass, and rubber.

Materials that are **conductors** contain electrons that can move more easily in the material. The electric wire in **Figure 7** is made from a conductor coated with an insulator such as plastic. Electrons move easily in the conductor but do not move easily through the plastic insulation. This prevents electrons from moving through the insulation and causing an electric shock if someone touches the wire.

Metals as Conductors The best conductors are metals such as copper, gold, and aluminum. In a metal atom, a few electrons are not attracted as strongly to the nucleus as the other electrons, and are loosely bound to the atom. When metal atoms form a solid, the metal atoms can move only short distances. However, the electrons that are loosely bound to the atoms can move easily in the solid piece of metal. In an insulator, the electrons are bound tightly in the atoms that make up the insulator and therefore cannot move easily.

Topic: Superconductors
Visit blue.msscience.com for Web links to information about materials that are superconductors.

Activity Make a table listing five materials that can become superconductors and the critical temperature for each material.

Induced Charge

Has this ever happened to you? You walk across a carpet and as you reach for a metal doorknob, you feel an electric shock. Maybe you even see a spark jump between your fingertip and the doorknob. To find out what happened, look at **Figure 8.**

As you walk, electrons are rubbed off the rug by your shoes. The electrons then spread over the surface of your skin. As you bring your hand close to the doorknob, the electric field around the excess electrons on your hand repels the electrons in the doorknob. Because the doorknob is a good conductor, its electrons move easily away from your hand. The part of the doorknob closest to your hand then becomes positively charged. This separation of positive and negative charges due to an electric field is called an induced charge.

If the electric field between your hand and the knob is strong enough, charge can be pulled from your hand to the doorknob, as shown in **Figure 8.** This rapid movement of excess charge from one place to another is an **electric discharge.** Lightning is an example of an electric discharge. In a storm cloud, air currents sometimes cause the bottom of the cloud to become negatively charged. This negative charge induces a positive charge in the ground below the cloud. A cloud-to-ground lightning stroke occurs when electric charge moves between the cloud and the ground.

Figure 8 A spark that jumps between your fingers and a metal doorknob starts at your feet. **Identify** *another example of an electric discharge.*

As you walk across the floor, you rub electrons from the carpet onto the bottom of your shoes. These electrons then spread out over your skin, including your hands.

As you bring your hand close to the metal doorknob, electrons on the doorknob move as far away from your hand as possible. The part of the doorknob closest to your hand is left with a positive charge.

The attractive electric force between the electrons on your hand and the induced positive charge on the doorknob might be strong enough to pull electrons from your hand to the doorknob. You might see this as a spark and feel a mild electric shock.

Figure 9 A lightning rod can protect a building from being damaged by a lightning strike.

Grounding

Lightning is an electric discharge that can cause damage and injury because a lightning bolt releases an extremely large amount of electric energy. Even electric discharges that release small amounts of energy can damage delicate circuitry in devices such as computers. One way to avoid the damage caused by electric discharges is to make the excess charges flow harmlessly into Earth's surface. Earth can be a conductor, and because it is so large, it can absorb an enormous quantity of excess charge.

The process of providing a pathway to drain excess charge into Earth is called grounding. The pathway is usually a conductor such as a wire or a pipe. You might have noticed lightning rods at the top of buildings and towers, as shown in **Figure 9**. These rods are made of metal and are connected to metal cables that conduct electric charge into the ground if the rod is struck by lightning.

---

section 1 review

Summary

Electric Charges
- There are two types of electric charge—positive charge and negative charge.
- The amount of negative charge on an electron is equal to the amount of positive charge on a proton.
- Objects that are electrically neutral become negatively charged when they gain electrons and positively charged when they lose electrons.

Electric Forces
- Like charges repel and unlike charges attract.
- The force between two charged objects increases as they get closer together.
- A charged object is surrounded by an electric field that exerts a force on other charged objects.

Insulators and Conductors
- Electrons cannot move easily in an insulator but can move easily in a conductor.

Self Check

1. **Explain** why when objects become charged it is electrons that are transferred from one object to another rather than protons.
2. **Compare and contrast** the movement of electric charge in a solution with the transfer of electric charge between solid objects.
3. **Explain** why metals are good conductors.
4. **Compare and contrast** the electric field around a negative charge and the electric field around a positive charge.
5. **Explain** why an electric discharge occurs.
6. **Think Critically** A cat becomes negatively charged when it is brushed. How does the electric charge on the brush compare to the charge on the cat?

Applying Skills

7. **Analyze** You slide out of a car seat and as you touch the metal car door, a spark jumps between your hand and the door. Describe how the spark was formed.

section 2

Electric Current

Flow of Charge

An electric discharge, such as a lightning bolt, can release a huge amount of energy in an instant. However, electric lights, refrigerators, TVs, and stereos need a steady source of electrical energy that can be controlled. This source of electrical energy comes from an **electric current**, which is the flow of electric charge. In solids, the flowing charges are electrons. In liquids, the flowing charges are ions, which can be positively or negatively charged. Electric current is measured in units of amperes (A). A model for electric current is flowing water. Water flows downhill because a gravitational force acts on it. Similarly, electrons flow because an electric force acts on them.

A Model for a Simple Circuit How does a flow of water provide energy? If the water is separated from Earth by using a pump, the higher water now has gravitational potential energy, as shown in **Figure 10**. As the water falls and does work on the waterwheel, the water loses potential energy and the waterwheel gains kinetic energy. For the water to flow continuously, it must flow through a closed loop. Electric charges will flow continuously only through a closed conducting loop called a **circuit**.

as you read

What You'll Learn
- **Relate** voltage to the electrical energy carried by an electric current.
- **Describe** a battery and how it produces an electric current.
- **Explain** electrical resistance.

Why It's Important
Electric current provides a steady source of electrical energy that powers the electric appliances you use every day.

Review Vocabulary
gravitational potential energy: the energy stored in an object due to its position above Earth's surface

New Vocabulary
- electric current
- voltage
- circuit
- resistance

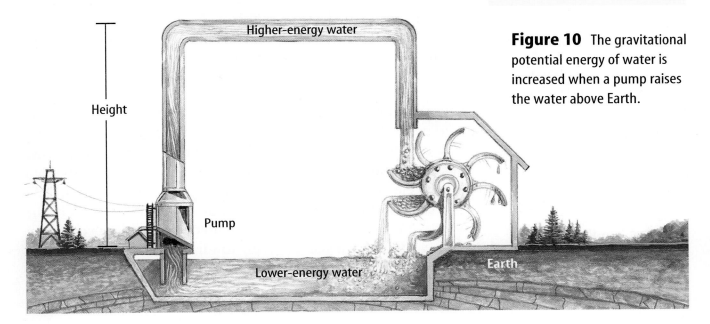

Figure 10 The gravitational potential energy of water is increased when a pump raises the water above Earth.

Figure 11 As long as there is a closed path for electrons to follow, electrons can flow in a circuit. They move away from the negative battery terminal and toward the positive terminal.

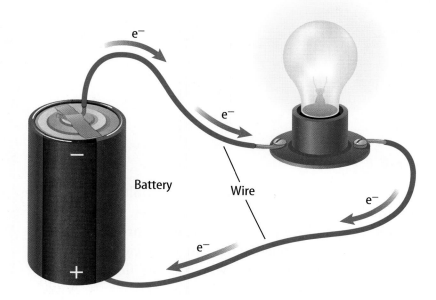

Electric Circuits The simplest electric circuit contains a source of electrical energy, such as a battery, and an electric conductor, such as a wire, connected to the battery. For the simple circuit shown in **Figure 11,** a closed path is formed by wires connected to a lightbulb and to a battery. Electric current flows in the circuit as long as none of the wires, including the glowing filament wire in the lightbulb, is disconnected or broken.

Voltage In a water circuit, a pump increases the gravitational potential energy of the water by raising the water from a lower level to a higher level. In an electric circuit, a battery increases the electrical potential energy of electrons. This electrical potential energy can be transformed into other forms of energy. The **voltage** of a battery is a measure of how much electrical potential energy each electron can gain. As voltage increases, more electrical potential energy is available to be transformed into other forms of energy. Voltage is measured in volts (V).

How a Current Flows You may think that when an electric current flows in a circuit, electrons travel completely around the circuit. Actually, individual electrons move slowly in an electric circuit. When the ends of a wire are connected to a battery, the battery produces an electric field in the wire. The electric field forces electrons to move toward the positive battery terminal. As an electron moves, it collides with other electric charges in the wire and is deflected in a different direction. After each collision, the electron again starts moving toward the positive terminal. A single electron may undergo more than ten trillion collisions each second. As a result, it may take several minutes for an electron in the wire to travel one centimeter.

Investigating the Electric Force

Procedure
1. Pour a layer of **salt** on a **plate.**
2. Sparingly sprinkle grains of **pepper** on top of the salt. Do not use too much pepper.
3. Rub a **rubber** or **plastic comb** on an article of **wool clothing.**
4. Slowly drag the comb through the salt and observe.

Analysis
1. How did the salt and pepper react to the comb?
2. Explain why the pepper reacted differently than the salt.

Batteries A battery supplies energy to an electric circuit. When the positive and negative terminals in a battery are connected in a circuit, the electric potential energy of the electrons in the circuit is increased. As these electrons move toward the positive battery terminal, this electric potential energy is transformed into other forms of energy, just as gravitational potential energy is converted into kinetic energy as water falls.

A battery supplies energy to an electric circuit by converting chemical energy to electric potential energy. For the alkaline battery shown in **Figure 12,** the two terminals are separated by a moist paste. Chemical reactions in the moist paste cause electrons to be transferred to the negative terminal from the atoms in the positive terminal. As a result, the negative terminal becomes negatively charged and the positive terminal becomes positively charged. This produces the electric field in the circuit that causes electrons to move away from the negative terminal and toward the positive terminal.

Battery Life Batteries don't supply energy forever. Maybe you know someone whose car wouldn't start after the lights had been left on overnight. Why do batteries run down? Batteries contain only a limited amount of the chemicals that react to produce chemical energy. These reactions go on as the battery is used and the chemicals are changed into other compounds. Once the original chemicals are used up, the chemical reactions stop and the battery is "dead."

Alkaline Batteries Several chemicals are used to make an alkaline battery. Zinc is a source of electrons at the negative terminal, and manganese dioxide combines with electrons at the positive terminal. The moist paste contains potassium hydroxide that helps transport electrons from the positive terminal to the negative terminal. Research dry-cell batteries and lead-acid batteries. Make a table listing the chemicals used in these batteries and their purpose.

Figure 12 When this alkaline battery is connected in an electric circuit, chemical reactions occur in the moist paste of the battery that move electrons from the positive terminal to the negative terminal.

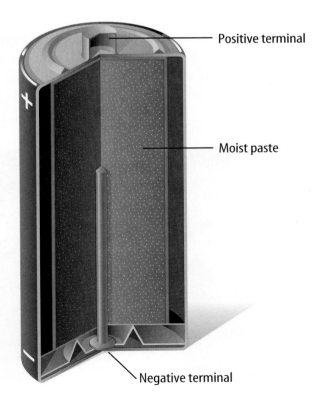

The Ohm The unit for electrical resistance was named in honor of the German physicist Georg Simon Ohm (1787–1854). Ohm is credited for discovering the relationship between current flow, voltage, and resistance. Research and find out more about Georg Ohm. Write a brief biography of him to share with the class.

Resistance

Electrons can move much more easily through conductors than through insulators, but even conductors interfere somewhat with the flow of electrons. The measure of how difficult it is for electrons to flow through a material is called **resistance.** The unit of resistance is the ohm (Ω). Insulators generally have much higher resistance than conductors.

As electrons flow through a circuit, they collide with the atoms and other electric charges in the materials that make up the circuit. Look at **Figure 13.** These collisions cause some of the electrons' electrical energy to be converted into thermal energy—heat—and sometimes into light. The amount of electrical energy that is converted into heat and light depends on the resistance of the materials in the circuit.

Buildings Use Copper Wires The amount of electrical energy that is converted into thermal energy increases as the resistance of the wire increases. Copper has low resistance and is one of the best electric conductors. Because copper is a good conductor, less heat is produced as electric current flows in copper wires, compared to wires made of other materials. As a result, copper wire is used in household wiring because the wires usually don't become hot enough to cause fires.

Resistance of Wires The electric resistance of a wire also depends on the length and thickness of the wire, as well as the material it is made from. When water flows through a hose, the water flow decreases as the hose becomes narrower or longer, as shown in **Figure 14** on the next page. The electric resistance of a wire increases as the wire becomes longer or as it becomes narrower.

Figure 13 As electrons flow through a wire, they travel in a zigzag path as they collide with atoms and other electrons. In these collisions, electrical energy is converted into other forms of energy.
Identify *the other forms of energy that electrical energy is converted into.*

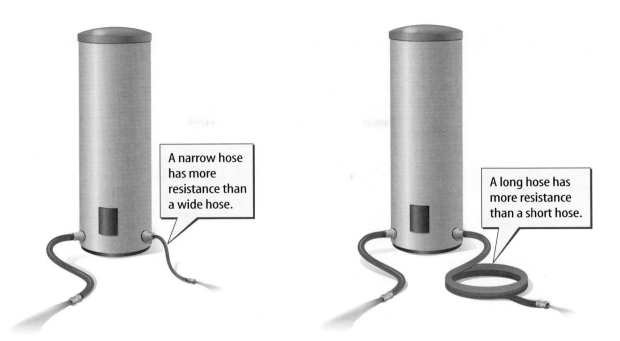

Figure 14 The resistance of a hose to the flow of water depends on the diameter and length of the hose used.
Compare and contrast *water flowing in a hose and electric current flowing in a wire.*

Lightbulb Filaments In a lightbulb, the filament is made of wire so narrow that it has a high resistance. When electric current flows in the filament, it becomes hot enough to emit light. The filament is made of tungsten metal, which has a much higher melting point than most other metals. This keeps the filament from melting at the high temperatures needed to produce light.

section 2 review

Flow of Charge

- Electric current is the flow of electric charges.
- Electric charges will flow continuously only through a closed conducting loop, called a circuit.
- The voltage in a circuit is a measure of the electrical potential energy of the electrons in the circuit.
- A battery supplies energy to an electric circuit by increasing the electric potential energy of electrons in the circuit.

Resistance

- Electric resistance is the measure of how difficult it is for electrons to flow through a material.
- Electric resistance is due to collisions between flowing electrons and the atoms in a material.
- Electric resistance in a circuit converts electrical energy into thermal energy and light.

Self Check

1. **Compare and contrast** an electric discharge with an electric current.
2. **Describe** how a battery causes electrons to move in a circuit.
3. **Describe** how the electric resistance of a wire changes as the wire becomes longer. How does the resistance change as the wire becomes thicker?
4. **Explain** why the electric wires in houses are usually made of copper.
5. **Think Critically** In an electric circuit, where do the electrons come from that flow in the circuit?

Applying Skills

6. **Infer** Find the voltage of various batteries such as a watch battery, a camera battery, a flashlight battery, and an automobile battery. Infer whether the voltage produced by a battery is related to its size.

section 3
Electric Circuits

as you read

What You'll Learn
- **Explain** how voltage, current, and resistance are related in an electric circuit.
- **Investigate** the difference between series and parallel circuits.
- **Determine** the electric power used in a circuit.
- **Describe** how to avoid dangerous electric shock.

Why It's Important
Electric circuits control the flow of electric current in all electrical devices.

Review Vocabulary
power: the rate at which energy is transferred; power equals the amount of energy transferred divided by the time over which the transfer occurs

New Vocabulary
- Ohm's law
- series circuit
- parallel circuit
- electric power

Controlling the Current

When you connect a conductor, such as a wire or a lightbulb, between the positive and negative terminals of a battery, electrons flow in the circuit. The amount of current is determined by the voltage supplied by the battery and the resistance of the conductor. To help understand this relationship, imagine a bucket with a hose at the bottom, as shown in **Figure 15.** If the bucket is raised, water will flow out of the hose faster than before. Increasing the height will increase the current.

Voltage and Resistance Think back to the pump and waterwheel in **Figure 10.** Recall that the raised water has energy that is lost when the water falls. Increasing the height from which the water falls increases the energy of the water. Increasing the height of the water is similar to increasing the voltage of the battery. Just as the water current increases when the height of the water increases, the electric current in a circuit increases as voltage increases.

If the diameter of the tube in **Figure 15** is decreased, resistance is greater and the flow of the water decreases. In the same way, as the resistance in an electric circuit increases, the current in the circuit decreases.

Figure 15 Raising the bucket higher increases the potential energy of the water in the bucket. This causes the water to flow out of the hose faster.

Ohm's Law A nineteenth-century German physicist, Georg Simon Ohm, carried out experiments that measured how changing the voltage in a circuit affected the current. He found a simple relationship among voltage, current, and resistance in a circuit that is now known as **Ohm's law.** In equation form, Ohm's law often is written as follows.

Ohm's Law
Voltage (in volts) = **current** (in amperes) × **resistance** (in ohms)
$V = IR$

According to Ohm's law, when the voltage in a circuit increases the current increases, just as water flows faster from a bucket that is raised higher. However, if the voltage in the circuit doesn't change, then the current in the circuit decreases when the resistance is increased.

Applying Math — Solving a Simple Equation

VOLTAGE FROM A WALL OUTLET A lightbulb is plugged into a wall outlet. If the lightbulb has a resistance of 220 Ω and the current in the lightbulb is 0.5 A, what is the voltage provided by the outlet?

Solution

1 *This is what you know:*
- current: $I = 0.5$ A
- resistance: $R = 220$ Ω

2 *This is what you need to find:* voltage: V

3 *This is the procedure you need to use:* Substitute the known values for current and resistance into Ohm's law to calculate the voltage:
$V = IR = (0.5 \text{ A})(220 \text{ Ω}) = 110$ V

4 *Check your answer:* Divide your answer by the resistance 220 Ω. The result should be the given current 0.5 A.

Practice Problems

1. An electric iron plugged into a wall socket has a resistance of 24 Ω. If the current in the iron is 5.0 A, what is the voltage provided by the wall socket?
2. What is the current in a flashlight bulb with a resistance of 30 Ω if the voltage provided by the flashlight batteries is 3.0 V?
3. What is the resistance of a lightbulb connected to a 110-V wall outlet if the current in the lightbulb is 1.0 A?

For more practice, visit blue.msscience.com/math_practice

Mini LAB

Identifying Simple Circuits

Procedure

1. The filament in a lightbulb is a piece of wire. For the bulb to light, an electric current must flow through the filament in a complete circuit. Examine the base of a **flashlight bulb** carefully. Where are the ends of the filament connected to the base?
2. Connect one piece of **wire,** a **battery,** and a flashlight bulb to make the bulb light. (There are four possible ways to do this.)

Analysis
Draw and label a diagram showing the path that is followed by the electrons in your circuit. Explain your diagram.

Series and Parallel Circuits

Circuits control the movement of electric current by providing paths for electrons to follow. For current to flow, the circuit must provide an unbroken path for current to follow. Have you ever been putting up holiday lights and had a string that would not light because a single bulb was missing or had burned out and you couldn't figure out which one it was? Maybe you've noticed that some strings of lights don't go out no matter how many bulbs burn out or are removed. These two strings of holiday lights are examples of the two kinds of basic circuits—series and parallel.

Wired in a Line A **series circuit** is a circuit that has only one path for the electric current to follow, as shown in **Figure 16.** If this path is broken, then the current no longer will flow and all the devices in the circuit stop working. If the entire string of lights went out when only one bulb burned out, then the lights in the string were wired as a series circuit. When the bulb burned out, the filament in the bulb broke and the current path through the entire string was broken.

Reading Check *How many different paths can electric current follow in a series circuit?*

In a series circuit, electrical devices are connected along the same current path. As a result, the current is the same through every device. However, each new device that is added to the circuit decreases the current throughout the circuit. This is because each device has electrical resistance, and in a series circuit, the total resistance to the flow of electrons increases as each additional device is added to the circuit. By Ohm's law, if the voltage doesn't change, the current decreases as the resistance increases.

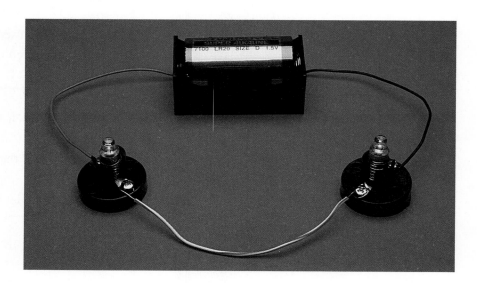

Figure 16 This circuit is an example of a series circuit. A series circuit has only one path for electric current to follow.
Predict *what will happen to the current in this circuit if any of the connecting wires are removed.*

Branched Wiring What if you wanted to watch TV and had to turn on all the lights, a hair dryer, and every other electrical appliance in the house to do so? That's what it would be like if all the electrical appliances in your house were connected in a series circuit.

Instead, houses, schools, and other buildings are wired using parallel circuits. A **parallel circuit** is a circuit that has more than one path for the electric current to follow, as shown in **Figure 17**. The current branches so that electrons flow through each of the paths. If one path is broken, electrons continue to flow through the other paths. Adding or removing additional devices in one branch does not break the current path in the other branches, so the devices on those branches continue to work normally.

In a parallel circuit, the resistance in each branch can be different, depending on the devices in the branch. The lower the resistance is in a branch, the more current flows in the branch. So the current in each branch of a parallel circuit can be different.

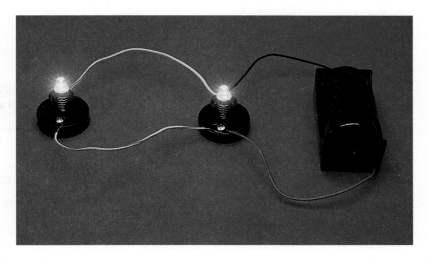

Figure 17 This circuit is an example of a parallel circuit. A parallel circuit has more than one path for electric current to follow. **Predict** what will happen to the current in the circuit if either of the wires connecting the two lightbulbs is removed.

Protecting Electric Circuits

In a parallel circuit, the current that flows out of the battery or electric outlet increases as more devices are added to the circuit. As the current through the circuit increases, the wires heat up.

To keep the wire from becoming hot enough to cause a fire, the circuits in houses and other buildings have fuses or circuit breakers like those shown in **Figure 18** that limit the amount of current in the wiring. When the current becomes larger than 15 A or 20 A, a piece of metal in the fuse melts or a switch in the circuit breaker opens, stopping the current. The cause of the overload can then be removed, and the circuit can be used again by replacing the fuse or resetting the circuit breaker.

Fuse
In some buildings, each circuit is connected to a fuse. The fuses are usually located in a fuse box.

Figure 18 You might have fuses in your home that prevent electric wires from overheating.

Wire
A fuse contains a piece of wire that melts and breaks when the current flowing through the fuse becomes too large.

Table 1 Power Used by Common Appliances	
Appliance	Power (W)
Computer	350
Color TV	200
Stereo	250
Refrigerator	450
Microwave	700–1,500
Hair dryer	1,000

Electric Power

When you use an appliance such as a toaster or a hair dryer, electrical energy is converted into other forms of energy. The rate at which electrical energy is converted into other forms of energy is **electric power**. In an electric appliance or in any electric circuit, the electric power that is used can be calculated from the electric power equation.

Electric Power Equation
Power (in watts) = **current** (in amperes) × **voltage** (in volts)
$$P = IV$$

The electric power is equal to the voltage provided to the appliance times the current that flows into the appliance. In the electric power equation, the SI unit of power is the watt. **Table 1** lists the electric power used by some common appliances.

Applying Math — Solving a Simple Equation

ELECTRIC POWER USED BY A LIGHTBULB A lightbulb is plugged into a 110-V wall outlet. How much electric power does the lightbulb use if the current in the bulb is 0.55 A?

Solution

1 *This is what you know:*
- voltage: $V = 110$ V
- current: $I = 0.55$ A

2 *This is what you need to find:* power: P

3 *This is the procedure you need to use:* To calculate electric power, substitute the known values for voltage and current into the equation for electric power:
$P = IV = (0.55\ \text{A})(110\ \text{V}) = 60$ W

4 *Check your answer:* Divide your answer by the current 0.55 A. The result should be the given voltage 110 V.

Practice Problems

1. The batteries in a portable CD player provide 6.0 V. If the current in the CD player is 0.5 A, how much power does the CD player use?

2. What is the current in a toaster if the toaster uses 1,100 W of power when plugged into a 110-V wall outlet?

3. An electric clothes dryer uses 4,400 W of electric power. If the current in the dryer is 20.0 A, what is the voltage?

For more practice, visit blue.msscience.com/math_practice

Cost of Electric Energy Power is the rate at which energy is used, or the amount of energy that is used per second. When you use a hair dryer, the amount of electrical energy that is used depends on the power of the hair dryer and the amount of time you use it. If you used it for 5 min yesterday and 10 min today, you used twice as much energy today as yesterday.

Using electrical energy costs money. Electric companies generate electrical energy and sell it in units of kilowatt-hours to homes, schools, and businesses. One kilowatt-hour, kWh, is an amount of electrical energy equal to using 1 kW of power continuously for 1 h. This would be the amount of energy needed to light ten 100-W lightbulbs for 1 h, or one 100-W lightbulb for 10 h.

 *What does kWh stand for and what does it measure?*

An electric company usually charges its customers for the number of kilowatt-hours they use every month. The number of kilowatt-hours used in a building such as a house or a school is measured by an electric meter, which usually is attached to the outside of the building, as shown in **Figure 19.**

Figure 19 Electric meters measure the amount of electrical energy used in kilowatt-hours. **Identify** *the electric meter attached to your house.*

Electrical Safety

 Have you ever had a mild electric shock? You probably felt only a mild tingling sensation, but electricity can have much more dangerous effects. In 1997, electric shocks killed an estimated 490 people in the United States. **Table 2** lists a few safety tips to help prevent electrical accidents.

Table 2 Preventing Electric Shock
Never use appliances with frayed or damaged electric cords.
Unplug appliances before working on them, such as when prying toast out of a jammed toaster.
Avoid all water when using plugged-in appliances.
Never touch power lines with anything, including kite string and ladders.
Always respect warning signs and labels.

Science Online

Topic: Cost of Electrical Energy
Visit blue.msscience.com for Web links to information about the cost of electrical energy in various parts of the world.

Activity Make a bar graph showing the cost of electrical energy for several countries on different continents.

SECTION 3 Electric Circuits

Current's Effects The scale below shows how the effect of electric current on the human body depends on the amount of current that flows into the body.

0.0005 A	Tingle
0.001 A	Pain threshold
0.01 A	Inability to let go
0.025 A	
0.05 A	Difficulty breathing
0.10 A	
0.25 A	
0.50 A	Heart failure
1.00 A	

Electric Shock You experience an electric shock when an electric current enters your body. In some ways your body is like a piece of insulated wire. The fluids inside your body are good conductors of current. The electrical resistance of dry skin is much higher. Skin insulates the body like the plastic insulation around a copper wire. Your skin helps keep electric current from entering your body.

A current can enter your body when you accidentally become part of an electric circuit. Whether you receive a deadly shock depends on the amount of current that flows into your body. The current that flows through the wires connected to a 60-W lightbulb is about 0.5 A. This amount of current entering your body could be deadly. Even a current as small as 0.001 A can be painful.

Lightning Safety On average, more people are killed every year by lightning in the United States than by hurricanes or tornadoes. Most lightning deaths and injuries occur outdoors. If you are outside and can see lightning or hear thunder, take shelter indoors immediately. If you cannot go indoors, you should take these precautions: avoid high places and open fields; stay away from tall objects such as trees, flag poles, or light towers; and avoid objects that conduct current such as bodies of water, metal fences, picnic shelters, and metal bleachers.

section 3 review

Summary

Electric Circuits

- In an electric circuit, voltage, resistance, and current are related. According to Ohm's law, this relationship can be written as $V = IR$.
- A series circuit has only one path for electric current to follow.
- A parallel circuit has more than one path for current to follow.

Electric Power and Energy

- The electric power used by an appliance is the rate at which the appliance converts electrical energy to other forms of energy.
- The electric power used by an appliance can be calculated using the equation $P = IV$.
- The electrical energy used by an appliance depends on the power of the appliance and the length of time it is used. Electrical energy usually is measured in kWh.

Self Check

1. **Compare** the current in two lightbulbs wired in a series circuit.
2. **Describe** how the current in a circuit changes if the resistance increases and the voltage remains constant.
3. **Explain** why buildings are wired using parallel circuits rather than series circuits.
4. **Identify** what determines the damage caused to the human body by an electric shock.
5. **Think Critically** What determines whether a 100-W lightbulb costs more to use than a 1,200-W hair dryer costs to use?

Applying Math

6. **Calculate Energy** A typical household uses 1,000 kWh of electrical energy every month. If a power company supplies electrical energy to 1,000 households, how much electrical energy must it supply every year?

Current in a Parallel Circuit

The brightness of a lightbulb increases as the current in the bulb increases. In this lab you'll use the brightness of a lightbulb to compare the amount of current that flows in parallel circuits.

Real-World Question

How does connecting devices in parallel affect the electric current in a circuit?

Goal
- **Observe** how the current in a parallel circuit changes as more devices are added.

Materials
1.5-V lightbulbs (4)
1.5-V batteries (2)
10-cm-long pieces of insulated wire (8)
battery holders (2)
minibulb sockets (4)

Safety Precautions

Procedure

1. Connect one lightbulb to the battery in a complete circuit. After you've made the bulb light, disconnect the bulb from the battery to keep the battery from running down. This circuit will be the brightness tester.

2. Make a parallel circuit by connecting two bulbs as shown in the diagram. Reconnect the bulb in the brightness tester and compare its brightness with the brightness of the two bulbs in the parallel circuit. Record your observations.

3. Add another bulb to the parallel circuit as shown in the figure. How does the brightness of the bulbs change?

4. Disconnect one bulb in the parallel circuit. Record your observations.

Conclude and Apply

1. **Describe** how the brightness of each bulb depends on the number of bulbs in the circuit.
2. **Infer** how the current in each bulb depends on the number of bulbs in the circuit.

Communicating Your Data

Compare your conclusions with those of other students in your class. **For more help, refer to the** Science Skill Handbook.

A Model for Voltage and Current

Goal
- **Model** the flow of current in a simple circuit.

Materials
plastic funnel
rubber or plastic tubing of different diameters (1 m each)
meterstick
ring stand with ring
stopwatch
*clock displaying seconds
hose clamp
*binder clip
500-mL beakers (2)
*Alternate materials

Safety Precautions

Real-World Question
The flow of electrons in an electric circuit is something like the flow of water in a tube connected to a water tank. By raising or lowering the height of the tank, you can increase or decrease the potential energy of the water. How does the flow of water in a tube depend on the diameter of the tube and the height the water falls?

Procedure

1. **Design** a data table in which to record your data. It should be similar to the table below.
2. Connect the tubing to the bottom of the funnel and place the funnel in the ring of the ring stand.
3. **Measure** the inside diameter of the rubber tubing. Record your data.
4. Place a 500-mL beaker at the bottom of the ring stand and lower the ring so the open end of the tubing is in the beaker.
5. Use the meterstick to measure the height from the top of the funnel to the bottom of the ring stand.

Flow Rate Data

Trial	Height (cm)	Diameter (mm)	Time (s)	Flow Rate (mL/s)
1				
2				
3		Do not write in this book.		
4				

656 CHAPTER 22 Electricity

6. Working with a classmate, pour water into the funnel fast enough to keep the funnel full but not overflowing. Measure and record the time needed for 100 mL of water to flow into the beaker. Use the hose clamp to start and stop the flow of water.

7. Connect tubing with a different diameter to the funnel and repeat steps 2 through 6.

8. Reconnect the original piece of tubing and repeat steps 4 through 6 for several lower positions of the funnel, lowering the height by 10 cm each time.

Analyze Your Data

1. **Calculate** the rate of flow for each trial by dividing 100 mL by the time measured for 100 mL of water to flow into the beaker.

2. **Make a graph** that shows how the rate of flow depends on the funnel height.

Conclude and Apply

1. **Infer** from your graph how the rate of flow depends on the height of the funnel.
2. **Explain** how the rate of flow depends on the diameter of the tubing. Is this what you expected to happen?
3. **Identify** which of the variables you changed in your trials that corresponds to the voltage in a circuit.
4. **Identify** which of the variables you changed in your trials that corresponds to the resistance in a circuit.
5. **Infer** from your results how the current in a circuit would depend on the voltage.
6. **Infer** from your results how the current in a circuit would depend on the resistance in the circuit.

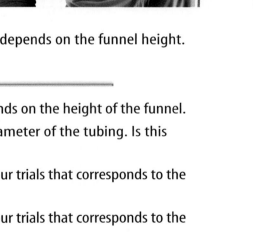

Communicating Your Data

Share your graph with other students in your class. Did other students draw the same conclusions as you? **For more help, refer to the** Science Skill Handbook.

LAB 657

TIME SCIENCE AND Society
SCIENCE ISSUES THAT AFFECT YOU!

Fire in the Forest

Plant life returns after a forest fire in Yellowstone National Park.

Fires started by lightning may not be all bad

When lightning strikes a tree, the intense heat of the lightning bolt can set the tree on fire. The fire then can spread to other trees in the forest. Though lightning is responsible for only about ten percent of forest fires, it causes about one-half of all fire damage. For example, in 2000, fires set by lightning raged in 12 states at the same time, burning a total area roughly the size of the state of Massachusetts.

Fires sparked by lightning often start in remote, difficult-to-reach areas, such as national parks and range lands. Burning undetected for days, these fires can spread out of control. In addition to threatening lives, the fires can destroy millions of dollars worth of homes and property. Smoke from forest fires also can have harmful effects on people, especially for those with preexisting conditions, such as asthma.

People aren't the only victims of forest fires. The fires kill animals as well. Those who survive the blaze often perish because their habitats have been destroyed. Monster blazes spew carbon dioxide and other gases into the atmosphere. Some of these gases may contribute to the greenhouse effect that warms the planet. Moreover, fires cause soil erosion and loss of water reserves.

But fires caused by lightning also have some positive effects. In old, dense forests, trees often become diseased and insect-ridden. By removing these unhealthy trees, fires allow healthy trees greater access to water and nutrients. Fires also clear away a forest's dead trees, underbrush, and needles, providing space for new vegetation. Nutrients are returned to the ground as dead organic matter decays, but it can take a century for dead logs to rot completely. A fire enables nutrients to be returned to soil much more quickly. Also, the removal of these highly combustible materials prevents more widespread fires from occurring.

Research Find out more about the job of putting out forest fires. What training is needed? What gear do firefighters wear? Why would people risk their lives to save a forest? Use the media center at your school to learn more about forest firefighters and their careers.

Science online
For more information, visit blue.msscience.com/time

chapter 22 Study Guide

Reviewing Main Ideas

Section 1 Electric Charge

1. The two types of electric charge are positive and negative. Like charges repel and unlike charges attract.

2. An object becomes negatively charged if it gains electrons and positively charged if it loses electrons.

3. Electrically charged objects have an electric field surrounding them and exert electric forces on one another.

4. Electrons can move easily in conductors, but not so easily in insulators.

Section 2 Electric Current

1. Electric current is the flow of charges—usually either electrons or ions.

2. The energy carried by the current in a circuit increases as the voltage in the circuit increases.

3. In a battery, chemical reactions provide the energy that causes electrons to flow in a circuit.

4. As electrons flow in a circuit, some of their electrical energy is lost due to resistance in the circuit.

Section 3 Electric Circuits

1. In an electric circuit, the voltage, current, and resistance are related by Ohm's law.

2. The two basic kinds of electric circuits are parallel circuits and series circuits.

3. The rate at which electric devices use electrical energy is the electric power used by the device.

Visualizing Main Ideas

Copy and complete the following concept map about electricity.

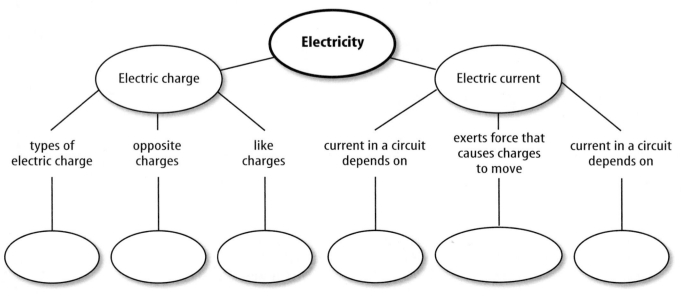

Chapter 22 Review

Using Vocabulary

circuit p. 643
conductor p. 640
electric current p. 643
electric discharge p. 641
electric field p. 639
electric force p. 639
electric power p. 652
insulator p. 640
ion p. 636
Ohm's law p. 649
parallel circuit p. 651
resistance p. 646
series circuit p. 650
static charge p. 637
voltage p. 644

Answer the following questions using complete sentences.

1. What is the term for the flow of electric charge?
2. What is the relationship among voltage, current, and resistance in a circuit?
3. In what type of material do electrons move easily?
4. What is the name for the unbroken path that current follows?
5. What is an excess of electric charge on an object?
6. What is an atom that has gained or lost electrons called?
7. Which type of circuit has more than one path for electrons to follow?
8. What is the rapid movement of excess charge known as?

Checking Concepts

Choose the word or phrase that best answers the question.

9. Which of the following describes an object that is positively charged?
   A) has more neutrons than protons
   B) has more protons than electrons
   C) has more electrons than protons
   D) has more electrons than neutrons

10. Which of the following is true about the electric field around an electric charge?
    A) It exerts a force on other charges.
    B) It increases the resistance of the charge.
    C) It increases farther from the charge.
    D) It produces protons.

11. What is the force between two electrons?
    A) frictional   C) attractive
    B) neutral   D) repulsive

12. What property of a wire increases when it is made thinner?
    A) resistance
    B) voltage
    C) current
    D) static charge

13. What property does Earth have that enables grounding to drain static charges?
    A) It has a high static charge.
    B) It has a high resistance.
    C) It is a large conductor.
    D) It is like a battery.

Use the graph below to answer question 14.

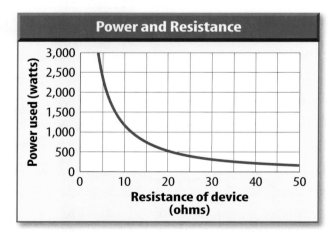

14. How does the resistance change if the power decreases from 2,500 W to 500 W?
    A) It increases four times.
    B) It decreases four times.
    C) It doubles.
    D) It doesn't change.

Chapter 22 Review

Thinking Critically

15. **Determine** A metal wire is made thinner. How would you change the length of the wire to keep the electric resistance of the wire from changing?

The tables below show how the voltage and current vary in a portable radio and a portable CD player. Use these tables to answer questions 16 through 19.

Portable Radio	
Voltage (V)	Current (A)
2.0	1.0
4.0	2.0
6.0	3.0

Portable CD Player	
Voltage (V)	Current (A)
2.0	0.5
4.0	1.0
6.0	1.5

16. **Make a graph** with current plotted on the horizontal axis and voltage plotted on the vertical axis. Plot the data in the above tables for both devices on your graph.

17. **Identify** from your graph which line is more horizontal—the line for the portable radio or the line for the portable CD player.

18. **Calculate** the electric resistance using Ohm's law for each value of the current and voltage in the tables above. What is the resistance of each device?

19. **Determine** For which device is the line plotted on your graph more horizontal—the device with higher or lower resistance?

20. **Explain** why a balloon that has a static electric charge will stick to a wall.

21. **Describe** how you can tell whether the type of charge on two charged objects is the same or different.

22. **Infer** Measurements show that Earth is surrounded by a weak electric field. If the direction of this field points toward Earth, what is the type of charge on Earth's surface?

Performance Activities

23. **Design a board game** about a series or parallel circuit. The rules of the game could be based on opening or closing the circuit, adding more devices to the circuit, blowing fuses or circuit breakers, replacing fuses, or resetting circuit breakers.

Applying Math

24. **Calculate Resistance** A toaster is plugged into a 110-V outlet. What is the resistance of the toaster if the current in the toaster is 10 A?

25. **Calculate Current** A hair dryer uses 1,000 W when it is plugged into a 110-V outlet. What is the current in the hair dryer?

26. **Calculate Voltage** A lightbulb with a resistance of 30 Ω is connected to a battery. If the current in the lightbulb is 0.10 A, what is the voltage of the battery?

Use the table below to answer question 27.

Average Standby Power Used	
Appliance	Power (W)
Computer	7.0
VCR	6.0
TV	5.0

27. **Calculate Cost** The table above shows the power used by several appliances when they are turned off. Calculate the cost of the electrical energy used by each appliance in a month if the cost of electrical energy is $0.08/kWh, and each appliance is in standby mode for 600 h each month.

Chapter 22 Standardized Test Practice

Part 1 Multiple Choice

Record your answers on the answer sheet provided by your teacher or on a sheet of paper.

1. What happens when two materials are charged by rubbing against each other?
   A. both lose electrons
   B. both gain electrons
   C. one loses electrons
   D. no movement of electrons

Use the table below to answer questions 2–4.

Power Ratings of Some Appliances	
Appliance	Power (W)
Computer	350
Color TV	200
Stereo	250
Toaster	1,100
Microwave	900
Hair dryer	1,000

2. Which appliance will use the most energy if it is run for 15 minutes?
   A. microwave   C. stereo
   B. computer   D. color TV

3. What is the current in the hair dryer if it is plugged into a 110-V outlet?
   A. 110 A   C. 9 A
   B. 130,000 A   D. 1,100 A

4. Suppose using 1,000 W for 1 h costs $0.10. How much would it cost to run the color TV for 8 hours?
   A. $1.00   C. $1.60
   B. $10.00   D. $0.16

5. How does the current in a circuit change if the voltage is doubled and the resistance remains unchanged?
   A. no change   C. doubles
   B. triples   D. reduced by half

6. Which statement does NOT describe how electric changes affect each other?
   A. positive and negative charges attract
   B. positive and negative charges repel
   C. two positive charges repel
   D. two negative charges repel

Use the illustration below to answer questions 7 and 8.

7. What is the device on the chimney called?
   A. circuit breaker   C. fuse
   B. lightning rod   D. circuit

8. What is the device designed to do?
   A. stop electricity from flowing
   B. repel an electric charge
   C. turn the chimney into an insulator
   D. to provide grounding for the house

9. Which of the following is a material through which charge cannot move easily?
   A. conductor   C. wire
   B. circuit   D. insulator

10. What property of a wire increases when it is made longer?
    A. charge   C. voltage
    B. resistance   D. current

11. Which of the following materials are good insulators?
    A. copper and gold
    B. wood and glass
    C. gold and aluminum
    D. plastic and copper

662 STANDARDIZED TEST PRACTICE

Standardized Test Practice

Part 2 Short Response/Grid In

Record your answers on the answer sheet provided by your teacher or on a sheet of paper.

Use the illustration below to answer questions 12 and 13.

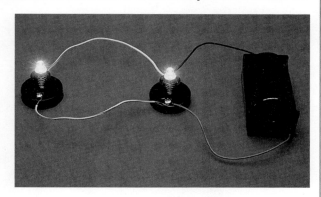

12. In this circuit, if one lightbulb is unscrewed, what happens to the current in the other lightbulb? Explain.

13. In this circuit, is the resistance and the current in each branch of the circuit always the same? Explain.

14. A 1,100-W toaster may be used for five minutes each day. A 400-W refrigerator runs all the time. Which appliance uses more electrical energy? Explain.

15. How much current does a 75-W bulb require in a 100-V circuit?

16. A series circuit containing mini-lightbulbs is opened and some of the lightbulbs are removed. What happens when the circuit is closed?

17. Suppose you plug an electric heater into the wall outlet. As soon as you turn it on, all the lights in the room go out. Explain what must have happened.

18. Explain why copper wires used in appliances or electric circuits are covered with plastic or rubber.

Part 3 Open Ended

Record your answers on a sheet of paper.

19. Why is it dangerous to use a fuse that is rated 30 A in a circuit calling for a 15-A fuse?

Use the illustration below to answer question 20.

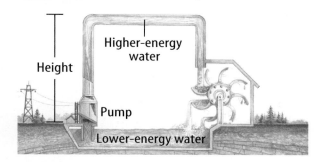

20. Compare the water pump in the water circuit above with the battery in an electric circuit.

21. Explain what causes the lightning that is associated with a thunderstorm.

22. Explain why two charged balloons push each other apart even if they are not touching.

23. Explain what can happen when you rub your feet on a carpet and then touch a metal doorknob.

24. Why does the fact that tungsten wire has a high melting point make it useful in the filaments of lightbulbs?

Test-Taking Tip

Recall Experiences Recall any hands-on experience as you read the question. Base your answer on the information given on the test.

Question 23 Recall from your personal experience the jolt you feel when you touch a doorknob after walking across a carpet.

chapter 23

Magnetism

The BIG Idea
Magnets exert forces on other magnets and on moving charges.

SECTION 1
What is magnetism?
Main Idea Moving electric charges produce magnetic fields.

SECTION 2
Electricity and Magnetism
Main Idea Magnetic fields can produce electric currents.

Magnetic Suspension
This experimental train can travel at speeds as high as 500 km/h—without even touching the track! It uses magnetic levitation, or maglev, to reach these high speeds. Magnetic forces lift the train above the track, and propel it forward at high speeds.

Science Journal List three ways that you have seen magnets used.

Start-Up Activities

Magnetic Forces

A maglev is moved along at high speeds by magnetic forces. How can a magnet get something to move? The following lab will demonstrate how a magnet is able to exert forces.

1. Place two bar magnets on opposite ends of a sheet of paper.
2. Slowly slide one magnet toward the other until it moves. Measure the distance between the magnets.
3. Turn one magnet around 180°. Repeat Step 2. Then turn the other magnet and repeat Step 2 again.
4. Repeat Step 2 with one magnet perpendicular to the other, in a T shape.
5. **Think Critically** In your Science Journal, record your results. In each case, how close did the magnets have to be to affect each other? Did the magnets move together or apart? How did the forces exerted by the magnets change as the magnets were moved closer together? Explain.

 **Preview this chapter's content and activities at blue.msscience.com**

 **Magnetic Forces and Fields** Make the following Foldable to help you see how magnetic forces and magnetic fields are similar and different.

STEP 1 Draw a mark at the midpoint of a vertical sheet of paper along the side edge.

STEP 2 Turn the paper horizontally and fold the outside edges in to touch at the midpoint mark.

STEP 3 Label the flaps *Magnetic Force* and *Magnetic Field*.

Compare and Contrast As you read the chapter, write information about each topic on the inside of the appropriate flap. After you read the chapter, compare and contrast the terms *magnetic force* and *magnetic field*. Write your observations under the flaps.

Get Ready to Read

Identify Cause and Effect

1 Learn It! A *cause* is the reason something happens. The result of what happens is called an *effect*. Learning to identify causes and effects helps you understand why things happen. By using graphic organizers, you can sort and analyze causes and effects as you read.

2 Practice It! Read the following paragraph. Then use the graphic organizer below to show what happened when the Sun ejects charged particles toward Earth.

> Sometimes the Sun ejects a large number of charged particles all at once. Most of these charged particles are deflected by Earth's magnetosphere. However, some of the ejected particles from the Sun produce other charged particles in Earth's outer atmosphere. These charged particles spiral along Earth's magnetic field lines toward Earth's magnetic poles. There they collide with atoms in the atmosphere. These collisions cause the atoms to emit light.
>
> —from page 677

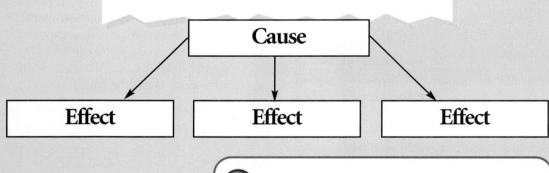

3 Apply It! As you read the chapter, be aware of causes and effects of charged particles moving in a magnetic field. Find three causes and their effects.

666 A CHAPTER 23 Magnetism

Target Your Reading

Reading Tip

Graphic organizers such as the Cause-Effect organizer help you organize what you are reading so you can remember it later.

Use this to focus on the main ideas as you read the chapter.

① Before you read the chapter, respond to the statements below on your worksheet or on a numbered sheet of paper.
- Write an **A** if you **agree** with the statement.
- Write a **D** if you **disagree** with the statement.

② After you read the chapter, look back to this page to see if you've changed your mind about any of the statements.
- If any of your answers changed, explain why.
- Change any false statements into true statements.
- Use your revised statements as a study guide.

Science Online
Print out a worksheet of this page at blue.msscience.com

Before You Read A or D		Statement	After You Read A or D
	1	Opposite poles of magnets attract each other.	
	2	An electric motor converts electrical energy into kinetic energy.	
	3	Earth's magnetic field has not changed since the Earth formed.	
	4	Magnetic fields get stronger as you move away from the magnet's poles.	
	5	A wire carrying electric current is surrounded by a magnetic field.	
	6	An electromagnet is wire wrapped around a magnet.	
	7	Magnetic fields have no effect on moving electric charges.	
	8	Earth's magnetic field affects only Earth's surface.	
	9	Magnetic fields are produced by moving masses.	
	10	Transformers convert kinetic energy to electrical energy.	

666 B

Section 1
What is magnetism?

as you read

What You'll Learn
- **Describe** the behavior of magnets.
- **Relate** the behavior of magnets to magnetic fields.
- **Explain** why some materials are magnetic.

Why It's Important
Magnetism is one of the basic forces of nature.

Review Vocabulary
compass: a device which uses a magnetic needle that can turn freely to determine direction

New Vocabulary
- magnetic field
- magnetic domain
- magnetosphere

Early Uses

Do you use magnets to attach papers to a metal surface such as a refrigerator? Have you ever wondered why magnets and some metals attract? Thousands of years ago, people noticed that a mineral called magnetite attracted other pieces of magnetite and bits of iron. They discovered that when they rubbed small pieces of iron with magnetite, the iron began to act like magnetite. When these pieces were free to turn, one end pointed north. These might have been the first compasses. The compass was an important development for navigation and exploration, especially at sea. Before compasses, sailors had to depend on the Sun or the stars to know in which direction they were going.

Magnets

A piece of magnetite is a magnet. Magnets attract objects made of iron or steel, such as nails and paper clips. Magnets also can attract or repel other magnets. Every magnet has two ends, or poles. One end is called the north pole and the other is the south pole. As shown in **Figure 1,** a north magnetic pole always repels other north poles and always attracts south poles. Likewise, a south pole always repels other south poles and attracts north poles.

Two north poles repel

Two south poles repel

Opposite poles attract

Figure 1 Two north poles or two south poles repel each other. North and south magnetic poles are attracted to each other.

The Magnetic Field You have to handle a pair of magnets for only a short time before you can feel that magnets attract or repel without touching each other. How can a magnet cause an object to move without touching it? Recall that a force is a push or a pull that can cause an object to move. Just like gravitational and electric forces, a magnetic force can be exerted even when objects are not touching. And like these forces, the magnetic force becomes weaker as the magnets get farther apart.

This magnetic force is exerted through a **magnetic field**. Magnetic fields surround all magnets. If you sprinkle iron filings near a magnet, the iron filings will outline the magnetic field around the magnet. Take a look at **Figure 2**. The iron filings form a pattern of curved lines that start on one pole and end on the other. These curved lines are called magnetic field lines. Magnetic field lines help show the direction of the magnetic field.

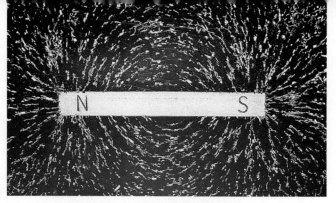

Iron filings show the magnetic field lines around a bar magnet.

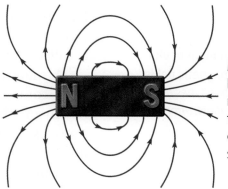

Magnetic field lines start at the north pole of the magnet and end on the south pole.

Reading Check *What is the evidence that a magnetic field exists?*

Magnetic field lines begin at a magnet's north pole and end on the south pole, as shown in **Figure 2**. The field lines are close together where the field is strong and get farther apart as the field gets weaker. As you can see in the figures, the magnetic field is strongest close to the magnetic poles and grows weaker farther from the poles.

Field lines that curve toward each other show attraction. Field lines that curve away from each other show repulsion. **Figure 3** illustrates the magnetic field lines between a north and a south pole and the field lines between two north poles.

Figure 2 A magnetic field surrounds a magnet. Where the magnetic field lines are close together, the field is strong.
Determine *for this magnet where the strongest field is.*

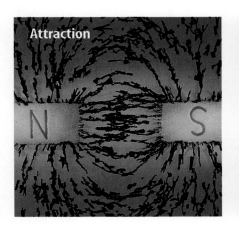

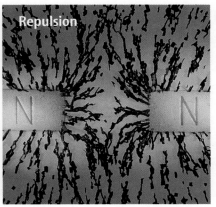

Figure 3 Magnetic field lines show attraction and repulsion.
Explain *what the field between two south poles would look like.*

SECTION 1 What is magnetism? **667**

Making Magnetic Fields Only certain materials, such as iron, can be made into magnets that are surrounded by a magnetic field. How are magnetic fields made? A moving electric charge, such as a moving electron, creates a magnetic field.

Inside every magnet are moving charges. All atoms contain negatively charged particles called electrons. Not only do these electrons swarm around the nucleus of an atom, they also spin, as shown in **Figure 4**. Because of its movement, each electron produces a magnetic field. The atoms that make up magnets have their electrons arranged so that each atom is like a small magnet. In a material such as iron, a large number of atoms will have their magnetic fields pointing in the same direction. This group of atoms, with their fields pointing in the same direction, is called a **magnetic domain.**

A material that can become magnetized, such as iron or steel, contains many magnetic domains. When the material is not magnetized, these domains are oriented in different directions, as shown in **Figure 5A.** The magnetic fields created by the domains cancel, so the material does not act like a magnet.

A magnet contains a large number of magnetic domains that are lined up and pointing in the same direction. Suppose a strong magnet is held close to a material such as iron or steel. The magnet causes the magnetic field in many magnetic domains to line up with the magnet's field, as shown in **Figure 5B.** As you can see in **Figure 5C** this process magnetizes paper clips.

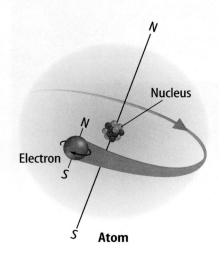

Figure 4 Movement of electrons produces magnetic fields. **Describe** what two types of motion are shown in the illustration.

Figure 5 Some materials can become temporary magnets.

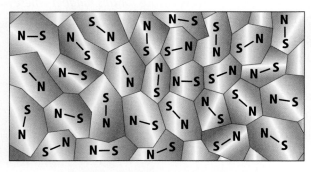

A Microscopic sections of iron and steel act as tiny magnets. Normally, these domains are oriented randomly and their magnetic fields cancel each other.

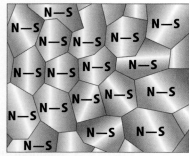

B When a strong magnet is brought near the material, the domains line up, and their magnetic fields add together.

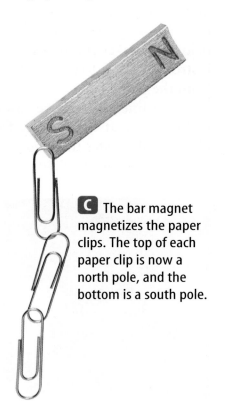

C The bar magnet magnetizes the paper clips. The top of each paper clip is now a north pole, and the bottom is a south pole.

Earth's Magnetic Field

Magnetism isn't limited to bar magnets. Earth has a magnetic field, as shown in **Figure 6.** The region of space affected by Earth's magnetic field is called the **magnetosphere** (mag NEE tuh sfihr). This deflects most of the charged particles from the Sun. The origin of Earth's magnetic field is thought to be deep within Earth in the outer core layer. One theory is that movement of molten iron in the outer core is responsible for generating Earth's magnetic field. The shape of Earth's magnetic field is similar to that of a huge bar magnet tilted about 11° from Earth's geographic north and south poles.

Figure 6 Earth has a magnetic field similar to the field of a bar magnet.

Applying Science

Finding the Magnetic Declination

The north pole of a compass points toward the magnetic pole, rather than true north. Imagine drawing a line between your location and the north pole, and a line between your location and the magnetic pole. The angle between these two lines is called the magnetic declination. Magnetic declination must be known if you need to know the direction to true north. However, the magnetic declination changes depending on your position.

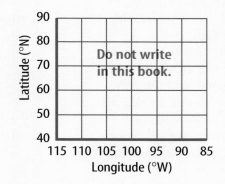

Identifying the Problem

Suppose your location is at 50° N and 110° W. The location of the north pole is at 90° N and 110° W, and the location of the magnetic pole is at about 80° N and 105° W. What is the magnetic declination angle at your location?

Solving the Problem
1. Draw and label a graph like the one shown above.
2. On the graph, plot your location, the location of the magnetic pole, and the location of the north pole.
3. Draw a line from your location to the north pole, and a line from your location to the magnetic pole.
4. Using a protractor, measure the angle between the two lines.

SECTION 1 What is magnetism? **669**

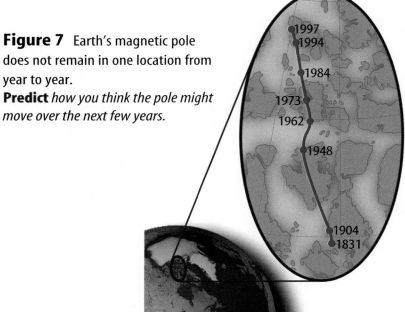

Figure 7 Earth's magnetic pole does not remain in one location from year to year.
Predict how you think the pole might move over the next few years.

Observing Magnetic Fields

Procedure
1. Place **iron filings** in a **plastic petri dish**. Cover the dish and seal it with **clear tape**.
2. Collect **several magnets**. Place the magnets on the table and hold the dish over each one. Draw a diagram of what happens to the filings in each case.
3. Arrange two or more magnets under the dish. Observe the pattern of the filings.

Analysis
1. What happens to the filings close to the poles? Far from the poles?
2. Compare the fields of the individual magnets. How can you tell which magnet is strongest? Weakest?

Nature's Magnets Honeybees, rainbow trout, and homing pigeons have something in common with sailors and hikers. They take advantage of magnetism to find their way. Instead of using compasses, these animals and others have tiny pieces of magnetite in their bodies. These pieces are so small that they may contain a single magnetic domain. Scientists have shown that some animals use these natural magnets to detect Earth's magnetic field. They appear to use Earth's magnetic field, along with other clues like the position of the Sun or stars, to help them navigate.

Earth's Changing Magnetic Field Earth's magnetic poles do not stay in one place. The magnetic pole in the north today, as shown in **Figure 7,** is in a different place from where it was 20 years ago. In fact, not only does the position of the magnetic poles move, but Earth's magnetic field sometimes reverses direction. For example, 700 thousand years ago, a compass needle that now points north would point south. During the past 20 million years, Earth's magnetic field has reversed direction more than 70 times. The magnetism of ancient rocks contains a record of these magnetic field changes. When some types of molten rock cool, magnetic domains of iron in the rock line up with Earth's magnetic field. After the rock cools, the orientation of these domains is frozen into position. Consequently, these old rocks preserve the orientation of Earth's magnetic field as it was long ago.

Figure 8 The compass needles align with the magnetic field lines around the magnet.
Explain what happens to the compass needles when the bar magnet is removed.

The Compass A compass needle is a small bar magnet with a north and south magnetic pole. In a magnetic field, a compass needle rotates until it is aligned with the magnetic field line at its location. **Figure 8** shows how the orientation of a compass needle depends on its location around a bar magnet.

Earth's magnetic field also causes a compass needle to rotate. The north pole of the compass needle points toward Earth's magnetic pole that is in the north. This magnetic pole is actually a magnetic south pole. Earth's magnetic field is like that of a bar magnet with the magnet's south pole near Earth's north pole.

Topic: Compasses
Visit blue.msscience.com for Web links to information about different types of compasses.

Activity Find out how far from true north a compass points in your location.

section 1 review

Summary

Magnets
- A magnet has a north pole and a south pole.
- Like magnetic poles repel each other; unlike poles attract each other.
- A magnet is surrounded by a magnetic field that exerts forces on other magnets.
- Some materials are magnetic because their atoms behave like magnets.

Earth's Magnetic Field
- Earth is surrounded by a magnetic field similar to the field around a bar magnet.
- Earth's magnetic poles move slowly, and sometimes change places. Earth's magnetic poles now are close to Earth's geographic poles.

Self Check

1. **Explain** why atoms behave like magnets.
2. **Explain** why magnets attract iron but do not attract paper.
3. **Describe** how the behavior of electric charges is similar to that of magnetic poles.
4. **Determine** where the field around a magnet is the strongest and where it is the weakest.
5. **Think Critically** A horseshoe magnet is a bar magnet bent into the shape of the letter U. When would two horseshoe magnets attract each other? Repel? Have little effect?

Applying Skills

6. **Communicate** Ancient sailors navigated by using the Sun, stars, and following a coastline. Explain how the development of the compass would affect the ability of sailors to navigate.

Make a Compass

A valuable tool for hikers and campers is a compass. Almost 1,000 years ago, Chinese inventors found a way to magnetize pieces of iron. They used this method to manufacture compasses. You can use the same procedure to make a compass.

Real-World Question

How do you construct a compass?

Goals
- **Observe** induced magnetism.
- **Build** a compass.

Materials
petri dish
*clear bowl
water
sewing needle
magnet
tape
marker
paper
plastic spoon
*Alternate material

Safety Precautions

Procedure

1. Reproduce the circular protractor shown. Tape it under the bottom of your dish so it can be seen but not get wet. Add water until the dish is half full.
2. Mark one end of the needle with a marker. Magnetize a needle by placing it on the magnet aligned north and south for 1 min.
3. Float the needle in the dish using a plastic spoon to lower the needle carefully onto the water. Turn the dish so the marked part of the needle is above the 0° mark. This is your compass.
4. Bring the magnet near your compass. Observe how the needle reacts. Measure the angle the needle turns.

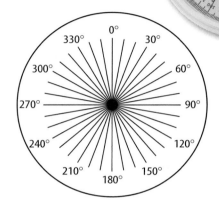

Conclude and Apply

1. **Explain** why the marked end of the needle always pointed the same way in step 3, even though you rotated the dish.
2. **Describe** the behavior of the compass when the magnet was brought close.
3. **Observe** the marked end of your needle. Does it point to the north or south pole of the bar magnet? **Infer** whether the marked end of your needle is a north or a south pole. How do you know?

Communicating Your Data

Make a half-page insert that will go into a wilderness survival guide to describe the procedure for making a compass. Share your half-page insert with your classmates. **For more help, refer to the Science Skill Handbook.**

672 CHAPTER 23 Magnetism

Section 2
Electricity and Magnetism

Current Can Make a Magnet

Magnetic fields are produced by moving electric charges. Electrons moving around the nuclei of atoms produce magnetic fields. The motion of these electrons causes some materials, such as iron, to be magnetic. You cause electric charges to move when you flip a light switch or turn on a portable CD player. When electric current flows in a wire, electric charges move in the wire. As a result, a wire that contains an electric current also is surrounded by a magnetic field. **Figure 9A** shows the magnetic field produced around a wire that carries an electric current.

Electromagnets Look at the magnetic field lines around the coils of wire in **Figure 9B**. The magnetic fields around each coil of wire add together to form a stronger magnetic field inside the coil. When the coils are wrapped around an iron core, the magnetic field of the coils magnetizes the iron. The iron then becomes a magnet, which adds to the strength of the magnetic field inside the coil. A current-carrying wire wrapped around an iron core is called an **electromagnet**, as shown in **Figure 9C**.

as you read

What **You'll Learn**
- **Explain** how electricity can produce motion.
- **Explain** how motion can produce electricity.

Why **It's Important**
Electricity and magnetism enable electric motors and generators to operate.

Review Vocabulary
electric current: the flow of electric charge

New Vocabulary
- electromagnet
- motor
- aurora
- generator
- alternating current
- direct current
- transformer

Figure 9 A current-carrying wire produces a magnetic field.

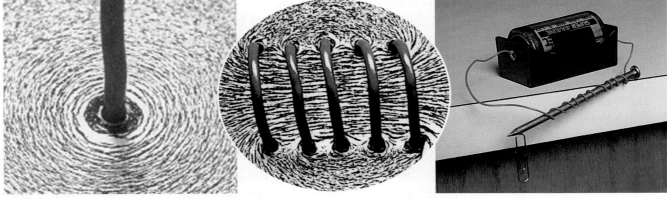

A Iron particles show the magnetic field lines around a current-carrying wire.

B When a wire is wrapped in a coil, the field inside the coil is made stronger.

C An iron core inside the coils increases the magnetic field because the core becomes magnetized.

Figure 10 An electric doorbell uses an electromagnet. Each time the electromagnet is turned on, the hammer strikes the bell. **Explain** how the electromagnet is turned off.

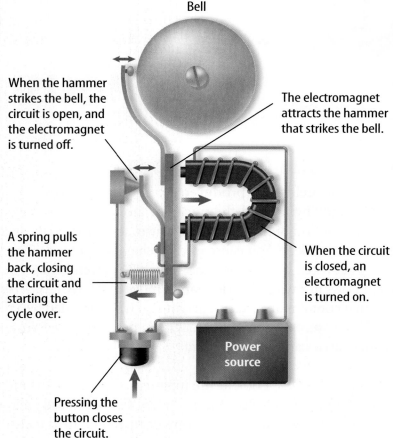

Bell

When the hammer strikes the bell, the circuit is open, and the electromagnet is turned off.

The electromagnet attracts the hammer that strikes the bell.

A spring pulls the hammer back, closing the circuit and starting the cycle over.

When the circuit is closed, an electromagnet is turned on.

Pressing the button closes the circuit.

Power source

Assembling an Electromagnet

Procedure

1. Wrap a **wire** around a **16-penny steel nail** ten times. Connect one end of the wire to a **D-cell battery,** as shown in **Figure 9C.** Leave the other end loose until you use the electromagnet. **WARNING: When current is flowing in the wire, it can become hot over time.**
2. Connect the wire. Observe how many **paper clips** you can pick up with the magnet.
3. Disconnect the wire and rewrap the nail with 20 coils. Connect the wire and observe how many paper clips you can pick up. Disconnect the wire again.

Analysis

1. How many paper clips did you pick up each time? Did more coils make the electromagnet stronger or weaker?
2. Graph the number of coils versus number of paper clips attracted. Predict how many paper clips would be picked up with five coils of wire. Check your prediction.

Try at Home

Using Electromagnets The magnetic field of an electromagnet is turned on or off when the electric current is turned on or off. By changing the current, the strength and direction of the magnetic field of an electromagnet can be changed. This has led to a number of practical uses for electromagnets. A doorbell, as shown in **Figure 10,** is a familiar use of an electromagnet. When you press the button by the door, you close a switch in a circuit that includes an electromagnet. The magnet attracts an iron bar attached to a hammer. The hammer strikes the bell. When the hammer strikes the bell, the hammer has moved far enough to open the circuit again. The electromagnet loses its magnetic field, and a spring pulls the iron bar and hammer back into place. This movement closes the circuit, and the cycle is repeated as long as the button is pushed.

Some gauges, such as the gas gauge in a car, use a galvanometer to move the gauge pointer. **Figure 11** shows how a galvanometer makes a pointer move. Ammeters and voltmeters used to measure current and voltage in electric circuits also use galvanometers, as shown in **Figure 11.**

674 **CHAPTER 23** Magnetism

NATIONAL GEOGRAPHIC VISUALIZING VOLTMETERS AND AMMETERS

Figure 11

The gas gauge in a car uses a device called a galvanometer to make the needle of the gauge move. Galvanometers are also used in other measuring devices. A voltmeter uses a galvanometer to measure the voltage in a electric circuit. An ammeter uses a galvanometer to measure electric current. Multimeters can be used as an ammeter or voltmeter by turning a switch.

A galvanometer has a pointer attached to a coil that can rotate between the poles of a permanent magnet. When a current flows through the coil, it becomes an electromagnet. Attraction and repulsion between the magnetic poles of the electromagnet and the poles of the permanent magnet makes the coil rotate. The amount of rotation depends on the amount of current in the coil.

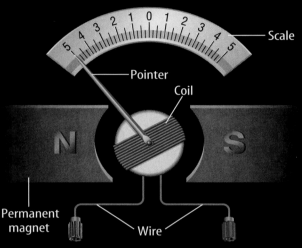

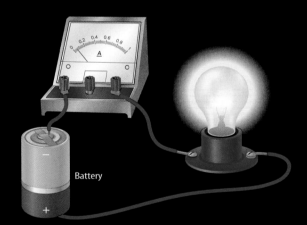

To measure the current in a circuit an ammeter is used. An ammeter contains a galvanometer and has low resistance. To measure current, an ammeter is connected in series in the circuit, so all the current in the circuit flows through it. The greater the current in the circuit, the more the needle moves.

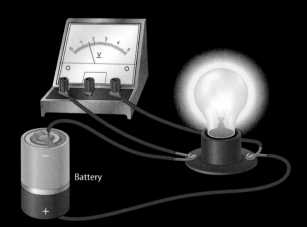

To measure the voltage in a circuit a voltmeter is used. A voltmeter also contains a galvanometer and has high resistance. To measure voltage, a voltmeter is connected in parallel in the circuit, so almost no current flows through it. The higher the voltage in the circuit, the more the needle moves.

SECTION 2 Electricity and Magnetism

Magnets Push and Pull Currents

Look around for electric appliances that produce motion, such as a fan. How does the electric energy entering the fan become transformed into the kinetic energy of the moving fan blades? Recall that current-carrying wires produce a magnetic field. This magnetic field behaves the same way as the magnetic field that a magnet produces. Two current-carrying wires can attract each other as if they were two magnets, as shown in **Figure 12.**

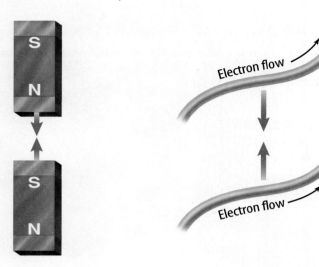

Figure 12 Two wires carrying current in the same direction attract each other, just as unlike magnetic poles do.

Electric Motor Just as two magnets exert a force on each other, a magnet and a current-carrying wire exert forces on each other. The magnetic field around a current-carrying wire will cause it to be pushed or pulled by a magnet, depending on the direction the current is flowing in the wire. As a result, some of the electric energy carried by the current is converted into kinetic energy of the moving wire, as shown on the left in **Figure 13.** Any device that converts electric energy into kinetic energy is a **motor.** To keep a motor running, the current-carrying wire is formed into a loop so the magnetic field can force the wire to spin continually, as shown on the right in **Figure 13.**

Figure 13 In an electric motor, the force a magnet exerts on a current-carrying wire transforms electric energy into kinetic energy.

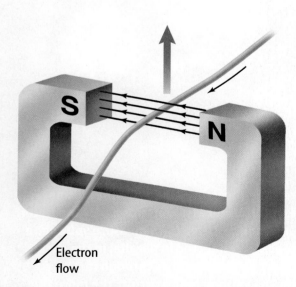

A magnetic field like the one shown will push a current-carrying wire upward.

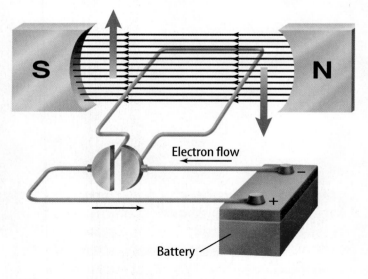

The magnetic field exerts a force on the wire loop, causing it to spin as long as current flows in the loop.

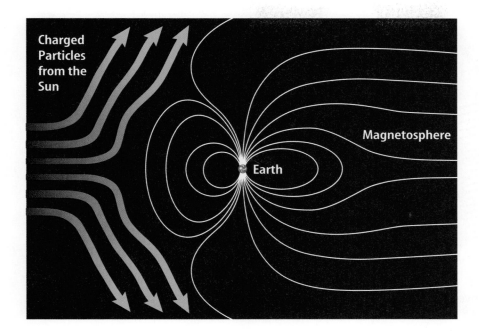

Figure 14 Earth's magnetosphere deflects most of the charged particles streaming from the Sun. **Explain** why the magnetosphere is stretched away from the Sun.

Earth's Magnetosphere The Sun emits charged particles that stream through the solar system like an enormous electric current. Just like a current-carrying wire is pushed or pulled by a magnetic field, Earth's magnetic field pushes and pulls on the electric current generated by the Sun. This causes most of the charged particles in this current to be deflected so they never strike Earth, as shown in **Figure 14.** As a result, living things on Earth are protected from damage that might be caused by these charged particles. At the same time, the solar current pushes on Earth's magnetosphere so it is stretched away from the Sun.

The Aurora Sometimes the Sun ejects a large number of charged particles all at once. Most of these charged particles are deflected by Earth's magnetosphere. However, some of the ejected particles from the Sun produce other charged particles in Earth's outer atmosphere. These charged particles spiral along Earth's magnetic field lines toward Earth's magnetic poles. There they collide with atoms in the atmosphere. These collisions cause the atoms to emit light. The light emitted causes a display known as the **aurora** (uh ROR uh), as shown in **Figure 15.** In northern latitudes, the aurora sometimes is called the northern lights.

Figure 15 An aurora is a natural light show that occurs far north and far south.

SECTION 2 Electricity and Magnetism **677**

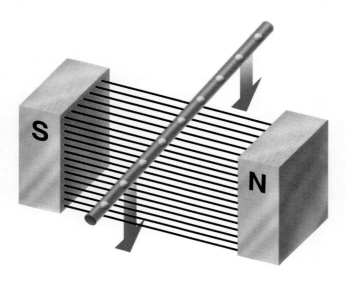
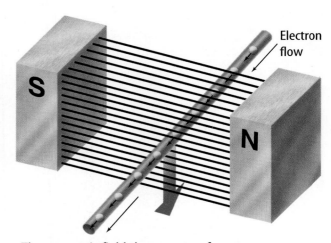

If a wire is pulled through a magnetic field, the electrons in the wire also move downward.

The magnetic field then exerts a force on the moving electrons, causing them to move along the wire.

Figure 16 When a wire is made to move through a magnetic field, an electric current can be produced in the wire.

Using Magnets to Create Current

In an electric motor, a magnetic field turns electricity into motion. A device called a **generator** uses a magnetic field to turn motion into electricity. Electric motors and electric generators both involve conversions between electric energy and kinetic energy. In a motor, electric energy is changed into kinetic energy. In a generator, kinetic energy is changed into electric energy. **Figure 16** shows how a current can be produced in a wire that moves in a magnetic field. As the wire moves, the electrons in the wire also move in the same direction, as shown on the left. The magnetic field exerts a force on the moving electrons that pushes them along the wire on the right, creating an electric current.

Figure 17 In a generator, a power source spins a wire loop in a magnetic field. Every half turn, the current will reverse direction. This type of generator supplies alternating current to the lightbulb.

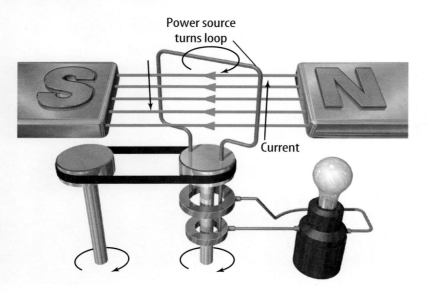

Electric Generators To produce electric current, the wire is fashioned into a loop, as in **Figure 17.** A power source provides the kinetic energy to spin the wire loop. With each half turn, the current in the loop changes direction. This causes the current to alternate from positive to negative. Such a current is called an **alternating current** (AC). In the United States, electric currents change from positive to negative to positive 60 times each second.

Types of Current A battery produces direct current instead of alternating current. In a **direct current** (DC) electrons flow in one direction. In an alternating current, electrons change their direction of movement many times each second. Some generators are built to produce direct current instead of alternating current.

 *What type of currents can be produced by a generator?*

Power Plants Electric generators produce almost all of the electric energy used all over the world. Small generators can produce energy for one household, and large generators in electric power plants can provide electric energy for thousands of homes. Different energy sources such as gas, coal, and water are used to provide the kinetic energy to rotate coils of wire in a magnetic field. Coal-burning power plants, like the one pictured in **Figure 18,** are the most common. More than half of the electric energy generated by power plants in the United States comes from burning coal.

Topic: Power Plants
Visit blue.msscience.com for Web links to more information about the different types of power plants used in your region of the country.

Activity Describe the different types of power plants.

Voltage The electric energy produced at a power plant is carried to your home in wires. Recall that voltage is a measure of how much energy the electric charges in a current are carrying. The electric transmission lines from electric power plants transmit electric energy at a high voltage of about 700,000 V. Transmitting electric energy at a low voltage is less efficient because more electric energy is converted into heat in the wires. However, high voltage is not safe for use in homes and businesses. A device is needed to reduce the voltage.

Figure 18 Coal-burning power plants supply much of the electric energy for the world.

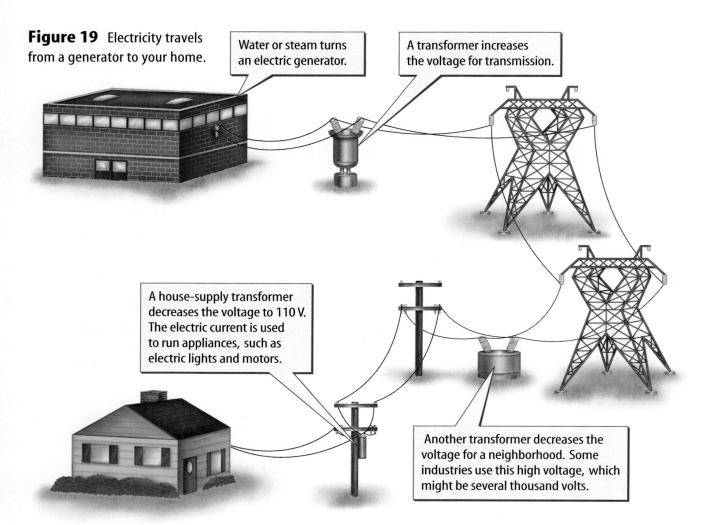

Figure 19 Electricity travels from a generator to your home.

Water or steam turns an electric generator.

A transformer increases the voltage for transmission.

A house-supply transformer decreases the voltage to 110 V. The electric current is used to run appliances, such as electric lights and motors.

Another transformer decreases the voltage for a neighborhood. Some industries use this high voltage, which might be several thousand volts.

Changing Voltage

A **transformer** is a device that changes the voltage of an alternating current with little loss of energy. Transformers are used to increase the voltage before transmitting an electric current through the power lines. Other transformers are used to decrease the voltage to the level needed for home or industrial use. Such a power system is shown in **Figure 19.** Transformers also are used in power adaptors. For battery-operated devices, a power adaptor must change the 120 V from the wall outlet to the same voltage produced by the device's batteries.

Reading Check *What does a transformer do?*

A transformer usually has two coils of wire wrapped around an iron core, as shown in **Figure 20.** One coil is connected to an alternating current source. The current creates a magnetic field in the iron core, just like in an electromagnet. Because the current is alternating, the magnetic field it produces also switches direction. This alternating magnetic field in the core then causes an alternating current in the other wire coil.

Figure 20 A transformer can increase or decrease voltage. The ratio of input coils to output coils equals the ratio of input voltage to output voltage.
Determine *the output voltage if the input voltage is 60 V.*

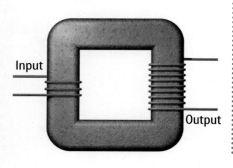

The Transformer Ratio Whether a transformer increases or decreases the input voltage depends on the number of coils on each side of the transformer. The ratio of the number of coils on the input side to the number of coils on the output side is the same as the ratio of the input voltage to the output voltage. For the transformer in **Figure 20,** the ratio of the number of coils on the input side to the number of coils on the output side is three to nine, or one to three. If the input voltage is 60 V, the output voltage will be 180 V.

In a transformer the voltage is greater on the side with more coils. If the number of coils on the input side is greater than the number of coils on the output side, the voltage is decreased. If the number of coils on the input side is less than the number on the output side, the voltage is increased.

Superconductors

Electric current can flow easily through materials, such as metals, that are electrical conductors. However, even in conductors, there is some resistance to this flow and heat is produced as electrons collide with atoms in the material.

Unlike an electrical conductor, a material known as a superconductor has no resistance to the flow of electrons. Superconductors are formed when certain materials are cooled to low temperatures. For example, aluminum becomes a superconductor at about −272°C. When an electric current flows through a superconductor, no heat is produced and no electric energy is converted into heat.

Superconductors and Magnets Superconductors also have other unusual properties. For example, a magnet is repelled by a superconductor. As the magnet gets close to the superconductor, the superconductor creates a magnetic field that is opposite to the field of the magnet. The field created by the superconductor can cause the magnet to float above it, as shown in **Figure 21.**

INTEGRATE History

The Currents War In the late 1800s, electric power was being transmitted using a direct-current transmission system developed by Thomas Edison. To preserve his monopoly, Edison launched a public-relations war against the use of alternating-current power transmission, developed by George Westinghouse and Nikola Tesla. However, by 1893, alternating current transmission had been shown to be more efficient and economical, and quickly became the standard.

Figure 21 A small magnet floats above a superconductor. The magnet causes the superconductor to produce a magnetic field that repels the magnet.

Figure 22 The particle accelerator at Fermi National Accelerator Laboratory near Batavia, Illinois, accelerates atomic particles to nearly the speed of light. The particles travel in a beam only a few millimeters in diameter. Magnets made of superconductors keep the beam moving in a circular path about 2 km in diameter.

Figure 23 A patient is being placed inside an MRI machine. The strong magnetic field inside the machine enables images of tissues inside the patient's body to be made.

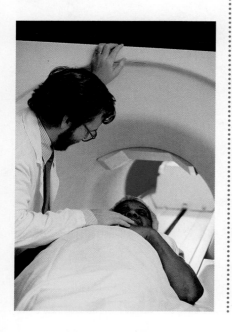

Using Superconductors Large electric currents can flow through electromagnets made from superconducting wire and can produce extremely strong magnetic fields. The particle accelerator shown in **Figure 22** uses more than 1,000 superconducting electromagnets to help accelerate subatomic particles to nearly the speed of light.

Other uses for superconductors are being developed. Transmission lines made from a superconductor could transmit electric power over long distances without having any electric energy converted to heat. It also may be possible to construct extremely fast computers using microchips made from superconductor materials.

Magnetic Resonance Imaging

INTEGRATE Health A method called magnetic resonance imaging, or MRI, uses magnetic fields to create images of the inside of a human body. MRI images can show if tissue is damaged or diseased, and can detect the presence of tumors.

Unlike X-ray imaging, which uses X-ray radiation that can damage tissue, MRI uses a strong magnetic field and radio waves. The patient is placed inside a machine like the one shown in **Figure 23**. Inside the machine an electromagnet made from superconductor materials produces a magnetic field more than 20,000 times stronger than Earth's magnetic field.

682 CHAPTER 23 Magnetism

Producing MRI Images About 63 percent of all the atoms in your body are hydrogen atoms. The nucleus of a hydrogen atom is a proton, which behaves like a tiny magnet. The strong magnetic field inside the MRI tube causes these protons to line up along the direction of the field. Radio waves are then applied to the part of the body being examined. The protons absorb some of the energy in the radio waves, and change the direction of their alignment.

When the radio waves are turned off, the protons realign themselves with the magnetic field and emit the energy they absorbed. The amount of energy emitted depends on the type of tissue in the body. This energy emitted is detected and a computer uses this information to form an image, like the one shown in **Figure 24.**

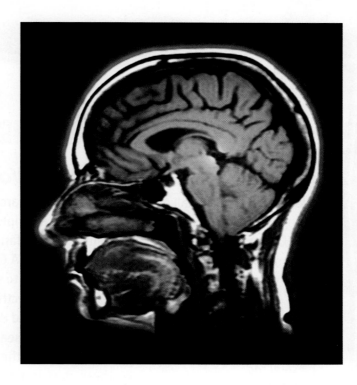

Figure 24 This MRI image shows a side view of the brain.

Connecting Electricity and Magnetism Electric charges and magnets are related to each other. Moving electric charges produce magnetic fields, and magnetic fields exert forces on moving electric charges. It is this connection that enables electric motors and generators to operate.

section 2 review

Summary

Electromagnets
- A current-carrying wire is surrounded by a magnetic field.
- An electromagnet is made by wrapping a current-carrying wire around an iron core.

Motors, Generators, and Transformers
- An electric motor transforms electrical energy into kinetic energy. An electric motor rotates when current flows in a wire loop that is surrounded by a magnetic field.
- An electric generator transforms kinetic energy into electrical energy. A generator produces current when a wire loop is rotated in a magnetic field.
- A transformer changes the voltage of an alternating current.

Self-Check

1. **Describe** how the magnetic field of an electromagnet depends on the current and the number of coils.
2. **Explain** how a transformer works.
3. **Describe** how a magnetic field affects a current-carrying wire.
4. **Describe** how alternating current is produced.
5. **Think Critically** What are some advantages and disadvantages to using superconductors as electric transmission lines?

Applying Math

6. **Calculate Ratios** A transformer has ten turns of wire on the input side and 50 turns of wire on the output side. If the input voltage is 120 V, what will the output voltage be?

How does an electric motor work?

Goals
- **Assemble** a small electric motor.
- **Observe** how the motor works.

Materials
22-gauge enameled wire (4 m)
steel knitting needle
*steel rod
nails (4)
hammer
ceramic magnets (2)
18-gauge insulated wire (60 cm)
masking tape
fine sandpaper
approximately 15-cm square wooden board
wooden blocks (2)
6-V battery
*1.5-V batteries connected in a series (4)
wire cutters
*scissors
*Alternate materials

Safety Precautions

WARNING: *Hold only the insulated part of a wire when it is attached to the battery. Use care when hammering nails. After cutting the wire, the ends will be sharp.*

Real-World Question

Electric motors are used in many appliances. For example, a computer contains a cooling fan and motors to spin the hard drive. A CD player contains electric motors to spin the CD. Some cars contain electric motors that move windows up and down, change the position of the seats, and blow warm or cold air into the car's interior. All these electric motors consist of an electromagnet and a permanent magnet. In this activity you will build a simple electric motor that will work for you. How can you change electric energy into motion?

684 **CHAPTER 23** Magnetism

Using Scientific Methods

Procedure

1. Use sandpaper to strip the enamel from about 4 cm of each end of the 22-gauge wire.

2. Leaving the stripped ends free, make this wire into a tight coil of at least 30 turns. A D-cell battery or a film canister will help in forming the coil. Tape the coil so it doesn't unravel.

3. Insert the knitting needle through the coil. Center the coil on the needle. Pull the wire's two ends to one end of the needle.

4. Near the ends of the wire, wrap masking tape around the needle to act as insulation. Then tape one bare wire to each side of the needle at the spot where the masking tape is.

5. Tape a ceramic magnet to each block so that a north pole extends from one and a south pole from the other.

6. Make the motor. Tap the nails into the wood block as shown in the figure. Try to cross the nails at the same height as the magnets so the coil will be suspended between them.

7. Place the needle on the nails. Use bits of wood or folded paper to adjust the positions of the magnets until the coil is directly between the magnets. The magnets should be as close to the coil as possible without touching it.

8. Cut two 30-cm lengths of 18-gauge wire. Use sandpaper to strip the ends of both wires. Attach one wire to each terminal of the battery. Holding only the insulated part of each wire, place one wire against each of the bare wires taped to the needle to close the circuit. Observe what happens.

Conclude and Apply

1. **Describe** what happens when you close the circuit by connecting the wires. Were the results expected?

2. **Describe** what happens when you open the circuit.

3. **Predict** what would happen if you used twice as many coils of wire.

Compare your conclusions with other students in your class. **For more help, refer to the** Science Skill Handbook.

LAB 685

Science and Language Arts

"Aagjuuk[1] and Sivulliit[2]"
from Intellectual Culture of the Copper Eskimos
by Knud Rasmussen, told by Tatilgak

The following are "magic words" that are spoken before the Inuit (IH noo wut) people go seal hunting. Inuit are native people that live in the arctic region. Because the Inuit live in relative darkness for much of the winter, they have learned to find their way by looking at the stars to guide them. The poem is about two constellations that are important to the Inuit people because their appearance marks the end of winter when the Sun begins to appear in the sky again.

By which way, I wonder the mornings—
You dear morning, get up!
See I am up!
By which way, I wonder,
the constellation *Aagjuuk* rises up in the sky?
By this way—perhaps—by the morning
It rises up!

Morning, you dear morning, get up!
See I am up!
By which way, I wonder,
the constellation *Sivulliit*
Has risen to the sky?
By this way—perhaps—by the morning.
It rises up!

[1] Inuit name for the constellation of stars called Aquila (A kwuh luh)
[2] Inuit name for the constellation of stars called Bootes (boh OH teez)

Understanding Literature

Ethnography Ethnography is a description of a culture. To write an ethnography, an ethnographer collects cultural stories, poems, or other oral tales from the culture that he or she is studying. Why must the Inuit be skilled in navigation?

Respond to the Reading

1. How can you tell the importance of constellations to the Inuit for telling direction?
2. How is it possible that the Inuit could see the constellations in the morning sky?
3. **Linking Science and Writing** Research the constellations in the summer sky in North America and write a paragraph describing the constellations that would help you navigate from south to north.

 Earth's magnetic field causes the north pole of a compass needle to point in a northerly direction. Using a compass helps a person to navigate and find his or her way. However, at the far northern latitudes where the Inuit live, a compass becomes more difficult to use. Some Inuit live north of Earth's northern magnetic pole. In these locations a compass needle points in a southerly direction. As a result, the Inuit developed other ways to navigate.

Chapter 23 Study Guide

Reviewing Main Ideas

Section 1 What is magnetism?

1. All magnets have two poles—north and south. Like poles repel each other and unlike poles attract.
2. A magnet is surrounded by a magnetic field that exerts forces on other magnets.
3. Atoms in magnetic materials are magnets. These materials contain magnetic domains which are groups of atoms whose magnetic poles are aligned.
4. Earth is surrounded by a magnetic field similar to the field around a bar magnet.

Section 2 Electricity and Magnetism

1. Electric current creates a magnetic field. Electromagnets are made from a coil of wire that carries a current, wrapped around an iron core.
2. A magnetic field exerts a force on a moving charge or a current-carrying wire.
3. Motors transform electric energy into kinetic energy. Generators transform kinetic energy into electric energy.
4. Transformers are used to increase and decrease voltage in AC circuits.

Visualizing Main Ideas

Copy and complete the following concept map on magnets.

- **Magnets**
  - are made from → **Magnetic materials**
    - in which → **Moving electrons in atoms**
    - produce → ( )
    - that line up to make → ( )
  - are used by → **Electric motors**
    - in which → ( )
    - generates → ( )
    - that produces → **Kinetic energy**
  - are used by → **Generators**
    - in which → ( )
    - causes → **A wire loop to rotate**
    - that generates → ( )

blue.msscience.com/interactive_tutor

chapter 23 Review

Using Vocabulary

alternating current p. 678
aurora p. 677
direct current p. 679
electromagnet p. 673
generator p. 678
magnetic domain p. 668
magnetic field p. 667
magnetosphere p. 669
motor p. 676
transformer p. 680

Explain the relationship that exists between each set of vocabulary words below.

1. generator—transformer
2. magnetic force—magnetic field
3. alternating current—direct current
4. current—electromagnet
5. motor—generator
6. electron—magnetism
7. magnetosphere—aurora
8. magnet—magnetic domain

Checking Concepts

Choose the word or phrase that best answers the question.

9. What can iron filings be used to show?
   A) magnetic field C) gravitational field
   B) electric field D) none of these

10. Why does the needle of a compass point to magnetic north?
    A) Earth's north pole is strongest.
    B) Earth's north pole is closest.
    C) Only the north pole attracts compasses.
    D) The compass needle aligns itself with Earth's magnetic field.

11. What will the north poles of two bar magnets do when brought together?
    A) attract
    B) create an electric current
    C) repel
    D) not interact

12. How many poles do all magnets have?
    A) one   C) three
    B) two   D) one or two

13. When a current-carrying wire is wrapped around an iron core, what can it create?
    A) an aurora   C) a generator
    B) a magnet   D) a motor

14. What does a transformer between utility wires and your house do?
    A) increases voltage
    B) decreases voltage
    C) leaves voltage the same
    D) changes DC to AC

Use the figure below to answer question 15.

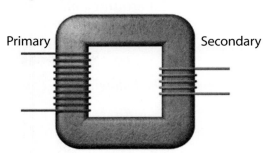

15. For this transformer which of the following describes how the output voltage compares with the input voltage?
    A) larger   C) smaller
    B) the same   D) zero voltage

16. Which energy transformation occurs in an electric motor?
    A) electrical to kinetic
    B) electrical to thermal
    C) potential to kinetic
    D) kinetic to electrical

17. What prevents most charged particles from the Sun from hitting Earth?
    A) the aurora
    B) Earth's magnetic field
    C) high-altitude electric fields
    D) Earth's atmosphere

chapter 23 Review

Thinking Critically

18. **Concept Map** Explain how a doorbell uses an electromagnet by placing the following phrases in the cycle concept map: *circuit open, circuit closed, electromagnet turned on, electromagnet turned off, hammer attracted to magnet and strikes bell,* and *hammer pulled back by a spring.*

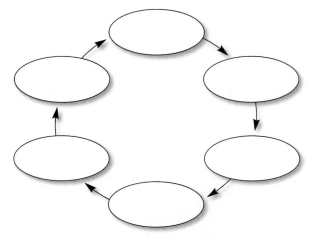

19. **Infer** A nail is magnetized by holding the south pole of a magnet against the head of the nail. Does the point of the nail become a north pole or a south pole? Include a diagram with your explanation.

20. **Explain** why an ordinary bar magnet doesn't rotate and align itself with Earth's magnetic field when you place it on a table.

21. **Determine** Suppose you were given two bar magnets. One magnet has the north and south poles labeled, and on the other magnet the magnetic poles are not labeled. Describe how you could use the labeled magnet to identify the poles of the unlabeled magnet.

22. **Explain** A bar magnet touches a paper clip that contains iron. Explain why the paper clip becomes a magnet that can attract other paper clips.

23. **Explain** why the magnetic field produced by an electromagnet becomes stronger when the wire coils are wrapped around an iron core.

24. **Predict** Magnet A has a magnetic field that is three times as strong as the field around magnet B. If magnet A repels magnet B with a force of 10 N, what is the force that magnet B exerts on magnet A?

25. **Predict** Two wires carrying electric current in the same direction are side by side and are attracted to each other. Predict how the force between the wires changes if the current in both wires changes direction.

Performance Activities

26. **Multimedia Presentation** Prepare a multimedia presentation to inform your classmates on the possible uses of superconductors.

Applying Math

Use the table below to answer questions 27 and 28.

Transformer Properties

Transformer	Number of Input Coils	Number of Output Coils
R	4	12
S	10	2
T	3	6
U	5	10

27. **Input and Output Coils** According to this table, what is the ratio of the number of input coils to the number of output coils on transformer T?

28. **Input and Output Voltage** If the input voltage is 60 V, which transformer gives an output voltage of 12 V?

Chapter 23 Standardized Test Practice

Part 1 Multiple Choice

Record your answers on the answer sheet provided by your teacher or on a sheet of paper.

Use the figure below to answer questions 1 and 2.

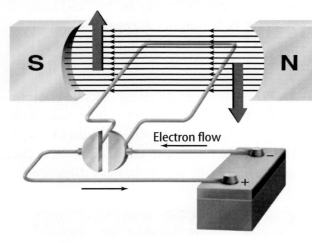

1. What is the device shown?
   A. electromagnet   C. electric motor
   B. generator   D. transformer

2. Which of the following best describes the function of this device?
   A. It transforms electrical energy into kinetic energy.
   B. It transforms kinetic energy into electrical energy.
   C. It increases voltage.
   D. It produces an alternating current.

3. How is an electromagnet different from a permanent magnet?
   A. It has north and south poles.
   B. It attracts magnetic substances.
   C. Its magnetic field can be turned off.
   D. Its poles cannot be reversed.

Test-Taking Tip

Check the Question Number For each question, double check that you are filling in the correct answer bubble for the question number you are working on.

4. Which of the following produces alternating current?
   A. electromagnet   C. generator
   B. superconductor D. motor

5. Which statement about the domains in a magnetized substance is true?
   A. Their poles are in random directions.
   B. Their poles cancel each other.
   C. Their poles point in one direction.
   D. Their orientation cannot change.

Use the figure below to answer questions 6–8.

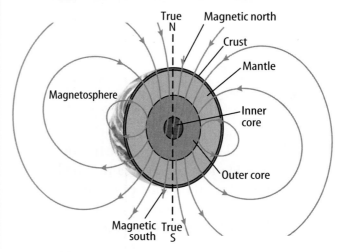

6. What is the region of space affected by Earth's magnetic field called?
   A. declination   C. aurora
   B. magnetosphere   D. outer core

7. What is the shape of Earth's magnetic field similar to?
   A. that of a horseshoe magnet
   B. that of a bar magnet
   C. that of a disk magnet
   D. that of a superconductor

8. In which of Earth's layers is Earth's magnetic field generated?
   A. crust   C. outer core
   B. mantle   D. inner core

Standardized Test Practice

Part 2 Short Response/Grid In

Record your answers on the answer sheet provided by your teacher or on a sheet of paper.

Use the figure below to answer questions 9 and 10.

9. Explain why the compass needles are pointed in different directions.

10. What will happen to the compass needles when the bar magnet is removed? Explain why this happens.

11. Describe the interaction between a compass needle and a wire in which an electric current is flowing.

12. What are two ways to make the magnetic field of an electromagnet stronger?

13. The input voltage in a transformer is 100 V and the output voltage is 50 V. Find the ratio of the number of wire turns on the input coil to the number of turns on the output coil.

14. Explain how you could magnetize a steel screwdriver.

15. Suppose you break a bar magnet in two. How many magnetic poles does each of the pieces have?

16. Alnico is a mixture of steel, aluminum, nickel, and cobalt. It is very hard to magnetize. However, once magnetized, it remains magnetic for a long time. Explain why it would not be a good choice for the core of an electromagnet.

Part 3 Open Ended

Record your answers on a sheet of paper.

17. Explain why the aurora occurs only near Earth's north and south poles.

18. Why does a magnet attract an iron nail to either of its poles, but attracts another magnet to only one of its poles?

19. A battery is connected to the input coil of a step-up transformer. Describe what happens when a lightbulb is connected to the output coil of the transformer.

20. Explain how electric forces and magnetic forces are similar.

Use the figure below to answer questions 21 and 22.

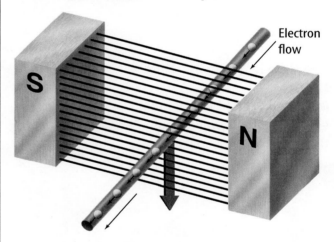

21. Describe the force that is causing the electrons to flow in the wire.

22. Infer how electrons would flow in the wire if the wire were pulled upward.

23. Explain why a nail containing iron can be magnetized, but a copper penny that contains no iron cannot be magnetized.

24. Every magnet has a north pole and a south pole. Where would the poles of a magnet that is in the shape of a disc be located?

chapter 24

Waves, Sound, and Light

The BIG Idea

Sound and light are waves that transfer energy from one place to another.

SECTION 1
Waves
Main Idea Waves transfer energy outward from a vibrating object.

SECTION 2
Sound Waves
Main Idea Sound is a compressional wave that travels only through matter.

SECTION 3
Light
Main Idea Light waves are electromagnetic waves that travel in matter or space.

Ups and Downs

This wind surfer is riding high for now, but that will change soon. The energy carried by ocean waves makes this a thrilling ride, but other waves carry energy, too. Sound waves and light waves carry energy that enable you to hear and see the world around you.

Science Journal Write a short paragraph describing water waves you have seen.

Start-Up Activities

Wave Properties

If you drop a pebble into a pool of water, you notice how the water rises and falls as waves spread out in all directions. How could you describe the waves? In this lab you'll make a model of one type of wave. By describing the model, you'll learn about some properties of all waves.

1. Make a model of a wave by forming a piece of thick string about 50-cm long into a series of *S* shapes with an up and down pattern.
2. Compare the wave you made with those of other students. Notice how many peaks you have in your wave.
3. Reform your wave so that you have a different number of peaks.
4. **Think Critically** Write a description of your wave model. How did the distance between the peaks change as the number of peaks increased?

Preview this chapter's content and activities at blue.msscience.com

FOLDABLES
Study Organizer

Waves Make the following Foldable to compare and contrast the characteristics of transverse and compressional waves.

STEP 1 **Fold** one sheet of lengthwise paper in half.

STEP 2 **Fold** into thirds.

STEP 3 **Unfold and draw** overlapping ovals. **Cut** the top sheet along the folds.

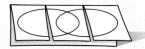

STEP 4 **Label** the ovals as shown.

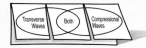

Construct a Venn Diagram As you read the chapter, list the characteristics unique to transverse waves under the left tab, those unique to compressional waves under the right tab, and those characteristics common to both under the middle tab.

Get Ready to Read

Make Connections

① Learn It! Make connections between what you read and what you already know. Connections can be based on personal experiences (text-to-self), what you have read before (text-to-text), or events in other places (text-to-world).

As you read, ask connecting questions. Are you reminded of a personal experience? Have you read about the topic before? Did you think of a person, a place, or an event in another part of the world?

② Practice It! Read the excerpt below and make connections to your own knowledge and experience.

> What caused the loudest sound you ever heard?

> According to the preceding section, what must an object do to produce waves?

> How does a guitar player make a guitar produce sound?

> How does the motion of a drummer's drumsticks produce sound waves? The impact of the sticks on the head of a drum causes the drum head to vibrate. These vibrations transfer energy to nearby air particles, producing sound waves in air. You can hear the sound because energy from the drums travels as sound waves to your ears. Every sound you hear is caused by something vibrating.
>
> —*from page 701*

③ Apply It! As you read this chapter, choose five words or phrases that make a connection to something you already know.

Target Your Reading

Reading Tip

Make connections with memorable events, places, or people in your life. The better the connection, the more likely you will remember.

Use this to focus on the main ideas as you read the chapter.

1 Before you read the chapter, respond to the statements below on your worksheet or on a numbered sheet of paper.
- Write an **A** if you **agree** with the statement.
- Write a **D** if you **disagree** with the statement.

2 After you read the chapter, look back to this page to see if you've changed your mind about any of the statements.
- If any of your answers changed, explain why.
- Change any false statements into true statements.
- Use your revised statements as a study guide.

Science Online
Print out a worksheet of this page at blue.msscience.com

Before You Read A or D		Statement	After You Read A or D
	1	As the wavelength of light waves increases, the frequency also increases.	
	2	Loud sounds can permanently damage your ears.	
	3	The energy carried by a wave depends on the speed of the wave.	
	4	Different colors of light have different wavelengths.	
	5	Sound waves with a low frequency have a low pitch.	
	6	Waves transfer matter from one place to another.	
	7	Light cannot travel through empty space.	
	8	The human eye can detect waves in most of the electromagnetic spectrum.	
	9	Sound waves travel faster in warmer air than in cooler air.	
	10	Refraction occurs when waves change speed going from one material to another.	

694 B

section 1
Waves

as you read

What You'll Learn
- **Explain** how waves transport energy.
- **Distinguish** among transverse, compressional, and electromagnetic waves.
- **Describe** the properties of waves.
- **Describe** reflection, refraction, and diffraction of waves.

Why It's Important
Devices such as televisions, radios, and cell phones receive and transmit information by waves.

Review Vocabulary
density: the mass per cubic meter of a substance

New Vocabulary
- wave
- transverse wave
- compressional wave
- wavelength
- frequency
- law of reflection
- refraction
- diffraction

What are waves?

When you float in the pool on a warm summer day, the up-and-down movement of the water tells you waves are moving past. Sometimes the waves are so strong they almost push you over. Other times, the waves just gently rock you. You know about water waves because you can see and feel their movement, but there are other types of waves, also. Different types of waves carry signals to televisions and radios. Sound and light waves move all around you and enable you to hear and see. Waves are even responsible for the damage caused by earthquakes.

Waves Carry Energy, not Matter A **wave** is a disturbance that moves through matter or space. Waves carry energy from one place to another. You can see that the waves in **Figure 1** carry energy by the way they crash against the rocks. In water waves, the energy is transferred by water molecules. When a wave moves, it may seem that the wave carries matter from place to place as the wave moves.

But that's not what really happens. When waves travel through solids, liquids, and gases, matter is not carried along with the waves. The movement of the fishing bob in **Figure 1** transfers energy to nearby water molecules. The energy is then passed from molecule to molecule as the wave spreads out. The wave disturbance moves outward, but the locations of the water molecules hardly change at all.

Figure 1 Waves carry energy from place to place without carrying matter.

The energy carried by ocean waves can break rocks.

The movement of the fishing bob produces water waves that carry energy through the water.

Types of Waves

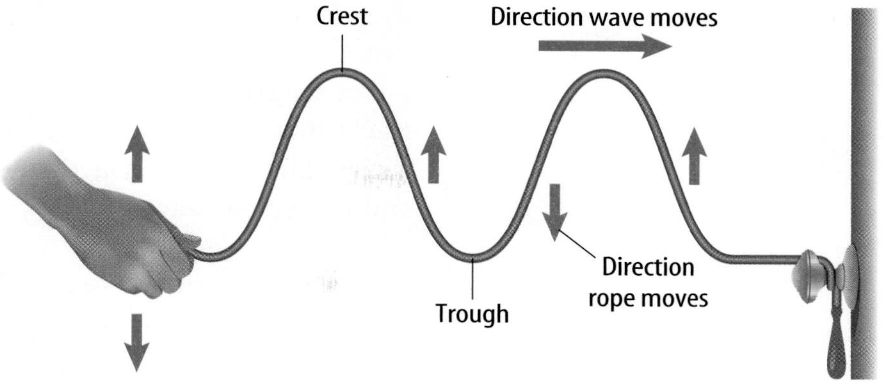

Waves usually are produced by something moving back and forth, or vibrating. It is the energy of the vibrating object that waves carry outward. This energy can spread out from the vibrating object in different types of waves. Some waves, known as mechanical waves, can travel only through matter. Other waves called electromagnetic waves can travel either through matter or through empty space.

Figure 2 You make a transverse wave when you shake the end of a rope up and down.

Transverse Waves One type of mechanical wave is a transverse wave, shown in **Figure 2**. A **transverse wave** causes particles in matter to move back and forth at right angles to the direction in which the wave travels. If you tie a rope to a door handle and shake the end of the rope up and down, transverse waves travel through the rope.

High points in the wave are called crests. Low points are called troughs. The series of crests and troughs forms a transverse wave. The crests and troughs travel along the rope, but the particles in the rope move only up and down.

Compressional Waves Another type of mechanical wave is a compressional wave. **Figure 3** shows a compressional wave traveling along a spring coil. A **compressional wave** causes particles in matter to move back and forth along the same direction in which the wave travels.

In **Figure 3** the places where the coils are squeezed together are called compressions. The places where the coils are spread apart are called rarefactions. The series of compressions and rarefactions forms a compressional wave. The compressions and rarefactions travel along the spring, but the coils move only back and forth.

Reading Check *How does matter move in a compressional wave?*

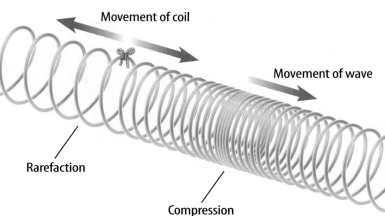

Figure 3 A wave on a spring coil is an example of a compressional wave.

SECTION 1 Waves **695**

INTEGRATE Earth Science

Seismic waves move through the ground during an earthquake. Some of these waves are compressional, and others are transverse. The seismic waves that cause most damage to buildings are a kind of rolling waves. These rolling waves are a combination of compressional and transverse waves.

Electromagnetic Waves Light, radio waves, and X rays are examples of electromagnetic waves. Just like waves on a rope, electromagnetic waves are transverse waves. However, electromagnetic waves contain electric and magnetic parts that vibrate up and down perpendicular to the direction the wave travels.

Properties of Waves

The properties that waves have depend on the vibrations that produce the waves. For example, if you move a pencil slowly up and down in a bowl of water, the waves produced by the pencil's motion will be small and spread apart. If you move the pencil rapidly, the waves will be larger and close together.

Wavelength The distance between one point on a wave and the nearest point moving with the same speed and direction is the **wavelength. Figure 4** shows how the wavelengths of transverse and compressional waves are measured. The wavelength of a transverse wave is the distance between two adjacent crests or two adjacent troughs. The wavelength of a compressional wave is the distance between two adjacent compressions or rarefactions.

Frequency The **frequency** of a wave is the number of wavelengths that pass by a point each second. If you were watching a transverse wave on a rope, the frequency of the wave would be the number of crests or troughs that pass you each second. In the same way, the frequency of a compressional wave is the number of compressions or rarefactions that would pass by each second.

Figure 4 The wavelength of a transverse wave is the distance from crest to crest or from trough to trough. The wavelength of a compressional wave is the distance from compression to compression or rarefaction to rarefaction.

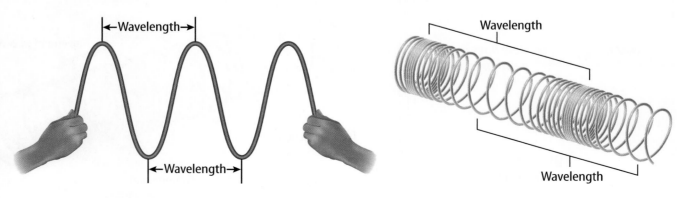

Amplitude of a Transverse Wave Waves have another property called amplitude. Suppose you shake the end of a rope by moving your hand up and down a large distance. Then you make a transverse wave with high crests and deep troughs. The wave you've made has a large amplitude. The amplitude of a transverse wave is half the distance between a crest and trough as shown in **Figure 5.** As the distance between crests and troughs increases, the amplitude of a transverse wave increases.

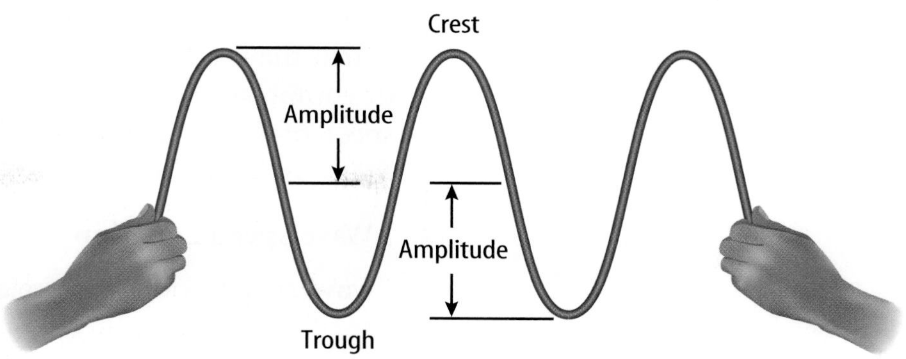

Figure 5 The amplitude of a transverse wave depends on the height of the crests or the depth of the troughs.

Amplitude of a Compressional Wave The amplitude of a compressional wave depends on the density of material in compressions and rarefactions as shown in **Figure 6.** Compressional waves with greater amplitude have compressions that are more squeezed together and rarefactions that are more spread apart. For example, in a spring, squeezing some coils together more tightly causes the nearby coils to be more spread apart.

Reading Check *What is the amplitude of a compressional wave?*

Amplitude and Energy The vibrations that produce a wave transfer energy to the wave. The more energy a wave carries, the larger its amplitude. By moving your hand up and down a larger distance in making a wave on a rope, you transfer more energy to the wave. Seismic waves are produced by vibrations in Earth's crust that cause earthquakes. The more energy these waves have, the larger their amplitudes and the more damage they cause as they travel along Earth's surface.

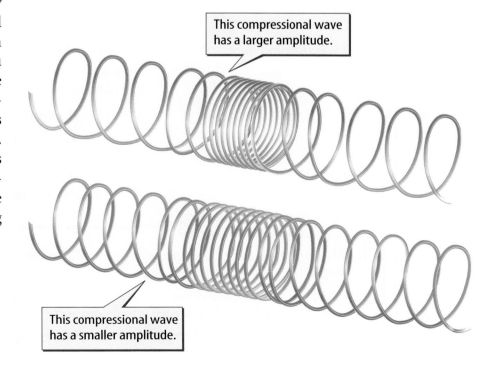

This compressional wave has a larger amplitude.

This compressional wave has a smaller amplitude.

Figure 6 The amplitude of a compressional wave depends on the density of the material in the compressions and rarefactions.

SECTION 1 Waves **697**

Wave Speed The speed of a wave depends on the medium in which the wave travels. The faster waves travel, the more crests or compressions pass by you each second. You can calculate the speed of a wave if you know its wavelength and frequency using the equation below.

Wave Speed Equation

wave speed (in m/s) = **wavelength** (in m) × **frequency** (in Hz)

$$v = \lambda f$$

In this equation, v is the symbol for wave speed and f is the symbol for frequency. The SI unit for frequency is the hertz, abbreviated Hz. One hertz equals one vibration per second, or one wavelength passing a point in one second. One hertz is equal to the unit 1/s. The wavelength is represented by the Greek letter lambda, λ, and is measured in meters.

Applying Math — Solve a Simple Equation

SPEED OF SOUND A sound wave produced by a lightning bolt has a frequency of 34 Hz and a wavelength of 10.0 m. What is the speed of the sound wave?

Solution

1 *This is what you know:*
- wavelength: $\lambda = 10$ m
- frequency: $f = 34$ Hz

2 *This is what you need to find:* wave speed: $v = ?$ m/s

3 *This is the procedure you need to use:* Substitute the known values for wavelength and frequency into the wave speed equation and calculate the wave speed:

$$v = \lambda f = (10.0 \text{ m})(34 \text{ Hz})$$
$$= 340 \text{ m} \times \text{Hz} = 340 \text{ m} \times (1/\text{s}) = 340 \text{ m/s}$$

4 *Check your answer:* Divide your answer by the wavelength 10.0 m. The result should be the given frequency 34 Hz.

Practice Problems

1. Waves on a string have a wavelength of 0.55 m. If the frequency of the waves is 6.0 Hz, what is the wave speed?
2. If the frequency of a sound wave in water is 15,000 Hz, and the sound wave travels through water at a speed of 1,500 m/s, what is the wavelength?

For more practice, visit blue.msscience.com/math_practice

Waves Can Change Direction

Waves don't always travel in a straight line. When you look into a mirror, you use the mirror to make light waves change direction. Waves can change direction when they travel from one material to another. The waves can reflect (bounce off a surface), refract (change direction), or diffract (bend around an obstacle).

The Law of Reflection When waves reflect off a surface, they always obey the law of reflection, as shown in **Figure 7**. A line that makes an angle of 90 degrees with a surface is called the normal to the surface. According to **law of reflection,** the angle that the incoming wave makes with the normal equals the angle that the outgoing wave makes with the normal.

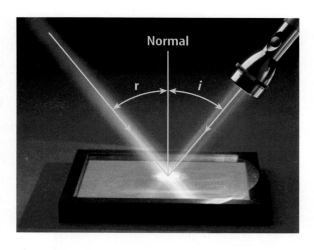

Figure 7 All waves obey the law of reflection. The angle of reflection, r, always equals the angle of incidence, i.

Refraction The speed of the wave depends on properties of the material through which it travels. A light wave, for example, travels faster through air than it does through water. **Figure 8** shows that a change in a wave's speed changes the direction in which the wave travels. When the light wave moves from air to water, it slows down. This change in speed causes the light wave to bend. **Refraction** is the change in direction of a wave when it changes speed as it travels from one material to another.

Figure 8 Refraction occurs when a wave changes speed. Light waves change direction when they slow down as they pass from air to water.

Refraction of Light

Procedure
1. Fill a **drinking glass** about half full with water.
2. Place a **pencil** in the glass. Describe the appearance of the pencil.
3. Slowly add water to the glass. Describe how the appearance of the pencil changes.

Analysis
1. How does the appearance of the pencil depend on the level of water in the glass?
2. Where do the light waves coming from the pencil change speed?
3. **Infer** how the appearance of the pencil and the change in speed of the light waves are related.

Diffraction Waves can change direction by **diffraction**, which is the bending of waves around an object. In **Figure 9**, the water waves are not completely blocked by the obstacle, but instead bend around the obstacle.

The amount of diffraction or bending of the wave depends on the size of the obstacle the wave encounters. If the size of the obstacle is much larger than the wavelength, very little diffraction occurs. Then there is a shadow behind the object where there are no waves.

As the wavelength increases compared with the size of the obstacle, the amount of diffraction increases. The amount of diffraction is greatest if the wavelength is much larger than the obstacle.

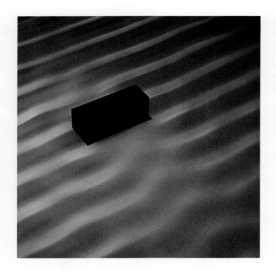

Figure 9 The amount of diffraction or bending around an obstacle depends on the size of the obstacle and the wavelength of the wave.

Diffraction of Sound and Light The wavelengths of sound waves are similar to the size of objects around you, but the wavelength of light waves are much shorter. As a result, you can hear people talking in a room with an open door even though you can't see them.

section 1 review

Summary

Wave Energy
- Waves transport energy without transporting matter.

Types of Waves
- Transverse waves cause particles in a material to move back and forth at right angles to the direction the waves travel.
- Compressional waves cause particles in a material to move back and forth along the same direction the waves travel.
- Electromagnetic waves are transverse waves that can travel through empty space.

Wave Properties
- A wave can be described by its wavelength, frequency, and amplitude.
- The energy carried by a wave increases as the amplitude of the wave increases.
- The speed of a wave, v, equals its wavelength, λ, multiplied by its frequency, f:
$$v = \lambda f$$
- Reflection, refraction, or diffraction can cause waves to change direction.

Self Check

1. **Analyze** How can waves transport energy without transporting matter from place to another?
2. **Explain** how the spacing between coils of a spring changes if the amplitude of compressional waves traveling along the spring increases.
3. **Predict** how the wavelength of waves traveling with the same speed would change if the frequency of the waves increases.
4. **Apply** Two similar-sized stones, one heavy and one light, are dropped from the same height into a pond. Explain why the impact of the heavy stone would produce waves with higher amplitude than the impact of the light stone would.
5. **Think Critically** Water waves produced by a speed boat strike a floating inner tube. Describe the motion of the inner tube as the waves pass by.

Applying Math

6. **Calculate Wave Speed** Find the speed of a wave with a wavelength of 0.2 m and a frequency of 1.5 Hz.
7. **Calculate Wavelength** Find the wavelength of a wave with a speed of 3.0 m/s and a frequency of 0.5 Hz.

section 2
Sound Waves

Making Sound Waves

How does the motion of a drummer's drumsticks produce sound waves? The impact of the sticks on the head of a drum causes the drum head to vibrate. These vibrations transfer energy to nearby air particles, producing sound waves in air. You can hear the sound because energy from the drums travels as sound waves to your ears. Every sound you hear is caused by something vibrating. For example, when you talk, tissues in your throat vibrate in different ways to form sounds.

Sound Waves are Compressional Waves Sound waves produced by a vibrating object are compressional waves. **Figure 10** shows how the vibrating drum produces compressional waves. When the drummer hits the drum, the head of the drum vibrates. Nearby air particles vibrate with the same frequency as the frequency of vibrations. The drum head moving outward compresses nearby air particles. The drum head moving inward causes rarefactions in nearby air particles. The inward and outward movement of the drum head produces the same pattern of compressions and rarefactions in the air particles.

Sound waves can only travel through matter. The energy carried by a sound wave is transferred by the collisions between the particles in the material the wave is traveling in. A spaceship traveling outside Earth's atmosphere, for example, does not make any sound outside the ship.

as you read

What You'll Learn
- **Describe** how sound waves are produced.
- **Explain** how sound waves travel through matter.
- **Describe** the relationship between loudness and sound intensity.
- **Explain** how humans hear sound.

Why It's Important
A knowledge of sound helps you understand how to protect your hearing.

Review Vocabulary
perception: a recognition, sense, or understanding of something

New Vocabulary
- intensity
- pitch
- reverberation

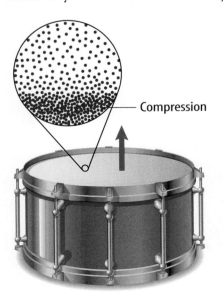
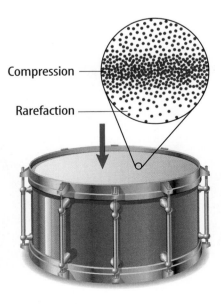

Figure 10 A vibrating drumhead produces a sound wave. The drum head produces a compression each time it moves upward and a rarefaction each time it moves downward.

SECTION 2 Sound Waves **701**

Table 1 Speed of Sound in Different Materials	
Material	Speed (m/s)
Air (20°C)	343
Glass	5,640
Steel	5,940
Water (25°C)	1,493
Seawater (25°C)	1,533
Rubber	1,600
Diamond	12,000
Iron	5,130

The Speed of Sound

Like all waves, the speed of sound depends on the matter through which it travels. Sound waves travel faster through solids and liquids. **Table 1** shows the speed of sound in different materials.

The speed of sound through a material increases as the temperature of the material increases. The effect of temperature is greatest in gases. For example, the speed of sound in air increases from about 330 m/s to about 350 m/s as the air temperature increases from 0° to 30°C.

Reading Check *How does temperature affect the speed of sound through a material?*

The Loudness of Sound

What makes a sound loud or soft? The girl in **Figure 11** can make a loud sound by clapping the cymbals together sharply. She can make a soft sound by clapping the cymbals together gently. The difference is the amount of energy the girl gives to the cymbals. Loud sounds have more energy than soft sounds.

Intensity The amount of energy that a wave carries past a certain area each second is the **intensity** of the sound. **Figure 12** shows how the intensity of sound from the cymbals decreases with distance. A person standing close when the girl claps the cymbals would hear an intense sound. The sound would be less intense for someone standing farther away. The intensity of sound waves is related to the amplitude. Sound with a greater amplitude also has a greater intensity.

Figure 11 The loudness of a sound depends on the amount of energy the sound waves carry.

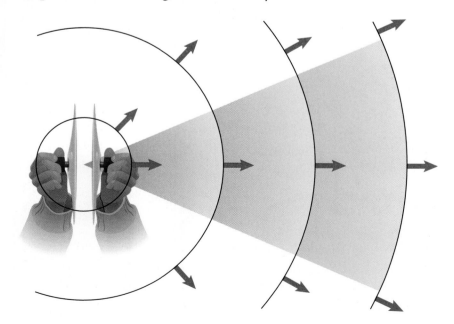

Figure 12 The intensity of a sound wave decreases as the wave spreads out from the source of the sound. The energy the wave carries is spread over a larger area.

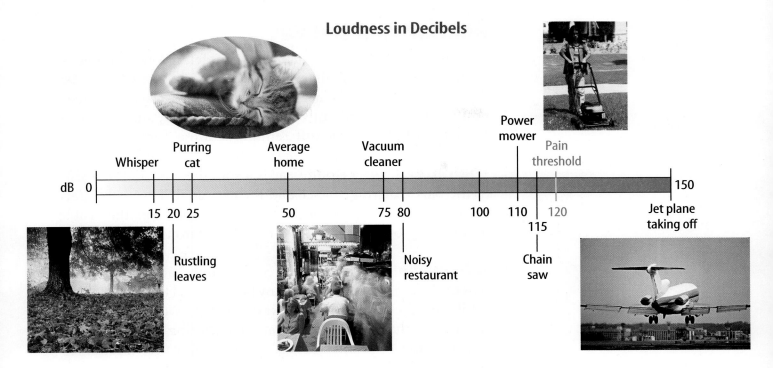

The Decibel Scale and Loudness The intensity of sound waves is measured in units of decibels (dB), as shown in **Figure 13**. The softest sound a person can hear has an intensity of 0 dB. Normal conversation has an intensity of about 50 dB. Sound with intensities of about 120 dB or higher are painful to people.

Loudness is the human perception of the intensity of sound waves. Each increase of 10 dB in intensity multiplies the energy of the sound waves ten times. Most people perceive this as a doubling of the loudness of the sound. An intensity increase of 20 dB corresponds to a hundred times the energy and an increase in loudness of about four times.

Figure 13 The intensity of sound is measured on the decibel scale.
Infer *how many times louder a power mower is compared to a noisy restaurant.*

 *How much has the energy of a sound wave changed if its intensity has increased by 30 dB?*

Frequency and Pitch

The frequency of sound waves is determined by the frequency of the vibrations that produce the sound. Recall that wave frequency is measured in units of hertz (Hz), which is the number of vibrations each second. On the musical scale, the note C has a frequency of 262 Hz. The note E has a frequency of 330 Hz. People are usually able to hear sounds with frequencies between about 20 Hz and 20,000 Hz.

Pitch is the human perception of the frequency of sound. The sounds from a tuba have a low pitch and the sounds from a flute have a high pitch. Sounds with low frequencies have low pitch and sounds with high frequencies have high pitch.

Hearing Damage
Prolonged exposure to sounds above 85 dB can damage your hearing. Research to find out the danger of noise levels you might experience at activities such as loud music concerts or basketball games.

Hearing and the Ear

The ear is a complex organ that can detect a wide range of sounds. You may think that the ear is just the structure that you see on the side of your head. However, the ear can be divided into three parts—the outer ear, the middle ear, and the inner ear. **Figure 14** shows the different parts of the human ear.

The Outer Ear The outer ear is a sound collector. It consists of the part that you can see and the ear canal. The visible part is shaped somewhat like a funnel. This shape helps the visible part collect sound waves and direct them into the ear canal.

The Middle Ear The middle ear is a sound amplifier. It consists of the ear drum and three tiny bones called the hammer, the anvil, and the stirrup. Sound waves that pass through the ear canal cause the eardrum to vibrate. Theses vibrations are transmitted to the three small bones, which amplify the vibrations.

The Inner Ear The inner ear contains the cochlea. The cochlea is filled with fluid and is lined with tiny hair-like cells. Vibrations of the stirrup bone are transmitted to the hair cells. The movement of the hair cells produce signals that travel to your brain, where they are interpreted as sound.

Figure 14 The human ear can be divided into three parts. The outer ear is the sound collector, the middle ear is the sound amplifier, and the inner ear is the sound interpreter.

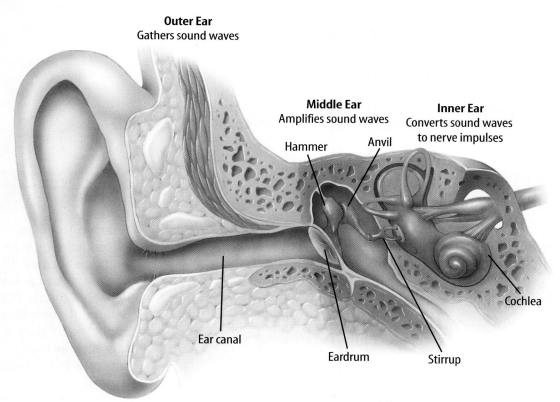

The Reflection of Sound

Have you ever stood in an empty room and heard echoes when you talked very loudly? Echoes are sounds that reflect off surfaces. Repeated echoes are called **reverberation.** Concert halls and auditoriums are designed with soft materials on the ceilings and walls to avoid too much reverberation. Theaters like the one in **Figure 15** often have curtains on the walls because sounds won't reflect off soft surfaces. The curtains absorb the energy of the sound waves.

The reflection of sound can be used to locate or identify objects. Echolocation is the process of locating objects by bouncing sounds off them. Bats, dolphins, and other animals emit short, high-frequency sound waves toward a certain area. By interpreting the reflected waves, the animals can locate and determine properties of other animals. Doctors use reflection of sound waves in medicine. Computers can analyze ultrasonic waves that reflect off body parts to produce an internal picture of the body. These pictures help doctors monitor pregnancies, heart problems, and other medical conditions.

Figure 15 A modern concert hall contains materials that absorb sound waves to control reverberation and other sound reflections.

section 2 review

Summary

Making Sound Waves
- Sound waves are compressional waves produced by something vibrating.
- The speed of sound waves depends on the material in which the waves travel and its temperature.

Loudness and Pitch
- The intensity of a wave is the amount of energy the wave transports each second across a unit surface.
- The intensity of sound waves is measured in units of decibels.
- Loudness is the human perception of sound intensity.
- Pitch is the human perception of the frequency of a sound.

Hearing Sound
- You hear a sound when a sound wave reaches your ear and causes structures in your ear to vibrate.

Self Check

1. **Explain** why you hear a sound when you clap your hands together.
2. **Predict** whether sound will would travel faster in air in the summer or in the winter.
3. **Compare and contrast** the sound waves produced by someone whispering and someone shouting.
4. **Describe** how vibrations produced in your ear by a sound wave enable you to hear the sound.
5. **Think Critically** Vibrations cause sounds, yet if you move your hand back and forth through the air, you don't hear a sound. Explain.

Applying Math

6. **Calculate a Ratio** How many times louder is a sound wave with an intensity of 50 dB than a sound wave with an intensity of 20 dB?
7. **Calculate Increase in Intensity** If the energy carried by a sound wave is multiplied by a thousand times, by what factor does the intensity of the sound wave increase?

Sound Waves in Matter

In this lab you can hear differences in sound when the sound waves travel through various materials.

Real-World Question

How does the movement of sound waves through different materials affect the sounds we hear?

Goals
- **Notice** the variations in sound when sound waves travel through different materials.
- **Infer** what property of the materials cause the sound waves to produce a different sound.

Materials
150-mL beakers (4)   corn syrup
water   pencil
vegetable oil

Safety Precautions

Procedure

1. Fill a beaker to the 140-mL line with water. Fill another beaker with 140 mL of vegetable oil. Fill a third beaker with 140 mL of corn syrup. Leave the fourth beaker empty.
2. Hold the pencil securely and tap the side of the beaker about halfway down from its rim. Use the metal band near the end of the pencil to make a clear sound.
3. Pay careful attention to the pitch of the sound. Notice whether the sound continues for a moment after the tap or if it stops suddenly. Write a description of the sound you hear in your data table.
4. Repeat steps 3 and 4 for the remaining beakers. You may wish to tap each beaker several times to be sure you hear the sound well.
5. **Compare** the sounds made by the beaker filled with air and the beaker filled with the different liquids.

Conclude and Apply

1. **List** the materials in the beakers in order of increasing density.
2. **Infer** how the pitch of the sound changes as the density of the material in the beaker increases.
3. How does the density of the material in the beaker affect how long the sound continued to be heard after the beaker was tapped?

Compare your results with other students in your class.

SECTION 3

Light

Waves in Empty Space

On a clear night you might see the Moon shining brightly, as in **Figure 16.** Like other waves, light waves can travel through matter, but light waves are different from water waves and sound waves. Light from the Moon has traveled through space that contains almost no matter. You can see light from the moon, distant stars, and galaxies because light is an electromagnetic wave. **Electromagnetic waves** are waves that can travel through matter or through empty space.

The Speed of Light Have you ever seen a movie where a spaceship travels faster than the speed of light? In reality, nothing travels faster than the speed of light. In empty space, light travels at a speed of about 300,000 km/s. Light travels so fast that light emitted from the Sun travels 150 million km to Earth in only about eight and a half minutes.

However, when light travels in matter, it interacts with the atoms and molecules in the material and slows down. As a result, light travels fastest in empty space, and travels slowest in solids. In glass, for example, light travels about 197,000 km/s.

Wavelength and Frequency of Light Can you guess how long a wavelength of light is? Wavelengths of light are usually expressed in units of nanometers (nm). One nanometer is equal to one billionth of a meter. For example, green light has a wavelength of about 500 nm, or 500 billionths of a meter. A light wave with this wavelength has a frequency of 600 trillion Hz.

as you read

What You'll Learn
- **Identify** the properties of light waves.
- **Describe** the electromagnetic spectrum.
- **Describe** the types of electromagnetic waves that travel from the Sun to Earth.
- **Explain** human vision and color perception.

Why It's Important

Light is necessary for vision. Other electromagnetic waves are used in devices such as cell phones and microwave ovens.

Review Vocabulary
spectrum: a range of values or properties

New Vocabulary
- electromagnetic waves
- electromagnetic spectrum
- infrared waves
- ultraviolet waves

Figure 16 The Moon reflects light from the Sun. These light waves travel through space to reach your eyes.
Infer *whether a sound wave could travel from the Moon to Earth.*

Figure 17 A light wave is a transverse wave that contains vibrating electric and magnetic fields. The fields vibrate at right angles to the direction the wave travels.

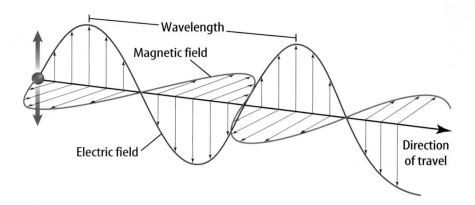

Properties of Light Waves

Light waves, and all electromagnetic waves, are transverse waves. Recall that a wave on a rope is a transverse wave that causes the rope to move at right angles to the direction the wave is traveling. An electromagnetic wave traveling through matter also can cause matter to move at right angles to the direction the wave is moving.

An electromagnetic wave contains an electric part and a magnetic part, as shown in **Figure 17.** Both parts are called fields and vibrate at right angles to the wave motion. The number of times the electric and magnetic parts vibrate each second is the frequency of the wave. The wavelength is the distance between the crests or troughs of the vibrating electric or magnetic parts.

Intensity of Light Waves The intensity of waves is a measure of the amount of energy that the waves carry. For light waves, the intensity determines the brightness of the light. A dim light has lower intensity because the waves carry less energy. However, as you move away from a light source, the energy spreads out and the intensity decreases.

 *What determines the intensity of light waves?*

Topic: Lasers
Visit blue.msscience.com for Web links to information about why the intensity of light emitted by lasers makes them useful.

Activity Write a short paragraph describing three uses for lasers.

The Electromagnetic Spectrum

Light waves aren't the only kind of electromagnetic waves. In fact, there is an entire spectrum of electromagnetic waves, as shown in **Figure 18.** The **electromagnetic spectrum** is the complete range of electromagnetic wave frequencies and wavelengths. At one end of the spectrum the waves have low frequency, long wavelength, and low energy. At the other end of the spectrum the waves have high frequency, short wavelength, and high energy. All of the waves—from radio waves to visible light to gamma rays—are the same kind of waves. They differ from each other only by their frequencies, wavelengths, and energy.

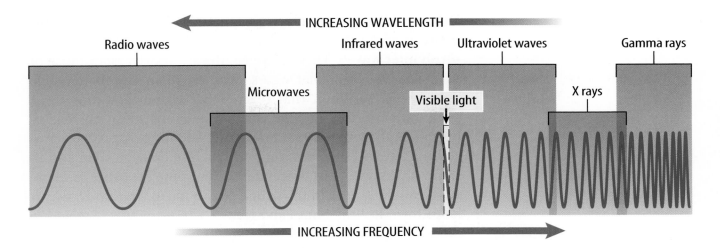

Figure 18 Electromagnetic waves have a range of frequencies and wavelengths called the electromagnetic spectrum.
Infer *how the frequency of electromagnetic waves change as their wavelength decreases.*

Radio Waves and Microwaves The waves that carry radio and television signals to your home are radio waves. The wavelengths of radio waves are greater than about 0.3 meters. Some are even thousands of meters long. The shortest radio waves are called microwaves. These waves have a wavelength between about 0.3 meters and 0.001 meters. You use these waves when you cook food in a microwave oven. Microwaves are also used to transmit information to and from cell phones.

Infrared Waves When you use a remote control, infrared waves travel from the remote to a receiver on your television. **Infrared waves** have wavelengths between 0.001 meters and 700 billionths of a meter. All warm bodies emit infrared waves. Because of this, law enforcement officials and military personnel sometimes use special night goggles that are sensitive to infrared waves. These goggles can be used to help locate people in the dark.

Visible Light and Color The range of electromagnetic waves between 700 and 400 billionths of a meter is special, because that is the range of wavelengths people can see. Electromagnetic waves in this range are called visible light. **Figure 19** shows how different wavelengths correspond to different colors of light. White light, like the light from the Sun or a flashlight, is really a combination of different colors. You can see this by using a prism to separate white light into different colors. When the light passes through the prism, the different wavelengths of light are bent different amounts. Violet light is bent the most because it has the shortest wavelength. Red light is bent the least.

Figure 19 Visible light waves are electromagnetic waves with a narrow range of wavelengths from about 700 to 400 billionths of a meter. The color of visible light waves depends on their wavelength.
Determine *the color of the visible light waves with the highest frequency.*

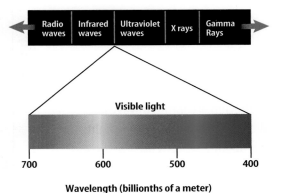

Reading Check *What range of wavelengths of electromagnetic waves can people see?*

SECTION 3 Light **709**

Ultraviolet Waves Electromagnetic waves with wavelengths between about 400 billionths and 10 billionths of a meter are **ultraviolet waves.** These wavelengths are shorter than those of visible light. Ultraviolet waves carry more energy than visible light waves. Sunlight that reaches Earth's surface contains a small fraction of ultraviolet waves. These waves can cause sunburn if skin is exposed to sunlight for too long. Excessive exposure to ultraviolet waves can permanently damage skin, and in some cases cause skin cancer. However, some exposure to ultraviolet waves is needed for your body to make vitamin D, which helps form healthy bones and teeth.

X Rays and Gamma Rays The electromagnetic waves with the highest energy, highest frequency, and shortest wavelengths are X rays and gamma rays. If you've ever broken a bone, the doctor probably took an X ray to examine the injured area. X rays are energetic enough to pass through the body. X rays pass through soft tissues, but are blocked by denser body parts, such as bones. This enables images to be made of internal body parts. Gamma rays are even more energetic than X rays. One use of gamma rays is in the food industry to kill bacteria that might increase the rate of spoilage of food.

Electromagnetic Waves from the Sun Most of the energy emitted by the Sun is in the form of ultraviolet, visible, and infrared waves, as shown in **Figure 20.** These waves carry energy away from the Sun and spread out in all directions. Only a tiny fraction of this energy reaches Earth. Most of the ultraviolet waves from the Sun are blocked by Earth's atmosphere. As a result, almost all energy from the Sun that reaches Earth's surface is carried by infrared and visible electromagnetic waves.

Mini LAB

Separating Wavelengths

Procedure

1. Place a **prism** in sunlight. Adjust its position until a color spectrum is produced.
2. Place the prism on a **desktop.** Dim the lights and shine a **flashlight** on the prism. Record your observations.
3. Shine a **laser pointer** on the prism. Record your observations.

WARNING: *Do not shine the laser pointer into anyone's eyes.*

Analysis

1. Determine whether sunlight and the light emitted from the flashlight contain light waves of more than one wavelength.
2. Determine whether the light emitted from the laser pointer contains light waves of more than one wavelength.

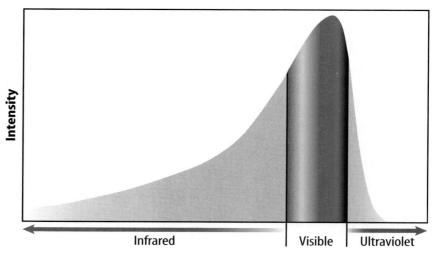

Figure 20 About 49 percent of the electromagnetic waves emitted by the Sun are infrared waves, about 43 percent are visible light, and about 7 percent are ultraviolet waves.

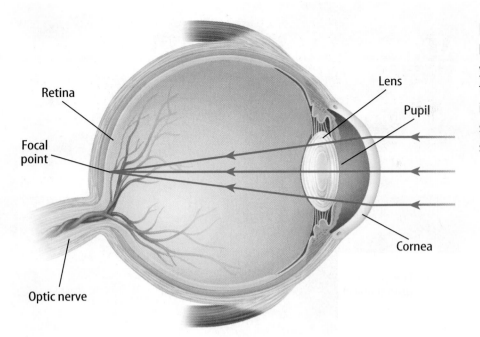

Figure 21 The cornea and the lens focus light waves that enter your eye so that a sharp image is formed on the retina. Special cells in the retina cause signals to be sent to the brain when they are struck by light.

The Eye and Seeing Light

You see an object when light emitted or reflected from the object enters your eye, as shown in **Figure 21.** Light waves first pass through a transparent layer called the cornea (KOR nee uh), and then the transparent lens. The lens is flexible and changes shape to enable you to focus on objects that are nearby and far away, as shown in **Figure 22.** However, sometimes the eye is unable to form sharp images of both nearby and distant objects, as shown in **Figure 23** on the next page.

Why do objects have color?
When light waves strike an object, some of the light waves are reflected. The wavelengths of the light waves that are reflected determine the object's color. For example, a red rose reflects light waves that have wavelengths in the red part of the visible spectrum. The color of objects that emit light is determined by the wavelengths of light that they emit. A neon sign appears to be red because it emits red light waves.

Figure 22 The shape of the lens changes when you focus on nearby and distant objects.

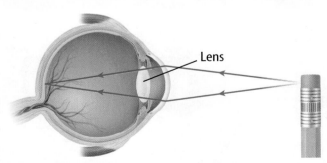

The lens becomes flatter when you focus on a distant object.

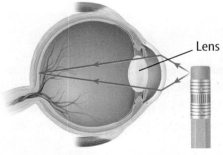

The lens becomes more curved when you focus on an object nearby.

SECTION 3 Light **711**

NATIONAL GEOGRAPHIC VISUALIZING COMMON VISION PROBLEMS

Figure 23

In a human eye, light waves pass through the transparent cornea and the lens of the eye. The cornea and the lens cause light waves from an object to be focused on the retina, forming a sharp image. However, vision problems result when a sharp image is not formed on the retina. The two most common vision problems are farsightedness and nearsightedness.

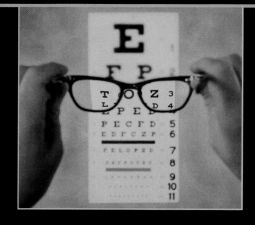

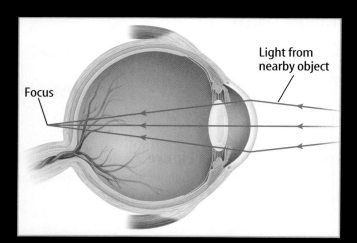

◀ **Nearsightedness** A person that is nearsighted can see nearby objects clearly, but distant objects seem blurry. Nearsightedness results if the eyeball is too long, so that light waves from far away objects are brought to a focus before they reach the retina. This vision problem usually is corrected by wearing glasses or contact lenses. Laser surgery also is used to correct nearsightedness by reshaping the cornea.

◀ **Farsightedness** A farsighted person can see distant objects clearly, but cannot focus clearly on nearby objects. Farsightedness results if the eyeball is too short, so light waves from nearby objects have not been brought to a focus when they strike the retina.

▶ Farsightedness also can be corrected by wearing glasses. People commonly become farsighted as they get older because of changes in the lens of the eye. Laser surgery sometimes is used to correct farsightedness.

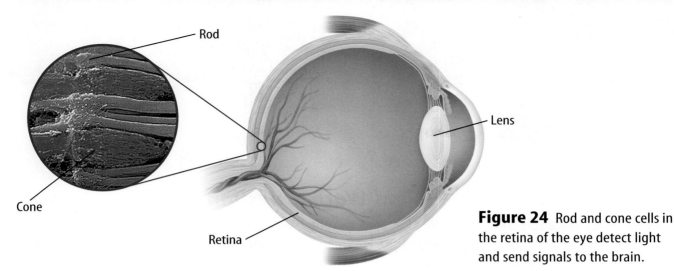

Figure 24 Rod and cone cells in the retina of the eye detect light and send signals to the brain.

Rod and Cone Cells The retina contains over a hundred million light-sensitive cells called rods and cones, shown in **Figure 24**. Rod cells are sensitive to dim light, and cone cells enable you to see colors. There are three types of cone cells. One type is sensitive to red and yellow light, another type is sensitive to green and yellow light, and the third type is sensitive to blue and violet light. The combination of the signals sent to the brain by all three types of cone cells forms the color image that you see.

section 3 review

Summary

Light and Electromagnetic Waves

- Light waves are electromagnetic waves. These waves travel through empty space at a speed of 300,000 km/s.
- Electromagnetic waves are transverse waves made of vibrating electric and magnetic fields.
- Radio waves, infrared waves, visible light, ultraviolet waves, X rays, and gamma rays form the electromagnetic spectrum.
- Most of the electromagnetic waves emitted by the Sun are infrared waves, visible light, and ultraviolet waves.

Color and Vision

- The color of an object is the color of the light the object emits or reflects.
- You see an object when light waves emitted or reflected by the object enter your eye and strike the retina.
- Rod cells and cone cells in the retina of the eye are light-sensitive cells that send signals to the brain when light strikes them.

Self Check

1. **Identify** the electromagnetic waves with the longest wavelengths and the electromagnetic waves with the shortest wavelengths.
2. **Describe** the difference between radio waves, visible light, and gamma rays.
3. **Compare and contrast** the rod cells and the cone cells in the retina of the human eye.
4. **Explain** why most of the electromagnetic waves emitted by the Sun that strike Earth's surface are infrared and visible light waves.
5. **Think Critically** Explain why the brightness of the light emitted by a flashlight decreases as the flashlight moves farther away from you.

Applying Skills

6. **Make a Concept Map** Design a concept map to show the sequence of events that occurs when you see a blue object.
7. **Recognize Cause and Effect** Why does light travel faster through empty space than it does through matter?

Bending Light

Real-World Question

What happens to light waves when they strike the boundary between two materials? Some of the light waves might be reflected from the boundary and some of the waves might travel into the second material. These light waves can change direction and be refracted in the second material. Transmission occurs when the light waves finally pass through the second material. What happens to light waves when they strike a boundary between air and other materials?

Goals
- **Compare and contrast** the reflection, refraction, and transmission of light.
- **Observe** how the refraction of white light can produce different colors of light.

Materials
small piece of cardboard
scissors
tape
flashlight
flat mirror
clear plastic CD case
250-mL beaker
prism

Safety Precautions

Procedure

1. Make a data table similar to the one shown below.

Bending of Light by Different Surfaces		
Surface	How Beam is Affected	Colors Formed
Mirror		
CD case	Do not write in this book.	
Water		
Prism		

2. Cut a slit about 3 cm long and 2 mm wide in a circular piece of cardboard. Tape the cardboard to the face of the flashlight.

3. In a darkened room, shine the flashlight at an angle toward the mirror. Determine whether the flashlight beam is reflected, refracted, or transmitted. Look at the color of the light beam after it strikes the mirror. Has the white light been changed into different colors of light? Record your observations on the chart.

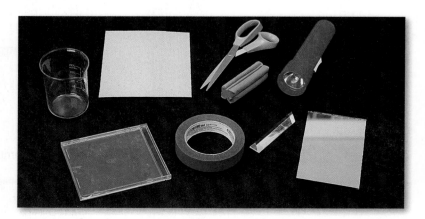

714

Using Scientific Methods

4. Remove the clear plastic front from an empty CD case. Shine the flashlight at an angle toward the plastic. Does transmission occur? Record your observations about how the direction of the beam changes and colors of the light.

5. Fill the beaker with water. Shine the flashlight toward the side of the beaker so that the light shines through the water. Move the light beam from side to side. Record your observations.

6. Shine the flashlight toward a side of the prism. Move the light beam around until you see the outgoing beam spread into different colors. Record your observations.

◉ Analyze Your Data

1. For which objects did reflection occur? For which objects did refraction occur? For which objects did transmission occur?

2. For which objects did refraction cause the flashlight beam to be separated into different colors?

◉ Conclude and Apply

1. **Compare and contrast** the behavior of light waves when they strike the mirror and the CD case.

2. **Explain** why the beam that passes through the CD case does or does not change direction.

3. **Describe** how the light beam changes after it passes through the prism.

Communicating Your Data

Create a sketch showing how light refracts in a prism and divides into different colors.

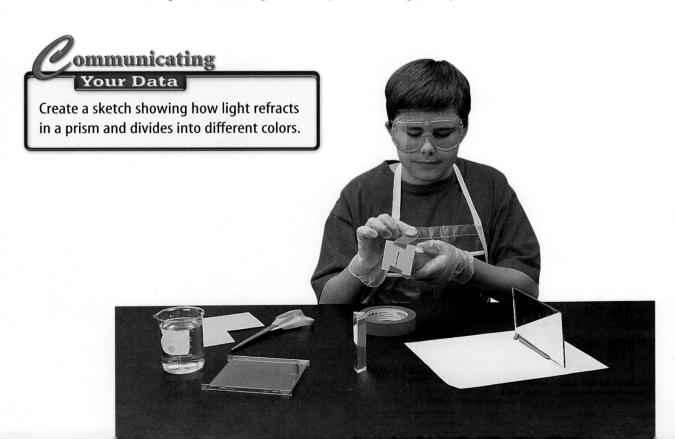

Oops! Accidents in SCIENCE

SOMETIMES GREAT DISCOVERIES HAPPEN BY ACCIDENT!

Jansky's Merry-Go-Round

Before the first radio signals were sent across the Atlantic Ocean in 1902, ships could only communicate if they could see one another. Being able to communicate using radio waves was a real breakthrough. But it wasn't without its problems—namely lots of static. Around 1930, Bell Labs was trying to improve radio communication by using radio waves with shorter wavelengths—between 10 and 20 m. They put Karl Jansky to work finding out what might be causing the static.

Karl Jansky built the first radiotelescope.

An Unexpected Discovery

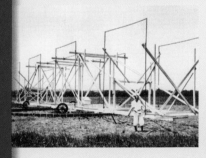

This antenna built by Janksy detected radio waves from the Milky Way galaxy.

Jansky built an antenna to receive radio waves with a wavelength of about 14.5 m. He mounted it on a turntable so that he could rotate it in any direction. His coworkers called it "Jansky's merry-go-round."

After recording signals for several months, Jansky found that there were three types of static. Two were caused by nearby and distant thunderstorms.

But the third was totally unexpected. It seemed to come from the center of our Milky Way galaxy! Jansky wanted to follow up on this unexpected discovery, but Bell Labs had the information it wanted. They were in the telephone business, not astronomy!

A New Branch of Astronomy

Fortunately, other scientists were fascinated with Jansky's find. Grote Reber built a "radiotelescope" in his Illinois backyard. He confirmed Jansky's discovery and did the first systematic survey of radio waves from space. The field of radioastronomy was born.

Previously, astronomers could observe distant galaxies only by gathering the light arriving from their stars. But they couldn't see past the clouds of gas and small particles surrounding the galaxies. Radio waves emitted by a galaxy can penetrate much of the gas and dust in space. This allows radio astronomers to make images of galaxies and other objects they can't see. As a result, Radio astronomy has revealed previously invisible objects such as quasars and pulsars.

The blue-white colors in this image are all you could see without radio waves.

Experiment Research how astronomers convert the radio waves received by radio telescopes into images of galaxies and stars.

For more information, visit blue.msscience.com/oops

chapter 24 Study Guide

Reviewing Main Ideas

Section 1 Waves

1. Waves carry energy from place to place without transporting matter.
2. Transverse waves move particles in matter at right angles to the direction in which the waves travel.
3. Compressional waves move particles back and forth along the same direction in which the waves travel.
4. The speed of a wave equals its wavelength multiplied by its frequency.

Section 2 Sound Waves

1. Sound waves are compressional waves produced by something vibrating.
2. The intensity of sound waves is measured in units of decibels.
3. You hear sound when sound waves reach your ear and cause parts of the ear to vibrate.

Section 3 Electromagnetic Waves and Light

1. Electromagnetic waves are transverse waves that can travel in matter or empty space.
2. Light waves are electromagnetic waves.
3. The range of frequencies and wavelengths of electromagnetic waves forms the electromagnetic spectrum.
4. You see an object when light waves emitted or reflected by the object enter your eye and strike light-sensitive cells inside the eye.

Visualizing Main Ideas

Copy and complete the following concept map on waves.

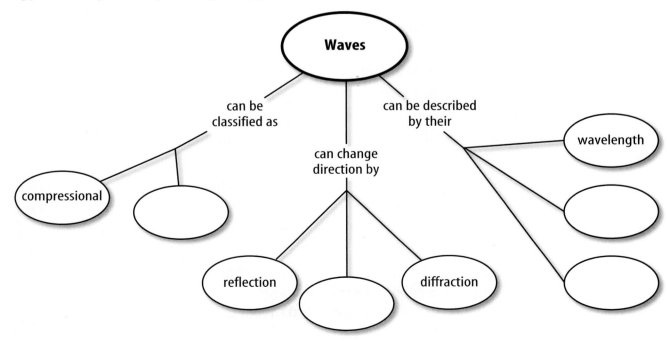

Chapter 24 Review

Using Vocabulary

compressional wave p. 695
diffraction p. 700
electromagnetic spectrum p. 708
electromagnetic waves p. 707
frequency p. 696
infrared waves p. 709
intensity p. 702
law of reflection p. 699
pitch p. 703
refraction p. 699
reverberation p. 705
transverse wave p. 695
ultraviolet waves p. 710
wave p. 694
wavelength p. 696

Complete each statement using a word(s) from the vocabulary list above.

1. The bending of a wave when it moves from one material into another is _____.
2. The bending of waves around an object is due to _____.
3. The _____ is the complete range of electromagnetic wave frequencies and wavelengths.
4. The amount of energy that a wave carries past a certain area each second is the _____.
5. In a(n) _____, the particles in the material move at right angles to the direction the wave moves.
6. The _____ of a wave is the number of wavelengths that pass a point each second.
7. In a _____, particles in the material move back and forth along the direction of wave motion.

Checking Concepts

Choose the word or phrase that best answers the question.

8. If the distance between the crest and trough of a wave is 0.6 m, what is the wave's amplitude?
   A) 0.3 m C) 0.6 m
   B) 1.2 m D) 2.4

9. Which of the following are units for measuring frequency?
   A) decibels C) meters
   B) hertz D) meters/second

10. Through which of these materials does sound travel fastest?
    A) empty space C) steel
    B) water D) air

11. An increase in a sound's pitch corresponds to an increase in what other property?
    A) intensity C) wavelength
    B) frequency D) loudness

12. Soft materials are sometimes used in concert halls to prevent what effect?
    A) refraction C) compression
    B) diffraction D) reverberation

13. Which of the following are not transverse waves?
    A) radio waves C) sound waves
    B) infrared waves D) visible light

14. Which of the following wave properties determines the energy carried by a wave?
    A) amplitude C) wavelength
    B) frequency D) wave speed

15. Which of the following best describes why refraction of a wave occurs when the wave travels from one material into another?
    A) The wavelength increases.
    B) The speed of the wave changes.
    C) The amplitude increases.
    D) The frequency decreases.

16. What produces waves?
    A) sound C) transfer of energy
    B) heat D) vibrations

17. Which of the following has wavelengths longer than the wavelengths of visible light?
    A) X rays C) radio waves
    B) gamma rays D) ultraviolet waves

chapter 24 Review

Thinking Critically

18. **Infer** Radio waves broadcast by a radio station strike your radio and your ear. Infer whether the human ear can hear radio waves. What evidence supports your conclusion?

19. **Solve an Equation** Robotic spacecraft on Mars have sent radio signals back to Earth. The distance from Mars to Earth, at its greatest, is about 401,300,000 km. About how many minutes would it take a signal to reach Earth from that distance?

20. **Recognize Cause and Effect** When a musician plucks a string on a guitar it produces sound with a certain pitch. If the musician then presses down on the string and plucks it, the sound produced has a shorter wavelength. How does the pitch of the sound change?

21. **Interpret Scientific Illustrations** One way that radio waves can carry signals to radios is by varying the amplitude of the wave. This is known as amplitude modulation (AM). Another way is by varying the frequency. This is called frequency modulation (FM). Which of the waves below shows AM, and which shows FM? Explain.

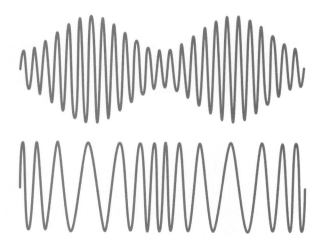

22. **Infer** When light passes through a prism, infer how the amount of bending of a light wave depends on the frequency of the light wave. How does the amount of bending depend on the wavelength of the light wave?

23. **Describe** how the lenses in your eyes change shape when you first look at your wristwatch to read the time, and then look at a mountain in the distance.

Performance Activities

24. **Poster** Investigate a musical instrument to find out how it produces sound. Make a poster showing the instrument and describing how it works.

25. **Model** Make an instrument out of common materials. Present the instrument to the class, and explain how it can produce different pitches.

Applying Math

26. **Noise Levels** A noisy restaurant has an intensity of about 80 dB, and a lawn mower has an intensity of about 110 dB. How many times louder does the lawn mower noise seem?

27. **Wavelength of Sound** Sound waves with a frequency of 150 Hz travel at a speed of 340 m/s. What is the wavelength of the sound waves?

28. **Ultrasound** Physicians sometimes use high-frequency sound waves to diagnose and monitor medical conditions. A typical frequency for the sound waves is about 5,000,000.0 Hz. Sound travels through soft body tissue at about 1500.0 m/s. What is the wavelength of the sound waves?

29. **Frequency of Radio Waves** Find the frequency of radio waves that have a wavelength of 15 m if they are traveling at a speed of 300,000,000 m/s.

Chapter 24 Standardized Test Practice

Part 1 Multiple Choice

Record your answers on the answer sheet provided by your teacher or on a sheet of paper.

1. Which of the following terms refers to the bending of waves around objects?
   A. diffraction
   B. reflection
   C. refraction
   D. transmission

Use the table below to answer questions 2 and 3.

Speed of Sound in Different Materials	
Material	Speed (m/s)
Air (20°C)	343
Glass	5,640
Steel	5,940
Water (25°C)	1,493
Seawater (25°C)	1,533

2. The table above shows the speed of sound through different materials. About how far can sound travel in air in 2.38 s if the air temperature is 20°C?
   A. 144 m
   B. 343 m
   C. 684 m
   D. 816 m

3. Sound travels 2,146 m through a material in 1.4 seconds. What is the material?
   A. air (20°C)
   B. glass
   C. water (25°C)
   D. seawater (25°C)

4. Which of the following are not able to travel through empty space?
   A. gamma rays
   B. ultraviolet waves
   C. sound waves
   D. light waves

Test-Taking Tip

Be Well-Rested Get plenty of sleep—at least eight hours every night—during test week and the week before the test.

5. Which of the following is a true statement?
   A. Waves do not transport the matter through which they travel.
   B. Waves can transport matter through solids, liquids, and gases, but not through empty space.
   C. Waves can transport matter through liquids and solids, but not through gases or empty space.
   D. Sound and water waves can transport matter, but light waves can't.

Use the following table to answer questions 6 and 7.

Decibel Scale	
Sound Source	Loudness (dB)
jet plane taking off	150
running lawn mower	100
average home	50
whisper	15

6. The table above shows typical sound intensity values on a decibel scale. Which of the following would you expect to be the approximate sound intensity of a noisy restaurant?
   A. 20 dB
   B. 40 dB
   C. 80 dB
   D. 120 dB

7. What sound intensity level would you expect to be painful to most humans?
   A. 30 dB
   B. 60 dB
   C. 90 dB
   D. 120 dB

8. What is the maximum range of sound frequencies that humans can hear?
   A. 0 to 150 Hz
   B. 0 to 200 Hz
   C. 20 to 5000 H4
   D. 20 to 20,000 Hz

Standardized Test Practice

Part 2 — Short Response/Grid In

Record your answers on the answer sheet provided by your teacher or on a sheet of paper.

9. If the intensity of a sound increases by 20 dB, by what factor is the energy carried by the sound wave increased?

10. Why do concert halls often have drapes or other soft material on the walls?

Use the illustration below to answer questions 11 and 12.

11. The photograph above shows flashes of lightning strikes during a thunderstorm. Why do you sometimes hear thunder at about the same time you see the flash of lightning, but other times you hear the thunder after the flash?

12. You hear thunder 3.0 seconds after you see a lightning flash. Later you hear thunder 2.5 seconds after you see a flash. If the sound travels at 340 m/s, how much closer was the second lightning strike than the first one?

13. The frequency of a sound is 37.5 Hz, and the sound travels through air at 343 m/s. What is the wavelength of the sound?

14. The speed of all electromagnetic waves through space is 300,000,000 m/s. What is the frequency of a radio wave that has a wavelength of 10 m?

Part 3 — Open Ended

Record your answers on a sheet of paper.

15. Compare and contrast light waves and sound waves.

16. Describe the process that occurs when light waves enter your eye and produce a signal in the optic nerve.

17. Name the different types of electromagnetic waves from longest to shortest wavelength. Give an example of each type.

18. Describe compressional and transverse waves. Explain the difference between them.

19. Explain why sound travels faster through some types of matter than through others. How does temperature affect the speed of sound through a material?

Use the photograph below to answer questions 20 and 21.

20. The girl in the photograph above produces sound by clapping cymbals together. Describe how the cymbals produce sound.

21. What determines the intensity of the sound that the girl produces with the cymbals? How does this affect whether the sound is loud or soft?

22. If you stand near a large tree, you can hear someone talking on the other side of the tree. Explain why you can hear the person, but can't see them.

Student Resources

Student Resources

CONTENTS

Science Skill Handbook 724

Scientific Methods 724
- Identify a Question 724
- Gather and Organize Information 724
- Form a Hypothesis 727
- Test the Hypothesis 728
- Collect Data 728
- Analyze the Data 731
- Draw Conclusions 732
- Communicate 732

Safety Symbols 733

Safety in the Science Laboratory 734
- General Safety Rules 734
- Prevent Accidents 734
- Laboratory Work 734
- Laboratory Cleanup 735
- Emergencies 735

Extra Try at Home Labs 736
- Animal Watch 736
- Guppies of All Colors 736
- Sports Drink Minerals 737
- Rock Creatures 737
- A Light in the Forest 738
- Immovable Echinoderms 738
- Measuring Movement 739
- Earth's Layers 739
- Making Burrows 740
- History in a Bottle 740
- Creating Craters 741
- Many Moons 741
- Big Stars 742
- Make An Electroscope 742
- Research Race 743
- Human Bonding 743
- Mini Fireworks 744
- Measuring Momentum 744
- Friction in Traffic 745
- Toolbox Simple Machines 745
- Estimating Temperature 746
- Bending Water 746
- Testing Magnets 747
- Disappearing Dots 747

Technology Skill Handbook ...748

Computer Skills 748
- Use a Word Processing Program ..748
- Use a Database 749
- Use the Internet 749
- Use a Spreadsheet 750
- Use Graphics Software 750

Presentation Skills 751
- Develop Multimedia Presentations 751
- Computer Presentations 751

Math Skill Handbook 752

Math Review 752
- Use Fractions 752
- Use Ratios 755
- Use Decimals 755
- Use Proportions 756
- Use Percentages 757
- Solve One-Step Equations 757
- Use Statistics 758
- Use Geometry 759

Science Applications 762
- Measure in SI 762
- Dimensional Analysis 762
- Precision and Significant Digits ...764
- Scientific Notation 764
- Make and Use Graphs 765

Reference Handbooks 767
Topographic Map Symbols 767
Physical Science Reference Tables 768
Periodic Table of the Elements 770

English/Spanish Glossary 772

Index 793

Credits 812

STUDENT RESOURCES 723

Scientific Methods

Scientists use an orderly approach called the scientific method to solve problems. This includes organizing and recording data so others can understand them. Scientists use many variations in this method when they solve problems.

Identify a Question

The first step in a scientific investigation or experiment is to identify a question to be answered or a problem to be solved. For example, you might ask which gasoline is the most efficient.

Gather and Organize Information

After you have identified your question, begin gathering and organizing information. There are many ways to gather information, such as researching in a library, interviewing those knowledgeable about the subject, testing and working in the laboratory and field. Fieldwork is investigations and observations done outside of a laboratory.

Researching Information Before moving in a new direction, it is important to gather the information that already is known about the subject. Start by asking yourself questions to determine exactly what you need to know. Then you will look for the information in various reference sources, like the student is doing in **Figure 1.** Some sources may include textbooks, encyclopedias, government documents, professional journals, science magazines, and the Internet. Always list the sources of your information.

Figure 1 The Internet can be a valuable research tool.

Evaluate Sources of Information Not all sources of information are reliable. You should evaluate all of your sources of information, and use only those you know to be dependable. For example, if you are researching ways to make homes more energy efficient, a site written by the U.S. Department of Energy would be more reliable than a site written by a company that is trying to sell a new type of weatherproofing material. Also, remember that research always is changing. Consult the most current resources available to you. For example, a 1985 resource about saving energy would not reflect the most recent findings.

Sometimes scientists use data that they did not collect themselves, or conclusions drawn by other researchers. This data must be evaluated carefully. Ask questions about how the data were obtained, if the investigation was carried out properly, and if it has been duplicated exactly with the same results. Would you reach the same conclusion from the data? Only when you have confidence in the data can you believe it is true and feel comfortable using it.

Interpret Scientific Illustrations As you research a topic in science, you will see drawings, diagrams, and photographs to help you understand what you read. Some illustrations are included to help you understand an idea that you can't see easily by yourself, like the tiny particles in an atom in **Figure 2.** A drawing helps many people to remember details more easily and provides examples that clarify difficult concepts or give additional information about the topic you are studying. Most illustrations have labels or a caption to identify or to provide more information.

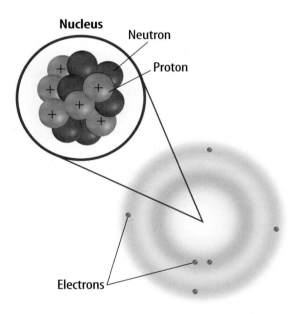

Figure 2 This drawing shows an atom of carbon with its six protons, six neutrons, and six electrons.

Concept Maps One way to organize data is to draw a diagram that shows relationships among ideas (or concepts). A concept map can help make the meanings of ideas and terms more clear, and help you understand and remember what you are studying. Concept maps are useful for breaking large concepts down into smaller parts, making learning easier.

Network Tree A type of concept map that not only shows a relationship, but how the concepts are related is a network tree, shown in **Figure 3.** In a network tree, the words are written in the ovals, while the description of the type of relationship is written across the connecting lines.

When constructing a network tree, write down the topic and all major topics on separate pieces of paper or notecards. Then arrange them in order from general to specific. Branch the related concepts from the major concept and describe the relationship on the connecting line. Continue to more specific concepts until finished.

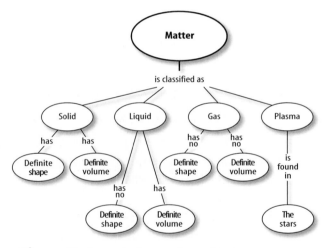

Figure 3 A network tree shows how concepts or objects are related.

Events Chain Another type of concept map is an events chain. Sometimes called a flow chart, it models the order or sequence of items. An events chain can be used to describe a sequence of events, the steps in a procedure, or the stages of a process.

When making an events chain, first find the one event that starts the chain. This event is called the initiating event. Then, find the next event and continue until the outcome is reached, as shown in **Figure 4.**

SCIENCE SKILL HANDBOOK 725

Science Skill Handbook

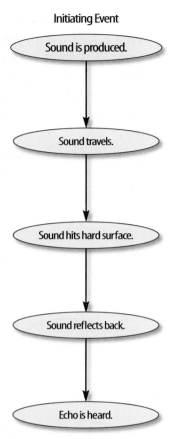

Figure 4 Events-chain concept maps show the order of steps in a process or event. This concept map shows how a sound makes an echo.

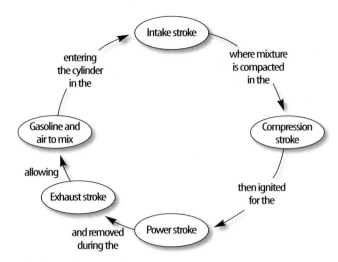

Figure 5 A cycle map shows events that occur in a cycle.

Cycle Map A specific type of events chain is a cycle map. It is used when the series of events do not produce a final outcome, but instead relate back to the beginning event, such as in **Figure 5.** Therefore, the cycle repeats itself.

To make a cycle map, first decide what event is the beginning event. This is also called the initiating event. Then list the next events in the order that they occur, with the last event relating back to the initiating event. Words can be written between the events that describe what happens from one event to the next. The number of events in a cycle map can vary, but usually contain three or more events.

Spider Map A type of concept map that you can use for brainstorming is the spider map. When you have a central idea, you might find that you have a jumble of ideas that relate to it but are not necessarily clearly related to each other. The spider map on sound in **Figure 6** shows that if you write these ideas outside the main concept, then you can begin to separate and group unrelated terms so they become more useful.

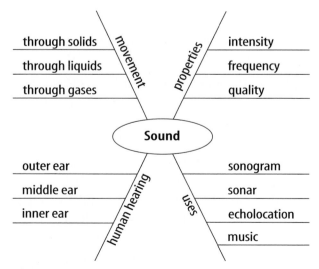

Figure 6 A spider map allows you to list ideas that relate to a central topic but not necessarily to one another.

726 STUDENT RESOURCES

Science Skill Handbook

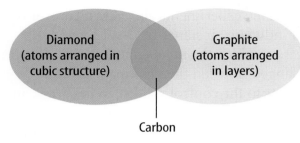

Figure 7 This Venn diagram compares and contrasts two substances made from carbon.

Venn Diagram To illustrate how two subjects compare and contrast you can use a Venn diagram. You can see the characteristics that the subjects have in common and those that they do not, shown in **Figure 7**.

To create a Venn diagram, draw two overlapping ovals that that are big enough to write in. List the characteristics unique to one subject in one oval, and the characteristics of the other subject in the other oval. The characteristics in common are listed in the overlapping section.

Make and Use Tables One way to organize information so it is easier to understand is to use a table. Tables can contain numbers, words, or both.

To make a table, list the items to be compared in the first column and the characteristics to be compared in the first row. The title should clearly indicate the content of the table, and the column or row heads should be clear. Notice that in **Table 1** the units are included.

Table 1 Recyclables Collected During Week			
Day of Week	Paper (kg)	Aluminum (kg)	Glass (kg)
Monday	5.0	4.0	12.0
Wednesday	4.0	1.0	10.0
Friday	2.5	2.0	10.0

Make a Model One way to help you better understand the parts of a structure, the way a process works, or to show things too large or small for viewing is to make a model. For example, an atomic model made of a plastic-ball nucleus and pipe-cleaner electron shells can help you visualize how the parts of an atom relate to each other. Other types of models can by devised on a computer or represented by equations.

Form a Hypothesis

A possible explanation based on previous knowledge and observations is called a hypothesis. After researching gasoline types and recalling previous experiences in your family's car you form a hypothesis—our car runs more efficiently because we use premium gasoline. To be valid, a hypothesis has to be something you can test by using an investigation.

Predict When you apply a hypothesis to a specific situation, you predict something about that situation. A prediction makes a statement in advance, based on prior observation, experience, or scientific reasoning. People use predictions to make everyday decisions. Scientists test predictions by performing investigations. Based on previous observations and experiences, you might form a prediction that cars are more efficient with premium gasoline. The prediction can be tested in an investigation.

Design an Experiment A scientist needs to make many decisions before beginning an investigation. Some of these include: how to carry out the investigation, what steps to follow, how to record the data, and how the investigation will answer the question. It also is important to address any safety concerns.

Science Skill Handbook

Test the Hypothesis

Now that you have formed your hypothesis, you need to test it. Using an investigation, you will make observations and collect data, or information. This data might either support or not support your hypothesis. Scientists collect and organize data as numbers and descriptions.

Follow a Procedure In order to know what materials to use, as well as how and in what order to use them, you must follow a procedure. **Figure 8** shows a procedure you might follow to test your hypothesis.

Procedure
1. Use regular gasoline for two weeks.
2. Record the number of kilometers between fill-ups and the amount of gasoline used.
3. Switch to premium gasoline for two weeks.
4. Record the number of kilometers between fill-ups and the amount of gasoline used.

Figure 8 A procedure tells you what to do step by step.

Identify and Manipulate Variables and Controls In any experiment, it is important to keep everything the same except for the item you are testing. The one factor you change is called the independent variable. The change that results is the dependent variable. Make sure you have only one independent variable, to assure yourself of the cause of the changes you observe in the dependent variable. For example, in your gasoline experiment the type of fuel is the independent variable. The dependent variable is the efficiency.

Many experiments also have a control—an individual instance or experimental subject for which the independent variable is not changed. You can then compare the test results to the control results. To design a control you can have two cars of the same type. The control car uses regular gasoline for four weeks. After you are done with the test, you can compare the experimental results to the control results.

Collect Data

Whether you are carrying out an investigation or a short observational experiment, you will collect data, as shown in **Figure 9**. Scientists collect data as numbers and descriptions and organize it in specific ways.

Observe Scientists observe items and events, then record what they see. When they use only words to describe an observation, it is called qualitative data. Scientists' observations also can describe how much there is of something. These observations use numbers, as well as words, in the description and are called quantitative data. For example, if a sample of the element gold is described as being "shiny and very dense" the data are qualitative. Quantitative data on this sample of gold might include "a mass of 30 g and a density of 19.3 g/cm^3."

Figure 9 Collecting data is one way to gather information directly.

Science Skill Handbook

Figure 10 Record data neatly and clearly so it is easy to understand.

When you make observations you should examine the entire object or situation first, and then look carefully for details. It is important to record observations accurately and completely. Always record your notes immediately as you make them, so you do not miss details or make a mistake when recording results from memory. Never put unidentified observations on scraps of paper. Instead they should be recorded in a notebook, like the one in **Figure 10.** Write your data neatly so you can easily read it later. At each point in the experiment, record your observations and label them. That way, you will not have to determine what the figures mean when you look at your notes later. Set up any tables that you will need to use ahead of time, so you can record any observations right away. Remember to avoid bias when collecting data by not including personal thoughts when you record observations. Record only what you observe.

Estimate Scientific work also involves estimating. To estimate is to make a judgment about the size or the number of something without measuring or counting. This is important when the number or size of an object or population is too large or too difficult to accurately count or measure.

Sample Scientists may use a sample or a portion of the total number as a type of estimation. To sample is to take a small, representative portion of the objects or organisms of a population for research. By making careful observations or manipulating variables within that portion of the group, information is discovered and conclusions are drawn that might apply to the whole population. A poorly chosen sample can be unrepresentative of the whole. If you were trying to determine the rainfall in an area, it would not be best to take a rainfall sample from under a tree.

Measure You use measurements everyday. Scientists also take measurements when collecting data. When taking measurements, it is important to know how to use measuring tools properly. Accuracy also is important.

Length To measure length, the distance between two points, scientists use meters. Smaller measurements might be measured in centimeters or millimeters.

Length is measured using a metric ruler or meter stick. When using a metric ruler, line up the 0-cm mark with the end of the object being measured and read the number of the unit where the object ends. Look at the metric ruler shown in **Figure 11.** The centimeter lines are the long, numbered lines, and the shorter lines are millimeter lines. In this instance, the length would be 4.50 cm.

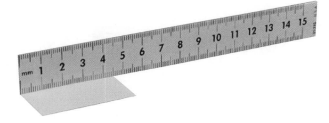

Figure 11 This metric ruler has centimeter and millimeter divisions.

Science Skill Handbook

Mass The SI unit for mass is the kilogram (kg). Scientists can measure mass using units formed by adding metric prefixes to the unit gram (g), such as milligram (mg). To measure mass, you might use a triple-beam balance similar to the one shown in **Figure 12**. The balance has a pan on one side and a set of beams on the other side. Each beam has a rider that slides on the beam.

When using a triple-beam balance, place an object on the pan. Slide the largest rider along its beam until the pointer drops below zero. Then move it back one notch. Repeat the process for each rider proceeding from the larger to smaller until the pointer swings an equal distance above and below the zero point. Sum the masses on each beam to find the mass of the object. Move all riders back to zero when finished.

Instead of putting materials directly on the balance, scientists often take a tare of a container. A tare is the mass of a container into which objects or substances are placed for measuring their masses. To mass objects or substances, find the mass of a clean container. Remove the container from the pan, and place the object or substances in the container. Find the mass of the container with the materials in it. Subtract the mass of the empty container from the mass of the filled container to find the mass of the materials you are using.

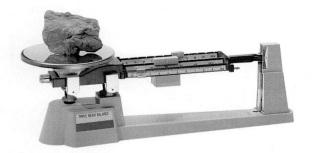

Figure 12 A triple-beam balance is used to determine the mass of an object.

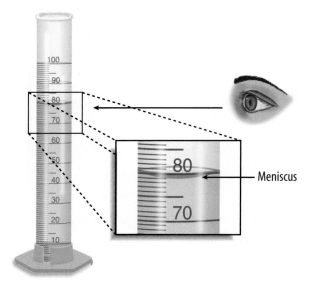

Figure 13 Graduated cylinders measure liquid volume.

Liquid Volume To measure liquids, the unit used is the liter. When a smaller unit is needed, scientists might use a milliliter. Because a milliliter takes up the volume of a cube measuring 1 cm on each side it also can be called a cubic centimeter (cm^3 = cm × cm × cm).

You can use beakers and graduated cylinders to measure liquid volume. A graduated cylinder, shown in **Figure 13,** is marked from bottom to top in milliliters. In lab, you might use a 10-mL graduated cylinder or a 100-mL graduated cylinder. When measuring liquids, notice that the liquid has a curved surface. Look at the surface at eye level, and measure the bottom of the curve. This is called the meniscus. The graduated cylinder in **Figure 13** contains 79.0 mL, or 79.0 cm^3, of a liquid.

Temperature Scientists often measure temperature using the Celsius scale. Pure water has a freezing point of 0°C and boiling point of 100°C. The unit of measurement is degrees Celsius. Two other scales often used are the Fahrenheit and Kelvin scales.

Science Skill Handbook

Figure 14 A thermometer measures the temperature of an object.

Scientists use a thermometer to measure temperature. Most thermometers in a laboratory are glass tubes with a bulb at the bottom end containing a liquid such as colored alcohol. The liquid rises or falls with a change in temperature. To read a glass thermometer like the thermometer in **Figure 14,** rotate it slowly until a red line appears. Read the temperature where the red line ends.

Form Operational Definitions An operational definition defines an object by how it functions, works, or behaves. For example, when you are playing hide and seek and a tree is home base, you have created an operational definition for a tree.

Objects can have more than one operational definition. For example, a ruler can be defined as a tool that measures the length of an object (how it is used). It can also be a tool with a series of marks used as a standard when measuring (how it works).

Analyze the Data

To determine the meaning of your observations and investigation results, you will need to look for patterns in the data. Then you must think critically to determine what the data mean. Scientists use several approaches when they analyze the data they have collected and recorded. Each approach is useful for identifying specific patterns.

Interpret Data The word *interpret* means "to explain the meaning of something." When analyzing data from an experiement, try to find out what the data show. Identify the control group and the test group to see whether or not changes in the independent variable have had an effect. Look for differences in the dependent variable between the control and test groups.

Classify Sorting objects or events into groups based on common features is called classifying. When classifying, first observe the objects or events to be classified. Then select one feature that is shared by some members in the group, but not by all. Place those members that share that feature in a subgroup. You can classify members into smaller and smaller subgroups based on characteristics. Remember that when you classify, you are grouping objects or events for a purpose. Keep your purpose in mind as you select the features to form groups and subgroups.

Compare and Contrast Observations can be analyzed by noting the similarities and differences between two more objects or events that you observe. When you look at objects or events to see how they are similar, you are comparing them. Contrasting is looking for differences in objects or events.

SCIENCE SKILL HANDBOOK 731

Science Skill Handbook

Recognize Cause and Effect A cause is a reason for an action or condition. The effect is that action or condition. When two events happen together, it is not necessarily true that one event caused the other. Scientists must design a controlled investigation to recognize the exact cause and effect.

Draw Conclusions

When scientists have analyzed the data they collected, they proceed to draw conclusions about the data. These conclusions are sometimes stated in words similar to the hypothesis that you formed earlier. They may confirm a hypothesis, or lead you to a new hypothesis.

Infer Scientists often make inferences based on their observations. An inference is an attempt to explain observations or to indicate a cause. An inference is not a fact, but a logical conclusion that needs further investigation. For example, you may infer that a fire has caused smoke. Until you investigate, however, you do not know for sure.

Apply When you draw a conclusion, you must apply those conclusions to determine whether the data supports the hypothesis. If your data do not support your hypothesis, it does not mean that the hypothesis is wrong. It means only that the result of the investigation did not support the hypothesis. Maybe the experiment needs to be redesigned, or some of the initial observations on which the hypothesis was based were incomplete or biased. Perhaps more observation or research is needed to refine your hypothesis. A successful investigation does not always come out the way you originally predicted.

Avoid Bias Sometimes a scientific investigation involves making judgments. When you make a judgment, you form an opinion. It is important to be honest and not to allow any expectations of results to bias your judgments. This is important throughout the entire investigation, from researching to collecting data to drawing conclusions.

Communicate

The communication of ideas is an important part of the work of scientists. A discovery that is not reported will not advance the scientific community's understanding or knowledge. Communication among scientists also is important as a way of improving their investigations.

Scientists communicate in many ways, from writing articles in journals and magazines that explain their investigations and experiments, to announcing important discoveries on television and radio. Scientists also share ideas with colleagues on the Internet or present them as lectures, like the student is doing in **Figure 15.**

Figure 15 A student communicates to his peers about his investigation.

Science Skill Handbook

SAFETY SYMBOLS

SAFETY SYMBOLS	HAZARD	EXAMPLES	PRECAUTION	REMEDY
DISPOSAL	Special disposal procedures need to be followed.	certain chemicals, living organisms	Do not dispose of these materials in the sink or trash can.	Dispose of wastes as directed by your teacher.
BIOLOGICAL	Organisms or other biological materials that might be harmful to humans	bacteria, fungi, blood, unpreserved tissues, plant materials	Avoid skin contact with these materials. Wear mask or gloves.	Notify your teacher if you suspect contact with material. Wash hands thoroughly.
EXTREME TEMPERATURE	Objects that can burn skin by being too cold or too hot	boiling liquids, hot plates, dry ice, liquid nitrogen	Use proper protection when handling.	Go to your teacher for first aid.
SHARP OBJECT	Use of tools or glassware that can easily puncture or slice skin	razor blades, pins, scalpels, pointed tools, dissecting probes, broken glass	Practice common-sense behavior and follow guidelines for use of the tool.	Go to your teacher for first aid.
FUME	Possible danger to respiratory tract from fumes	ammonia, acetone, nail polish remover, heated sulfur, moth balls	Make sure there is good ventilation. Never smell fumes directly. Wear a mask.	Leave foul area and notify your teacher immediately.
ELECTRICAL	Possible danger from electrical shock or burn	improper grounding, liquid spills, short circuits, exposed wires	Double-check setup with teacher. Check condition of wires and apparatus.	Do not attempt to fix electrical problems. Notify your teacher immediately.
IRRITANT	Substances that can irritate the skin or mucous membranes of the respiratory tract	pollen, moth balls, steel wool, fiberglass, potassium permanganate	Wear dust mask and gloves. Practice extra care when handling these materials.	Go to your teacher for first aid.
CHEMICAL	Chemicals can react with and destroy tissue and other materials	bleaches such as hydrogen peroxide; acids such as sulfuric acid, hydrochloric acid; bases such as ammonia, sodium hydroxide	Wear goggles, gloves, and an apron.	Immediately flush the affected area with water and notify your teacher.
TOXIC	Substance may be poisonous if touched, inhaled, or swallowed.	mercury, many metal compounds, iodine, poinsettia plant parts	Follow your teacher's instructions.	Always wash hands thoroughly after use. Go to your teacher for first aid.
FLAMMABLE	Flammable chemicals may be ignited by open flame, spark, or exposed heat.	alcohol, kerosene, potassium permanganate	Avoid open flames and heat when using flammable chemicals.	Notify your teacher immediately. Use fire safety equipment if applicable.
OPEN FLAME	Open flame in use, may cause fire.	hair, clothing, paper, synthetic materials	Tie back hair and loose clothing. Follow teacher's instruction on lighting and extinguishing flames.	Notify your teacher immediately. Use fire safety equipment if applicable.

 Eye Safety Proper eye protection should be worn at all times by anyone performing or observing science activities.

 Clothing Protection This symbol appears when substances could stain or burn clothing.

 Animal Safety This symbol appears when safety of animals and students must be ensured.

 Handwashing After the lab, wash hands with soap and water before removing goggles.

Safety in the Science Laboratory

The science laboratory is a safe place to work if you follow standard safety procedures. Being responsible for your own safety helps to make the entire laboratory a safer place for everyone. When performing any lab, read and apply the caution statements and safety symbol listed at the beginning of the lab.

General Safety Rules

1. Obtain your teacher's permission to begin all investigations and use laboratory equipment.

2. Study the procedure. Ask your teacher any questions. Be sure you understand safety symbols shown on the page.

3. Notify your teacher about allergies or other health conditions which can affect your participation in a lab.

4. Learn and follow use and safety procedures for your equipment. If unsure, ask your teacher.

5. Never eat, drink, chew gum, apply cosmetics, or do any personal grooming in the lab. Never use lab glassware as food or drink containers. Keep your hands away from your face and mouth.

6. Know the location and proper use of the safety shower, eye wash, fire blanket, and fire alarm.

Prevent Accidents

1. Use the safety equipment provided to you. Goggles and a safety apron should be worn during investigations.

2. Do NOT use hair spray, mousse, or other flammable hair products. Tie back long hair and tie down loose clothing.

3. Do NOT wear sandals or other open-toed shoes in the lab.

4. Remove jewelry on hands and wrists. Loose jewelry, such as chains and long necklaces, should be removed to prevent them from getting caught in equipment.

5. Do not taste any substances or draw any material into a tube with your mouth.

6. Proper behavior is expected in the lab. Practical jokes and fooling around can lead to accidents and injury.

7. Keep your work area uncluttered.

Laboratory Work

1. Collect and carry all equipment and materials to your work area before beginning a lab.

2. Remain in your own work area unless given permission by your teacher to leave it.

734 STUDENT RESOURCES

Science Skill Handbook

3. Always slant test tubes away from yourself and others when heating them, adding substances to them, or rinsing them.

4. If instructed to smell a substance in a container, hold the container a short distance away and fan vapors towards your nose.

5. Do NOT substitute other chemicals/substances for those in the materials list unless instructed to do so by your teacher.

6. Do NOT take any materials or chemicals outside of the laboratory.

7. Stay out of storage areas unless instructed to be there and supervised by your teacher.

Laboratory Cleanup

1. Turn off all burners, water, and gas, and disconnect all electrical devices.

2. Clean all pieces of equipment and return all materials to their proper places.

3. Dispose of chemicals and other materials as directed by your teacher. Place broken glass and solid substances in the proper containers. Never discard materials in the sink.

4. Clean your work area.

5. Wash your hands with soap and water thoroughly BEFORE removing your goggles.

Emergencies

1. Report any fire, electrical shock, glassware breakage, spill, or injury, no matter how small, to your teacher immediately. Follow his or her instructions.

2. If your clothing should catch fire, STOP, DROP, and ROLL. If possible, smother it with the fire blanket or get under a safety shower. NEVER RUN.

3. If a fire should occur, turn off all gas and leave the room according to established procedures.

4. In most instances, your teacher will clean up spills. Do NOT attempt to clean up spills unless you are given permission and instructions to do so.

5. If chemicals come into contact with your eyes or skin, notify your teacher immediately. Use the eyewash or flush your skin or eyes with large quantities of water.

6. The fire extinguisher and first-aid kit should only be used by your teacher unless it is an extreme emergency and you have been given permission.

7. If someone is injured or becomes ill, only a professional medical provider or someone certified in first aid should perform first-aid procedures.

EXTRA  Labs

From Your Kitchen, Junk Drawer, or Yard

1 Animal Watch

Real-World Question
What does your favorite wild animal do?

Possible Materials
- meterstick
- metric ruler
- binoculars
- hand lens
- microscope slide
- aquatic net
- insect net
- collecting jar
- hiking equipment
- waders or boots

Procedure
1. Choose a wild animal to observe. You may choose a common animal such as an ant, squirrel, or backyard bird, or you may choose an animal that only lives in a forest or stream.
2. Create a data chart to record your observations and measurements about your animal. Consider using an electronically generated chart.
3. Observe the physical characteristics of your animal including its approximate size, color, and distinct features.
4. Observe the behavior of your animal. If possible, observe what it eats and how it behaves. Try to measure the distance it travels.
5. Record all your observations and measurements in your data chart.
6. Compare your data with the data collected by your classmates.

Conclude and Apply
1. Describe several new facts you learned about your animal.
2. Explain why careful observations are a vital skill for life scientists.

2 Guppies of All Colors

Real-World Question
How can the effects of selective breeding be observed?

Possible Materials
- metric ruler
- Science Journal
- pencil
- access to a pet store

Procedure
1. Go to a pet store with an adult and ask an employee if you would be able to observe the store's guppies for a school assignment.
2. Ask the employee to show you the aquaria housing plain guppies, which are sold as "feeder fish," and fancy tail guppies.
3. Observe the different varieties of plain guppies. Estimate their average body length and tail size. Describe their colors.
4. Observe all the fancy tail guppies. Estimate their average body length and tail size. Describe their colors.
5. Record any other differences in the traits of plain and fancy tail guppies you observe.

Conclude and Apply
1. Compare the traits of the plain guppies to the fancy tail guppies.
2. Infer why the fancy tail guppies have such a large variety of colors.

Adult supervision required for all labs.

Extra Try at Home Labs

3 Sports Drink Minerals

Real-World Question
How do the minerals compare in sports drinks and natural foods?

Possible Materials
- sport drink bottle labels (several different brands)
- orange juice bottle or carton

Procedure
1. Create a data chart in your Science Journal to record your data. Make a *Food and Drinks* column, and columns for two important minerals needed by athletes, *Potassium* and *Sodium*.
2. Read the Nutrition Facts labels of the sport drinks and the orange juice. For a large variety of drinks, visit your local grocery store with an adult and record the information you need. In your data chart, record the amounts of sodium and potassium found in each.
3. Research the amount of potassium and sodium in a banana or other type of fruit. Record these numbers on your data chart.

Conclude and Apply
1. Compare the amount of potassium and sodium found in the different drinks and foods.
2. Infer why a person may not need sodium in a sports drink.
3. The minerals in a sports drink are not needed unless a person exercises continually for at least an hour. Interview several of your friends who drink sports drinks and evaluate whether or not they need the drink.

4 Rock Creatures

Real-World Question
What types of organisms live under stream rocks?

Possible Materials
- waterproof boots
- ice cube tray (white)
- aquarium net
- bucket
- collecting jars
- guidebook to pond life

Procedure
1. With permission, search under the rocks of a local stream. Look for aquatic organisms under the rocks and leaves of the stream. Compare what you find in fast- and slow-moving water.
2. With permission, carefully pull organisms you find off the rocks and put them into separate compartments of your ice cube tray. Take care not to injure the creatures you find.
3. Use your net and bucket to collect larger organisms.
4. Use your guidebook to pond life to identify the organisms you find.
5. Release the organisms back into the stream once you identify them.

Conclude and Apply
1. Identify and list the organisms you found under the stream rocks.
2. Infer why so many aquatic organisms make their habitats beneath stream rocks.

Adult supervision required for all labs.

Extra Try at Home Labs

5 A Light in the Forest

Real-World Question
Does the amount of sunlight vary in a forest?

Possible Materials
- empty toilet paper or paper towel roll
- Science Journal

Procedure
1. Copy the data table into your Science Journal.
2. Go with an adult to a nearby forest or large grove of trees.
3. Stand near the edge of the forest and look straight up through your cardboard tube. Estimate the percentage of blue sky and clouds you can see in the circle. This percentage is the amount of sunlight reaching the forest floor.
4. Record your location and estimated percentage of sunlight in your data table.
5. Test several other locations in the forest. Choose places where the trees completely cover the forest floor and where sunlight is partially coming through.

Data Table

Location	% of Sunlight

Conclude and Apply
1. Explain how the amount of sunlight reaching the forest floor changed from place to place.
2. Infer why it is important for leaves and branches to stop sunlight from reaching much of the forest floor.

6 Echinoderm Hold

Real-World Question
How do echinoderms living in intertidal ecosystems hold on to rocks?

Possible Materials
- plastic suction cup
- water
- paper towel or sponge

Procedure
1. Moisten a paper towel or sponge with water.
2. Press a plastic suction cup on the moist towel or sponge until the entire bottom surface of the cup is wet.
3. Firmly press the suction cup down on a kitchen counter for 10 s.
4. Grab the top handle of the suction cup and try removing the cup from the counter by pulling it straight up.

Conclude and Apply
1. Describe what happened when you tried to remove the cup from the counter.
2. Infer how echinoderms living in intertidal ecosystems withstand the constant pull of ocean waves and currents.

Extra Try at Home Labs

7 Measuring Movement

Real-World Question
How can we model continental drift?

Possible Materials
- flashlight, nail, rubber band or tape, thick circle of paper
- protractor
- mirror
- stick-on notepad paper
- marker
- metric ruler
- calculator

Procedure
1. Cut a circle of paper to fit around the lens of the flashlight. Use a nail to make a hole in the paper. Fasten the paper with the rubber band or tape. You should now have a flashlight that shines a focused beam of light.
2. Direct the light beam of the flashlight on a protractor held horizontally so that the beam lines up to the 90° mark.
3. Darken a room and aim the light beam at a mirror from an angle. Measure the angle. Observe where the reflected beam hits the wall.
4. Have a partner place a stick-on note on the wall and mark the location of the beam on the paper with a marker.
5. Move the flashlight to a 100° angle and mark the beam's location on the wall with a second note.
6. Measure the distance between the two points on the wall and divide by ten to determine the distance per degree.

Conclude and Apply
1. What was the distance per degree of your measurements?
2. Calculate what the distance would be between the first spot and a third spot marking the location of the flashlight at a 40° angle. Test your calculations.
3. Explain how this lab models measuring continental drift.

8 Earth's Layers

Real-World Question
What is the relative thickness of Earth's different layers?

Possible Materials
- meterstick
- masking tape

Procedure
1. Use a piece of masking tape to mark a spot on the floor. This spot represents the center of Earth.
2. Measure a distance of 1.22 m from the first tape mark and place a second piece of tape.
3. From the second piece of tape, measure a distance of 2.27 m and place a third piece of tape.
4. From the third piece of tape, measure a distance of 2.89 m and place a fourth piece of tape.
5. From the fourth piece of tape, make two measurements. Measure a distance of 0.005 m and a distance of 0.06 m. Place two more pieces of tape to mark these two distances.

Conclude and Apply
1. Identify the name of each of the levels you drew.
2. Calculate the scale you used for the thickness of your earth layers.

Adult supervision required for all labs.

Extra Try at Home Labs

9 Making Burrows

Real-World Question
How does burrowing affect sediment layers?

Possible Materials
- clear-glass bowl
- white flour
- colored gelatin powder (3 packages)
- paintbrush
- pencil

Procedure
1. Add 3 cm of white flour to the bowl. Flatten the top of the flour layer.
2. Carefully sprinkle gelatin powder over the flour to form a colored layer about 0.25 cm thick.
3. The two layers represent two different layers of sediment.
4. Use a paintbrush or pencil to make "burrows" in the "sediment."
5. Make sure to make some of the burrows at the edge of the bowl so that you can see how it affects the sediment.
6. Continue to make more burrows and observe the effect on the two layers.

Conclude and Apply
1. How did the two layers of powder change as you continued to make burrows?
2. Were the "trace fossils" easy to recognize at first? How about after a lot of burrowing?
3. How do you think burrowing animals affect layers of sediment on the ocean floor? How could this burrowing be recognized in rock?

10 History in a Bottle

Real-World Question
What does the geologic column look like?

Possible Materials
- clear-plastic 2-liter bottle
- scissors
- 3-in × 5-in index cards
- colored markers
- permanent marker
- transparent tape
- metric ruler
- sand (3 different colors)
- aquarium gravel (3 different colors)

Procedure
1. Cut the top 5 cm off a clear-plastic 2-L soda bottle. Remove the label.
2. Cut 12 square cards measuring 2 cm × 2 cm.
3. Draw a picture of a trilobite, coral, fish, amphibian, insect, reptile, mouse, coniferous tree, dinosaur, bird, flower, large mammal, and human on the 12 cards.
4. Starting at the bottom, inside of the bottle and working up, tape the trilobite, coral, fish, amphibian, insect, and reptile pictures face out in that order. The reptile picture should be about half way up the bottle.
5. Pour red sand into the bottle until it covers your reptile picture.
6. Tape the mouse, conifer tree, dinosaur, bird, and flower pictures above the red sand in that order. Pour in blue sand until it covers the flower picture.
7. Tape the large mammal and human pictures above the blue sand. Pour green sand into the bottle until it covers the person.

Conclude and Apply
1. Research what era each color of sand represents.
2. Infer why few fossils of organisms living before the Paleozoic Era are found.

Extra Try at Home Labs

11 Creating Craters

Real-World Question
Why does the Moon have craters?

Possible Materials
- drink mix or powdered baby formula
- black pepper or paprika
- large, deep cooking tray or large bowl
- marbles
- small, round candies
- aquarium gravel
- tweezers
- bag of cotton balls

Procedure
1. Pour a 3-cm layer of powder over the bottom of a large, deep cooking tray.
2. Sprinkle a fine layer of black pepper over the powder.
3. Lay a 2–3 cm layer of cotton over half of the powder.
4. Drop marbles and other small objects into the powder not covered by the cotton. Carefully remove the objects with tweezers and observe the craters and impact patterns they make.
5. Drop objects on to the half of the tray covered by cotton.
6. Remove the objects and cotton and observe the marks made by objects in the powder.

Conclude and Apply
1. Compare the impacts made by the objects in the powder not covered by cotton with the impacts in the powder covered by cotton.
2. Infer why the Moon has many craters on its surface but Earth does not.

12 Many Moons

Real-World Question
How do the number of moons of the nine planets compare?

Possible Materials
- golf balls (5)
- softballs (4)
- colored construction paper
- hole puncher
- pennies (10)
- quarters (8)
- meterstick

Procedure
1. Lay the golf balls and softballs on the floor in a row to represent the nine planets. The golf balls should represent the terrestrial planets and the softballs the gas planets.
2. Next to the golf ball representing Earth place one quarter. A quarter represents a moon with a diameter greater than 1,000 km. Research which planets have moons this size and place quarters next to them.
3. Use pennies to represent moons with a diameter between 200–1,000 km. Place pennies next to the planets with moons this size.
4. Use a hole punch to punch out holes from colored construction paper. These holes represent moons smaller than 200 km in diameter. Research which planets have moons this size and place the holes next to them.

Conclude and Apply
1. Infer why terrestrial planets have fewer moons than gas planets.
2. Infer why astronomers do not believe all the moons in the solar system have been discovered.

Adult supervision required for all labs.

Extra Try at Home Labs

13 Big Stars

Real-World Question
How does the size of Earth compare to the size of stars?

Possible Materials
- metric ruler
- meterstick
- tape measure
- masking tape
- white paper
- black marker

Procedure
1. Tape a sheet of white paper to the floor.
2. Draw a dot in the center to the paper. Measure a 1-mm distance from the dot and draw a second dot. This distance represents the diameter of Earth.
3. Measure a distance of 10.9 cm from the first dot and draw a third dot. This distance represents the diameter of the Sun.
4. Measure a distance of 5 m from the first dot and mark the location on the floor with a piece of masking tape. This distance represents the average diameter of a red giant star.
5. Measure a distance of 30 m from the first dot and mark the location on the floor with a piece of masking tape. This distance represents the diameter of the supergiant star Antares.

Conclude and Apply
1. The diameter of Earth is 12,756 km. What is the diameter of the Sun?
2. What is the diameter of an average red giant?

14 Make an Electroscope

Real-World Question
How can you test the radiation coming from an old television or your smoke detector?

Materials
- glass jar
- thin cardboard
- paper clip
- aluminum foil
- hammer and nail
- tape
- plastic rod
- fur, wool, or cotton cloth

Procedure
1. Cut two identical pieces of foil about 1.25 cm by 2.5 cm.
2. Hang the two pieces of foil side by side on one loop of the paper clip.
3. Straighten the other loop of the paper clip. Use the hammer and nail to tap a small hole in the cardboard lid, and push the paper clip through the hole so that the straightened portion of the paper clip sticks out. If necessary, tape the clip in place at right angles to the card. Put the lid on the jar.
4. Charge the plastic rod by rubbing with the cloth or fur. Touch it to the straightened portion of the paper clip. The leaves of foil will get equal charges and repel each other.
5. Observe how much time it takes for the leaves to lose their charge and fall back together. Now, recharge your electroscope and bring it near an old television or smoke detector. If radiation is present, the leaves will fall back together much more quickly than they did without the radiation present.

Conclude and Apply
1. Why do you think the foil leaves discharge faster when there is ionizing radiation present?
2. If you touch the top of the electroscope with your finger, the leaves fall back together. Why is this?

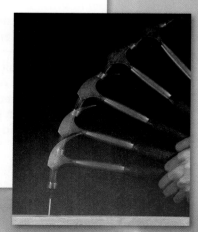

Extra Try at Home Labs

15 Research Race

▶ Real-World Question
How many secrets of the periodic table's elements can you find by research?

Possible Materials
- reference materials
- access to library

▶ Procedure
1. Get together with a team of your friends. Look at the Race Questions and divide them between you.
2. Try to get as many answers as you can in a certain amount of time.
3. Be sure to keep a record of each resource. You get a point for each correct answer. You also get a point for each properly listed book, magazine, or Web site that you list.

Race Questions:
- List colored compounds of transition metals.
- List uses of colored transition metal compounds.
- List elements that are dangerous to human health, especially heavy metals. Where are they found in society?
- List elements that are needed for human health. What food sources are each found in?
- List any other interesting information about elements that show up as you do your research.

▶ Conclude and Apply
1. Which resources did you find most helpful?
2. Name an interesting fact you found.

16 Human Bonding

▶ Real-World Question
How can humans model atoms bonding together?

Possible Materials
- family members or friends
- sheets of blank paper
- markers
- large safety pins
- large colored rubber bands

▶ Procedure
1. Draw a large electron dot diagram of an element you choose. Have other activity participants do that too.
2. Pin the diagram to your shirt.
3. How many electrons does your element have? Gather that many rubber bands.
4. Place about half of the rubber bands on one wrist and half on the other.
5. Form bonds by finding someone who has the number of rubber bands you need to total eight. Try to form as many different compounds with different elements as you can. (You may need two or three of another element's atoms to make a compound.) Record the compounds you make in your Science Journal. Label each compound as ionic or covalent.

▶ Conclude and Apply
1. Which elements don't form any bonds?
2. Which elements form four bonds?

Adult supervision required for all labs.

Extra Try at Home Labs

17 Mini Fireworks

Real-World Question
Where do the colors in fireworks come from?

Possible Materials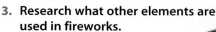
- candle
- lighter
- wooden chopsticks (or a fork or tongs)
- penny
- water in an old cup
- steel wool

Procedure
1. Light the candle.
2. Use the chopsticks to get a firm grip on the penny.
3. Hold the penny in the flame until you observe a change. *(Hint: this experiment is more fun in the bathroom with the lights off!)*
4. Drop the penny in the water when you are finished and plunge the burning end of the chopsticks or hot part of the fork into the water as well.
5. Repeat the procedure using steel wool.

Conclude and Apply
1. What color did you see?
2. Infer why copper and iron are used in fireworks.
3. Research what other elements are used in fireworks.

18 Measuring Momentum

Real-World Question
How much momentum do rolling balls have?

Possible Materials
- meterstick
- orange cones or tape
- scale
- stopwatch
- bucket
- bowling ball
- plastic baseball
- golf ball
- tennis ball
- calculator

Procedure
1. Use a balance to measure the masses of the tennis ball, golf ball, and plastic baseball. Convert their masses from grams to kilograms.
2. Find the weight of the bowling ball in pounds. The weight should be written on the ball. Divide the ball's weight by 2.2 to calculate its mass in kilograms.
3. Go outside and measure a 10-m distance on a blacktop or concrete surface. Mark the distance with orange cones or tape.
4. Have a partner roll each ball the 10-m distance. Measure the time it takes each ball to roll 10 m.
5. Use the formula: velocity = $\frac{\text{distance}}{\text{time}}$ to calculate each ball's velocity.

Conclude and Apply
1. Calculate the momentum of each ball.
2. Infer why the momentums of the balls differed so greatly.

Extra Try at Home Labs

19 Friction in Traffic

Real-World Question
How do the various kinds of friction affect the operation of vehicles?

Possible Materials
- erasers taken from the ends of pencils (4)
- needles (2)
- small match box
- toy car

Procedure
1. Build a match box car with the materials listed, or use a toy car.
2. Invent ways to demonstrate the effects of static friction, sliding friction, and rolling friction on the car. Think of hills, ice or rain conditions, graveled roads and paved roads, etc.
3. Make drawings of how friction is acting on the car, or how the car uses friction to work.

Conclude and Apply
1. In what ways are static, sliding, and rolling friction helpful to drivers?
2. In what ways are static, sliding, and rolling friction unfavorable to car safety and operation?
3. Explain what your experiment taught you about driving in icy conditions.

20 Simple Machines

Real-World Question
What types of simple machines are found in a toolbox?

Possible Materials
- box of tools

Procedure
1. Obtain a box of tools and lay all the tools and other hardware from the box on a table.
2. Carefully examine all the tools and hardware, and separate all the items that are a type of inclined plane.
3. Carefully examine all the tools and hardware, and separate all the items that are a type of lever.
4. Identify and separate all the items that are a wheel and axle.
5. Identify any pulleys in the toolbox.
6. Identify any tools that are a combination of two or more simple machines.

Conclude and Apply
1. List all the tools you found that were a type of inclined plane, lever, wheel and axle, or pulley.
2. List all the tools that were a combination of two or more simple machines.
3. Infer how a hammer could be used as both a first class lever and a third class lever.

Adult supervision required for all labs.

Extra Try at Home Labs

21 Estimate Temperature

Real-World Question
How can we learn to estimate temperatures?

Possible Materials
- thermometer
- bowl
- water
- ice

Procedure
1. If you have a dual-scale weather thermometer, you can learn twice as much by trying to do your estimation in degrees Fahrenheit and Celsius each time.
2. Fill a bowl with ice water. Submerge your fingers in the water and estimate the water temperature.
3. Place the thermometer in the bowl and observe the temperature.
4. Place a bowl of warm water in direct sunlight for 20 min. Submerge your fingers in the water and estimate the water temperature.
5. Place the thermometer in the bowl and observe the temperature.
6. Place the thermometer outside in a location where you can see it each day.
7. Each day for a month, step outside and estimate the temperature. Check the accuracy of your estimates with the thermometer. Record the weather conditions as well.

Conclude and Apply
1. Describe how well you can estimate air temperatures after estimating the temperature each day for a month. Did the cloudiness of the day affect your estimation skills?
2. Infer why understanding the Celsius scale might be helpful to you in the future.

22 Bending Water

Real-World Question
How can a plastic rod bend water without touching it?

Possible Materials
- plastic rod
- plastic clothes hanger
- 100% wool clothing
- water faucet

Procedure
1. Turn on a faucet until a narrow, smooth stream of water is flowing out of it. The stream of water cannot be too wide, and it cannot flow in a broken pattern.
2. Vigorously rub a plastic rod on a piece of 100% wool clothing for about 15 s.
3. Immediately hold the rod near the center of the stream of water. Move the rod close to the stream. Do not touch the water.
4. Observe what happens to the water.

Conclude and Apply
1. Describe how the plastic rod affected the stream of water.
2. Explain why the plastic rod affected the water.

Adult supervision required for all labs.

Extra Try at Home Labs

23 Testing Magnets

Real-World Question
How do the strengths of kitchen magnets compare?

Possible Materials
- several kitchen magnets
- metric ruler
- small pin or paper clip

Procedure
1. Place a small pin or paper clip on a flat, nonmetallic surface such as a wooden table.
2. Holding your metric ruler vertically, place it next to the pin with the 0 cm mark on the tabletop.
3. Hold a kitchen magnet at the 10 cm mark on the ruler.
4. Slowly lower the magnet toward the pin. At the point where the pin is attracted to the magnet, measure the height of the magnet from the table. Record the height in your Science Journal.
5. Repeat steps 2–4 to test your other kitchen magnets.

Conclude and Apply
1. Describe the results of your experiment.
2. Infer how the kitchen magnets should be used based on the results of your experiment.

24 Disappearing Dots

Real-World Question
Do your eyes have a blind spot?

Possible Materials
- white paper
- metric ruler
- colored pencils

Procedure
1. Hold a sheet of white paper horizontally. Near the left edge of the paper, draw a black dot about 0.5 cm in diameter.
2. Draw a red dot 5 cm to the right of the black dot.
3. Hold the paper out in front of you, close your left eye, and look at the black dot with your right eye. Slowly move the paper toward you and observe what happens to the red dot.
4. Draw a blue dot 10 cm to the right of the black dot and a green dot 15 cm from the black dot.
5. Hold the paper out at arm's length, close your left eye, and look at the black dot with your right eye. Slowly move the paper toward you and observe what happens to the dots.

Conclude and Apply
1. Describe what happened to the red, blue, and green dots as you moved the paper toward you.
2. The optic nerve carries visual images to the brain, and it is attached to the retina in your eye. Infer why the dots disappeared.

Adult supervision required for all labs.

Computer Skills

People who study science rely on computers, like the one in **Figure 16,** to record and store data and to analyze results from investigations. Whether you work in a laboratory or just need to write a lab report with tables, good computer skills are a necessity.

Using the computer comes with responsibility. Issues of ownership, security, and privacy can arise. Remember, if you did not author the information you are using, you must provide a source for your information. Also, anything on a computer can be accessed by others. Do not put anything on the computer that you would not want everyone to know. To add more security to your work, use a password.

Use a Word Processing Program

A computer program that allows you to type your information, change it as many times as you need to, and then print it out is called a word processing program. Word processing programs also can be used to make tables.

Learn the Skill To start your word processing program, a blank document, sometimes called "Document 1," appears on the screen. To begin, start typing. To create a new document, click the *New* button on the standard tool bar. These tips will help you format the document.

- The program will automatically move to the next line; press *Enter* if you wish to start a new paragraph.
- Symbols, called non-printing characters, can be hidden by clicking the *Show/Hide* button on your toolbar.
- To insert text, move the cursor to the point where you want the insertion to go, click on the mouse once, and type the text.
- To move several lines of text, select the text and click the *Cut* button on your toolbar. Then position your cursor in the location that you want to move the cut text and click *Paste*. If you move to the wrong place, click *Undo*.
- The spell check feature does not catch words that are misspelled to look like other words, like "cold" instead of "gold." Always reread your document to catch all spelling mistakes.
- To learn about other word processing methods, read the user's manual or click on the *Help* button.
- You can integrate databases, graphics, and spreadsheets into documents by copying from another program and pasting it into your document, or by using desktop publishing (DTP). DTP software allows you to put text and graphics together to finish your document with a professional look. This software varies in how it is used and its capabilities.

Figure 16 A computer will make reports neater and more professional looking.

Technology Skill Handbook

Use a Database

A collection of facts stored in a computer and sorted into different fields is called a database. A database can be reorganized in any way that suits your needs.

Learn the Skill A computer program that allows you to create your own database is a database management system (DBMS). It allows you to add, delete, or change information. Take time to get to know the features of your database software.

- Determine what facts you would like to include and research to collect your information.
- Determine how you want to organize the information.
- Follow the instructions for your particular DBMS to set up fields. Then enter each item of data in the appropriate field.
- Follow the instructions to sort the information in order of importance.
- Evaluate the information in your database, and add, delete, or change as necessary.

Use the Internet

The Internet is a global network of computers where information is stored and shared. To use the Internet, like the students in **Figure 17**, you need a modem to connect your computer to a phone line and an Internet Service Provider account.

Learn the Skill To access internet sites and information, use a "Web browser," which lets you view and explore pages on the World Wide Web. Each page is its own site, and each site has its own address, called a URL. Once you have found a Web browser, follow these steps for a search (this also is how you search a database).

Figure 17 The Internet allows you to search a global network for a variety of information.

- Be as specific as possible. If you know you want to research "gold," don't type in "elements." Keep narrowing your search until you find what you want.
- Web sites that end in *.com* are commercial Web sites; *.org, .edu,* and *.gov* are nonprofit, educational, or government Web sites.
- Electronic encyclopedias, almanacs, indexes, and catalogs will help locate and select relevant information.
- Develop a "home page" with relative ease. When developing a Web site, NEVER post pictures or disclose personal information such as location, names, or phone numbers. Your school or community usually can host your Web site. A basic understanding of HTML (hypertext mark-up language), the language of Web sites, is necessary. Software that creates HTML code is called authoring software, and can be downloaded free from many Web sites. This software allows text and pictures to be arranged as the software is writing the HTML code.

TECHNOLOGY SKILL HANDBOOK 749

Technology Skill Handbook

Use a Spreadsheet

A spreadsheet, shown in **Figure 18,** can perform mathematical functions with any data arranged in columns and rows. By entering a simple equation into a cell, the program can perform operations in specific cells, rows, or columns.

Learn the Skill Each column (vertical) is assigned a letter, and each row (horizontal) is assigned a number. Each point where a row and column intersect is called a cell, and is labeled according to where it is located—Column A, Row 1 (A1).

- Decide how to organize the data, and enter it in the correct row or column.
- Spreadsheets can use standard formulas or formulas can be customized to calculate cells.
- To make a change, click on a cell to make it activate, and enter the edited data or formula.
- Spreadsheets also can display your results in graphs. Choose the style of graph that best represents the data.

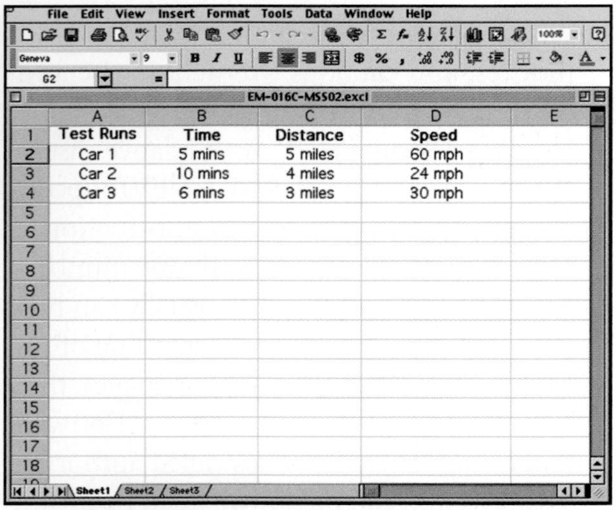

	A	B	C	D
1	Test Runs	Time	Distance	Speed
2	Car 1	5 mins	5 miles	60 mph
3	Car 2	10 mins	4 miles	24 mph
4	Car 3	6 mins	3 miles	30 mph

Figure 18 A spreadsheet allows you to perform mathematical operations on your data.

Use Graphics Software

Adding pictures, called graphics, to your documents is one way to make your documents more meaningful and exciting. This software adds, edits, and even constructs graphics. There is a variety of graphics software programs. The tools used for drawing can be a mouse, keyboard, or other specialized devices. Some graphics programs are simple. Others are complicated, called computer-aided design (CAD) software.

Learn the Skill It is important to have an understanding of the graphics software being used before starting. The better the software is understood, the better the results. The graphics can be placed in a word-processing document.

- Clip art can be found on a variety of internet sites, and on CDs. These images can be copied and pasted into your document.
- When beginning, try editing existing drawings, then work up to creating drawings.
- The images are made of tiny rectangles of color called pixels. Each pixel can be altered.
- Digital photography is another way to add images. The photographs in the memory of a digital camera can be downloaded into a computer, then edited and added to the document.
- Graphics software also can allow animation. The software allows drawings to have the appearance of movement by connecting basic drawings automatically. This is called in-betweening, or tweening.
- Remember to save often.

Presentation Skills

Develop Multimedia Presentations

Most presentations are more dynamic if they include diagrams, photographs, videos, or sound recordings, like the one shown in **Figure 19.** A multimedia presentation involves using stereos, overhead projectors, televisions, computers, and more.

Learn the Skill Decide the main points of your presentation, and what types of media would best illustrate those points.

- Make sure you know how to use the equipment you are working with.
- Practice the presentation using the equipment several times.
- Enlist the help of a classmate to push play or turn lights out for you. Be sure to practice your presentation with him or her.
- If possible, set up all of the equipment ahead of time, and make sure everything is working properly.

Figure 19 These students are engaging the audience using a variety of tools.

Computer Presentations

There are many different interactive computer programs that you can use to enhance your presentation. Most computers have a compact disc (CD) drive that can play both CDs and digital video discs (DVDs). Also, there is hardware to connect a regular CD, DVD, or VCR. These tools will enhance your presentation.

Another method of using the computer to aid in your presentation is to develop a slide show using a computer program. This can allow movement of visuals at the presenter's pace, and can allow for visuals to build on one another.

Learn the Skill In order to create multimedia presentations on a computer, you need to have certain tools. These may include traditional graphic tools and drawing programs, animation programs, and authoring systems that tie everything together. Your computer will tell you which tools it supports. The most important step is to learn about the tools that you will be using.

- Often, color and strong images will convey a point better than words alone. Use the best methods available to convey your point.
- As with other presentations, practice many times.
- Practice your presentation with the tools you and any assistants will be using.
- Maintain eye contact with the audience. The purpose of using the computer is not to prompt the presenter, but to help the audience understand the points of the presentation.

Math Skill Handbook

Math Review

Use Fractions

A fraction compares a part to a whole. In the fraction $\frac{2}{3}$, the 2 represents the part and is the numerator. The 3 represents the whole and is the denominator.

Reduce Fractions To reduce a fraction, you must find the largest factor that is common to both the numerator and the denominator, the greatest common factor (GCF). Divide both numbers by the GCF. The fraction has then been reduced, or it is in its simplest form.

Example Twelve of the 20 chemicals in the science lab are in powder form. What fraction of the chemicals used in the lab are in powder form?

Step 1 Write the fraction.
$\frac{\text{part}}{\text{whole}} = \frac{12}{20}$

Step 2 To find the GCF of the numerator and denominator, list all of the factors of each number.
Factors of 12: 1, 2, 3, 4, 6, 12 (the numbers that divide evenly into 12)
Factors of 20: 1, 2, 4, 5, 10, 20 (the numbers that divide evenly into 20)

Step 3 List the common factors.
1, 2, 4.

Step 4 Choose the greatest factor in the list.
The GCF of 12 and 20 is 4.

Step 5 Divide the numerator and denominator by the GCF.
$\frac{12 \div 4}{20 \div 4} = \frac{3}{5}$

In the lab, $\frac{3}{5}$ of the chemicals are in powder form.

Practice Problem At an amusement park, 66 of 90 rides have a height restriction. What fraction of the rides, in its simplest form, has a height restriction?

Add and Subtract Fractions To add or subtract fractions with the same denominator, add or subtract the numerators and write the sum or difference over the denominator. After finding the sum or difference, find the simplest form for your fraction.

Example 1 In the forest outside your house, $\frac{1}{8}$ of the animals are rabbits, $\frac{3}{8}$ are squirrels, and the remainder are birds and insects. How many are mammals?

Step 1 Add the numerators.
$\frac{1}{8} + \frac{3}{8} = \frac{(1+3)}{8} = \frac{4}{8}$

Step 2 Find the GCF.
$\frac{4}{8}$ (GCF, 4)

Step 3 Divide the numerator and denominator by the GCF.
$\frac{4}{4} = 1, \frac{8}{4} = 2$

$\frac{1}{2}$ of the animals are mammals.

Example 2 If $\frac{7}{16}$ of the Earth is covered by freshwater, and $\frac{1}{16}$ of that is in glaciers, how much freshwater is not frozen?

Step 1 Subtract the numerators.
$\frac{7}{16} - \frac{1}{16} = \frac{(7-1)}{16} = \frac{6}{16}$

Step 2 Find the GCF.
$\frac{6}{16}$ (GCF, 2)

Step 3 Divide the numerator and denominator by the GCF.
$\frac{6}{2} = 3, \frac{16}{2} = 8$

$\frac{3}{8}$ of the freshwater is not frozen.

Practice Problem A bicycle rider is going 15 km/h for $\frac{4}{9}$ of his ride, 10 km/h for $\frac{2}{9}$ of his ride, and 8 km/h for the remainder of the ride. How much of his ride is he going over 8 km/h?

Math Skill Handbook

Unlike Denominators To add or subtract fractions with unlike denominators, first find the least common denominator (LCD). This is the smallest number that is a common multiple of both denominators. Rename each fraction with the LCD, and then add or subtract. Find the simplest form if necessary.

Example 1 A chemist makes a paste that is $\frac{1}{2}$ table salt (NaCl), $\frac{1}{3}$ sugar ($C_6H_{12}O_6$), and the rest water (H_2O). How much of the paste is a solid?

Step 1 Find the LCD of the fractions.
$\frac{1}{2} + \frac{1}{3}$ (LCD, 6)

Step 2 Rename each numerator and each denominator with the LCD.
$1 \times 3 = 3, \quad 2 \times 3 = 6$
$1 \times 2 = 2, \quad 3 \times 2 = 6$

Step 3 Add the numerators.
$\frac{3}{6} + \frac{2}{6} = \frac{(3+2)}{6} = \frac{5}{6}$

$\frac{5}{6}$ of the paste is a solid.

Example 2 The average precipitation in Grand Junction, CO, is $\frac{7}{10}$ inch in November, and $\frac{3}{5}$ inch in December. What is the total average precipitation?

Step 1 Find the LCD of the fractions.
$\frac{7}{10} + \frac{3}{5}$ (LCD, 10)

Step 2 Rename each numerator and each denominator with the LCD.
$7 \times 1 = 7, \quad 10 \times 1 = 10$
$3 \times 2 = 6, \quad 5 \times 2 = 10$

Step 3 Add the numerators.
$\frac{7}{10} + \frac{6}{10} = \frac{(7+6)}{10} = \frac{13}{10}$

$\frac{13}{10}$ inches total precipitation, or $1\frac{3}{10}$ inches.

Practice Problem On an electric bill, about $\frac{1}{8}$ of the energy is from solar energy and about $\frac{1}{10}$ is from wind power. How much of the total bill is from solar energy and wind power combined?

Example 3 In your body, $\frac{7}{10}$ of your muscle contractions are involuntary (cardiac and smooth muscle tissue). Smooth muscle makes $\frac{3}{15}$ of your muscle contractions. How many of your muscle contractions are made by cardiac muscle?

Step 1 Find the LCD of the fractions.
$\frac{7}{10} - \frac{3}{15}$ (LCD, 30)

Step 2 Rename each numerator and each denominator with the LCD.
$7 \times 3 = 21, \quad 10 \times 3 = 30$
$3 \times 2 = 6, \quad 15 \times 2 = 30$

Step 3 Subtract the numerators.
$\frac{21}{30} - \frac{6}{30} = \frac{(21-6)}{30} = \frac{15}{30}$

Step 4 Find the GCF.
$\frac{15}{30}$ (GCF, 15)
$\frac{1}{2}$

$\frac{1}{2}$ of all muscle contractions are cardiac muscle.

Example 4 Tony wants to make cookies that call for $\frac{3}{4}$ of a cup of flour, but he only has $\frac{1}{3}$ of a cup. How much more flour does he need?

Step 1 Find the LCD of the fractions.
$\frac{3}{4} - \frac{1}{3}$ (LCD, 12)

Step 2 Rename each numerator and each denominator with the LCD.
$3 \times 3 = 9, \quad 4 \times 3 = 12$
$1 \times 4 = 4, \quad 3 \times 4 = 12$

Step 3 Subtract the numerators.
$\frac{9}{12} - \frac{4}{12} = \frac{(9-4)}{12} = \frac{5}{12}$

$\frac{5}{12}$ of a cup of flour.

Practice Problem Using the information provided to you in Example 3 above, determine how many muscle contractions are voluntary (skeletal muscle).

Math Skill Handbook

Multiply Fractions To multiply with fractions, multiply the numerators and multiply the denominators. Find the simplest form if necessary.

Example Multiply $\frac{3}{5}$ by $\frac{1}{3}$.

Step 1 Multiply the numerators and denominators.
$$\frac{3}{5} \times \frac{1}{3} = \frac{(3 \times 1)}{(5 \times 3)} = \frac{3}{15}$$

Step 2 Find the GCF.
$$\frac{3}{15} \quad (\text{GCF, 3})$$

Step 3 Divide the numerator and denominator by the GCF.
$$\frac{3}{3} = 1, \quad \frac{15}{3} = 5$$
$$\frac{1}{5}$$

$\frac{3}{5}$ multiplied by $\frac{1}{3}$ is $\frac{1}{5}$.

Practice Problem Multiply $\frac{3}{14}$ by $\frac{5}{16}$.

Find a Reciprocal Two numbers whose product is 1 are called multiplicative inverses, or reciprocals.

Example Find the reciprocal of $\frac{3}{8}$.

Step 1 Inverse the fraction by putting the denominator on top and the numerator on the bottom.
$$\frac{8}{3}$$

The reciprocal of $\frac{3}{8}$ is $\frac{8}{3}$.

Practice Problem Find the reciprocal of $\frac{4}{9}$.

Divide Fractions To divide one fraction by another fraction, multiply the dividend by the reciprocal of the divisor. Find the simplest form if necessary.

Example 1 Divide $\frac{1}{9}$ by $\frac{1}{3}$.

Step 1 Find the reciprocal of the divisor.
The reciprocal of $\frac{1}{3}$ is $\frac{3}{1}$.

Step 2 Multiply the dividend by the reciprocal of the divisor.
$$\frac{\frac{1}{9}}{\frac{1}{3}} = \frac{1}{9} \times \frac{3}{1} = \frac{(1 \times 3)}{(9 \times 1)} = \frac{3}{9}$$

Step 3 Find the GCF.
$$\frac{3}{9} \quad (\text{GCF, 3})$$

Step 4 Divide the numerator and denominator by the GCF.
$$\frac{3}{3} = 1, \quad \frac{9}{3} = 3$$
$$\frac{1}{3}$$

$\frac{1}{9}$ divided by $\frac{1}{3}$ is $\frac{1}{3}$.

Example 2 Divide $\frac{3}{5}$ by $\frac{1}{4}$.

Step 1 Find the reciprocal of the divisor.
The reciprocal of $\frac{1}{4}$ is $\frac{4}{1}$.

Step 2 Multiply the dividend by the reciprocal of the divisor.
$$\frac{\frac{3}{5}}{\frac{1}{4}} = \frac{3}{5} \times \frac{4}{1} = \frac{(3 \times 4)}{(5 \times 1)} = \frac{12}{5}$$

$\frac{3}{5}$ divided by $\frac{1}{4}$ is $\frac{12}{5}$ or $2\frac{2}{5}$.

Practice Problem Divide $\frac{3}{11}$ by $\frac{7}{10}$.

Math Skill Handbook

Use Ratios

When you compare two numbers by division, you are using a ratio. Ratios can be written 3 to 5, 3:5, or $\frac{3}{5}$. Ratios, like fractions, also can be written in simplest form.

Ratios can represent probabilities, also called odds. This is a ratio that compares the number of ways a certain outcome occurs to the number of outcomes. For example, if you flip a coin 100 times, what are the odds that it will come up heads? There are two possible outcomes, heads or tails, so the odds of coming up heads are 50:100. Another way to say this is that 50 out of 100 times the coin will come up heads. In its simplest form, the ratio is 1:2.

Example 1 A chemical solution contains 40 g of salt and 64 g of baking soda. What is the ratio of salt to baking soda as a fraction in simplest form?

Step 1 Write the ratio as a fraction.
$$\frac{\text{salt}}{\text{baking soda}} = \frac{40}{64}$$

Step 2 Express the fraction in simplest form.
The GCF of 40 and 64 is 8.
$$\frac{40}{64} = \frac{40 \div 8}{64 \div 8} = \frac{5}{8}$$

The ratio of salt to baking soda in the sample is 5:8.

Example 2 Sean rolls a 6-sided die 6 times. What are the odds that the side with a 3 will show?

Step 1 Write the ratio as a fraction.
$$\frac{\text{number of sides with a 3}}{\text{number of sides}} = \frac{1}{6}$$

Step 2 Multiply by the number of attempts.
$$\frac{1}{6} \times 6 \text{ attempts} = \frac{6}{6} \text{ attempts} = 1 \text{ attempt}$$

1 attempt out of 6 will show a 3.

Practice Problem Two metal rods measure 100 cm and 144 cm in length. What is the ratio of their lengths in simplest form?

Use Decimals

A fraction with a denominator that is a power of ten can be written as a decimal. For example, 0.27 means $\frac{27}{100}$. The decimal point separates the ones place from the tenths place.

Any fraction can be written as a decimal using division. For example, the fraction $\frac{5}{8}$ can be written as a decimal by dividing 5 by 8. Written as a decimal, it is 0.625.

Add or Subtract Decimals When adding and subtracting decimals, line up the decimal points before carrying out the operation.

Example 1 Find the sum of 47.68 and 7.80.

Step 1 Line up the decimal places when you write the numbers.
$$\begin{array}{r} 47.68 \\ + 7.80 \end{array}$$

Step 2 Add the decimals.
$$\begin{array}{r} 47.68 \\ + 7.80 \\ \hline 55.48 \end{array}$$

The sum of 47.68 and 7.80 is 55.48.

Example 2 Find the difference of 42.17 and 15.85.

Step 1 Line up the decimal places when you write the number.
$$\begin{array}{r} 42.17 \\ -15.85 \end{array}$$

Step 2 Subtract the decimals.
$$\begin{array}{r} 42.17 \\ -15.85 \\ \hline 26.32 \end{array}$$

The difference of 42.17 and 15.85 is 26.32.

Practice Problem Find the sum of 1.245 and 3.842.

Math Skill Handbook

Multiply Decimals To multiply decimals, multiply the numbers like any other number, ignoring the decimal point. Count the decimal places in each factor. The product will have the same number of decimal places as the sum of the decimal places in the factors.

Example Multiply 2.4 by 5.9.

Step 1 Multiply the factors like two whole numbers.
$24 \times 59 = 1416$

Step 2 Find the sum of the number of decimal places in the factors. Each factor has one decimal place, for a sum of two decimal places.

Step 3 The product will have two decimal places.
14.16

The product of 2.4 and 5.9 is 14.16.

Practice Problem Multiply 4.6 by 2.2.

Divide Decimals When dividing decimals, change the divisor to a whole number. To do this, multiply both the divisor and the dividend by the same power of ten. Then place the decimal point in the quotient directly above the decimal point in the dividend. Then divide as you do with whole numbers.

Example Divide 8.84 by 3.4.

Step 1 Multiply both factors by 10.
$3.4 \times 10 = 34$, $8.84 \times 10 = 88.4$

Step 2 Divide 88.4 by 34.

$$\begin{array}{r} 2.6 \\ 34\overline{)88.4} \\ -68 \\ \hline 204 \\ -204 \\ \hline 0 \end{array}$$

8.84 divided by 3.4 is 2.6.

Practice Problem Divide 75.6 by 3.6.

Use Proportions

An equation that shows that two ratios are equivalent is a proportion. The ratios $\frac{2}{4}$ and $\frac{5}{10}$ are equivalent, so they can be written as $\frac{2}{4} = \frac{5}{10}$. This equation is a proportion.

When two ratios form a proportion, the cross products are equal. To find the cross products in the proportion $\frac{2}{4} = \frac{5}{10}$, multiply the 2 and the 10, and the 4 and the 5. Therefore $2 \times 10 = 4 \times 5$, or $20 = 20$.

Because you know that both proportions are equal, you can use cross products to find a missing term in a proportion. This is known as solving the proportion.

Example The heights of a tree and a pole are proportional to the lengths of their shadows. The tree casts a shadow of 24 m when a 6-m pole casts a shadow of 4 m. What is the height of the tree?

Step 1 Write a proportion.
$$\frac{\text{height of tree}}{\text{height of pole}} = \frac{\text{length of tree's shadow}}{\text{length of pole's shadow}}$$

Step 2 Substitute the known values into the proportion. Let h represent the unknown value, the height of the tree.
$$\frac{h}{6} = \frac{24}{4}$$

Step 3 Find the cross products.
$h \times 4 = 6 \times 24$

Step 4 Simplify the equation.
$4h = 144$

Step 5 Divide each side by 4.
$$\frac{4h}{4} = \frac{144}{4}$$
$h = 36$

The height of the tree is 36 m.

Practice Problem The ratios of the weights of two objects on the Moon and on Earth are in proportion. A rock weighing 3 N on the Moon weighs 18 N on Earth. How much would a rock that weighs 5 N on the Moon weigh on Earth?

Math Skill Handbook

Use Percentages

The word *percent* means "out of one hundred." It is a ratio that compares a number to 100. Suppose you read that 77 percent of the Earth's surface is covered by water. That is the same as reading that the fraction of the Earth's surface covered by water is $\frac{77}{100}$. To express a fraction as a percent, first find the equivalent decimal for the fraction. Then, multiply the decimal by 100 and add the percent symbol.

Example Express $\frac{13}{20}$ as a percent.

Step 1 Find the equivalent decimal for the fraction.

$$\begin{array}{r} 0.65 \\ 20\overline{)13.00} \\ \underline{12\ 0} \\ 1\ 00 \\ \underline{1\ 00} \\ 0 \end{array}$$

Step 2 Rewrite the fraction $\frac{13}{20}$ as 0.65.

Step 3 Multiply 0.65 by 100 and add the % sign.
$0.65 \times 100 = 65 = 65\%$

So, $\frac{13}{20} = 65\%$.

This also can be solved as a proportion.

Example Express $\frac{13}{20}$ as a percent.

Step 1 Write a proportion.
$\frac{13}{20} = \frac{x}{100}$

Step 2 Find the cross products.
$1300 = 20x$

Step 3 Divide each side by 20.
$\frac{1300}{20} = \frac{20x}{20}$
$65\% = x$

Practice Problem In one year, 73 of 365 days were rainy in one city. What percent of the days in that city were rainy?

Solve One-Step Equations

A statement that two things are equal is an equation. For example, $A = B$ is an equation that states that A is equal to B.

An equation is solved when a variable is replaced with a value that makes both sides of the equation equal. To make both sides equal the inverse operation is used. Addition and subtraction are inverses, and multiplication and division are inverses.

Example 1 Solve the equation $x - 10 = 35$.

Step 1 Find the solution by adding 10 to each side of the equation.
$x - 10 = 35$
$x - 10 + 10 = 35 + 10$
$x = 45$

Step 2 Check the solution.
$x - 10 = 35$
$45 - 10 = 35$
$35 = 35$

Both sides of the equation are equal, so $x = 45$.

Example 2 In the formula $a = bc$, find the value of c if $a = 20$ and $b = 2$.

Step 1 Rearrange the formula so the unknown value is by itself on one side of the equation by dividing both sides by b.
$a = bc$
$\frac{a}{b} = \frac{bc}{b}$
$\frac{a}{b} = c$

Step 2 Replace the variables a and b with the values that are given.
$\frac{a}{b} = c$
$\frac{20}{2} = c$
$10 = c$

Step 3 Check the solution.
$a = bc$
$20 = 2 \times 10$
$20 = 20$

Both sides of the equation are equal, so $c = 10$ is the solution when $a = 20$ and $b = 2$.

Practice Problem In the formula $h = gd$, find the value of d if $g = 12.3$ and $h = 17.4$.

Math Skill Handbook

Use Statistics

The branch of mathematics that deals with collecting, analyzing, and presenting data is statistics. In statistics, there are three common ways to summarize data with a single number—the mean, the median, and the mode.

The **mean** of a set of data is the arithmetic average. It is found by adding the numbers in the data set and dividing by the number of items in the set.

The **median** is the middle number in a set of data when the data are arranged in numerical order. If there were an even number of data points, the median would be the mean of the two middle numbers.

The **mode** of a set of data is the number or item that appears most often.

Another number that often is used to describe a set of data is the range. The **range** is the difference between the largest number and the smallest number in a set of data.

A **frequency table** shows how many times each piece of data occurs, usually in a survey. **Table 2** below shows the results of a student survey on favorite color.

Table 2 Student Color Choice		
Color	**Tally**	**Frequency**
red	IIII	4
blue	HHI	5
black	II	2
green	III	3
purple	HHI II	7
yellow	HHI I	6

Based on the frequency table data, which color is the favorite?

Example The speeds (in m/s) for a race car during five different time trials are 39, 37, 44, 36, and 44.

To find the mean:

Step 1 Find the sum of the numbers.
$39 + 37 + 44 + 36 + 44 = 200$

Step 2 Divide the sum by the number of items, which is 5.
$200 \div 5 = 40$

The mean is 40 m/s.

To find the median:

Step 1 Arrange the measures from least to greatest.
36, 37, 39, 44, 44

Step 2 Determine the middle measure.
36, 37, <u>39</u>, 44, 44

The median is 39 m/s.

To find the mode:

Step 1 Group the numbers that are the same together.
44, 44, 36, 37, 39

Step 2 Determine the number that occurs most in the set.
<u>44, 44</u>, 36, 37, 39

The mode is 44 m/s.

To find the range:

Step 1 Arrange the measures from largest to smallest.
44, 44, 39, 37, 36

Step 2 Determine the largest and smallest measures in the set.
<u>44</u>, 44, 39, 37, <u>36</u>

Step 3 Find the difference between the largest and smallest measures.
$44 - 36 = 8$

The range is 8 m/s.

Practice Problem Find the mean, median, mode, and range for the data set 8, 4, 12, 8, 11, 14, 16.

Math Skill Handbook

Use Geometry

The branch of mathematics that deals with the measurement, properties, and relationships of points, lines, angles, surfaces, and solids is called geometry.

Perimeter The **perimeter** (P) is the distance around a geometric figure. To find the perimeter of a rectangle, add the length and width and multiply that sum by two, or $2(l + w)$. To find perimeters of irregular figures, add the length of the sides.

Example 1 Find the perimeter of a rectangle that is 3 m long and 5 m wide.

Step 1 You know that the perimeter is 2 times the sum of the width and length.
$P = 2(3\text{ m} + 5\text{ m})$

Step 2 Find the sum of the width and length.
$P = 2(8\text{ m})$

Step 3 Multiply by 2.
$P = 16\text{ m}$

The perimeter is 16 m.

Example 2 Find the perimeter of a shape with sides measuring 2 cm, 5 cm, 6 cm, 3 cm.

Step 1 You know that the perimeter is the sum of all the sides.
$P = 2 + 5 + 6 + 3$

Step 2 Find the sum of the sides.
$P = 2 + 5 + 6 + 3$
$P = 16$

The perimeter is 16 cm.

Practice Problem Find the perimeter of a rectangle with a length of 18 m and a width of 7 m.

Practice Problem Find the perimeter of a triangle measuring 1.6 cm by 2.4 cm by 2.4 cm.

Area of a Rectangle The **area** (A) is the number of square units needed to cover a surface. To find the area of a rectangle, multiply the length times the width, or $l \times w$. When finding area, the units also are multiplied. Area is given in square units.

Example Find the area of a rectangle with a length of 1 cm and a width of 10 cm.

Step 1 You know that the area is the length multiplied by the width.
$A = (1\text{ cm} \times 10\text{ cm})$

Step 2 Multiply the length by the width. Also multiply the units.
$A = 10\text{ cm}^2$

The area is 10 cm^2.

Practice Problem Find the area of a square whose sides measure 4 m.

Area of a Triangle To find the area of a triangle, use the formula:

$A = \frac{1}{2}(\text{base} \times \text{height})$

The base of a triangle can be any of its sides. The height is the perpendicular distance from a base to the opposite endpoint, or vertex.

Example Find the area of a triangle with a base of 18 m and a height of 7 m.

Step 1 You know that the area is $\frac{1}{2}$ the base times the height.
$A = \frac{1}{2}(18\text{ m} \times 7\text{ m})$

Step 2 Multiply $\frac{1}{2}$ by the product of 18×7. Multiply the units.
$A = \frac{1}{2}(126\text{ m}^2)$
$A = 63\text{ m}^2$

The area is 63 m^2.

Practice Problem Find the area of a triangle with a base of 27 cm and a height of 17 cm.

Math Skill Handbook

Circumference of a Circle The **diameter** (*d*) of a circle is the distance across the circle through its center, and the **radius** (*r*) is the distance from the center to any point on the circle. The radius is half of the diameter. The distance around the circle is called the **circumference** (C). The formula for finding the circumference is:

$$C = 2\pi r \quad or \quad C = \pi d$$

The circumference divided by the diameter is always equal to 3.1415926... This nonterminating and nonrepeating number is represented by the Greek letter π (pi). An approximation often used for π is 3.14.

Example 1 Find the circumference of a circle with a radius of 3 m.

Step 1 You know the formula for the circumference is 2 times the radius times π.
$C = 2\pi(3)$

Step 2 Multiply 2 times the radius.
$C = 6\pi$

Step 3 Multiply by π.
$C = 19$ m

The circumference is 19 m.

Example 2 Find the circumference of a circle with a diameter of 24.0 cm.

Step 1 You know the formula for the circumference is the diameter times π.
$C = \pi(24.0)$

Step 2 Multiply the diameter by π.
$C = 75.4$ cm

The circumference is 75.4 cm.

Practice Problem Find the circumference of a circle with a radius of 19 cm.

Area of a Circle The formula for the area of a circle is:
$A = \pi r^2$

Example 1 Find the area of a circle with a radius of 4.0 cm.

Step 1 $A = \pi(4.0)^2$

Step 2 Find the square of the radius.
$A = 16\pi$

Step 3 Multiply the square of the radius by π.
$A = 50$ cm^2

The area of the circle is 50 cm^2.

Example 2 Find the area of a circle with a radius of 225 m.

Step 1 $A = \pi(225)^2$

Step 2 Find the square of the radius.
$A = 50625\pi$

Step 3 Multiply the square of the radius by π.
$A = 158962.5$

The area of the circle is 158,962 m^2.

Example 3 Find the area of a circle whose diameter is 20.0 mm.

Step 1 You know the formula for the area of a circle is the square of the radius times π, and that the radius is half of the diameter.
$A = \pi\left(\frac{20.0}{2}\right)^2$

Step 2 Find the radius.
$A = \pi(10.0)^2$

Step 3 Find the square of the radius.
$A = 100\pi$

Step 4 Multiply the square of the radius by π.
$A = 314$ mm^2

The area is 314 mm^2.

Practice Problem Find the area of a circle with a radius of 16 m.

Math Skill Handbook

Volume The measure of space occupied by a solid is the **volume** (V). To find the volume of a rectangular solid multiply the length times width times height, or $V = l \times w \times h$. It is measured in cubic units, such as cubic centimeters (cm^3).

Example Find the volume of a rectangular solid with a length of 2.0 m, a width of 4.0 m, and a height of 3.0 m.

Step 1 You know the formula for volume is the length times the width times the height.
$V = 2.0 \text{ m} \times 4.0 \text{ m} \times 3.0 \text{ m}$

Step 2 Multiply the length times the width times the height.
$V = 24 \text{ m}^3$

The volume is 24 m^3.

Practice Problem Find the volume of a rectangular solid that is 8 m long, 4 m wide, and 4 m high.

To find the volume of other solids, multiply the area of the base times the height.

Example 1 Find the volume of a solid that has a triangular base with a length of 8.0 m and a height of 7.0 m. The height of the entire solid is 15.0 m.

Step 1 You know that the base is a triangle, and the area of a triangle is $\frac{1}{2}$ the base times the height, and the volume is the area of the base times the height.
$V = \left[\frac{1}{2}(b \times h)\right] \times 15$

Step 2 Find the area of the base.
$V = \left[\frac{1}{2}(8 \times 7)\right] \times 15$
$V = \left(\frac{1}{2} \times 56\right) \times 15$

Step 3 Multiply the area of the base by the height of the solid.
$V = 28 \times 15$
$V = 420 \text{ m}^3$

The volume is 420 m^3.

Example 2 Find the volume of a cylinder that has a base with a radius of 12.0 cm, and a height of 21.0 cm.

Step 1 You know that the base is a circle, and the area of a circle is the square of the radius times π, and the volume is the area of the base times the height.
$V = (\pi r^2) \times 21$
$V = (\pi 12^2) \times 21$

Step 2 Find the area of the base.
$V = 144\pi \times 21$
$V = 452 \times 21$

Step 3 Multiply the area of the base by the height of the solid.
$V = 9490 \text{ cm}^3$

The volume is 9490 cm^3.

Example 3 Find the volume of a cylinder that has a diameter of 15 mm and a height of 4.8 mm.

Step 1 You know that the base is a circle with an area equal to the square of the radius times π. The radius is one-half the diameter. The volume is the area of the base times the height.
$V = (\pi r^2) \times 4.8$
$V = \left[\pi\left(\frac{1}{2} \times 15\right)^2\right] \times 4.8$
$V = (\pi 7.5^2) \times 4.8$

Step 2 Find the area of the base.
$V = 56.25\pi \times 4.8$
$V = 176.63 \times 4.8$

Step 3 Multiply the area of the base by the height of the solid.
$V = 847.8$

The volume is 847.8 mm^3.

Practice Problem Find the volume of a cylinder with a diameter of 7 cm in the base and a height of 16 cm.

MATH SKILL HANDBOOK 761

Math Skill Handbook

Science Applications

Measure in SI

The metric system of measurement was developed in 1795. A modern form of the metric system, called the International System (SI), was adopted in 1960 and provides the standard measurements that all scientists around the world can understand.

The SI system is convenient because unit sizes vary by powers of 10. Prefixes are used to name units. Look at **Table 3** for some common SI prefixes and their meanings.

Table 3 Common SI Prefixes			
Prefix	Symbol	Meaning	
kilo-	k	1,000	thousand
hecto-	h	100	hundred
deka-	da	10	ten
deci-	d	0.1	tenth
centi-	c	0.01	hundredth
milli-	m	0.001	thousandth

Example How many grams equal one kilogram?

Step 1 Find the prefix *kilo* in **Table 3**.

Step 2 Using **Table 3**, determine the meaning of *kilo*. According to the table, it means 1,000. When the prefix *kilo* is added to a unit, it means that there are 1,000 of the units in a "*kilo*unit."

Step 3 Apply the prefix to the units in the question. The units in the question are grams. There are 1,000 grams in a kilogram.

Practice Problem Is a milligram larger or smaller than a gram? How many of the smaller units equal one larger unit? What fraction of the larger unit does one smaller unit represent?

Dimensional Analysis

Convert SI Units In science, quantities such as length, mass, and time sometimes are measured using different units. A process called dimensional analysis can be used to change one unit of measure to another. This process involves multiplying your starting quantity and units by one or more conversion factors. A conversion factor is a ratio equal to one and can be made from any two equal quantities with different units. If 1,000 mL equal 1 L then two ratios can be made.

$$\frac{1,000 \text{ mL}}{1 \text{ L}} = \frac{1 \text{ L}}{1,000 \text{ mL}} = 1$$

One can covert between units in the SI system by using the equivalents in **Table 3** to make conversion factors.

Example 1 How many cm are in 4 m?

Step 1 Write conversion factors for the units given. From **Table 3**, you know that 100 cm = 1 m. The conversion factors are

$$\frac{100 \text{ cm}}{1 \text{ m}} \text{ and } \frac{1 \text{ m}}{100 \text{ cm}}$$

Step 2 Decide which conversion factor to use. Select the factor that has the units you are converting from (m) in the denominator and the units you are converting to (cm) in the numerator.

$$\frac{100 \text{ cm}}{1 \text{ m}}$$

Step 3 Multiply the starting quantity and units by the conversion factor. Cancel the starting units with the units in the denominator. There are 400 cm in 4 m.

$$4 \text{ m} \times \frac{100 \text{ cm}}{1 \text{ m}} = 400 \text{ cm}$$

Practice Problem How many milligrams are in one kilogram? (Hint: You will need to use two conversion factors from **Table 3**.)

Math Skill Handbook

Table 4 Unit System Equivalents	
Type of Measurement	Equivalent
Length	1 in = 2.54 cm
	1 yd = 0.91 m
	1 mi = 1.61 km
Mass and Weight*	1 oz = 28.35 g
	1 lb = 0.45 kg
	1 ton (short) = 0.91 tonnes (metric tons)
	1 lb = 4.45 N
Volume	1 in^3 = 16.39 cm^3
	1 qt = 0.95 L
	1 gal = 3.78 L
Area	1 in^2 = 6.45 cm^2
	1 yd^2 = 0.83 m^2
	1 mi^2 = 2.59 km^2
	1 acre = 0.40 hectares
Temperature	°C = $\frac{(°F - 32)}{1.8}$
	K = °C + 273

*Weight is measured in standard Earth gravity.

Convert Between Unit Systems Table 4 gives a list of equivalents that can be used to convert between English and SI units.

Example If a meterstick has a length of 100 cm, how long is the meterstick in inches?

Step 1 Write the conversion factors for the units given. From **Table 4,** 1 in = 2.54 cm.

$$\frac{1 \text{ in}}{2.54 \text{ cm}} \text{ and } \frac{2.54 \text{ cm}}{1 \text{ in}}$$

Step 2 Determine which conversion factor to use. You are converting from cm to in. Use the conversion factor with cm on the bottom.

$$\frac{1 \text{ in}}{2.54 \text{ cm}}$$

Step 3 Multiply the starting quantity and units by the conversion factor. Cancel the starting units with the units in the denominator. Round your answer based on the number of significant figures in the conversion factor.

$$100 \text{ cm} \times \frac{1 \text{ in}}{2.54 \text{ cm}} = 39.37 \text{ in}$$

The meterstick is 39.4 in long.

Practice Problem A book has a mass of 5 lbs. What is the mass of the book in kg?

Practice Problem Use the equivalent for in and cm (1 in = 2.54 cm) to show how 1 in^3 = 16.39 cm^3.

Math Skill Handbook

Precision and Significant Digits

When you make a measurement, the value you record depends on the precision of the measuring instrument. This precision is represented by the number of significant digits recorded in the measurement. When counting the number of significant digits, all digits are counted except zeros at the end of a number with no decimal point such as 2,050, and zeros at the beginning of a decimal such as 0.03020. When adding or subtracting numbers with different precision, round the answer to the smallest number of decimal places of any number in the sum or difference. When multiplying or dividing, the answer is rounded to the smallest number of significant digits of any number being multiplied or divided.

Example The lengths 5.28 and 5.2 are measured in meters. Find the sum of these lengths and record your answer using the correct number of significant digits.

Step 1 Find the sum.

5.28 m	2 digits after the decimal
+ 5.2 m	1 digit after the decimal
10.48 m	

Step 2 Round to one digit after the decimal because the least number of digits after the decimal of the numbers being added is 1.

The sum is 10.5 m.

Practice Problem How many significant digits are in the measurement 7,071,301 m? How many significant digits are in the measurement 0.003010 g?

Practice Problem Multiply 5.28 and 5.2 using the rule for multiplying and dividing. Record the answer using the correct number of significant digits.

Scientific Notation

Many times numbers used in science are very small or very large. Because these numbers are difficult to work with scientists use scientific notation. To write numbers in scientific notation, move the decimal point until only one non-zero digit remains on the left. Then count the number of places you moved the decimal point and use that number as a power of ten. For example, the average distance from the Sun to Mars is 227,800,000,000 m. In scientific notation, this distance is 2.278×10^{11} m. Because you moved the decimal point to the left, the number is a positive power of ten.

The mass of an electron is about 0.000 000 000 000 000 000 000 000 000 000 911 kg. Expressed in scientific notation, this mass is 9.11×10^{-31} kg. Because the decimal point was moved to the right, the number is a negative power of ten.

Example Earth is 149,600,000 km from the Sun. Express this in scientific notation.

Step 1 Move the decimal point until one non-zero digit remains on the left.
1.496 000 00

Step 2 Count the number of decimal places you have moved. In this case, eight.

Step 3 Show that number as a power of ten, 10^8.

The Earth is 1.496×10^8 km from the Sun.

Practice Problem How many significant digits are in 149,600,000 km? How many significant digits are in 1.496×10^8 km?

Practice Problem Parts used in a high performance car must be measured to 7×10^{-6} m. Express this number as a decimal.

Practice Problem A CD is spinning at 539 revolutions per minute. Express this number in scientific notation.

Math Skill Handbook

Make and Use Graphs

Data in tables can be displayed in a graph—a visual representation of data. Common graph types include line graphs, bar graphs, and circle graphs.

Line Graph A line graph shows a relationship between two variables that change continuously. The independent variable is changed and is plotted on the *x*-axis. The dependent variable is observed, and is plotted on the *y*-axis.

Example Draw a line graph of the data below from a cyclist in a long-distance race.

Table 5 Bicycle Race Data	
Time (h)	Distance (km)
0	0
1	8
2	16
3	24
4	32
5	40

Step 1 Determine the *x*-axis and *y*-axis variables. Time varies independently of distance and is plotted on the *x*-axis. Distance is dependent on time and is plotted on the *y*-axis.

Step 2 Determine the scale of each axis. The *x*-axis data ranges from 0 to 5. The *y*-axis data ranges from 0 to 40.

Step 3 Using graph paper, draw and label the axes. Include units in the labels.

Step 4 Draw a point at the intersection of the time value on the *x*-axis and corresponding distance value on the *y*-axis. Connect the points and label the graph with a title, as shown in **Figure 20**.

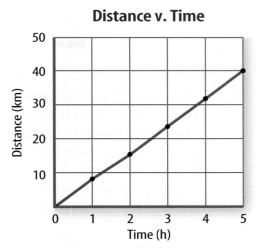

Figure 20 This line graph shows the relationship between distance and time during a bicycle ride.

Practice Problem A puppy's shoulder height is measured during the first year of her life. The following measurements were collected: (3 mo, 52 cm), (6 mo, 72 cm), (9 mo, 83 cm), (12 mo, 86 cm). Graph this data.

Find a Slope The slope of a straight line is the ratio of the vertical change, rise, to the horizontal change, run.

$$\text{Slope} = \frac{\text{vertical change (rise)}}{\text{horizontal change (run)}} = \frac{\text{change in } y}{\text{change in } x}$$

Example Find the slope of the graph in **Figure 20**.

Step 1 You know that the slope is the change in *y* divided by the change in *x*.

$$\text{Slope} = \frac{\text{change in } y}{\text{change in } x}$$

Step 2 Determine the data points you will be using. For a straight line, choose the two sets of points that are the farthest apart.

$$\text{Slope} = \frac{(40-0) \text{ km}}{(5-0) \text{ hr}}$$

Step 3 Find the change in *y* and *x*.

$$\text{Slope} = \frac{40 \text{ km}}{5 \text{ h}}$$

Step 4 Divide the change in *y* by the change in *x*.

$$\text{Slope} = \frac{8 \text{ km}}{\text{h}}$$

The slope of the graph is 8 km/h.

MATH SKILL HANDBOOK 765

Math Skill Handbook

Bar Graph To compare data that does not change continuously you might choose a bar graph. A bar graph uses bars to show the relationships between variables. The *x*-axis variable is divided into parts. The parts can be numbers such as years, or a category such as a type of animal. The *y*-axis is a number and increases continuously along the axis.

Example A recycling center collects 4.0 kg of aluminum on Monday, 1.0 kg on Wednesday, and 2.0 kg on Friday. Create a bar graph of this data.

Step 1 Select the *x*-axis and *y*-axis variables. The measured numbers (the masses of aluminum) should be placed on the *y*-axis. The variable divided into parts (collection days) is placed on the *x*-axis.

Step 2 Create a graph grid like you would for a line graph. Include labels and units.

Step 3 For each measured number, draw a vertical bar above the *x*-axis value up to the *y*-axis value. For the first data point, draw a vertical bar above Monday up to 4.0 kg.

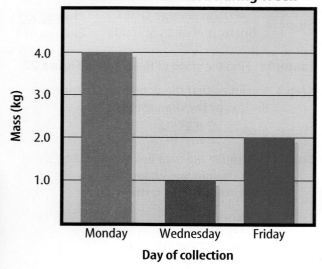

Practice Problem Draw a bar graph of the gases in air: 78% nitrogen, 21% oxygen, 1% other gases.

Circle Graph To display data as parts of a whole, you might use a circle graph. A circle graph is a circle divided into sections that represent the relative size of each piece of data. The entire circle represents 100%, half represents 50%, and so on.

Example Air is made up of 78% nitrogen, 21% oxygen, and 1% other gases. Display the composition of air in a circle graph.

Step 1 Multiply each percent by 360° and divide by 100 to find the angle of each section in the circle.

$$78\% \times \frac{360°}{100} = 280.8°$$

$$21\% \times \frac{360°}{100} = 75.6°$$

$$1\% \times \frac{360°}{100} = 3.6°$$

Step 2 Use a compass to draw a circle and to mark the center of the circle. Draw a straight line from the center to the edge of the circle.

Step 3 Use a protractor and the angles you calculated to divide the circle into parts. Place the center of the protractor over the center of the circle and line the base of the protractor over the straight line.

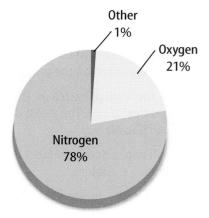

Practice Problem Draw a circle graph to represent the amount of aluminum collected during the week shown in the bar graph to the left.

766 STUDENT RESOURCES

Topographic Map Symbols

Topographic Map Symbols

	Primary highway, hard surface		Index contour
	Secondary highway, hard surface		Supplementary contour
	Light-duty road, hard or improved surface		Intermediate contour
	Unimproved road		Depression contours
	Railroad: single track		
	Railroad: multiple track		Boundaries: national
	Railroads in juxtaposition		State
			County, parish, municipal
	Buildings		Civil township, precinct, town, barrio
	Schools, church, and cemetery		Incorporated city, village, town, hamlet
	Buildings (barn, warehouse, etc.)		Reservation, national or state
	Wells other than water (labeled as to type)		Small park, cemetery, airport, etc.
	Tanks: oil, water, etc. (labeled only if water)		Land grant
	Located or landmark object; windmill		Township or range line, U.S. land survey
	Open pit, mine, or quarry; prospect		Township or range line, approximate location
	Marsh (swamp)		
	Wooded marsh		Perennial streams
	Woods or brushwood		Elevated aqueduct
	Vineyard		Water well and spring
	Land subject to controlled inundation		Small rapids
	Submerged marsh		Large rapids
	Mangrove		Intermittent lake
	Orchard		Intermittent stream
	Scrub		Aqueduct tunnel
	Urban area		Glacier
			Small falls
x7369	Spot elevation		Large falls
670	Water elevation		Dry lake bed

Physical Science Reference Tables

Standard Units

Symbol	Name	Quantity
m	meter	length
kg	kilogram	mass
Pa	pascal	pressure
K	kelvin	temperature
mol	mole	amount of a substance
J	joule	energy, work, quantity of heat
s	second	time
C	coulomb	electric charge
V	volt	electric potential
A	ampere	electric current
Ω	ohm	resistance

Physical Constants and Conversion Factors

Acceleration due to gravity	g	9.8 m/s/s or m/s^2
Avogadro's Number	N_A	6.02×10^{23} particles per mole
Electron charge	e	1.6×10^{-19} C
Electron rest mass	m_e	9.11×10^{-31} kg
Gravitation constant	G	6.67×10^{-11} N $\times$ m^2/kg^2
Mass-energy relationship		1 u (amu) = 9.3×10^2 MeV
Speed of light in a vacuum	c	3.00×10^8 m/s
Speed of sound at STP		331 m/s
Standard Pressure		1 atmosphere
		101.3 kPa
		760 Torr or mmHg
		14.7 lb/in.^{2}

Wavelengths of Light in a Vacuum

Violet	$4.0 - 4.2 \times 10^{-7}$ m
Blue	$4.2 - 4.9 \times 10^{-7}$ m
Green	$4.9 - 5.7 \times 10^{-7}$ m
Yellow	$5.7 - 5.9 \times 10^{-7}$ m
Orange	$5.9 - 6.5 \times 10^{-7}$ m
Red	$6.5 - 7.0 \times 10^{-7}$ m

The Index of Refraction for Common Substances
($\lambda = 5.9 \times 10^{-7}$ m)

Air	1.00
Alcohol	1.36
Canada Balsam	1.53
Corn Oil	1.47
Diamond	2.42
Glass, Crown	1.52
Glass, Flint	1.61
Glycerol	1.47
Lucite	1.50
Quartz, Fused	1.46
Water	1.33

Heat Constants

	Specific Heat (average) (kJ/kg $\times$ °C) (J/g $\times$ °C)	Melting Point (°C)	Boiling Point (°C)	Heat of Fusion (kJ/kg) (J/g)	Heat of Vaporization (kJ/kg) (J/g)
Alcohol (ethyl)	2.43 (liq.)	−117	79	109	855
Aluminum	0.90 (sol.)	660	2467	396	10500
Ammonia	4.71 (liq.)	−78	−33	332	1370
Copper	0.39 (sol.)	1083	2567	205	4790
Iron	0.45 (sol.)	1535	2750	267	6290
Lead	0.13 (sol.)	328	1740	25	866
Mercury	0.14 (liq.)	−39	357	11	295
Platinum	0.13 (sol.)	1772	3827	101	229
Silver	0.24 (sol.)	962	2212	105	2370
Tungsten	0.13 (sol.)	3410	5660	192	4350
Water (solid)	2.05 (sol.)	0	–	334	–
Water (liquid)	4.18 (liq.)	–	100	–	–
Water (vapor)	2.01 (gas)	–	–	–	2260
Zinc	0.39 (sol.)	420	907	113	1770

Reference Handbooks

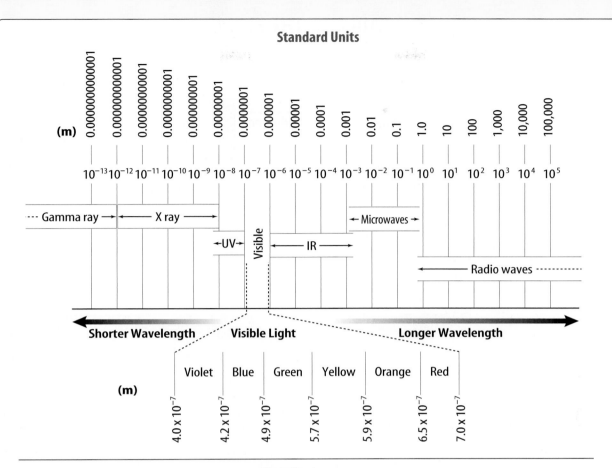

Heat Constants

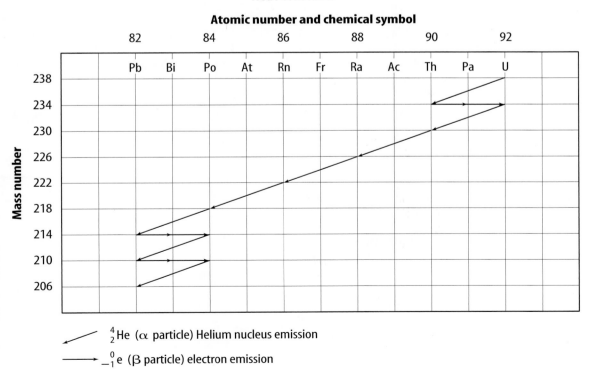

Reference Handbooks

PERIODIC TABLE OF THE ELEMENTS

Columns of elements are called groups. Elements in the same group have similar chemical properties.

- Gas
- Liquid
- Solid
- Synthetic

Element — Hydrogen
Atomic number — 1
Symbol — H
Atomic mass — 1.008
State of matter

The first three symbols tell you the state of matter of the element at room temperature. The fourth symbol identifies elements that are not present in significant amounts on Earth. Useful amounts are made synthetically.

Group	1	2	3	4	5	6	7	8	9
1	Hydrogen 1 H 1.008								
2	Lithium 3 Li 6.941	Beryllium 4 Be 9.012							
3	Sodium 11 Na 22.990	Magnesium 12 Mg 24.305							
4	Potassium 19 K 39.098	Calcium 20 Ca 40.078	Scandium 21 Sc 44.956	Titanium 22 Ti 47.867	Vanadium 23 V 50.942	Chromium 24 Cr 51.996	Manganese 25 Mn 54.938	Iron 26 Fe 55.845	Cobalt 27 Co 58.933
5	Rubidium 37 Rb 85.468	Strontium 38 Sr 87.62	Yttrium 39 Y 88.906	Zirconium 40 Zr 91.224	Niobium 41 Nb 92.906	Molybdenum 42 Mo 95.94	Technetium 43 Tc (98)	Ruthenium 44 Ru 101.07	Rhodium 45 Rh 102.906
6	Cesium 55 Cs 132.905	Barium 56 Ba 137.327	Lanthanum 57 La 138.906	Hafnium 72 Hf 178.49	Tantalum 73 Ta 180.948	Tungsten 74 W 183.84	Rhenium 75 Re 186.207	Osmium 76 Os 190.23	Iridium 77 Ir 192.217
7	Francium 87 Fr (223)	Radium 88 Ra (226)	Actinium 89 Ac (227)	Rutherfordium 104 Rf (261)	Dubnium 105 Db (262)	Seaborgium 106 Sg (266)	Bohrium 107 Bh (264)	Hassium 108 Hs (277)	Meitnerium 109 Mt (268)

The number in parentheses is the mass number of the longest-lived isotope for that element.

Rows of elements are called periods. Atomic number increases across a period.

The arrow shows where these elements would fit into the periodic table. They are moved to the bottom of the table to save space.

Lanthanide series

Cerium 58 Ce 140.116	Praseodymium 59 Pr 140.908	Neodymium 60 Nd 144.24	Promethium 61 Pm (145)	Samarium 62 Sm 150.36

Actinide series

Thorium 90 Th 232.038	Protactinium 91 Pa 231.036	Uranium 92 U 238.029	Neptunium 93 Np (237)	Plutonium 94 Pu (244)

Reference Handbooks

Metal
Metalloid
Nonmetal

The color of an element's block tells you if the element is a metal, nonmetal, or metalloid.

Science Online
Visit blue.msscience.com for updates to the periodic table.

13	14	15	16	17	18
					Helium 2 **He** 4.003
Boron 5 **B** 10.811	Carbon 6 **C** 12.011	Nitrogen 7 **N** 14.007	Oxygen 8 **O** 15.999	Fluorine 9 **F** 18.998	Neon 10 **Ne** 20.180

10	11	12						
			Aluminum 13 **Al** 26.982	Silicon 14 **Si** 28.086	Phosphorus 15 **P** 30.974	Sulfur 16 **S** 32.065	Chlorine 17 **Cl** 35.453	Argon 18 **Ar** 39.948
Nickel 28 **Ni** 58.693	Copper 29 **Cu** 63.546	Zinc 30 **Zn** 65.409	Gallium 31 **Ga** 69.723	Germanium 32 **Ge** 72.64	Arsenic 33 **As** 74.922	Selenium 34 **Se** 78.96	Bromine 35 **Br** 79.904	Krypton 36 **Kr** 83.798
Palladium 46 **Pd** 106.42	Silver 47 **Ag** 107.868	Cadmium 48 **Cd** 112.411	Indium 49 **In** 114.818	Tin 50 **Sn** 118.710	Antimony 51 **Sb** 121.760	Tellurium 52 **Te** 127.60	Iodine 53 **I** 126.904	Xenon 54 **Xe** 131.293
Platinum 78 **Pt** 195.078	Gold 79 **Au** 196.967	Mercury 80 **Hg** 200.59	Thallium 81 **Tl** 204.383	Lead 82 **Pb** 207.2	Bismuth 83 **Bi** 208.980	Polonium 84 **Po** (209)	Astatine 85 **At** (210)	Radon 86 **Rn** (222)
Darmstadtium 110 **Ds** (281)	Roentgenium 111 **Rg** (272)	Ununbium * 112 **Uub** (285)		Ununquadium * 114 **Uuq** (289)				

* The names and symbols for elements 112 and 114 are temporary. Final names will be selected when the elements' discoveries are verified.

Europium 63 **Eu** 151.964	Gadolinium 64 **Gd** 157.25	Terbium 65 **Tb** 158.925	Dysprosium 66 **Dy** 162.500	Holmium 67 **Ho** 164.930	Erbium 68 **Er** 167.259	Thulium 69 **Tm** 168.934	Ytterbium 70 **Yb** 173.04	Lutetium 71 **Lu** 174.967
Americium 95 **Am** (243)	Curium 96 **Cm** (247)	Berkelium 97 **Bk** (247)	Californium 98 **Cf** (251)	Einsteinium 99 **Es** (252)	Fermium 100 **Fm** (257)	Mendelevium 101 **Md** (258)	Nobelium 102 **No** (259)	Lawrencium 103 **Lr** (262)

Glossary/Glosario

Cómo usar el glosario en español:
1. Busca el término en inglés que desees encontrar.
2. El término en español, junto con la definición, se encuentran en la columna de la derecha.

Pronunciation Key

Use the following key to help you sound out words in the glossary.

a	back (BAK)	ew	food (FEWD)
ay	day (DAY)	yoo	pure (PYOOR)
ah	father (FAH thur)	yew	few (FYEW)
ow	flower (FLOW ur)	uh	comma (CAH muh)
ar	car (CAR)	u (+ con)	rub (RUB)
e	less (LES)	sh	shelf (SHELF)
ee	leaf (LEEF)	ch	nature (NAY chur)
ih	trip (TRIHP)	g	gift (GIHFT)
i (i + con + e)	idea (i DEE uh)	j	gem (JEM)
oh	go (GOH)	ing	sing (SING)
aw	soft (SAWFT)	zh	vision (VIH zhun)
or	orbit (OR buht)	k	cake (KAYK)
oy	coin (COYN)	s	seed, cent (SEED, SENT)
oo	foot (FOOT)	z	zone, raise (ZOHN, RAYZ)

English — A — Español

abiotic: nonliving, physical features of the environment, including air, water, sunlight, soil, temperature, and climate. (p. 122)

abiótico: características inertes y físicas del medio ambiente, incluyendo el aire, el agua, la luz solar, el suelo, la temperatura y el clima. (p. 122)

absolute age: age, in years, of a rock or other object; can be determined by using properties of the atoms that make up materials. (p. 257)

edad absoluta: edad, en años, de una roca u otro objeto; puede determinarse utilizando las propiedades de los átomos de los materiales. (p. 257)

absolute magnitude: measure of the amount of light a star actually gives off. (p. 372)

magnitud absoluta: medida de la cantidad real de luz que genera una estrella. (p. 372)

absorption: process in which food molecules pass through the walls of the villi and into the bloodstream. (p. 75)

absorción: proceso mediante el cual las moléculas de los alimentos pasan a través de las paredes de las vellosidades intestinales y entran al torrente sanguíneo. (p. 75)

acceleration: equals the change in velocity divided by the time for the change to take place; occurs when an object speeds up, slows down, or turns. (p. 528)

aceleración: es igual al cambio de velocidad dividido por el tiempo que toma en realizarse dicho cambio; sucede cuando un objeto aumenta su velocidad, la disminuye o gira. (p. 528)

actinide: the second series of inner transition elements which goes from thorium to lawrencium. (p. 450)

actínido: la segunda serie de los elementos de transición interna que abarca desde el torio hasta el laurencio. (p. 450)

activation energy: minimum amount of energy needed to start a chemical reaction. (p. 503)

energía de activación: cantidad mínima de energía necesaria para iniciar una reacción química. (p. 503)

adaptive radiation: production of several species from one ancestral species. (p. 52)

radiación adaptable: producción de varias especies a partir de una especie ancestral. (p. 52)

alkali metals: elements in group 1 of the periodic table. (p. 441)

metales alcalinos: elementos en el grupo 1 de la tabla periódica. (p. 441)

Glossary/Glosario

alkaline earth metals: elements in group 2 of the periodic table. (p. 442)

alleles (uh LEEL): an alternate form that a gene may have for a single trait; can be dominant or recessive. (p. 45)

alpha particle: consists of a particle containing two protons and two neutrons. (p. 417)

alternating current (AC): electric current that changes its direction repeatedly. (p. 678)

alveoli (al VEE uh li): in the lungs, the tiny, thin-walled sacs surrounded by capillaries where exchange of oxygen and carbon dioxide takes place. (p. 76)

amplitude: for a transverse wave, one half the distance between a crest and a trough. (p. 697)

apparent magnitude: measure of the amount of light from a star that is received on Earth. (p. 372)

asteroid: a piece of rock or metal made up of material similar to that which formed the planets; mostly found in the asteroid belt between the orbits of Mars and Jupiter. (p. 358)

asthenosphere (as THE nuh sfihr): plasticlike layer of Earth on which the lithospheric plates float and move around. (p. 190)

atmosphere: air surrounding Earth; is made up of gases, including 78 percent nitrogen, 21 percent oxygen, and 0.03 percent carbon dioxide. (p. 123)

atomic number: number of protons in the nucleus of an atom of a given element. (p. 415)

aurora: light display that occurs when charged particles trapped in the magnetosphere collide with Earth's atmosphere above the poles. (p. 677)

average speed: equals the total distance traveled divided by the total time taken to travel the distance. (p. 525)

axis: imaginary vertical line that cuts through the center of Earth and around which Earth spins. (p. 307)

metales alcalinotérreos: elementos en el grupo 2 de la tabla periódica. (p. 442)

alelos: forma alterna que puede tener un gen para un solo rasgo; un alelo puede ser dominante o recesivo. (p. 45)

partícula alfa: partícula que contiene dos protones y dos neutrones. (p. 417)

corriente alterna (CA): corriente eléctrica que cambia de dirección repetidamente. (p. 678)

alvéolos: en los pulmones, los pequeños sacos de paredes delgadas rodeados por capilares en donde se realiza el intercambio de oxígeno y dióxido de carbono. (p. 76)

amplitud: para una onda transversal, es la mitad de la distancia entre la cresta y la depresión. (p. 697)

magnitud aparente: medida de la cantidad de luz recibida en la Tierra desde una estrella. (p. 372)

asteroide: pedazo de roca o metal formado de material similar al que forma los planetas; se encuentran principalmente en el cinturón de asteroides entre las órbitas de Marte y Júpiter. (p. 358)

astenosfera: capa flexible de la Tierra en la que las placas litosféricas flotan y se mueven de un lugar a otro. (p. 190)

atmósfera: aire que rodea a la Tierra; está compuesta de gases, incluyendo 78% de nitrógeno, 21% de oxígeno y 0.03% de dióxido de carbono. (p. 123)

número atómico: número de protones en el núcleo de un átomo de un elemento determinado. (p. 415)

aurora: despliegue de luz que se produce cuando partículas cargadas atrapadas en la magnetosfera chocan contra la atmósfera terrestre por encima de los polos. (p. 677)

velocidad promedio: es igual al total de la distancia recorrida dividida por el tiempo total necesario para recorrer dicha distancia. (p. 525)

eje: línea vertical imaginaria que atraviesa el centro de la Tierra y alrededor de la cual gira ésta. (p. 307)

B

balanced forces: two or more forces whose effects cancel each other out and do not change the motion of an object. (p. 551)

beta particle: a high-energy electron that comes from the nucleus. (p. 418)

big bang theory: states that about 13.7 billion years ago, the universe began with a huge, fiery explosion. (p. 391)

fuerzas balanceadas: dos o más fuerzas cuyos efectos se cancelan mutuamente sin cambiar el movimiento de un objeto. (p. 551)

partícula beta: electrón de alta energía que proviene del núcleo. (p. 418)

teoría de la gran explosión: establece que hace aproximadamente 13.7 billones de años el universo se originó con una enorme explosión. (p. 391)

Glossary/Glosario

biomes/chemical bond — **biomas/enlace químico**

biomes (BI ohmz): large geographic areas with similar climates and ecosystems; includes tundra, taiga, desert, temperate deciduous forest, temperate rain forest, tropical rain forest, and grassland. (p. 154)

biosphere: part of Earth that supports life, including the top portion of Earth's crust, the atmosphere, and all the water on Earth's surface. (p. 93)

biotic (bi AH tihk): features of the environment that are alive or were once alive. (p. 122)

black hole: final stage in the evolution of a very massive star, where the core's mass collapses to a point that its gravity is so strong that not even light can escape. (p. 384)

biomas: grandes áreas geográficas con climas y ecosistemas similares; incluyen la tundra, la taiga, el desierto, el bosque caducifolio templado, el bosque lluvioso templado, la selva húmeda tropical y los pastizales. (p. 154)

biosfera: capa de la Tierra que alberga la vida, incluyendo la porción superior de la corteza terrestre, la atmósfera y toda el agua de la superficie terrestre. (p. 93)

biótico: características del ambiente que tienen o alguna vez tuvieron vida. (p. 122)

agujero negro: etapa final en la evolución de una estrella masiva, en donde la masa del núcleo se colapsa hasta el punto de que su gravedad es tan fuerte que ni siquiera la luz puede escapar. (p. 384)

C

carbon cycle: model describing how carbon molecules move between the living and nonliving world. (p. 135)

carbon film: thin film of carbon residue preserved as a fossil. (p. 244)

carrying capacity: largest number of individuals of a particular species that an ecosystem can support over time. (p. 101)

cast: a type of body fossil that forms when crystals fill a mold or sediments wash into a mold and harden into rock. (p. 245)

catalyst: substance that speeds up a chemical reaction but is not used up itself or permanently changed. (pp. 449, 507)

cell: smallest functional unit of an organism. (p. 68)

cellular respiration: series of chemical processes in which oxygen combines with food molecules, energy is released, and carbon dioxide and water are given off as wastes. (p. 76)

Cenozoic (sen uh ZOH ihk) Era: era of recent life that began about 66 million years ago and continues today; includes the first appearance of *Homo sapiens* about 400,000 years ago. (p. 292)

center of mass: point in a object that moves as if all of the object's mass were concentrated at that point. (p. 562)

chemical bond: force that holds two atoms together. (p. 471)

ciclo del carbono: modelo que describe cómo se mueven las moléculas de carbono entre el mundo vivo y el mundo inerte. (p. 135)

película de carbono: capa delgada de residuos de carbono preservada como un fósil. (p. 244)

capacidad de carga: el mayor número de individuos de una especie en particular que un ecosistema puede albergar en un periodo de tiempo. (p. 101)

vaciado: tipo de cuerpo fósil que se forma cuando los cristales llenan un molde o los sedimentos son lavados hacia un molde y se endurecen convirtiéndose en roca. (p. 245)

catalizador: sustancia que acelera una reacción química pero que ella misma ni se agota ni sufre cambios permanentes. (pp. 449, 507)

célula: la unidad funcional más pequeña de un organismo. (p. 68)

respiración celular: serie de procesos químicos en los que el oxígeno se mezcla con las moléculas de los alimentos, se libera energía y el dióxido de carbono y el agua son desechados como residuos. (p. 76)

Era Cenozoica: era de vida reciente que comenzó hace aproximadamente 66 millones de años y continúa hasta hoy; incluye la aparición del *Homo sapiens* cerca de 400,000 años atrás. (p. 292)

centro de masa: punto en un objeto que se mueve como si toda la masa del objeto estuviera concentrada en ese punto. (p. 562)

enlace químico: fuerza que mantiene a dos átomos unidos. (p. 471)

chemical equation: shorthand form for writing what reactants are used and what products are formed in a chemical reaction; sometimes shows whether energy is produced or absorbed. (p. 494)

chemical formula: combination of chemical symbols and numbers that indicates which elements and how many atoms of each element are present in a molecule. (p. 480)

chemical reaction: process that produces chemical change, resulting in new substances that have properties different from those of the original substances. (p. 492)

chemosynthesis (kee moh SIN thuh sus): process in which producers make energy-rich nutrient molecules from chemicals. (p. 137)

chromosphere: layer of the Sun's atmosphere above the photosphere. (p. 375)

cinder cone volcano: relatively small volcano formed by moderate to explosive eruptions of tephra. (p. 222)

circuit: closed conducting loop in which electric current can flow continually. (p. 643)

climate: average weather conditions of an area over time, including wind, temperature, and rainfall or other types of precipitation such as snow or sleet. (p. 127)

climax community: stable, end stage of ecological succession in which balance is in the absence of disturbance. (p. 153)

comet: space object made of dust and rock particles mixed with frozen water, methane, and ammonia that forms a bright coma as it approaches the Sun. (p. 356)

commensalism: a type of symbiotic relationship in which one organism benefits and the other organism is not affected. (p. 108)

community: all the populations of different species that live in an ecosystem. (p. 96)

composite volcano: steep-sided volcano formed from alternating layers of violent eruptions of tephra and quieter eruptions of lava. (p. 223)

compound: pure substance that contains two or more elements. (p. 473)

compound machine: machine made up of a combination of two or more simple machines. (p. 591)

compressional wave: mechanical wave that causes particles in matter to move back and forth along the direction the wave travels. (p. 695)

ecuación química: forma breve para representar los reactivos utilizados y los productos que se forman en una reacción química; algunas veces muestra si se produce o absorbe energía. (p. 494)

fórmula química: combinación de símbolos y números químicos que indican cuáles elementos y cuántos átomos de cada elemento están presentes en una molécula. (p. 480)

reacción química: proceso que produce cambios químicos que dan como resultado nuevas sustancias cuyas propiedades son diferentes a aquellas de las sustancias originales. (p. 492)

quimiosíntesis: proceso a través del cual los productores fabrican moléculas ricas en energía a partir de agentes químicos. (p. 137)

cromosfera: capa de la atmósfera del sol que se encuentra sobre la fotosfera. (p. 375)

volcán de cono de cenizas: volcán relativamente pequeño formado por erupciones moderadas o explosivas de tefra. (p. 222)

circuito: circuito conductor cerrado en el cual la energía puede fluir continuamente. (p. 643)

clima: condiciones meteorológicas promedio de un área durante un periodo de tiempo; incluye viento, temperatura y precipitación pluvial u otros tipos de precipitación como la nieve o el granizo. (p. 127)

clímax comunitario: etapa final estable de la sucesión ecológica en la cual se da un equilibrio en ausencia de alteraciones. (p. 153)

cometa: objeto espacial formado por partículas de polvo y roca mezcladas con agua congelada, metano y amoníaco que forman una cola brillante cuando se aproxima al sol. (p. 356)

comensalismo: tipo de relación simbiótica en la que un organismo se beneficia sin afectar al otro. (p. 108)

comunidad: todas las poblaciones de diferentes especies que viven en un mismo ecosistema. (p. 96)

volcán compuesto: volcán de costados inclinados formado por capas alternas producto de erupciones violentas de tefra y erupciones silenciosas de lava. (p. 223)

compuesto: sustancia pura que contiene dos o más elementos. (p. 473)

máquina compuesta: máquina compuesta por la combinación de dos o más máquinas. (p. 591)

onda de compresión: onda mecánica que hace que las partículas de materia se muevan hacia adelante y hacia atrás en la dirección en que viaja la onda. (p. 695)

concentration/cyanobacteria

concentration: describes how much solute is present in a solution compared to the amount of solvent. (p. 505)

condensation: process that takes place when a gas changes to a liquid. (p. 131)

conduction: transfer of thermal energy by direct contact; occurs when energy is transferred by collisions between particles. (p. 613)

conductor: material in which electrons can move easily (p. 612); material that transfers heat easily. (p. 640)

constant: variable that stays the same during an experiment. (p. 21)

constellation: group of stars that forms a pattern in the sky that looks like a familiar object (Libra), animal (Pegasus), or character (Orion). (p. 370)

consumer: organism that cannot create energy-rich molecules but obtains its food by eating other organisms. (p. 107)

continental drift: Wegener's hypothesis that all continents were once connected in a single large landmass that broke apart about 200 million years ago and drifted slowly to their current positions. (p. 182)

control: sample that is treated like other experimental groups except that the independent variable is not applied to it. (p. 22)

convection: transfer of thermal energy by the movement of particles from one place to another in a gas or liquid. (p. 614)

convection current: current in Earth's mantle that transfers heat in Earth's interior and is the driving force for plate tectonics. (p. 195)

coral reef: diverse ecosystem formed from the calcium carbonate shells secreted by corals. (p. 167)

corona: outermost, largest layer of the Sun's atmosphere; extends millions of kilometers into space and has temperatures up to 2 million K. (p. 375)

covalent bond: chemical bond formed when atoms share electrons. (p. 475)

cyanobacteria: chlorophyll-containing, photosynthetic bacteria thought to be one of Earth's earliest life-forms. (p. 281)

concentración/cianobacteria

concentración: describe la cantidad de soluto presente en una solución en relación con la cantidad de solvente. (p. 505)

condensación: proceso que tiene lugar cuando un gas cambia a estado líquido. (p. 131)

conducción: transferencia de energía térmica por contacto directo; se produce cuando la energía se transfiere mediante colisiones entre las partículas. (p. 613)

conductor: material en el cual los electrones se pueden mover fácilmente (p. 612); material que transfiere calor fácilmente. (p. 640)

constante: variable que permanece igual durante un experimento. (p. 21)

constelación: grupo de estrellas que forma un patrón en el cielo y que semeja un objeto (Libra), un animal (Pegaso) o un personaje familiar (Orión). (p. 370)

consumidor: organismo que no puede fabricar moléculas ricas en energía por lo que debe obtener su alimento ingiriendo otros organismos. (p. 107)

deriva continental: hipótesis de Wegener respecto a que todos los continentes estuvieron alguna vez conectados en una gran masa terrestre única que se fraccionó cerca de 200 millones de años atrás y sus trozos se han movilizado lentamente a la deriva hasta sus posiciones actuales. (p. 182)

control: muestra que es tratada de igual manera que otro grupo de experimentos, con la excepción de que no se le aplica la variable independiente. (p. 22)

convección: transferencia de energía térmica por el movimiento de partículas de un sitio a otro en un líquido o un gas. (p. 614)

corriente de convección: corriente en el manto de la Tierra que transfiere calor en el interior de la Tierra y es la causa de la tectónica de placas. (p. 195)

arrecife de coral: ecosistema diverso conformado de caparazones de carbonato de calcio secretados por los corales. (p. 167)

corona: capa más externa y más grande de la atmósfera solar; se extiende millones de kilómetros dentro del espacio y tiene una temperatura hasta de 2 millones de grados Kelvin. (p. 375)

enlace covalente: enlace químico que se forma cuando los átomos comparten electrones. (p. 475)

cianobacteria: bacteria fotosintética que contiene clorofila; se cree que es una de las primeras formas de vida que surgió en la tierra. (p. 281)

D

dependent variable: factor that is being measured during an experiment. (p. 21)

descriptive research: answers scientific questions through observation. (p. 13)

desert: driest biome on Earth with less than 25 cm of rain each year; has dunes or thin soil with little organic matter, where plants and animals are adapted to survive extreme conditions. (p. 160)

diffraction: bending of waves around an object. (p. 700)

digestion: breakdown of foods into smaller and simpler molecules that can be used by the cells of the body. (p. 74)

direct current (DC): electric current that flows only in one direction. (p. 679)

dominant (DAH muh nunt): describes a trait that covers over, or dominates, another form of that trait. (p. 45)

variable dependiente: factor que es medido durante un experimento. (p. 21)

investigación descriptiva: responde a preguntas científicas por medio de la observación. (p. 13)

desierto: el bioma más seco sobre la Tierra con menos de 25 centímetros cúbicos de lluvia al año; tiene dunas o un suelo delgado con muy poca materia orgánica y aquí las plantas y animales están adaptados para sobrevivir en condiciones extremosas. (p. 160)

difracción: curvatura de las ondas alrededor de un objeto. (p. 700)

digestión: desdoblamiento de los alimentos en moléculas más simples y pequeñas que pueden ser utilizadas por las células del cuerpo. (p. 74)

corriente directa (CD): corriente eléctrica que fluye solamente en una dirección. (p. 679)

dominante: describe un rasgo que encubre por completo o domina a otra forma de dicho rasgo. (p. 45)

E

Earth: third planet from the Sun; has an atmosphere that protects life and surface temperatures that allow water to exist as a solid, liquid, and gas. (p. 344)

earthquake: movement of the ground that occurs when rocks inside Earth pass their elastic limit, break suddenly, and experience elastic rebound. (p. 210)

ecology: study of the interactions that take place among organisms and their environment. (p. 95)

ecosystem: all the living organisms that live in an area and the nonliving features of their environment. (p. 95)

efficiency: equals the output work divided by the input work; expressed as a percentage. (p. 589)

electric current: the flow of electric charge, measured in amperes (A). (p. 643)

electric discharge: rapid movement of excess charge from one place to another. (p. 641)

electric field: surrounds every electric charge and exerts forces on other electric charges. (p. 639)

Tierra: tercer planeta más cercano al sol; tiene una atmósfera que protege la vida y temperaturas en su superficie que permiten la presencia de agua en estado sólido, líquido y gaseoso. (p. 344)

terremoto: movimiento del suelo que ocurre cuando las rocas del interior de la Tierra sobrepasan su límite de elasticidad, se rompen súbitamente y experimentan rebotes elásticos. (p. 210)

ecología: estudio de las interacciones que se dan entre los organismos y su medio ambiente. (p. 95)

ecosistema: conjunto de organismos vivos que habitan en un área y las características de su medio ambiente. (p. 95)

eficiencia: equivale al trabajo aplicado dividido el trabajo generado y se expresa en porcentaje. (p. 589)

corriente eléctrica: flujo de carga eléctrica, el cual se mide en amperios (A). (p. 643)

descarga eléctrica: movimiento rápido de carga excesiva de un lugar a otro. (p. 641)

campo eléctrico: campo que rodea a todas las cargas eléctricas y que ejerce fuerzas sobre otras cargas eléctricas. (p. 639)

electric force: attractive or repulsive force exerted by all charged objects on each other. (p. 639)

electric power: rate at which electrical energy is converted into other forms of energy, measured in watts (W) or kilowatts (kW). (p. 652)

electromagnet: magnet created by wrapping a current-carrying wire around an iron core. (p. 673)

electromagnetic spectrum: complete range of electromagnetic wave frequencies and wavelengths. (p. 708)

electromagnetic waves: waves that can travel through matter or empty space, includes radio waves, infrared waves, visible light waves, ultraviolet waves, X rays, and gamma rays. (p. 707)

electron: negatively-charged particle that exists in an electron cloud formation around an atom's nucleus. (p. 407)

electron cloud: region surrounding the nucleus of an atom, where electrons are most likely to be found. (pp. 404, 473)

electron dot diagram: chemical symbol for an element, surrounded by as many dots as there are electrons in its outer energy level. (p. 470)

element: substance that cannot be broken down into simpler substances. (p. 405)

ellipse (ee LIHPS): elongated, closed curve that describes Earth's yearlong orbit around the Sun. (p. 309)

endothermic (en duh THUR mihk) reaction: chemical reaction in which heat energy is absorbed. (p. 499)

energy level: the different positions for an electron in an atom. (p. 465)

energy pyramid: model that shows the amount of energy available at each feeding level in an ecosystem. (p. 139)

enzyme (EN zime): a type of protein that regulates chemical reactions in cells without being changed or used up itself. (pp. 75, 508)

eon: longest subdivision in the geologic time scale that is based on the abundance of certain types of fossils and is subdivided into eras, periods, and epochs. (p. 273)

epicenter: point on Earth's surface directly above an earthquake's focus. (p. 212)

epoch: next-smaller division of geologic time after the period; is characterized by differences in life-forms that may vary regionally. (p. 273)

fuerza eléctrica: fuerza de atracción o de repulsión que ejercen todos los objetos cargados entre ellos mismos. (p. 639)

potencia eléctrica: tasa a la cual la energía eléctrica se convierte en otras formas de energía, la cual se mide en vatios (W) o en kilovatios (kW). (p. 652)

electroimán: imán que se crea al enrollar un cable transportador de corriente alrededor de un centro de hierro. (p. 673)

espectro electromagnético: rango total de las frecuencias y longitudes de onda de las ondas electromagnéticas. (p. 708)

ondas electromagnéticas: ondas que pueden viajar a través de la materia o del espacio vacío; incluyen las ondas de radio, las infrarrojas, las de luz visible, las ultravioleta y los rayos X y gama. (p. 707)

electrón: partícula cargada negativamente que existe en una nube de electrones alrededor del núcleo de un átomo. (p. 407)

nube de electrones: región que rodea al núcleo de un átomo en la cual es muy probable que se encuentren los electrones. (pp. 404, 473)

diagrama de punto de electrones: símbolo químico para un elemento, rodeado de tantos puntos como electrones se encuentran en su nivel exterior de energía. (p. 470)

elemento: sustancia que no se puede dividir en sustancias más simples. (p. 405)

elipse: curva cerrada y elongada que describe la órbita anual de la Tierra alrededor del sol. (p. 309)

reacción endotérmica: reacción química en la cual se absorbe energía calórica. (p. 499)

nivel de energía: las diferentes posiciones de un electrón en un átomo. (p. 465)

pirámide de energía: modelo que muestra la cantidad de energía disponible en cada nivel alimenticio de un ecosistema. (p. 139)

enzima: tipo de proteína que regula las reacciones químicas en las células sin que ésta sufra modificaciones o se agote. (pp. 75, 508)

eón: la más grande subdivisión en la escala del tiempo geológico; se basa en la abundancia de cierto tipo de fósiles y está dividida en eras, periodos y épocas. (p. 273)

epicentro: punto de la superficie terrestre directamente encima del foco del terremoto. (p. 212)

época: la siguiente división más pequeña del tiempo geológico después del periodo; está caracterizada por diferencias en las formas de vida que pueden variar regionalmente. (p. 273)

Glossary/Glosario

equinoccio/fósiles

equinoccio: dos veces al año—en primavera y otoño—cuando el sol está posicionado directamente sobre el ecuador y el número de horas del día y de la noche son iguales en todo el mundo. (p. 311)

era: la segunda división más grande del tiempo geológico; está subdividida en periodos y se basa en cambios mayores en todo el mundo con respecto a los tipos de fósiles. (p. 273)

estuario: área extremadamente fértil donde un río desemboca en el océano; contiene una mezcla de agua dulce y salada y sirve como vivero para muchas especies de peces. (p. 168)

evaporación: proceso que tiene lugar cuando un líquido cambia a estado gaseoso. (p. 130)

evolución: cambio genético de una especie a través del tiempo. (p. 50)

excreción: en los humanos, la eliminación de productos de desecho del organismo por medio de esfuerzos combinados de los sistemas circulatorio, respiratorio y excretorio. (p. 78)

reacción exotérmica: reacción química en la cual se libera energía calórica. (p. 499)

diseño de investigación experimental: se utiliza para responder preguntas científicas comprobando una hipótesis a través del uso de una serie de cuidadosos pasos controlados. (p. 13)

extinción: ocurre cuando muere el último miembro de una especie y produce pérdida de diversidad entre seres vivos. (p. 53)

---

equinox (EE kwuh nahks): twice-yearly time—each spring and fall—when the sun is directly over the equator and the number of daylight and nighttime hours are equal worldwide. (p. 311)

era: second-longest division of geologic time; is subdivided into periods and is based on major worldwide changes in types of fossils. (p. 273)

estuary: extremely fertile area where a river meets an ocean; contains a mixture of freshwater and saltwater and serves as a nursery for many species of fish. (p. 168)

evaporation: process that takes place when a liquid changes to a gas. (p. 130)

evolution: change in the genetics of a species over time. (p. 50)

excretion: in humans, the removal of waste products from the body through combined efforts of the circulatory, respiratory, and excretory systems. (p. 78)

exothermic (ek soh THUR mihk) reaction: chemical reaction in which heat energy is released. (p. 499)

experimental research design: used to answer scientific questions by testing a hypothesis through the use of a series of carefully controlled steps. (p. 13)

extinction: occurs when the last member of a species dies; causes a loss of diversity among living things. (p. 53)

---

fault: fracture that occurs when rocks break and the results in relative movement of opposing sides; form as a result of compression (reverse fault), pulled apart (normal fault), or shear (strike-slip fault). (p. 211)

focus: point deep inside Earth where energy is released causing an earthquake. (p. 212)

food web: model that shows the complex feeding relationships among organisms in a community. (p. 138)

force: a push or a pull. (p. 550)

fossils: remains, imprints, or traces of preserved organisms that can tell when and where such organisms once lived and how they lived. (p. 243)

---

falla: fractura que ocurre cuando al romperse una roca presentan relativos movimientos de los lados opuestos; se pueden formar como resultado de compresión (falla reversa), al separarse (falla normal) o al deslizarse (falla por desplazamiento). (p. 211)

foco: punto profundo de la Tierra donde se genera energía causando un terremoto. (p. 212)

red alimenticia: modelo que muestra las complejas relaciones alimenticias entre los organismos de una comunidad. (p. 138)

fuerza: presión o tracción. (p. 550)

fósiles: restos, huellas o trazas de organismos preservados que pueden informar cuándo, dónde y cómo vivieron tales organismos. (p. 243)

GLOSSARY/GLOSARIO 779

Glossary/Glosario

frequency/half-life

frequency: number of wavelengths that pass a given point in one second, measured in hertz (Hz). (p. 696)

friction: force that acts to oppose sliding between two surfaces that are touching. (p. 552)

full moon: phase that occurs when all of the Moon's surface facing Earth reflects light. (p. 313)

G

galaxy: large group of stars, dust, and gas held together by gravity; can be elliptical, spiral, or irregular. (p. 386)

gene: segment of DNA on a chromosome. (p. 40)

generator: device that uses a magnetic field to turn kinetic energy into electrical energy. (p. 6)

genetics: study of heredity. (p. 44)

genotype (JEE nuh tipe): genes that an organism has—its genetic makeup. (p. 40)

geologic time scale: division of Earth's history into units based largely on the types of life that lived only during certain periods. (p. 272)

giant: late stage in the life of comparatively less main sequence star in which hydrogen is depleted, the core contracts and temperature of the star increase, causing its outer layers to expand and cool. (p. 383)

grasslands: temperate and tropical regions to 75 cm of precipitation each year, dominated by climax communities of grasses, growing crops and raising cattle and sheep.

Great Red Spot: giant, high-pressure storm in atmosphere. (p. 348)

group: family of elements in the periodic table that have similar physical or chemical properties.

habitat: place where an organism lives that provides the types of food, shelter, moisture, and temperature needed for survival.

half-life: time needed for one-half of a sample of a radioactive isotope to decay.

---

frecuencia/vida media

frecuencia: número de longitudes de onda que pasan un punto determinado en un segundo; se mide en hertz (Hz). (p. 696)

fricción: fuerza que actúa para oponerse al deslizamiento entre dos superficies que se tocan. (p. 552)

luna llena: fase que ocurre cuando la superficie de la luna frente a la Tierra refleja la luz. (p. 313)

G

galaxia: grupo grande de estrellas, polvo y gas en donde todo está unido por gravedad; puede ser elíptica, espiral o irregular. (p. 386)

gen: segmento de ADN en un cromosoma. (p. 40)

generador: dispositivo que utiliza un campo magnético para convertir energía cinética en energía eléctrica. (p. 678)

genética: estudio de la herencia. (p. 44)

genotipo: genes que posee un organismo: su composición genética. (p. 40)

escala del tiempo geológico: división de la historia de la Tierra en unidades de tiempo; se basa en los tipos de formas de vida que vivieron sólo durante ciertos periodos. (p. 272)

gigante: etapa tardía en la vida de una estrella de secuencia principal, de relativamente poca masa, en la que el hidrógeno en el núcleo está agotado, el núcleo se contrae y la temperatura en el interior de la estrella aumenta, causando que las capas externas se expandan y enfríen. (p. 383)

pastizales: regiones tropicales y templadas con 25 a 75 centímetros cúbicos de lluvia al año; son dominadas por el clímax comunitario de los pastos e ideales para la cría de ganado y ovejas. (p. 161)

La Gran Mancha Roja: tormenta gigante de alta presión en la atmósfera de Júpiter. (p. 348)

grupo: familia de elementos en la tabla periódica que tienen propiedades físicas o químicas similares. (p. 435)

---

hábitat: lugar donde vive un organismo y que le proporciona los tipos de alimento, refugio, humedad y temperatura necesarios para su supervivencia. (p. 97)

vida media: tiempo necesario para que la mitad de la masa de una muestra de un isótopo radiactivo se descomponga. (pp. 256, 420)

equinox (EE kwuh nahks): twice-yearly time—each spring and fall—when the Sun is directly over the equator and the number of daylight and nighttime hours are equal worldwide. (p. 311)

era: second-longest division of geologic time; is subdivided into periods and is based on major worldwide changes in types of fossils. (p. 273)

estuary: extremely fertile area where a river meets an ocean; contains a mixture of freshwater and saltwater and serves as a nursery for many species of fish. (p. 168)

evaporation: process that takes place when a liquid changes to a gas. (p. 130)

evolution: change in the genetics of a species over time. (p. 50)

excretion: in humans, the removal of waste products from the body through combined efforts of the circulatory, respiratory, and excretory systems. (p. 78)

exothermic (ek soh THUR mihk) reaction: chemical reaction in which heat energy is released. (p. 499)

experimental research design: used to answer scientific questions by testing a hypothesis through the use of a series of carefully controlled steps. (p. 13)

extinction: occurs when the last member of a species dies; causes a loss of diversity among living things. (p. 53)

equinoccio: dos veces al año—en primavera y otoño—cuando el sol está posicionado directamente sobre el ecuador y el número de horas del día y de la noche son iguales en todo el mundo. (p. 311)

era: la segunda división más grande del tiempo geológico; está subdividida en periodos y se basa en cambios mayores en todo el mundo con respecto a los tipos de fósiles. (p. 273)

estuario: área extremadamente fértil donde un río desemboca en el océano; contiene una mezcla de agua dulce y salada y sirve como vivero para muchas especies de peces. (p. 168)

evaporación: proceso que tiene lugar cuando un líquido cambia a estado gaseoso. (p. 130)

evolución: cambio genético de una especie a través del tiempo. (p. 50)

excreción: en los humanos, la eliminación de productos de desecho del organismo por medio de esfuerzos combinados de los sistemas circulatorio, respiratorio y excretorio. (p. 78)

reacción exotérmica: reacción química en la cual se libera energía calórica. (p. 499)

diseño de investigación experimental: se utiliza para responder preguntas científicas comprobando una hipótesis a través del uso de una serie de cuidadosos pasos controlados. (p. 13)

extinción: ocurre cuando muere el último miembro de una especie y produce pérdida de diversidad entre los seres vivos. (p. 53)

F

fault: fracture that occurs when rocks break and that results in relative movement of opposing sides; can form as a result of compression (reverse fault), being pulled apart (normal fault), or shear (strike-slip fault). (p. 211)

focus: point deep inside Earth where energy is released, causing an earthquake. (p. 212)

food web: model that shows the complex feeding relationships among organisms in a community. (p. 138)

force: a push or a pull. (p. 550)

fossils: remains, imprints, or traces of prehistoric organisms that can tell when and where organisms once lived and how they lived. (p. 243)

falla: fractura que ocurre cuando al romperse una roca se presentan relativos movimientos de los lados opuestos; se pueden formar como resultado de una compresión (falla reversa), al separarse (falla normal) o al deslizarse (falla por desplazamiento). (p. 211)

foco: punto profundo de la Tierra donde se genera energía causando un terremoto. (p. 212)

cadena alimenticia: modelo que muestra las complejas relaciones alimenticias entre los organismos de una comunidad. (p. 138)

fuerza: presión o tracción. (p. 550)

fósiles: restos, huellas o trazas de organismos prehistóricos que pueden informar cuándo, dónde y cómo vivieron tales organismos. (p. 243)

frequency/half-life　　　　　　　　　　　　　　　　　　　　　　　　　　　　　　　　　**frecuencia/vida media**

frequency: number of wavelengths that pass a given point in one second, measured in hertz (Hz). (p. 696)

friction: force that acts to oppose sliding between two surfaces that are touching. (p. 552)

full moon: phase that occurs when all of the Moon's surface facing Earth reflects light. (p. 313)

frecuencia: número de longitudes de onda que pasan un punto determinado en un segundo; se mide en hertz (Hz). (p. 696)

fricción: fuerza que actúa para oponerse al deslizamiento entre dos superficies que se tocan. (p. 552)

luna llena: fase que ocurre cuando toda la superficie de la luna frente a la Tierra refleja la luz del sol. (p. 313)

G

galaxy: large group of stars, dust, and gas held together by gravity; can be elliptical, spiral, or irregular. (p. 386)

gene: segment of DNA on a chromosome. (p. 40)

generator: device that uses a magnetic field to turn kinetic energy into electrical energy. (p. 678)

genetics: study of heredity. (p. 44)

genotype (JEE nuh tipe): genes that an organism has—its genetic makeup. (p. 40)

geologic time scale: division of Earth's history into time units based largely on the types of life-forms that lived only during certain periods. (p. 272)

giant: late stage in the life of comparatively low-mass main sequence star in which hydrogen in the core is deleted, the core contracts and temperatures inside the star increase, causing its outer layers to expand and cool. (p. 383)

grasslands: temperate and tropical regions with 25 cm to 75 cm of precipitation each year that are dominated by climax communities of grasses; ideal for growing crops and raising cattle and sheep. (p. 161)

Great Red Spot: giant, high-pressure storm in Jupiter's atmosphere. (p. 348)

group: family of elements in the periodic table that have similar physical or chemical properties. (p. 435)

galaxia: grupo grande de estrellas, polvo y gas en donde todo está unido por gravedad; puede ser elíptica, espiral o irregular. (p. 386)

gen: segmento de ADN en un cromosoma. (p. 40)

generador: dispositivo que utiliza un campo magnético para convertir energía cinética en energía eléctrica. (p. 678)

genética: estudio de la herencia. (p. 44)

genotipo: genes que posee un organismo: su composición genética. (p. 40)

escala del tiempo geológico: división de la historia de la Tierra en unidades de tiempo; se basa en los tipos de formas de vida que vivieron sólo durante ciertos periodos. (p. 272)

gigante: etapa tardía en la vida de una estrella de secuencia principal, de relativamente poca masa, en la que el hidrógeno en el núcleo está agotado, el núcleo se contrae y la temperatura en el interior de la estrella aumenta, causando que las capas externas se expandan y enfríen. (p. 383)

pastizales: regiones tropicales y templadas con 25 a 75 centímetros cúbicos de lluvia al año; son dominadas por el clímax comunitario de los pastos e ideales para la cría de ganado y ovejas. (p. 161)

La Gran Mancha Roja: tormenta gigante de alta presión en la atmósfera de Júpiter. (p. 348)

grupo: familia de elementos en la tabla periódica que tienen propiedades físicas o químicas similares. (p. 435)

H

habitat: place where an organism lives and that provides the types of food, shelter, moisture, and temperature needed for survival. (p. 97)

half-life: time needed for one half of the mass of a sample of a radioactive isotope to decay. (pp. 256, 420)

hábitat: lugar donde vive un organismo y que le proporciona los tipos de alimento, refugio, humedad y temperatura necesarios para su supervivencia. (p. 97)

vida media: tiempo necesario para que la mitad de la masa de una muestra de un isótopo radiactivo se descomponga. (pp. 256, 420)

halogen/intensity   **halógenos/intensidad**

halogen: any element in group 17 of the periodic table. (p. 446)

heat engine: device that converts thermal energy into mechanical energy. (p. 619)

heat: thermal energy transferred from a warmer object to a cooler object. (p. 612)

homeostasis: process by which the body maintains a stable internal environment. (p. 79)

hot spot: hot, molten rock material that has been forced upward from deep inside Earth, which may cause magma to break through Earth's mantle and crust and may form volcanoes. (p. 228)

hypothesis (hi PAH thuh sus): prediction or statement that can be tested and may be formed by prior knowledge, any previous observations, and new information. (p. 21)

halógenos: elementos en el grupo 17 de la tabla periódica. (p. 446)

motor de calor: motor que transforma la energía térmica en energía mecánica. (p. 619)

calor: energía térmica transferida de un objeto con más calor a uno con menos calor. (p. 612)

homeostasis: proceso mediante el cual el cuerpo mantiene un entorno interno estable. (p. 79)

punto caliente: material de roca fundida, caliente, que ha sido lanzado hacia arriba desde lo más profundo de la Tierra y que puede producir que el magma se rompa a través del manto y la corteza pudiendo formar volcanes. (p. 228)

hipótesis: predicción o enunciado que puede ser probado y que se formula con base a previos conocimientos, previas observaciones y nuevas informaciones. (p. 21)

I

impact basin: a hollow left on the surface of the Moon caused by an object striking its surface. (p. 323)

inclined plane: simple machine that is a flat surface, sloped surface, or ramp. (p. 591)

independent variable: variable that can be changed during an experiment. (p. 21)

index fossils: remains of species that existed on Earth for a relatively short period of time, were abundant and widespread geographically, and can be used by geologists to assign the ages of rock layers. (p. 247)

inertia: tendency of an object to resist a change in its motion. (p. 533)

infrared waves: electromagnetic waves with wavelengths between about one thousandth of a meter and 700 billionths of a meter. (p. 709)

inhibitor: substance that slows down a chemical reaction, making the formation of a certain amount of product take longer. (p. 506)

input force: force exerted on a machine. (p. 586)

instantaneous speed: the speed of an object at one instant of time. (p. 525)

insulator: material in which electrons cannot move easily. (p. 640)

intensity: amount of energy a wave carries past a certain area each second. (p. 702)

cráter de impacto: un hueco dejado en la superficie de la luna causada por un objeto que chocó contra su superficie. (p. 323)

plano inclinado: máquina simple que consiste en una superficie plana, inclinada, o una rampa. (p. 591)

variable independiente: variable que se puede cambiar durante un experimento. (p. 21)

fósiles índice: restos de especies que existieron sobre la Tierra durante un periodo de tiempo relativamente corto y que fueron abundantes y ampliamente diseminadas geográficamente; los geólogos pueden usarlos para inferir las edades de las capas rocosas. (p. 247)

inercia: tendencia de un objeto a resistirse a un cambio de movimiento. (p. 533)

ondas infrarrojas: ondas electromagnéticas con longitudes de onda entre aproximadamente una milésima y 700 billonésimas de metro. (p. 709)

inhibidor: sustancia que reduce la velocidad de una reacción química, haciendo que la formación de una determinada cantidad de producto tarde más tiempo. (p. 506)

fuerza aplicada: fuerza que se ejerce sobre una máquina. (p. 586)

velocidad instantánea: la velocidad de un objeto en un instante de tiempo. (p. 525)

aislante: material en el cual los electrones no se pueden mover fácilmente. (p. 640)

intensidad: cantidad de energía que transporta una onda al pasar por un área determinada en un segundo. (p. 702)

Glossary/Glosario

internal combustion engine/limiting factor   **motor de combustión interna/factor limitante**

internal combustion engine: heat engine in which fuel is burned in a combustion chamber inside the engine. (p. 620)

intertidal zone: part of the shoreline that is under water at high tide and exposed to the air at low tide. (p. 168)

ion (I ahn): atom that is positively or negatively charged because it has gained or lost one or more electrons. (pp. 473, 636)

ionic bond: attraction that holds oppositely charged ions close together. (p. 473)

isotopes (I suh tohps): atoms of the same element that have different numbers of neutrons. (p. 415)

motor de combustión interna: motor de calor en el cual el combustible es quemado en una cámara de combustión dentro del motor. (p. 620)

zona litoral: parte de la línea costera que está bajo el agua durante la marea alta y expuesta al aire durante la marea baja. (p. 168)

ion: átomo cargado positiva o negativamente debido a que ha ganado o perdido uno o más electrones. (pp. 473, 636)

enlace iónico: atracción que mantiene unidos a iones con cargas opuestas. (p. 473)

isótopos: átomos del mismo elemento que tienen diferente número de neutrones. (p. 415)

J

Jupiter: largest and fifth planet from the Sun; contains more mass than all the other planets combined, has continuous storms of high-pressure gas, and an atmosphere mostly of hydrogen and helium. (p. 348)

Júpiter: el quinto planeta más cercano al sol, y también el más grande; contiene más masa que todos los otros planetas en conjunto, tiene tormentas continuas de gas a alta presión y una atmósfera compuesta principalmente por hidrógeno y helio. (p. 348)

L

lanthanide: the first series of inner transition elements which goes from cerium to lutetium. (p. 450)

lava: molten rock flowing onto Earth's surface. (p. 219)

law of conservation of momentum: states that the total momentum of objects that collide with each other is the same before and after the collision. (p. 535)

law of reflection: states that the angle the incoming wave makes with the normal to the reflecting surface equals the angle the reflected wave makes with the surface. (p. 699)

lever: simple machine consisting of a rigid rod or plank that pivots or rotates about a fixed point called the fulcrum. (p. 594)

light-year: unit representing the distance light travels in one year—about 9.5 trillion km—used to record distances between stars and galaxies. (p. 373)

limiting factor: anything that can restrict the size of a population, including living and nonliving features of an ecosystem, such as predators or drought. (p. 100)

lantánidos: la primera serie de los elementos de transición interna que va desde el cerio hasta el lutecio. (p. 450)

lava: roca fundida que fluye en la superficie terrestre. (p. 219)

ley de conservación de momento: establece que el momento total de los objetos que chocan entre sí es el mismo antes y después de la colisión. (p. 535)

ley de reflexión: establece que el ángulo que forman la onda que llega y la normal hacia la superficie reflejante es igual al ángulo que la onda reflejada forma con la superficie. (p. 699)

palanca: máquina simple que consiste en una barra rígida que puede girar sobre un punto fijo llamado punto de apoyo. (p. 594)

año luz: unidad que representa la distancia que la luz viaja en un año—cerca de 9.5 trillones de kilómetros—usada para registrar las distancias entre las estrellas y las galaxias. (p. 373)

factor limitante: cualquier factor que pueda restringir el tamaño de una población, incluyendo las características biológicas y no biológicas de un ecosistema, tales como los depredadores o las sequías. (p. 100)

lithosphere (LIH thuh sfihr): rigid layer of Earth about 100 km thick, made of the crust and a part of the upper mantle. (p. 190)

lunar eclipse: occurs when Earth's shadow falls on the Moon. (p. 316)

litosfera: capa rígida de la Tierra de unos 100 kilómetros de profundidad, comprende la corteza y una parte del manto superior. (p. 190)

eclipse lunar: ocurre cuando la sombra de la Tierra cubre la luna. (p. 316)

M

magnetic domain: group of atoms whose fields point in the same direction. (p. 668)

magnetic field: surrounds a magnet and exerts a magnetic force on other magnets. (p. 667)

magnetosphere: region of space affected by Earth's magnetic field. (p. 669)

magnitude: a measure of the energy released by an earthquake. (p. 213)

maria (MAHR ee uh): dark-colored, relatively flat regions of the Moon formed when ancient lava reached the surface and filled craters on the Moon's surface. (p. 317)

Mars: fourth planet from the Sun; has polar ice caps, a thin atmosphere, and a reddish appearance caused by iron oxide in weathered rocks and soil. (p. 344)

mass: amount of matter in an object. (p. 533)

mass number: the sum of neutrons and protons in the nucleus of an atom. (p. 416)

mechanical advantage: number of times the input force is multiplied by a machine; equal to the output force divided by the input force. (p. 587)

Mercury: smallest planet, closest to the Sun; does not have a true atmosphere; has a surface with many craters and high cliffs. (p. 342)

Mesozoic (mez uh ZOH ihk) Era: middle era of Earth's history, during which Pangaea broke apart, dinosaurs appeared, and reptiles and gymnosperms were the dominant land life-forms. (p. 288)

metal: element that has luster, is malleable, ductile, and a good conductor of heat and electricity. (p. 438)

metallic bond: bond formed when metal atoms share their pooled electrons. (p. 474)

metalloid (MET ul oyd): element that shares some properties with both metals and nonmetals. (p. 438)

meteor: a meteoroid that burns up in Earth's atmosphere. (p. 357)

meteorite: a meteoroid that strikes the surface of a moon or planet. (p. 358)

dominio magnético: grupo de átomos cuyos campos apuntan en la misma dirección. (p. 668)

campo magnético: campo que rodea a un imán y ejerce fuerza magnética sobre otros imanes. (p. 667)

magnetosfera: región del espacio afectada por el campo magnético de la Tierra. (p. 669)

magnitud: medida de la energía generada por un terremoto. (p. 213)

mares: regiones relativamente planas, de color oscuro, que se encuentran en la luna y que fueron formadas cuando la lava antigua alcanzó la superficie y llenó los cráteres sobre la superficie lunar. (p. 317)

Marte: cuarto planeta más cercano al sol; tiene casquetes de hielo polar, una atmósfera delgada y una apariencia rojiza causada por el óxido de hierro presente en las rocas y suelo de su superficie. (p. 344)

masa: cantidad de materia en un objeto. (p. 533)

número de masa: la suma de neutrones y protones en el núcleo de un átomo. (p. 416)

ventaja mecánica: número de veces que la fuerza aplicada es multiplicada por una máquina; equivale a la fuerza producida dividida por la fuerza aplicada. (p. 587)

Mercurio: el planeta más pequeño y más cercano al sol; no tiene una atmósfera verdadera; tiene una superficie con muchos cráteres y grandes acantilados. (p. 342)

Era Mesozoica: era media de la historia de la Tierra durante la cual se escindió la Pangea y aparecieron los dinosaurios; los reptiles y gimnospermas fueron las formas de vida que dominaron la tierra. (p. 288)

metal: elemento que tiene brillo, es maleable, dúctil y buen conductor de calor y electricidad. (p. 438)

enlace metálico: enlace que se forma cuando átomos metálicos comparten sus electrones agrupados. (p. 474)

metaloide: elemento que comparte algunas propiedades de los metales y de los no metales. (p. 438)

meteoro: un meteoroide que se incinera en la atmósfera de la Tierra. (p. 357)

meteorito: un meteoroide que choca contra la superficie de la luna o de algún planeta. (p. 358)

Glossary/Glosario

mineral/neutron

mineral: inorganic substance required in small amounts that is involved in many of the chemical reactions that occur in the body and is needed to promote good health and fight disease. (p. 65)

model: represents something that is too big, too small, too dangerous, too time consuming, or too expensive to observe directly. (p. 16)

mold: a type of body fossil that forms in rock when an organism with hard parts is buried, decays or dissolves, and leaves a cavity in the rock. (p. 245)

molecule (MAH lih kewl): neutral particle formed when atoms share electrons. (p. 475)

momentum: a measure of how difficult it is to stop a moving object; equals the product of mass and velocity. (p. 534)

moon phase: change in appearance of the Moon as viewed from the Earth, due to the relative positions of the Moon, Earth, and Sun. (p. 313)

motor: device that transforms electrical energy into kinetic energy. (p. 676)

mutation: process that changes DNA to form new alleles. (p. 52)

mutualism: a type of symbiotic relationship in which both organisms benefit. (p. 108)

mineral/neutrón

mineral: sustancia inorgánica requerida en pequeñas cantidades, la cual participa en muchas de las reacciones químicas que ocurren en el cuerpo y que es necesaria para mantener la salud y combatir enfermedades. (p. 65)

modelo: representa algo que es muy grande, muy pequeño, muy peligroso, muy tardado o muy costoso para ser observado directamente. (p. 16)

moldura: tipo de cuerpo fósil que se formó en la roca cuando un organismo con partes duras fue enterrado, descompuesto o disuelto, dejando una cavidad en la roca. (p. 245)

molécula: partícula neutra que se forma cuando los átomos comparten electrones. (p. 475)

momento: medida de la dificultad para detener un objeto en movimiento; es igual al producto de la masa por la velocidad. (p. 534)

fase lunar: cambio en la apariencia de la luna según es vista desde la Tierra; se debe a las posiciones relativas de la luna, la Tierra y el sol. (p. 313)

motor: dispositivo que transforma energía eléctrica en energía cinética. (p. 676)

mutación: proceso que modifica al ADN para formar nuevos alelos. (p. 52)

mutualismo: tipo de relación simbiótica en la que ambos organismos se benefician. (p. 108)

N

natural selection: process by which organisms that are suited to a particular environment are better able to survive and reproduce than organisms that are not. (pp. 50, 275)

nebula: large cloud of gas and dust that contracts under gravitational force and breaks apart into smaller pieces, each of which might collapse to form a star. (p. 382)

negative feedback: mechanism that helps the body change an internal condition and return to its normal state. (p. 80)

Neptune: usually the eighth planet from the Sun; is large and gaseous, has rings that vary in thickness, and is bluish-green in color. (p. 352)

net force: combination of all forces acting on an object. (p. 551)

neutron (NEW trahn): electrically-neutral particle that has the same mass as a proton and is found in an atom's nucleus. (p. 411)

selección natural: proceso mediante el cual los organismos que están adaptados a un ambiente particular están mejor capacitados para sobrevivir y reproducirse que los organismos que no están adaptados. (pp. 50, 275)

nebulosa: nube grande de polvo y gas que se contrae bajo la fuerza gravitacional y se descompone en pedazos más pequeños, cada uno de los cuales se puede colapsar para formar una estrella. (p. 382)

retroalimentación negativa: mecanismo que ayuda a que el organismo cambie una condición interna para volver a su estado normal. (p. 80)

Neptuno: el octavo planeta desde el sol; es grande y gaseoso, tiene anillos que varían en espesor y tiene un color verde-azulado. (p. 352)

fuerza neta: la combinación de todas las fuerzas que actúan sobre un objeto. (p. 551)

neutrón: partícula con carga eléctrica neutra que tiene la misma masa que un protón y que se encuentra en el núcleo de un átomo. (p. 411)

Glossary/Glosario

neutron star: collapsed core of a supernova that can shrink to about 20 km in diameter and contains only neutrons in the dense core. (p. 384)

new moon: moon phase that occurs when the Moon is between Earth and the Sun, at which point the Moon cannot be seen because its lighted half is facing the Sun and its dark side faces Earth. (p. 313)

Newton's first law of motion: states that if the net force acting on an object is zero, the object will remain at rest or move in a straight line with a constant speed. (p. 552)

Newton's second law of motion: states that an object acted upon by a net force will accelerate in the direction of the force, and that the acceleration equals the net force divided by the object's mass. (p. 556)

Newton's third law of motion: states that forces always act in equal but opposite pairs. (p. 563)

niche: in an ecosystem, refers to the unique ways an organism survives, obtains food and shelter, and avoids danger. (p. 109)

nitrogen cycle: model describing how nitrogen moves from the atmosphere to the soil, to living organisms, and then back to the atmosphere. (p. 132)

nitrogen fixation: process in which some types of bacteria in the soil change nitrogen gas into a form of nitrogen that plants can use. (p. 132)

noble gases: elements in group 18 of the periodic table. (p. 446)

nonmetal: element that is usually a gas or brittle solid at room temperature and is a poor conductor of heat and electricity. (p. 438)

estrella de neutrones: núcleo colapsado de una supernova que puede contraerse hasta tener un diámetro de 20 kilómetros y contiene sólo neutrones en su denso núcleo. (p. 384)

luna nueva: fase lunar que ocurre cuando la luna se encuentra entre la Tierra y el sol, punto en el cual la luna no puede verse porque su mitad iluminada está frente al sol y su lado oscuro frente a la Tierra. (p. 313)

primera ley de movimiento de Newton: establece que si la fuerza neta que actúa sobre un objeto es igual a cero, el objeto se mantendrá en reposo o se moverá en línea recta a una velocidad constante. (p. 552)

segunda ley de movimiento de Newton: establece que si una fuerza neta se ejerce sobre un objeto, éste se acelerará en la dirección de la fuerza y la aceleración es igual a la fuerza neta dividida por la masa del objeto. (p. 556)

tercera ley de movimiento de Newton: establece que las fuerzas siempre actúan en pares iguales pero opuestos. (p. 563)

nicho: en un ecosistema, se refiere a las formas únicas en las que un organismo sobrevive, obtiene alimento, refugio y evita el peligro. (p. 109)

ciclo del nitrógeno: modelo que describe cómo se mueve el nitrógeno de la atmósfera al suelo, a los organismos vivos y de nuevo a la atmósfera. (p. 132)

fijación del nitrógeno: proceso en el cual algunos tipos de bacterias en el suelo transforman el nitrógeno gaseoso en una forma de nitrógeno que las plantas pueden usar. (p. 132)

gases inertes: elementos en el grupo 18 de la tabla periódica. (p. 446)

no metal: elemento que por lo general es un gas o un sólido frágil a temperatura ambiente y mal conductor de calor y electricidad. (p. 438)

O

Ohm's law: states that the current in a circuit equals the voltage divided by the resistance in the circuit. (p. 649)

organ: structure, such as the heart, that is made up of several different types of tissues that work together. (p. 70)

organ system: group of organs that are interdependent and work together to do a certain job. (p. 71)

ley de Ohm: establece que la corriente en un circuito es igual al voltaje dividido por la resistencia en el circuito. (p. 649)

órgano: estructura, como el corazón, compuesta por diferentes tipos de tejidos que trabajan conjuntamente. (p. 70)

sistema orgánico: grupo de órganos interdependientes que trabajan conjuntamente para desempeñar una función determinada. (p. 71)

Glossary/Glosario

organic compounds/plate tectonics   **compuestos orgánicos/tectónica de placas**

organic compounds: most compounds that contain carbon, including nucleic acids, proteins, carbohydrates, and lipids. (p. 66)

organic evolution: change of organisms over geologic time. (p. 274)

output force: force exerted by a machine. (p. 586)

compuestos orgánicos: la mayoría de los compuestos que contienen carbono, incluyendo los ácidos nucleicos, las proteínas, los carbohidratos y los lípidos. (p. 66)

evolución orgánica: cambio de los organismos a través del tiempo geológico. (p. 274)

fuerza generada: fuerza producida por una máquina. (p. 586)

P

Paleozoic Era: era of ancient life, which began about 544 million years ago, when organisms developed hard parts, and ended with mass extinctions about 245 million years ago. (p. 282)

Pangaea (pan JEE uh): large, ancient landmass that was composed of all the continents joined together. (pp. 182, 279)

parallel circuit: circuit that has more than one path for electric current to follow. (p. 651)

parasitism: a type of symbiotic relationship in which one organism benefits and the other organism is harmed. (p. 108)

period: horizontal row of elements in the periodic table whose properties change gradually and predictably (p. 273); third-longest division of geologic time; is subdivided into epochs and is characterized by the types of life that existed worldwide. (p. 435)

permineralized remains: fossils in which the spaces inside are filled with minerals from groundwater. (p. 244)

phenotype (FEE nuh tipe): observable product of genetic makeup and the environment's influences on that genetic makeup. (p. 40)

photosphere: lowest layer of the Sun's atmosphere; gives off light and has temperatures of about 6,000 K. (p. 375)

pioneer species: first organisms to grow in new or disturbed areas. (p. 150)

pitch: human perception of the frequency of sound. (p. 703)

plate: a large section of Earth's oceanic or continental crust and rigid upper mantle that moves around on the asthenosphere. (p. 190)

plate tectonics: theory that Earth's crust and upper mantle are broken into plates that float and move around on a plasticlike layer of the mantle. (p. 190)

Era Paleozoica: era de la vida antigua que comenzó hace 544 millones de años, cuando los organismos desarrollaron partes duras; terminó con extinciones en masa hace unos 245 millones de años. (p. 282)

Pangea: masa terrestre extensa y antigua que estaba compuesta por todos los continentes unidos. (pp. 182, 279)

circuito paralelo: circuito en el cual la corriente eléctrica puede seguir más de una trayectoria. (p. 651)

parasitismo: tipo de relación simbiótica en la que un organismo se beneficia y el otro es perjudicado. (p. 108)

período: fila horizontal de elementos en la tabla periódica cuyas propiedades cambian gradualmente y en forma predecible (p. 273); la tercera división más grande del tiempo geológico; está subdividido en épocas y se caracteriza por los tipos de vida que existieron en todo el mundo. (p. 435)

restos permineralizados: fósiles en los que los espacios interiores son llenados con minerales de aguas subterráneas. (p. 244)

fenotipo: producto de la composición genética que puede ser observado, así como la influencia del medio ambiente ejercida sobre dicha composición genética. (p. 40)

fotosfera: capa más interna de la atmósfera del sol; emite luz y tiene temperaturas de cerca de 6,000 grados Kelvin. (p. 375)

especies pioneras: primeros organismos que crecen en áreas nuevas o alteradas. (p. 150)

tono: percepción humana de la frecuencia del sonido. (p. 703)

placa: gran sección de la corteza terrestre u oceánica y del manto rígido superior que se mueve sobre la astenosfera. (p. 190)

tectónica de placas: teoría respecto a que la corteza terrestre y el manto superior están fraccionados en placas que flotan y se mueven sobre una capa plástica del manto. (p. 190)

Glossary/Glosario

Pluto: considered to be the ninth planet from the Sun; has a solid icy-rock surface and a single moon, Charon. (p. 353)

polar bond: bond resulting from the unequal sharing of electrons. (p. 476)

population: all the organisms that belong to the same species living in a community. (p. 96)

power: rate at which work is done; equal to the work done divided by the time it takes to do the work; measured in watts (W). (p. 583)

Precambrian (pree KAM bree un) time: longest part of Earth's history, lasting from 4.0 billion to about 544 million years ago. (p. 280)

principle of superposition: states that in undisturbed rock layers, the oldest rocks are on the bottom and the rocks become progressively younger toward the top. (p. 250)

producer: organism, such as a green plant or alga, that uses an outside source of energy like the Sun to create energy-rich food molecules. (p. 106)

product: substance that forms as a result of a chemical reaction. (p. 494)

proton: positively-charged particle in the nucleus of an atom. (p. 410)

pulley: simple machine made from a grooved wheel with a rope or cable wrapped around the groove. (p. 596)

Punnett square: tool used to predict the probability of certain traits in offspring that shows the different ways alleles can combine. (p. 47)

Plutón: considerado como el noveno planeta desde el sol; tiene una superficie sólida de roca congelada y una luna, Caronte. (p. 353)

enlace polar: enlace que resulta de compartir electrones en forma desigual. (p. 476)

población: todos los organismos que pertenecen a la misma especie dentro de una comunidad. (p. 96)

potencia: velocidad a la que se realiza un trabajo y que equivale al trabajo realizado dividido por el tiempo que toma realizar el trabajo; se mide en vatios (W). (p. 583)

tiempo precámbrico: la parte más duradera de la historia de la Tierra; duró desde hace 4.0 billones de años hasta hace aproximadamente 544 millones de años. (p. 280)

principio de superposición: establece que en las capas rocosas no perturbadas, las rocas más antiguas están en la parte inferior y las rocas son más jóvenes conforme están más cerca de la superficie. (p. 250)

productor: organismo, como una planta o un alga verde, que utiliza una fuente externa de energía, como la luz solar, para producir moléculas de nutrientes ricas en energía. (p. 106)

producto: sustancia que se forma como resultado de una reacción química. (p. 494)

protón: partícula cargada positivamente en el núcleo de un átomo. (p. 410)

polea: máquina simple que consiste en una rueda acanalada con una cuerda o cable que corre alrededor del canal. (p. 596)

cuadrado de Punnett: herramienta utilizada para predecir la probabilidad de ciertos rasgos en la descendencia y que muestra las diferentes formas en que pueden combinarse los alelos. (p. 47)

R

radiation: transfer of energy by electromagnetic waves. (p. 613)

radioactive decay: release of nuclear particles and energy from unstable atomic nuclei. (pp. 257, 416)

radiometric dating: process used to calculate the absolute age of rock by measuring the ratio of parent isotope to daughter product in a mineral and knowing the half-life of the parent. (p. 259)

rate of reaction: measure of how fast a chemical reaction occurs. (p. 504)

reactant: substance that exists before a chemical reaction begins. (p. 494)

radiación: transferencia de energía mediante ondas electromagnéticas. (p. 613)

descomposición radioactiva: liberación de partículas nucleares y energía de un núcleo atómico inestable. (pp. 257, 416)

fechado radiométrico: proceso utilizado para calcular la edad absoluta de las rocas midiendo la relación isótopo parental a producto derivado en un mineral y conociendo la vida media del parental. (p. 259)

velocidad de reacción: medida de la rapidez con que se produce una reacción química. (p. 504)

reactivo: sustancia que existe antes de que comience una reacción química. (p. 494)

recessive/seismic waves

recessive (rih SE sihv)**:** describes a trait that is covered over, or dominated, by another form of that trait. (p. 45)

refraction: change in direction of a wave when it changes speed as it travels from one material into another. (p. 699)

relative age: the age of something compared with other things. (p. 251)

representative elements: elements in groups 1 and 2 and 13–18 in the periodic table that include metals, metalloids, and nonmetals. (p. 435)

resistance: a measure of how difficult it is for electrons to flow in a material; unit is the ohm (Ω). (p. 646)

reverberation: repeated echoes of sound waves. (p. 705)

revolution: Earth's yearlong elliptical orbit around the Sun. (p. 309)

rift: long crack, fissure, or trough that forms between tectonic plates moving apart at plate boundaries. (p. 227)

rotation: spinning of Earth on its imaginary axis, which takes about 24 hours to complete and causes day and night to occur. (p. 307)

recesivo/ondas sísmicas

recesivo: describe un rasgo que es encubierto por completo o dominado por otra forma de dicho rasgo. (p. 45)

refracción: cambio de dirección de una onda al cambiar su velocidad cuando pasa de un material a otro. (p. 699)

edad relativa: la edad de algo comparado con otras cosas. (p. 251)

elementos representativos: elementos en los grupos 1 y 2 y 13–18 en la tabla periódica; incluyen metales, metaloides y no metales. (p. 435)

resistencia: medida de la dificultad que tienen los electrones para fluir en un material; se mide en ohmios (Ω). (p. 646)

reverberación: ecos repetidos de ondas sonoras. (p. 705)

revolución: órbita elíptica de un año de duración que la Tierra recorre alrededor del sol. (p. 309)

ruptura: grieta larga, fisura o hueco que se forma entre placas tectónicas que se separan en los límites de las placas. (p. 227)

rotación: rotación de la Tierra sobre su eje imaginario, lo cual toma cerca de 24 horas para completarse y causa la alternancia entre el día y la noche. (p. 307)

S

Saturn: second-largest and sixth planet from the Sun; has a complex ring system, at least 31 moons, and a thick atmosphere made mostly of hydrogen and helium. (p. 350)

science: process used to investigate what is happening around us in order to solve problems or answer questions; part of everyday life. (p. 6)

scientific methods: ways to solve problems that can include step-by-step plans, making models, and carefully thought-out experiments. (p. 13)

screw: simple machine that is an inclined plane wrapped around a cylinder or post. (p. 593)

seafloor spreading: Hess's theory that new seafloor is formed when magma is forced upward toward the surface at a mid-ocean ridge. (p. 187)

seismic safe: describes the ability of structures to stand up against the vibrations caused by an earthquake. (p. 217)

seismic waves: earthquake waves, including primary waves, secondary waves, and surface waves. (p. 212)

Saturno: además de ser el sexto planeta más cercano al sol, también es el segundo en tamaño; tiene un sistema de anillos complejo, por lo menos 31 lunas y una atmósfera gruesa compuesta principalmente de hidrógeno y helio. (p. 350)

ciencia: proceso usado para investigar lo que sucede a nuestro alrededor con el fin de solucionar problemas o despejar dudas como parte de la vida diaria. (p. 6)

métodos científicos: formas de solucionar problemas, las cuales pueden incluir planes paso a paso, creación de modelos y elaboración cuidadosa de experimentos. (p. 13)

tornillo: máquina simple que consiste en un plano inclinado envuelto en espiral alrededor de un cilindro o poste. (p. 593)

expansión del suelo oceánico: teoría de Hess respecto a que se forma un nuevo suelo oceánico cuando el magma es empujado hacia la superficie a través de un surco en la mitad del océano. (p. 187)

seguridad antisísmica: describe la capacidad de las estructuras de resistir las vibraciones producidas por los terremotos. (p. 217)

ondas sísmicas: ondas producidas durante los terremotos, las cuales pu ser primarias, secundarias y superficiales. (p. 212)

seismograph: instrument used to record seismic waves. (p. 213)

semiconductor: element that does not conduct electricity as well as a metal but conducts it better than a nonmetal. (p. 443)

series circuit: circuit that has only one path for electric current to follow. (p. 650)

shield volcano: large, broad volcano with gently sloping sides that is formed by the buildup of basaltic layers. (p. 222)

simple machine: a machine that does work with only one movement; includes the inclined plane, wedge, screw, lever, wheel and axle, and pulley. (p. 591)

soil: mixture of mineral and rock particles, the remains of dead organisms, air, and water that forms the topmost layer of Earth's crust and supports plant growth. (p. 124)

solar eclipse: occurs when the Moon passes directly between the Sun and Earth and casts a shadow over part of Earth. (p. 315)

solar system: system of nine planets, including Earth, and other objects that revolve around the Sun. (p. 337)

solstice: twice-yearly point at which the Sun reaches its greatest distance north or south of the equator. (p. 310)

species: group of organisms that reproduces only with other members of their own group. (p. 274)

specific heat: amount of heat needed to raise the temperature of 1 kg of a substance by 1°C. (p. 616)

speed: equals the distance traveled divided by the time it takes to travel that distance. (p. 524)

sphere (SFIHR): a round, three-dimensional object whose surface is the same distance from its center at all points; Earth is a sphere that bulges somewhat at the equator and is slightly flattened at the poles. (p. 306)

static charge: imbalance of electric charge on an object. (p. 637)

succession: natural, gradual changes in the types of species that live in an area; can be primary or secondary. (p. 150)

sunspots: areas on the Sun's surface that are cooler and less bright than surrounding areas, are caused by the Sun's magnetic field, and occur in cycles. (p. 376)

supergiant: late stage in the life cycle of a massive star in which the core heats up, heavy elements form by fusion, and the star expands; can eventually explode to form a supernova. (p. 385)

sismógrafo: instrumento usado para registrar las ondas sísmicas. (p. 213)

semiconductor: elemento que no conduce electricidad tan bien como un metal pero que la conduce mejor que un no metal. (p. 443)

circuito en serie: circuito en el cual la corriente eléctrica sólo puede seguir una trayectoria. (p. 650)

volcán escudo: volcán grande y ancho con lados ligeramente inclinados que se forma por la aparición de capas basálticas. (p. 222)

máquina simple: máquina que ejecuta el trabajo con un solo movimiento; incluye el plano inclinado, la palanca, el tornillo, la rueda y el eje y la polea. (p. 591)

suelo: mezcla de partículas minerales y rocas, restos de organismos muertos, aire y del agua que forma la capa superior de la corteza terrestre y favorece el crecimiento de las plantas. (p. 124)

eclipse solar: ocurre cuando la luna pasa directamente entre el sol y la Tierra y se genera una sombra sobre una parte de la Tierra. (p. 315)

sistema solar: sistema de nueve planetas, incluyendo a la Tierra y otros objetos que giran alrededor del sol. (p. 337)

solsticio: punto en el cual dos veces al año el sol alcanza su mayor distancia al norte o al sur del ecuador. (p. 310)

especie: grupo de organismos que se reproduce sólo entre los miembros de su mismo grupo. (p. 274)

calor específico: cantidad de calor necesario para elevar la temperatura de 1 kilogramo de una sustancia en 1 grado centígrado. (p. 616)

rapidez: equivale a dividir la distancia recorrida por el tiempo que toma recorrer dicha distancia. (p. 524)

esfera: un objeto tridimensional y redondo donde cualquier punto de su superficie está a la misma distancia del centro; la Tierra es una esfera algo abultada en el ecuador y ligeramente achatada en los polos. (p. 306)

carga estática: desequilibrio de la carga eléctrica en un objeto. (p. 637)

sucesión: cambios graduales y naturales en los tipos de especies que viven en un área; puede ser primaria o secundaria. (p. 150)

manchas solares: áreas en la superficie solar que son más frías y menos brillantes que las áreas circundantes, son causadas por el campo magnético solar y ocurren en ciclos. (p. 376)

supergigante: etapa tardía en el ciclo de vida de una estrella masiva en la que el núcleo se calienta, se forman elementos pesados por fusión y la estrella se expande; eventualmente puede explotar para formar una supernova. (p. 385)

Glossary/Glosario

symbiosis/trilobite

symbiosis: any close relationship between species, including mutualism, commensalism, and parasitism. (p. 108)

simbiosis/trilobite

simbiosis: cualquier relación estrecha entre especies, incluyendo mutualismo, comensalismo y parasitismo. (p. 108)

T

taiga (TI guh): world's largest biome, located south of the tundra between 50°N and 60°N latitude; has long, cold winters, precipitation between 35 cm and 100 cm each year, cone-bearing evergreen trees, and dense forests. (p. 156)

technology: application of science to make useful products and tools, such as computers. (p. 9)

temperate deciduous forest: biome usually having four distinct seasons, annual precipitation between 75 cm and 150 cm, and climax communities of deciduous trees. (p. 157)

temperate rain forest: biome with 200 cm to 400 cm of precipitation each year, average temperatures between 9°C and 12°C, and forests dominated by trees with needlelike leaves. (p. 157)

temperature: a measure of the average value of the kinetic energy of the particles in a material. (p. 608)

thermal energy: the sum of the kinetic and potential energy of the particles in a material. (p. 611)

thermal pollution: increase in temperature of a natural body of water; caused by adding warmer water. (p. 617)

tissue: group of similar cells, such as nerve cells, that work together to do a job. (p. 70)

trait: feature that an organism inherits from its parents, such as eye color, that is coded for by DNA. (p. 39)

transformer: device used to increase or decrease the voltage of an alternating current. (p. 680)

transition elements: elements in groups 3–12 in the periodic table, all of which are metals. (p. 435)

transmutation: the change of one element into another through radioactive decay. (p. 416)

transverse wave: mechanical wave that causes particles in matter to move at right angles to the direction the wave travels. (p. 695)

trilobite (TRI luh bite): organism with a three-lobed exoskeleton that was abundant in Paleozoic oceans and is considered to be an index fossil. (p. 273)

taiga: el bioma más grande del mundo, localizado al sur de la tundra entre 50° y 60° de latitud norte; tiene inviernos prolongados y fríos, una precipitación que alcanza entre 35 y 100 centímetros cúbicos al año, coníferas perennifolias y bosques espesos. (p. 156)

tecnología: aplicación de la ciencia para producir bienes y herramientas útiles, tales como los computadoras. (p. 9)

bosque caducifolio templado: bioma que generalmente tiene cuatro estaciones distintas, con una precipitación anual entre 75 y 150 centímetros cúbicos y un clímax comunitario de árboles caducifolios. (p. 157)

bosque lluvioso templado: bioma con 200 a 400 centímetros cúbicos de precipitación al año; tiene una temperatura promedio entre 9 y 12°C y bosques dominados por árboles de hojas aciculares. (p. 157)

temperatura: medida del valor promedio de energía cinética de las partículas en un material. (p. 608)

energía térmica: la suma de la energía cinética y potencial de las partículas en un material. (p. 611)

polución térmica: incremento de la temperatura de una masa natural de agua producido al agregarle agua a mayor temperatura. (p. 617)

tejido: grupo de células, como las células nerviosas, que trabajan conjuntamente para realizar una función determinada. (p. 70)

rasgo: característica que un organismo hereda de sus progenitores, tal como el color de los ojos, que es regulado por el ADN. (p. 39)

transformador: dispositivo utilizado para aumentar o disminuir el voltaje de una corriente alterna. (p. 680)

elementos de transición: elementos en los grupos 3–12 en la tabla periódica, todos los cuales son metales. (p. 435)

transmutación: cambio de un elemento a otro a través de la descomposición radiactiva. (p. 416)

onda transversal: onda mecánica que hace que las partículas de materia se muevan en ángulos rectos respecto a la dirección en que viaja la onda. (p. 695)

trilobite: organismo con un exoesqueleto trilobulado que fue abundante en los océanos del Paleozoico y es considerado como un fósil índice. (p. 273)

tropical rain forest: most biologically diverse biome; has an average temperature of 25°C and receives between 200 cm and 600 cm of precipitation each year. (p. 158)

tsunami: powerful seismic sea wave that begins over an ocean-floor earthquake, can reach 30 m in height when approaching land, and can cause destruction in coastal areas. (p. 215)

tundra: cold, dry, treeless biome with less than 25 cm of precipitation each year, a short growing season, permafrost, and winters that can be six to nine months long. Tundra is separated into two types: arctic tundra and alpine tundra. (p. 155)

selva húmeda tropical: el bioma más diverso biológicamente; tiene una temperatura promedio de 25°C y recibe entre 200 y 600 centímetros cúbicos de precipitación al año. (p. 158)

tsunami: poderosa onda sísmica marina que comienza en un terremoto en el lecho oceánico, pudiendo alcanzar 30 metros de altura al acercarse a la tierra y causar gran destrucción en las áreas costeras. (p. 215)

tundra: bioma sin árboles, frío y seco, con menos de 25 centímetros cúbicos de precipitación al año; tiene una estación corta de crecimiento y permafrost e inviernos que pueden durar entre 6 y 9 meses. La tundra se divide en dos tipos: tundra ártica y tundra alpina. (p. 155)

ultraviolet waves: electromagnetic waves with wavelengths between about 400 billionths and 10 billionths of a meter. (p. 710)

unbalanced forces: two or more forces acting on an object that do not cancel, and cause the object to accelerate. (p. 551)

unconformity (un kun FOR mih tee): gap in the rock layer that is due to erosion or periods without any deposition. (p. 252)

uniformitarianism: principle stating that Earth processes occurring today are similar to those that occurred in the past. (p. 261)

Uranus (YOOR uh nus): seventh planet from the Sun; is large and gaseous, has a distinct bluish-green color, and rotates on an axis nearly parallel to the plane of its orbit. (p. 351)

ondas ultravioleta: ondas electromagnéticas con longitudes de onda entre aproximadamente 10 y 400 billonésimas de metro. (p. 710)

fuerzas no balanceadas: dos o más fuerzas que actúan sobre un objeto sin anularse y que hacen que el objeto se acelere. (p. 551)

discordancia: brecha en la capa rocosa que es debida a la erosión o a periodos sin deposición. (p. 252)

uniformitarianismo: principio que establece que los procesos de la Tierra que ocurren actualmente son similares a los que ocurrieron en el pasado. (p. 261)

Urano: séptimo planeta desde el sol; es grande y gaseoso, tiene un color verde-azulado distintivo y gira sobre un eje casi paralelo al plano de su órbita. (p. 351)

velocity: speed and direction of a moving object. (p. 527)

Venus: second planet from the Sun; similar to Earth in mass and size; has a thick atmosphere and a surface with craters, faultlike cracks, and volcanoes. (p. 343)

villi: tiny fingerlike projections that line the small intestine, contain many blood vessels, and increase the intestine's surface area where food molecules are absorbed. (p. 75)

velocidad: rapidez y dirección de un objeto en movimiento. (p. 527)

Venus: segundo planeta más cercano al sol; similar a la Tierra en masa y tamaño; tiene una atmósfera gruesa y una superficie con cráteres, grietas similares a fallas y volcanes. (p. 343)

vellosidades intestinales: proyecciones en forma de dedo que cubren las paredes del intestino delgado, contienen muchos vasos sanguíneos y aumentan el área de superficie para la absorción de las moléculas alimenticias. (p. 75)

Glossary/Glosario

volcano/work   **volcán/trabajo**

volcano: cone-shaped hill or mountain formed when hot magma, solids, and gas erupt onto Earth's surface through a vent. (p. 219)

voltage: a measure of the amount of electrical potential energy an electron flowing in a circuit can gain; measured in volts (V). (p. 644)

volcán: colina o montaña cónica que se forma cuando el magma caliente, sólidos y gases, hacen erupción en la superficie terrestre a través de una abertura. (p. 219)

voltaje: medida de la cantidad de energía eléctrica potencial que puede adquirir un electrón que fluye en un circuito; se mide en voltios (V). (p. 644)

W

waning: describes phases that occur after a full moon, as the visible lighted side of the Moon grows smaller. (p. 314)

water cycle: model describing how water moves from Earth's surface to the atmosphere and back to the surface again through evaporation, condensation, and precipitation. (p. 131)

wave: disturbance that moves through matter and space and carries energy. (p. 694)

wavelength: distance between one point on a wave and the nearest point moving with the same speed and direction. (p. 696)

waxing: describes phases following a new moon, as more of the Moon's lighted side becomes visible. (p. 314)

wedge: simple machine consisting of an inclined plane that moves; can have one or two sloping sides. (p. 592)

weight: gravitational force between an object and Earth. (p. 557)

wetland: a land region that is wet most or all of the year. (p. 165)

wheel and axle: simple machine made from two circular objects of different sizes that are attached and rotate together. (p. 594)

white dwarf: late stage in the life cycle of a comparatively low-mass main sequence star; formed when its core depletes its helium and its outer layers escape into space, leaving behind a hot, dense core. (p. 384)

work: is done when a force exerted on an object causes that object to move some distance; equal to force times distance; measured in joules (J). (p. 580)

menguante: describe las fases posteriores a la luna llena, de manera que el lado iluminado de la luna es cada vez menos visible. (p. 314)

ciclo del agua: modelo que describe cómo se mueve el agua de la superficie de la Tierra hacia la atmósfera y nuevamente hacia la superficie terrestre a través de la evaporación, la condensación y la precipitación. (p. 131)

onda: perturbación que se mueve a través de la materia y el espacio y que transporta energía. (p. 694)

longitud de onda: distancia entre un punto en una onda y el punto más cercano, moviéndose con la misma rapidez y dirección. (p. 696)

creciente: describe las fases posteriores a la luna nueva, de manera que el lado iluminado de la luna es cada vez más visible. (p. 314)

cuña: máquina simple que consiste en un plano inclinado que se mueve; puede tener uno o dos lados inclinados. (p. 592)

peso: fuerza gravitacional entre un objeto y la Tierra. (p. 557)

zona húmeda: región lluviosa la mayor parte del año. (p. 165)

rueda y eje: máquina simple compuesta por dos objetos circulares de diferentes tamaños que están interconectados y giran. (p. 594)

enana blanca: etapa tardía en el ciclo de vida de una estrella de secuencia principal, de relativamente poca masa, formada cuando el núcleo agota su helio y sus capas externas escapan al espacio, dejando atrás un núcleo denso y caliente. (p. 384)

trabajo: se realiza cuando la fuerza ejercida sobre un objeto hace que el objeto se mueva determinada distancia; es igual a la fuerza multiplicada por la distancia y se mide en julios (J). (p. 580)

Index

Abiotic factors / **Applying Math**

Italic numbers = illustration/photo **Bold numbers** = vocabulary term
lab = a page on which the entry is used in a lab
act = a page on which the entry is used in an activity

A

Abiotic factors, *122,* **122**–129; air, *122,* 123, 127; climate, *127,* 127–128, *128;* soil, 124, 124 *lab,* 129 *lab;* sunlight, 124, *124;* temperature, *125,* 125–126, *126;* water, *122,* 123, *123*
Absolute ages, 257–261
Absolute magnitude, 372
Absorption, 75, *75*
Acceleration, 528–532; calculating, 529–530, 530 *act,* 559, 559 *act;* equation for, 530; and force, *556,* 556–557, 560; graph of, 532, *532;* modeling, 531, 531 *lab;* and motion, 528–529; negative, 531, *531;* positive, 531; and speed, *528,* 528–529; unit of measurement with, 557; and velocity, *528,* 528–529, *529*
Acetic acid, 494, 495, 496
Actinides, 450, *451*
Action and reaction, 563–566, *564, 566*
Activation energy, 503, *503*
Activities, Applying Math, 47, 77, 101, 126, 166, 230, 291, 296, 419, 498, 524, 530, 534, 559, 582, 583, 587, 589, 610, 649, 652, 698; Applying Science, 14, 192, 260, 319, 372, 439, 469, 669; Integrate, 7, 15, 21, 30, 39, 49, 50, 67, 75, 107, 109, 127, 128, 137, 160, 165, 168, 187, 197, 198, 202, 221, 228, 229, 245, 248, 257, 274, 281, 307, 308, 317, 338, 340, 352, 382, 383, 384, 412, 420, 423, 444, 445, 450, 452, 456, 467, 473, 484, 495, 505, 507, 534, 551, 557, 564, 589, 593, 616, 617, 622, 645, 646, 681, 682, 696, 703; Science Online, 8, 25, 42, 52, 66, 80, 96, 102, 127, 135, 151, 167, 183, 192, 216, 221, 251, 254, 284, 289, 309, 311, 315, 323, 337, 346, 377, 382, 416, 421, 438, 452, 466, 477, 497, 503, 526, 536, 553, 564, 584, 587, 620, 640, 653, 671, 679, 708; Standardized Test Practice, 34–35, 60–61, 88–89, 118–119, 146–147, 176–177, 206–207, 238–239, 268–269, 300–301, 332–333, 366–367, 398–399, 430–431, 460–461, 488–489, 516–517, 546–547, 576–577, 604–605, 630–631, 662–663, 690–691, 720–721
Adaptive radiation, 52
Advertising, science in, 7
Africa, savannas of, 161, *161*
Age, absolute, **257**–261; relative, **251**–256, 256 *lab*
Agriculture, on grasslands, 161; isotopes in, 421 *act,* 423; and nitrogen fixation, *132,* 133, *133;* traits in corn, 38, *38*
Air, as abiotic factor in environment, *122,* 123, 127; early, 281
Air conditioners, 623
Air resistance, 561
Alchemist, 479
Algae, and mutualism, 108, *108;* as producers, *106*
Alkali metals, 441, *441,* 469, *469*
Alkaline earth metals, 442, *442*
Allele(s), 45, *45*
Alligator(s), *110*
Alpha Centauri, 378
Alpha decay, 258, *258,* 417, *417*
Alpha particle(s), 408, 409, *409,* 410, 412, **417**
Alternating current (AC), 678
Aluminum, 442
Alveoli, 76, *77*
Amalgam, 452
Americium, 417, *417,* 450
Ammeter, 674, *675*
Ammonia, 444, *444*
Amphibians, 284
Amplitude, 697; of compressional wave, 697, *697;* and energy, 697; of transverse wave, 697, *697*
Angiosperms, 290
Angular unconformities, 252, *252, 253*
Animal(s). *See also* Invertebrate animals; Vertebrate animals, competition among, 98, *98;* cooperation among, 110; in desert, 160, *160;* Ediacaran, 282, *283;* effect of momentum on motion of, 534; in energy flow, 137, *137,* 138, *138;* and food chain, 137, *137;* in grasslands, 161, *161;* habitats of, 97, *97,* 98, *98,* 109, *109;* insulation of, 616; invertebrate, 281; migration of, 103; speed of, 524; on taiga, 156, *156;* in temperate deciduous forest, *156,* 157; in temperate rain forest, 157, *157;* and temperature, 125, *125;* in tropical rain forest, 158, *159;* on tundra, 155, *155;* warm-blooded v. cold-blooded, 289 *act*
Anode, 406, *406*
Antares, 381, *381*
Antibiotics, 75
Anvil (of ear), 704, *704*
Appalachian Mountains, 285, *285*
Apparent magnitude, 372
Applying Math, Acceleration of a Bus, 530; Acceleration of a Car, 559; Calculating Efficiency, 589; Calculating Extinction by Using Percentages, 291; Calculating Mechanical Advantage, 587; Calculating Power, 583; Calculating Work, 582;

Index

Applying Science, Chapter Reviews, 33, 59, 87, 117, 145, 175, 205, 237, 267, 299, 331, 365, 397, 429, 459, 487, 515, 545, 575, 603, 629, 661, 689, 719; Conserving Mass, 498; Converting to Celsius, 610; Diameter of Mars, 346; Electric Power Used by a Lightbulb, 652; Find Half-Lives, 419; Lung Volume, 77; Momentum of a Bicycle, 534; Percent of Offspring with Certain Traits, 47; P-Wave Travel Time, 230; Section Reviews, 23, 48, 71, 128, 139, 153, 188, 224, 261, 293, 320, 347, 359, 385, 413, 440, 471, 501, 508, 527, 532, 538, 562, 568, 584, 590, 597, 611, 654, 683, 700, 705; Speed of Sound, 698; Speed of a Swimmer, 524; Temperature, 166; Temperature Changes, 126; Voltage from a Wall Outlet, 649

Applying Science, Are distance and brightness related?, 372; Do you have too many crickets?, 101; Finding Magnetic Declination, 669; How does the periodic table help you identify properties of elements?, 469; How well do the continents fit together?, 192; Problem-Solving Skills, 14; What does *periodic* mean in the periodic table?, 439; What will you use to survive on the Moon?, 319; When did the Iceman die?, 260

Applying Skills, 11, 27, 42, 53, 81, 97, 105, 110, 135, 161, 169, 185, 199, 218, 249, 255, 279, 286, 311, 325, 340, 353, 374, 378, 391, 423, 447, 452, 480, 555, 617, 623, 642, 647, 671, 713

Aquatic ecosystem(s), 163–171; freshwater, *163*, 163–165, *164*, 164 *lab*, *165*, 170–171 *lab*, *172*, *172*; saltwater, 166–169, *167*, 167 *act*, *168*, *169*

Archean Eon, 280
Arctic, 94, *94*, 96
Argon, 447, *447*
Aristotle, 25, 306
Arsenic, 445

Arteries, modeling blood flow in, 63 *lab*
Artificial body parts, 600, *600*
Artificial selection, 276, *276*
Astatine, 446
Asteroid, 358–359, *359*
Asteroid belt, 358, *358*
Asthenosphere, 190
Astronauts, *318*
Astronomical unit (AU), 344
Atlantic Ocean, 292 *lab*
Atmosphere, as abiotic factor in environment, *122*, 123; early, 281; and gravity, 127 *act*; of Jupiter, 348, *348*; of Mercury, 343; of Neptune, 352, *352*; of Saturn, 350; of Sun, 375, *375*
Atom(s), 257, 402–425; components of, 636, *636*; history of research on, 404–413; model of, 405–414, 414 *lab*, *415*, *418*, *420*, *421*; nucleus of, *410*, 410–412, *411*, 411 *lab*, *412*, *414*, 415–416; structure of, *464*, 464–465, *465*, 482–483 *lab*; symbols for, 479, *479*
Atomic number, 415
Aurora, 677, *677*
Aurora borealis, 377, *377*
Automobiles, air bags in, 572, *572*; internal combustion engines in, 620, *620*, 620 *act*, *621*; safety in, 540–541 *lab*, 572, *572*
Average speed, 525, *525*, 525 *lab*
Awiakta, Marilou, 484
Axis, 307, *307*, *309*; magnetic, 308, *308*; tilt of, 326–327 *lab*
Axle, See Wheel and axle

B

Bacteria, and digestion, 75; early, 281, *281*; and food, 505
Baking soda, 494, 495, 496
Balanced chemical equations, 497, 498 *act*
Balanced forces, 551, *551*
Balloon races, 569 *lab*
Basaltic lava, 222, *222*
Basin(s), impact, *323*, *323*, 323 *lab*, 324

Batteries, chemical energy in, 645, *645*; in electrical circuit, 644, *644*; life of, 645; lithium, 441
Begay, Fred, 26
Beginning growth, *104*
Beryllium, 442, 467
Beta decay, 258, *258*, 418, *418*
Beta particle, 418
Betelgeuse, 370, *370*, 394, *394*
Bias, eliminating in research, 15
Bicycles, 591, *591*
Big bang theory, 388, *390*, **391**
Big Dipper, 371, *371*
Binary stars, 378
Biological organization, 96, *96*
Biomechanics, 551
Biomes, 154–161. *See also* Land biomes
Bionics, 600
Biosphere, 94, **94**–95
Biotic factors, 122
Biotic potential, 102, 103 *lab*
Bird(s), and competition, 98, *98*; habitats of, 97, *97*, 98, *98*; how birds fly, 564 *act*; interactions with other animals, 92, *92*, 95, *95*; migration of, 103; and natural selection, 51; origin of, 290, *290*; selective breeding of, 52, *52*
Birth, 81, *81*
Birthrates, 102, 102 *act*
Bison, 95, *95*
Black hole(s), 384, *384*, 557
Blood, clotting of, 81
Blood cells, red, 69; white, 69
Blood vessels, arteries, 63 *lab*; capillaries, 76, 77, 84; dilation of, 79; modeling blood flow in, 63 *lab*; veins, 63 *lab*
Blue shift, 389, *389*
Body, cells in. *See* Cell(s); elements in, 67; energy for, 76, 76–79, *77*; feeding, 73–75, *74*, *75*; interdependence of systems in, 79–81; organization in. *See* Organization
Body parts, artificial, 600, *600*
Body systems, 71; interactions of, *73*, 73–83, 82–83 *lab*
Body temperature, 79, *79*, 589
Bohr, Niels, 412

Index

Bond(s), 472–481; chemical, **471**, *471*; covalent, *475*, **475**–476, *476*; double, 476, *476*; ionic, *472*, 472–474, **473**, *473*, *474*; metallic, **474**, *474*; polar, **476**, *476*; triple, 476, *476*, 477
Bone(s), 65, 84
Boomerangs, 542, *542*
Boron, 442
Boron family, 442, *442*
Brain, mass of, 84
Breeding, selective, 52, *52*
Bromine, 446, 468
Bronchi, 76, *77*
Bryce Canyon National Park, 254, *255*
Bubbles, 494, *494*
Building materials, insulators, 616, *616*
Burning, 492, *492*, 499, *499*, 503, *503*
Butterflies, 109
Butyl hydroxytoluene (BHT), 506

C

Cactus, 98, *98*
Calcium, in body, 65, 67
Calcium carbonate, 245
Calcium phosphate, 245
Calendar, Mayan, 328, *328*
Californium-252, 450
Callisto (moon of Jupiter), 349, *349*
Camels, 125, *125*
Camouflage, 43 *lab*
Cancer, treatment of, 423
Canis Major, 370
Canyonlands National Park, 254, *255*
Capillaries, 76, *77*, 84
Carbohydrates, 66, *66*, 67
Carbon, 443, *443*; isotopes of, *415*, 415–416
Carbon cycle, *134*, **135**
Carbon dioxide, in carbon cycle, *134*, 135; chemical formula for, 496; covalent bond in, 476, *476*; in photosynthesis, 123; as waste, 76, *77*, 78, 79
Carbon films, 243, *244*
Carbon-14 dating, 258, *258*, 258 *lab*, 259, *259*, 260 *act*, 420, *420*

Carbon group, 443, *443*
Cardiac muscles, 69
Carnivores, 107, *107*, 137, *137*, 593
Carroll, Lewis, 296
Carrying capacity, 101, *104*, 105
Car safety testing, 540–541 *lab*
Carson, Rachel, 165
Cascade Mountains, *128*
***Cassini* space probe,** 350
Cassiopeia, 371
Cast, 245, *245*
Catalysts, 449, *449*, 507, **507**–508, *508*
Catalytic converters, 507, *507*
Cathode, 406, *406*
Cathode-ray tube (CRT), 406, *406*, 407, *407*
Cat(s), 41, *41*, 276, *276*, 296, *296*
Cell(s), 68, **68**–72, *69*, *70*. *See also* Blood cells; epithelial, *69*; feeding, 73–75, *74*, *75*; nerve, *637*, *638*; observing, 72 *lab*
Cellular respiration, 76, *76*
Celsius scale, *609*, 609–610, 610 *act*
Cenozoic Era, 292–293, *293*
Census, 99, 114
Ceres (asteroid), 359
Cerium, 450, *450*
Charge, electric, *636*, 636–642, *637*, *640*; flow of, *643*, 643–645, *644*, *645*; induced, 641, *641*; static, **637**, *637*
Charon (moon of Pluto), 353, *353*
Chart(s), 18, *18*
Chemical basis of life, 64–67, *66*, *67*
Chemical bonds, 471, *471*
Chemical changes, 491 *lab*, 492, *492*, 494 *lab*, 509 *lab*
Chemical energy, 645, *645*
Chemical equations, 494–496, 497, 497 *act*; balanced, 497, 498 *act*; energy in, 501, *501*
Chemical formulas, *479*, **480**, 496
Chemical names, 494, 495
Chemical plant, 490, *490*
Chemical reactions, 76 *lab*, 490–511, **492,** *493*, 589; describing, 494–495; endothermic, **499**, 510–511 *lab*; energy in, 498–501, *499*, *500*; exothermic, **499**, *500*, 510–511 *lab*; heat absorbed in, 500, *500*; heat released in, 499–500; identifying, 491 *lab*, 494;

rates of, *504*, 504–508, *505*, *506*; slowing down, 506, *506*; speeding up, *507*, 507–508, *508*; and surface area, 506, *506*
Chemosynthesis, 106, *136*, **136**–137
Chesapeake Bay, *169*
Childbirth, 81, *81*
Chlorine, 65, 446, *446*, *472*, 472–473, *473*, 476, *476*
Chlorophyll, 106, 108, 495
Chromium, 449
Chromosome(s), 39, *39*, 46, *46*
Chromosphere, 375, *375*
Cinder cone volcano, 222, *222*
Circuit, 643, 648–657; electric energy in, 644, *644*; parallel, **651,** *651*, 655 *lab*; protecting, 651, *651*; resistance in, 646, 646–647, *647*, 648, *648*, 649; series, **650,** *650*; simple, *643*, 643–644, *644*, 650 *lab*
Circuit breakers, 651, *651*
Circular motion, 560–561, *561*
Circulatory system, 71. *See also* Blood vessels; and digestion, 75, *75*; and respiratory system, 73, *73*
Circumference, of Earth, 307
Circumpolar constellations, 371
Cities, heat in, 626, *626*
Classification, of stars, 380–381
Clean Water Act of 1987, 15
Climate, 127; as abiotic factor in environment, *127*, 127–128, *128*; change of, 248, *248*; as evidence of continental drift, 184; extreme, 142, *142*; fossils as indicators of, 248, *248*, 249, *249*; and land, 154; and mountains, 292
Climax community, 153, *153*, 154
Clotting, 81
Clouds of Magellan, 387, *387*
Clown fish, 42, *42*, 108, *108*
CMEs (coronal mass ejections), 377, *377*
Coal, 245
Cobalt, 65, 448
Coelacanth, 264, *264*
Collisions, 521 *lab*, *535*, 535–538, *536*, *537*, *538*, 539 *lab*, 540–541 *lab*
Color, seeing, 713, *713*
Columbia River Plateau, 223, *223*
Coma, 357, *357*

Index

Comet(s), *356*, **356**–357; Kuiper Belt of, 353; structure of, 357, *357*
Commensalism, 108, *108*
Communicating Your Data, 12, 19, 29, 43, 55, 72, 83, 111, 113, 129, 141, 162, 171, 195, 201, 225, 233, 256, 263, 321, 327, 341, 361, 379, 393, 414, 425, 455, 481, 483, 509, 511, 539, 541, 569, 571, 585, 599, 618, 625, 655, 657, 672, 685, 706, 715
Communication, in science, *10*, 10–11
Communities, 96; climax **153**, *153*, 154; interactions within, 96, 106–110; symbiosis in, 108, *108*
Compass, 308, 308 *lab*, 666, 671, *671*, 671 *act*, 672 *lab*
Competition, 98, *98*, 99, 99 *lab*
Composite volcano, 223, *223*
Compound(s), 473; as chemical basis of life, 64; ionic, 472–474, 481 *lab;* organic, **66**–17; symbols for, 479, *479*
Compound machines, 591, *591*
Compression, 211, *211*
Compressional waves, 695, *695*, 696, *696;* amplitude of, 697, *697*; sound waves as, 701, *701*
Compression forces, 197
Compressor, 622, *622*
Computers, *9*, 16, 27; and semiconductors, 443, *443*
Concentration, 505; and rate of reaction, 505, *505*
Conclusions, 14 *act*, 19, *19*
Condensation, 131, *131*
Conduction, 613, *613*
Conductor, 615, 640, *640*, 640 *act*
Cone(s), of eye, 713, *713*
Connective tissue, 69
Conservation, of energy, 619; of mass, 496, *496*, 496 *lab*, 498; of momentum, 535, 535–538, *536, 537*
Constant, 17
Constant speed, 525, *525*
Constellation, *370*, **370**–371, *371*
Consumers, 107, *107*, 120, *136*, 137, *137*
Continent(s), fitting together, 181 *lab*, 192, 192 *act*

Continental drift, 182–185, 183 *act;* course of, 185, *185;* evidence for, 181 *lab*, 182, *183*, 184, 184 *lab*
Control, 22
Convection, *614*, **614**–615, *615*, 615 *lab*
Convection current, 195, *195*, 195 *act*
Convergent plate boundaries, 192, 193, *193*, 194, 197, 228
Coolant, 622, *622*, 623, *623*
Cooling, 618 *lab*
Cooperation, 110
Copernicus, Nicholas, 337, 340
Copper, 502, *502*; in body, 65
Copper wire, 646
Coral reef, 94, *94*, **167**, *167*, 167 *lab*
Corn, 38, *38*
Cornea, 711, *711*
Corona, 375, *375*
Covalent bond, *475*, **475**–476, *476*
Coyotes, 104
Crankshaft, 620
Cretaceous Period, 289, 291
Crickets, 98, 99, 100
Crinoid, 249, *249*
Crookes, William, 405, 406
Crystal, ionic, 477, *478;* molecular, 477, *478;* structure of, *478*
Curie point, 187
Current. *See* Electric current
Cyanobacteria, 281, *281*
Cycles, 130–135; carbon, *134*, **135**; nitrogen, *132*, **132**–133, *133*; water, *130*, 130–**131**, *131*
Cylinders, 620, *621*

Dalton, John, 405, *405*
Dark energy, 391
Darwin, Charles, 50, 51, 52, 275, *275*
Data Source, 28, 170, 200, 262, 454
Data tables, 14 *act*, 18
Dating, carbon-14, 254, 258, *258*, 259, *259*, 260 *act;* radiocarbon, 420, *420;* radiometric, 259, **259**–260, *260;* relative, 251 *act;* of rocks, 247, *247*, 249, 259–260, *260*, 282 *lab*
Days, length of, 309, *309*

Death rates, 102, 102 *act*
Decibel scale, 703, *703*
Decision making, 6, *6*
Decomposers, 107, *107*
Decompression melting, 228
Deer, 156
Deformation, 211 *lab*
Deimos (moon of Mars), 347
Dentistry, elements used in, 452
Deoxyribonucleic acid (DNA), 39, *39*, 67
Dependent variable, 17
Descriptive research, 13–15, *15*, 20
Desert(s), 94, *94*, **160**, *160;* competition in, 98, *98;* water in, *123*
Desertification, 160
Design Your Own, Car Safety Testing, 540–541; Comparing Thermal Insulators, 624–625; Does exercise affect respiration?, 82–83; Exothermic or Endothermic?, 510–511; Half-Life, 424–425; Measuring Parallax, 392–393; Modeling Motion in Two Directions, 570–571; Population Growth in Fruit Flies, 112–113; Pulley Power, 598–599
Devonian Period, 282, *282*
Diamond, 443, 512, *512*
Diffraction, 700; of waves, 700, *700*
Digestion, 74, **74**–75, *75*
Digestive system, 71, 74, 74–75, 75
Dinosaurs, era of, 288; extinction of, 286, 291 *act*, 296, *296*; fossils of, 242, *242*, 243, 244, 270, 289, 289–290, *290;* tracks of, 246, *246;* warm-blooded v. cold-blooded, 289 *act*
Direct current (DC), 679
Direction, changing, 588, *588;* of force, 581, *581*, 588, *588*
Disconformity, 252, *253*
Diseases, and bacteria, 75; fighting, 8 *act*
Displacement, and distance, 523, *523*
Distance, changing, 588, *588;* and displacement, 523, *523;* in space, 372 *act*, 373, 388 *lab;* and work, 582, 588, *588*
Distance-time graph, 526, *526*

Divergent plate boundaries, 191, 193, *193,* 227
DNA (deoxyribonucleic acid), 39, *39,* 67
Dodo, 296, *296*
Dolphin, 293
Domain, magnetic, **668,** *668*
Dominant traits, 45, *45*
Doppler shift, 370–389, *389*
Double bonds, 476, *476*

Ear, 704, *704*
Earth, 344, *344,* 354, *354;* axis of, 307, *307,* 309, 326–327 *lab;* biosphere of, *94,* 94–95; as center of solar system, 336; circumference of, 307; density of, 307; diameter of, 307; distance from Sun, 307, 309, 360–361 *lab;* ecosystems of, 121 *lab;* life on, 95; lithosphere of, 190, *190, 191,* 226, *227;* magnetic axis of, 308, *308;* magnetic field of, 181–188, *188,* 308, *308,* 669, 669–671, *670;* magnetosphere of, 669, *669,* 677, *677;* mantle of, 195, *195;* Mars compared to, *345;* mass of, 307; moving plates of, 219, *219,* 226, *226, 227, 231;* orbital speed of, 340; orbit of, 309; revolution of, 305 *lab,* 307, 309; rotation of, 305 *lab,* 307, *307;* spherical shape of, *306,* 306–307
Earth history, 240–263, 262–263 *lab,* 279–295; and absolute age, 257–261; Cenozoic Era, *292,* 292–293; discovering, 294–295 *lab;* and fossils, 242–249; Mesozoic Era, 288, 288–291, *289, 290;* Paleozoic Era, *282,* 282–286, *284, 284 act, 285, 286;* and plate tectonics, 279, *279,* 285, *285;* Precambrian time, *280,* 280–282, *281;* and relative age, 251–256, 256 *lab*
Earthquakes, 192 *act,* **210**–218, 317, 697; building for, 209 *lab,* 234; causes of, *210,* 210–211; damage caused by, 216, 234, *234;* and Earth's plates, 229–231, *231;* epicenter of, 212, 214, *214;* and faults, 211, *211,* 212, *212,* 218; focus of, 212, *212;* locations of, 229, *229;* magnitude of, 213, 214, 216, 216 *act;* measuring, 213, *213,* 214, 216; predicting, 218, *218;* preparation for, 209 *lab,* 217, 217–218, 234; and seismic waves, *212,* 212–213, *214,* 230, *230,* 230 *act,* 232–233 *lab*
East African Rift Valley, 180
East Pacific Rise, 196
Echolocation, 705
Eclipses, *314,* 314–316, 315 *act;* causes of, 315; lunar, **316,** *316,* 321 *lab;* solar, 314, *314,* **315,** *315*
Ecological succession, 150–153, *152*
Ecology, 95, 248
Ecosystem(s), 95, *95,* 121 *lab,* 148–171; aquatic, 163–171, 164 *lab,* 170–171 *lab;* carrying capacity of, 101, *104,* 105; changes in, *150,* 150–153, *151, 152;* competition in, 98, *98;* habitats in, 97, *97,* 98, *98,* 109, *109;* land, *154,* 154–162, 162 *lab;* limiting factors in, 100; populations in, 96, 99–105, *109, 110,* 112–113 *lab*
Ediacaran fauna, 282, *283*
Efficiency, 589–590, *590;* calculating, 589 *act;* equation for, 589; and friction, 590
Einstein, Albert, 381
Elastic limit, 210, *210*
Elastic rebound, 210
Electrical energy, 619
Electric charge, *636,* 636–645, *637,* 640
Electric circuit. *See* Circuit
Electric current, 643–649, 655 *lab;* controlling, *648,* 648–649; effect on body, 654; generating, *678,* 678–679, *679;* and magnetism, 673–681, 683; model for, 656–657 *lab;* in a parallel circuit, 655 *lab;* and resistance, 646, 646–647, *647,* 648, *648,* 649; types of, 678, 679

Electric discharge, 641, *641*
Electric energy, in circuit, 644, *644;* cost of, 653, 653 *act;* and resistance, 646, 646–647, *647,* 648, *648,* 649
Electric field, 639, *639*
Electric forces, 635 *lab,* **639,** *639,* 644 *lab*
Electricity, 634–657; connecting with magnetism, 683; generating, *678,* 678–679, *679;* safety with, 653–654
Electric meter, 653, *653*
Electric motors, 676, *676,* 684–685 *lab*
Electric power, 652–653
Electric shock, 653–654
Electric wire, 640, 646, *646*
Electromagnet(s), 673, 673–674, *674,* 674 *lab*
Electromagnetic spectrum, 708–709, *709*
Electromagnetic waves, 696, 707, 708, *708,* 710, *710*
Electron(s), 257, **407,** 412–413, *413,* 466 *act,* 636, 636–637, *637,* 644; arrangement of, 465, 465–466, *466;* energy levels of, 465, 465–467, *466, 467;* in magnetic fields, 668, *668;* model of energy of, 463 *lab;* movement of, 464, *464*
Electron cloud, 413, *413,* **464,** *464*
Electron dot diagrams, *470,* 470 *lab,* **470**–471
Element(s), 405, 438 *act;* atomic number of, 415; atomic structure of, 465, *465,* 482–483 *lab;* boron family of, 442, *442;* carbon group of, 443, *443;* as chemical basis of life, 64; halogen family of, 468, *468;* halogens, 446, *446;* in human body, 67; identifying properties of, 469 *act;* isotopes of, 257, *415,* 415–416, 421 *act,* 421–423, *422;* metalloids, 438, *438,* 442, 443, *443,* 445, 446; metals, 438, *438,* 441, 441–442, *442,* 443, *443,* 448–452; nitrogen group of, 444, *444;* noble gases, 446, 446–447, *447,* 468, *468;*

nonmetals, 438, *438*, 443, *443*, 444, *444*, 445, 446, *446;* oxygen family of, 445, *445;* periodic table of, 466–467, *467*, 468, *468*. *See* Periodic table; radioactive, 450, *451;* representative, **435;** symbols for, 440; synthetic, 421–423, *422*, **450**, *451;* tracer, 421–423, *422;* transition, **435**, *448*, 448–452, *449*, *450*, *451*

Element keys, 439, *439*

Elevation, and temperature, 126, *126*, 126 *act*

Ellipse, 309, 309 *act*

Elliptical galaxy, 387, *387*

Endocrine system, 71

Endothermic reactions, 499, 510–511 *lab*

Energy, activation, **503,** *503;* and amplitude, 697; chemical, 645, *645;* in chemical reactions, 498–501, *499*, *500;* conservation of, 619; converting, *136*, 136–137, 420; dark, 391; electrical, 619; in equations, 501, *501;* flow of, 136–139; in food chain, 137, *137;* forms of, 619; from fusion, 381–382, *382;* for human body, *76*, 76–79, *77;* loss of, 139, *139;* and mass, 381; mechanical, 619, *619*, *621;* nuclear, 619; obtaining, *106*, 106–107, *107;* and photosynthesis, 106, 136; and power, 584; radiant, 619; from Sun, 120; thermal. *See* Thermal energy; transfer of, *137*, 137–138, *138;* and waves, 694, *694;* and work, 584

Energy levels, of electrons, *465*, 465–467, *466*, *467*

Energy pyramids, 138–139, *139*

Engines, *619*, 619–621, *620*, 620 *act*, *621*

Environment, abiotic factors in, *122*, 122–129, 129 *lab;* biotic factors in, 122; and fossils, *248*, 248–249, *249;* freshwater, modeling, 164 *lab;* for houseplants, 149 *lab;* model of, 271 *lab;* and species, 50–53, *51*, *52*, *53*, 54–55 *lab;* and survival, 49, 49–50, *50;* and traits, 40, 40–42, *41*, *42*, 49–53

Environmental Protection Agency, 167

Enzymes, 75, 508; as catalysts, 507–508, *508*

Eon, 273, *273*

Epicenter, 212, 214, *214*

Epithelial cells, 69

Epoch, 273, *273*

Equation(s). *See* Chemical equations; acceleration, 530; for efficiency, 589; for mechanical advantage, 587; one-step, 582 *act*, 583 *act*, 587 *act*, 589 *act;* for power, 583; simple, 559 *act*, 610, 649 *act*, 652 *act*, 698 *act;* for wave speed, 698; for work, 582

Equinox, 310, 311

Era, 273, *273*

Eros (asteroid), 359

Eruptions, 220, *220*, 220 *lab*, 225 *lab;* fissure, 223, *223;* largest, 224; quiet, 221, 223; violent, 221, 223

Estuaries, 168–169, *169*

Europa (moon of Jupiter), 349, *349*

Europium oxide, 450

Evaporation, 130, *131*

Event horizon, 383

Everglades, 165

Evolution, 50–52, *51;* of mammals, 293, *293;* organic, 274, **274**–276, *275*, *276;* of stars, 382 *act*, 382–385

Excretion, *78*, 78–79

Exercise, and respiration, 82–83 *lab*

Exhaling, 74 *lab*

Exhaust valve, 621

Exothermic reactions, 499, *500*, 510–511 *lab*

Expansion, thermal, 609, *609*

Experiment(s), *21*, 21–23

Experimental research design, 13, *20*, *21*, 21–23, *22*, *23*

Exponential growth, 104, 105, *105*

Extinction, 53, *53*, 286, *286*, 288, 291 *act*, 296, *296*

Eye, *711*, 711–713, *712*, *713*

Fahrenheit scale, *609*, 609–610, 610 *act*

Fat(s), dietary, 66, *66*, 67

Fault(s), 194, *194*, **211,** *212;* measuring movement along, 218, *218;* normal, 196; strike-slip, 198, *198;* types of, 211, *211*

Fault-block mountains, 196, *196*

Feedback, negative, **80,** *80*, 80 *act;* positive, 81, *81*

Fertilizer(s), 133 *lab*, 423, 444, *444*

Field(s), electric, **639,** *639;* magnetic. *See* Magnetic field(s)

Filaments, 468, 647

Filtration, in kidneys, 78, *78*

Finches, 51

Fingerprinting, 37 *lab*

Fire. *See* Wildfires; chemical changes caused by, 492, *492*, *493*

Firefighting foam, 445, *445*

Fireworks, 502, *502*

First-class lever, 595

Fish, early, 264, *264*, 282, *282*, 284, *284;* and environment, 54–55 *lab;* gender of, 42, *42;* with lungs, 284, *284*

Fissure eruptions, 223, *223*

Fixed pulleys, 596, *597*

Fleming, Alexander, 75

Flight, 564 *act*

Flint, 450, *450*

Flood basalts, 223, *223*

Florida Everglades, 165

Fluoride, 452

Fluorine, 65, 446, 468, *468*

Foam, for firefighting, 445, *445*

Focus, 212, *212*

Foldables, 5, 37, 63, 93, 121, 149, 181, 209, 241, 271, 305, 335, 369, 403, 433, 463, 491, 521, 549, 579, 607, 635, 665, 693

Food, and bacteria, 505; getting to cells, 73–75, *74*, *75;* irradiated, 426, *426;* reaction rates in, 504, *504*, 505, 506

Food chain, 107, *107;* energy in, 137, *137*

Food web, 138, *138*

Force(s), *550*, **550**–551;

and acceleration, *556*, 556–557, 560; action and reaction, 563–566, *564*, *566*; balanced, **551**, *551*; changing, 587; combining, 551; comparing, 579 *lab*; compression, 197; direction of, 198, 581, *581*, 588, *588*; effects of, 549 *lab*; electric, 635 *lab*, **639**, *639*, 644 *lab*; input, **586**, *586*; magnetic, 655 *lab*; net, **551**, 560; output, **586**, *586*; shear, 211; strong nuclear, 416; unbalanced, **551**; unit of measurement with, 557; and work, 579 *lab*, 581, *581*, 585 *lab*, 587

Force pairs, 567 *lab*

Forests. See also Rain forests; as climax community, 153, *153*, 154; temperate deciduous, 154, *156*, 156–**157**; and wildfires, 148, *148*, 151 *act*, 152, 658, *658*

Formulas, chemical, *479*, **480**, 496

Fossil(s), *242*, 242–249, *243*; and ancient environments, 248, 248–249, *249*; changes shown by, 278; and climate, 248, *248*, 249, *249*; of dinosaurs, *270*, 289, 289–290, *290*; Ediacaran, 282, *283*; formation of, 243, *243*; index, **247**, *247*, 254 *act*; making model of, 241 *lab*; minerals in, 244, *244*; organic remains, 246, *246*; Paleozoic, 282; Precambrian, 280; preservation of, 243 *lab*, 243–247, *244*, *245*, *246*; in rocks, 282 *lab*; trace, 246, *246*, 262–263 *lab*

Fossil record, as evidence of continental drift, 183, *183*, 184, 184 *lab*

Four-stroke cycle, 620, *621*

Fox, 41

Free fall, 567, *567*, 568

Frequency, 696; of light, 707; of sound waves, 703; unit of, 698

Freshwater ecosystems, 163–165; lakes and ponds, *164*, 164 *lab*, 164–165; rivers and streams, *163*, 163–164; wetlands, 165, *165*, 170–171 *lab*, 172, *172*

Friction, 229, *552*, **552**–555, 554 *lab*, 590, *590*; rolling, 555, *555*; sliding, 553, 554, *554*, 555, 559, 562; static, 554

Fruit flies, phenotypes of, 46 *lab*; population growth in, 112–113 *lab*

Fulcrum, 594, *595*

Full moon, *314*, **314**

Fungi, and mutualism, 108, *108*

Fuses, 651, *651*

Fusion, 338, *339*, 381–382, *382*, 451

G

Galaxies, *368*, **386**–387; clusters of, 369 *lab*, 386; elliptical, 387, *387*; irregular, 387, *387*; spiral, *386*, 386–387

Galilei, Galileo, 320, 337, 349, 376, 552, 553 *act*

Galileo space probe, 348, 359

Gallium, 442

Galvanometer, 674, *675*

Gamma rays, 710

Ganymede (moon of Jupiter), 349, *349*

Gas(es), exhalation of, 74 *lab*; noble, 468, *468*

Gaspra (asteroid), 359, *359*

Gender, 42, *42*, 42 *act*

Gene, 39, *39*

Generator, 678, 678–679

Genetics, 44, **44**–48; dominant and recessive factors in, 45, *45*; early study of, 45, 45–46, *46*; and mutations, 52; and traits, 44–48

Genotype, 39

Geologic time scale, 272–273, *273*. See also Earth history

Germanium, 443

Giants, 381, *381*, **383**, *383*, 384, 394, *394*

Giraffes, 276, *276*

Glaciers, as evidence of continental drift, 184

Glass, 443

Global Surveyor space probe, 345, 346

Glomar Challenger (research ship), 187

Glucagon, 80

Glucose, 107; regulation of, 80, *80*

Glycogen, 80

Gondwanaland, 288, *288*

Goodall, Jane, 30, *30*

Grand Canyon National Park, 254

Graph(s), 18, *18*; of accelerated motion, 532, *532*; distance-time, 526, *526*; of motion, 526, *526*; speed-time, 532, *532*

Graphite, 443

Graptolites, 244, *244*

Grass, life in, 93 *lab*

Grasslands, 161, *161*

Gravity, 557–558; and air resistance, 561; and atmosphere, 127 *act*; effects of, 345 *lab*; and motion, 557, 560, *560*, 561, *561*; and stem growth, 40 *lab*

Great Barrier Reef, 167

Great Dark Spot (Neptune), 352

Great Red Spot (Jupiter), 348, *348*

Great Rift Valley, 191, 196

Greenhouse effect, 343

Grounding, 642, *642*

Group, 435

Growth, beginning, *104*; and environment, 40, *40*; exponential, *104*, 105, *105*; of plants, 40, *40*, 40 *lab*, 140–141 *lab*; of population, 102–105, *103*, *104*, *105*, 112–113 *lab*

Gymnosperms, 290

H

Habitats, 97, *97*, 98, *98*, 109, *109*

Hadean Eon, 280

Hale-Bopp comet, 356, *356*

Half-life, 258, 259, *418*, 418 *lab*, 418–419, 419 *act*, 424–425 *lab*

Halley, Edmund, 356

Halogens, 446, *446*, 468, *468*

Hammer (of ear), 704, *704*

Hawaiian Islands, volcanoes in, 222, *222*, 224, 228, *228*

Hawking, Stephen, 26

Health, and heavy metals, 454–455 *lab*; and mercury, 452 *act*

Hearing, *493*, 703, 704

Heart, interaction with lungs, 73, *73*
Heat, 612–618, in chemical reactions, 499–500, *500;* conduction of, 613, *613;* convection of, *614,* 614–615, *615,* 615 *lab;* radiation of, 613; specific, **616;** and thermal energy, 612–615; transfer of, 612, *612*
Heat engines, *619,* 619–621, *620,* 620 *act, 621*
Heating, 618 *lab*
Heat island, 626, *626*
Heat pumps, 623, *623*
Heavy metals, 443, *443,* 449, 454–455 *lab*
Helium, 446, 447, *447,* 467, 468
Hemoglobin, 448
Herbivores, 107, *107,* 137, *137,* 593
Hertz (Hz), 698
Hertzsprung, Ejnar, 380
Hertzsprung-Russell (H-R) diagram, 380, *380,* 381, 382
Hess, Harry, 187
Himalaya, 197, *197,* 292, *292*
History. *See* Earth history
Homeostasis, 79, *79*
Homo sapiens, 293
Hopper, Grace Murray, 26
Horse, 293
Hot spots, 228, *228,* 231
Hubble, Edwin, 389
Hubble Space Telescope, 352, 353, 357, 384, 391
Human(s), origin of, 293
Human Genome Project, 44
Humus, 124, 129 *lab*
Hurricanes, on Neptune, 352
Hutton, James, 261
Hybrid, 48
Hydrogen, v. helium, 446; isotopes of, 257
Hydrogen chloride, 476, *476*
Hydrothermal vents, 137
Hyoliths, *286*
Hypothesis, 21

Iceman, 260
Impact basin, *323, 323,* 323, 324 *lab*

Impact theory, 319, *319*
Inclined plane, 591–593, *592*
Independent variable, 17
Index fossils, 247, *247,* 254 *act*
Induced charge, 641, *641*
Inertia, 533, *533*
Inference, 9 *lab*
Infrared waves, 709
Inhibitor, 506, *506,* 506 *lab*
Inner planets, 338, 342–347, 344, *344. See also* Earth; Mars, 340, *344,* 344–347, 354, *354,* 360–361 *lab;* Mercury, 340, *342,* 342–343, 354, *354,* 360–361 *lab;* Venus, *336,* 337, 340, 343, 353, 354, *354,* 360–361 *lab*
Inner transition elements, 450, *450,* 451
Inorganic substances, 65–66
Input force, 586, *586*
Insect(s), and competition, 98; counting population of, 99; niches of, 109, *109,* 110
Instantaneous speed, 525, *525*
Insulator(s), 616, *616,* 624–625 *lab,* **640,** *640*
Insulin, 80
Integrate Career, Environmental Author, 165; Farmer, 127, 444; Mechanical Engineering, 622; Nobel Prize Winner, 467; Seismology, 317; Volcanologist, 50, 197
Integrate Chemistry, Alkaline Batteries, 645; Classifying Elements, 456; Curie Point, 187; DNA Structure, 39; Earth's First Air, 281; Glucose, 107; Melting Point, 228; What determines how a volcano erupts?, 221; White Dwarf Matter, 383
Integrate Earth Science, Body Elements, 67; Carbon Dating, 420; Desertification, 160; Hydrothermal Vents, 137; rain shadow effect, 128; seismic waves, 696; volcanoes, 202
Integrate Environment, The Clean Water Act, 15; Energy Conversion, 420; nonliving influences, 49; seashores, 168;
Integrate Health, dentistry and

dental materials, 452; Hearing Damage, 703; magnetic resonance imaging (MRI), 682; reaction rates in food, 505; variables, 21
Integrate History, Antibiotics, 75; Breathe Easy, 507; The Currents War, 681; James Prescott Joule, 589; Newton and Gravity, 557; The Ohm, 646; Plant Poisons, 109; Protons, 412; 386 Supernova, 385
Integrate Language Arts, Friction, 229; Name of Planets, 352
Integrate Life Science, Ancient Ecology, 248; Animal Insulation, 616; Autumn Leaves, 495; Biomechanics, 551; Body Temperature, 589; Cell Division in Tumors, 423; Earth's Rotation, 307; flight, 564; Ions, 473, Poison Buildup, 445; research for writing, 30; species, 274; Thermal Pollution, 617; wedges in your body, 593
Integrate Physics, Bright Lights, 450; Direction of Forces, 198; evolution of stars, 382; magnetic clues, 187; magnetic field, 308; motion of planets, 340; nuclear fission, 484; radioactive decay, 257; Rotational Motion, 338
Integrate Social Studies, Coal Mining, 245; Forensics and Momentum, 534; Science in Advertising, 7
Integumentary system, 71
Intensity, **702;** of sound, 702, *702,* 703
Interactions, of body systems, 73, 73–83, 82–83 *lab;* and survival, 50, *50*
Internal combustion engines, 620, *620,* 620 *act, 621*
International System of Units (SI), 17
International Union of Pure and Applied Chemistry (IUPAC), 440
Internet, 28–29 *lab. See* Use the Internet
Intertidal zone, 168, *168*

**Invertebrate animals, ** 281
Iodine, 65, 446, *470*
Iodine-131, *422,* 423
Io (moon of Jupiter), 349, *349*
Ion(s), 473, *473,* **636,** 637, *637*
Ionic bond, 472, 472–474, **473,** *473,* 474
Ionic compounds, 472–474, 481 *lab*
Ionic crystal, 477, *478*
Iridium, 449
Iron, 448, 450; in body, 65
Iron triad, 448, *448*
Irradiated food, 426, *426*
Irregular galaxy, 387, *387*
Isotopes, 257, *415,* **415**–416; radioactive, 421 *act,* 421–423, *422*

James, Sarita M., *26*
Jellyfish, 282
Joule, James Prescott, 589
Journal, 4, 36, 62, 92, 120, 148, 180, 208, 240, 270, 304, 334, 368, 402, 432, 462, 490, 520, 548, 578, 606, 634, 664, 692
Jupiter, *348,* 348–349, 355, *355;* distance from Sun, 360–361 *lab;* exploration of, 348; Great Red Spot on, 348, *348;* moons of, 349, *349;* orbital speed of, 340
Jurassic Period, 289

Kelvin scale, 610
Kepler, Johannes, 340
Kidney(s), filtration in, 78, *78;* and waste elimination, *78,* 78–79
Kilogram (kg), 557
Kilowatt-hour (unit of electric energy), 653
Krypton, 447
Kuiper Belt, 353

Lab(s), Balloon Races, 569; Battle of the Beverages Mixes, 12; Bending Light, 710, 714–715; Building the Pyramids, 585; Changing Species, 287; Collisions, 539; Current in a Parallel Circuit, 655; Design Your Own, 82–83, 112–113, 392–393, 424–425, 510–511, 540–541, 570–571, 598–599, 624–625; Discovering the Past, 294–295; Disruptive Eruptions, 225; Exothermic or Endothermic?, 510–511; Feeding Habits of Planaria, 111; Half-Life, 418; Heating Up and Cooling Down, 618; How does an electric motor work?, 684–685; Humus Farm, 129; Ionic Compounds, 481; Jelly Bean Hunt, 43; Launch Labs, 5, 37, 63, 93, 121, 149, 181, 209, 241, 271, 305, 335, 369, 403, 433, 463, 491, 521, 549, 579, 607, 635, 665, 693; Make a Compass, 672; Making a Model of the Invisible, 414; Mini Labs, 9, 46, 76, 103, 133, 164, 195, 211, 258, 282, 323, 345, 388, 418, 435, 470, 494, 496, 531, 567, 596, 615, 650, 670, 710; Model and Invent, 262–263, 360–361, 482–483; Model for Voltage and Current, 656–657; Moon Phases and Eclipses, 321; Observing Cells, 72; Physical or Chemical Change?, 509; Planetary Orbits, 341; Relative Ages, 256; Seafloor Spreading Rates, 189; Seismic Waves, 232–233; Sound Waves in Matter, 706; Studying a Land Ecosystem, 162; Sunspots, 379; Tilt and Temperature, 326–327; Toothpick Fish, 54–55; Try at Home Mini Labs, 18, 40, 74, 99, 124, 158, 184, 220, 243, 292, 308, 313, 350, 371, 411, 475, 506, 525, 554, 583, 614, 644, 674, 699; Use the Internet, 28–29, 170–171, 200–201, 294–295, 454–455; Where does the mass of a plant come from?, 140–141

Lakes, *164,* 164–165
Land biomes, *154,* 154–162, 162 *lab;* deserts, 160, *160;* grasslands, 161, *161;* taiga, 156, *156;* temperate deciduous forests, 154, *156,* 156–157; temperate rain forests, 157, *157;* tropical rain forests, 154, *158,* 158–159, *159;* tundra, 155, *155*
Land speed, 526 *act*
Lanthanides, 450, *450*
Lanthanum, 450, *450*
Large Magellanic Cloud, 387, *387*
Lasers, 708 *act*
Latitude, and temperature, 125, *125*
Launch Labs, Clues to Life's Past, 241; Compare Forces, 579; Construct with Strength, 209; Earth Has Many Ecosystems, 121; Forces and Motion, 549; How are people different?, 37; How can you tour the solar system?, 335; How do lawn organisms survive?, 93; Identify a Chemical Change, 491; Magnetic Forces, 665; Make a Model of a Periodic Pattern, 433; Measuring Temperature, 607; Measure Using Tools, 5; Model Blood Flow in Arteries and Veins, 63; Model Crater Formation, 335; Model Rotation and Revolution, 305; Model the Energy of Electrons, 463; Model the Unseen, 403; Motion After a Collision, 521; Observing Electric Forces, 635; Reassemble an Image, 181; Survival Through Time, 271; Wave Properties, 693; Were the continents connected?, 181; What are some properties of waves?, 693; What environment do houseplants need?, 149; Why do clusters of galaxies move apart?, 369
Laurasia, 288, *288*
Lava, 219, 221, 222, *222,* 223
Lava plateaus, 223, *223*
Lavoisier, Antoine, 496
Law(s), on clean water, 15;

Lawrencium   **Mercury**

of conservation of energy, 619; of conservation of mass, 496, *496*, 496 *lab*, 498; of conservation of momentum, *535*, **535**–538, *536*, *537*; Newton's first law of motion, **552**–555, *555*; Newton's second law of motion, **556**–562, *565*; Newton's third law of motion, **563**–568, *565*, 569 *lab*; Ohm's, **649**; of reflection, **699**, *699*
Lawrencium, 450
Lead, 443, *443*, 450
Leaves, 158 *lab*; changing colors of, 495; and genetics, 41, *41*
Lenses, of eye, 711, *711*
Lever, 594, *594*, 595, *595*
Lichens, and mutualism, 108, *108*; as pioneer species, 150, *151*
Life, chemical basis of, 64–67, *66*, *67*; origins of, 56, *281*, 281–282; unusual forms of, 282, *283*
Life processes, 135
Light, 707–715; as abiotic factor in environment, 124, *124*; bending, 710, 714–715 *lab*; frequency of, 707; seeing, 711; speed of, 707; visible, 709, *709*; wavelength of, 707
Lightbulb, 449, *449*, 468
Lightning, 658
Lightning rod, 642, *642*
Light waves, in empty space, 707, *707*; properties of, 708, *708*
Light-year, 373
Limestone, 254, *254*
Limiting factors, 100
Lipids, 66, *66*, 67
Lithium, 441, 465, *465*, 467, 469
Lithosphere, 190, *190*, *191*, 226, 227
Liver, 80
Lizards, *274*
Local Group, 386, 389
Loudness, *702*, 702–703, *703*
Lunar eclipse, 316, *316*, 321 *lab*
Lunar Orbiter, 322
Lunar Prospector, 324, *324*
Lung(s), calculating volume of air held by, 77 *act*; interaction with heart, 73, *73*; and respiration, 76, *77*
Lutetium, 450

Lymphatic system, 71
Lynx, *156*

Machine(s), 586–599; compound, **591,** *591*; and efficiency, 589–590, *590*; and friction, 590, *590*; and mechanical advantage, 586, 586–588, 594, *594*; simple. *See* Simple machines
Magellan space probe, 343, *343*
Maglev, 664, *664*, 665
Magma, 187, 221, 223; silica-rich, 221
Magnesium, 65, 442, 474, *474*
Magnet(s), 666–668; electromagnets, 673, 673–674, *674*, 674 *lab*; poles of, 666, *666*, 667, 669 *act*; superconductors, 681, 681–682, *682*
Magnetic axis of Earth, 308, *308*
Magnetic declination, 669 *act*
Magnetic domain, 668, *668*
Magnetic field(s), 667, *667*–671; of Earth, 187–188, *188*, 308, *308*, 669, 669–671, *670*; making, 668, *668*; observing, 670 *lab*; and seafloor spreading, 187–188, *188*
Magnetic field lines, 667, *667*
Magnetic force, 655 *lab*
Magnetic properties, 448
Magnetic resonance imaging (MRI), 682, 682–683, *683*
Magnetic time scale, 188
Magnetism, 664–683; early uses of, 666; and electric current, 673–681, *683*
Magnetite, 188, 666
Magnetometer, 188
Magnetosphere, 669, *669*, 677, *677*
Magnitude, 213, 214, 216, 216 *act*; absolute, **372;** apparent, **372**
Main sequence, *380,* 380–381, 382–383
Maize, 38, *38*
Mammal(s), evolution of, 293, *293*; marsupials, 293, *293*; origin of, 290, *290*
Mammoth, 246, 296, *296*
Manganese, 65
Mantle, of Earth, 195, *195*

Map(s), 527, *527*; of Moon, 324, 324–325
Maria, 317, *318*, 320
Mariner **space probes,** 342, 343
Mars, 95, *344*, **344**–347, 354, *354*; distance from Sun, 360–361 *lab*; Earth compared to, 345; exploration of, 344–346, 346 *act*; moons of, 347, *347*; orbital speed of, 340; polar ice caps on, 344, 346; seasons on, 346; surface features of, 344, *344*
Mars Odyssey, 345
Mars Pathfinder, 345
Marsupials, 293, *293*
Mass, 533, *533*; conservation of, 496, *496*, 496 *lab*, 498; and energy, 381; unit of measurement with, 557; and weight, 558
Mass number, 416
Materials, semiconductors, 443, *443*
Matter, cycles of, 130–135; and motion, 522; recycling, 385
Mauna Loa, Hawaii, 222, *222*
Mayan calendar, 328, *328*
Measurement, of average speed, 525 *lab*; of distances in solar system, 336–337, 344, 360–361 *lab*; of earthquakes, 213, *213*, 214, 216; of force pairs, 567 *lab*; of movement along faults, 218, *218*; of parallax, 392–393 *lab*; scientific, 17; in space, 373, 388 *lab*; of temperature, 609, 609–610; units of, 17, 530, 557, 583, 646, 653; using tools, 5 *lab*; of weight, 567, *567*; of work, 584
Mechanical advantage, *586,* 586–588, **587,** 594, *594*
Mechanical energy, 619, *619*, 621
Medicine, isotopes in, 421, 421 *act*, 422, 423; magnetic resonance imaging (MRI) in, 682, 682–683, *683*; research in, 27, *27*
Meitner, Lise, 440
Melting point, 228
Melting rates, 614 *lab*
Mendel, Gregor, 45–46, 47, 48
Mendeleev, Dmitri, 434, *434*, 435, 468
Mercury, 449, 452, 452 *act*

802   **STUDENT RESOURCES**

Index

Mercury (planet)

Mercury (planet), 95, 340, *342,* **342**–343, 354, *354,* 360–361 *lab*
Mesozoic Era, *288,* **288**–291, *289, 290*
Metal(s), 438; alkali, **441,** *441,* 469, *469;* alkaline earth, **442,** *442;* as catalysts, 449, *449;* as conductors, 640; heavy, 443, *443,* 449, 454–455 *lab;* iron triad, 448, *448;* misch, 450, *450;* on periodic table, 438, *438,* 441, 441–442, *442,* 443, *443,* 448–452; transition, **435,** 448, 448–452, *449, 450, 451*
Metallic bond, 474, *474*
Metalloids, 438, *438,* 442, 443, *443,* 445, 446
Meteor, *357,* **357**–358
Meteorite, 56, 260, *260,* 358, *358,* 362
Meteoroid, 357, 358
Meteor shower, 358
Meter, electric, 653, *653*
Methane, 475 *lab*
Microwaves, 709
Mid-Atlantic Ridge, 191, 192, 196
Mid-ocean ridges, 186, *186,* 191, 192, 196
Migration, 103
Milkweed plants, 109
Milky Way Galaxy, 368, 386, *386,* 387
Millipedes, 109
Mineral(s), 65; in fossils, 244, *244*
Mini Labs, Comparing Biotic Potential, 103; Comparing Fertilizers, 133; Dating Rock Layers with Fossils, 282; Designing a Periodic Table, 435; Drawing Electron Dot Diagrams, 470; Graphing Half-Life, 418; Identifying Simple Circuits, 650; Inferring Effects of Gravity, 345; Inferring from Pictures, 9; Making Your Own Compass, 308; Measuring Distance in Space, 388; Measuring Force Pairs, 567; Modeling Acceleration, 531; Modeling Carbon-14 Dating, 258; Modeling Convection Currents, 195; Modeling Freshwater Environments, 164; Observing a Chemical Change, 494; Observing a Chemical Reaction, 76; Observing Convection, 615; Observing Deformation, 211; Observing Fruit Fly Phenotypes, 46; Observing the Law of Conservation of Mass, 496; Observing Magnetic Fields, 670; Observing Pulleys, 596
Misch metal, 450, *450*
Model(s), 16; of atom, *405,* 405–414, *408, 410, 411,* 414 *lab;* of unseen, 403 *lab,* 414 *lab*
Model and Invent, Atomic Structure, 482–483; Solar System Distance Model, 360–361; Trace Fossils, 262–263
Molds, of organic remains, **245,** *245*
Molecular crystal, 477, *478*
Molecules, 475, *475;* nonpolar, 477, *477;* polar, 476, 476–477, *477,* 477 *act*
Momentum, 534–**538;** calculating, 534 *act;* and collisions, 521 *lab,* 535, 535–538, *536, 537, 538,* 539 *lab,* 540–541 *lab;* conservation of, *535,* 535–538, *536, 537*
Montserrat volcano, 219, *219,* 220, *220,* 221, 221 *act,* 224, 227
Moon(s), 312–325, 319 *lab;* craters on, 317, *318,* 320; eclipse of, 316, *316,* 321 *lab;* exploration of, 322, 322–325; ice on, 324, *325;* interior of, 317, *317;* of Jupiter, 349, *349;* mapping, *324,* 324–325; of Mars, 347, *347;* movement of, 305 *lab,* 312, *312, 316, 322;* of Neptune, 352, *352;* origin of, 319, *319;* of Pluto, 353, *353;* poles of, 325; and reflection of Sun, 313; rocks on, *320,* 323; of Saturn, 350; surface of, *304,* 317, *317, 318, 323,* 323–325, *324;* of Uranus, 351
Moon phases, 313, *314,* 321 *lab*
Moonquakes, 317, *317*
Moseley, Henry, 435
Motion, 520, 522 *act,* 522–527, 548–571; and acceleration, 528–529, *556,* 556–557, 560; after a collision, 521 *lab;* and air resistance, 561; and changing position, *522,* 522–523; circular, 560–561, *561;* and friction, *552,* 552–555; graphing, 526, *526,* 532, *532;* and gravity, 557, 560, *560,* 561, *561;* and matter, 522; modeling in two directions, 570–571 *lab;* and momentum, 534–538; Newton's first law of, **552**–**555,** *565;* Newton's second law of, **556**–**562,** *565;* Newton's third law of, **563**–**568,** *565,* 569 *lab;* on a ramp, 549 *lab;* relative, 523, *523;* and speed, 524–525, *525;* and work, *580,* 580–581, *581*
Motors, electric, **676,** *676,* 684–685 *lab*
Mountains, and climate, 292; as evidence of continental drift, 184; fault-block, 196, *196;* formation of, 197, *197,* 285, *285,* 286, 292, *292;* rain shadow effect in, 128, *128;* and temperature, 126, *126*
Mount St. Helens eruption (Washington state), 223, 224
Movable pulleys, 597, *597*
Movement, of populations, 103, *103*
MRI (magnetic resonance imaging), 682, 682–683, *683*
Muscle(s), cardiac, 69; skeletal, 69; smooth, 69
Muscle tissue, 69
Muscular system, 71
Mutation, 52
Mutualism, 108, *108*

Names, chemical, 494, 495
Nanometer, 707
National Aeronautics and Space Administration (NASA), 318, 324
National Geographic Unit Openers, How are Electricity and DNA Connected?, 2; How are Beverages and Wildlife Connected?, 90; How are Volcanoes and Fish Connected?, 178; How are Thunderstorms and Neutron Stars Connected?, 302;

Index

National Geographic Visualizing, How are Charcoal and Celebrations Connected?, 400; How are City Streets and Zebra Mussels Connected?, 518; How are Radar and Popcorn Connected?, 632

National Geographic Visualizing, The Big Bang Theory, 390; The Carbon Cycle, 134; Chemical Reactions, 493; Common Vision Problems, 712; The Conservation of Momentum, 537; Crystal Structure, 478; Descriptive and Experimental Research, 20; The Four-Stroke Cycle, 621; Human Cells, 69; Levers, 595; The Moon's Surface, 318; Natural Selection, 51; Nerve Impulses, 638; Newton's Laws in Sports, 565; Plate Boundaries, 193; Population Growth, 104; Secondary Succession, 152; The Solar System's Formation, 339; Synthetic Elements, 451; Tracer Elements, 422; Tsunamis, 215; Unconformities, 253; Unusual Life Forms, 283; Vision Defects, 712; Voltmeters and Ammeters, 675

Natural selection, 43 *lab*, 50, 51, 275, **275**–276, 276

NEAR spacecraft, 359

Nebula, 338, *338*, **382**–383, 385, *385*

Negative acceleration, 531, *531*

Negative charge, 636, *636*

Negative feedback, **80**, *80*, 80 *act*

Neodymium, 450, *450*

Neon, 447, *447*, 468, *468*

Nephron, 79

Neptune, 340, **342**, *342*, 355, *355*, 360–361 *lab*

Neptunium, 417, *417*

Nerve cells, 637, *638*

Nervous system, 71

Net force, **551**, 560

Neurotransmitters, *638*

Neutron(s), 257, **411**, 415, *415*, 636, *636*

Neutron star, 384

New moon, 313, *314*

Newton, Isaac, 552, 553 *act*

Newton (unit of force), 557

Newton's first law of motion, **552**–555, *565*

Newton's second law of motion, **556**–562, *565*; and air resistance, 561; and gravity, 557–558; using, *558*, 558–560, *560*

Newton's third law of motion, **563**–568, *565*, 569 *lab*

Niche, *109*, **109**–110

Nickel, 448

Nitrogen, 444, *444*; electron dot diagram of, 470

Nitrogen cycle, *132*, **132**–133, *133*

Nitrogen fixation, **132**, *132*

Nitrogen group, 444, *444*

Noble gases, **446**, *446*–447, *447*, 468, *468*

Nonconformity, 252, *253*

Nonmetals, **438**, *438*, 443, *443*, 444, *444*, 445, 446, *446*; noble gases, 468, *468*

Nonpolar molecules, 477, *477*

Normal fault, 196, 211, *211*

Northern lights, *377*, 377, 677, *677*

North Star (Polaris), 371, *371*

Nuclear energy, 619

Nuclear fusion, 338, *339*, 381–382, *382*

Nuclear radiation, 417

Nucleic acid, 66, *66*, 67, *67*

Nucleus, 257, **410**, 410–412, *411*, 411 *lab*, 412, 415, 415–416

Numbers, using, 419 *act*

Nutrient(s), 66, *66*, 67; carbohydrates, 66, *66*, 67; fats, 66, *66*, 67; minerals, 65; proteins, 66, *66*, 67; water, 66, 66 *act*

Ocean(s), age of, 292 *lab*

Ocean floor, mapping, 186; spreading of, *186*, 187–188, 189 *lab*

Ocean water, 94, *94*

Ochoa, Ellen, 26

Ohm (unit of resistance), 646

Ohm's law, 649

Olympic torch, 503, *503*

Omnivores, *107*, 107, 137, *137*

One-step equations, 582 *act*, 583 *act*, 587 *act*, 589 *act*

Oops! Accidents in Science, It Came from Outer Space, 362; What Goes Around Comes Around, 542; The World's Oldest Fish Story, 264

Oort, Jan, 356

Oort Cloud, 356

Orbit, of Earth, 309; of planets, 340, 341 *lab*; of satellite, 561; weightlessness in, *568*, 568

Organ(s), *70*, **70**–71

Organic compounds, **66**–67

Organic evolution, **274**, 274–276, *275*, *276*

Organism(s), *70*, 70–71

Organization, *64*, 64; cells, *68*, 68–71, *69*, *70*, 72 *lab*; chemical substances in, 64–67, *66*, *67*; organs, *70*, 70–71; organ systems, 71; tissues, 70, *70*

Organ systems, 71; interactions of, *73*, 73–83, 82–83 *lab*

Orion, 370, *370*, 371

Ortelius, Abraham, 182

Oscillating model of universe, 388

Osmium, 449

Outer planets, 338, 348–353, *355*; Jupiter, 340, *348*, 348–349, 355, *355*, 360–361 *lab*; Neptune, 340, 352, *352*, 355, *355*, 360–361 *lab*; Pluto, 340, 352, 353, *353*, 355, *355*, 360–361 *lab*; Saturn, 340, 350, *350*, 355, *355*, 360–361 *lab*; Uranus, 340, 351, *351*, 355, *355*, 360–361 *lab*

Output force, **586**, *586*

Oxygen, on periodic table, 445, *445*; and respiration, 76, *77*, 123; use in body, 67

Oxygen family, 445, *445*

Ozone, 445

Pacific Ring of Fire, 228, 229

Paleozoic Era, *282*, **282**–286, *284*, 284 *act*, 285, 286

Palladium, 449

Pancreas, 80, *80*

Pangaea, *182*, **182**, 183, **279**, *279*, 286, 288, *288*

Parallax

Parallax, 373, *373*, 392–393 *lab*
Parallel circuit, **651**, *651*, 655 *lab*
Parasitism, **108**, *108*
Particle(s), alpha, *408*, 409, *409*, 410, 412, **417**; beta, **418**; charged, 407–409
Particle accelerator, 421, *421*, 451, *451*
Particle size, and rate of reaction, 506, *506*
Penguins, 125, *125*
Penicillin, 75
Pennsylvanian Period, 284
Penumbra, 315, 316
Percentages, 47 *act*, 291 *act*, 346 *act*
Period, 273, *273*, **435**. *See also* names of individual periods
Periodic pattern, making models of, 433 *lab*
Periodic table, 432–455, *436–437*; boron family on, 442, *442*; carbon group on, 443, *443*; designing, 435 *lab*; development of, *434*, 434–435; element keys on, 439, *439*; and energy levels of electrons, 466–467, *467*; halogen family on, 468, *468*; halogens on, 446, *446*; in identifying properties of elements, 469 *act*; metalloids on, 438, *438*, 442, 443, *443*, 445, 446; metals on, 438, *438*, 441, 441–442, *442*, 443, 448–452; nitrogen group on, 444, *444*; noble gases on, 446, 446–447, *447*, 468, *468*; nonmetals on, 438, *438*, 443, *443*, 444, *444*, 445, 446, *446*; oxygen family on, 445, *445*; symbols for elements on, 440; zones on, 435, *435*, *438*, 438–440, *439*
Permafrost, 155, *155*
Permian Period, 286
Permineralized remains, **244**, *244*
Perspiration. *See* Sweat
Phases of Moon, **313**, *314*, 321 *lab*
Phenotype, **39**; and environment, 40, 40–42, *41*, *42*; of fruit flies, 46 *lab*
Phobos (moon of Mars), 347, *347*
Phosphorus, 444, *444*; in bones, 65
Phosphorus-32, 423

Photosphere, **375**, *375*
Photosynthesis, 106, 108, 123, 124, *124;* and energy, 136; and respiration, 135 *lab*
Physical changes, **492**, *492*, 509 *lab*
Physical properties, magnetic, 448
Pictures, inferring from, 9 *lab*
Pigeons, 52, *52*
Pioneer species, **150**, *151*
Pistons, 620, *621*
Pitch, **703**
Plains, 161
Planaria, 111 *lab*
Planet(s). *See also* individual planets; distances between, 336–337, *344*, 360–361 *lab*; formation of, 338; inner, 338, 342–347, *354*, *354;* modeling, 350 *lab*; moons of. *See* Moon(s); motions of, 340, 341 *lab*, 351, *351;* orbital speed of, 340; orbits of, 340, 341 *lab*; outer, 338, 348–353, *355*, 360–361 *lab*; ring systems of, 348, *348*, 350, *350*, 351
Planetariums, 335 *lab*
Plant(s), chlorophyll in, 495; and competition, 99 *lab*; as evidence of continental drift, *183*, 184, *184;* growth of, 40, *40*, 40 *lab*, 140–141 *lab*; houseplants, 149 *lab*; leaves of, 41, *41*, 495; movement of, 103, *103;* and nitrogen fixation, 132, *132;* photosynthesis in, 106, 123, 124, *124*, 135 *act*, 136; and poison, 109; seed, 290; stems of, 40 *lab*
Plate(s), **190**, *191;* collisions of, *193*, 194, 285, *285;* composition of, 190, *190;* and earthquakes, 229–231, *231;* movement of, 226, *226*, *227*, *231;* and volcanoes, 219, *219*, *226*, 226–228, *227*
Plate boundaries, 191, *191;* convergent, 192, 193, *193*, 194, 197, 228; divergent, 191, 193, *193*, 227; transform, 194, *194*
Platelets, 69, 81
Plate tectonics, 180, **190**–201; causes of, 195, *195;* and Earth history, 279, *279*, 285, *285;*

features caused by, *196*, 196–198, *197*, *198;* predicting activity, 200–201 *lab*; testing for, *198*, 198–199
Platinum, 449
Platinum group, 449
Pluto, 340, 352, 353, *353*, 355, *355*, 360–361 *lab*
Plutonium, 450
Poisons, 109, 435
Polar bears, 616
Polar bond, **476**, *476*
Polaris (North Star), 371, *371*
Polar molecules, **476**, 476–477, *477*, 477 *act*
Polar regions, 94, *94*, 96
Pole(s), of Earth, 125; magnetic, **666**, *666*, 667, 669 *act*; of Moon, 325; South, 142
Pollution, thermal, **617**, *617;* of water, 165, *165*, 172, *172*
Polonium, 445
Ponds, *164*, 164–165
Population(s), **96**; biotic potential of, 102, 103 *lab*; data on, 96 *act*; growth of, 102–105, *103*, *104*, *105*, 112–113 *lab*; movement of, 103, *103;* size of, 99, 99–102, *100*
Population density, **99**, *99*
Position, changing, *522*, 522–523
Positive acceleration, **531**
Positive charge, **636**, *636*
Positive feedback, 81, *81*
Potassium, 65, 441, 469, *469;* use in body, 67
Power, **583**–584; calculating, 583, 583 *act*; electric, **652**–653; and energy, 584; equation for, 583; of pulley, 598–599 *lab*; and work, 583 *lab*
Power plants, 679, *679*, 679 *act*
Prairies, 161; life on, 96
Precambrian time, **280**, 280–282, *281*
Precipitation, extreme amounts of, 142; and land, 154
Predators, 43 *lab*, 50, *50*, 110, *110*
Prey, 43 *lab*, 50, *50*, 110, *110*
Primary succession, **150**, 150–151, 153
Primordial soup, 56
Principle of superposition, **250**, *250*

Index

Principle of uniformitarianism, 261, *261*
Problem solving, 6, *6*, 8, *8*, 12 *lab*, 13, *13*, 14 *act*
Producers, 106, *106*, 120, 123, 137, *137*
Product, 494, 495, *496*, 497
Project Apollo, 322
Prominences, 376, *377*
Properties, of light waves, 708, *708*; magnetic, 448; of waves, 693 *lab*, *696*, 696–698, *697*
Prostheses, 600, *600*
Protactinium, 450
Proteases, 508, *508*
Proteins, 66, *66*, 67
Proterozoic Eon, 280
Proton(s), 257, **410,** 636, *636*
Proxima Centauri, 373, 378
Ptarmigan, 155
Pulley, 596 *lab,* **596**–599; fixed, 596, *597*; movable, 597, *597*; power of, 598–599 *lab*
Pulley system, 597, *597*, 598–599 *lab*
Punnett square, 47, *48*
Pyramids, 579 *lab*, 585 *lab*
Pyroclastic flows, 220, *220*, 221

Quartz, 443
Quasars, 384
Quaternary Period, 292

Rabbits, 99, 100
Raccoons, 50, *50*
Radiant energy, 619
Radiation, 613; adaptive, **52;** and food, 426, *426*; nuclear, 417; from Sun, 310
Radiation therapy, 423
Radioactive decay, 257–258, *258*, 416 *act,* **416**–420, *417*, *418*, 424–425 *lab*
Radioactive elements, 450, *451*
Radioactive isotopes, 421 *act,* 421–423, *422*
Radioactive wastes, 420

Radiocarbon dating, 258, *258*, 258 *lab,* 259, *259*, 260 *act*, 420, *420*
Radiometric dating, 259, **259**–260, *260*
Radio waves, 709
Radon, 447
Rain, extreme amounts of, 142
Rain forests, leaves in, 158 *lab;* life in, 94, *94;* temperate, **157,** *157;* tropical, 154, *158,* **158**–159, *159;* water in, 123
Rainier, Mount (Washington state), 223
Rain shadow effect, 128, *128*
Rasmussen, Knud, 686
Rate of reaction, 504, **504**–508, *505, 506*
Ratio, input coils/output coils, 681
Reactant, 494, 495, *496*, 497
Reaction, and action, 563–566, *564, 566;* chemical, 76 *lab*, 589. *See* Chemical reactions
Reaction rate, 504, **504**–508, *505, 506*
Reading Check, 10, 11, 15, 19, 22, 23, 25, 39, 41, 44, 46, 50, 66, 67, 68, 74, 79, 80, 94, 95, 99, 101, 107, 109, 125, 132, 135, 137, 139, 151, 158, 160, 165, 169, 182, 183, 186, 187, 191, 196, 197, 210, 212, 222, 223, 227, 228, 245, 247, 252, 254, 257, 260, 273, 276, 281, 282, 284, 289, 292, 307, 309, 313, 314, 320, 323, 324, 338, 345, 352, 358, 372, 376, 381, 383, 384, 387, 406, 408, 410, 411, 416, 418, 438, 442, 444, 446, 447, 448, 450, 452, 466, 467, 469, 471, 475, 476, 480, 494, 499, 503, 504, 523, 525, 532, 553, 554, 556, 580, 582, 587, 589, 593, 596, 609, 613, 615, 637, 642, 646, 650, 653, 667, 679, 680, 695, 697, 702, 703, 708, 709
Reading Strategies, 94A, 122A, 150A, 182A, 210A, 242A, 272A, 306A, 336A, 370A, 404A, 434A, 464A, 492A, 522A, 550A, 580A, 608A, 636A, 666A, 694A
Real-World Questions, 12, 28, 43, 54, 72, 82, 111, 112, 129, 140, 162, 170, 189, 200, 225, 232, 256, 262, 287, 294, 321, 326, 341, 360, 379, 392, 414, 424, 454, 481, 482,

509, 510, 539, 540, 569, 570, 585, 598, 618, 624, 655, 656, 672, 684, 706, 714
Recessive traits, 45
Recycling, 385
Red blood cells, 69
Red giants, 381, *381,* **383,** *383,* 384, *394, 394*
Red shift, 389, *389*
Reef, 94, *94,* 167, *167,* 167 *act*
Reflection, law of, **699,** *699;* of sound, 705; of waves, 699, *699,* 699 *lab*
Refraction, 699; of waves, 699, *699*
Refrigerators, 622, *622*
Regulation, of glucose levels, 80, *80*
Relative ages, 251–256, 256 *lab*
Relative dating, 251 *act*
Relative motion, 523, *523*
Representative elements, 435
Reproduction, artificial selection, 276, *276*
Reproductive system, 71
Reptiles, 284, *284,* 288, 296, *296*
Research, *7;* conclusions in, 19; data in, 18; descriptive, **13,** 14–15, *15,* 20; design of, 13, 15, *15, 21,* 21–23, *22, 23;* eliminating bias in, 15; equipment used in, 16, *17;* experimental, **13,** 20, 21, 21–23, *22, 23;* in medicine, 27, *27;* objective of, 14; scientific measurement in, 17; selecting materials for, 16, *16;* using models in, 16
Resistance, *646,* **646**–647, *647,* 648, *648,* 649
Respiration, cellular, **76,** *76;* and exercise, 82–83 *lab;* and oxygen, 76, *77,* 123; and photosynthesis, 135 *lab*
Respiratory system, 71, 76, *76;* and circulatory system, 73, *73*
Retina, 711, *711,* 713, *713*
Reverberation, 705
Reverse fault, 211, *211*
Revolution, 305 *lab,* 307, *309,* 312
Rhinoceros, 92, *92*
Rhodium, 449
Ribonucleic acid (RNA), 56, 67
Richter scale, 214, 216
Rift, 227

Rift valleys, 180, 191, *193*, 196
Rift zones, 228
Rigel, 372
Ring of Fire, 228, 229
River(s), 163–164
RNA (ribonucleic acid), 56, 67
Rock(s), absolute ages of, 257–261; dating, 247, *247*, 249, 259–260, *260*, 282 *lab*; as evidence of continental drift, 184; fossils in, 247, *247*, 248, 249, *249*, 282 *lab*; Moon, *320*, 323; and principle of superposition, 250, *250*; relative ages of, 251–256, 256 *lab*
Rockets, balloon, 569 *lab*; launching, 566, *566*
Rock layers, matching up, *254*, 254–255, *255*; unconformities in, 252, *252*, *253*
Rodriguez, Eloy, 26
Rods, of eye, 713, *713*
Rolling friction, 555, *555*
Rotation, 338; of Earth, 305 *lab*, *307*, **307**; of Moon, 312, *322*; of Uranus, 351, *351*
Roundworm, as parasite, 108, *108*
Rubidium, 469
Russell, Henry, 380
Rust(s), 500
Ruthenium, 449
Rutherford, Ernest, 408–409, 410

Saber-toothed cat, 296, *296*
Safe Drinking Water Act of 1986, 15
Safety, and air bags, 572, *572*; in automobiles, 540–541 *lab*, 572, *572*; in earthquakes, 209 *lab*, *217*, 217–218, 234; with electricity, 653–654; in experiments, 22
Salamanders, 157
Salt(s), 441, 446, *446*; in body, 66; bonding in, *472*, 472–473, *473*; movement of ions in, 637, *637*
Saltwater ecosystems, 166–169; coral reefs, 167, *167*, 167 *act*; estuaries, 168–169, *169*; oceans, 167; seashores, 168, *168*

San Andreas Fault, 194, *194*, 198, 234
Sand, 443
San Francisco earthquake (1906), 234, *234*
Satellite(s), 561
Satellite Laser Ranging System, 198, *198*
Saturn, 340, **350**, *350*, 355, *355*, 360–361 *lab*
Savannas, 161, *161*
Science, 6–29; in advertising, 7; communication in, *10*, 10–11; and problem solving, 6, *6*, 8, *8*, 12 *lab*, 13, *13*, 14 *act*; and technology, 9, *24*, 24–27, *25*, 28–29 *lab*
Science and History, Quake, 234; Synthetic Diamonds, 512; The Census Measures a Human Population, 114; The Mayan Calendar, 328; Pioneers in Radioactivity, 426
Science and Language Arts, "Aagjuuk and Sivulliit" (Rasmussen), 686; "Anansi Tries to Steal All the Wisdom in the World", 456; "Baring the Atom's Mother Heart" (Awiakta), 484; The Everglades: River of Grass, 30; Listening In, 202
Science and Society, Air Bag Safety, 572; Bionic People, 600; Creating Wetlands to Purify Wastewater, 172; Fire in the Forest, 658; Food for Thought, 426; The Heat Is On, 626; How Did Life Begin?, 56
Science Online, Automobile Engines, 620; Birth and Death Rates, 102; Changing Gender, 42; Chemical Equations, 497; Collisions, 536; Compasses, 671; Continental Drift, 183; Coral Reefs, 167; Correlating with Index Fossils, 254; Cost of Electrical Energy, 653; Different Species, 52; Disease Control, 8; Earthquake Magnitude, 216; Earthquakes and Volcanoes, 192; Eclipses, 315; Electrons, 466; Elements, 438; Ellipses, 309; Eutrophication, 151; Evolution of Stars, 382; The Far Side, 323;

Galileo and Newton, 553; Health Risks, 452; Historical Tools, 587; How Birds Fly, 564; Human Population Data, 96; Isotopes in Ice Cores, 260; Isotopes in Medicine and Agriculture, 421; James Watt, 584; Land Speed Record, 526; Lasers, 708; Life Processes, 135; Mars Exploration, 346; Montserrat Volcano, 221; Negative Feedback, 80; Olympic Torch, 503, *503*; Paleozoic Life, 284; Polar Molecules, 477; Power Plants, 679; Radioactive Decay, 416; Relative Dating, 251; Seasons, 311; Solar System, 337; Space Weather, 377; Student Scientists, 25; Superconductors, 640; Warm Versus Cold, 289; Water, 66; Weather Data, 127
Science Stats, Astonishing Human Systems, 84; Extinct!, 296; Extreme Climates, 142; Stars and Galaxies, 394
Science writer, 165
Scientific information, use of, 27
Scientific measurement, 17
Scientific Methods, 12, **13**–23, *14*, 28–29, 43, 54–55, 72, 82–83, 111, 112–113, 129, 140–141, 162, 170–171, 189, 200–201, 225, 232–233, 256, 262–263, 287, 294–295, 321, 326–327, 341, 360–361, 379, 392–393, 414, 424–425, 453, 454–455, 481, 482–483, 509, 510–511, 539, 540–541, 569, 570–571, 585, 598–599, 618, 624–625, 655, 656–657, 672, 684–685, 706, 714–715; Analyze Your Data, 6, 18, 23, 55, 83, 113, 171, 201, 263, 295, 393, 425, 455, 483, 510, 541, 569, 571, 599, 625, 715; Conclude and Apply, 19, 29, 43, 55, 72, 83, 111, 113, 129, 141, 162, 171, 189, 201, 225, 233, 256, 287, 295, 321, 327, 341, 361, 379, 393, 414, 425, 455, 481, 483, 509, 511, 539, 541, 569, 571, 585, 599, 618, 625, 655, 657, 672, 685, 706, 715; Follow Your Plan, 29, 83,

425, 455, 510, 571, 599, 625; Form a Hypothesis, 21, 82, 112, 200, 392, 424, 540, 570, 598, 624; Make a Plan, 29, 83, 425, 455, 510, 571, 599, 625; Make the Model, 361, 483; Making the Model, 263; Planning the Model, 263; Plan the Model, 361, 483; Test Your Hypothesis, 83, 113, 201, 295, 361, 392–393, 425, 541, 570, 625
Scorpion, 160
Screw, 593, *593*
Sea anemone, 108, *108*
Seafloor spreading, *186,* **187**–188, 189 *lab*
Seashores, 168, *168*
Seasons, *309,* 309–310, *310,* 311 *act,* 326–327 *lab;* on Mars, 346
Sea stars, *168*
Secondary succession, 151, *152,* 153
Second-class lever, *595*
Secretion, 78
Seed(s), movement of, 103, *103*
Seedling competition, 99 *lab*
Seed plants, angiosperms, 290; gymnosperms, 290
Seismic-safe structures, 209 *lab,* **217,** 217–218, 234
Seismic waves, *212,* **212**–213, *214, 230, 230,* 230 *act,* 232–233 *lab,* 317, 696, 697
Seismograph, 213, *213,* 214, *214*
Selection, artificial, 276, *276;* natural, *275,* **275**–276, *276*
Selective breeding, 52, *52*
Selenium, 445, *445*
Semiconductors, 443, *443*
Sense(s), hearing, *493,* 703, 704; sight, *493;* smell, *493;* taste, *493;* touch, *493;* vision, 711–713, *712*
Series circuit, 650, *650*
Shark, *243*
Shear forces, 211
Sheep, 103
Shield volcano, 222, *222*
Shock, electric, 653–654
SI (International System of Units), 17
Sight. *See* Vision
Silicon, 443, *443*
Silver tarnish, *480,* 480, 497

Simple machines, 591–599; inclined plane, 591–593, *592;* lever, 594, *594,* 595, *595;* pulley, 596 *lab,* 596–599, *597,* 598–599 *lab;* screw, 593, *593;* wedge, *592,* 592–593, *593;* wheel and axle, 594, *594,* 596, *596*
Sirius, 370, 372
Skeletal muscles, 69
Skeletal system, 71, 84
Sliding friction, 553, 554, *554, 555,* 559, 562
Smell, *493*
Smoke detectors, 417, *417*
Smooth muscles, 69
Sodium, 65, 469, *472,* 472–473, *473*
Sodium bicarbonate, 494, 495, 496
Sodium chloride, 66, 441, 446, *446*
Soil, 124; as abiotic factor in environment, 124, 124 *lab,* 129 *act;* building, 150–151, *151;* determining makeup of, 124 *lab;* nitrogen in, 133, *133;* in tropical rain forests, 158–159
Sojourner robot rover, 345
Solar eclipse, 314, *314,* 315, *315*
Solar flares, 376, *377*
Solar system, 334–362, **337,** 337 *act;* asteroids in, *358,* 358–359, *359;* comets in, 353, *356,* 356–357, *357;* distances in, 336–337, *344,* 360–361 *lab;* formation of, *338,* 338–339, *339;* inner planets of, 338, 342–347, 354, *354;* meteors in, *357,* 357–358; models of, 336–337, 360–361 *lab;* outer planets of, 338, 348–353, *355,* 360 *lab;* planetary motions in, 340, 341 *lab,* 351, *351*
Solid(s), movement of electrons in, 637, *637*
Solstice, 310, *310*
Solution(s), movement of ions in, 637, *637*
Soufrière Hills volcano (Montserrat), 219, *219,* 220, *220,* 221, 221 *act,* 224, 227
Sound, intensity of, 702, *702,* 703; loudness of, *702,* 702–703, *703;* pitch of, 703; reflection of, 705; speed of, 698 *act,* 702

Sound waves, 692, 698 *act,* 701–706; as compressional waves, 701, *701;* frequency of, 703; making, 701, *701;* in matter, 706 *lab*
South Pole, 142
Space, distance in, 372 *act,* 373, 388 *lab;* light waves in, 707, *707;* measurement in, 373, 388 *lab;* weather in, 375 *act*
Space exploration, of Jupiter, 348; of Mars, 344–346, 346 *lab;* of Mercury, 342; of Moon, *322,* 322–325; of Neptune, 352; of Pluto, 353; of Saturn, 350; of Uranus, 351; of Venus, 343
Space probes, *Cassini,* 350; *Galileo,* 348, 359; *Global Surveyor,* 345, 346; *Magellan,* 343, *343; Mariner,* 342, 343; *Mars Odyssey,* 345; *Mars Pathfinder,* 345; *NEAR,* 359; *Stardust,* 357; *Viking,* 345, 346; *Voyager,* 348, 349, *349,* 350, 351, 352
Space shuttle, 568, *568*
Species, 274, *274;* changing, 287 *lab;* differences within, *51,* 52 *act;* and environment, 50–53, *51, 52, 53,* 54–55 *lab;* extinction of, 53, *53, 286,* 286, 288, 291 *act, 296,* 296; natural selection within, 276, *276;* new, 276; pioneer, **150,** *151*
Specific heat, 616
Spectroscope, 374
Spectrum, electromagnetic, **708**–709, *709;* of star, 374, *374, 389, 389*
Speed, 524–525; and acceleration, *528,* 528–529; of animals, 524; average, **525,** *525,* 525 *lab;* calculating, 524 *act;* constant, 525, *525;* and distance-time graphs, 526, *526;* of heating and cooling, 618 *lab;* instantaneous, **525,** *525;* land, 526 *act;* of light, 707; and motion, 524–525, *525;* of sound, 698 *act,* 702; and velocity, 527; of waves, 698
Speed-time graph, 532, *532*
Sphere, *306,* **306**–307

Index

Spiders, 109
Spiral galaxy, *386*, 386–387
Sports, Newton's laws in, *565*
Standardized Test Practice, 34–35, 60–61, 88–89, 118–119, 146–147, 176–177, 206–207, 238–239, 268–269, 300–301, 332–333, 366–367, 398–399, 430–431, 460–461, 488–489, 516–517, 546–547, 604–605, 630–631, 662–663, 690–691, 720–721
Star(s), 370–374; absolute magnitude of, 372; apparent magnitude of, 372; binary, 378; classifying, 380–381; constellations of, *370*, 370–371, *371*; evolution of, 382 *act*, 382–385, *383*; fusion reaction in, 381–382, *382*; life cycle of, 382–385, *383*; main sequence, *380*, 380–381, 382, 383; neutron, **384**; patterns of, 371 *lab*; properties of, 374, *374*; spectrum of, 374, *374*, 389, *389*; Sun as, 375, 378; triple, 378
Star cluster, 378, *378*
Stardust **spacecraft,** 357
Static charge, 637, *637*
Static friction, 554
Steady state theory, 388
Steel, 448, *448*
Stem(s), 40 *lab*
Stirrup (of ear), 704, *704*
Stream(s), *163*, 163–164
Strike-slip fault, *198*, 198, 211, *211*
Stromatolites, *270*, 270, 281, *281*
Strong nuclear force, 416
Student scientists, 25 *act*
Study Guide, 31, 57, 85, 115, 143, 173, 203, 235, 265, 297, 329, 363, 395, 427, 457, 485, 513, 543, 573, 601, 627, 659, 687, 717
Subduction zones, 192, 194, 228
Subscripts, 479, 496
Succession, 150–153, *152*; primary, 150–151, *151*, 153; secondary, 151, *152*, 153
Sugars, in blood, 80, *80*
Sulfur, 65, 445
Sulfuric acid, 445
Sun, 375–379, 385; atmosphere of, 375, *375*; as center of solar system, 337; corona of, 375, *375*; distance from Earth, 307, 309, 360–361 *lab*; and Earth's rotation, 307; eclipse of, 314, *314*, 315, *315*; electromagnetic waves from, 710, *710*; energy from, 120; layers of, 375, *375*; origin of, 338, *339*; radiation from, 310; as star, 375, 378; surface features of, *376*, 376–377, *377*; temperature of, 375
Sunlight, ultraviolet waves in, 710
Sunset Crater, Arizona, 222, *222*
Sunspots, **376**, *376*, 379 *lab*
Superconductors, 640 *act*, 681, 681–682, *682*
Supergiants, 381, *381*, **384**
Supernova, 384
Superposition principle, **250**, *250*
Survival, and environment, 49, 49–50, *50*
Sweat, 78, 79, *79*
Symbiosis, **108**, *108*
Symbols, for atoms, 479, *479*; for compounds, 479, *479*; for elements, 440
Synthetic elements, 421–423, *422*, 450, *451*
System(s), *See* Body systems

Taiga, **156**, *156*
Tarnish, 480, *480*, 497
Taste, 493
Technetium-99, 423
Technology, 9, *24*, 24–27, *25*, 28–29 *lab*. *See also* Space probes; advances in, 25, *25*; air conditioners, 623; ammeter, 674, *675*; bicycle, 591, *591*; catalytic converters, 507, *507*; cathode-ray tube (CRT), 406, *406*, 407, *407*; circuit breakers, 651, *651*; compass, 308, 308 *lab*, 666, 671, 671 *act*, 672 *lab*; computers, 443, *443*; in dentistry, 452; electric meter, 653, *653*; electric motors, 676, *676*, 684–685 *lab*; electromagnets, 673, 673–674, 674, 674 *lab*; fingerprinting, 37 *lab*; fireworks, 502, *502*; fuses, 651, *651*; galvanometer, 674, *675*; generators, 678, 678–679; *Glomar Challenger* (research ship), 187; heat pumps, 623, *623*; Hubble Space Telescope, 352, 353, 357, 384, 391; internal combustion engines, 620, *620*, 620 *act*, 621; Internet, 28–29 *lab*; lasers, 708 *act*; lightbulb, 449, *449*, 468; lightning rod, 642, *642*; maglev, 664, *664*, 665; magnetic resonance imaging (MRI), 682, 682–683, *683*; magnetometer, 188; particle accelerator, 421, *421*, 451, *451*; power plants, 679, *679*, 679 *act*; for predicting earthquakes, 218, *218*; pyramids, 579 *lab*, 585 *lab*; radiation therapy, 423; refrigerators, 622, *622*; rockets, 566, *566*, 569 *lab*; Satellite Laser Ranging System, 198, *198*; satellites, 561; seismograph, 213, *213*, 214, *214*; semiconductors, 443, *443*; smoke detector, 417, *417*; space shuttle, 568, *568*; spectroscope, 374; superconductors, 681, 681–682, *682*; synthetic elements, 421–423, *422*, 450, *451*; testing for plate tectonics, 198–199, *199*; thermometers, 609, *609*; transformers, 680, 680–681; Tsunami Warning System, 215; voltmeter, 674, *675*; welding torch, 499
Tectonic plates. *See* Plate(s); Plate tectonics
Tectonics. *See* Plate tectonics
Teeth, dentistry, 452; of herbivores and carnivores, 593, *593*
Telescopes, *Hubble*, 352, 353, 357, 384, 391
Tellurium, 445
Temperate deciduous forests, 154, 156, **156**–157
Temperate rain forests, **157**, *157*

Index

Temperature, 607 lab, 608, **608**–610; as abiotic factor in environment, 125, 125–126, 126; of body, 589; converting measures of, 166 act; and elevation, 126, 126, 126 act; extreme, 142, 142; of human body, 79, 79; and land, 154; measuring, 609, 609–610; of oceans, 166; and rate of reaction, 504, 504–505, 505; of Sun, 375; and thermal energy, 611, 611; and tilt of axis, 326–327 lab
Temperature scales, Celsius, 609, 609–610, 610 act; converting, 610, 610 act; Fahrenheit, 609, 609–610, 610 act; Kelvin, 610
Teosinte, 38, 38
Tephra, 219, 223
Termites, 109, 109
Thermal conductors, 615
Thermal energy, 606–626, **609**, 611, 619, 619, 621; and heat, 612–615; and temperature, 611, 611; transfer of, 612–614, 613, 614
Thermal expansion, 609, 609
Thermal insulators, 616, 616, 624–625 lab
Thermal pollution, 617, 617
Thermometer, 5 lab, 609, 609
Third-class lever, 595
Thomson, J. J., 407–408, 410
Thorium, 450
Thyroid gland, 422, 423
TIME, Science and History, 114, 234, 328, 426, 512; Science and Society, 56, 172, 426, 572, 600, 626, 658
Tin, 443
Tissues, 69, 70, 70
Titan (moon of Saturn), 350
Titania (moon of Uranus), 351
Tools, historical, 587 act
Tornadoes, 142
Touch, 493
Trace fossils, 246, 246, 262–263 lab
Tracer elements, 421–423, 422
Tracks, of dinosaurs, 246, 246
Traits, 36–55, **38**; dominant, **45**, 45; and environment, 40, 40–41, 41, 42, 49–53; and genetics, 44–48; and mutations, 52; observing, 38, 38; predicting, 44–48; recessive, **45**
Transformer, 680, 680–681
Transform plate boundaries, 194, 194
Transition elements, 435, 448, 448–452, 449; in dentistry, 452; inner, 450, 450, 451
Transmutation, 416–418, 421, 421
Transpiration, 130
Transverse waves, 695, 695, 696, 697, 697
Trials, 22
Triassic Period, 288, 288, 289
Triceratops, 242
Trilobites, 272, 272, 277, 277–278, 278
Triple bonds, 476, 476, 477
Triple stars, 378
Triton (moon of Neptune), 352, 352
Tropical rain forests, 154, 158, **158**–159, 159; life in, 94, 94
Try at Home Mini Labs, Assembling an Electromagnet, 674; Calculating the Age of the Atlantic Ocean, 292; Comparing Paper Towels, 18; Comparing Rates of Melting, 614; Comparing the Sun and the Moon, 313; Constructing a Model of Methane, 475; Determining Soil Makeup, 124; Identifying Inhibitors, 506; Interpreting Fossil Data, 184; Investigating the Electric Force, 644; Measuring Average Speed, 525; Modeling an Eruption, 220; Modeling the Nuclear Atom, 411; Modeling Planets, 350; Modeling Rain Forest Leaves, 158; Modeling a Shaded-Impact Basin, 323; Observing Friction, 554; Observing Gravity and Stem Growth, 40; Observing Seedling Competition, 99; Observing Star Patterns, 371; Predicting Fossil Preservation, 243; Refraction of Light, 699; Work and Power, 583
Tsunami, 215, 216
Tube worms, 187
Tundra, 155, 155
Tungsten, 449, 449
Tyrannosaurus rex, 243

Ultraviolet waves, 710
Umbra, 315, 316, 316
Unbalanced forces, 551
Unconformities, 252, 252, 253
Uniformitarianism, 261, 261
Universe, expansion of, 369 lab, 388–389, 388–389, 391; origin of, 388, 390, 391
Uranium, 415, 416, 420, 450, 451
Uranus, 340, **351,** 351, 355, 355, 360–361 lab
Urinary system, 71, 78, 78–79
Ursa Major, 371, 371
Use the Internet, Discovering the Past, 294–295; Exploring Wetlands, 170–171; Health Risks from Heavy Metals, 454–455; Predicting Tectonic Activity, 200–201; When is the Internet the busiest?, 28–29
Uterus, 81, 81

Variables, dependent, **17**; independent, **17**
Veins, modeling blood flow in, 63 lab
Velocity, 527; and acceleration, 528, 528–529, 529; and speed, 527
Venera space probe, 343
Venus, 95, 336, 337, 340, **343,** 354, 354, 360–361 lab
Vertebrate animal(s), amphibians, 284; early, 282, 282, 284; mammals, 290, 290, 293, 293; reptiles, 284, 284, 288, 296, 296
Viking space probes, 345, 346
Villi, 75, 75
Vinegar, 494, 495, 496

Visible light, 709, *709*
Vision, *493,* 711–713, *712*
Vision defects, *712*
Volcano(es), 219–225, in early Earth history, *280,* 285, *285;* and Earth's plates, 219, *219,* 226, 226–228, *227;* eruptions of, 220, *220,* 220 *lab,* 221, 223, 224, 225 *lab;* formation of, *219,* 219–221, 227–228, *228;* forms of, 221–223, *222, 223;* on other planets, 344, *344,* 349, *349;* and plate tectonics, 192 *act,* 194, 197; risks of, 221, *221*
Volcanologist, 202
Voltage, 644, *644,* 648, 649, 649 *act,* 656–657 *lab,* 679; changing, *680,* 680–681
Voltmeter, 674, *675*
Volume, calculating, 77 *act*
***Voyager* space probes,** 348, 349, *349,* 350, 351, 352

Wallace, Alfred Russell, 50
Waning, 314, *314*
Waste(s), eliminating, *78,* 78–79; production of, 76, 79; radioactive, 420
Wastewater, purifying, 172, *172*
Water. *See also* Aquatic ecosystems; as abiotic factor in environment, *122,* 123, *123;* from hydrothermal vents, 137; as limiting factor in ecosystem, 100; in living things, 66, 66 *act;* molecules of, 477, *477;* pollution of, 165, *165,* 170–171 *lab,* 172, *172;* use of, 131

Water cycle, *130,* **130**–131, *131*
Water waves, 692
Watt (W), 583
Watt, James, 583, 584 *act*
Wave(s), 694–700; amplitude of, 697, *697;* changing direction of, *699,* 699–700, *700;* compressional, **695,** *695,* 696, *696,* 697, *697,* 701, *701;* diffraction of, 700, *700;* electromagnetic, 696, 707, 708, *708,* 710, *710;* electrons as, 413; and energy, 694, *694;* frequency of, 696; infrared, 709; microwaves, 709; properties of, 693 *lab,* 696, 696–698, *697;* radio, 709; reflection of, 699, *699,* 699 *lab;* refraction of, 699, *699;* seismic, 212, **212**–213, *214,* 230, 230 *act,* 231, 232–233 *lab,* 317, 696, 697; sound. *See* Sound waves; speed of, 698; transverse, **695,** *695,* 696, 697, *697;* tsunami, *215,* **216;** types of, 694–696, *695;* ultraviolet, **710;** visible light, 709, *709;* water, 692
Wavelength, 696, *696;* of light, 707
Waxing, 314, *314*
Weather, 127 *act;* in space, 375 *act*
Wedge, *592,* **592**–593, *593*
Wegener, Alfred, 182, 183, 184, 185
Weight, 557–558; and mass, 558; measuring, 567, *567*
Weightlessness, *567,* **567**–568, *568*
Welding, 499
Wetlands, 165, *165,* 170–171 *lab,* 172, *172*
Whales, 293
Wheel and axle, 594, *594,* 596, *596*
White blood cells, 69
White dwarf, 381, **383,** *383*
Wildebeests, *100*

Wildfires, *148,* 151 *act;* benefits of, 148, *152;* in forests, 658, *658*
Williams, Daniel Hale, 26
Wind, 127, *127,* 142
Wire, copper, 646; electric, 640, 646, *646*
Wombat, 293, *293*
Woodpeckers, 97, *97,* 98, *98*
Work, 580–582; calculating, 582, 582 *act;* and distance, 582, 588, *588;* and energy, 584; equation for, 582; and force, 579 *lab,* 581, *581,* 585 *lab,* 587; measuring, 584; and mechanical advantage, *586,* 586–588; and motion, *580,* 580–581, *581;* and power, 583 *lab*

Xenon, 447
X rays, 710

Yttrium oxide, 450

Zewail, Ahmed H., 467, 484
Zinc, 65, 645

Credits

Magnification Key: Magnifications listed are the magnifications at which images were originally photographed.
LM–Light Microscope
SEM–Scanning Electron Microscope
TEM–Transmission Electron Microscope

Acknowledgments: Glencoe would like to acknowledge the artists and agencies who participated in illustrating this program: Absolute Science Illustration; Andrew Evansen; Argosy; Articulate Graphics; Craig Attebery represented by Frank & Jeff Lavaty; CHK America; John Edwards and Associates; Gagliano Graphics; Pedro Julio Gonzalez represented by Melissa Turk & The Artist Network; Robert Hynes represented by Mendola Ltd.; Morgan Cain & Associates; JTH Illustration; Laurie O'Keefe; Matthew Pippin represented by Beranbaum Artist's Representative; Precision Graphics; Publisher's Art; Rolin Graphics, Inc.; Wendy Smith represented by Melissa Turk & The Artist Network; Kevin Torline represented by Berendsen and Associates, Inc.; WILDlife ART; Phil Wilson represented by Cliff Knecht Artist Representative; Zoo Botanica.

Photo Credits

Cover PhotoDisc; **i** (bkgd)Arnulf Husmo/Getty Images, (l)Georgette Douwma/Getty Images, (r)John Lawrence/Getty Images; **ii** (bkgd)Arnulf Husmo/Getty Images, (l)Georgette Douwma/Getty Images, (r)John Lawrence/Getty Images; **vii** Aaron Haupt; **viii** John Evans; **ix** (t)PhotoDisc, (b)John Evans; **x** (l)John Evans, (r)Geoff Butler; **xi** (l)John Evans, (r)PhotoDisc; **xii** PhotoDisc; **xiii** Marian Bacon/Animals Animals; **xiv** (t)Dwight Kuhn, (b)John Kaprielian/Photo Researchers; **xv** Kevin West/AP/Wide World Photos; **xvi** (t)Pat O'Hara/CORBIS, (b)JPL; **xvii** Peter Menzel/Stock Boston; **xviii** Richard Megna/Fundamental Photographs/Photo Researchers; **xix** Philip Bailey/The Stock Market/CORBIS; **xx** AFP Photo/Hector Mata/CORBIS; **xxi** Matt Meadows; **xxii** Jeff J. Daly/Visuals Unlimited; **xxiii** Alexis Duclos/Liaison/Getty Images; **xxvi** Otto Hahn/Peter Arnold, Inc.; **xxvii** Bob Daemmrich; **1** R. Arndt/Visuals Unlimited; **2** (l)Sinclair Stammers/Science Photo Library/Photo Researchers, (r)James D. Wilson/Woodfin Camp & Assoc.; **2–3** (bkgd)WT Sullivan III/Science Photo Library/Photo Researchers; **4–5** TEK Image/Science Photo Library/Photo Researchers; **6** (tr)Stephen Webster, (others)KS Studios; **7 8** Aaron Haupt; **9** (t)Bob Daemmrich, (bl)Paul A. Souders/CORBIS, (br)KS Studios; **10** Aaron Haupt; **11** Geoff Butler; **14** KS Studios; **15** (l)Icon Images, (r)F. Fernandes/Washington Stock Photo; **16** Aaron Haupt; **17** Matt Meadows; **18** Doug Martin; **19** Aaron Haupt; **20** (tr)courtesy IWA Publishing, (others)Patricia Lanza; **21** Amanita Pictures; **22** John Evans; **23** Jeff Greenberg/Visuals Unlimited; **24** KS Studios; **25** The Image Bank/Getty Images; **26** (tl)Sarita M. James, (tc)AFP/CORBIS, (tr)James D. Wilson/Liaison Agency/Getty Images, (c)Bob Rowan/Progressive Image/CORBIS, (cr)Fred Begay, (bl)NASA, (bcl)CORBIS, (bcr)Provident Foundation/CED Photographic Service, (cl br)Bettmann/CORBIS; **27** The Image Bank; **28** (tr)Dominic Oldershaw, (bl)Richard Hutchings; **29** Dominic Oldershaw; **30** Gary Retherford/Photo Researchers; **33** Amanita Pictures; **34** (l)Aaron Haupt, (r)Geoff Butler; **35** (l)Doug Martin, (r)Aaron Haupt; **36–37** Tim Davis/CORBIS; **38** (l)Runk/Schoenberger from Grant Heilman, (r)Alan Pitcairn from Grant Heilman; **40** Hermann Eisenbeiss/Photo Researchers; **41** (t)Alan & Sandy Carey/Photo Researchers, (b)Runk/Schoenberger from Grant Heilman; **42** Marian Bacon/Animals Animals; **44** David Parker/Science Photo Library/Photo Researchers; **45** unknown; **48** Getty Images; **49** Ken Lucas/Visuals Unlimited; **50** Karl H. Maslowski/Photo Researchers; **51** (t)Mickey Gibson/Animals Animals, (others)Tui DeRoy/Bruce Coleman, Inc.; **52** (tl)Mark Stouffer/Animals Animals, (tc)Kenneth W. Fink/Photo Researchers, (tr)George F. Godfrey/Animals Animals, (bl)Cathy & Gordan/ILLG/Animals Animals; **54** Ian Adams; **55** David Woodfall/Stone/Getty Images; **56** (t) Keven Laubacher/FPG, (b) Michael Black/Bruce Coleman, Inc.; **57** Kenneth W. Fink/Photo Researchers; **58** Alan Pitcairn from Grant Heilman; **60** Runk/Schoenberger from Grant Heilman; **61** Bios (Klein & Hubert)/Peter Arnold, Inc.; **62–63** Paul A. Souders/CORBIS; **64** Barry L. Runk from Grant Heilman; **66** Laura Sifferlin; **67** Prof. Oscar Miller/Science Photo Library/Photo Researchers; **69** (cw from top)Ken Eward/Science Source/Photo Researchers, Cabisco/Visuals Unlimited, Eric Grave/Science Source/Photo Researchers, Robert Knauft/Biology Media/Photo Researchers, Quest/Science Photo Library/Photo Researchers, Science Vu/Visuals Unlimited, Stan Elems/Visuals Unlimited, Cabisco/Visuals Unlimited; **70 72 76** Bob Daemmrich; **78** Laura Sifferlin; **79** Rolf Bruderer/The Stock Market/CORBIS; **82** (t)Bob Daemmrich, (b)KS Studios; **83** KS Studios; **89** National Cancer Institute/Science Photo Library/Photo Researchers; **90–91** (bkgd)Lynn M. Stone; **91** (inset)Mark Burnett; **92–93** Joe McDonald/Visuals Unlimited; **94** (tr)Richard Kolar/Animals Animals, (l)Adam Jones/Photo Researchers, (c)Tom Van Sant/Geosphere Project, Santa Monica/Science Photo Library/Photo Researchers, (br)G. Carleton Ray/Photo Researchers; **95** (t)John W. Bova/Photo Researchers, (b)David Young/Tom Stack & Assoc.; **97** (l)Zig Leszczynski/Animals Animals, (r)Gary W. Carter/Visuals Unlimited; **100** Mitsuaki Iwago/Minden Pictures; **101** Joel Sartore from Grant Heilman; **103** (t)Norm Thomas/Photo Researchers, (b)Maresa Pryor/Earth Scenes; **104** (r)Bud Neilson/Words & Pictures/PictureQuest, (others)Wyman P. Meinzer; **106** (l)Michael Abbey/Photo Researchers, (r)OSF/Animals Animals, (b)Michael P. Gadomski/Photo Researchers; **107** (tcr)Lynn M. Stone, (bl)Larry Kimball/Visuals Unlimited, (bcl)George D. Lepp/Photo Researchers, (bcr)Stephen J. Krasemann/Peter Arnold, Inc., (br)Mark Steinmetz, (others)William J. Weber; **108** (t)Milton Rand/Tom Stack & Assoc., (c)Marian Bacon/Animals Animals, (b)Sinclair Stammers/Science Photo Library/Photo Researchers; **109** (tl)Raymond A. Mendez/Animals Animals, (bl)Donald Specker/Animals Animals, (br)Joe McDonald/Animals Animals; **110** Ted Levin/Animals Animals; **111** Richard L. Carlton/Photo Researchers; **112** (t)Jean Claude Revy/PhotoTake, NYC, (b)OSF/Animals Animals; **113** Runk/Schoenberger from Grant Heilman; **114** Eric Larravadieu/Stone/Getty Images; **115** (l)C.K. Lorenz/Photo Researchers, (r)Hans Pfletschinger/Peter Arnold, Inc.; **116** CORBIS; **118** (l)Michael P. Gadomski/Photo Researchers, (r)William J. Weber; **120–121** Ron Thomas/Getty Images; **122** Kenneth Murray/Photo Researchers; **123** (t)Jerry L. Ferrara/Photo Researchers, (b)Art Wolfe/Photo Researchers; **124** (t)Telegraph Colour Library/FPG/Getty Images, (b)Hal Beral/Visuals Unlimited; **125** (l)Fritz Polking/Visuals Unlimited, (r)R. Arndt/Visuals Unlimited; **126** Tom Uhlman/Visuals Unlimited; **130** Jim Grattan; **133** (l)Runk/Schoenberger from Grant Heilman, (r)Rob & Ann Simpson/Visuals Unlimited; **136** WHOI/Visuals Unlimited; **140** Gerald and Buff Corsi/Visuals Unlimited; **141** Jeff J. Daly/Visuals Unlimited; **142** Gordon Wiltsie/Peter Arnold, Inc.; **143** (l)Soames Summerhay/Photo Researchers, (r)Tom Uhlman/Visuals Unlimited; **148–149** William Campbell/CORBIS Sygma; **150** Jeff Greenberg/Visuals Unlimited; **151** Larry Ulrich/DRK Photo; **152** (bkgd)Craig Fujii/Seattle Times, (l)Kevin R. Morris/CORBIS, (tr br)Jeff Henry; **153** Rod Planck/Photo Researchers; **155** (t)Steve McCutcheon/Visuals Unlimited, (bl)Pat O'Hara/DRK Photo, (br)Erwin & Peggy Bauer/Tom Stack & Assoc.; **156** (tl)Peter Ziminski/Visuals Unlimited, (c)Leonard Rue III/Visuals Unlimited, (bl)C.C. Lockwood/DRK Photo, (br)Larry Ulrich/DRK Photo; **157** (t)Fritz Polking/Visuals Unlimited, (b)William Grenfell/Visuals Unlimited; **158** Lynn M. Stone/DRK Photo; **160** (l)Joe McDonald/DRK Photo, (r)Steve Solum/Bruce Coleman, Inc.; **161** Kevin Schafer; **163** W. Banaszewski/Visuals Unlimited; **164** (l)Dwight Kuhn, (r)Mark E. Gibson/Visuals Unlimited; **165** James R. Fisher/DRK Photo; **166** D. Foster/WHOI/Visuals Unlimited; **167** (l)C.C. Lockwood/Bruce Coleman, Inc., (r)Steve Wolper/DRK Photo; **168** (tl)Dwight Kuhn, (tr)Glenn Oliver/Visuals Unlimited, (b)Stephen J. Krasemann/DRK Photo; **169** (l)John Kaprielian/Photo Researchers, (r)Jerry Sarapochiello/Bruce Coleman, Inc.; **170** (t)Dwight Kuhn, (b)John Gerlach/DRK Photo; **171** Fritz Polking/Bruce Coleman, Inc.; **172** courtesy Albuquerque Public Schools; **173** (l)James P. Rowan/DRK Photo, (r)John Shaw/Tom Stack & Assoc.; **177** (l)Leonard Rue III/Visuals Unlimited, (r)Joe McDonald/DRK Photo; **178** (l)Ken Lucas/TCL/Masterfile, (r)Ken Lucas/TCL/Masterfile; **178–179** James Watt/Earth Scenes; **179** (l)Patrice Ceisel/Stock Boston/PictureQuest, (r)Hal Beral/Photo Network/PictureQuest; **180–181** Bourseiller/Durieux/Photo Researchers; **184** Martin Land/Science Source/Photo Researchers; **187** Ralph White/CORBIS; **193** Davis Meltzer; **194** Craig Aurness/CORBIS; **196** Craig Brown/Index Stock; **197** Ric Ergenbright/CORBIS; **198** Roger Ressmeyer/CORBIS; **200** AP/Wide World Photos; **202** L. Lauber/Earth Scenes; **208–209** Reuters NewMedia Inc./CORBIS; **210** KS Studios; **213** (t)Krafft/Explorer/Photo Researchers, (b)Jean Miele/The Stock Market/CORBIS; **216** (bkgd)Galen Rowell/CORBIS, (others)NOAA; **217** (t)KS Studios, (b)Pacific Seismic Products, Inc.; **218** Roger Ressmeyer/CORBIS; **220** AP/Wide World Photos; **222** (t)Breck P. Kent/Earth Scenes, (b)Dewitt Jones/CORBIS; **223** (t)Lynn Gerig/Tom Stack & Assoc., (b)Milton Rand/Tom Stack & Assoc.; **225** Otto Hahn/Peter Arnold, Inc.; **226** Spencer Grant/PhotoEdit, Inc.; **228** Image courtesy NASA/GSFC/JPL, MISR Team; **232 233** Aaron Haupt; **234** (t)Ted Streshinky/CORBIS, (b)Underwood & Underwood/CORBIS; **235** (t)James L. Amos/CORBIS, (c)Michael Collier, (b)Phillip Wallick/The Stock Market/CORBIS; **240–241** Hugh Sitton/Getty Images; **242** (t)Mark E. Gibson/Visuals Unlimited, (b)D.E. Hurlbert & James DiLoreto/Smithsonian Institution; **243** Jeffrey Rotman/CORBIS; **244** (t)Dr. John A. Long, (b)A.J. Copley/Visuals Unlimited; **246** (t)PhotoTake, NYC/PictureQuest, (b)Louis Psihoyos/Matrix; **248** David M. Dennis; **249** (l)Gary Retherford/Photo Researchers, (r)Lawson Wood/CORBIS; **250** Aaron Haupt; **253** (bkgd)Lyle Rosbotham, (l)IPR/12–18 T. Bain, British Geological Survey/NERC. All rights reserved, (r)Tom Bean/CORBIS; **254** Jim Hughes/PhotoVenture/Visuals Unlimited; **255** (l)Michael T. Sedam/CORBIS, (r)Pat O'Hara/CORBIS; **257** Aaron Haupt; **259** James King-Holmes/Science Photo Library/Photo Researchers; **260** Kenneth Garrett; **261** WildCountry/CORBIS; **262** (t)A.J. Copley/Visuals Unlimited, (b)Lawson Wood/CORBIS; **263** Matt Meadows; **264** Jacques Bredy; **265** (tl)François Gohier/Photo Researchers, (tr)Sinclair Stammers/Photo Researchers, (b)Mark E. Gibson/DRK Photo; **268** Tom Bean/CORBIS; **270–271** Roger Garwood & Trish Ainslie/CORBIS; **271** KS Studios; **272** Tom & Therisa Stack/Tom Stack & Assoc.; **274** (l)Gerald & Buff Corsi/Visuals Unlimited, (r)John Gerlach/Animals Animals; **276** (tl)Mark Boulron/Photo Researchers, (others)Walter Chandoha; **277** Jeff Lepore/Photo Researchers; **281** (l)Mitsuaki Iwago/Minden Pictures, (r)R. Calentine/Visuals Unlimited; **283** J.G. Gehling/Discover Magazine; **284** Gerry Ellis/ENP Images; **287** Matt Meadows; **290** (l)David Burnham/Fossilworks, Inc., (r)François Gohier/Photo Researchers; **292** Michael Andrews/Earth Scenes; **293** Tom J. Ulrich/Visuals Unlimited; **294** David M. Dennis; **295** Mark Burnett; **297** (l)E. Webber/Visuals Unlimited, (r)Len Rue, Jr./Animals Animals; **299** John Cancalosi/Stock Boston; **302–303** Steve Murray/PictureQuest; **303** (inset) Davis Meltzer; **304–305** Chad Ehlers/Stone/Getty Images; **314** (bl)Richard J. Wainscoat/Peter Arnold, Inc., (others)Lick Observatory; **316** Dr. Fred Espenak/Science Photo Library/Photo Researchers; **317** Bettmann/CORBIS; **318** NASA; **320** Roger Ressmeyer/CORBIS; **323** BMDO/NRL/LLNL/Science Photo Library/Photo Researchers; **324** (t)Zuber et al/Johns Hopkins University/NASA/Photo Researchers, (b)NASA; **325** NASA; **327** Matt Meadows; **328** Cosmo Condina/Stone; **330** Lick Observatory; **331** NASA; **334–335** Roger Ressmeyer/CORBIS; **335** Matt Meadows; **338** European Southern Observatory/Photo Researchers; **340** Bettmann/CORBIS; **342** USGS/Science Photo Library/Photo Researchers; **343** (t)NASA/Photo Researchers, (inset) JPL/TSADO/Tom Stack & Assoc.; **344** (t)Science Photo Library/Photo Researchers, (bl)USGS/TSADO/Tom Stack & Assoc., (inset)USGS/Tom Stack & Assoc., (br)USGS/Tom Stack & Assoc.; **345** NASA/JPL/Malin Space Science Systems; **347** Science Photo Library/Photo Researchers; **348** (l)NASA/Science Photo Library/Photo Researchers, (r)CORBIS; **349** (t to b)USGS/TSADO/Tom Stack & Assoc., NASA/JPL/Photo Researchers, NASA/TSADO/Tom Stack & Assoc., JPL, NASA; **350** JPL; **351** Heidi Hammel/NASA; **352** (inset) NASA/Science Source/Photo Researchers, (r)NASA/JPL/TSADO/Tom Stack & Assoc.; **353** NASA, ESA, H. Weaver (JHU/APL), A. Stern (SWRI), and the HST Pluto Companion Search Team; **354** (t to b) NASA/JPL/TSADO/Tom Stack & Assoc., NASA/Science Source/Photo Researchers, CORBIS, NASA/USGS/TSADO/Tom Stack & Assoc.; **355** (t to b)NASA/Science Photo Library/Photo Researchers, NASA/Science Source/Photo Researchers, ASP/Science Source/Photo Researchers, W. Kaufmann/JPL/Science Source/Photo Researchers, CORBIS; **356** Pekka Parviainen/Science Photo Library/Photo Researchers; **357** Pekka Parviainen/Science Photo Library/Photo Researchers; **358** Georg Gerster/Photo Researchers; **359** JPL/TSADO/Tom Stack & Assoc.; **361** Bettmann/CORBIS; **362** (t) Museum of Natural History/Smithsonian Institution, (b)Museum of Natural History/Smithsonian Institution; **363** (l)NASA, (r)NASA, (cl)JPL/NASA, (cr)file photo; **364** NASA/Science Source/Photo Researchers; **366** John R. Foster/Photo Researchers; **368–369** TSADO/ESO/Tom Stack & Assoc.; **373** Bob Daemmrich;

Credits

376 (t)Carnegie Institution of Washington, (b)NSO/SEL/Roger Ressmeyer/CORBIS; 377 (l)NASA, (r)Picture Press/CORBIS, (b)Bryan & Cherry Alexander/Photo Researchers; 378 Celestial Image Co./Science Photo Library/Photo Researchers; 379 Tim Courlas; 381 Luke Dodd/Science Photo Library/Photo Researchers; 384 AFP/CORBIS; 385 NASA; 387 (t)Kitt Peak National Observatory, (b)CORBIS; 391 R. Williams (ST ScI)/NASA; 392 Matt Meadows; 394 Dennis Di Cicco/Peter Arnold, Inc.; 395 (l)file photo, (r)AFP/CORBIS; 400–401 (bkgd)PhotoDisc; 401 (inset)Stephen Frisch/Stock Boston/PictureQuest; 402–403 courtesy IBM; 404 EyeWire; 405 (tl)Culver Pictures/PictureQuest, (tr)E.A. Heiniger/Photo Researchers, (b)Andy Roberts/Stone/Getty Images; 406 Elena Rooraid/PhotoEdit, Inc.; 407 (t)L.S. Stepanowicz/Panographics, (b)Skip Comer; 408 Aaron Haupt; 412 Fraser Hall/Robert Harding Picture Library; 414 Aaron Haupt; 417 Timothy Fuller; 421 Peter Menzel/Stock Boston; 422 (tl)T. Youssef/Custom Medical Stock Photo, (tr)Alfred Pasieka/Science Photo Library/Custom Medical Stock Photo, (cl)CNRI/PhotoTake, NYC, (bl)Voker Steger/Science Photo Library/Photo Researchers, (br)Bob Daemmrich/Stock Boston/PictureQuest; 424 425 Mark Burnett; 426 (l r) Bettmann/CORBIS, (b)Astrid & Hanns-Frieder Michler/Photo Researchers; 427 Aaron Haupt; 429 Dr. Paul Zahl/Photo Researchers; 432–433 Jim Corwin/Index Stock; 434 Stamp from the collection of Prof. C.M. Lang, photo by Gary Shulfer, University of WI Stevens Point; 438 (tl)Tom Pantages, (tr)Elaine Shay, (bl)Paul Silverman/Fundamental Photographs; 441 Amanita Pictures; 442 (l)Joail Hans Stern/Liaison Agency/Getty Images, (r)Leonard Freed/Magnum/PictureQuest; 443 (l)David Young-Wolff/PhotoEdit/PictureQuest, (c)Jane Sapinsky/The Stock Market/CORBIS, (r)Dan McCoy/Rainbow/PictureQuest; 444 (t)George Hall/CORBIS, (b)Aaron Haupt; 445 SuperStock; 446 (t)Don Farrall/PhotoDisc, (b)Matt Meadows; 447 (l)file photo, (r)Bill Freund/CORBIS; 448 CORBIS; 449 (t)Geoff Butler, (b)Royalty-Free/CORBIS; 450 Amanita Pictures; 451 (l)Achim Zschau, (r)Ted Streshinsky/CORBIS; 454 Robert Essel NYC/CORBIS; 455 Mark Burnett; 456 Tim Flach/Stone/Getty Images; 457 (l)Yoav Levy/PhotoTake NYC/PictureQuest, (r)Louvre, Paris/Bridgeman Art Library, London/New York; 458 Matt Meadows; 462–463 Christian Michel; 471 Laura Sifferlin; 472 (l)Lester V. Bergman/CORBIS, (r)Doug Martin; 477 Matt Meadows; 478 (tr cr) Kenneth Libbrecht/Caltech, (cl)Albert J. Copley/Visuals Unlimited, (bl)E.R. Degginger/Color-Pic; 480 James L. Amos/Photo Researchers; 481–483 Aaron Haupt; 484 Fulcrum Publishing; 489 Matt Meadows; 490–491 Simon Fraser/Science Photo Library/Photo Researchers; 492 (l)Aaron Haupt, (r)Doug Martin; 493 (tl)Patricia Lanza, (tc)Jeff J. Daly/Visuals Unlimited, (tr)Susan T. McElhinney, (bl)Craig Fuji/Seattle Times, (br)Sovfoto/Eastfoto/PictureQuest; 494 Amanita Pictures; 497 Sovfoto/Eastfoto/PictureQuest; 499 Christopher Swann/Peter Arnold, Inc.; 500 (tl)Frank Balthis, (tr)Lois Ellen Frank/CORBIS, (b)Matt Meadows; 501 David Young-Wolff/PhotoEdit/PictureQuest; 502 (l)Amanita Pictures, (r)Richard Megna/Fundamental Photographs/Photo Researchers; 503 Victoria Arocho/AP/Wide World Photos; 504 (t)Aaron Haupt, (bl)Kevin Schafer/CORBIS, (br)Icon Images; 505 SuperStock; 506 (tl)Chris Arend/Alaska Stock Images/PictureQuest, (tr)Aaron Haupt, (b)Bryan F. Peterson/CORBIS; 507 courtesy General Motors; 508 509 Matt Meadows; 510 Amanita Pictures; 511 Bob Daemmrich; 512 (l)Tino Hammid Photography, (r)Joe Richard/UF News & Public Affairs; 513 David Young-Wolff/PhotoEdit, Inc.; 516 Lester V. Bergman/CORBIS; 517 Peter Walton/Index Stock; 518–519 (bkgd)Museum of the City of New York/CORBIS; 519 (l)Lee Snider/CORBIS, (r)Scott Camazine/Photo Researchers; 520–521 William Dow/CORBIS; 522 Telegraph Colour Library/FPG/Getty Images; 523 Geoff Butler; 526 Richard Hutchings; 529 Runk/Schoenberger from Grant Heilman; 531 Mark Doolittle/Outside Images/Picturequest; 533 (l)Ed Bock/The Stock Market/CORBIS, (r)Will Hart/PhotoEdit, Inc.; 535 (t)Tom & DeeAnn McCarthy/The Stock Market/CORBIS, (bl)Jodi Jacobson/Peter Arnold, Inc., (br)Jules Frazier/Photo Disc; 536 Mark Burnett; 538 Robert Brenner/PhotoEdit, Inc.; 539 Laura Sifferlin; 540 541 Icon Images; 542 Alexis Duclos/Liaison/Getty Images; 543 (l r)Rudi Von Briel/PhotoEdit, Inc., (c)PhotoDisc; 543 Rudi von Briel/Photo Edit, Inc.; 547 (l)Jodi Jacobson/Peter Arnold, Inc., (r)Runk/Schoenberger from Grant Heilman; 548–549 Wendell Metzen/Index Stock; 549 Richard Hutchings; 550 (l)Globus Brothers Studios, NYC, (r)Stock Boston; 551 Bob Daemmrich; 552 (t)Beth Wald/ImageState, (b)David Madison; 553 Rhoda Sidney/Stock Boston/PictureQuest; 555 (l)Myrleen Cate/PhotoEdit, Inc., (r)David Young-Wolff/PhotoEdit, Inc.; 556 Bob Daemmrich; 558 (t)Stone/Getty Images, (b)Myrleen Cate/PhotoEdit, Inc.; 560 David Madison; 562 Richard Megna/Fundamental Photographs; 563 Mary M. Steinbacher/PhotoEdit, Inc.; 564 (t)Betty Sederquist/Visuals Unlimited, (b)Jim Cummins/FPG/Getty Images; 565 (tl)Denis Boulanger/Allsport, (tr)Donald Miralle/Allsport, (b)Tony Freeman/PhotoEdit/PictureQuest; 566 (t)David Madison, (b)NASA; 568 NASA; 569 Richard Hutchings; 570 571 Mark Burnett; 572 (t)Tom Wright/CORBIS, (b)Didier Charre/Image Bank; 573 (tl)Philip Bailey/The Stock Market/CORBIS, (tr)Romilly Lockyer/Image Bank/Getty Images, (bl)Tony Freeman/PhotoEdit, Inc.; 577 Betty Sederquist/Visuals Unlimited; 578–579 Rich Iwasaki/Getty Images; 579 Mark Burnett; 580 Mary Kate Denny/PhotoEdit, Inc.; 588 (l)David Young-Wolff/PhotoEdit, Inc., (r)Frank Siteman/Stock Boston; 591 Duomo; 592 Robert Brenner/PhotoEdit, Inc.; 593 (t)Tom McHugh/Photo Researchers, (b)Amanita Pictures; 594 Amanita Pictures; 595 (t)Dorling Kindersley, (bl br)Bob Daemmrich; 596 (l)Wernher Krutein/Liaison Agency/Getty Images, (r)Siegfried Layda/Stone/Getty Images; 598 Tony Freeman/PhotoEdit, Inc.; 599 Aaron Haupt; 600 (t)Ed Kashi/CORBIS, (b)James Balog; 601 (l)Inc. Janeart/The Image Bank/Getty Images, (r)Ryan McVay/PhotoDisc; 605 Comstock Images, PhotoDisc; 606–607 Peter Walton/Index Stock; 608 John Evans; 609 (t)Nancy P. Alexander/Visuals Unlimited, (b)Morton & White; 611 Tom Stack & Assoc.; 612 Doug Martin; 613 Matt Meadows; 614 Jeremy Hoare/PhotoDisc; 615 Donnie Kamin/PhotoEdit, Inc.; 616 SuperStock; 617 Colin Raw/Stone/Getty Images; 618 Aaron Haupt; 619 PhotoDisc; 620 (l)Barbara Stitzer/PhotoEdit, Inc., (c)Doug Menuez/PhotoDisc, (r)Addison Geary/Stock Boston; 622 C. Squared Studios/PhotoDisc; 624 625 Morton & White; 626 (bkgd)Chip Simons/FPG/Getty Images, (inset)Joseph Sohm/CORBIS; 627 SuperStock; 630 John Evans; 631 Michael Newman/Photo Edit, Inc.; 632–633 (bkgd)Matthew Borkoski/Stock Boston/PictureQuest; 633 (inset)L. Fritz/H. Armstrong Roberts; 634–635 V.C.L./Getty Images; 637 (t)Richard Hutchings, (b)KS Studios; 640 Royalty Free/CORBIS; 642 J. Tinning/Photo Researchers; 645 Gary Rhijnsburger/Masterfile;

Credits

650 Doug Martin; **651** (t)Doug Martin, (b)Geoff Butler; **653** Bonnie Freer/Photo Researchers; **655** Matt Meadows; **656 657** Richard Hutchings; **658** (bkgd)Tom & Pat Leeson/Photo Researchers, (inset)William Munoz/Photo Researchers; **662** J. Tinning/Photo Researchers; **663** Doug Martin; **664–665** James Leynse/CORBIS; **667** Richard Megna/Fundamental Photographs; **668** Amanita Pictures; **671** John Evans; **672** Amanita Pictures; **673** (l)Kodansha, (c)Manfred Kage/Peter Arnold, Inc., (r)Doug Martin; **677** Bjorn Backe/Papilio/CORBIS; **679** Norbert Schafer/The Stock Market/CORBIS; **681** AT&T Bell Labs/Science Photo Library/Photo Researchers, (t)Richard Hutchings, (b)Tony Freeman/PhotoEdit, Inc.; **682** (t)Science Photo Library/Photo Researchers, (c)Fermilab/Science Photo Library/Photo Researchers, (b)SuperStock; **683** PhotoDisc; **684** (t)file photo, (b)Aaron Haupt; **685** Aaron Haupt; **686** John MacDonald; **687** (l)SIU/Peter Arnold, Inc., (r)Latent Image; **691** John Evans; **692–693** Mark A. Johnson/CORBIS; **694** (l)David W. Hamilton/Getty Images, (r)Ray Massey/Getty Images; **699 700** Richard Megna/Fundamental Photographs; **702** David Young-Wolff/Photo Edit, Inc.; **703** (tl)Ian O'Leary/Stone/Getty Images, (tr)David Young-Wolff/PhotoEdit, Inc., (bl)Mark A. Schneider/Visuals Unlimited, (bc)Rafael Macia/Photo Researchers, (br)SuperStock; **705** AFP Photo/Hector Mata/CORBIS; **706** Matt Meadows; **707** James Blank/Getty Images; **712** (t)Nation Wong/CORBIS, (b)Jon Feingersh/CORBIS; **713** Ralph C. Eagle Jr./Photo Researchers; **714 715** Matt Meadows; **716** (t)Bettmann/CORBIS, (bl br)image courtesy of NRAO/AUI; **721** (l)Edward Burchard/Index Stock, (r)David Young-Wolff/Photo Edit, Inc.; **722** PhotoDisc; **724** Tom Pantages; **728** Michell D. Bridwell/PhotoEdit, Inc.; **729** (t)Mark Burnett, (b)Dominic Oldershaw; **730** StudiOhio; **731** Timothy Fuller; **732** Aaron Haupt; **734** KS Studios; **735** Matt Meadows; **738** Rod Planck/Photo Researchers; **742** Runk/Schoenberger from Grant Heilman; **744** (t)Amanita Pictures, (b)Mark Burnett; **747** Icon Images; **748** Amanita Pictures; **749** Bob Daemmrich; **751** Davis Barber/PhotoEdit, Inc.